AF183085

MACROCYCLIC CHEMISTRY

Macrocyclic Chemistry

Current Trends and Future Perspectives

Edited by

KARSTEN GLOE
Technische Universität Dresden, Germany

Springer

A C.I.P. Catalogue record for this book is available from the Library of Congress.

ISBN-10 1-4020-3364-8 (HB) Springer Dordrecht, Berlin, Heidelberg, New York
ISBN-10 1-4020-3687-6 (e-book) Springer Dordrecht, Berlin, Heidelberg, New York
ISBN-13 978-1-4020-3364-3 (HB) Springer Dordrecht, Berlin, Heidelberg, New York
ISBN-13 978-1-4020-3687-3 (e-book) Springer Dordrecht, Berlin, Heidelberg, New York

---

Published by Springer,
P.O. Box 17, 3300 AA Dordrecht, The Netherlands.

The front cover illustration depicts the X-ray structure of 5,10,15,20-(4-bromophenyl)-21,23-dithiaporphyrin obtained by S. Stute and T. Söhnel, TU Dresden.

The image on the back cover and in the preface shows the logo of the ISMC 2005 in Dresden designed by Cornelia Kreiß, TU Dresden.

Printed on acid-free paper

All Rights Reserved
© 2005 Springer
No part of this work may be reproduced, stored in a retrieval system, or transmitted
in any form or by any means, electronic, mechanical, photocopying, microfilming, recording
or otherwise, without written permission from the Publisher, with the exception
of any material supplied specifically for the purpose of being entered
and executed on a computer system, for exclusive use by the purchaser of the work.

Printed in the Netherlands.

CONTENTS

PREFACE

Macrocyclic and supramolecular chemistry have developed into one of the most active and promising research areas of chemical science – located at the interface between chemistry, physics and biology. Based on the pioneering and stimulating work of the 1987 Nobel Prize Winners Charles J. Pedersen, Donald J. Cram and Jean-Marie Lehn, this field encompasses manifold fundamental aspects of molecular recognition and self-organization and uses these biological principles for the design of smart artificial systems. Therefore an enormous potential exists (some of which is currently being exploited) for future applications in such fields as bio- and nanotechnology, environmental protection, catalysis, molecular electronics and photonics as well as medicine.

The interdisciplinary character of supramolecular chemistry thus provides a platform for the mixing and cross-pollination of ideas between chemists, physicists, biologists and engineers. Since the first International Symposium on Macrocyclic Chemistry (ISMC) was held at Brigham Young University in Provo, USA in 1977, the yearly conferences have been a forum for experts in supramolecular chemistry, bringing together researchers from around the globe in a pleasant and creative atmosphere. In July 2005 the XXX ISMC conference will be held in Dresden, Germany. In order to celebrate the 30th meeting, it was the idea of Emma Roberts from Kluwer Academic Publishers to produce a book in which selected aspects of current macrocyclic and supramolecular chemistry would be highlighted. Owing to the enormous number of publications that continue to appear, it has been a difficult venture to cover all important facets of this expanding field. Nevertheless, the result of our efforts and of the hard work of 25 leading groups in the field is now presented to you. Starting with a discussion of the history of ISMC, the following chapters serve to characterize a representative number of topics in macrocyclic chemistry and to indicate its future developments. Thus, the synthesis, structure and properties of functional macrocycles, their modeling and an outlook towards their possible application, such as for monitoring, environmental clean-up and recovery, in medicine or as models for biological processes, are all presented.

I am indebted to all the authors for their stimulating contributions and for their commitment and patience during the very short preparation time for the chapters between April and October 2004.

I should like to thank my coworkers, especially Marco Wenzel, Christoph Naumann and my wife, Kerstin, who have helped me complete this book by reading and correcting the manuscripts. In particular, I am very grateful to Bianca Antonioli who fostered close contact with the authors and who had the tremendous burden of coming to grips with the formatting of the individual chapters.

Karsten Gloe (ed.), Macrocyclic Chemistry: Current Trends and Future Perspectives, ix–x.
© 2005 *Springer. Printed in the Netherlands.*

Finally, I sincerely thank Emma Roberts and Vaska Krabbe from Springer (formerly Kluwer Academic Publishers) for their considerable assistance during preparation of the book.

Hopefully this book will find a timely and useful place in the supramolecular chemistry community.

Dresden, Germany
October 2004

Karsten Gloe

1.
CONTRIBUTIONS OF THE INTERNATIONAL SYMPOSIUM ON MACROCYCLIC CHEMISTRY TO THE DEVELOPMENT OF MACROCYCLIC CHEMISTRY

REED M. IZATT, KRYSTYNA PAWLAK
AND JERALD S. BRADSHAW
Department of Chemistry and Biochemistry, Brigham Young University
Provo, UT 84602, U.S.A.

1. Introduction

In the early 1960s, Charles J. Pedersen at du Pont serendipitously discovered the compound that later came to be known as dibenzo-18-crown-6 (DB18C6). He isolated DB18C6 in a 0.4% yield from a "brownish goo" while attempting to prepare a completely different compound [1]. The decision to expend the effort needed to isolate, purify, and characterize the compound that became known as DB18C6 represents a true example of scientific creativity and luck. Many organic chemists have isolated pure compounds from various shades of „goo", but few have had the experience of seeing a new field of chemistry arise from their „goo" as Pedersen did. Pedersen knew he was on to something important when his subsequent characterization of the colorless, crystalline by-product revealed that its solubilization in methanol was due to the presence of Na^+ and that its molecular weight was double that of benzo-9-crown-3 (B9C3). Pedersen [1] remarks that with the "realization that I had something very unusual and with the utmost curiosity and anticipation, I devoted all my energies to the study of this fascinating class of ligands by synthesizing a great variety of macrocyclic polyethers and determining their interaction with inorganic cations." The culmination of this effort was the publication of a monumental paper by him in 1967 [2].

The publication of Pedersen's work attracted instant and increasing attention from a variety of scientists worldwide. During the ensuing decade, research in an increasing number of laboratories led to the synthesis and characterization of many novel macrocyclic chemical structures, and to the application of macrocycles in a wide variety of fields. This effort cut across many areas including organic and inorganic synthesis, biochemistry, ion transport in membranes, phase transfer catalysis, and structure analysis.

Karsten Gloe (ed.), Macrocyclic Chemistry: Current Trends and Future Perspectives, 1–14.
© 2005 *Springer. Printed in the Netherlands.*

One of the identifying characteristics of this new developing field of macrocyclic chemistry was the focus on the high selectivity that **host** macrocycles had for a variety of **guest** compounds. As the principles governing this selectivity were discovered, the intelligent design of hosts with pre-determined selectivity for specific guests became possible. During this period, terms with new meanings appeared in the chemical literature including macrocyclic chemistry, host-guest chemistry, crown compounds, molecular recognition, nanochemistry, selective cation and anion complexation, phase transfer catalysis, and supramolecular chemistry.

Dr. James J. Christensen of Brigham Young University (BYU) and one of us (RMI) were two of the first scientists to visit Pedersen in his DuPont laboratory following publication of his 1967 paper [2]. Our interest at that time was in the highly selective carrier-mediated transport of K^+ and Na^+ in biological membranes. Pedersen's results offered the possibility of using cyclic polyethers to study selective cation transport [3]. Indeed,Pedersen's work resulted in a resurgence of interest in alkali metal ion chemistry. The ability of macrocycles, natural and synthetic, to selectively interact with these cations provided the opportunity to investigate their chemistry in ways that had not been possible before. By the late 1970s, the growth of macrocyclic chemistry was impressive and the field had attracted persons from many disciplines.

2. Origins of the International Symposium on Macrocyclic Chemistry

A *Chemical Reviews* article [4] in the mid-1970s reflected the growth in the macrocyclic chemistry field. About this time, Professor Christensen and one of us (RMI) envisioned the value of an annual symposium to provide a forum for the presentation and discussion of research activities in the field of macrocyclic chemistry. A need was recognized by them to bring together persons from a variety of chemical and non-chemical fields who had an interest in macrocyclic chemistry, but who were not personally acquainted with each other. It was apparent that the number of workers in the field was increasing rapidly. An annual symposium could be the means to catalyze growth in the field and lead to the exploration of new areas of chemistry. It was visualized that both theoretical and experimental aspects of the properties and behavior of synthetic and naturally occurring macrocyclic compounds would be covered in a series of invited lectures as well as accepted contributed papers. The First Symposium on Macrocyclic Compounds was held 15-17 August 1977 at BYU in Provo, Utah. Seventy-nine persons attended. Thirteen of these were from ten countries outside the U.S.A. Of those attending from the U.S.A., 23 were from BYU. Sixteen of the attendees came from 13 industrial companies. Twenty-eight universities were represented. The expenses for the Symposium totaled $9,500. The Provo symposia were held annually through 1981.

In 1980, the First European Symposium on Macrocyclic Compounds was held in Basel, Switzerland with Thomas Kaden as Chair. In 1982, the Second European Symposium on Macrocyclic Compounds was held in Strasbourg, France with Marie-José Schwing as Chair. No symposium was held in Provo in 1982, but one of us (RMI) attended the Strasbourg meeting where informal discussions were held on the possibility of combining these two meetings into an annual symposium, which would be international in nature. It was agreed that the 1983 Symposium on Macrocyclic Compounds in Provo and the 1984 European Symposium in Stirling, Scotland would be held as scheduled. The 1985 meeting in Provo would be the first to be held under the new title of International Symposium on Macrocyclic Chemistry (ISMC). The ISMC meetings have been held on an annual basis since 1985. In Table 1, the symposia held from 1977 through 2005 are listed together with their locations, dates, and chairs.

At the Strasbourg meeting, an International Advisory Committee was formed at the suggestion of Professor Jean-Marie Lehn consisting of A. V. Bogatsky (USSR), R. Hay (UK), R. Izatt (USA), T. Kaden (Switzerland), E. Kimura (Japan), P. Paoletti (Italy), A. Sargeson (Australia), and M.J. Schwing (France). The untimely death of Professor Bogatsky resulted in N.G. Lukyanenko (USSR) being added to the Advisory Committee at the Stirling meeting. R. Izatt was the Chair of the Committee. He continued in this position until 1998. Professor John D. Lamb of BYU has served as chair since 1998.

The symposia have afforded a unique opportunity for workers in macrocyclic chemistry to interact with each other, plan collaborative efforts, and be exposed to new ideas and concepts. The symposia have attracted people from academia, industry, and government. There has been a good mix of older and younger scientists including many students. The appeal of the symposia to students and younger people is especially noteworthy. As the symposia have moved from country to country, large numbers of students and younger investigators have taken advantage of the opportunity to attend and participate in an international meeting in their field.

The venues of the symposia have provided unique opportunities to visit scenic and cultural sites throughout the world. These sites have included the national parks in Utah; cathedrals and palaces in Europe, Korea, and Japan; dinner cruises on the Inland Sea from Hiroshima to Miyajima and on the canals of the Netherlands; the Edinburgh Tattoo at Edinburgh Castle; historic sites in Jerusalem; beautiful beaches of Hawaii; Jean-Marie Lehn presenting an organ recital in the Sheffield cathedral; the Great Barrier Reef in Australia, and ancient Tuscany villages in Italy. These visits provided opportunity for informal contact and increased the appreciation of the participants for the cultural and scenic values in other locations.

The Chairs of the various symposia are the individuals who have been responsible for the success of the meetings. Each of the organizers has also had an active part in the development of macrocyclic chemistry. Their interest, scientific background, and organizational skills have resulted in excellent scientific and social programs, involvement of many student participants, and the active participation of scientists from many nations.

Several of the organizers have received Izatt-Christensen (I-C) Awards and their scientific accomplishments are mentioned in Section 3. The remaining organizers have also had distinguished careers in macrocyclic chemistry and the scientific work of all of the chairs has brought recognition to them and to their respective countries. The nations represented by the organizers are: Australia (Lindoy, Keene), France, (Schwing), Germany (Knöchel, Gloe), Israel (Shanzer, Meyerstein), Italy (Paoletti, Fabbrizzi, Bianchi), Japan (Kimura, Shinkai), South Korea (Kim), Spain (Casabó, García-España), Switzerland (Kaden), The Netherlands (Reinhoudt), UK (Hay, Clay, Fenton, Schröder, Stoddart), Ukraine (Andronati, Lukyanenko, Kuklar), USA (Izatt, Christensen, Bradshaw, Anslyn, Lamb, Busch, Bowman-James, Sessler). These organizers were supported by numerous others who ensured the success of the symposia.

TABLE 1. Symposia on Macrocyclic Chemistry, 1977-2005

	Title	Location	Dates	Chair(s)
1	First Symposium on Macrocyclic Compounds	Provo, Utah	15-17 August 1977	R. M. Izatt, J. J. Christensen
2	Second Symposium on Macrocyclic Compounds	Provo, Utah	14-16 August 1978	R. M. Izatt, J. J. Christensen
3	Third Symposium on Macrocylic Compounds	Provo, Utah	6-8 August 1979	R. M. Izatt, J. J. Christensen
4	First European Symposium on Macrocyclic Compounds	Basel, Switzerland	2-4 July 1980	Th. Kaden
5	Fourth Symposium on Macrocyclic Compounds	Provo, Utah	11-13 August 1980	R. M Izatt, J. J. Christensen
6	Fifth Symposium on Macrocyclic Compounds	Provo, Utah	10-12 August 1981	R. M. Izatt, J. J. Christensen
7	Second European Symposium on Macrocyclic Compounds	Strasbourg, France	30 August-1 September 1982	M. -J. Schwing
8	Seventh Symposium on Macrocyclic Compounds	Provo, Utah	8-10 August 1983	R. M. Izatt, J. J. Christensen
9	Third European Symposium on Macrocyclic Compounds	Stirling, Scotland	29-31 August 1984	R. W. Hay, R. M. Clay
10	X International Symposium on Macrocyclic Chemistry	Provo, Utah	5-7 August 1985	R. M. Izatt, J. J. Christensen
11	XI International Symposium on Macrocyclic Chemistry	Florence, Italy	1-4 September 1986	P. Paoletti L. Fabbrizzi
12	XII International Symposium on Macrocyclic Chemistry	Hiroshima, Japan	20-23 July 1987	E. Kimura
13	XIII International Symposium on Macrocyclic Chemistry	Hamburg, Germany	4-8 September 1988	A. Knöchel
14	XIV International Symposium on Macrocyclic Chemistry	Townsville, Australia	25-28 June 1989	L. F. Lindoy
15	XV International Symposium on Macrocyclic Chemistry	Odessa, Ukraine	3-8 September 1990	S.A. Andronati, V.P. Kukhar, N.G. Lukyanenko
16	XVI International Symposium on Macrocyclic Chemistry	Sheffield, United Kingdom	1-6 September 1991	F. Stoddart, D. Fenton
17	XVII International Symposium on Macrocyclic Chemistry	Provo, Utah	9-14 August 1992	R. M. Izatt, J. S. Bradshaw
18	XVIII International Symposium on Macrocyclic Chemistry	Enschede, The Netherlands	27 June-2 July 1993	D.N. Reinhoudt
19	XIX International Symposium on Macrocyclic Chemistry	Lawrence, Kansas	12-17 June 1994	D. H. Busch K. Bowman-James
20	XX International Symposium on Macrocyclic Chemistry	Jerusalem, Israel	2-7 July 1995	A. Shanzer, D. Meyerstein
21	XXI International Symposium on Macrocyclic Chemistry	Montecatini Terme, Italy	23-28 June 1996	P. Paoletti, A. Bianchi
22	XXII International Symposium on Macrocyclic Chemistry	Seoul, Korea	3-8 August 1997	S. J. Kim
23	XXIII International Symposium on Macrocyclic Chemistry	Turtle Bay, Oahu, Hawaii	7-12 June 1998	J.L. Sessler, E.V. Anslyn

24	XXIV International Symposium on Macrocyclic Chemistry	Barcelona, Spain	18-23 July 1999	J. Casabo, E. Garcia-España
25	XXV International Symposium on Macrocyclic Chemistry	St. Andrews, United Kingdom	2-7 July 2000	R. W. Hay (posthumous), M. Schröder
26	XXVI International Symposium on Macrocyclic Chemistry	Fukuoka, Japan	15-20 July 2001	S. Shinkai
27	XXVII International Symposium on Macrocyclic Chemistry	Park City, Utah	23-27 June 2002	J. D. Lamb
28	XXVIII International Symposium on Macrocyclic Chemistry	Gdansk, Poland	13-18 July 2003	J. F. Biernat
29	XXIX International Symposium on Macrocyclic Chemistry	Cairns, Australia	4-8 July 2004	L.F. Lindoy, F.R. Keene
30	XXX International Symposium on Macrocyclic Chemistry	Dresden, Germany	17-21 July 2005	K. Gloe

Significant scientific breakthroughs involving macrocyclic compounds have occurred during the past three decades. The ISMC meetings have had a positive influence on this activity. In the material that follows, three examples of this influence are presented, i.e., the Izatt-Christensen Award; IBC Advanced Technologies, Inc. (IBC); and the awarding of the 1987 Nobel Prize in Chemistry to three individuals with close ties to the Symposium.

3. The Izatt-Christensen Award

The papers presented at the ISMC and its predecessors since 1977 reflect the developments in the field over nearly thirty years. During this time, the field has broadened resulting in the design, synthesis and characterization of increasingly more complex organic ligands and their application to new fields of chemistry that were scarcely envisioned three decades ago. This trend is illustrated by the titles of the lectures presented by the I-C awardees. The I-C Award was instituted in 1991 by IBC (American Fork, Utah). This competitive annual award recognizes excellence in macrocyclic chemistry and is given to individuals who have not received a major award in chemistry. The awardee receives a small honorarium and is expected to present an invited lecture at the ISMC meeting in the year of the award. In Table 2, the recipients of the I-C Award are listed together with the locations and titles of their Award lectures.

It is informative to examine the present research in macrocyclic chemistry being carried out by the I-C awardees. Their research interests reflect the variety of fields in which macrocyclic chemistry is involved.

TABLE 2. The Izatt-Christensen Award in Macrocyclic Chemistry

Year	Recipient	Location	Title
1991	Jean-Pierre Sauvage	Sheffield, United Kingdom	Synthetic Molecular Knots
1992	Eiichi Kimura	Provo, Utah	Roles of Zinc (II) in Zinc Enzymes
1993	J. Fraser Stoddart	Enschede, The Netherlands	Self-Assembly in Unnatural Product Synthesis
1994	Daryle H. Busch	Lawrence, Kansas	A Sampling of Multi-receptor Supramolecular Systems and Beginnings in the Chemistry of Orderly Entanglements
1995	David N. Reinhoudt	Jerusalem, Israel	Synthesis and Self-Assembly of Supramolecular Structures for Switches and Sensors
1996	George W. Gokel	Montecatini, Terme, Italy	Synthetic Models for Cation Channel Function
1997	Alan M. Sargeson	Seoul, Korea	Outer-Sphere Electron Transfer Reactions of Macro-Bicyclic Complexes
1998	Seiji Shinkai	Turtle Bay, Oahu, Hawaii	Dynamic Control of Ion and Molecule Recognition Processes in Macrocyclic Host-Guest Systems
1999	Fritz Vögtle	Barcelona, Spain	Rotaxanes, Catenanes. Pretzelanes- Template Synthesis and Chirality
2000	Jerry L. Atwood	St. Andrews, United Kingdom	Macrocycles as Building Blocks for Large Supramolecular Assemblies
2001	Jonathan L. Sessler	Fukuoka, Japan	Novel Polypyrrole Macrocycles
2002	C. David Gutsche	Park City, Utah	The Cornucopia of Calixarene Chemistry
2003	Jeremy K.M. Sanders	Gdansk, Poland	The Ins and Outs of Templating: A Dynamic Future for Macrocyclic Chemistry
2004	Makoto Fujita	Cairns, Australia	Self-assembly and Function of Metal-linked Macrocyclic and Cage-like Molecular Frameworks
2005	Kenneth N. Raymond	Dresden, Germany	Chemistry in Chiral, Nanoscale Flasks

Jean-Pierre Sauvage is a CNRS director of research and is located at the Université Louis Pasteur in Strasbourg, France. His current research interests include the development of models of the photosynthetic reaction centre using transition metals and porphyrins [5], topology (synthetic catenanes and knots) [6], and molecular machines [7].

Eiichi Kimura is retired from the Department of Medicinal Chemistry at Hiroshima University in Japan. His recent research interests have included the supramolecular chemistry of macrocyclic polyamines and their use in molecular recognition and as zinc-enzyme models. These interests have led to the development of fluorophore sensors for Zn(II) [8] use of macrocycles to effect selective recognition of anions [9], nucleobases in polynucleotides [10], thymidine mono- and diphosphate nucleotides (11), carbonic anhydrase and carboxypeptidase [12], and development of Zn(II)-macrocycle anti-HIV agents [13]. In May 2004, he received a Purple Ribbon Award from the Emperor of Japan.

J. Fraser Stoddart is the Saul Winstein Professor of Organic Chemistry at the University of California at Los Angeles. His current research interests involve the application of molecular recognition to the development of molecular self-assembly processes [14], template-directed protocols in noncovalent and covalent synthesis [15], and artifical molecular machines [16]. Applications lie in the construction of artificial molecular electronic devices on the nanoscale level [16, 17]. Stoddart and Atwood (also an I-C awardee) collaborated recently in the novel synthesis and characterization of molecular borromean rings [18]. These objects have interest in knot theory and are composed of three rings interlocked in such a manner that scission of any one ring leads to the other two falling apart. The symbol describing these rings can be traced back to early Christian iconography and Norse mythology and is found on crests and statues dating to the 15th century Borromeo family in Tuscany. In the synthesis, Zn(II) was used as a template. The authors suggest [18] that these compounds could be used as highly organized nanoclusters in a materials setting such as spintronics or in a biological context such as medical imaging.

Daryle H. Busch is the Roy A. Roberts Distinguished Professor of Chemistry at the University of Kansas. His research focuses on three subjects. First, the dynamics of tight binding ligands [19]. Second, the application of tight-binding ligands to transition metal ion control of the chemistry of O_2 and its reduction products HO_2^- and H_2O_2 [20]. Third, orderly molecular entanglements [21, 22]. Busch was a recent president of the American Chemical Society. He was an early pioneer in the 1950s and 1960s in the study of transition metal ion interactions with azamacrocycles.

David N. Reinhoudt is scientific director of the Laboratory of Supramolecular Chemistry and Technology, MESA Research Institute at the University of Twente in The Netherlands. His research is focused on supramolecular chemistry and technology including nanofabrication, molecular recognition, and non-covalent combinatorial synthesis [23, 24]

George W. Gokel is Professor of Molecular Biology and Pharmacology and Director of the Chemical Biology Program at Washington University, St. Louis, Missouri. His research interests lie in attempting to understand better the weak chemical interactions that permit and control the binding of molecular cations to other molecules. A major recent focus has been the development of models for cation and anion channels in biological systems. Long-range goals are to develop systems that will selectively transport cations, anions, and small molecular species through membranes having specific compositions [25-28].

Alan M. Sargeson is Professor Emeritus at the Research School of Chemistry, Canberra, Australia. His work has involved the synthesis, structure determination and characterization of metal-macrocycle complexes [29-31]. Potential applications of his studies are in catalysis, organic synthesis and biology.

Seiji Shinkai is Professor of Chemistry in Kyushu University, Fukuoka, Japan. He is currently serving as a leader at Kyushu University in the 21st Century Project entitled „Functional Innovation of Molecular Informatics". His research activities focus on host-guest chemistry and molecular recognition [32-34]. A particular interest has been the design of molecular nanomachines [33, 35]. Shinkai and Reinhoudt (another I-C awardee) have worked together on an international collaborative research project aimed at building host molecules capable of recognizing guest molecules of any size or shape [32, 36].

Fritz Vögtle is Professor and Director at the Kekulé-Institute for Organic Chemistry and Biochemistry at the University of Bonn, Germany. His research interests are supramolecular chemistry; deformed helical molecules and their chiroptical properties; and compounds with appealing architectures such as rotaxanes, catenanes, knots, and dendrimers [37-40].

Jerry L. Atwood is Curator's Professor of Chemistry at the University of Missouri-Columbia. His research has focused on supramolecular chemistry. His research group has synthesized and examined a broad array of host-guest chemical systems (e.g., liquid clathrates, macromolecular hosts). A principal method for characterization of these systems has been single crystal X-ray structure determination [41-43].

Jonathan L. Sessler is the Rowland K. Pettit Professor of Chemistry at the University of Texas-Austin. His research involves the design and construction of molecules carefully tailored so as to accomplish specific objectives. These molecules often have architectural elegance and interesting chemical, physical, and/or biological properties. He has made important contributions to the synthetic chemistry of porphyrins and related compounds [44-46]. New drugs have resulted from his efforts [47].

C. David Gutsche is retired from the position of Robert A. Welch Professor at Texas Christian University. He is a pioneer in the use of calixarenes in systems designed to mimic enzymes. He has prepared numerous calixarene-type molecules, evaluated their host-guest binding abilities, and studied their catalytic potentials [48, 49].

Jeremy K.M. Sanders is Head of the Department of Chemistry at Cambridge University, United Kingdom. His interests are in supramolecular chemistry with special emphasis on molecular recognition. The aim of his work is to uncover and exploit the rules which govern non-covalent interactions and to understand events that occur at interfaces and on surfaces using spectroscopies and microscopies [50-53]. He has collaborated with J. Fraser Stoddart, another I-C awardee [54].

Makoto Fujita is a professor in the Department of Applied Chemistry at The University of Tokyo. His research interests have involved the design and preparation of self-assembling molecular systems. For example, the combination of transition metal geometry with well-designed bridging ligands can result in the quantitative self-assembly of nano-sized, discrete organic frameworks [55-58]. He has also achieved chemical syntheses in the nanospace present within cavities of self-assembled cages [59].

Three of the I-C awardees (Atwood, Gokel, and Sessler) have made important contributions to macrocyclic chemistry by serving as editor or co-editor of several journals of interest to workers in the macrocyclic chemistry field. The editorial work of these individuals has provided an important means for the publication of macrocyclic chemistry research.

4. IBC Advanced Technologies, Inc.

IBC has made an important contribution to the ISMC by providing financial support for the I-C Award. This award, its recipients, and their contributions are described in Section 3. Significant contributions to the commercialization of macrocyclic chemistry have been made by IBC since it was founded in 1988 by Reed M. Izatt, Jerald S. Bradshaw and James J. Christensen, three early workers in the field. These individuals received much stimulus from contacts in the ISMC and made use of the ideas evolving in macrocyclic chemistry during the 1970s and 1980s to design and prepare ligands with high affinity and high selectivity for specific metal ions in the presence of other competing metal ions that often have chemical properties very similar to those of the target metal ion. The resulting high selectivity of the ligands enabled IBC to develop a series of products (termed SuperLig) using solid-supported macrocycles that could perform difficult and usually large scale metal ion separations. Descriptions of the separation process and its applications are available [60-62]. SuperLig products are used [61] in the separation of palladium and rhodium from other precious metals, of bismuth and antimony from copper in electrorefineries, and of cesium from sodium, potassium and other constituents of nuclear waste.

IBC has also developed a number of products (termed AnaLig) for use in analytical scale separations. 3M Corporation markets disks (termed Rad Disks) for rapid and selective testing of environmental samples featuring IBC AnaLig products. These disks have been developed for analysis of Sr^{2+} and Ra^{2+} [62]. The high selectivity of the IBC product results in major advantages over conventional materials in terms of time and high selectivity. For example, the published EPA Method 905 for radioactive Sr^{2+} determination requires over 50 steps and considerable time. The Sr Rad Disk method requires only six steps and approximately 20 minutes to complete the analysis [62]. The introduction of high selectivity and other improvements inherent in the IBC method results in appreciable savings in cost, time, and reagents [62].

5. 1987 Nobel Prize Awardees

An important indication of the importance of the effect on chemistry of the remarkable early growth of macrocyclic chemistry was the awarding of the 1987 Nobel Prize to three prominent workers in the field, Charles J. Pedersen, Donald J.Cram, and Jean-Marie Lehn (Table 3). The award was given "for their development and use of molecules with structure-specific interactions of high selectivity," [63]. All of these individuals have had close association with the ISMC either by personal contact or through their students. The Symposium has honored them in special sessions at the 1987 Hiroshima meeting (Pedersen) and the 1993 Sheffield meeting (Lehn and Cram). A special volume edited by two of us (RMI, JSB) honoring Pedersen has been published [64]. This volume contains a first-hand account of the discovery of crown ethers by

Pedersen, a description of the industrial scientific career of Pedersen by Herman E. Schroeder of DuPont, and contributed papers by colleagues of Pedersen worldwide.

TABLE 3. Nobel Prize in Chemistry Awardees -- 1987

Awardee	Affiliation	Nobel Lecture Title
Donald J. Cram	Department of Chemistry and Biochemistry; University of California; Los Angeles, California. U.S.A.	The Design of Molecular Hosts, Guests, and Their Complexes
Jean-Marie Lehn	Institut Le Bel; Université Louis Pasteur, Strasbourg, France	Supramolecular Chemistry — Scope and Perspectives — Molecules — Supermolecules — Molecular Devices
Charles J. Pedersen	E.I. du Pont de Nemours and Company, Wilmington, Delaware, U.S.A.	The Discovery of Crown Ethers

The seminal contributions to macrocyclic chemistry of these three individuals at the time they received the Nobel Prize in 1987 are summarized in the official record of the event [63]. In papers presented by each person, the events leading to their decisions to explore macrocyclic chemistry are given. In presenting the Nobel Prize recipients to the King of Sweden, Professor Salo Gronowitz made several salient points [63]. First, he pointed out the importance of specific and selective biological recognition to all life processes. He suggested that the cyclic polyether molecules prepared by Pedersen hold the promise of mimicking biomolecules through their common property of specific recognition and high selectivity for metal ions. Second, the important contributions of Cram and Lehn in carrying out extremely creative organic syntheses were recognized. Gronowitz indicated that these individuals built on the explosive development of the art of organic synthesis during the preceding decades and on the specific and selective recognition capabilities of the new compounds prepared by Pederson. He mentions the innovative way in which Cram and Lehn and their associates skillfully designed, prepared, and studied complex macrocyclic molecules containing holes and clefts which enabled them to selectively bind inorganic and organic anions, cations, and neutral molecules. He also pointed out that some of these new host molecules could even differentiate between the enantiomeric forms of a guest compound. Third, he emphasized that these three individuals had laid the foundation for what, in 1987, was one of the most expansive chemical research areas. He noted that in the twenty year period between 1967, when Pedersen's seminal paper was published [2], and the awarding of the Nobel Prize in 1987, the research of these three scientists had been of great importance in the development of coordination chemistry, organic synthesis, analytical chemistry, bioinorganic chemistry, and bioorganic chemistry.

The continuing impact of the work of Pedersen, Cram, and Lehn on macrocyclic chemistry is evident in the titles of the lectures delivered by the I-C awardees beginning in 1991 (Table 2). There are close connections between a number of the first awardees and the three Nobel Prize winners. The first I-C awardee, Jean-Pierre Sauvage received his Ph.D. degree under the direction of Professor Lehn, Eichii Kimura had a close connection with and interest in Pedersen whose mother was Japanese. George Gokel was a post-doctoral student with Cram. In addition, one of us (JSB) was a graduate student with Cram in his pre-macrocycle years.

Professor Lehn has had a particularly close relationship with the IMSC. He has presented lectures at several of the symposia. He was also instrumental in discussions at the 2nd European Symposium on Macrocyclic Compounds in Strasbourg in 1982 that led to the formation of the ISMC meetings beginning in 1985 (see Table 1).

6. Summary

The ISMC and its predecessor symposia have made significant contributions to the development of the field of macrocyclic chemistry in the past nearly three decades. These symposia have provided a venue for the presentation of new results, for the discussion of new ideas, and for the development of new collaborations among researchers young and old. As the meetings move from country to country, opportunity was afforded for younger scientists to meet and interact with senior scientists in the field.

The scientific programs of the Symposium have changed over the years and reflect the changing nature of the field. A strength of macrocyclic chemistry that has emerged over the decades has been its use in the hands of creative and skilled scientists to explore new areas of chemistry. For example, the creation of host molecules of predesigned shapes enables one to mimic on the molecular scale components of simple machines such as molecular on-off switches, molecular axles, and molecular wires. Macrocycles pre-designed to interact selectively with target inorganic or organic guests have also had an important impact in separations chemistry. The future of the macrocyclic chemistry field is limited only by the imagination and creativity of its practitioners. It is expected that the ISMC will continue to play an important role in facilitating personal interactions, exchanges of ideas, and the discovery of new chemistry.

7. References

1. Pedersen, C.J. (1992) The discovery of crown ethers, *J. Incl. Phenom. Mol. Recognit. Chem.* **12**, 7-10.
2. Pedersen, C.J. (1967) Cyclic polyethers and their complexes with metal salts, *J. Am. Chem. Soc.* **89**, 7017-7036.
3. Izatt, R.M., Rytting, J.H., Nelson, D.P., Haymore, B.L., and Christensen, J.J. (1969) Binding of alkali metal ions by cyclic polyethers: significance in ion transport processes, *Science* **164**, 443-444.
4. Christensen, J.J., Eatough, D.J., and Izatt, R.M. (1974) The synthesis and ion binding of synthetic multidentate macrocyclic compounds, *Chem. Rev.* **74**, 351-384.
5. Baranoff, E., Collin, J.-P., Flamigni, L., and Sauvage, J.-P. (2004) From ruthenium(II) to iridium(III): 15 years of triads based on bis-terpyridine complexes, *Chem. Soc. Rev.* **33**, 147-155.
6. Chambron, J. C., Collin, J.-P., Heitz, V., Jouvenot, D., Kern, J.-M., Moblan, P., Pomeranc, D., and Sauvage, J.-P. (2004) Rotaxanes and catenanes built around octahedral transition metals, *Eur. J. Org. Chem.* **8**, 1627-1638.
7. Dietrich-Buchecker, C., Jimenez-Molero, M.C., Sartor, V., and Sauvage, J.-P. (2003) Rotaxanes and catenanes as prototypes of molecular machines and motors, *Pure Appl. Chem.* **75**, 1383-1393.
8. Kimura, E. and Aoki, S. (2001) Chemistry of zinc(II) fluorophone sensors, *BioMetals*, **14**, 191-204.
9. Kimura, E. and Koike, T. (1998) Dynamic anion recognition by macrocyclic polyamines in neutral pH aqueous solution: development from static anion complexes to an enolate complex, *Chem. Commun.* **15**, 1495-1500.
10. Kimura, E. and Kikuta, E. (2000) Macrocyclic zinc(II) complexes for selective recognition of nucleobases in single- and double- stranded polynucleotides, *Prog. React. Kinetics Mech.* **25**, 1-64

11. Aoki, S. and Kimura, E. (2000) Highly selective recognition of thymidine mono- and diphosphate nucleotides in aqueous solution by ditopic receptors zinc(II)-bis(cyclen) complexes (cyclen = 1, 4, 7, 10-tetraazacyclododecane), *J. Am. Chem. Soc.* **122**, 4542-4548.
12. Kimura, E. (2001) Model studies for molecular recognition of carbonic anhydrase and carboxypeptidase, *Acc. Chem. Res.* **34**, 171-179.
13. Kikuta, E., Aoki, S., and Kimura, E. (2001) A new type of potent inhibitors of HIV-1 TAR RNA-Tat peptide binding by zinc(II)-macrocyclic tetraamine complexes, *J. Am. Chem. Soc.* **123**, 7911-7912.
14. Rowan, S.J. and Stoddart, J.F. (2002) Surrogate-stoppered [2]rotaxanes: a new route to larger interlocked architectures, *Polym. Adv. Tech.* **13**, 777-787.
15. Raymo, F.M. and Stoddart, J.R. (1999) Interlocked macromolecules, *Chem. Rev.* **99**, 1643-1663.
16. Balzani, V., Credi, A., Raymo, F.M. and Stoddart, J.F. (2000) Artificial molecular machines, *Angew. Chem., Int. Ed.* **39**, 3348-3391.
17. Flood, A.H., Ramirez, R.J.A., Deng, W.Q., Muller, R.P., Goddard, W.A., III and Stoddart, J.F. (2004) Meccano on the nanoscale - a blueprint for making some of the world"s tiniest machines, *Aust. J. Chem.* **57**, 301-322.
18. Chichack, K.S., Cantrill, S.J., Pease, A.R., Chiu, S.H., Cave, G.W.V., Atwood, J.L. and Stoddart, J.F. (2004) Molecular borromean rings, *Science* **304**, 1308-1312.
19. Hubin, T.J., McCormick, J.M., Alcock, N.W. and Busch, D.H. (2001) Topologically constrained manganese (III) and iron (III) complexes of two cross-bridged tetraazamacrocycles, *Inorg. Chem* **40** 435-444.
20. Busch, D.H. and Alcock, N.W. (1994) Iron and cobalt „lacunar" complexes as dioxygen carriers, *Chem. Rev.* **94**, 585-623.
21. Clifford, T., Abushamleh, A. and Busch, D.H. (2002) Factors affecting the threading of axle molecules through macrocycles: binding constants for semirotaxane formation, *Proc. Natl. Acad. Sci.* **99**, 4830-4836.
22. Hubin, T.J. and Busch, D.H. (2000) Template routes to interlocked molecular structures and orderly molecular entanglements, *Coord. Chem. Rev.* **200-202**, 5-52.
23. Mateos-Timoneda, M.A., Crego-Calama, M. and Reinhoudt, D.N. (2004) Supramolecular chirality of self-assembled systems in solution, *Chem. Soc. Rev.* **33**, 363-372.
24. Speets, E.A., Ravoo, B.J., Roesthius, F.J.G., Vroegindeweij, F, Blank, D.H.A and Reinhoudt, D.N. (2004) Fabrication of arrays of gold islands on self-assembled monolayers using pulsed laser deposition through nanosieves, *Nano Lett.* **4** 841-844.
25. Gokel, G.W., Leevy, W.M. and Weber, M.E. (2004) Crown ethers: sensors for ions and molecular scaffolds for materials and biological models, *Chem. Rev.* **104**, 2723-2750.
26. Gokel, G.W., Abel, E., De Wall, S.L., Jin, T., Maguire, G.E.M., Meadows, E.S., Murray, C.L., Murillo, O. and Suzuki, I. (1997) Syntheic organic models for transmembrane, cation-conducting channels, *Anales Quim. Int. Ed.* **93**, 347-354.
27. Schlesinger, P.H., Djedovic, N.K., Ferdani, R., Pajewska, J., Pajewski, R. and Gokel, G.W. (2003) Anchor chain length alters the apparent mechanism of chloride channel functions in SCMTR derivatives, *Chem. Commun.* 308-309.
28. Schlesinger, P.H., Ferdani, R., Pajewska, J., Pajewski, R. and Gokel, G.W. (2003) Replacing proline at the apex of heptapeptide-based chloride ion transporters alters their properties and their ionophoretic efficacy, *New J. Chem.* **27**, 60-67.
29. Walker, G.W., Geue, R.J., Sargeson, A.M. and Behm, C.A. (2003) Surface-active cobalt cage complexes: synthesis, surface chemistry, biological activity, and redox properties, *J. Chem. Soc., Dalton Trans.* 2992-3001.
30. Walker, G.W., Geue, R.J., Haller, K.J., David Rae, A. and Sargeson, A.M. (2003) New synthetic routes to hexa-aza cages using cobalt(III) tris(1,2-diamine) templates, *J. Chem. Soc., Dalton Trans.* 279-281.
31. Brown, K.N., Geue, R.J., Hambley, T.W., Hockless, D.C.R., Rae, A.D. and Sargeson, A.M. (2003) Specificity in template syntheses of hexaaza-macrobicyclic cages: [Pt(Me5-tricosatrieneN6)]4+ and [Pt(Me5-tricosaneN6)]4+, *Org. Biomol. Chem.* **1**, 1598-1608.
32. http://www.cstm.kyushu-u.ac.jp/shinkai/seiji.shinkai.html
33. Shinkai, S. and James, T.D. (2001) Molecular recognition events controllable by photochemical triggers or readable by photochemical outputs, *Mol. Supramol. Photochem.* **7**, 429-456.
34. Shinkai, S. (2001) Switchable molecular receptors and recognition processes: from photoresponsive crown ethers to allosteric sugar sensing systems, in Feringa, B.L. (ed.), *Molecular Switches*, Wiley-VCH, Weinheim, Germany, pp. 281-307.
35. Shinkai, S., Takeuchi, M. and Ikeda, A. (2000) Molecular machines useful for the design of chemosensor, in Osada, Y. and De Rossi, D.E. (eds.), *Polymer Sensors and Actuators,* Springer, New York, pp. 183-206.

36. Ishi-i, T., Mateos-Timoneda, M.A., Timmerman, P., Crego-Calama, M., Reinhoudt, D.N., Shinkai, S. (2003) Self-assembled receptors that stereoselectively recognize a saccharide, *Agnew. Chem., Int. Ed.* **42**, 2300-2305.
37. Reuter, C., Schmeider, R. and Vögtle, F. (2000) From rotaxanes to knots. Templating, hydrogen bond patterns, and cyclochirality, *Pure Appl. Chem.* **72**, 2233-2241.
38. Schalley, C.A., Beizai, K. and Voegtle, F. (2001) On the way to rotaxane-based molecular motors: studies in molecular mobility and topological chirality, *Acc. Chem. Res.* **34**, 465-476.
39. Voegtle, F. and Pawlitzki, G. (2002) Cyclophanes: from planar chirality and helicity to cyclochirality, in Takemura, H. (ed.), *Cyclophane Chemistry for the 21st Century*, Research Signpost, Trivandrum, India, pp. 55-90.
40. Saudan, C., Balzani, V., Gorka, M., Lee, S.-K., Maestri, M., Vicinelli, V. and Vögtle, F. (2003) Dendrimers as ligands. Formation of a 2:1 luminescent complex between a dendrimer with a 1, 4, 8, 11-tetraazacyclotetradecane (cyclam) core and Zn^{2+}, *J. Am. Chem. Soc*. **125**, 4424-4425.
41. Atwood, J.L., Barbour, L.J., Hardie, M.J. and Raston, C.L. (2001) Metal sulfonatocalix[4,5]arene complexes: bi-layers, capsules, spheres, tubular arrays and beyond, *Coord. Chem. Rev.* **222,** 3-32.
42. MacGillivray, L.R. and Atwood, J.L. (2000) Cavity-containing materials based upon resorcin[4]arenes by discovery and design, *J. Solid State Chem.* **152**, 199-210.
43. MacGillivray, L.R. and Atwood, J.L. (2000) Spherical molecular containers: from discovery to design, *Adv. Supramol. Chem.* **6**, 157-183.
44. Sessler, J.L. and Seidel, D. (2003) Synthetic expanded porphyrin chemistry, *Angew. Chem. Int. Ed.* ***42***, 5134-5175.
45. Sessler, J.L., Zimmerman, R.S., Bucher, C., Kral, V. and Andrioletti, B. (2001) Calixphyrins. Hybrid macrocycles at the structural crossroads between porphyrins and calixpyrroles, *Pure Appl. Chem.* **73**, 1041-1057.
46. Sessler, J.L. and Davis, J.M. (2001) Sapphyrins: versatile anion binding agents, *Acc. Chem. Res.* **34**, 989-997.
47. Sessler, J.L. and Miller, R.A. (2000) Texaphyrins: new drugs with diverse clinical applications in radiation and photodynamic therapy, *Biochem. Pharmacol.* **59**, 733-739.
48. Hanna, T.A., Liu, L., Angeles-Boza, A.M., Kou, X., Gutsche, C.D., Ejsmont, K., Watson, W.H., Zakharov, L.N., Incarvito, C.D. and Rheingold, A.L. (2003) Synthesis, structures, and conformational characteristics of calixarene monoanions and dianions, *J. Am. Chem. Soc.* **125**, 6228-6238.
49. Wang, J. and Gutsche, C.D. (2002) Tricalixarenes and pentacalixarenes: synthesis and complexation studies, *J. Org. Chem.* **67**, 4423-4429.
50. Sanders, J.K.M. (2004) Self-assembly using dynamic combinatorial chemistry, *Philosoph. Trans. Royal Soc. London, Series A: Math., Phys., Eng. Sci.* **362**, 1239-1245.
51. Brisig, B., Sanders, J.K.M. and Otto, S. (2003) Selection and amplification of a catalyst from a dynamic combinatorial library, *Angew. Chem., Int. Ed.* **42**, 1270-1273.
52. Johnstone, K.D., Bampos, N., Sanders, J.K.M. and Gunter, M.J. (2003) A self-assembling polymer-bound rotaxane under thermodynamic control, *Chem. Commun.*, 1396-1397.
53. Corbett, P.T., Otto, S. and Sanders, J.K.M. (2004) Correlation between host-guest binding and host amplification in simulated dynamic combinatorial libraries, *Chem. - A Eur. J.* **10**, 3139-3143.
54. Vignon, S.A., Jarrosson, T., Iijima, T., Tseng, H., Sanders, J.K.M. and Stoddart, J.F. (2004) Switchable neutral bistable rotaxanes, *J. Am. Chem. Soc.* **126**, 9884-9885.
55. Tominaga, M., Kusukawa, T., Sakamoto, S., Yamaguchi, K. and Fujita, M. (2004) Complementary multicomplexation of desymmetrized 2,2"-bipyridine ligands on square planar Pd(II) centers, *Chem. Lett.* **33**, 794-795.
56. Hori, A., Kataoka, H., Akasaka, A., Okano, T. and Fujita, M. (2003) Reversible catenation of coordination rings with Pd(II) and/or Pt(II) metal centers: chirality induction through catenation, *J. Polymer Sci., Part A: Polymer Chem.* **41**, 3478-3485.
57. Tominaga, M., Kato, M., Okano, T., Sakamoto, S., Yamaguchi, K. and Fujita, M. (2003) Stabilization of a self-assembled coordination nanotube by covalent link, *Chem. Lett.* **32**, 1012-1013.
58. Chand, D.K., Fujita, M., Biradha, K., Sakamoto, S. and Yamaguchi, K. (2003) Metal driven self-assembly of pyridine appended ligands with cis protected/naked Pd(II) ion: a comparative study, *J. Chem. Soc., Dalton Trans.* 2750-2756.
59 Yoshizawa, M. and Fujita, M. (2004) Cavity-directed syntheses within self-assembled nano-space, *Yuki Gosei Kagaku Kyokaishi* **62**, 416-423.
60. Izatt, R.M., Bradshaw, J.S., Bruening, R.L. and Bruening, M.L. (1994) Solid phase extraction of ions of analytical interest using molecular recognition technology, *Am. Lab.*, **26**, no. 18, 28c-28m.
61. Izatt, N.E., Bruening, R.L., Krakowiak, K.E. and Izatt, S.R. (2000) Contributions of Professor Reed M. Izatt to molecular recognition technology: from laboratory to commercial application, *Ind. Eng. Chem. Res.* **39**, 3405-3411.

62. Goken, G.L., Bruening, R.L., Krakowiak, K.E. and Izatt, R.M. (1999) Metal- ion separations using SuperLig or AnaLig materials encased in Empore cartridges and disks; in Bond, A.H.; Dietz, M.L.; Rogers, R.D., (eds.), *Metal-Ion Separation and Preconcentration; Progress and Opportunities*; ACS Symposium Series 716; American Chemical Society: Washington, D.C., Chapter 17, pp 251-259.
63. Nobel Lectures, Chemistry 1981-1990 (1992) Malmström, B.G. (ed.), World Scientific, London, UK, pp. 411-511.
64. J. Incl. Phenom. Mol. Recognit. Chem. (1992) **12**, 1-395. The Pedersen memorial issue, Izatt, R.M. and Bradshaw, J.S. (eds.)

2.
FROM FUNCTIONALISED CATENANES, ROTAXANES AND KNOTS TO HIGHER INTERTWINED ASSEMBLIES

OLEG LUKIN AND FRITZ VÖGTLE
Kekulé-Institut für Organische Chemie und Biochemie der Rheinischen Friedrich-Wilhelms-Universität Bonn, Gerhard-Domagk-Str. 1, 53121 Bonn, Germany

1. Introduction

Remarkable progress in carrying out diverse macrocyclisation reactions in 1950th prompted chemists to attempt threading of molecular chains through the macrorings to get unprecedented by then interlocked species. Initial synthetic approaches to such threading were of purely statistical character. For example, the first interlocked rings reported in 1960 by Wasserman [1] resulted from statistical threading of a macrocyclic hydrocarbon with linear 1,ω-diester with following macrocyclisation of the latter by means of acyloin condensation. Therefore, the first [2]catenane **1** was isolated in about 1% yield. Despite the small yield, the synthesis of the first catenane introduced a highly important concept of mechanical bonding in chemistry and defined the field of 'chemical topology' [2, 3]. Research in the field of interlocked structures was significantly boosted in 1967 when the discovery of catenated DNA was reported [4].

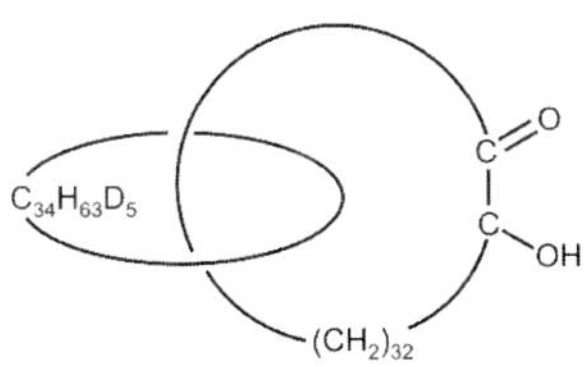

1

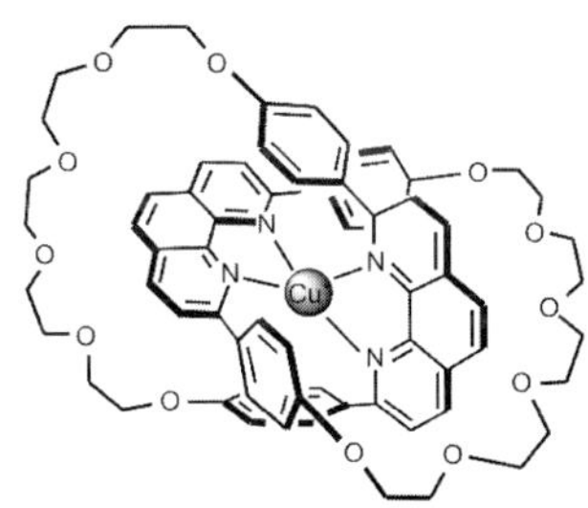

2

Preparation of the first rotaxanes was also accomplished by means of statistical threading of a chain through a cycle with following attaching of bulky substituents to the loose ends of the threaded chain [5]. Statistical means of molecular threading were pushed out by much more efficient covalent templation procedures developed by Schill et al. [6, 7].

Karsten Gloe (ed.), Macrocyclic Chemistry: Current Trends and Future Perspectives, 15–36.
© 2005 *Springer. Printed in the Netherlands.*

Since late 1960th the Schill group reported a number of more complex interlocked structures such as [3]catenanes [7, 8]. Development of supramolecular templation techniques [9], which exploit non-covalent interactions to preorganise reacting species in a necessary manner, made a breakthrough in the field of mechanically interlocked assemblies. For instance, the Sauvage group used for the first time copper(I) ions as an external template in a high-yield assembly of a bis-phenanthroline [2]catenane **2** [3, 10]. This success was followed by numerous reports on a variety of interlocked structures of Stoddart and coworkers [11] who made use of complexes between neutral π-donor and positively charged π-acceptor units involved in macrocyclic architectures, e.g., one shown in Figure 1.

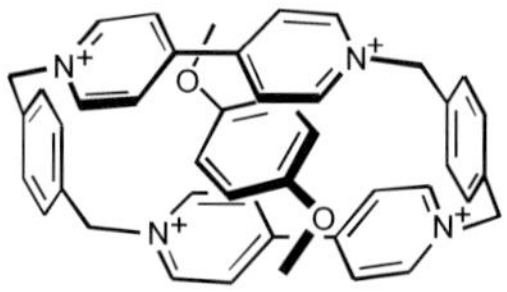

Figure 1. Threading of 1,4-dimethoxybenzene (π-donor) through a tetraquat macrocycle (π-acceptor).

The successful utilization of hydrogen-bonding in cyclic oligo-amides led groups of Hunter, [12] Vögtle [13] (both 1992) and Leigh [14] (1995) to the discovery of high-yielding syntheses of amide-based catenanes **3-5** depicted in Figure 2.

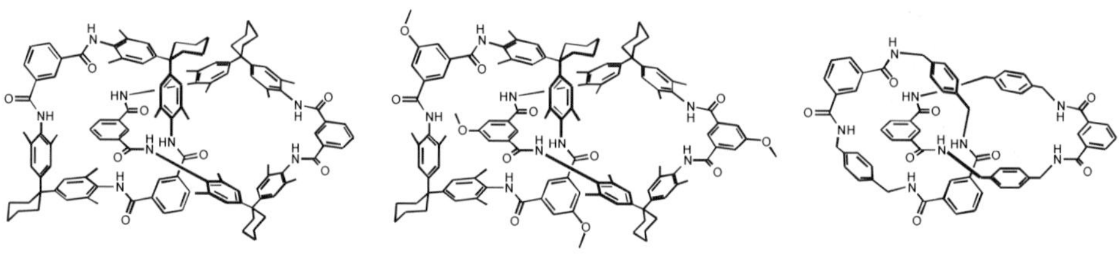

3 **4** **5**

Figure 2. Amide-based catenanes prepared by Hunter (catenane 3), Vögtle et al. (catenane 4), and Leigh et al. (catenane 5).

Further investigations of the supramolecular templation procedures led to preparation of even more complex topologies involving [4]-, [5]- and [6]catenanes [15, 16], doubly intertwined [2]catenane [17], [n]rotaxanes [18], and finally molecular knots **6-9** (Figure 3) [19-22]. Significant achievements in the synthesis of diverse intertwined species stimulated an increase in communication among chemists of diverse sub-disciplines as well as biochemists and mathematicians [23-25]. For instance, the growing number of molecular catenanes, rotaxanes and knots has required a revision of certain aspects of stereochemistry, especially with regard to the definition of chirality [26, 27]. Topologically chiral catenanes and knots are testing the performance of the latest chiral stationary phases [28, 29], while molecular switches [30] made up on the basis of catenanes and rotaxanes are currently of high interest for molecular electronics, photonics and nanotechnology. Furthermore, polymerised catenanes and rotaxanes have intrigued polymer researchers [31]. It is now well recognised that functions of intertwined species and their potential applications are dependent on the possibilities of their specific functionalisation.

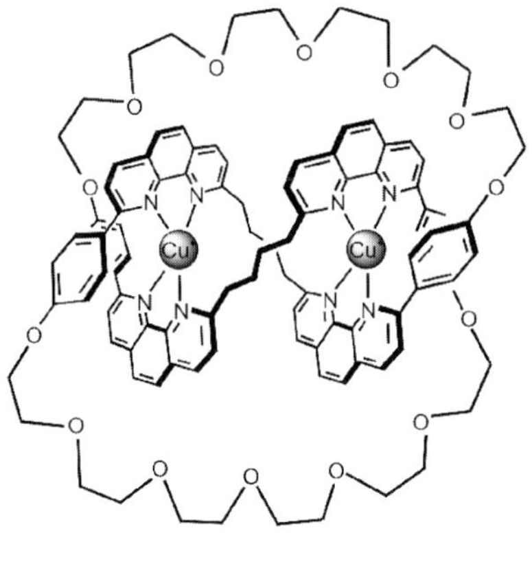

6

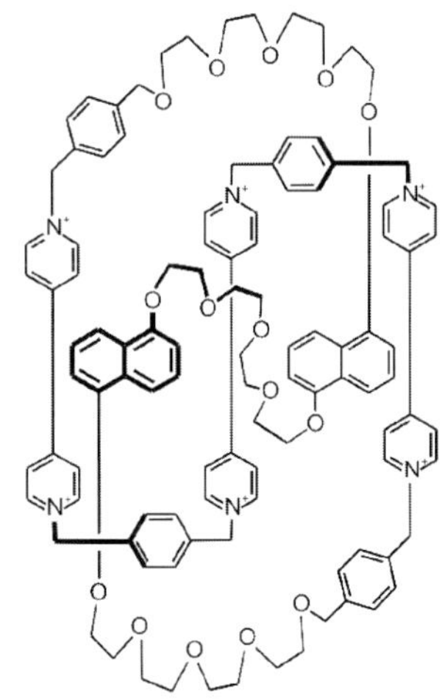

7

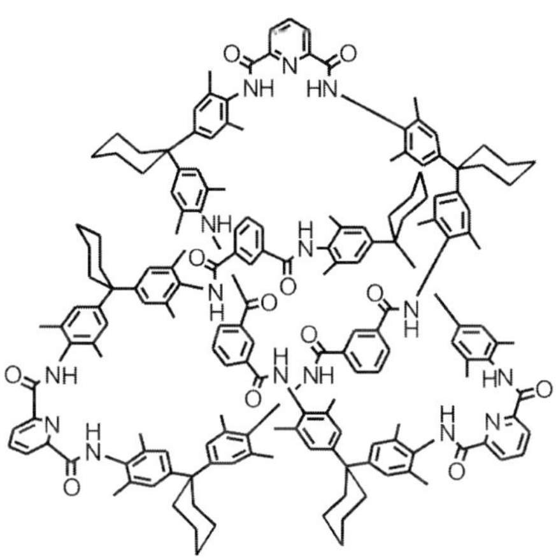

8

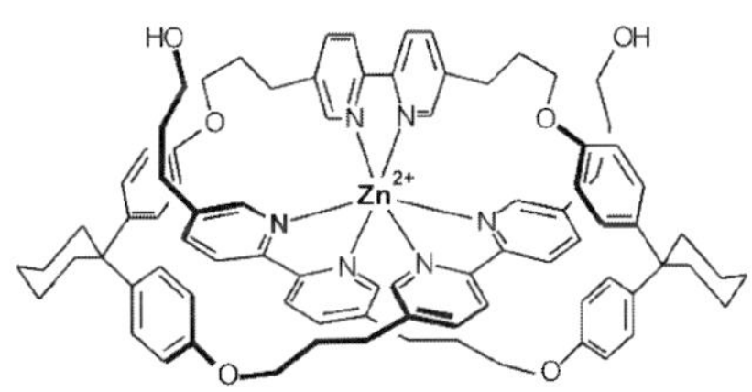

9

Figure 3. Molecular knots (knotanes) prepared by Sauvage et al. (knotane 6), Stoddart et al. (knotane 7), Vögtle et al. (knotane 8) and Hunter et al. (open knot 9).

The functionalisation also allows extending the complexity of intertwined molecular assemblies involving molecular catenanes, rotaxanes and knots. Elaborate interlocked assemblies constructed by means of metal-templation techniques and π-π-stacking preorganisation were reviewed [3, 11]. Our last survey was devoted to the hydrogen bond templated synthesis of amide-based catenanes and rotaxanes [32]. Since then a considerable advancement in elucidation of mechanisms of templation and derivatisation of the amide-based interlocked structures has been reached. Moreover, in 2000 we reported a one pot synthesis of amide-based knots such as **8** [21], which is so far the easiest preparation of molecular knots. In the following, specific possibilities of functionalisation of amide-based catenanes, rotaxanes and knots will be discussed.

2. Template Synthesis of Amide-based Catenanes, Rotaxanes and Knots

Although the hydrogen-bond templated preparation of intertwined species of the amide-type has already been reviewed [32], it seems essential for this survey to outline the template technique that is used for the preparation of amide-based catenanes and rotaxanes as well as its limits of application recognised by very recent efforts of the Schalley group [33-36]. As mentioned in the preceding section, amide-based intertwined species involving catenanes, rotaxanes and knots are obtained in a reaction that is templated by numerous complementary hydrogen bonds between reacting species. Figure 4 illustrates the principle of the hydrogen-bond templated preparation that leads to catenanes and rotaxanes of the amide-type.

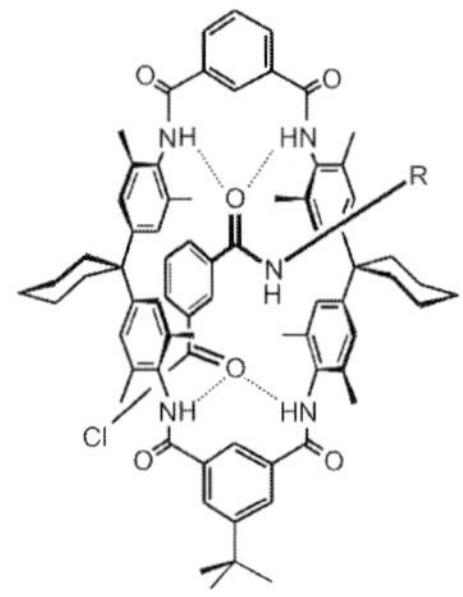

Figure 4. Hydrogen bond templated threading of a tetraamide ring in synthesis of amide-based catenanes and rotaxanes.

The template synthesis of amide-based catenanes and rotaxanes can be combined with iterative approach of interlocked chain growth or multiple ring threading. This iterative template strategy led to unprecedented amide-based oligocatenanes and [n]rotaxanes [16, 18]. Unlike the well understood templation procedures for the preparation of the latter topologies, the mechanism of formation of amide-based molecular knots (*knotanes*) [21] still requires explanation. From available single crystal X-ray structures of amide-knotanes (Figure 5) [21, 37] one can conclude that, most probably, the specific folding of the linear precursor **10** is programmed by the pattern of hydrogen bonds which should be strong in a non-competitive solvent such as dichloromethane.

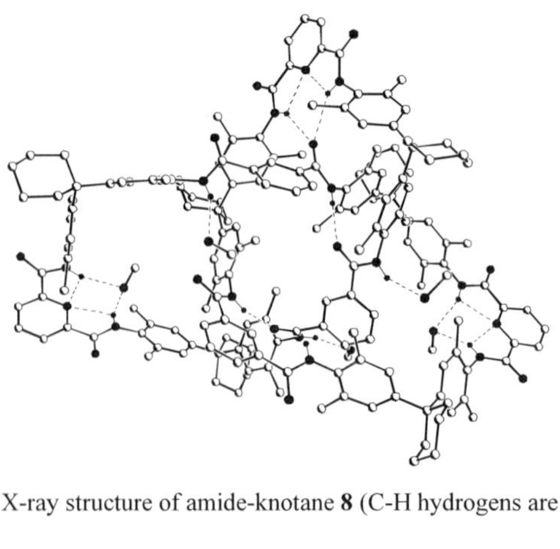

Figure 5. X-ray structure of amide-knotane **8** (C-H hydrogens are omitted).

Figure 6 shows schematically our suggested mechanism of folding-and-knotting of the linear diamine **10**. Although our synthetic [29, 38] and theoretical studies [39] support the latter conclusion, the analysis of folding of the linear precursor **10** has to be carried out in order to verify the 'knotting' mechanism.

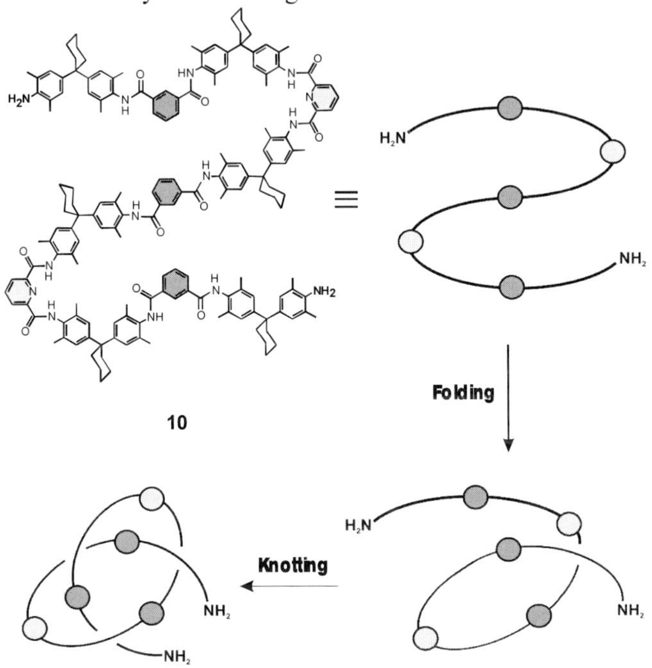

Figure 6. Proposed mechanism of the amide-knotane formation.

In 1999 we developed another template technique of threading called 'trapping' [40]. As outlined in Figure 7, it consists in the efficient binding of a phenolate anion by tetraamide macrorings (the anion is "trapped" in the macrocyclic cavity) followed by its through-ring nucleophilic attack on benzyl bromide leading to a rotaxane formation. This method attains up to 95% yield and is so far the most efficient procedure for the preparation of rotaxanes.

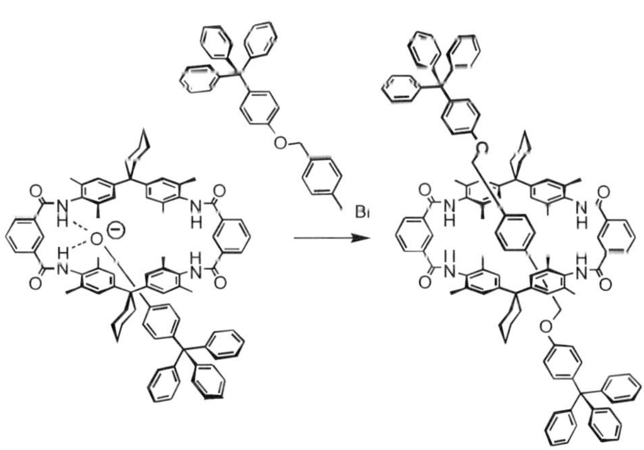

Figure 7. An anion-assisted assembly of rotaxanes - "trapping" method.

The procedure of the anion-templated synthesis was later used for the preparation of various rotaxanes by the groups of Smith [41] and Schalley [34-36]. The anion-templation technique was carefully tested by Schalley et al. Altering the length of the axle-center pieces and the size of stopper groups they synthesized a variety of rotaxanes and analysed their yields and de-slipping rates [34, 35].

It was found that sometimes subtle changes in the axle length and the stopper size affect significantly the rotaxane yield. While larger trityl phenol stoppers give high yields of rotaxanes (40–95% depending on the axle centre piece), smaller di-*tert*-butyl phenol stoppers generate the corresponding rotaxanes in much lower yields (with some centre pieces even < 5%). The most spectacular demonstration of such subtle effects was a 10% increased de-slipping rate of the rotaxane **11a** bearing fully deuterated *t*-Bu groups compared to its non-deuterated analogue **11b** [36]. Moreover, the Schalley group developed a novel anion-templated technique of rotaxane synthesis ionising 2,6-substituted phenol bearing two terminal amine groups [42]. Acylation of the latter with bulky acyl chlorides gives rise to rotaxanes such as **12**. An interesting feature of **12** is a high chemical inertness of the hydroxyl group of the axle as it is hidden inside the cavity of the ring.

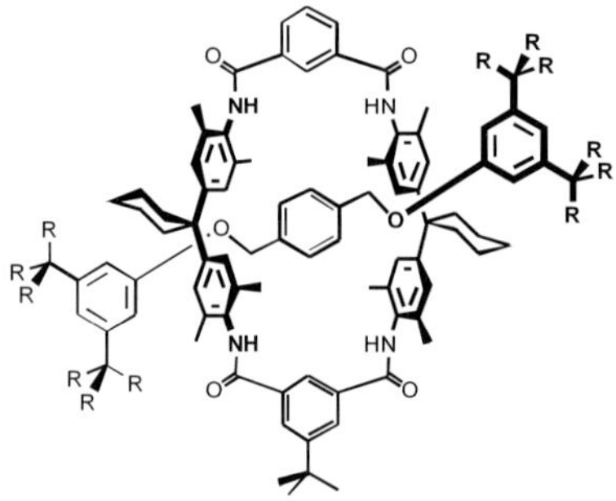

11a: R = CD_3

11b: R = CH_3

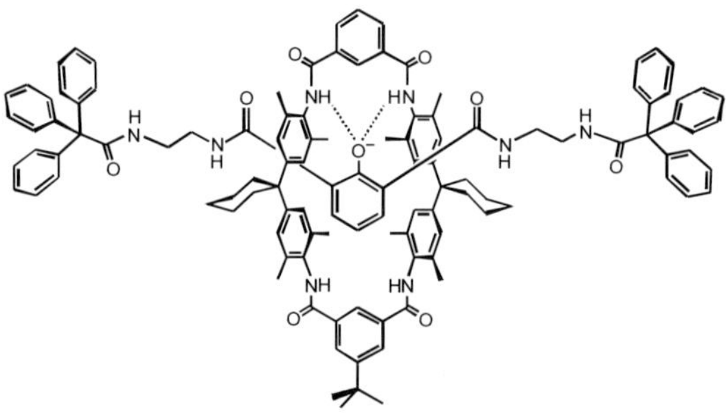

12

The advantages of all above amide-based assemblies include the electro-neutral character of the compounds, their relative inertness and, last but not least, possibility of their further functionalisation. The latter can be performed both by wise selection of reagents prior to the assembly of an intertwined compound and by its post-assembly derivatisation or by combination of these two methods. The next section discusses the state-of-the-art in derivatising of the amide-based topologies.

3. Functionalisation of the Amide-based Catenanes, Rotaxanes and Knots

The functionalisation of interlocked structures is highly desirable (i) to improve and to control the solubility, which is important if chiral HPLC resolution is necessary, (ii) to study the influence of the topological chirality of catenanes, rotaxanes and knots combining them with already known functional compounds, and (iii) to prepare higher interlocked assemblies introducing new bridges between topological [43] and Euclidean [44] stereochemistries. First functionalised catenanes and rotaxanes of the amide-type were obtained via integration of one sulfonamide unit into their wheel and/or axle components [45].

As shown in Figure 8, sulfonamide centers render catenanes such as **13** and rotaxanes such as **14** chiral, because these groups introduce the directionality into both constituent parts.

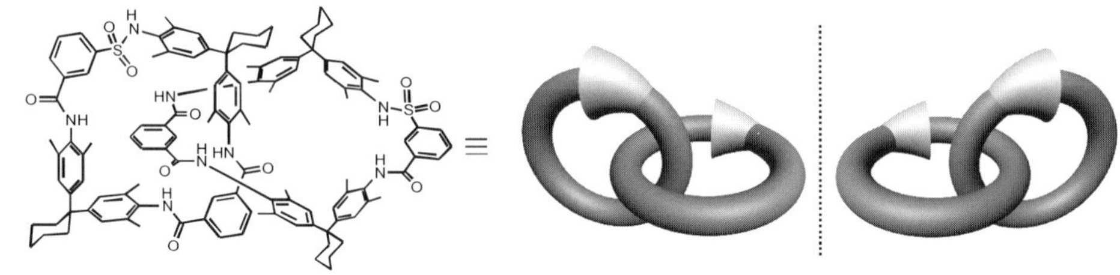

13

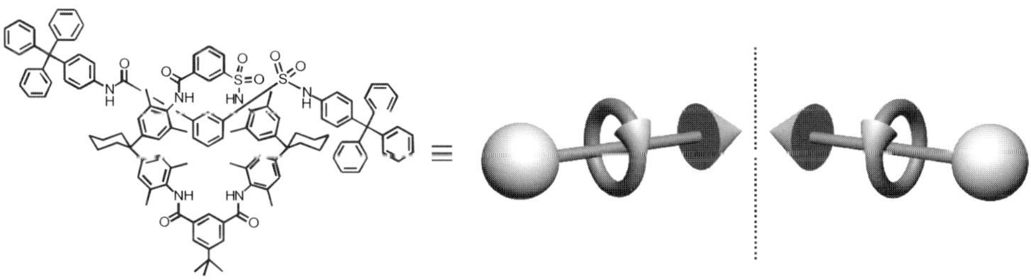

14

Figure 8. Chirality of sulfonamide catenanes and rotaxanes.

In addition, sulfonamides bear more acidic, compared to carboxamides, protons that can be selectively *N*-alkylated or *N*-acylated. Thus, alkylation of sulfonamide catenane and rotaxane with Frechet-dendrons of 2nd generation led to the first representatives of the dendrocatenane **15** and dendrorotaxane **16** [46], respectively. These compounds were of especial interest, since they allowed for the first time to study chiral induction of topologically chiral cores on appended dendrons and to compare to the analogous centrochiral dendrimers for which phenomenon of the crypto-optical activity was postulated [47].

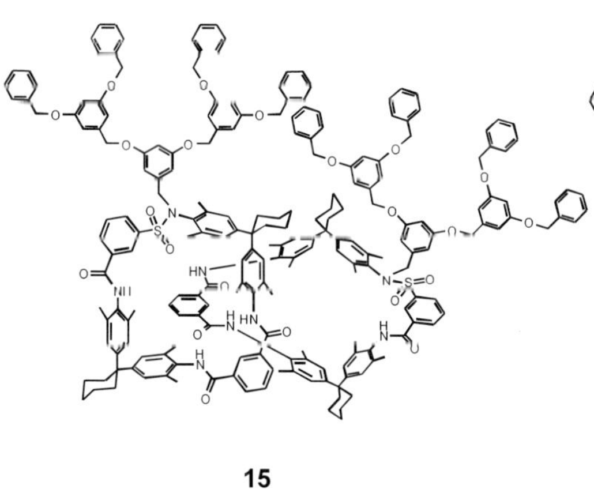

15

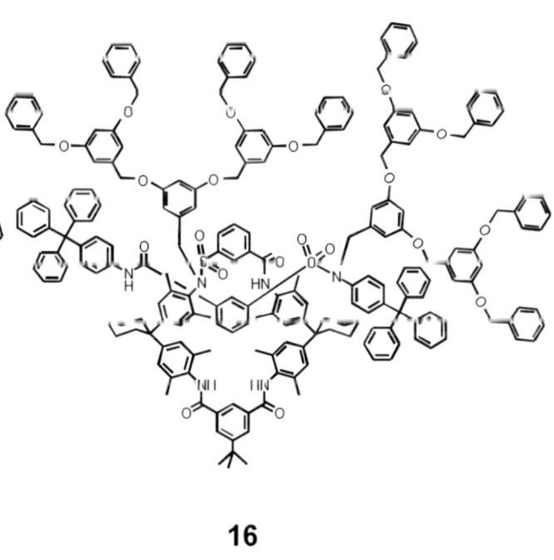

16

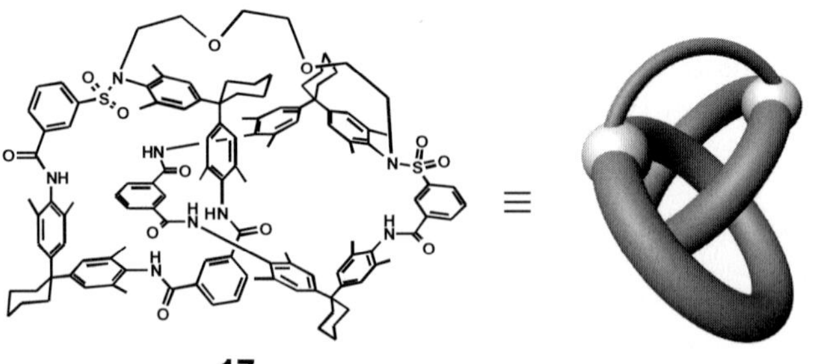

17

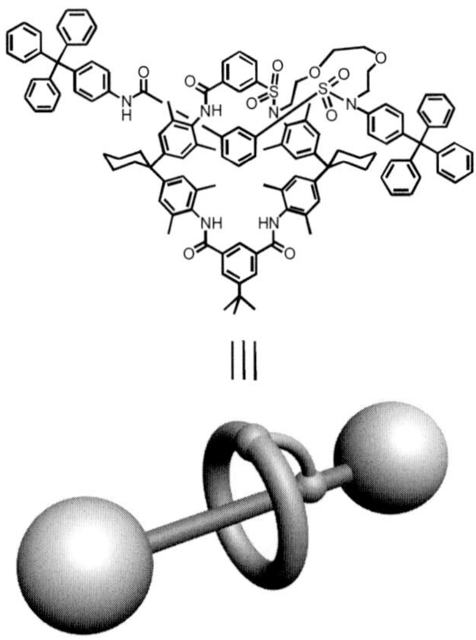

18

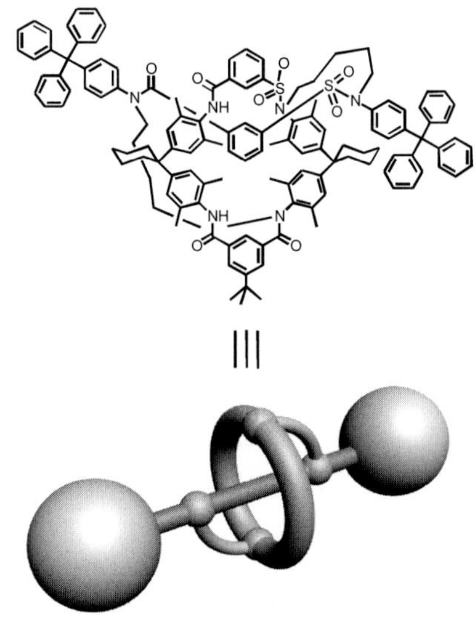

19

Mechanically connected components of sulfonamide catenanes and rotaxanes can also be connected in an intramolecular manner by linking their sulfonamide moieties and producing the first pretzel-shaped compound **17** [48] and [1]rotaxane **18** [49], respectively. Interestingly, the [1]rotaxane can be bridged intramolecularly giving rise to a 'molecular 8' compound **19** [50].

Functionalisation of amide-knotanes, such as **8**, is rather different. It is very difficult to obtain and, even impossible, to purify knotanes containing one sulfonamide unit [51]. Additionally, all amide units in knotanes are remarkably deactivated on account of their highly entangled structure [38]. Nevertheless, it turned out that amide-knotanes can easily be functionalised at 4-positions of 2,6-pyridine dicarboxamide units constituting their three peripheral edges. This makes amide-knotanes perfect, readily available nano-sized scaffolds which can be modified in different ways [29, 38]. We were able to show that amide knotanes could be equipped with various small substituents at isophthaloyl and 2,6-pyridine dicarboxamide fragments [29] by means of *direct* selection of suitably substituted 2,6-pyridindicarboxylic acid dichloride and elongated diamine prior the knot assembly. Therefore, we termed the latter method 'direct' [52] approach to functionalised knotanes. Despite the value of its one-pot procedure, the method of direct functionalisation of amide-knotanes has significant drawbacks. Firstly, large substituents are not tolerated and second, no rational selective derivatisation is possible.

These limitations prompted us to look for other possible ways of the amide-knotane functionalisation. The seminal concept that we call 'indirect' approach [52] to derivatisation of amide-knotanes has been to take advantage of the protecting group chemistry at 4-positions of their 2,6-pyridine dicarboxamide units. The first synthesis utilizing the indirect approach was the complete and partial deprotection of the tris(benzyloxy)knotane **20** [52] via Pd-catalyzed hydration followed by the alkylation with Fréchet-type dendrons. The isolated dendronised knotanes **21-24** (or simply *dendroknots*) which bear one to three dendritic wedges constituted the first examples of *selectively* derivatised knotanes. The main disadvantage of the latter synthesis is the step of deprotection that does not proceed completely giving rise to a mixture of mono-, di-, and trihydroxyknotanes that could not be separated.

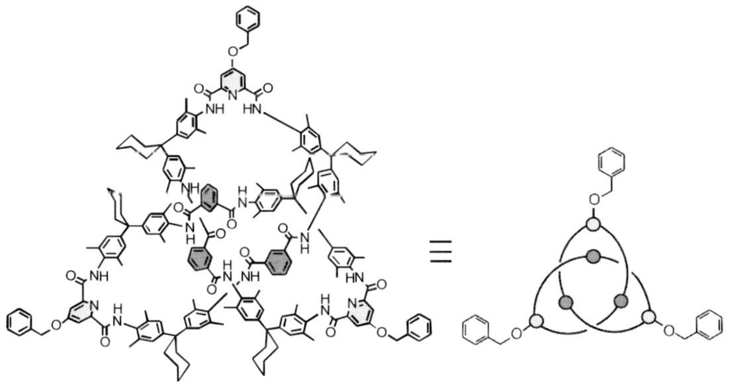

20

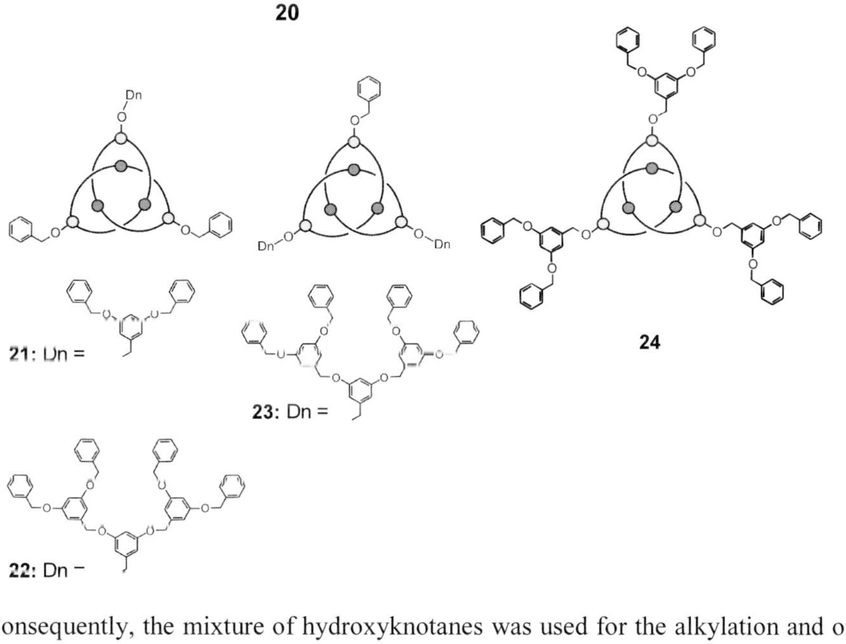

Consequently, the mixture of hydroxyknotanes was used for the alkylation and only the resulting mixtures of dendroknots were later resolved on HPLC. The difficulty can be overcome by selecting a better protecting group whose expected reactivity will neither affect the amide-knotane synthesis nor its further derivatisation. 4-allyloxypyridine 2,6-dicarboxylic acid dichloride seemed to be a suitable protected building block for this purpose. All three allyl-groups can be completely or selectively removed from the periphery of the tris(allyloxy)knotane **25** with the aid of tributyltin hydride and a Pd-catalyst resulting in the corresponding tri-, di-, and mono-hydroxyknotanes **26-28** [37, 38]. Unlike the above case with the tris(benzyloxy)knotane **20** [52], the latter synthesis proceeds smoothly and cleanly, yielding the products that can be purified by means of conventional column chromatography on a gram-scale.

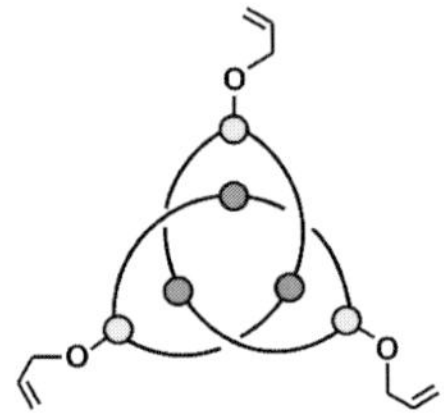

25

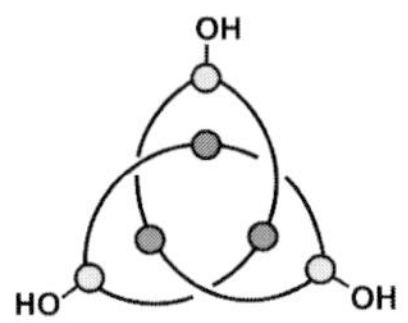

26

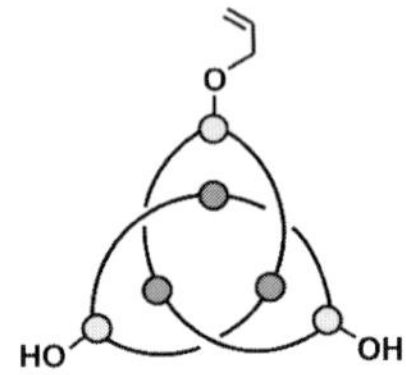

27

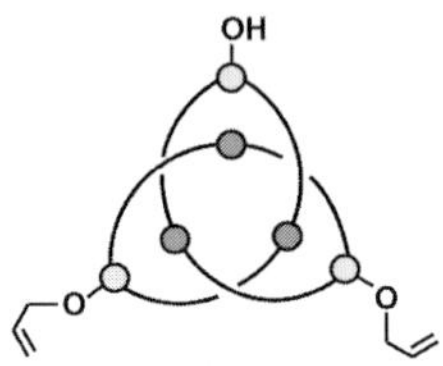

28

Having developed a simple and reliable synthetic strategy for the complete and selective removal of the outer protecting groups we have topologically chiral building blocks with unprecedented reactivity for a whole array of conceivable synthetic transformation. Subtle tuning and monitoring the properties of the amide-knotanes can now be carried out in a rational manner.

The hydroxyknotanes **26-28** can further be e.g., acylated with diethylchlorophosphate giving rise to tri-, di-, and monophosphorylated knotanes [38]. Sulfonylation of the bis(allyloxy)hydroxyknotane **27** with *p*-toluenesulfonyl chloride in the presence of triethylamine in acetonitrile gives rise to monosulfonate **29** in 95% yield. Allyl groups, in turn, can be completely or selectively removed from the periphery of **29** resulting in dihydroxy- and monohydroxy-knotanes **30** and **31**, respectively. The preparation of the knotane **31** with three different peripheral substituents constitutes a remarkable synthetic breakthrough, because it allows for the preparation of amide-knotanes with a wide range of substitution patterns and affords exceptional opportunities for further synthetic variations.

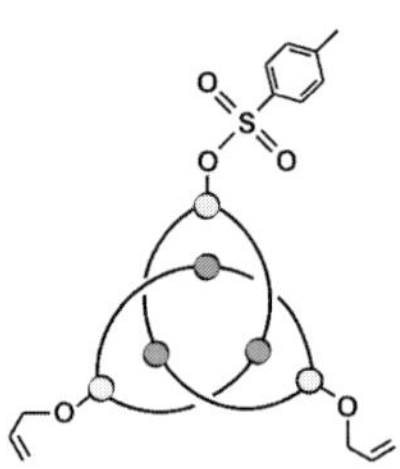

29

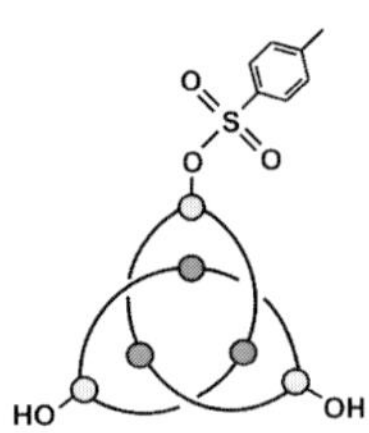

30

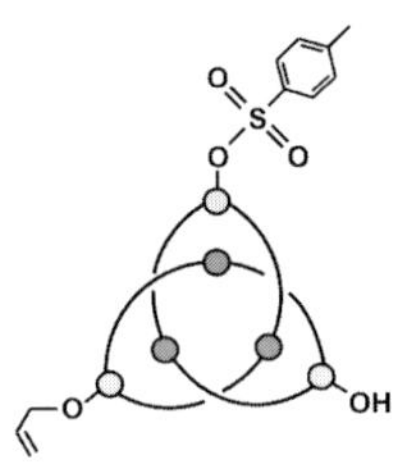

31

4. Construction of Assemblies of Amide Catenanes, Rotaxanes and Knots

In the introduction we mentioned extravagant interlocked structures of higher complexity such as doubly intertwined catenane and molecular composite knots of Sauvage et al. and multicatenanes made up of 4 to 7 interlocked rings obtained by Stoddart et al. In this section, we will discuss assemblies made up of amide-based catenanes, rotaxanes and knots. Here we use the term ‘assembly’ to describe covalent or non-covalent molecular architectures in which interlocked and intertwined compounds are used as subunits. We tend to apply this term to well-defined mono-disperse species, aside from known polymerised catenanes and rotaxanes [31].

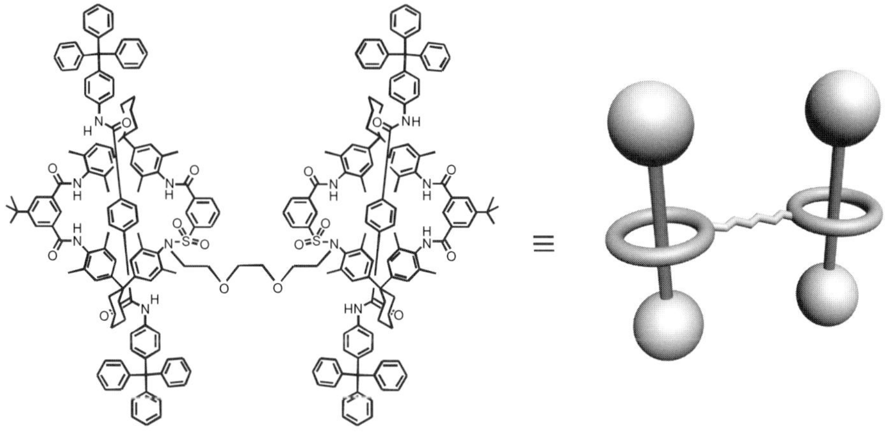

32

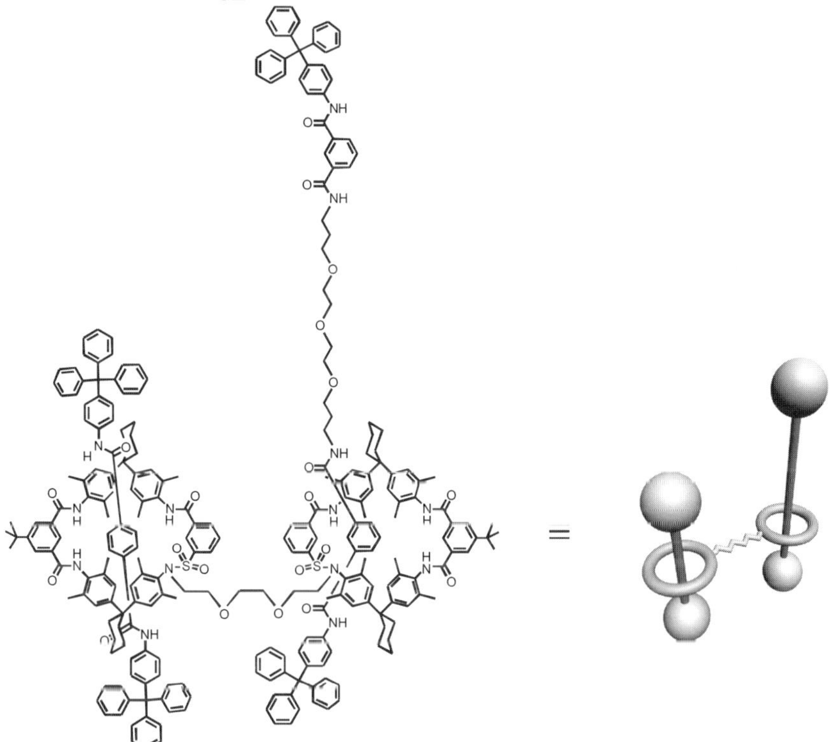

33

In 1996 we reported on the first covalent dimerization of the [2]rotaxane with sulfonamide function in its wheel component to give a [3]rotaxane **32** (bis[2]rotaxane) [49]. Subsequently, we expanded the concept of covalent bridging and produced more elaborate rotaxane assemblies such as an unsymmetrical bis[2]rotaxane **33**, and **34**, which involves two axle-linked [2]rotaxanes, and the first tris[2]rotaxane **35** possessing a dendritic structure [53]. Rotaxane assemblies with dendritic architecture e.g., **36** were also synthesized in our group taking advantage of both convergent and divergent methods [54].

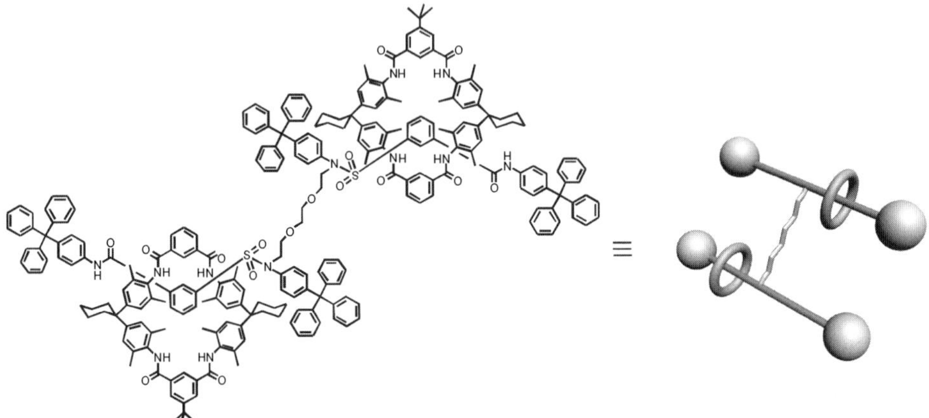

34

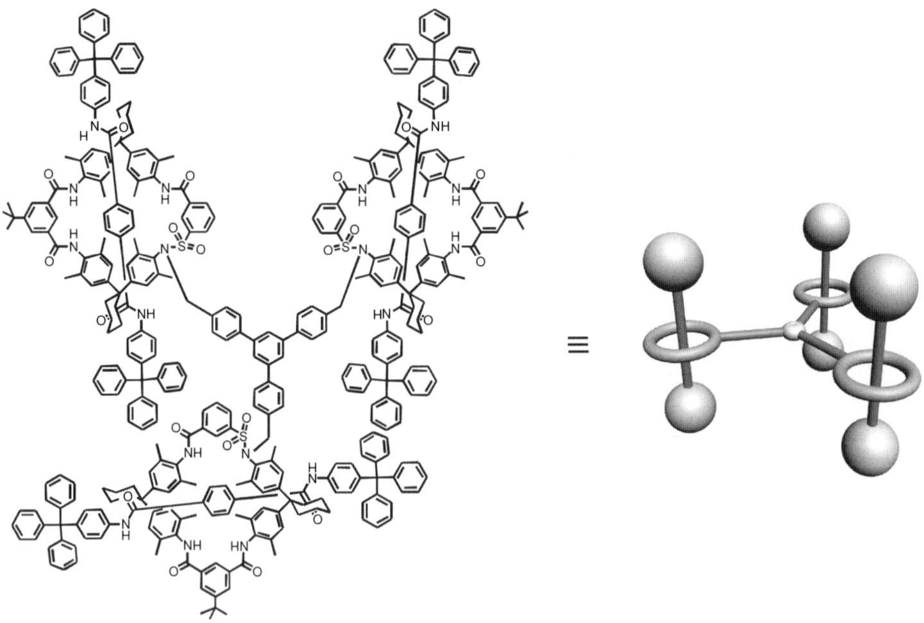

35

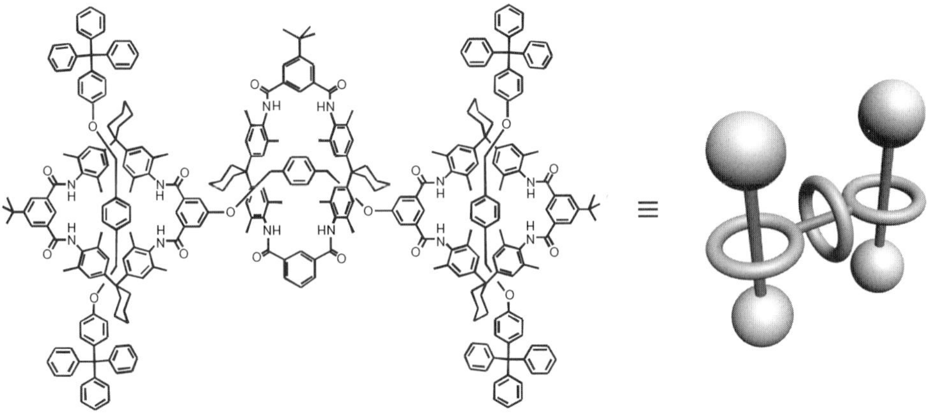

36

These assemblies are unique, since not covalent but mechanical bonds are involved in the branching of dendrimers. We also synthesized an oligomacrocycle **37** containing four units of sulfonamide wheel. Furthermore, a tetrakis(rotaxane) **38** and a bis(pretzelane) **39** [55] were prepared starting from the bis(sulfonamide) catenane and the monosulfonamide rotaxane, respectively. The durene-based spacered tetrabromide was used as the core unit in all three cases. It is noteworthy, the oligomacrocycle **37** and the bis(pretzelane) **39** are *residual topological* isomers [27]. The latter type of isomerism implies the physical impossibility of disentangling of two interlocked moieties of the bis(pretzelane) without bond breaking, while the rules of mathematical topology allow such a disentanglement. Catenanes and knotanes, in turn, belong to the truly topological structures, since both chemical and topological concepts agree about the impossibility of their disentangling.

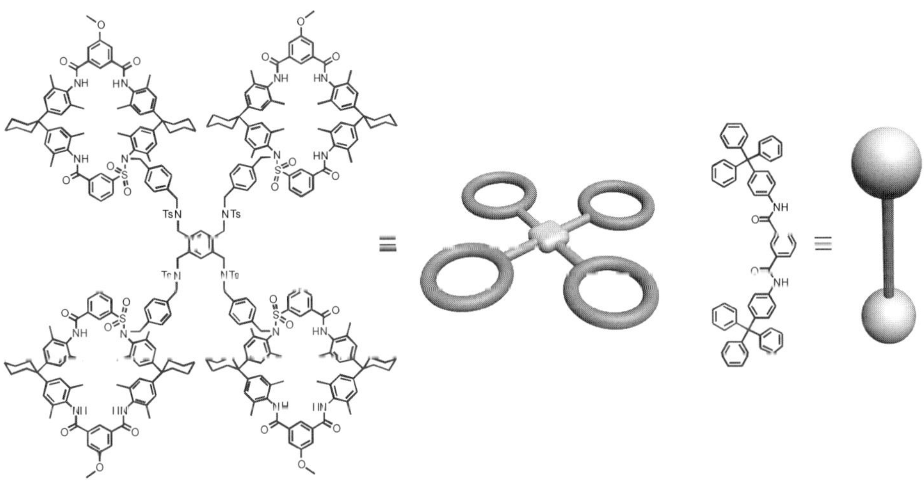

37

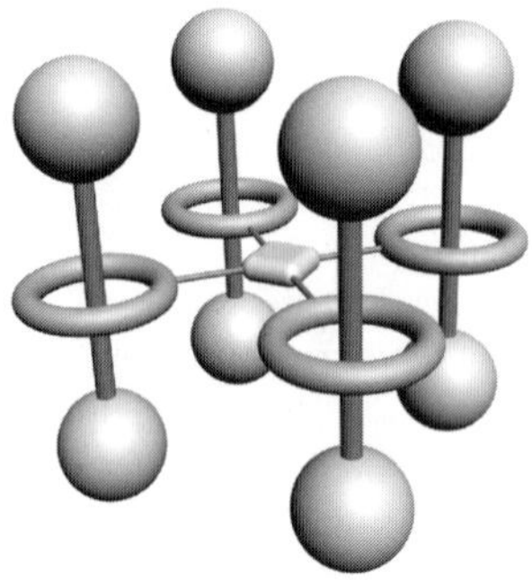

38

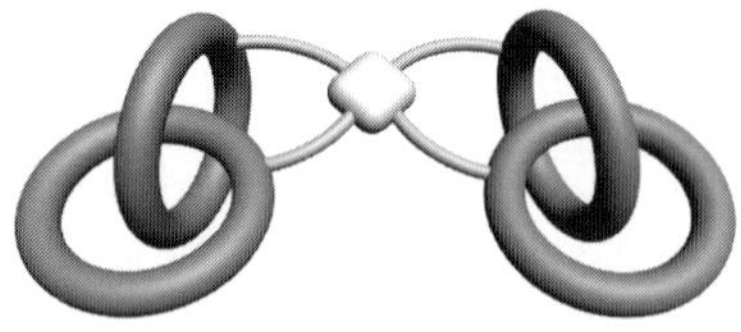

39

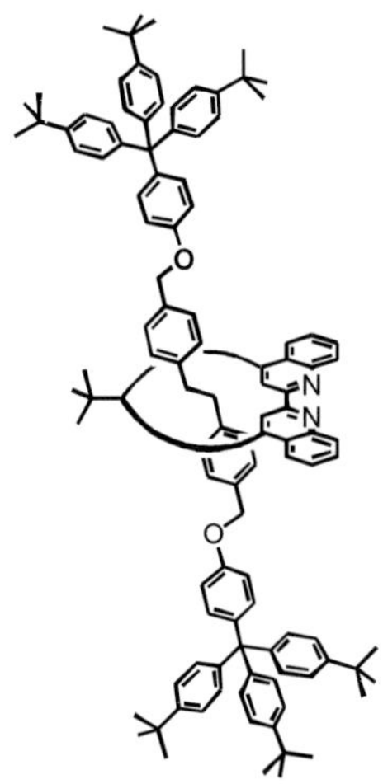

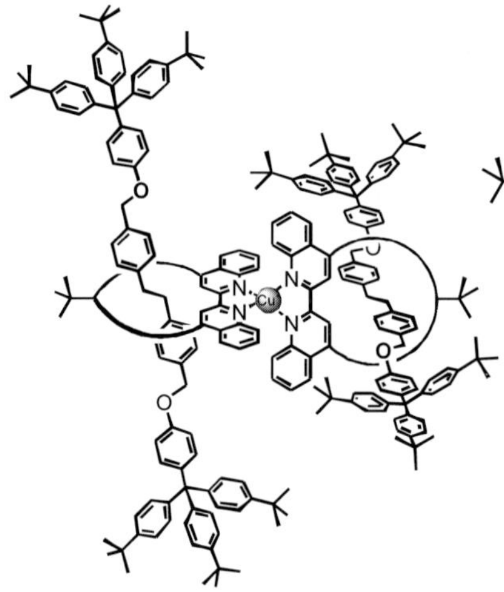

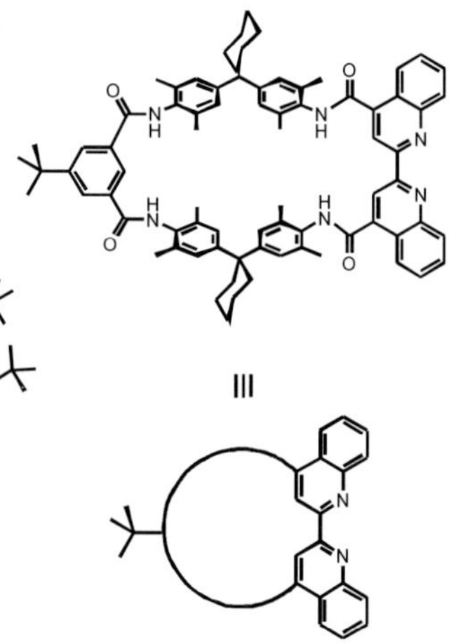

40

41

An original approach to assemblies of rotaxanes was suggested by Schalley et al. [56]. The authors synthesized a rotaxane **40** containing the bis-quinoline unit in its wheel component. Subsequent exo-coordination of copper cations leads to a copper-bridged bis(rotaxane) **41** with perpendicularly oriented axles.

Compared to the above assemblies which are composed of amide-catenanes and rotaxanes, construction of assemblies of molecular knots represents a considerable challenge. Thus, our previous efforts highlighted in the previous section showed that significant achievements were reached in the synthesis and derivatisation of knotanes. As in the case of catenanes and rotaxanes, the preparation of higher covalently linked knotanes implies the availability of selectively functionalised molecular knots e.g., the monohydroxy-knotane **28**. Reaction of **28** with biphenyl-4,4'-disulfonylchloride **42** in the presence of Et_3N yielded covalently linked molecular knots **43** which we termed *topologically chiral molecular dumbbell* [37]. The dumbbell **43** contains two topologically chiral stereogenic units meaning that similarly to centrochiral species with two asymmetric carbon atoms (e.g., tartaric acid) it should consist of two enantiomers and a *meso*-form. Chiral resolution of the dumbbell **43** carried out on a non-commercial Chiralpak AD HPLC column confirmed its expected isomeric composition [37].

The dumbbell **43** is a prerequisite for the construction of more elaborate assemblies of molecular knots which would have more complex isomeric compositions. We decided to use the rotaxane platform in which knotanes would play the role of nano-sized stoppers. The rotaxane concept makes this assembly, which we call "*knotaxane*" [57], a particularly attractive architecture with an option to control the directionality of rotation or shuttling of mechanically linked constituent parts with the aid of topologically chiral knotted stoppers. Therefore, we designed an elongated axle **44** that according to the preliminary molecular modeling can efficiently thread the monosulfonamide macrocycle **45** and prevent noticeable overcrowding of the mechanically bound parts both in the transition state and the final assembly. The reaction of the monohydroxy-knotane **28** with disulfonyl chloride **44** affords the desired *knotaxane* **46** in a yield of 19%. Structure of the knotaxane **46** constitutes the first example of three topological stereogenic units which are held together, both in covalent and an interlocked manner. Isomeric composition of the knotaxane **46** is more complex than that of the dumbbell **43**. The isomerism of knotaxanes resembles that of trihydroxyglutaric acid which contains two enantiomers and two meso-forms. The enantiomeric pair of **46** was successfully separated using, as in the case of **43**, the non-commercial Chiralpak AD HPLC column.

Generally, we have shown that the suggested strategy towards linear knotane assemblies consists in the selective removal of allyl groups with following linking of intermediate hydroxy-compounds with a disulfonyl chloride. Further growth of the knotted backbone can therefore be achieved in an iterative way. Consequently, the selective removal of one allyl group from **43** gives rise to a monohydroxy-dumbbell **47**, which is sulfonylated with 4,4'-biphenyldisulfonyl chloride yielding 55% of linear tetraknotane **48** [58].

This synthetic strategy can be altered for the preparation of branched oligo-knotanes, which necessitate a multifunctional core and monofunctional branching units. Reaction of the monohydroxy-knotane **28** with an excess of biphenyl-4,4'-disulfonyl chloride **42** readily gives a sulfonylated knotane **49** containing one reactive sulfonyl chloride unit. **49** is converted by reaction with the trihydroxy-knotane **26** to yield a branched tetraknotane **50** [58]. As we mentioned above, the isomerism of the topological dumbbell **43** and the knotaxane **46** bears a close analogy with that of open-chain sugar acids and can be described with the aid of Fischer-type schemes. Thus, the chirality designations of the topologically chiral non-symmetrical dumbbell **47** and the linear tetraknotane **48** are reminiscent of the Fischer projections [44] for erythrose/threose and for hexaric acid, respectively. The isomeric composition of the branched tetraknotane **50** is completely unique, because no centrochiral analogues with this constitution can exist. However, the chiral resolution of these new tetraknotanes attempted on chemically bonded Chiralpak AD stationary phases met severe difficulties on account of their highly complex isomeric compositions and a significant overlap of the isomer fractions was observed in most cases.

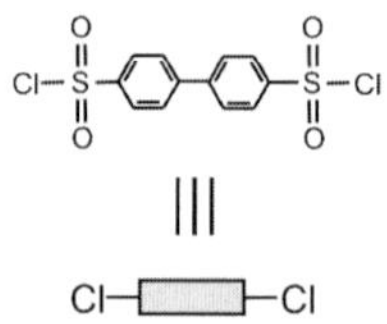

42

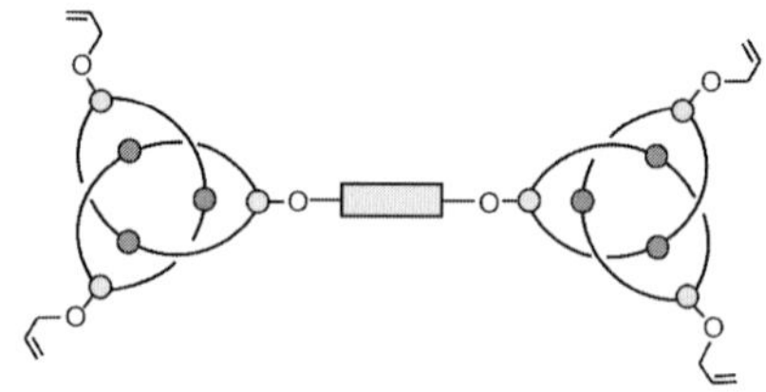

43

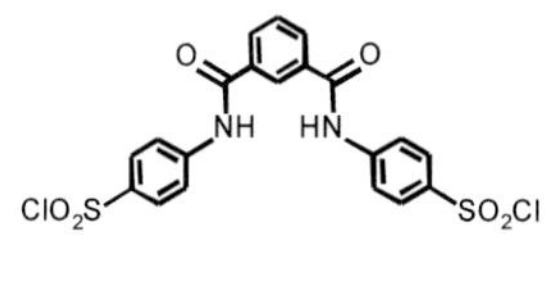

44

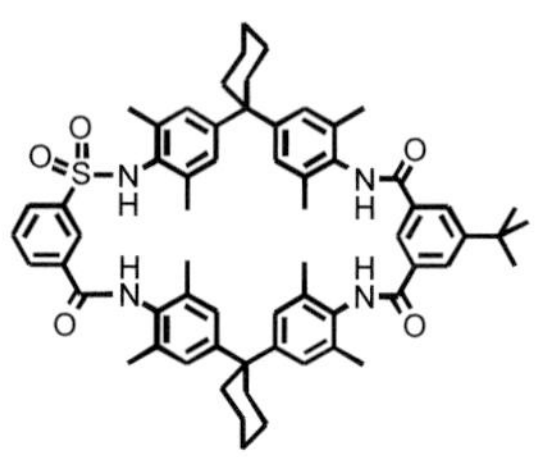

45

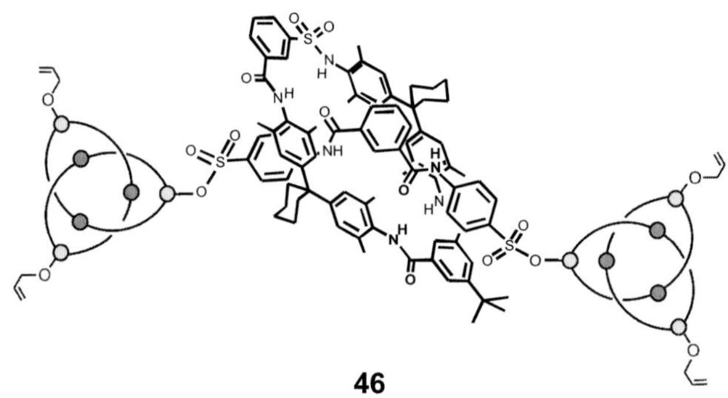

46

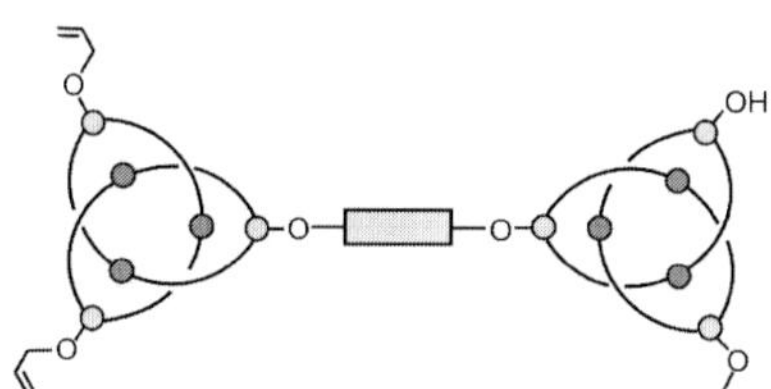

47

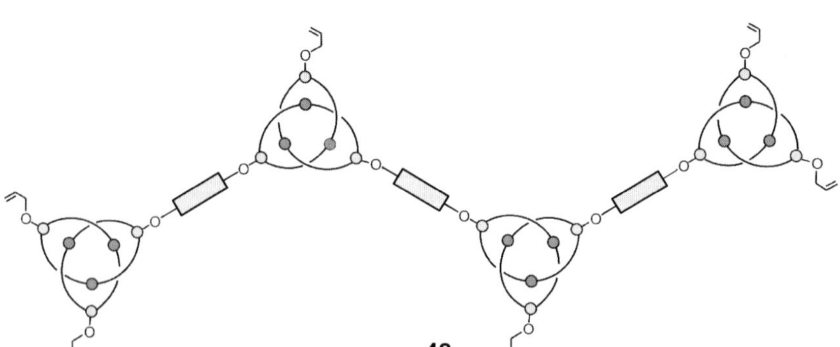

48

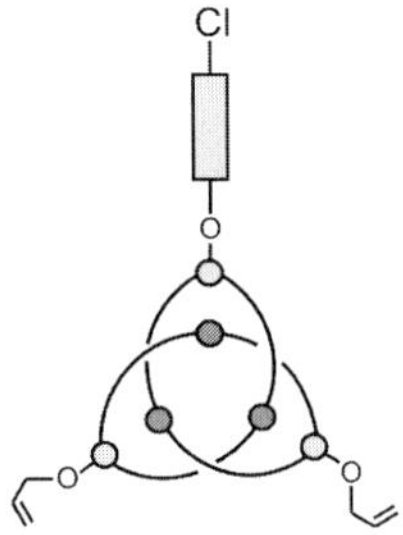

49

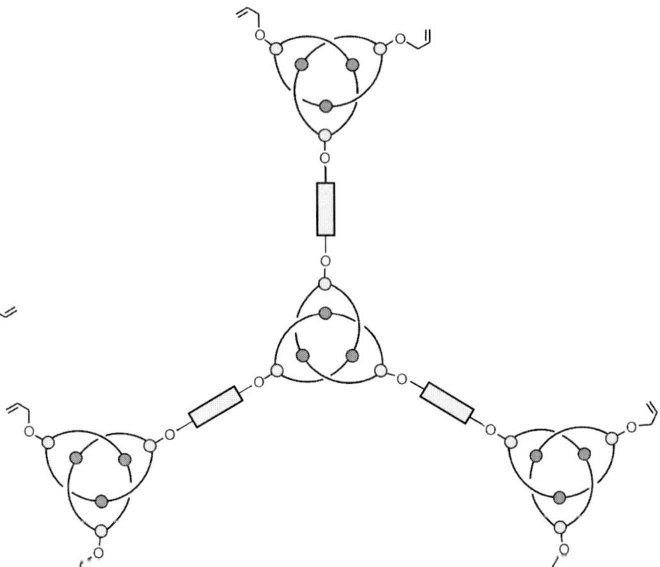

50

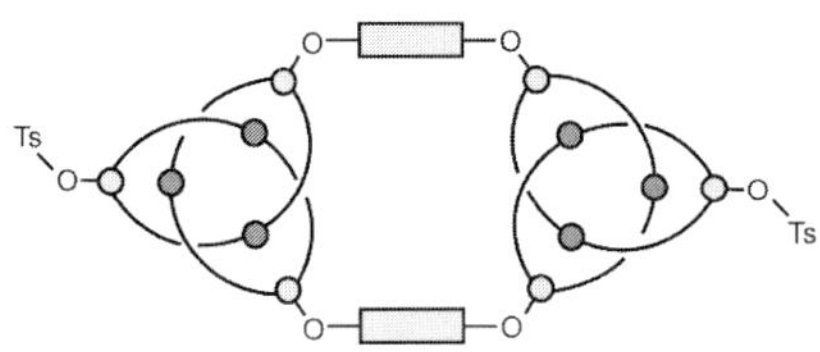

51

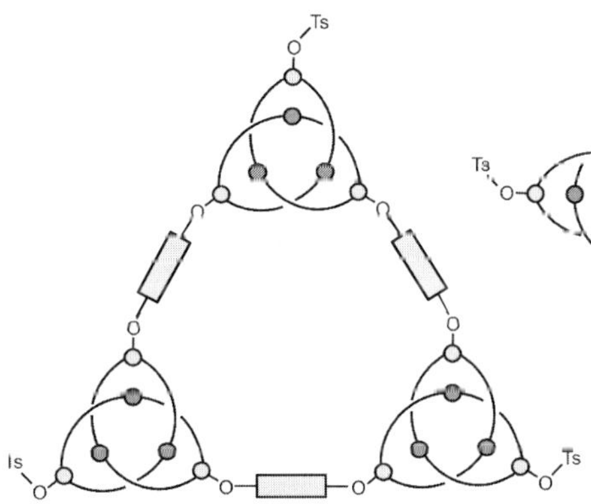

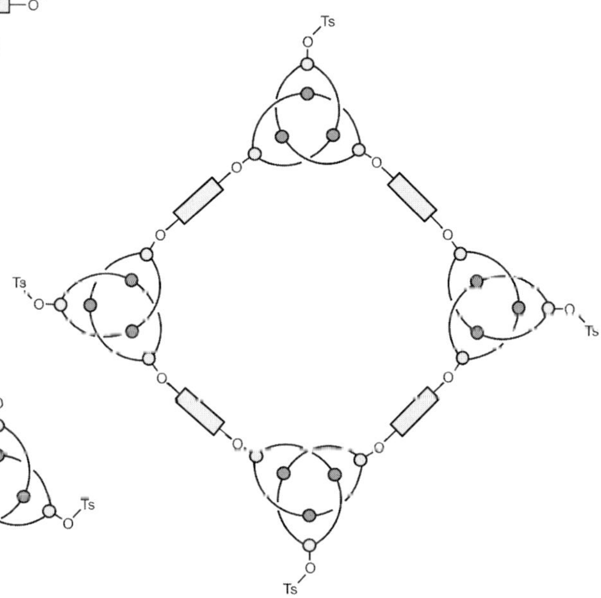

52

53

Finally, we aimed at the preparation of cyclic molecular architectures involving amide-knotanes. The preparation of macrocyclic knotane oligomers implies the availability of a selectively bifunctionalised knotane such as the dihydroxy-knotane **30**. Reaction of **30** with an equivalent amount of 4,4'-biphenyldisulfonyl dichloride **42** under high dilution conditions results in a mixture of the oligomeric macrocycles composed of two (**51**), three (**52**), and four (**53**) amide-knotane moieties in an overall yield of 65% [58]. Following the rules of the cyclophane nomenclature [59], we termed the latter macrocyclic knotane oligomers "*knotanophanes*" [58]. The preparative isolation of the individual compounds **51**–**53** from their mixture was afforded by using a standard silica gel HPLC column. As in the case of the tetraknotanes **48** and **50**, further attempts of chiral resolution of the knotanophanes proved difficult. Only enantiomers of the simplest knotanophane **51**, composed of two doubly bridged knots, were resolved.

5. Conclusions and Outlook

The successful synthesis of the oligomeric rotaxanes, catenanes and knotanes possessing linear, branched as well as cyclic architectures highlights the advance in synthetic chemistry of such complex topological compounds. We have shown that multiple covalently linked topological stereogenic units can be arranged in a desired manner. Topologically chiral covalent assemblies of amide-knotanes seem to be of especial significance. Taking into account their sizes and masses reaching 10 nm and 12000 Da, respectively, these assemblies define a new class of artificial macromolecules beyond polymers and dendritic species, yet perfect in shape and dispersity. Despite the limitations of the modern chiral separation science which do not allow for a complete isolation of all isomers of the synthesized topologies, they have been shown to be of high fundamental value in developing new knowledge about isomerism and chirality. For example, the chirality designations of the topologically chiral non-symmetrical dumbbell **47** and the linear tetraknotane **48** are analogous with the Fischer projections of erythrose/threose and hexaric acid respectively, while the isomeric composition of the branched tetraknotane **50** is original since no centrochiral analogues with this constitution can exist. Chirality of the knotanophanes represents, in turn, analogies with known cyclic forms of peptides or sugars having chiral centers.

Our vision is to use the knotanophanes and the linear oligomeric knotanes as chiral wheel and axle components, respectively, in future giant rotaxanes which would mimic naturally occurring enzyme complexes [60].

54

55

56

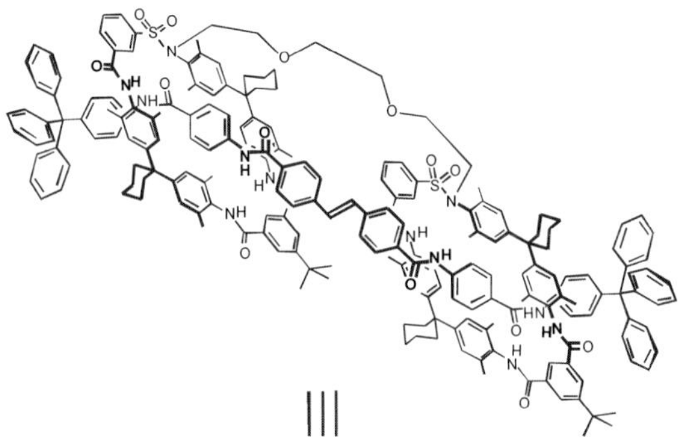

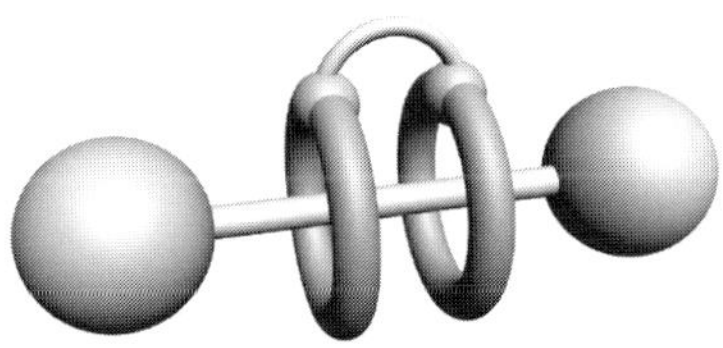

57

58

At the outlook, we hope that the studies of the mechanisms of template-controlled molecular threading and entangling will lead to functional topological species with nanotechnological applications in the future. Indeed, the number of more and more complex intertwined structures steadily grows. For instance, the research group of Stoddart [61] has recently accomplished a supramolecular template-assisted assembly of molecular Borromean rings **54** while the group of Sauvage is currently completing the preparation of five-star knotane **55** and a David star catenane **56** [62] which can be assembled with the aid of metal templated entangling of oligophenanthrolines. We, in turn, have recently been successful in the milligram-scale preparation of an intramolecularly bridged diastereoisomeric [3]rotaxane **57** that is the simplest member of a new topological family which we call *bonnanes* [63]. Bonnanes are expected to expand to the field of the rotaxane-based molecular motors in which rotary motion is controlled by link design. Structure **58** is a cartoon representation of such future bonnane in which two covalently connected terminal rings can reversibly jam the middle ring. Further investigations on the use of intertwined assemblies as molecular switches or motors will hopefully point to their concrete applications, most of which are still in the realm of the imagination.

6. Acknowledgements

Financial assistance by the Deutsche Forschungsgemeinschaft (SFB 624) and the Fonds der Chemischen Industrie, Frankfurt/Main is gratefully acknowledged. O. L. thanks the Alexander von Humboldt Foundation for the fellowship. We warmly thank Dr. C. A. Schalley and Mr. V. Lukin for sharing their graphical material with us.

7. References

1. Wasserman, E.: The preparation of interlocking rings: a catenane, *J. Am. Chem. Soc.* **82** (1960), 4433-4434.
2. Frisch, H.L., Wasserman, E.: Chemical topology, *J. Am. Chem. Soc.* **83** (1961), 3789-3795.
3. Sauvage, J.-P., Dietrich-Buchecker, C. (eds.), *Molecular Catenanes, Rotaxanes and Knots, A Journey Through the World of Molecular Topology*, Wiley-VCH, Weinheim, 1999.
4. a) Hudson, B., Vinograd, J.: Catenated circular DNA molecules in HeLa cell mitochondria, *Nature*, **216** (1967), 647-652; b) Clayton, D.A., Vinograd, J.: Circular dimer and catenate forms of mitochondrial DNA in human leukaemic leucocytes, *Nature* **216** (1967), 652-657.
5. Harrison, I.T., Harrison, S.: Synthesis of a stable complex of a macrocycle and a threaded chain, *J. Am. Chem. Soc.* **89** (1967), 5723-5724.
6. Schill, G., Zollenkopf, H.J.: Rotaxane, *Nachr. Chem. Tech.* **15** (1967), 149.
7. Schill, G. *Catenanes, Rotaxanes and Knots*, Academic Press, New York, 1971.
8. Schill, G., Zürcher, C.: [3]-catenane, *Angew. Chem.* **89** (1969), 996-997; *Angew. Chem. Int. Ed.* **8** (1969), 988-989.
9. a) Diederich, F., Stang, P.J. (eds.), *Templated Organic Synthesis*, VCH-Wiley, Weinheim, 2000; b) Sanders, J.K.M.: Adventures in molecular recognition. The ins and outs of templating, *Pure Appl. Chem.* **72** (2000), 2265-2274; c) Greig, L.M., Philip, D.: Applying biological principles to the assembly and selection of synthetic superstructures, *Chem. Soc. Rev.* **30** (2001), 287-302; d) Gerbeleu, N.V., Arion, V.B., Burgess, J. *Templated Synthesis of Macrocyclic Compounds*, Wiley-VCH, Weinheim, 1999; e) Hubin, T.J., Bush, D.H.: Template routes to interlocked molecular structures and orderly molecular entanglements, *Coord. Chem. Rev.* **200-202** (2000), 5-52; f) Vögtle, F., Hoss, R., Händel, M. *Template or Host/Guest Relations* in B. Cornils, W. A. Herrmann (eds.), *Applied Homogeneous Catalysis with Organometallic Compounds*, Wiley-VCH, Weinheim, 1996; g) Hoss, R., Vögtle, F.: Template syntheses, *Angew. Chem.* **106** (1994), 389-398; *Angew. Chem. Int. Ed.* **33** (1994), 375-384; h) Cacciapaglia, R., Mandolini, L.: Catalysis by metal ions in reactions of crown-ether substrates, *Chem. Soc. Rev.* **22** (1993), 221-231; i) Anderson, S., Anderson, H.L., Sanders, J.K.M.: Expanding roles of templates in synthesis, *Acc. Chem. Res.* **26** (1993), 469-475; j) Dietrich, B., Viout, P., Lehn, J.-M. *Macrocyclic Chemistry: Aspects of Organic and Inorganic Supramolecular Chemistry*, VCH, Weinheim, 1992.
10. Dietrich-Buchecker, C.O., Sauvage, J.P.: Une nouvelle famille de molecules: les metallo-catenanes, *Tetrahedron Lett.* **24** (1983), 5091-5094.
11. Amabilino, D.B., Stoddart, J.F.: Interlocked and intertwined structures and superstructures, *Chem. Rev.* **95** (1995), 2725-2828.
12. Hunter, C.A.: Synthesis and structure elucidation of a new [2]catenane, *J. Am. Chem. Soc.* **114** (1992), 5303-5311.
13. Hoss, R., Meier, S., Vögtle, F.: One-step synthesis of a fourfold functionalised catenane, *Angew. Chem.* **104** (1992), 1628-1631; *Angew. Chem. Int. Ed.* **31** (1992), 1619-1622.
14. Johnston, A.G., Leigh, D.A., Pritchard, R.J., Deegan, M.D.: Facile synthesis and solid-state structure of a benzylic amide [2]catenane, *Angew. Chem.* **107** (1995), 1324-1327; *Angew. Chem. Int. Ed.* **34** (1995), 1209-1212.
15. Amabilino, D.B., Ashton, P.R., Reder, A.S., Spencer, N., Stoddart, J.F.: Olympiadane, *Angew. Chem.* **106** (1994), 1316-1319; *Angew. Chem. Int. Ed.* **33** (1994), 1286-1290.
16. Schwanke, F., Safarowsky, O., Heim, C., Silva, G., Vögtle, F.: Amide-based oligocatenanes by an iterative template strategy, *Helv. Chim. Acta* **83** (2000), 3279-3290.
17. Nierengarten, J.-F., Dietrich-Buchecker, C.O., Sauvage, J.-P.: Synthesis of a doubly interlocked [2]catenane, *J. Am. Chem. Soc.* **116** (1994), 375-376.
18. Parham, A.H., Schmieder, R., Vögtle, F.: Iterative synthesis of [n]rotaxanes, *Synlett* **12** (1999), 1887-1890.
19. Dietrich-Buchecker, C.O., Sauvage, J.-P.: A trefoil knot compound, *Angew. Chem.* **101** (1989), 192-194; *Angew. Chem. Int. Ed.* **28** (1989), 189-192.
20. Ashton, P.R., Matthews, O.A., Menzer, S., Raymo, F.M., Spencer, N., Stoddart, J.F., Williams, D.J.: A template-directed synthesis of a molecular trefoil knot, *Liebigs Ann. Chem.* (1997), 2485-2494.
21. Safarowsky, O., Nieger, M., Fröhlich, R., Vögtle, F.: A molecular knot with twelve amide groups – one-step synthesis, crystal structure, chirality, *Angew. Chem.* **112** (2000), 1699-1701; *Angew. Chem. Int. Ed.* **39** (2000), 1616-1618.
22. Adams, H., Ashworth, E., Breault, G.A., Guo, J., Hunter, C.A., Mayers, P.C.: Knot tied around an octahedral metal centre, *Nature* **411** (2001), 763.
23. a) Taylor, W.R.: A deeply knotted protein structure and how it might fold, *Nature* **406** (2000), 916-919; b) Laurie, B., Katritch, V., Sogo, J., Koller, T., Dubochet, J., Stasiak, A.: Geometry and physics of catenanes applied to the study of DNA replication, *Biophys. J.* **74** (1998), 2815-2822.

24. Adams, C.C. *The Knot Book*, W.H. Freeman and Co., New York, 1994.
25. Flapan, E. *When Topology Meets Chemistry: A Topological Look at Molecular Chirality*, Cambridge University Press, 2000.
26. Dodziuk, H., Nowinski, K.S.: Topological isomerism: should rotaxanes, endohedral fullerene complexes and in-out isomers of perhydrogenated fullerenes be considered as such? *Tetrahedron* **54** (1998), 2917-2930.
27. Lukin, O., Godt, A., Vögtle, F.: Residual tolopogical isomerism of intertwined molecules, *Chem. Eur. J.* **10** (2004), 1878-1883.
28. Yamamoto, C., Okamoto, Y., Schmidt. T., Jäger, R., Vögtle, F.: Enantiomeric resolution of cycloenantiomeric rotaxane, topologically chiral catenane, and pretzel-shaped molecules: observation of pronounced circular dichroism, *J. Am. Chem. Soc.* **119** (1997), 10547-10548.
29. Vögtle, F., Hünten, A., Vogel, E., Buschbeck, S., Safarowsky, O., Recker, J., Parham, A., Knott, M., Müller, W.M., Müller, U., Okamoto, Y., Kubota, T., Lindner, W., Francotte, E., Grimme, S.: Novel amide-based knots: complete enantiomeric separation, chiroptical properties, and absolute configuration, *Angew. Chem.* **113** (2001), 2534-2537; *Angew. Chem. Int. Ed.* **40** (2001), 2468-2471.
30. Balzani, V., Venturi, M., Credi, A. *Molecular Devices and Machines. A Jorney into the Nanoworld*, Wiley-VCH, Weinheim, 2003; b) Feringa, B.L. (ed.), *Molecular Switches*, Wiley-VCH, Weinheim, 2001; c) Raehm, L., Sauvage, J.-P.: Molecular machines and motors based on transition metal-containing catenanes and rotaxanes, *Struct. Bond.* **99** (2001), 55-78.
31. a) Gibson, H.W., Bheda, M.C., Engen, P.T.: Rotaxanes, catenanes, polyrotaxanes, polycatenanes and related materials, *Prog. Polym. Sci.* **19** (1994), 843–945; b) Gibson, H.W. in J.A. Semlyen (ed.), *Large Ring Molecules*, Wiley, New York, 1996, pp. 191–262; c) Harada, A.: Polyrotaxanes, *Acta Polym.* **49** (1998), 3–17; d) Raymo, F.M., Stoddart, J.F.: Interlocked macromolecules, *Chem. Rev.* **99** (1999), 1643–1663; e) Sauvage, J.-P., Kern, J.M., Bidan, G., Divisia-Blohorn, B., Vidal, P.L.: Improvement of the polymer properties by using sterically hindered coordinating units, *New J. Chem.* **26** (2002), 1287-1290; f) Fustin, C.-A., Bailly, C., Clarkson, G.J., De Groote, P., Galow, T.H., Leigh, D.A., Robertson, D., Slawin, A.M.Z., Wong, J.K.Y.: Mechanically linked polycarbonate, *J. Am. Chem. Soc.* **125** (2003), 2200-2207; g) Kidd, T.J., Loontjens, T.J.A., Leigh, D.A., Wong, J.K.Y.: Rotaxane building blocks bearing blocked isocyanate stoppers: polyrotaxanes through post-assembly chain extension, *Angew. Chem.* **115** (2003), 3501-3505; *Angew. Chem. Int. Ed.* **42** (2003), 3379-3383; h) Panova, I.G., Topchieva, I.N.: Rotaxanes and polyrotaxanes. Their synthesis and supramolecular devices based on them, *Russ. Chem. Rev.* **70** (2001), 23-44; i) Mahan, E., Gibson, H.W. in A.J. Semlyen (ed.), *Cyclic Polymers, 2nd ed.*, Kluwer, Dordrecht, 2000, pp. 415-560.
32. Heim, C., Udelhofen, D., Vögtle, F. *Amide-Based Catenanes, Rotaxanes and Pretzelanes* in J.-P. Sauvage, C. Dietrich-Buchecker (eds.), *Molecular Catenanes, Rotaxanes and Knots. A Journey Through the World of Molecular Topology*, Wiley-VCH, 1999, pp. 177-222.
33. Affeld, A., Hübner, G.M., Seel, C., Schalley, C.A.: Rotaxane or pseudorotaxane? Effects of small structural variations on the deslipping kinetics of rotaxanes with stopper groups of intermediate size, *Eur. J. Org. Chem.* (2001), 2877-2890.
34. Schalley, C.A., Silva, G., Nising, C.F., Linnartz, P.: Analysis and improvement of an anion-templated rotaxane synthesis, *Helv. Chim. Acta*, **85** (2002), 1578-1596.
35. Linnartz, P., Bitter, S., Schalley, C.A.: Deslipping of ester rotaxanes: a kooperative interplay of hydrogen bonding with rotational barriers, *Eur. J. Org. Chem.* (2003), 4819-4829.
36. Felder, T., Schalley, C.A.: Secondary isotope effects on the deslipping reaction of rotaxanes: high-precision measurement of steric size, *Angew. Chem.* **115** (2003), 2360-2363; *Angew.Chem. Int. Ed.* **42**, (2003), 2258-2260.
37. Lukin, O., Recker. J., Böhmer, A., Müller, W.M., Kubota, T., Okamoto, Y., Nieger, M., Vögtle, F.: A topologically ciral molecular dumbbell, *Angew. Chem.* **115** (2003), 458-461; *Angew. Chem. Int. Ed.* **42** (2003), 442-445.
38. Lukin, O., Müller, W.M., Müller, U., Kaufmann, A., Schmidt, C., Leszczynski, J., Vögtle, F.: Covalent chemistry and conformational dynamics of topologically chiral amide-based molecular knots, *Chem. Eur. J.* **9** (2003), 3507-3517.
39. Schalley, C.A., Reckien, W., Peyerimhoff, S., Vögtle, F.: Theory and experiment in concert: templated synthesis of amide rotaxanes, catenanes and knots, *Chem. Eur. J.* in press.
40. a) Hübner, G.M., Gläser, J., Seel, C., Vögtle, F.: High-yielding rotaxane synthesis with an anion template, *Angew. Chem.* **111** (1999), 395-398; *Angew. Chem. Int. Ed.* **37** (1999), 3031-3034; b) Seel, C., Vögtle, F.: Templates, „wheeled reagents", and a new route to rotaxanes by anion complexation: the trapping method, *Chem. Eur. J.* **6** (2000), 21-24.
41. Shukla, R., Deetz, M.J., Smith, B.D.: [2]Rotaxane with a cation-binding wheel, *Chem. Commun.* (2000), 2397-2398.
42. Ghosh, P., Mermagen, O., Schalley, C.A.: Novel template effects for the preparation of [2]rotaxanes with functionalised centre pieces, *Chem. Commun.* (2002), 2628-2629.

43. Walba, D.M.: Topological stereochemistry, *Tetrahedron* **41** (1985), 3161-3212.
44. a) Helmchen, G. *Nomenclature and vocabulary of organic stereochemistry* in *Houben-Weyl, Methods in Organic Chemistry, 4th Edition*, Vol. E 21, Thieme, Stuttgart, New-York, 1995; b) Eliel, E.L., Wilen, S.H., Mander, L.N. *Stereochemistry of Organic Compounds*, Wiley, New-York, 1994.
45. a) Vögtle, F., Ottens-Hildebrandt, S., Schmidt, T., Harren, J.: Sulfonamide-based catenanes – regioselective template synthesis, *Liebigs Ann.* (1995), 1855-1860; b) Vögtle, F., Jäger, R., Händel, M., Ottens-Hildebrandt, S., Schmidt, W.: Amide-based rotaxanes with terephthal, furan, thiophene and sulfonamide subunits, *Synthesis* (1996), 353-356.
46. Reuter, C., Pawlitzki, G., Wörsdörfer, U., Plevoets, M., Mohry, A., Kubota, T., Okamoto, Y., Vögtle, F.: Chiral dendrophanes, dendro[2]catenanes, and dendro[2]rotaxanes: synthesis and chiroptical properties, *Eur. J. Org. Chem.* (2000), 3059-3067.
47. a) Peerlings, H.W., Struijk, M.P., Meijer, E.W.: Chiral objects with a dendritic architecture, *Chirality* **10** (1998), 46-52; b) Peerlings, H.W., Trimbach, D.C., Meijer, E.W.: Chiral dendrimers with backfolding wedges, *Chem. Commun.* (1998), 497-498.
48. Jäger, R., Schmidt, T., Karbach, D., Vögtle, F.: The first pretzel-shaped molecules via catenane precursors, *Synlett*, (1996), 723-725.
49. a) Jäger, R., Händel, M., Harren, J., Rissanen, K., Vögtle, F.: chemistry with rotaxanes: intra- and intermolecularly covalently linked rotaxanes, *Liebigs Ann.* (1996), 1201-1207; b) Reuter, C., Mohry, A., Sobanski, A., Vögtle, F.: [1]Rotaxanes and pretzelanes: synthesis, chirality, and absolute configuration, *Chem. Eur. J.* **6** (2000), 1674-1682.
50. Reuter, C., Wienand, W., Schmuck, C., Vögtle, F.: A self-threaded „molecular 8", *Chem. Eur. J.* **7** (2001), 1728-1733.
51. Buschbeck, S. *Ph. D. thesis*, University of Bonn, 2002.
52. Recker, J., Müller, W.M., Müller, U., Kubota, T., Okamoto, Y., Nieger, M., Vögtle, F.: Dendronized molecular knots: selective synthesis of various generations, enantiomer separation, circular dichroism, *Chem. Eur. J.* **8** (2002), 4434-4442.
53. Dünnwald, T., Jäger, R., Vögtle, F.: Synthesis of rotaxane assemblies, *Chem. Eur. J.* **3** (1997), 2043-2051.
54. Osswald, F., Vogel, E., Safarowsky, O., Schwanke, F., Vögtle, F.: Rotaxane assemblies with dendritic architecture, *Adv. Synth. Catal.* **343** (2001), 303-309.
55. Mohry, A., Schwierz, H., Vögtle, F.: Supramolecular assemblies: a bis(pretzelane) and a tetrakis(rotaxane), *Synthesis* (1999), 1753-1758.
56. Li, X., Illigen, J., Nieger, M., Michel, S., Schalley, C.A.: Tetra- and octalactam macrocycles and catenanes with exocyclic metal-coordination sites: versatile building blocks for supramolecular chemistry, *Chem. Eur. J.* **9** (2003), 1332-1347.
57. Lukin, O., Kubota, T., Okamoto, Y., Schelhase, F., Yoneva, A., Müller, W.M., Müller, U., Vögtle, F.: Knotaxanes – rotaxanes with knots as stoppers, *Angew. Chem.* **115** (2003), 4681-4684; *Angew. Chem. Int. Ed.* **42** (2003), 4542-4545.
58. Lukin, O., Kubota, T., Okamoto, Y., Kaufmann, A., Vögtle, F.: Topologically chiral covalent assemblies of molecular knots with linear, branched, and cyclic architectures, *Chem. Eur. J.* **10** (2004), 2804-2810.
59. a) Vögtle, F. *Cyclophan-Chemie*, Teubner, Stuttgart, 1990; *Cyclophane Chemistry*, Wiley, Chichester, 1993; b) Diederich, F. *Cyclophanes* in J.F. Stoddart (ed.), *Monographs in Supramolecular Chemistry, Volume 3.* Royal Society of Chemistry: Cambridge, U.K., 1991; c) Takemura, H. (ed.), *Cyclophane Chemistry for the 21st Century*, Research Singpost, 2002.
60. a) Schalley, C.A., Beizai, K., Vögtle, F.: On the way to rotaxane-based molecular motors: studies in molecular mobility and topological chirality, *Acc. Chem. Res.* **34** (2001), 465-476; b) Breyer, W.A., Matthews, B.W.: A structural basis of processivity, *Protein. Sci.* **10** (2001), 1699-1711; c) Benkovic, S.J., Valentine, A.M., Salinas, F.: Replisome-mediated DNA replication, *Annu. Rev. Biochem.* **70** (2001), 181-208; d) Kool, E.T., Morales, J.C., Guckian, K.M.: Mimicking the structure and function of DNA: insights into DNA stability and replication, *Angew. Chem.* **112** (2000), 1026-1044; *Angew. Chem. Int. Ed.* **39** (2000), 990-1009; e) Coumans, R.G.E., Elemans, J.A.A.W., Thordarson, P., Nolte, R.J.M., Rowan, A.E.: Synthesis of porphyrin-containing [3]rotaxanes by olefin metathesis, *Angew. Chem.* **115** (2003), 674-678; *Angew. Chem. Int. Ed.* **42** (2003), 650-654.
61. Chichak, K.S., Cantrill, S.J., Pease, A.R., Chiu, S.-H., Cave, G.W.V., Atwood, J.L., Stoddart, J.F.: Molecular borromean rings, *Science* **304** (2004), 1308-1312.
62. Sauvage, J.-P. : Private communication.
63. The name *Bonnane* is in honor of the former West-German capital Bonn: Schelhase, F., Vögtle, F. unpublished results.

3.
TOWARDS FUNCTIONAL MACROCYCLES: SELF-ASSEMBLY AND TEMPLATE STRATEGIES

C. A. SCHALLEY, H. T. BAYTEKIN AND B. BAYTEKIN
Kekulé-Institut für Organische Chemie und Biochemie der Rheinischen Friedrich-Wilhelms-Universität Bonn, Gerhard-Domagk-Str. 1, 53121 Bonn, Germany

1. Introduction

As the area of supramolecular chemistry [1] has expanded from the studies of simple crown ethers to larger molecules with functions, the so-called supramolecular machines, the employed synthetic strategies have also become more elaborate. In the world of supramolecules where weak intermolecular interactions play a pivotal role in governing the overall structures, tactics in building the desired supramolecules are not limited to basic synthetic programming. Thus, self-assembly [2] of the parts building the supramolecular systems and the use of sophisticated template strategies [3] emerged as two new approaches in this area.

2. Template Strategies in the Synthesis of Functional Macrocycle Assemblies

It was not only for the basic scientific knowledge but also for the new challenges in the synthesis and the beauty of the final structures that nearly half a century ago, chemists started to investigate intertwined macrocyclic supramolecules such as rotaxanes and catenanes [4]. Earlier, when the syntheses of such structures were at their infancy, the routes to such systems were troublesome. The statistical methods [5] proved to be low-yield processes. Multistep procedures [6] involving a covalent junction which is formed between two parts that are needed to stay together until the structure is complete were not convenient as well. The use of non-covalent templates thus provided a more straightforward and high-yield approach to the problem.

Rotaxanes and catenanes are two of such interlocked systems which can be described simply as a rod threaded in a macrocycle and captured by two stoppers at each end (rotaxane), and two or more rings which can not be separated unless one or more covalent bonds are broken (catenane). Specific non-covalent interactions are not required to make these systems stable. They do not break apart because of the presence of mechanical bonds [4] in these structures. The fact that they are interlocked does not mean that they are "frozen". Conversely, the motions within the final supramolecule can be tailored and controlled by external stimuli making them supramolecular machines

Karsten Gloe (ed.), Macrocyclic Chemistry: Current Trends and Future Perspectives, 37–52.
© 2005 *Springer. Printed in the Netherlands.*

[7]. This is one of the most important features of these architectures, which make them not only aesthetically attractive species, but also molecules that have functions which could be incorporated in more complex operations.

Although structurally and functionally appealing, the synthetic ways to build these species remained a hard task until mid 80s. Then, a strategy was developed by Sauvage and his co-workers that used coordination to a Cu(I) ion prior to the macrocyclization in the catenane synthesis (Figure 1a). The metal that served as a template with tetrahedral coordination geometry was then removed by demetallation with cyanide [8].

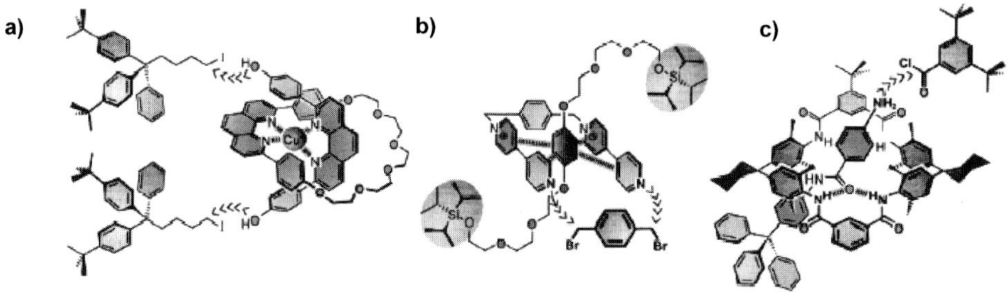

Figure 1. Different template strategies: a) Metal ions, b) π-acceptor π-donor interactions and c) hydrogen bonding mediate the template synthesis of rotaxanes.

Later, other template strategies employed different kinds of non-covalent interactions to improve the yield of the desired mechanically bound product and to avoid side products. To preorganize the reactands, Stoddart and his co-workers used π-acceptor π-donor interactions [9]. As an example the interaction between the electron-poor paraquat (*N,N*-dialkyl 4,4'-bipyridinium) macrocycle precursors (Figure 1b) and electron-rich hydroquinones or naphthoquinones have been used to achieve a suitable preorganization. This precursor can then be successfully converted into a rotaxane by a macrocyclization of the wheel in a clipping synthesis. The inverse way can also be chosen: Threading a paraquat axle into a *bis*-hydroquinone wheel and locking the centerpiece by stoppers or macrocyclization is also successful [10].

Because of the advantage of having directionality in hydrogen bonds, hydrogen bonding is also used in template synthesis of interlocked molecules [11] (Figure 1c). It is also advantageous because it allows us to use non-ionic templates. One example for hydrogen-bond-mediated templates is the use of amide binding inside a tetralactam macrocycle which turned out to permit a huge structural diversity of rotaxanes to be synthesized.

The most recently discovered template effect is the one that makes use of anions. Vögtle and coworkers [12] have found that a phenolate equipped with one stopper can bind in the cavity of the tetralactam macrocycle by two strong hydrogen bonds. Then this nucleophile complex is reacted with an electrophilic semi-axle to obtain rotaxane in very high yields up to 95%.

2.1 AMIDE BASED SYNTHESIS OF INTERLOCKED MOLECULES

The amide-based template synthesis of rotaxanes, catenanes and knots were first discovered by Vögtle and coworkers when they tried to synthesize macrocycle **3** [13]. The cyclization to yield the tetralactam ring was carried out under high-dilution conditions where intramolecular reactions are more favored over intermolecular reactions. This should prevent polymerization and provides higher yields of macrocycles. Three different products: Tetralactam wheel **3**, and octalactam macrocycle **5** as well as catenane **4** (Figure 2). This result was verified simultaneously by Hunter [14].

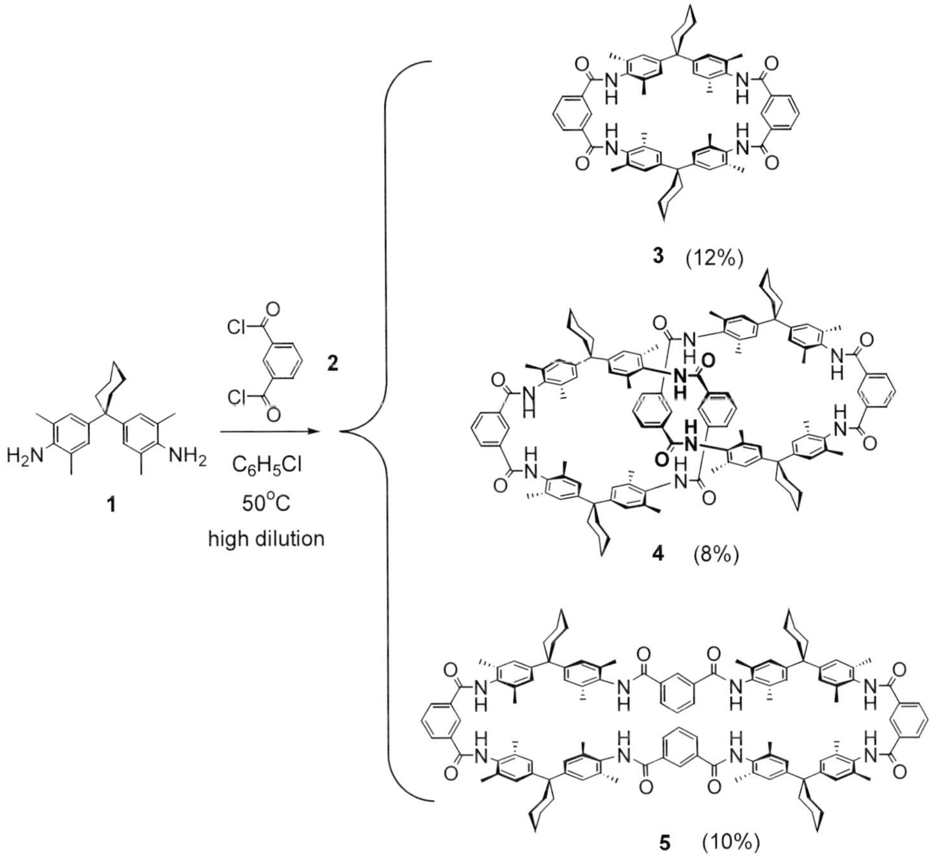

Figure 2. Under high-dilution conditions, the reaction of **1** with isophthaloyl chloride **2** yields tetralactam macrocycle **3**, the catenane **4**, and octalactam ring **5**.

When other dicarboxylic acids were used to investigate this result, it was found that the catenane could not be isolated with 2,6-pyridine dicarboxylic acid dichloride, whereas the yield of the tetralactam macrocycle was increased significantly [15]. This was attributed to the special feature of pyridine to bind nearby amide hydrogens by intramolecular two hydrogen bonds [16] (Figure 3). Thus, the open ring precursor is preorganized for the formation of the tetralactam macrocycle. It also induces a narrower cavity as shown by calculations, which likely disfavors the catenane threading [17].

In another reaction, the *para*-substitution pattern in terephthaloyl chloride that is used instead of isophthaloyl chloride induces the formation of larger octalactam macrocycles which provide two suitable binding sites for threading two semicycles and thus yield higher chain-like catenanes [18].

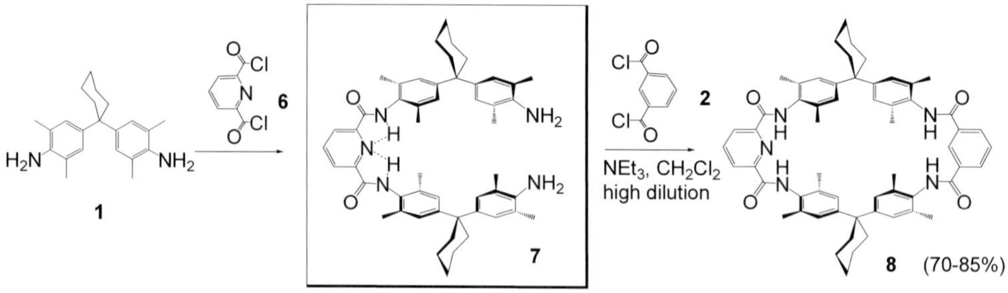

Figure 3. Preorganization of intermediate **7** through bifurcated hydrogen bonding with the central pyridine favors the formation of the tetralactam wheel at the expense of the corresponding catenane.

These results obtained from experiment with structurally different building blocks are meaningful in a geometric sense. Theoretical calculations provide a more detailed insight into the features of these systems. Figure 4 shows that two amide hydrogens of the isophthaloyl amide bind the carbonyl oxygen of the guest, and also, one amide hydrogen of the guest is coordinated a to carbonyl oxygen of the host [19]. The latter of these main interactions is proved to be of significant importance since ketones, esters or tertiary amides exhibit much smaller binding constants [20]. Single crystal X-ray structures show the presence of all these hydrogen bonds in the final structure providing evidence for a good agreement of theory and experiment [21].

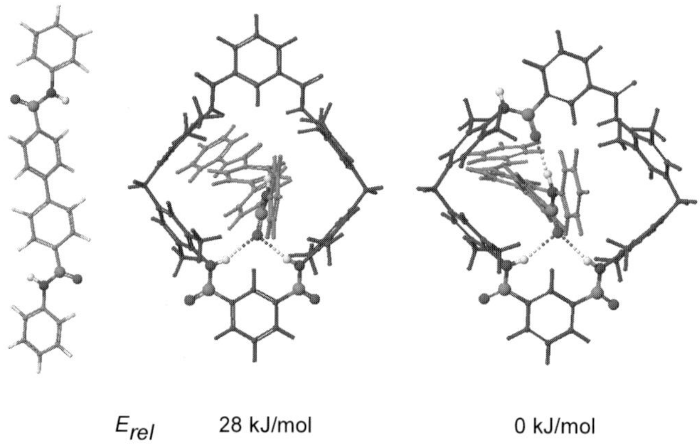

Figure 4. Structures of an axle (left), its complex with "oxygens out" conformer of the tetralactam ring (middle) and the complex with the most stable conformer. The third hydrogen bond plays an important role in stabilization of the complex. All structures were optimized by DFT methods.

A successful synthesis of a rotaxane of this type is shown in Figure 5. First, one trityl aniline stopper is reacted with the terephthaloyl chloride to from semiaxle **9**. This semiaxle threads into the tetralactam macrocycle **3** and is held by the amide template. Then, the preorganized complex is reacted with the second stopper **11** to yield rotaxane **12**. Figure 5 shows three hydrogen bonds to form, which is in accord with AM1 calculations on this system [19] and X-Ray crystal structure analysis [22].

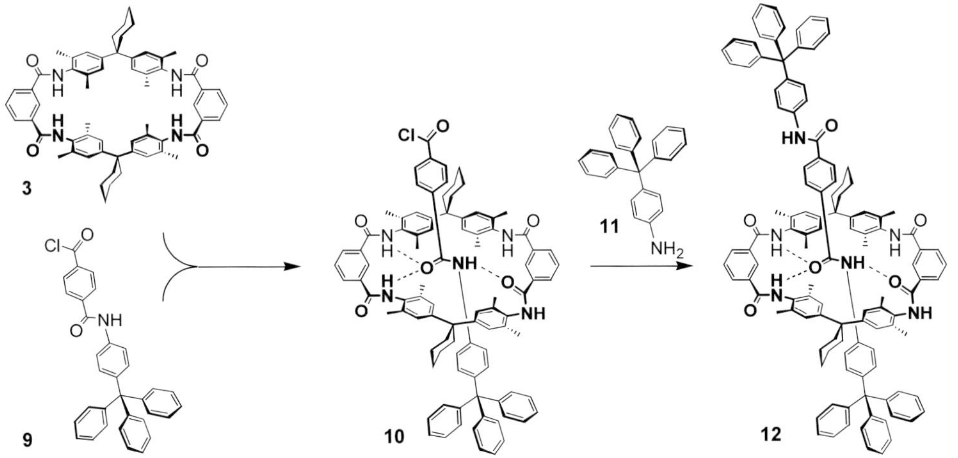

Figure 5. The amide template reaction leading to rotaxane **12**.

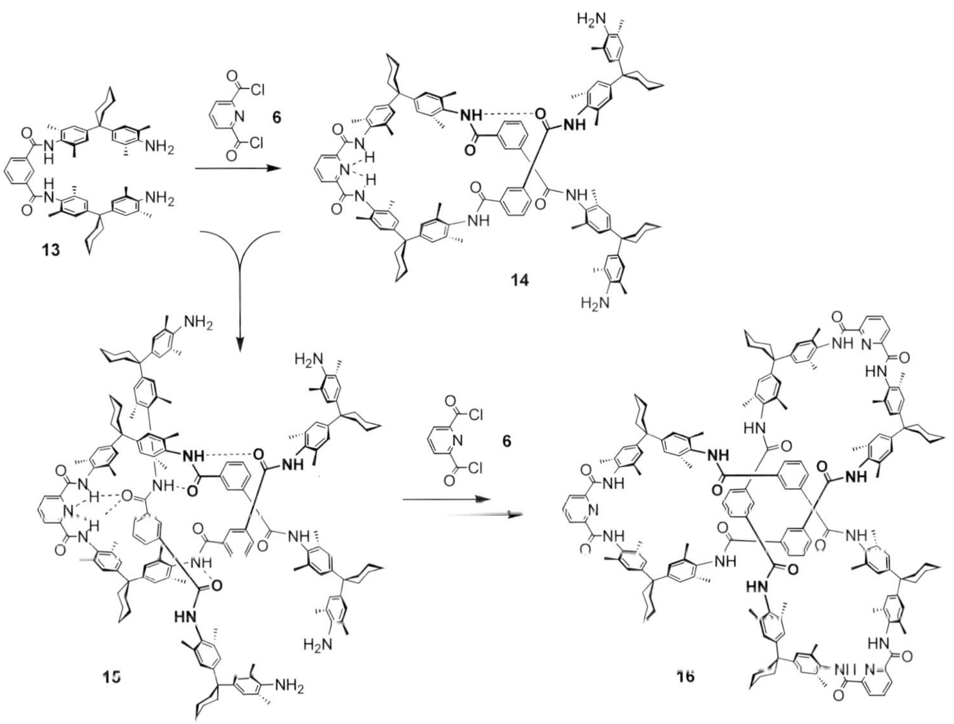

Figure 6. The formation of knot from a helical intermediate preorganized through hydrogen bonds.

A molecule with an even more spectacular topology was obtained when extended diamine **13** was reacted with 2,6-pyridine dicarboxylic acid dichloride. This was just a reversal of the order of the reaction steps in Figure 3. As well as the expected tetralactam macrocycle, its larger analogue octalactam wheel and a knot were formed [15]. It was very surprising that a knot which bears three crossing points in its molecular graph, and whose synthesis is thus a challenge to template design [23], has formed with a surprisingly high yield up to 20%. The structure was verified by X-ray analysis which showed that the knot has three different loops. The formation of the knot was also evident from CD spectra: Even though it contains no chiral centers in the structure, a knot is topologically chiral [24]. After enantiomeric separation it was straightforward to differentiate a knot from the achiral macrocycles and catenanes formed as side products [25]. The third evidence came from a so-called tandem-MS experiment by which a mono-macrocycle, a catenane and a knot can be differentiated from each other [26]. The mechanism of knot formation was analyzed by theory and likely follows the pathway indicated in Figure 6. Formation of a helical loop which just like a rotaxane binds the third extended diamine as an axle-like thread is finally followed by adding the two missing pyridine moieties which close the knot structure.

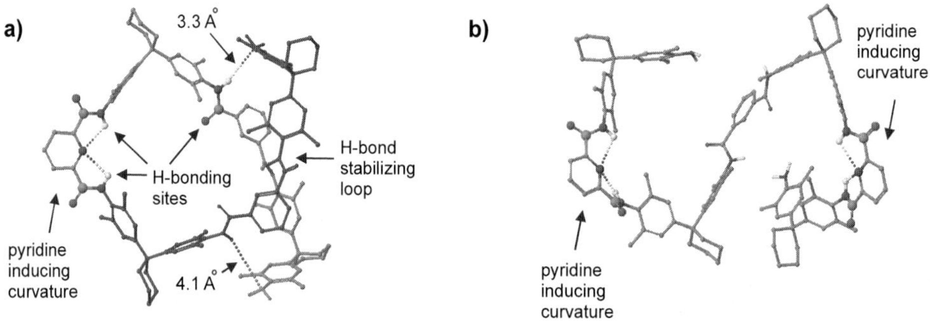

Figure 7. a) The lowest energy conformation of the intermediate **14** which promotes knot formation. b) The reversed sequence of adding the diacyl chlorides **2** and **6** into the reaction does not produce the helical intermediate; knot formation does not occur.

The reason why the knot was formed in this reaction but not in the reaction summarized in Figure 3 was cleared again by looking at the hydrogen bond patterns. When 2,6-pyridine diacylchloride reacts at both arms, it forms a helix-shaped, preorganized structure **14** by intramolecular hydrogen bond formation. Into this loop one of the reactant diamines **13** is threaded as expected from the template behavior of amides. Then the open ends of the threaded axle and the helical loop, which are in close vicinity of each other are tied with two 2,6-pyridine dicarboxylic acid dichlorides to end up with knot **16**. This is the main reason why the reversed sequence does not yield a knot: pyridine is not at the central position of the intermediate analogous to **14** and thus induces curvature in the wrong parts of the intermediate (Figure 7).

This mechanism was also supported by an experiment where isophthalic acid moiety is substituted by its bulkier t-butyl- analogue. No knot formation was observed suggesting that the isophthalic acid units are inside the knot and pyridines are located at the periphery (Figure 8).

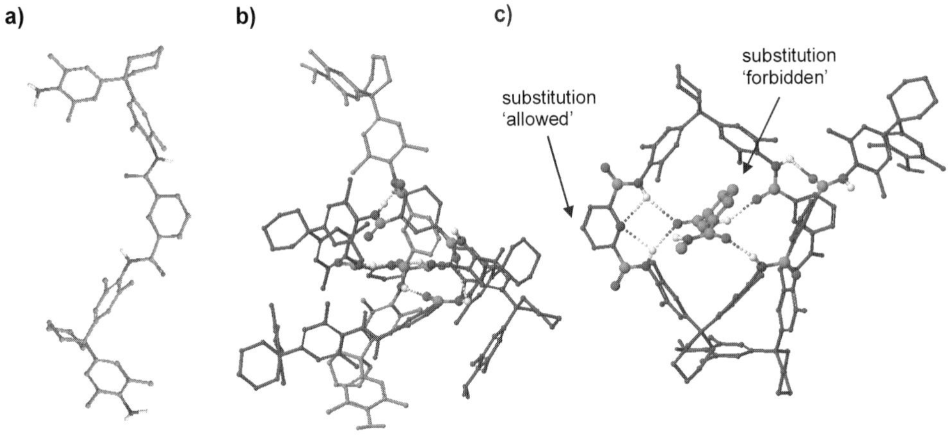

Figure 8. a) Lowest energy conformation of the extended diamine axle **13** that threads into the helix shaped intermediate **14**. b) Side view of the intermediate complex: the ends that will react are in close vicinity of each other. c) The "inner" part of the intermediate can not stand any bulkiness that is created by using bulkier isophthaloyl chlorides.

2.2 ANION TEMPLATE SYNTHESIS OF ROTAXANES

A very surprising and fruitful result was obtained when a control experiment related to an amide templated synthesis was made. Dibromo compound **19** utilized in the reaction looked similar to the axle centerpiece used in the amide template synthesis but lacked the amide in the middle which was crucial for this purpose. However, when the reaction was complete, it was found that rotaxane **24** was formed with 80-95% yield [12] (Figure 9). It seemed reasonable to assume that this time not the axle but the stopper coordinated to the macrocycle [27]. This suggestion was supported by the high binding constant of the deprotonated stopper-wheel complex **21•18** (> 10^5 M^{-1}) derived from ^1H NMR titrations. In the rotaxane synthesis, this complex reacts with the semiaxle **23** producing a rotaxane.

The anion template effect proved to be successful not only for ether linkages between the stopper and the axle, but also for the ester and acetal analogues [28]. Michael addition with a thiophenolate stopper could also be accomplished with yields up to 53% [29]. The yields of the rotaxane synthesis by anion template strategy depend strongly on the nucleophilicity of the stopper and structures of both the stoppers and the centerpieces. This fact was clearly seen when 3,5-di-t-butyl phenol stopper was used in the reaction [17,30]. The yield of the rotaxane was only 2% to 5% which was independent of the length of the axle centerpieces used. The reason for this behavior could be uncovered by a conformational search which yielded different families of favorable conformations [31] (Figure 10). This shows the larger trityl phenolate stopper complexes the wheel such that the axle centerpiece is attacked from the opposite side to form rotaxane with high yield. However, the 3,5-di-t-butyl phenolate stopper is bound to the wheel in such a conformation that it can not be attacked by the semiaxle electrophile from the opposite side but reacts more favorably from the same side of the wheel. After the reaction, a weakly bound, non-threaded axle-wheel complex is formed which easily dissociates into the two compounds.

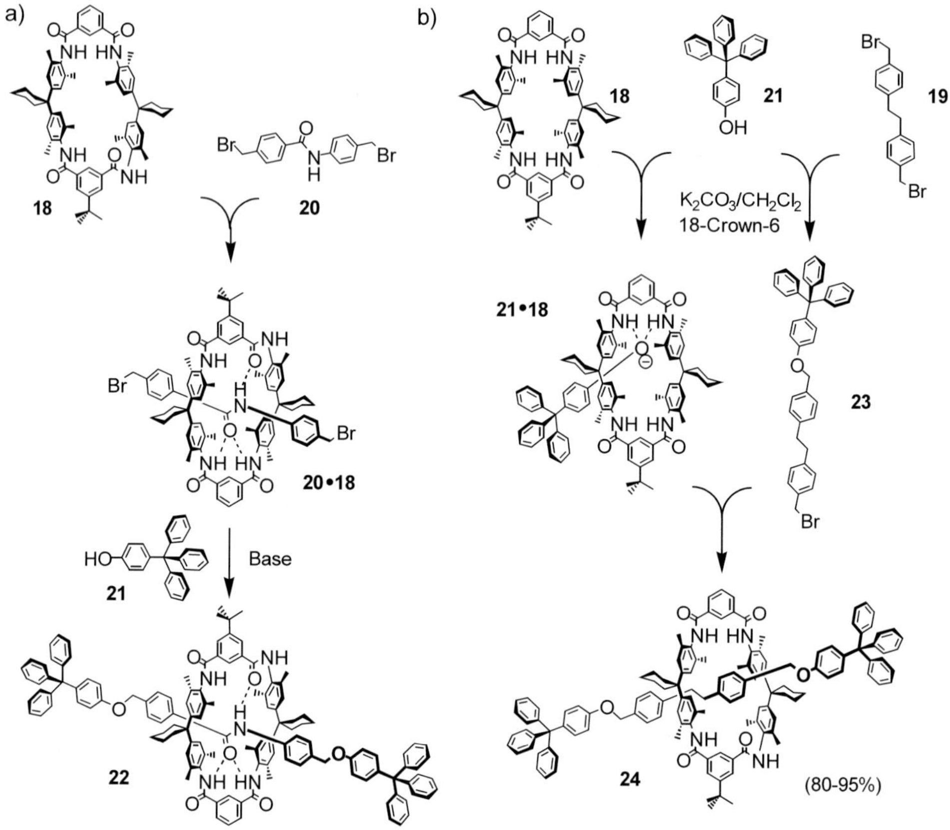

Figure 9. a) The amide-mediated template which was hypothesized to provide access to rotaxane **22**. b) A control reaction which was thought not to yield any rotaxane, but led to almost quantitative formation of rotaxane **24**.

To overcome this difficulty, a new solution is developed that involves spatial separation of the functional group for stopper attachment from the phenolate that provides the basis for anion template. Rotaxanes can be synthesized with yields up to 45% [32] (Figure 11).

One highly interesting feature of this synthetic approach is that rotaxanes are generated that contain a functional group at their axle centers. This permits to control the rotaxanes' properties by external stimuli, e.g. by protonation and deprotonation.

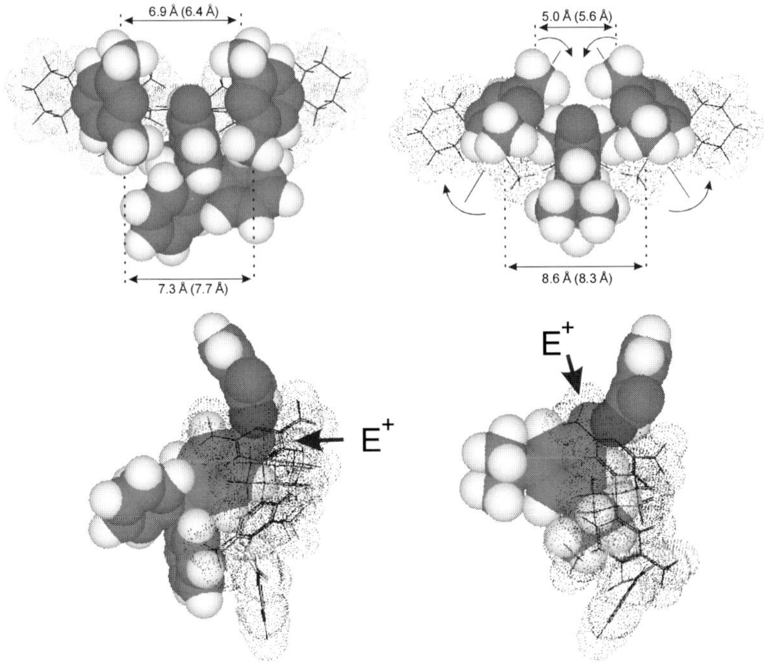

Figure 10. Top view and side view of the minimal energy conformations of the stopper-wheel complexes with trityl phenolate (left) and 3,5-di-*t*-butyl phenolate (right) as obtained from a 3000 step Monte Carlo conformational search with the AMBER* force field. The arrows show the most favorable attack paths for the electrophile. (E^+ = approaching electrophile)

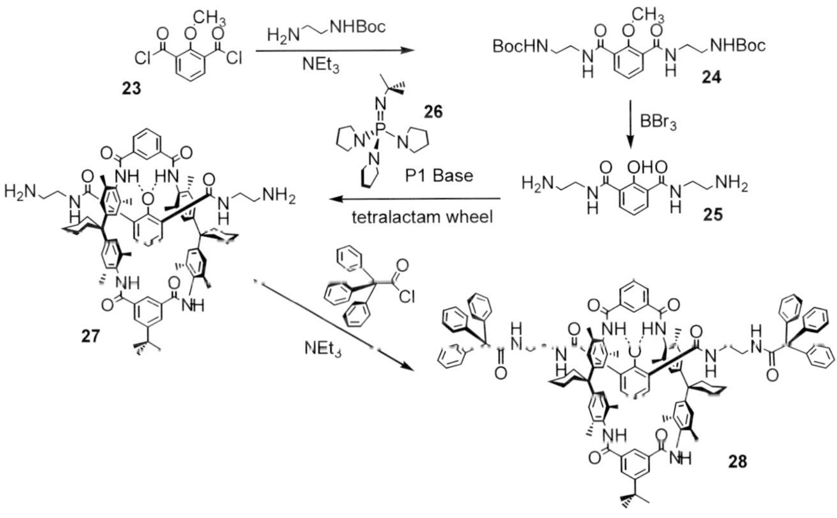

Figure 11. Anion-mediated rotaxane synthesis generating rotaxanes with functionalized axle center-pieces.

3. Self-Assembly in the Synthesis of Functional Macrocycle Assemblies

The synthetic approach of "molecular self assembly" combines features of most widely used organic synthetic strategies to constitute huge assemblies [2]. Among these synthetic strategies, the strategy referred as "self-organizing synthesis" aims to organize atoms, ions or molecules into of two- and three dimensional discrete supramolecular architectures, and it uses weaker and less directional bonds, such as ionic bonds, hydrogen bonding, and van-der-Waals forces, coulombic interactions, and dipole-dipole interactions instead of covalent bonds. Molecular triangles, squares and rectangles, pentagons and hexagons, catenanes, nanotubes and polytubes are prominent examples of transition metal-directed self assembled macrocycles [33].

3.1 METAL-DIRECTED SELF-ASSEMBLY OF MACROCYCLIC COMPLEXES

Up to now, metallo-supramolecular squares and rectangles have been extensively studied with respect to self-assembly strategies. A fascinating aspect of these macrocycles is their behavior as "inorganic cyclophanes" that is, they have an inner cavity surrounded by aromatic rings and have the ability for molecular recognition in aqueous solution. Both charge transfer interactions and hydrophobic interactions between electron-deficient pyridine nuclei and electron-rich aromatic guest molecules is attributed to complexation [34].

Initial work in this area by Fujita et al. [35] includes the synthesis of a series of ligand systems and the investigation of the dynamic behavior of the macrocycles in solution.

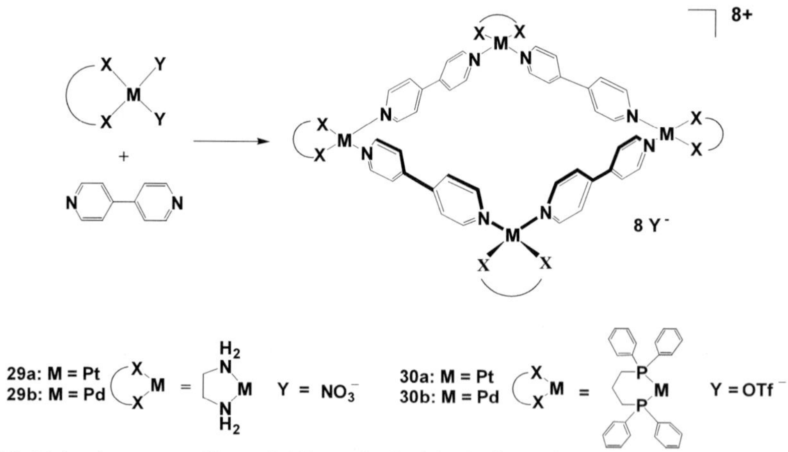

Figure 12. Molecular squares **29a** and **29b** synthesized by Fujita and coworkers [35], assembling from ethylenediamine complex of M(II) dinitrate [M(en)$(NO_3)_2$] in which M= Pt(II) and Pd(II) and 4,4'-bipyridine in water, and macrocyclic tetranuclear complexes **30a** and **30b** prepared by Stang and coworkers [38] assembled by treating (dppp)M$(OTf)_2$ (M= Pt(II) and Pd(II)) with 4,4'-bipyridine in dichloromethane.

An attempt to synthesize **29a**, a combination of [Pt(en)$(NO_3)_2$] and 4,4'-bipyridine, afforded a mixture in D_2O which was not straightforward to analyze at first [36]. For the transformation of kinetically distributed mixture of products into the thermodynamically most stable square, this oligomeric mixture needed to be heated at 100°C and then led to simple NMR spectra. Unlike the Pt containing analogue, the preparation of molecular square **29b** using [Pd(en)$(NO_3)_2$] and 4,4'-bipyridine is straightforward. This effect is a consequence of the metal-nitrogen bond strenghts which are considerably higher for the Pt compound (Figure 12).

In contrast, Stang and co-workers [37] used lipophilic phosphane-substituted metal centers **30a** and **30b** [38]. Due to stacking of the phenyl groups with the pyridine rings, the N-Pt-N bond angle decreased to 84° and the entire molecule was slightly puckered. The crystal structure showed an almost perfect square at room temperature with the pyridine rings slightly twisted.

Hydrophobic forces are also important in the assemblies of metallo-supramolecular catenanes. One of the most interesting examples is formed when one of the unpolar bipyridine ligands of one macrocycle is included spontaneously in the other macrocycle's internal cavity [39]. Here, the benzene unit of the one macrocycle serves as a guest molecule for the other macrocycle, and the cyclization is favored by π-π interactions. In addition, the minimization of hydrophobic surfaces in polar medium constitutes the second driving force for the catenane formation. The quantitative formation of the [2]catenanes **31a** and **31b** based on this principle are depicted in Figure 13. Formation of catenane **31b** was found to be reversible. Even at room temperature, two monomeric ring structures equilibrate quickly due to the labile nature of Pd-N bond and interlocked molecular ring system **31b** is formed.

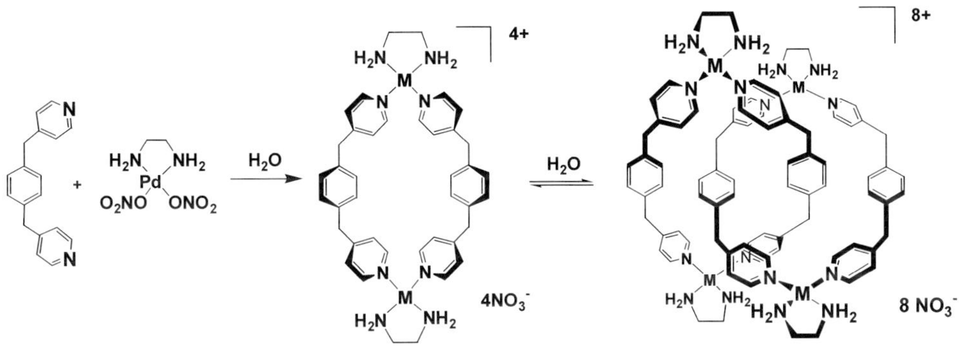

31a: M=Pt
31b: M=Pd

Figure 13. [2]Catenanes **31a** and **31b** self-assemble in water through hydrophobic interactions.

More elaborate catenanes were reported by Fujita in the following years [40]. It was discovered that modified ditopic ligands were quantitatively self-assembled with platinum and palladium complexes when the rectangles have ideal van-der-Waals interplane separation about 3.5 Å. In fact, it was revealed from the X-ray analysis that the aromatic rings stack on each other with distances of 3.2 – 3.6 Å, which provide the most efficient aromatic stacking. The building blocks and resulting catenanes are shown in Figure 14. Unlike the catenanes **31a** and **31b** shown in Figure 13, these catenanes were found to be stable in less polar media, (D_2O-CD_3OD 1:1). In addition to this, dissociation of the catenanes into their components were not observed even at low concentration. Formation of these structures was found to be reversible only when Pd(II) was used as a metal cation and even at room temperature, the catenane was in equilibrium with its precursors. The catenated structures **32a** and **32b** were confirmed by CSI-MS and they were unambiguously characterized by NMR spectroscopy and X-ray crystallography [41].

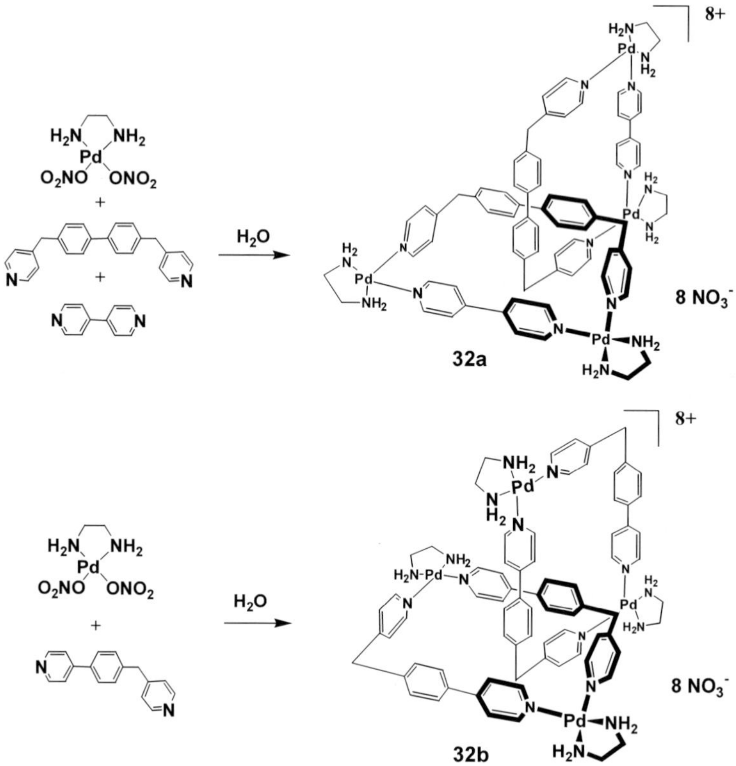

Figure 14. [2]Catenanes **32a** and **32b** which self-assemble in water through hydrophobic interactions.

3.2 COMBINATION OF TEMPLATE AND SELF-ASSEMBLY SYNTHETIC STRATEGIES

The combination of template and self-assembly strategies constitutes a new approach for the preparation of higher order structures. Interlocked architectures composed of transition metal coordinating ligands such as disubstituted bidentate ligands, e.g. [1,10]-phenanthrolines and 2,2'-bipyridines, or a tridentate ligand such as 2,2',6',2''-terpyridines, in their macrocycles are promising candidates for the development of new synthetic methods [42]. Metal coordination to copper(I) has been used extensively as templating metal center, allowing the preparation of simple to complex interlocking or knotted topologies. This phenomenon was exemplified for [2]catenane **34** [43]. Two building blocks **33** of the [2]catenane, i.e. the unclosed rings, are held together by attractive interactions between a copper(I) ion and two [1,10]-phenanthroline moieties to form a precursor of the desired [2]catenane. Furthermore, the organic building blocks of the [2]catenane are equipped with two pyridine ends, which can be "closed" with palladium(II) ethylenediamine [Pd(II)(en)] moieties quantitatively (Figure 15).

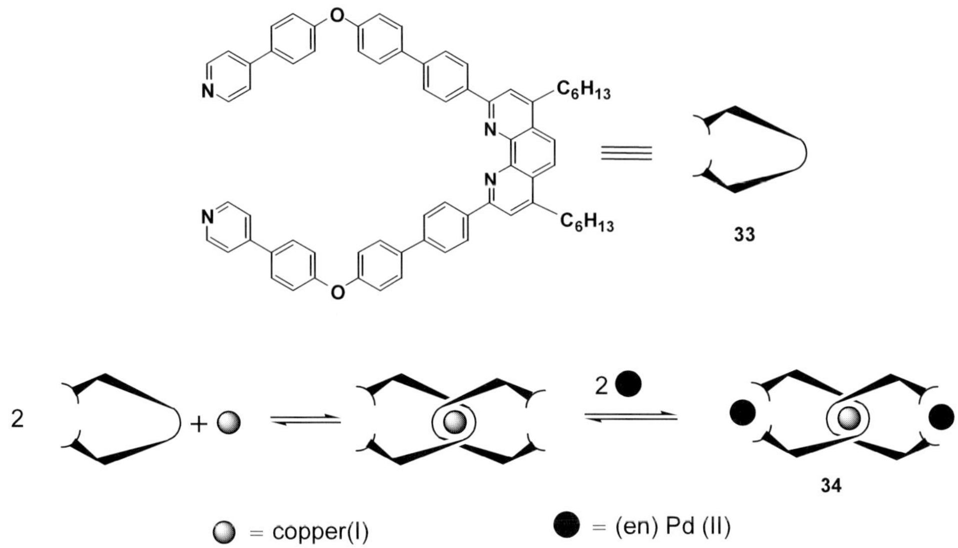

Figure 15. Self assembly of a [2]catenane using a metal-ion template effect.

3.3 SURFACE TEMPLATION OF MOLECULAR SQUARES

In order to utilize self-assembled macrocycles in the construction of nanoscale devices, their ordered deposition on surfaces seems a necessity. A surface-induced templating effect could be a very useful approach when the ordered deposition of supramolecular entities on a surface is desired. Especially, charged species are expected to organize on an oppositely charged surface. Also, a particular orientation of the square could be achieved by enhancing its interaction. When a highly-ordered pyrolytic graphite (HOPG) which was used by Stoddart et al. [44] to deposit a cationic cyclophane was incorporated in deposition of the molecular squares, substrate-induced template effect is not observed. Rather, strong intersquare forces determine the order of the deposited layer.

To overcome this problem, we used second-order templation in order to deposit square **29** on a surface [45]: First, chloride ions are deposited on a Cu(100) surface in the cell of an electrochemical STM cell (Figure 16). Cu(100) is chosen to meet the fourfold symmetry requirement. Since the chloride retains practically its full charge upon adsorption it was the preferred anion for the first layer deposition. Chloride ions are adsorbed from electrolyte containing e.g. 10 mM HCl. After preparation and characterization of the negatively charged layer, the pre supporting electrolyte (10 mM HCl, 5 mM KCl) was changed under potential control by a solution containing 10 mM HCl, 5 mM KCl and 0.1 mM of the water-soluble metallosupramolecular square **29**. Due to strong Coulomb interactions with the surface, cationic squares can then be deposited flat on top of the layer of anions.

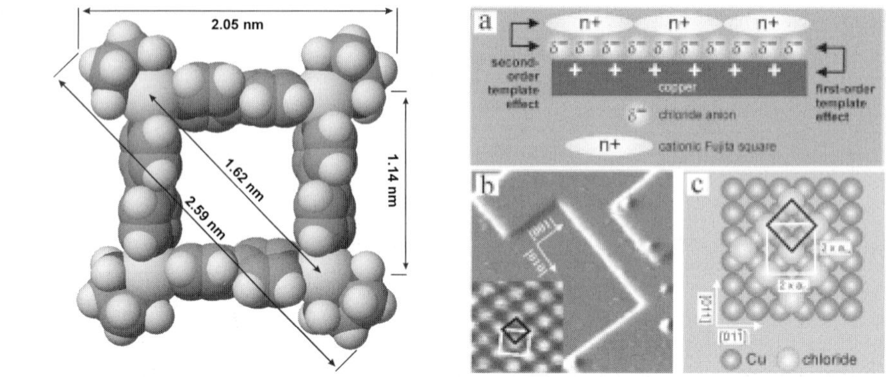

Figure 16. Left: MM2 minimized space filling molecular model of square **29** of the squares used in surface templation studies and. The dimensions along the edge and along the diagonal are given as metal-metal distances and as total width including Van-der-Waals radii. Right: (a) The principle of second-order templation: First, chloride ions are deposited on a Cu(100) surface in the cell of an electrochemical STM cell. Due to strong attractive interactions with the surface, cationic squares can then be deposited flat on top of the layer of anions. (b) STM image of the chloride-modified Cu(100) surface with the direction of the steps visible (inset: same surface at atomic resolution; Cl^- ions are oriented along the direction of the steps in a tetragonal grid, 1.7 nm × 1.7 nm). (c) Schematic representation of the chloride orientation on the Cu substrate.

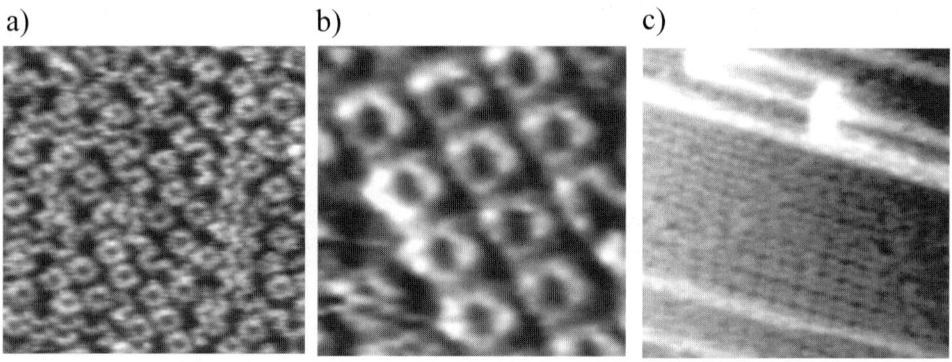

Figure 17. (a) Squares **29a** deposited onto the anion-covered surface. Quite many oligomer chains can be seen due to the acidic conditions of sample preparation; 20.5 nm × 20.5 nm. (b) Lateral ordering of squares **29a**, 7.6 nm × 7.6 nm.(c) Long-range order of squares **29a** to yield large perfect domains under optimized sample preparation conditions (see text); 35.6 nm × 35.6 nm.

It is obvious from the Figure 17a that the local ordering of molecular assemblies reveals a pronounced cavity in the center. This feature points to supramolecular square lying flat on the electrode surface. The distance of 1.9 nm to the nearest neighbor (Figure 17b) agrees well with the calculated size of the square in Figure 16. Most likely, the adsorbate-substrate interactions are mainly governed by attractive electrostatic interactions between the negatively charged chloride layer and the positively charged Pt(II) cations, while the lateral ordering is apparently dominated by van-der-Waals interactions. Note that the presence of chainlike oligomers strongly hinders the lateral ordering of molecular squares on the electrode surface (Figure 17a). Oligomeric degradation is one of the most serious problems on the electrode surface. Degradation was found to be time dependent and it takes place within 30 minutes. Therefore, application of aqueous solution of square molecules in droplets into the electrochemical cell to the pure supporting electrolyte was preferred to keep the presence of **29** in the acidic medium as short as possible. With these optimized sample preparation conditions, large perfectly ordered domains of the squares can be obtained, for which the amount of oligomer is reduced to a minimum (Figure 17c).

4. Conclusions

Macrocyclic chemistry spans the whole range from molecular recognition to template effects to STM experiments. The fundamentals for the development of macrocyclic molecules with more complex functions are laid. This basic knowledge that has been compiled over years, allows us now to more elaborate systems with diversity of functionalities. We discussed two of the many methods used in this respect, template and self-assembly strategies, and their combination. Current research in this area includes construction of systems bearing functions that allow incorporation into high-order architectures.

5. Acknowledgements

We thank our colleagues Prof. Fritz Vögtle, Prof. Sigrid Peyerimhoff, Prof. Bianca Hermann, Dr. Peter Broekmann, Dr. Arne Lützen, and Dr. Patrick Weis for their excellent and valuable contributions to joined projects. Our work was funded by the Deutsche Forschungsgemeinschaft (DFG) and the Fonds der Chemischen Industrie (FCI). C.A.S. is indebted to these institutions for a Heisenberg fellowship (DFG) and a Dozentenstipendium (FCI).

6. References

1) a) Vögtle, F. (1992) *Supramolekulare Chemie*, Teubner, Stuttgart. b) Lehn, J.M. (1995) *Supramolecular Chemistry*, Verlag Chemie, Weinheim. c) Schneider, H.J., Yatsimirsky, A. (2000) *Principles and Methods in Supramolecular Chemistry*, Wiley, New York. d) Steed, J.W., Atwood, J.L. (2000) *Supramolecular Chemistry*, Wiley, New York.
2) Whitesides, G.M., Mathias J.P., Seto, C.T. (1991) *Science* 254, 1312.
3) Schalley, C.A., Vögtle, F., Dötz, K.H. (2005) *Top. Curr. Chem.*, Vol. 248
4) a) Schill, G. (1971) *Catenanes, Rotaxanes and Knots*, Academic Press, New York. b) Sauvage, J.-P. Dietrich-Buchecker, C. (eds) (1999) *Molecular Catenanes, Rotaxanes and Knots* Wiley-VCH, Weinheim.
5) a) Wasserman, E. (1960) *J. Am. Chem. Soc.* 82, 4433. b) Harrison, I.T., Harrison, S. (1967) *J. Am. Chem. Soc.* 89, 5723.
6) a) Schill, G., Lüttringhaus, A. (1964) *Angew. Chem.* 76, 567; *Angew. Chem. Int. Ed.* 3, 564. b) Schill, G Zollenkopf, H. (1969) *Liebigs Ann.* 721, 53.
7) Balzani, V., Venturi, M., Credi, A. (2003) *Molecular Devices and Machines*, Wiley-VCH, Weinheim.
8) a) Dietrich-Buchecker, C., Sauvage, J.-P. (1987) *Chem. Rev.* 87, 795. b) Sauvage, J.-P. (1990) *Acc. Chem. Res.* 23, 319.
9) Raymo, F.M., Stoddart, J.F. *Organic Template Directed Synthesis of Catenanes, Rotaxanes and Knots*. In ref. 4b, p.143.
10) Ashton, P.R., Goodnow, T.T., Kaifer, A.E., Reddington, M.V., Slawin, A.M.Z., Spencer, N., Stoddart, J.F., Vicent, C., Williams, D.J., (1989) *Angew. Chem.* 101, 1404; *Angew. Chem. Int. Ed.* 28, 1396..
11) a) Kolchinski, A.G., Busch, D.H., Alcock, N.W. (1995) *Chem. Comm.* 1289. b) Ashton, P.R., Campell, P.J., Chrystal, E.J.T., Glink, P.T., Menzer, S., Philp, D., Spencer, N., Stoddart, J.F., Tasker, P.A., Williams, D.J. (1995) *Angew. Chem.* 107, 1997; *Angew. Chem. Int. Ed.* 34, 1865. c) Ashton, P.R., Chrystal, E.J.T., Glink, P.T., Menzer, S., Schiavo, C., Stoddart, J.F., Tasker, P.A., Williams, D.J. (1995) *Angew. Chem.* 107, 2001; *Angew. Chem. Int. Ed.* 34, 1869. d) Glink, P.T., Schiavo, C., Stoddart, J.F. Williams, D.J. (1996) *Chem. Comm.* 1483. e) Ashton, P.R., Collins, A.N., Fyfe, M.C.T., Menzer, S., Stoddart, J.F., Williams, D.J. (1997) *Angew. Chem.* 109, 760; *Angew. Chem. Int. Ed.* 36, 760.
12) Hübner, G.M., Gläser, J., Seel, C., Vögtle, F. (1999) *Angew. Chem.* 111, 395; *Angew. Chem. Int. Ed.* 38, 383.

13) a) Seel, C., Vögtle, F. (1992) *Angew. Chem.* 104, 542; *Angew. Chem. Int. Ed.* 31, 528. b) Vögtle, F., Hoss, R., Meier, S. (1992) *Angew. Chem.* 104, 1628; *Angew. Chem. Int. Ed.*, 31, 1619.
14) a) Hunter, C.A. (1991) *Chem. Comm.* 749. b) Hunter, C.A. (1992) *J. Am. Chem. Soc.* 114, 5303.
15) Safarowsky, O., Nieger, M., Fröhlich, R., Vögtle, F. (2000) *Angew. Chem.* 112, 1699; *Angew. Chem. Int. Ed.* 39, 1616.
16) a) Johnston, A.G., Leigh, D.A., Nehzat, L., Smart, J.P., Deegan, M.D. (1995) *Angew. Chem.* 107, 1327; *Angew. Chem. Int. Ed.* 34, 1202. b) Hamuro, Y., Geib, S.J., Hamilton, A.D. (1994) *Angew. Chem.* 106, 465; *Angew. Chem. Int. Ed.* 33, 446. c) Hunter, C.A. (1994) *Angew. Chem.* 106, 2424; *Angew. Chem. Int. Ed.* 33, 2313.
17) Affeld, A., Hübner, G.M., Seel, C., Schalley, C.A. (2001) *Eur. J. Org. Chem.* 2877.
18) a) Hunter, C.A., Purvis, D.H. (1992) *Angew. Chem.* 104, 779; *Angew. Chem. Int. Ed.* 31, 792. b) Carver, F.J., Hunter, C.A., Shannon, R.J. (1994) *Chem.Comm.* 1277. c) Adams, H., Carver, F.J., Hunter, C.A., Osborne, N.J. (1996) *Chem. Comm.* 2529. d) Allott, C., Adams, H., Bernad, Jr., P.L., Hunter, C.A., Rotger, C., Thomas, J.A. (1998) *Chem. Comm.* 2449.
19) Schalley, C.A., Reckien, W., Peyerimhoff, S., Baytekin B., Vögtle F. (2004) *Chem. Eur. J.* 10, 4777.
20) Seel, C., Parham, A.H., Safarowsky, O., Hübner, G.M., Vögtle, F. (1999) *J. Org. Chem.* 64, 7236.
21) Mohry, A., Vögtle., F., Nieger, M., Hupfer, H. (2000) Chirality 12, 76.
22) a) Reuter, C., Seel, C., Nieger, M., Vögtle, F. (2000) *Helv. Chim. Acta* 83, 630. b) Reuter, C., Mohry, A., Sobanski, A., Vögtle, F. (2000) *Chem. Eur. J.* 6, 1674.
23) a) Rapenne, G., Dietrich-Buchecker, C., Sauvage, J.-P. (1994) *J. Am. Chem. Soc.* 121, 994. b) Ashton, P.R., Matthews, O.A., Menzer, S., Raymo, F.M., Spencer, N., Stoddart, J.F., Williams, D.J. (1997) *Liebigs Ann./Recueil* 2485. c) Adams, H., Ashworth, E., Breault, G.A., Guo, J., Hunter, C.A., Mayers, P.C. (2001) *Nature* 411, 763.
24) a) Chambron, J.-C., Dietrich-Buchecker, C., Sauvage, J.-P. (1993) *Top. Curr. Chem.* 165, 131. b) Mislow, K. (1999) *Top. Stereochem.* 1. c) Reuter, C., Schmieder, R., Vögtle, F. (2000) *Pure Appl. Chem.* 72, 2233.
25) a) Vögtle, F., Hünten, A., Vogel, E., Buschbeck, S., Safarowsky, O., Recker, J., Parham, A., Knott, M., Müller, W.M., Müller, U., Okamoto, Y., Kubota, T., Lindner, W., Francotte, E., Grimme, S. (2001) *Angew. Chem. Int. Ed.* 40, 2468.; (2001) *Angew. Chem.* 113, 2534.
26) a) Schalley, C.A, Hoernschemeyer, J., Li. X., Silva, G., Weis, P. (2003) *Int. J. Mass Spectrom.* 228, 373. b) Schalley, C.A., Ghosh, P., Engeser, M. (2004) *Int. J. Mass Spectrom.* 232-233, 249.
27) Seel, C., Vögtle, F. (2000) *Chem. Eur. J.* 6, 21.
28) a) Reuter, C., Wienand, W., Hübner, G.M., Seel, C., Vögtle, F. (1999) *Chem.Eur. J.* 5, 2692. b) Li, X.-Y., Illigen, J., Nieger, M., Michel, S., Schalley, C.A. (2003) *Chem. Eur. J.* 9, 1332.
29) Reuter, C., Vögtle, F. (2000) *Org. Lett.* 2, 593.
30) a) Felder, T., Schalley, C.A. (2003) *Angew. Chem.* 115, 2360; *Angew. Chem. Int. Ed.* 42, 2258. b) Linnartz, P., Bitter, S., Schalley, C.A. (2003) *Eur. J. Org. Chem.* 4819.
31) Schalley, C.A., Silva, G., Nising, C.-F., Linnartz, P. (2002) *Helv. Chim. Acta* 85, 1578.
32) Ghosh, P., Mermagen, O., Schalley, C.A. (2002) *Chem. Comm.* 2628.
33) a) Schalley, C.A., Müller, T., Linnartz, P., Witt, M., Schäfer, M., Lützen, A. (2002) *Chem. Eur. J.* 8, 15, 3538. b) Holliday, J.H., Mirkin, C.A. (2001) *Angew. Chem.* 113, 2076; *Angew. Chem. Int. Ed.* 40, 2022.
34) Fujita. M., Yazaki, J., Ogura, K. (1991) *Tetrahedron Letters* 32, 5589.
35) Fujita, M. (1998) *Chem. Soc. Rev.* 27, 417.
36) Fujita, M., Yazaki, J., Ogura, K. (1991) *Chem. Lett.* 1031.
37) Leininger, S., Olenyuk. B., Stang, P.J. (2000) *Chem. Rev.* 100, 853.
38) Stang, P.J., Cao, D.H. (1994) *J. Am. Chem. Soc.* 116, 4981.
39) Ibukuro F., Fujita, M., Yamaguchi, K., Ogura K. (1995) *J. Am. Chem. Soc.* 117, 4175.
40) Fujita, M., Aoyagi, M., Ibukuro F., Ogura K., Yamaguchi, K. (1998) *J. Am. Chem. Soc.* 120, 611.
41) Hori, A., Kumazawa, K., Kusukawa, T., Chand, D.K., Fujuta M., Sakamoto, S., Yamaguchi, K. (2001) *Chem. Eur. J.* 7, 4142.
42) a) Hamann, C., Kern, J-M., Sauvage, J.-P. (2003) *Inorg. Chem.* 42, 1877. b) Dietrich-Buchecker, Colasson, B., Fujita, M., Hori, A., Geum., Sakamoto, S., Yamaguchi, K., Sauvage, J.-P. (2003) *J. Am. Chem. Soc.* 125, 5717.
43) Dietrich-Buchecker, C., Geum, N., Hori, A., Fujita, M., Sakamoto, S., Yamaguchi, K., Sauvage, J.-P. (2001) *Chem. Commun.* 1182.
44) Laitenberger, P., Claessens, C.G., Kuipers, L., Raymo, F.M., Palmer, R.E., Stoddart, J.F. (1997) *Chem. Phys. Lett.* 279, 209.
45) Safarowsky, C., Merz, L., Rang, A., Broekmann P., Hermann B.A., Schalley, C.A. (2004) *Angew. Chem.* 116, 1311; *Angew. Chem. Int. Ed.* 43, 1291.

4.

RECENT DEVELOPMENTS IN THE SYNTHESIS AND d-BLOCK CHEMISTRY OF LINKED MULTI-RING MACROCYCLIC LIGANDS

LEONARD F. LINDOY
School of Chemistry, University of Sydney, NSW, Australia, 2006

1. Introduction

Over recent years there has been a trend in super- and supramolecular chemistry towards the synthesis of extended molecular entities that are multi-component in nature. Such a trend has also occurred in macrocyclic ligand chemistry.

In this article the design, synthesis and d-block metal ion chemistry of some more recent examples of covalently-linked, macrocyclic ligand systems are discussed. The use of macrocyclic rings in such systems is not surprising given that the resulting macrocyclic complexes often exhibit both enhanced kinetic and thermodynamic stabilities and hence tend to retain their integrity under a variety of conditions - a lesson that nature knows well.

There have been a number of motivations for synthesising such ligands and investigating their metal ion binding properties. First, multi-metal ion complexes (including homo- and heterometallic systems) may yield unusual electronic, magnetic and/or redox properties that may arise as a consequence of the presence of cooperativity between the metal binding ion sites. Nevertheless, even when no cooperativity occurs there is the prospect that the metallo species may act as a multi-electron redox reagents. In addition, such metal-bound linked rings may serve as simple models for related metal-containing biochemical systems reflecting, at least in part, the relatively defined spatial and electronic environments associated with the metal binding sites in these structures. Further, linked systems are of considerable intrinsic interest since a number have now been demonstrated to give rise to new metal derivatives whose properties may be somewhat more than the 'sum of the parts'. Many examples of linked macrocyclic rings together with their metal ion chemistry have been reported. In particular, linked two-ring tetraaza macrocycles have been widely studied and interest in this ligand category has been heightened by the finding that individual compounds (and selected metal-ion derivatives), incorporating either alkyl or aromatic linking groups between the rings, are effective in inhibiting several strains of human immunodeficiency virus type 1 (HIV-1) and type 2 (HIV-2) [1].

Karsten Gloe (ed.), Macrocyclic Chemistry: Current Trends and Future Perspectives, 53–66.
© 2005 *Springer. Printed in the Netherlands.*

The metal ion chemistry of such linked bis-ring macrocycles has been extensively reviewed over recent years [2-6]. In view of this, the present discussion is restricted to linked tri- and higher-ring systems.

2. Linked Aza Macrocyclic Ligand Systems

A range of three-ring, linked aza macrocycles incorporating 9-membered 1,4,7-triazacyclononane (tacn), 12-membered cyclen and 14-membered cyclam and its derivatives have been reported but the number is much less than for the corresponding systems incorporating two linked macrocycles; systems containing four or more linked rings are quite rare.

Although 1,4,7-triazacyclononane was first incorporated into linked macrocyclic systems around three decades ago [7], it was only in 1997 that the corresponding three-ring analogue **1** was synthesised and investigated. The synthesis of **1** [8] (Scheme 1) proceeds from the tricyclic orthoamide derivative of 1,4,7-triazacyclononane **2** [9] and involves reaction with 1,3,5-tris(bromomethyl)benzene in acetonitrile in a 3:1 molar ratio followed by base hydrolytic work-up of the product. The addition of excess HBr to the reaction mixture led to isolation of **1** as its nonahydrobromide salt in 76% yield.

Reaction of **1** with Cu(II) yielded a magnetically dilute, trinuclear complex of type $[Cu_3(\mathbf{1})(H_2O)_6](ClO_4)_6 \cdot 6H_2O$. This species reacts with phosphate ion in the presence of hexafluorophosphate to yield the polymeric species $\{[Cu_3(\mathbf{1})(\mu\text{-}OH)(\mu_3\text{-}HPO_4)(OH_2)](PF_6)_3 \cdot 3H_2O\}_n$ whose X-ray structure shows that the cationic unit contains three Cu(II) ions coordinated to each of the tacn rings in a tetragonal distorted square pyramidal arrangement defined by the three nitrogens of a tacn ring and two oxygen atoms. The oxygen donors for two of these metal centres are provided by a bridging hydroxide ion together with an oxygen to each from the μ_3-phosphate anion. The phosphate also binds to a third Cu(II) belonging to a neighbouring complex, giving rise to the polymeric structure, with the coordination sphere of this latter copper ion being completed by a water molecule. A variable temperature magnetic study of the complex suggested that anti-ferromagnetic coupling occurs between the $Cu(\mu\text{-}OH)(\mu\text{-}HPO_4)Cu$ bridged centres as well as between these centres and the phosphate bridged centre; the complex has a $S = ½$ molecular ground state.

In a subsequent study two further hydroxo-bridged copper complexes of **1** were prepared [10]. The trinuclear species $[Cu_3(\mathbf{1})(\mu\text{-}OH)_2(H_2O)_2](ClO_4)_6 \cdot 3.2H_2O$ was isolated from an aqueous solution of $[Cu_3(\mathbf{1})(H_2O)_6](ClO_4)_6 \cdot 6H_2O$ that had been adjusted to pH 6. The former species was shown to contain a $[Cu_2(\mu\text{-}OH)_2]^{2+}$ core as well as an isolated Cu(II) centre. The magnetic properties of this species are in accord with the presence of an antiferromagnetically coupled Cu(II) pair and a magnetically isolated Cu(II) centre - overall the complex has an $S = ½$ ground state. When the pH of the above aqueous solution was adjusted to 9.5 a new species of type $[Cu_6(\mathbf{1})(\mu\text{-}OH)_6](ClO_4)_6 \cdot 2H_2O$ was isolated whose structure incorporates two trinuclear $[Cu_3(\mathbf{1})(\mu\text{-}OH)_2]$ units that are linked by two μ-hydroxo groups.

The synthesis and interaction of the related tetra-macrocyclic ligand **3** with Ni(II) and Cu(II) has been reported [11]. The synthesis of **3** follows a similar procedure to that employed for **1** starting from the tacn derivative **2** and 1,2,4,5-tetrakis(bromomethyl)benzene.

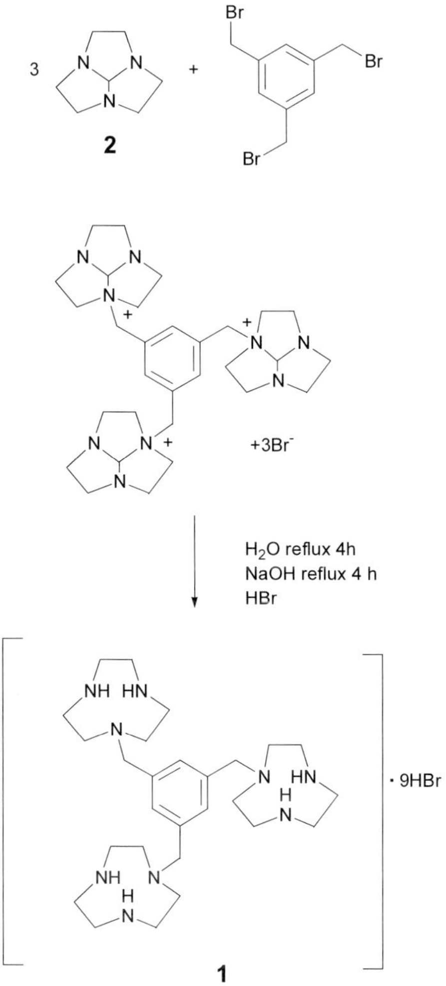

1

Scheme 1

Reaction of four equivalents of Ni(II) with **3** resulted in a mixture of di- and trinuclear complexes, $[Ni_2(\mathbf{3})](ClO_4)_4$ and $[Ni_3(\mathbf{3})(OH_2)_6](ClO_4)_6$, which were separated by cation exchange chromatography. A tetranuclear complex, $[Ni_4(\mathbf{3})(OH_2)_{12}](ClO_4)_8$, was also obtained when the reaction mixture contained a fifty-fold excess of Ni(II).

The X-ray structure of $[Ni_2(\mathbf{3})](ClO_4)_4$ shows that two pairs of *ortho*-attached tacn macrocycles each form a sandwich arrangement about a Ni(II) ion such that the coordination of each nickel is distorted octahedral. The two metal centres are separated by 9.09 Å and lie on opposite sides of the plane of the aromatic linker. A sandwich arrangement involving one pair of *ortho*-related tacn rings again occurs in the above trinuclear species; the other two rings bind one Ni(II) each with water ligands occupying the remaining sites to yield distorted octahedral arrangements.

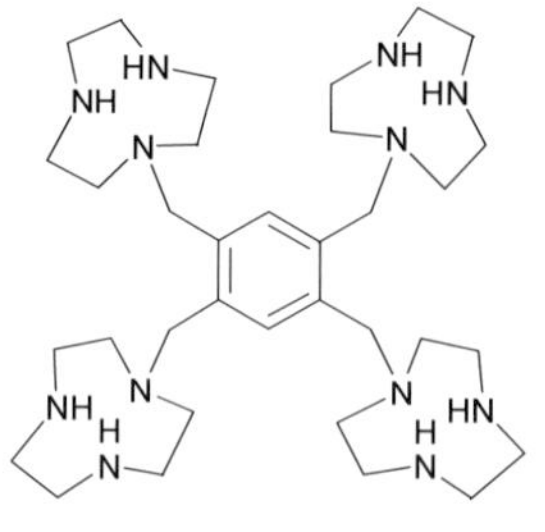

3

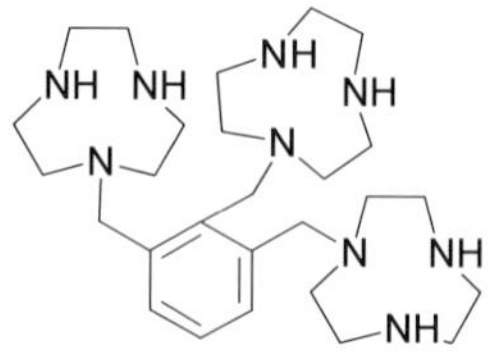

4

Di- and tetranuclear Cu(II) complexes were obtained on reaction of **3** with two and four equivalents of Cu(II), respectively. The former complex has a similar structure to the dinuclear nickel complex mentioned above, although the presence of Jahn-Teller distortion is also evident in the copper complex.

In the tetranuclear species, $[Cu_4(\mathbf{3})(OH_2)_8](ClO_4)_8$, each copper is bound to the four nitrogens of a separate macrocyclic ring as well as to two water molecules. The secondary amines and the waters around each copper form the basal plane of a square pyramid while the axial position is occupied by the tertiary amine. The arrangement of the ligand is such that *ortho*- and *para*-related macrocyclic compartments are *anti*, the shortest metal-metal distance being 7.04 Å between *ortho*-related pairs. Magnetic susceptibility measurements and EPR spectra confirmed the absence of coupling between metal centres in this complex.

The above study has been extended to the synthesis of **4** which was prepared in a similar fashion to **1** and **3** starting from tris(1,4,7-triazacyclonon-1-ylmethyl)benzene [12]. The X-ray structure of the $[Ni_2(\mathbf{4})(H_2O)_3]^{4+}$ cation shows that both nickel ions adopt distorted octahedral coordination geometries, with one nickel being sandwiched between two tacn residues while the other is coordinated to the third tacn residue with the coordination sphere being completed by three water ligands. Electrochemical studies indicate that the sandwiched Ni(II) centres in this and the related nickel complexes mentioned above may the be reversibly oxidised to the Ni(III) state. In the case of the dinuclear bis-sandwich complex, $[Ni_2(\mathbf{4})](ClO_4)_4$, the electrochemical results indicate that the two nickel centres behave in an essentially independent manner.

The two dendritic structures, **5** and **6**, incorporating tetra- and deca-macrocyclic sites, respectively, were both synthesised by a divergent procedure. This involved the use of (mono)tosylated N-protected tacn as the precursor for generation of the branching synthons and tacn itself for the core [13]. Spectrophotometric titrations in acetonitrile and DMF demonstrated that these ligands take up the expected four and ten Cu(II) ions, respectively.

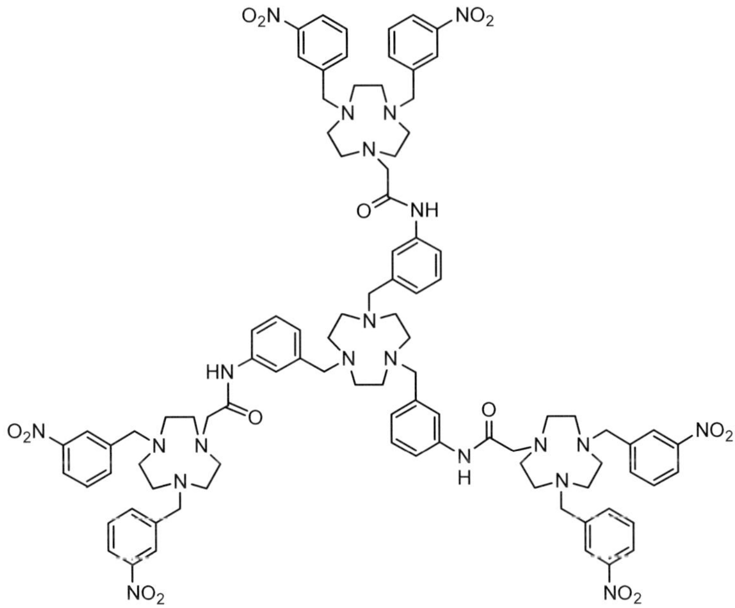

O2N
NO2
N
N
N
O
NH
N
N
N
HN
O
O
O2N
N
N
N
N
H
N
N
N
NO2
NO2
NO2

5

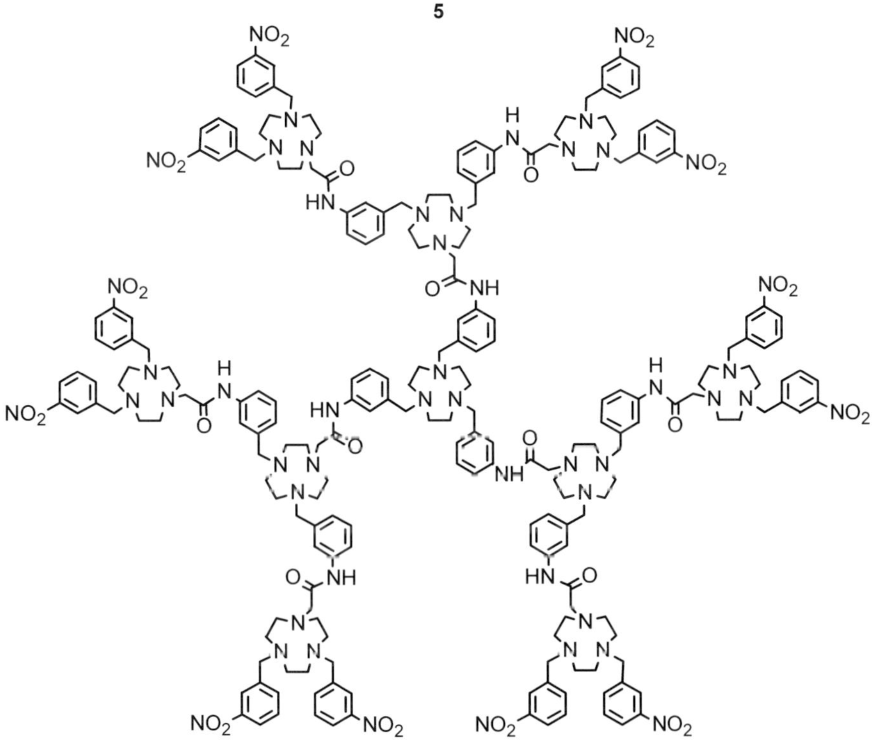

NO2
NO2
NO2
NO2
HN
O
O
NH
NO2
NO2
HN
O
O
NH
NO2
NO2
NO2
NO2
NO2
NO2

6

The trinuclear zinc complex of the tris(cyclen) derivative **7** linked through a 1,3,5-trimethylbenzene spacer has been shown to be an excellent receptor for organic phosphate dianions in aqueous solution relative to the parent Zn(cyclen)$^{2+}$ complex [14]. The tritopic ligand was synthesised by reaction between tris-Boc (*tert*-butyloxycarbonyl) protected cyclen and 1,3,5,-tris(bromomethyl)benzene in a sodium carbonate/acetonitrile mixture. Treatment of the tri-linked, *N*-protected product with concentrated HBr led to removal of the Boc substituents to yield the required tris(cyclen) derivative as its 9·HBr salt. An X-ray structure of $[Zn_3(\mathbf{7})(NO_3)_2(OH_2)](NO_3)_4$, shows that each zinc ion is bound to the four donors of a single cyclen ring with each ion residing on an equivalent face of the ring. An apical O-donor is also bound to each zinc (from nitrate ions in two cases and a water molecule for the third) to yield distorted tetragonal-pyramidal coordination spheres in each case.

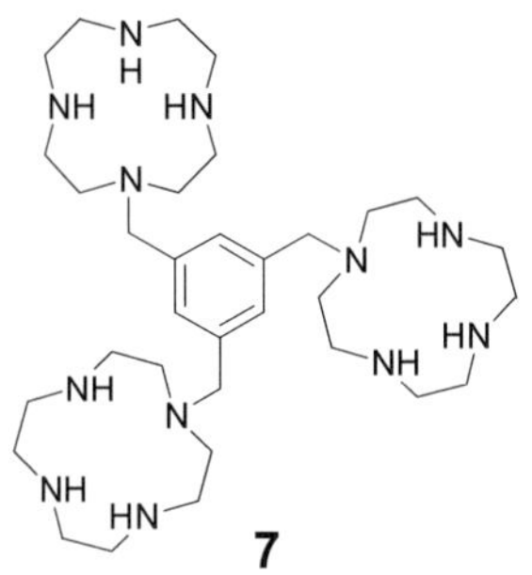

7

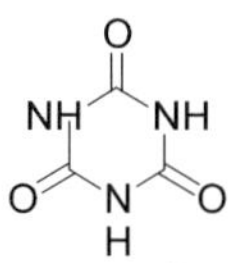

8

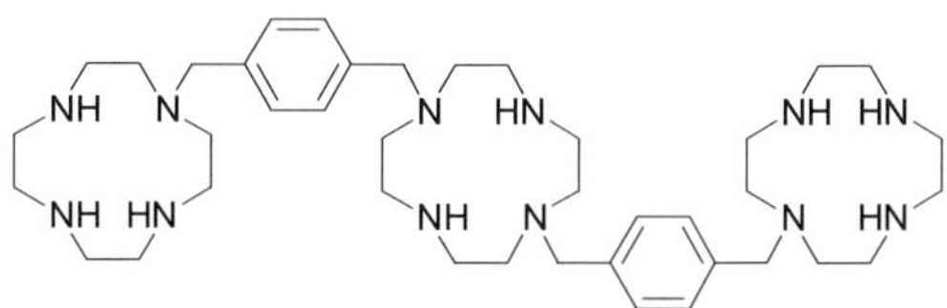

9

A 2:3 mixture of $[Zn_3(\mathbf{7})(H_2O)_3]^{6+}$ and cyanuric acid (**8**, CA) self-assemble in aqueous solution above pH 6 to yield a sandwich-like 2:3 supramolecular complex [15]. An X-ray diffraction study showed a hollow framework structure in which each doubly deprotonated CA group is sandwiched between two Zn(cyclen)$^{2+}$ moieties extending from two trinuclear zinc units. Unexpectedly, at pH 11.5 a new, highly symmetric 4:4 product was observed to form between the trinuclear zinc complex and CA^{3-}. This product may be represented as a truncated tetrahedron formed by binding four equilateral triangles and four scalene hexagons to each other through CA^{3-} to Zn(II) bonds. The tris-zinc species also interacts with thiocyanuric acid (TCA) to yield a self-assembled supramolecular species of type $(ZnL)_4(TCA^{3-})_4$. Bonding features of this assembly include the presence of Zn-S bonds as well as the presence of hydrogen bonds between the 1,3,5-triazine nitrogens and NH groups on cyclen [16]. An X-ray structure determination showed this species to be capsule-like: it corresponds to a twisted cubo-octahedral framework containing a truncated tetrahedral cavity.

The linear, xylyl-linked, tris-cyclen derivative **9** has also been prepared using a protecting group strategy [17]. Once again this species forms a trinuclear zinc complex which was demonstrated to exhibit selective and efficient binding of thymidylylthmidylylthymidine. In other work, the above "linear" tris-zinc complex has also been shown to recognise selectively an uridine-rich single strand in a biologically important RNA [18] as well as showing high binding affinity for AT-rich sequences in double-stranded DNA [19].

It has been well established that metal-directed condensation between metal-coordinated primary amines with aldehydes and dibasic acids such as nitroalkanes or primary amines provide a ready route to selected pendant arm macrocyclic ligands and their metal complexes [20]. In such a reaction a new six-membered chelate ring is generated between a pair of adjacent (cis-disposed) primary amines, with the pendant group attached to an uncoordinated nitrogen atom at the 'apex' of the 6-membered ring. A reaction of this type has been employed to generate the trinuclear copper complex of the tris-ring species **10** via reaction of three molar equivalents of the 1:1 copper complex of *N,N*-bis(2-aminoethyl)-1,3-propanediamine with melamine and formaldehyde [21]. The X-ray structure of the complex cation shows that each ring adopts a trans-III configuration, as is often observed in aliphatic 14-membered macrocyclic complexes [22]. The relative dispositions of the macrocyclic rings are different; two rings are located on the same side of the triazine core while the other is on the opposite side. The axial positions of each coordinated copper ion are filled by oxygens from either water or a perchlorato ligand. The presence of weak (three-way) dipole-dipole interactions was detected in the EPR spectrum of this complex. In other studies [23], a spectrophotometric investigation confirmed that a solution of the above complex interacts with its complementary 1,3,5-benzene-tricarboxylate anion. The 2:1 adduct of this anion with the trinuclear complex was isolated and its structure confirmed by X-ray diffraction.

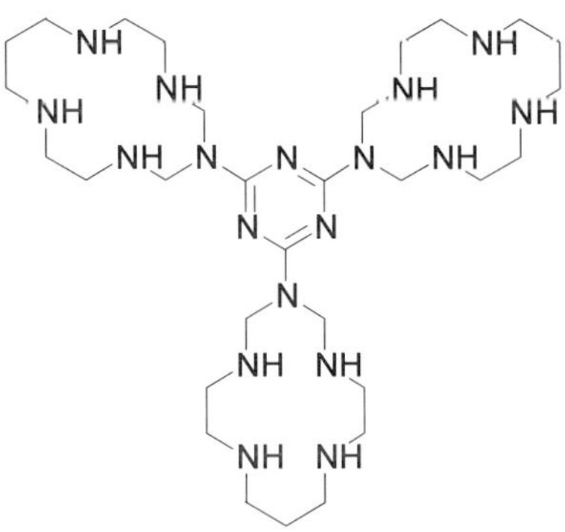

10

The synthesis of **11** (an analogue of **7** in which its 12-membered cyclen had been replaced by 14-membered cyclam rings) was reported in 1998 using the tri-N-tosylamide derivative of cyclam and 1,3,5-tris(bromomethyl)benzene as precursors [24]. The trinuclear Hg(II) derivative was shown to selectively recognise and bind a tripodal tris(histidine) ligand in which the three (S)-histidines are 12 Å apart. The system was seen as a model for peptide recognition.

Cyclam, with its 14-membered ring, has been demonstrated to complex with a very wide range of metal ions (and especially d-block metal ions) and tends to generate metal complexes that are especially kinetically and thermodynamically stable [22]. In particular instances it is also known to aid the stabilisation of less common oxidation states such as Ni(III), Cu(III), Ag(II) and Ag(III) [25,26].

Based on published procedures for preparing both di- and tri-N-protected *tert*-butoxycarbonyl (Boc) cyclam derivatives [27], we developed the alternate procedure shown in Scheme 2 for obtaining the tri-branched species **11** [28] as well as related procedures for the new tri-branched derivative **12** (see Scheme 3) [28], the linearly linked derivative **13** [28] and the linked cyclic derivative **14** [29]. In the case of the phloroglucinol derivative, tri-Boc protected cyclam was first acylated with chloroacetyl chloride; the resulting chloromethylamide was then used to trialkylate phloroglucinol in DMF at 70 °C over caesium carbonate as outlined in Scheme 3. Subsequent deprotection of this product followed by reduction with $BH_3 \cdot Me_2S$ afforded the corresponding deprotected species **12** [28].

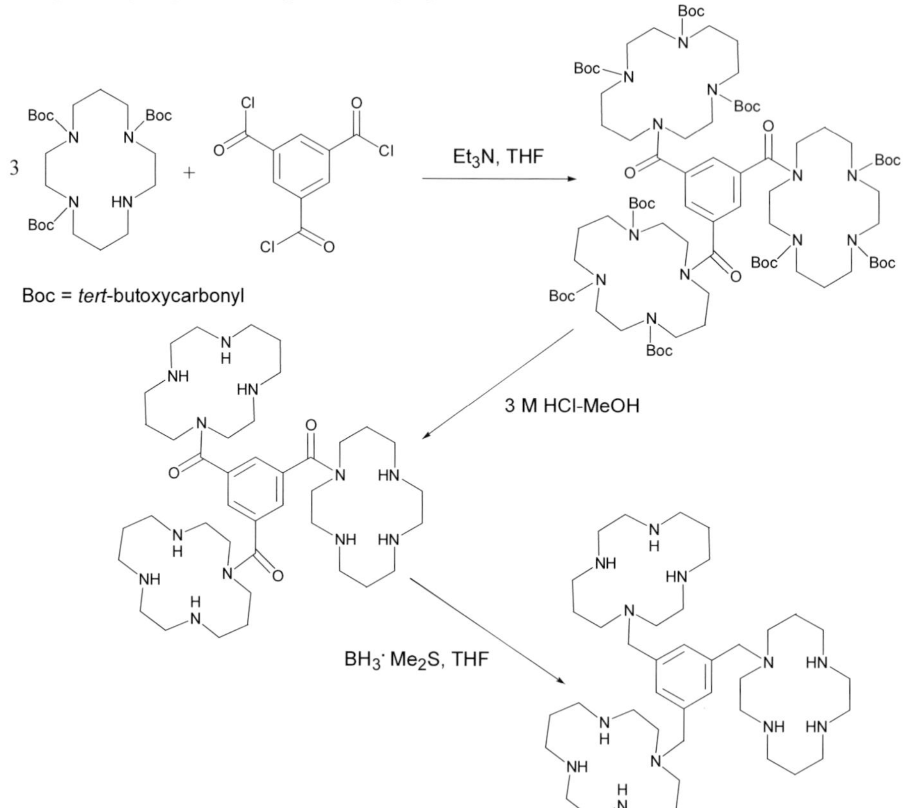

11

Scheme 2

The interaction of above two symmetrically branched tris-cyclam derivatives **11** and **12** with Ni(II), Cu(II), Zn(II) and Cd(II) has been investigated [30]. All four metal ions yield solid complexes for which the metal:ligand ratio is 3:1. For both ligand types, spectrophotometric titrations also confirm the formation of Ni(II) and Cu(II) complexes of similar 3:1 stoichiometry in dimethylsulfoxide. Where possible, visible spectral, electrochemical, magnetic moment, ESR and NMR studies have been performed to probe the nature of the respective complexes and the results compared with those from parallel investigations in which the corresponding mono-substituted cyclam analogues were employed as the ligands. In general, the metal ion chemistries of the latter ligands are similar to those of the corresponding tri-branched ligand systems. Little communication between individual metal centres in the complexes of **11** and **12** was evident. A structural determination employing a small crystal of the trinuclear Ni(II) complex of **11** was successfully carried out with the aid of a synchrotron radiation source [30]. A nickel ion was shown to occupy each cyclam ring in a square-planar (low-spin) coordination arrangement, with each cyclam ring adopting the stable *trans*-III configuration [31].

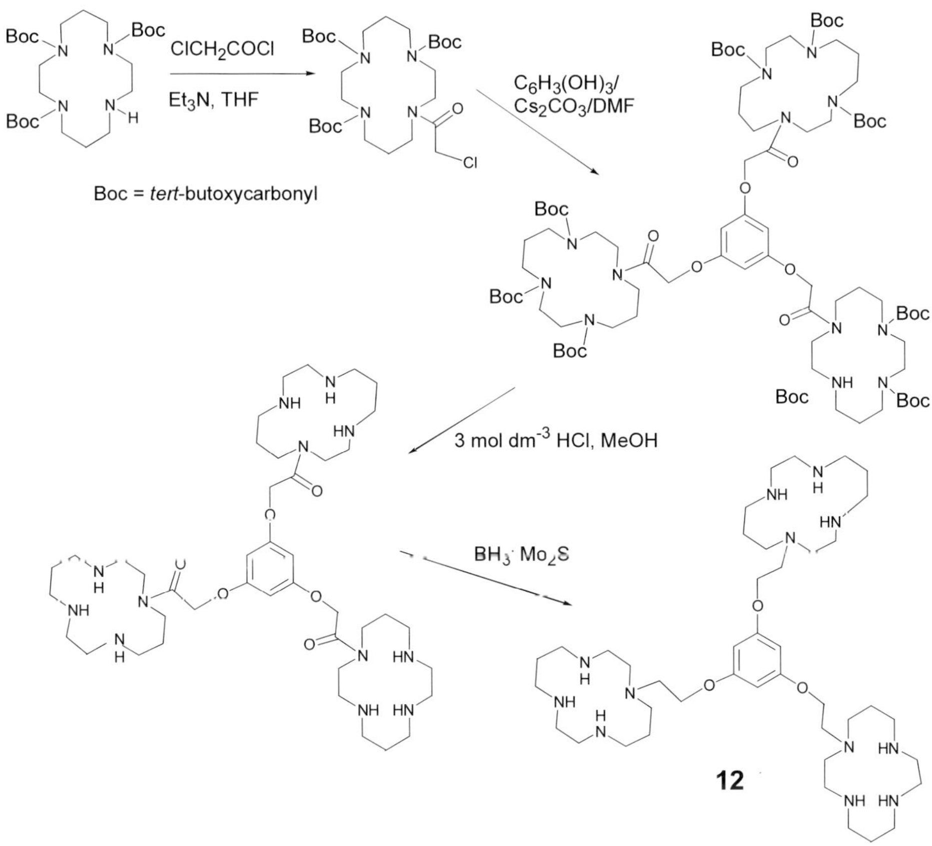

Scheme 3

The interaction of the linearly linked tris-cyclam derivative **13** with Ni(II), Cu(II), Zn(II), Cd(II), and Pd(II) has been investigated [32]. As for the above tri-branched systems, all five metals yield solid complexes in which the metal:ligand stoichiometry is 3:1 with, for Cu(II), a spectrophotometric titration also confirming the formation of a complex of this stoichiometry in acetonitrile. Cyclic voltammograms of both the Ni(II) (low-spin) and Cu(II) complexes both yield evidence for the presence of M(II)/M(III) as well as M(I)/M(II) couples in acetonitrile.

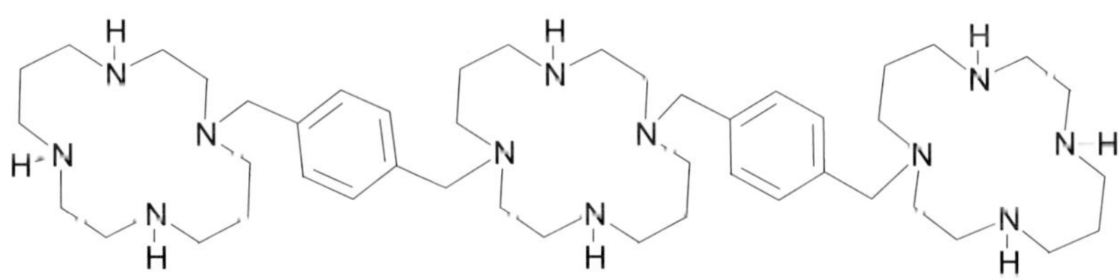

13

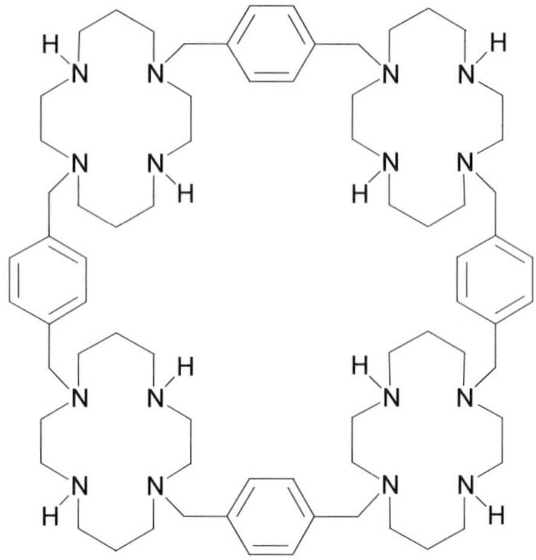

14

The novel tetra-linked derivative **14**, which forms 4:1 (metal:ligand) complexes with Ni(II) and Cu(II), [29] incorporates a large central cavity that has the potential to act as a large receptor for inclusion of a suitable guest. Variation of the coordinated metal in such a system provides the potential for 'tuning' the electronic nature of this cavity and hence influence the electronic environment of an included guest.

3. Linked mixed-donor Ring Systems

In an early study, the author's group synthesised the lipophilic, tri-branched ligand **15** for use in two-phase solvent extraction as well as three phase membrane transport studies [33]; the parent ring system in this case has been well documented to interact with a range of first-row transition and post transition ions [34].

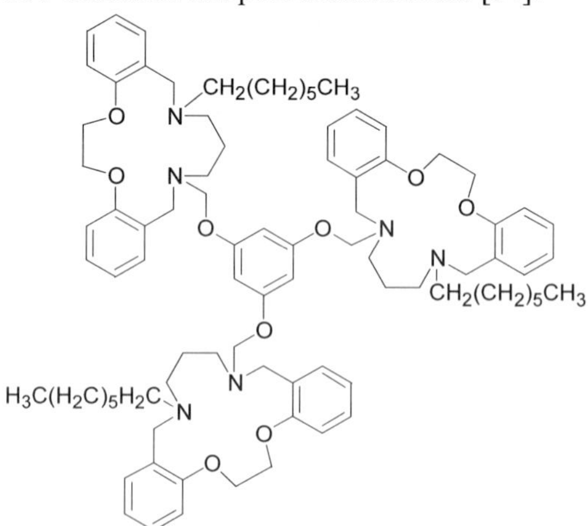

15

Based in part on parallel studies during which strategies for obtaining linked two-ring, mixed donor systems were developed [35], successful procedures for the preparation of the linked systems **16** and **17** incorporating 16-membered, N_2S_2-donor macrocycles and 1,3,5-tribenzyl or phloroglucinol cores were carried out [36]. The latter procedures are also related in broad detail to those outlined above for preparing the linked tetraaza macrocyclic systems **11 - 14**. The procedure for **16** (R = H), outlined in Scheme 4, is representative. It involves the use of both *tert*-butoxycarbonyl (Boc) and 2,2,2-trichloroethoxycarbonyl (Troc) N-protecting groups in order to provide differential deprotection at different stages in the step-wise synthesis.

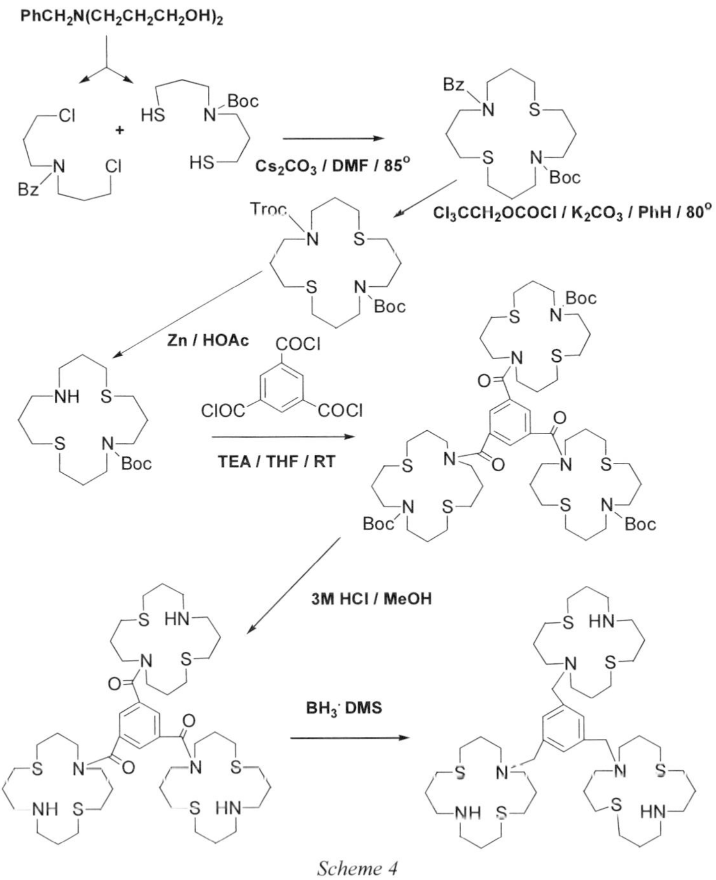

Scheme 4

The metal ion binding properties of **16** and **17** (with R= H and benzyl) towards the soft metal ions Ag(I) [37], Pd(II) [38] and Pt(II) [38] have been investigated. A number of complexes of type $[Ag_3L](NO_3)_3$ (L = **16**, **17**) and $[M_3L](PF_6)_6$ (M = Pd or Pt, L = **16**, **17**) were isolated and their properties compared with the corresponding complexes of the analogous (similarly substituted) single-ring macrocycles. Both the microanalytical and the mass spectrometric results confirmed the 3:1 (metal:ligand) stoichiometries of the tris-ring complexes. Undoubtedly the metal ions occupy the macrocyclic cavities in each of these complexes. In this context it is noted that the X-ray structure of mononuclear $[PdL](PF_6)_2$ (where L is a mono *N*-benzylated derivative of the corresponding single-ring, N_2S_2-donor macrocycle) showed that the palladium coordinates to all four donor atoms of the macrocycle in a square planar fashion. NMR titrations in the case of Ag(I) [37], and spectrophotometric titrations for Pd(II) and Pt(II) [38], all confirm that 3:1 stoichiometries once again occur for the complexes of the tri-linked ligands in non-aqueous solution. Clear step-wise coordination was detected in the case of Ag(I) interacting with both **16** and **17**, with the individual step-wise stability constants showing the expected decrease for formation of the respective 1:1, 2:1 and 3:1 metal complex species.

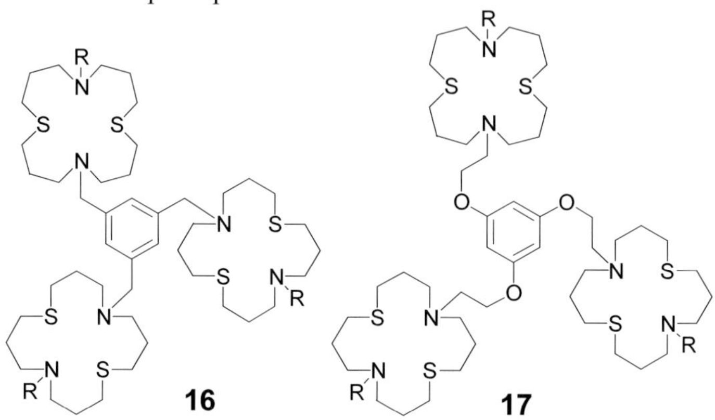

Competitive (seven-metal) solvent extraction experiments (water/chloroform) and related bulk membrane transport (water/chloroform/water) experiments have been performed in which each of the four tri-branched ligands as well as their single ring analogues were employed as the extractant/ionophore in the respective chloroform phases [37]. In both sets of experiments the aqueous source phases contained an equimolar mixture of Co(II), Ni(II), Cu(II), Zn(II), Cd(II), Ag(I) and Pb(II) nitrates and were buffered at pH 4.9. For membrane transport the aqueous receiving phase was buffered at pH 3; under these conditions any transport will be driven by the 'back' transfer of protons. Under the conditions employed, the results from the solvent extraction and the bulk membrane transport experiments clearly paralleled each other - for each ligand system high extraction/transport selectivity for Ag(I) was observed over the other six metal ions present in the respective source phases.

In an extension of the above studies the tri-branched system **16** (R = H) was employed as a first generation core for the synthesis of a second generation dendritic system [39]. Tris N-alkylation of **16** (R = H) with three equivalents of the bis macrocycle-containing fragment **18** yielded the nine-ring dendritic species **19** which was readily characterised using normal spectroscopic techniques (including high resolution mass spectrometry). This product reacts with Pd(II) and Pt(II) to yield yellow complexes that, however, proved more difficult to characterise. For example, both resisted mass spectral characterisation despite the use of a variety of ionisation methods (EI, FAB, MALDI or ES); this may reflect the presence of an 18+ charge arising from a Pd(II) or Pt(II) occupying each of the macrocyclic sites. Nevertheless, a spectrophotometric titration involving the addition of the nine-ring dendrimer to Pd(II) chloride in acetonitrile yielded a sharp 9:1 endpoint, confirming the uptake of nine Pd(II) ions per dendrimer molecule in this case.

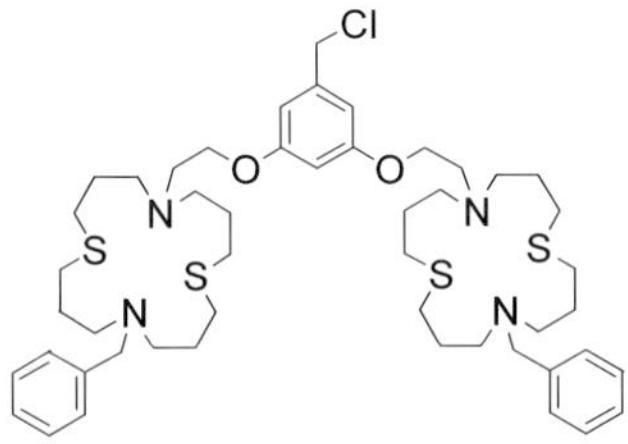

18

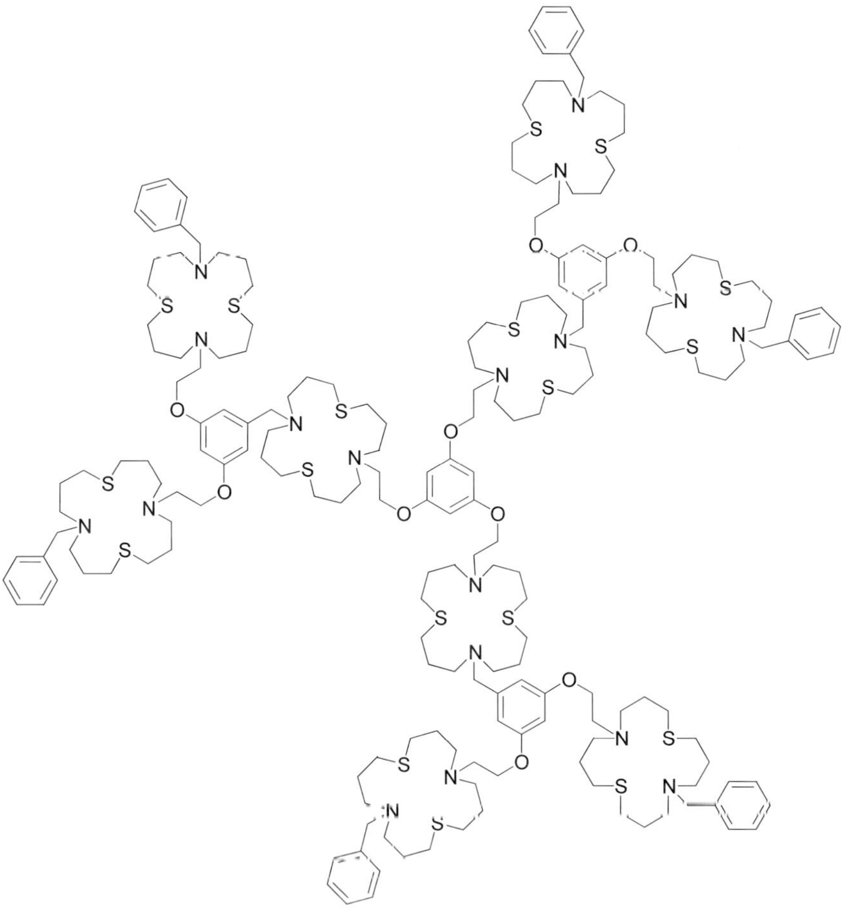

19

4. Acknowledgements

The author thanks his students and collaborators, named in the references, who performed many of the investigations described in this chapter; he especially acknowledges the major contribution of Assoc. Professor G. V. Meehan of James Cook University to several of the studies. The support of the Australian Research Council is also acknowledged.

5. References

1. G. J. Bridger, R. T. Skerlj, S. Padmanabhan, S. A. Martellucci, G. W. Henson, M. J. Abrams, H. C. Joao, M. Witvrouw, K. D. Vreese, R. Pauwels and E. De Clercq, *J. Med. Chem.*, 1996, **39**, 109; X. Liang, J. A. Parkinson, M. Weishaeupl, R. O. Gould, S. J. Paisey, H.-S. Park, T. M. Hunter, G. A. Blindauer, S. Parsons and P. J. Sadler, *J. Am. Chem. Soc.*, 2002, **124**, 9105 and refs. therein.
2. G. De Santis, L. Fabbrizzi, M. Licchelli, C. Mangano, P. Pallavicini, A. Poggi, D. Sacchi and A. Taglietti, *Supramol. Chem.*, 1996, **6**, 239.
3. L. F. Lindoy, *Adv. Inorg. Chem.*, 1998, **45**, 75.
4. L. F. Lindoy, *Coord. Chem. Rev.*, 1998, **174**, 327.
5. J. D. Chartres, L. F. Lindoy and G. V. Meehan, *Coord. Chem. Rev.*, 2001, **216-217**, 249.
6. A. McAuley and S. Subramanian, *Coord. Chem. Rev.*, 2000, **200-202**, 75.
7. N. Tanaka, Y. Kobayashi and S. Takamoto, *Chem. Lett.*, 1977, 107.
8. L. Spiccia, B. Graham, M. T. W. Hearn, G. Lazarev, B. Moubaraki, K. S. Murray and E. R. T. Tiekink, *Dalton Trans.*, 1997, 4089.
9. X. Zhang, W.-Y. Hsieh, T. N. Margulis and L. J. Zompa, *Inorg. Chem.*, 1995, **34**, 2883.
10. B. Graham, L. Spiccia, G. D. Fallon, M. T. W. Hearn, F. E. Mabbs, B. Moubaraki and K. S. Murray, *Dalton Trans.*, 2001, 1226.
11. B. Graham, M. J. Grannas, M. T. W. Hearn, C. M. Kepert, L. Spiccia, B. W. Skelton and A. H. White, *Inorg. Chem.*, 2000, ***39***, 1092.
12. B. Graham. L. Spiccia, A. M. Bond, M. T. W. Hearn and C. M. Kepert., *Dalton Trans.*, 2001, 2232.
13. P. D. Beer and D. Gao, *Chem. Commun.*, 2000, 443.
14. E. Kimura, S. Aoki, T. Koike and M. Shiro, *J. Am. Chem. Soc.*, 1997, **119**, 3068.
15. S. Aoki, M. Shiro, T. Koike and E. Kimura, *J. Am. Chem. Soc.*, 2000, **122**, 576.
16. S Aoki, M. Shiro and E. Kimura, *Chem. Eur. J.*, 2002, **8**, 929; S. Aoki, M. Zulkefeli, M. Shiro and E. Kimura, *Proc. Nat. Acad. Sci., USA.* 2002, **99**, 4894.
17. E. Kimura, M. Kikuchi, H. Kitamura and T. Koike, *Chem. Eur. J.*, 1999, **5**, 3113.
18. E. Kikuta, S. Aoki and E. Kimura, *J. Am. Chem. Soc.*, 2001, **123**, 7911.
19. E. Kikuta, S. Aoki and E. Kimura, *J. Biol. Inorg. Chem.*, 2002, **7**, 473.
20. P. V. Bernhardt and G. A. Lawrance, *Coord. Chem. Rev.,* 1990, **104**, 297.
21. P. V. Bernhardt and E. J. Hayes, *Dalton Trans.*, 1998, 3539.
22. L. F. Lindoy, *"The Chemistry of Macrocyclic Ligand Complexes"*, Cambridge University Press, 1989.
23. Comba, Y. D. Lampeka, A. I. Prikhod'ko and H. Pritzkow, *Theo. Exper. Chem.*, 2002, **37**, 236.
24. S. Sun, J. Saltmarsh, S. Mallik and K. Thomasson, *Chem. Commun.*, 1998, 519.
25. G. De Santis, L. Fabbrizzi, M. Licchelli and P. Pallavicini, *Coord. Chem. Rev.*, 1992, **120**, 237.
26. A. De Blas, G. De Santis, L. Fabbrizzi, M. Licchelli, P. Pallavicini, *Pure Appl. Chem.*, 1993, **65**, 455.
27. B. Boitrel, B. Andrioletti, M. Lachkar and R. Guilard, *Tetrahedron Lett.*, 1995, **36**, 4995; S. Brandes, C. Gros, F. Denat, P. Pullumbi and R. Guilard, *Bull. Soc. Chim. Fr.*, 1996, **133**, 65; F. Denat, S. Brandes and R. Guilard, *Synlett,* 2000, 561.
28. Y. Dong and L. F. Lindoy, *Aust. J. Chem.*, 2001, 54 291.
29. Y. Dong and L. F. Lindoy, unpublished work.
30. Y. Dong, L. F. Lindoy, P. Turner and G. Wei, *Dalton Trans.*, 2004, 1264.
31. M. A. Donnelly and M. Zimmer, *Inorg. Chem.*, 1999, **38**, 1650.
32. Y. Dong and L. F. Lindoy, *Coord. Chem. Rev.*, 2003, **245**, 11.
33. I. M. Atkinson, D. M. Boghai, B. Ghanbari, L. F. Lindoy, G. V. Meehan and V. Saini, *Aust. J. Chem.*, 1999, ***52***, 351.
34. K. R. Adam, A. J. Leong, L. F. Lindoy and G. Anderegg., *Dalton Trans.*, 1988, 1733.
35 J. D. Chartres, A. M. Groth, L. F. Lindoy, M. P. Lowe and G. V. Meehan, *Perkin Trans.1*, 2000, 3444.
36. A. M. Groth, L. F. Lindoy and G. V. Meehan, *Perkin Trans.1*, 1996, 1553; I. M. Atkinson, A. M. Groth, L. F. Lindoy, O. A. Matthews and G. V. Meehan, *Pure & Appl. Chem.*, 1996, **68**, 1231.
37. J. D. Chartres, A. M. Groth, L. F. Lindoy and G. V. Meehan, *J. Chem. Soc., Dalton Trans.*, 2002, 371.
38. I. M. Atkinson, J. D. Chartres, A. M. Groth, L. F. Lindoy, M. P. Lowe, G. V. Meehan, B. W. Skelton and A. H. White, *Dalton Trans.*, 2001, 2801.
39. I. M. Atkinson, J. D. Chartres, A. M. Groth, L. F. Lindoy, M. P. Lowe and G. V. Meehan, *Chem. Commun.*, 2002, 2428.

5.
BINDING AND STRUCTURAL ASPECTS OF NITRILE- AND AMINO-FUNCTIONALISED PENDANT ARM DERIVATIVES OF 1,4,7-TRIAZACYCLONONANE ([9]aneN_3)

MARTIN SCHRÖDER[a] AND VITO LIPPOLIS[b]
[a] *School of Chemistry, The University of Nottingham, University Park, Nottingham NG7 2RD, U.K.*
[b] *Dipartimento di Chimica Inorganica ed Analitica, Università degli Studi di Cagliari, S.S. 554 Bivio per Sestu, 09042 Monserrato (CA), Italy*

1. Introduction

The coordination chemistry of the tridentate macrocycle 1,4,7-triazacyclononane ([9]aneN_3) and its *N*-alkylated analogues (R_3-[9]aneN_3) with main group and transition metal ions has been comprehensively studied over the last thirty years [1-7]. In addition to kinetically inert and thermodynamically stable sandwich complexes, a series of polynuclear complexes have also been prepared in the presence of suitable bridging ligands. In these species, the [9]aneN_3 macrocyclic framework serves as a face-capping group, allowing variation of the donor array at the remaining coordination sites of a coordinated metal ion.

The binding properties of [9]aneN_3 to metal cations can be finely adapted through sequential functionalisation of the secondary amines with pendant arms bearing additional coordinating groups to generate ligands with increasing number of donor atoms. Symmetric *N*-functionalisation of [9]aneN_3 *via* incorporation of three identical pendant arms terminated with neutral or anionic N-, O-, S-, or P-donor groups has afforded a great variety of effective hexadentate ligands. These often confer remarkable stability upon metal centres in low nuclearity complexes, and can adapt to their binding mode to the preferred coordination geometries and oxidation states of metal centre(s) [8-15]. However, much less work has been reported on selectively *N*-functionalised derivatives of [9]aneN_3 bearing only one or two pendant arms having identical coordinating donor groups or "innocent" alkyl substituents [8-10, 16-21]. Even less work has been reported on the synthesis of asymmetric derivatives of [9]aneN_3 having different *N*-attached pendant-donors groups, reflecting the increased and significant synthetic difficulty encountered in the preparation of these types of ligands using a multistep approach [22, 23].

Karsten Gloe (ed.), Macrocyclic Chemistry: Current Trends and Future Perspectives, 67–86.
© 2005 *Springer. Printed in the Netherlands.*

Nevertheless, pendant arm derivatives of [9]aneN_3 and other polyaza-macrocycles have garnered great interest in recent years for their use in many different chemical applications such as catalysis [24-30], selective cation binding [30-32], surfactants [33, 34], mimicry of enzymes and siderophores [35-39], tumor-directed radioisotope carriers, and use in magnetic resonance imaging reagents [40].

The synthesis, structure, and coordination properties of [9]aneN_3 and other polyaza-macrocycles bearing *N*-attached coordinating pendant groups have been reviewed [1, 8-10, 40]. The present review covers the last four years of our work in this field and is only focussed on synthetic aspects and coordination properties of nitrile- and amino-functionalised pendant arms derivatives of [9]aneN_3. Their use in developing new high-yield synthetic routes for the preparation of asymmetric derivatives of [9]aneN_3 with more than one type of pendant donors and featuring amino-functionalised pendant arms are also discussed.

2. Nitrile-Functionalised Pendant arm Derivatives of [9]aneN_3

N-Functionalised derivatives of [9]aneN_3 bearing three nitrile pendant arms have been successfully used as precursors in the synthesis of the corresponding amino derivatives (see next paragraph)[41-45]. However, their coordination chemistry has not been studied extensively. To date few examples have been reported for the synthesis of mono- and di-nitrilealkyl pendant arm derivatives of [9]aneN_3 [42], although an efficient synthetic procedure has been developed for the ligands illustrated in Scheme 1 [41, 43-45]. Thus, addition of chloroacetonitrile to [9]aneN_3·3HBr in EtOH in the presence of an excess of Et_3N affords $\mathbf{L^1}$ [43, 45], while Michael addition of acrylonitrile to the appropriate [9]aneN_3 precursor gives $\mathbf{L^2}$-$\mathbf{L^4}$ in high yield [41, 44, 45].

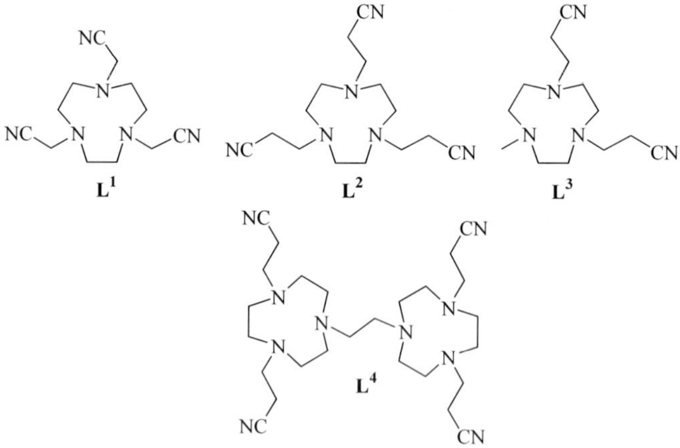

Scheme 1. Summary of nitrile-functionalised pendant arm derivatives of [9]aneN_3.

2.1 REACTIVITY TOWARDS Ag^I

In principle, $\mathbf{L^1}$-$\mathbf{L^4}$ should not form sandwich complexes or encapsulate a metal centre due to the steric hindrance of the nitrile-functionalised pendant arms and to the inability of the nitrile group to bind *via* σ M←:N≡C–R donation in a chelate fashion to a metal ion sitting within the macrocyclic cavity. These ligands should instead promote the formation of multinuclear or polymeric complexes *via exo* coordination of the nitrile functionality. In fact, reaction of one molar equivalent of $AgPF_6$ with $\mathbf{L^1}$ gives the polymeric complex $\{[Ag(\mathbf{L^1})]PF_6\}_\infty$ in which the Ag^I ion is bound to *N*-donors is a distorted octahedral geometry [45, 46]. One face is taken up by the three *N*-donors of the triaza ring [Ag–N([9]aneN_3) 2.523(4)-2.547(4) Å], with the remaining three positions occupied by the *N*-donors of nitrile groups [Ag–N(C≡N) 2.311(4)-2.486(4) Å] belonging to three different $[Ag(\mathbf{L^1})]^+$ complex cation (Fig. 1a). An uncommon 3D inorganic network is, therefore, formed in which each molecule of $\mathbf{L^1}$ is a node linked to four different Ag^I centres and each Ag^I ion represents a six-connected junction via NCH_2CN linkers to six other Ag^I ions in an overall 3D single network (Fig. 1b). Interestingly, the structure of this inorganic network does not depend upon whether BF_4^- or PF_6^- is the counter anion since the channels within the polymer can accommodate both anions.

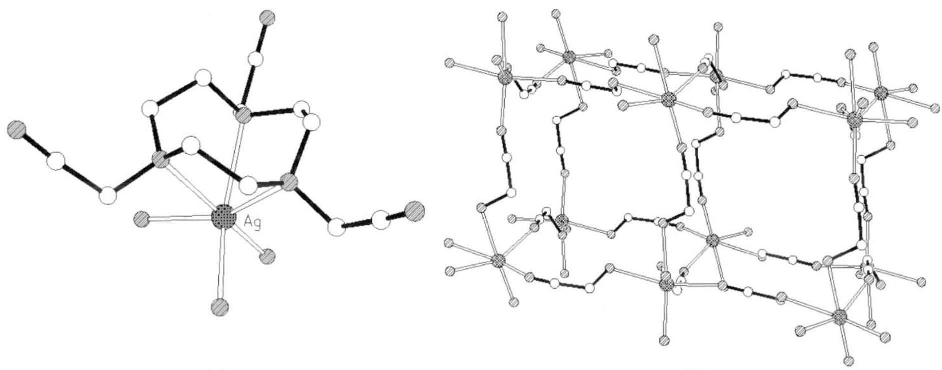

(a) (b)

Figure 1. (a) View of the coordination sphere in the complex cation $[Ag(\mathbf{L^1})]^+$; (b) partial view of the $\{[Ag(\mathbf{L^1})]^+\}_\infty$ 3D polymer: the ethylene units belonging to [9]aneN_3 frameworks are omitted for clarity to better show the six-connected single network at Ag^I centres.

The ligand $\mathbf{L^2}$, which differs from $\mathbf{L^1}$ in the length of each pendant arm, gives the polymeric complex $\{[Ag_2(\mathbf{L^2})_2][BF_4]_2\}_\infty$ upon reaction with $AgBF_4$ in a 1:1 molar ratio in MeCN [45, 46]. In this compound the repeating unit is the binuclear complex cation $[Ag_2(\mathbf{L^2})_2]^{2+}$ lying across a crystallographic inversion centre in which each Ag^I centre is bound to four *N*-donors in a distorted tetrahedral geometry: three are provided by the [9]aneN_3 framework of $\mathbf{L^2}$ [Ag–N([9]aneN_3) 2.412(5)-2.504(5) Å] and the fourth comes from the nitrile group [Ag–N(C≡N) 2.192(6) Å] of a pendant arm from a symmetry related $[Ag(\mathbf{L^2})]^+$ unit (Fig. 2). Each Ag^I ion interacts also with one of the two remaining pendant arms of $\mathbf{L^2}$ [Ag–N(C≡N) 2.779(7) Å] of an inversion-related $[Ag_2(\mathbf{L^2})_2]^{2+}$ binuclear fragment to give an infinite zigzag polymer along the [100] direction (Fig. 2).

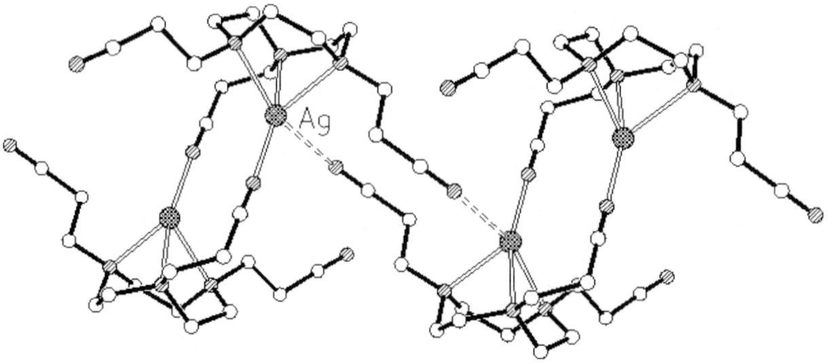

Figure 2. Packing diagram showing part of an $\{[Ag_2(\mathbf{L}^2)_2]^{2+}\}_\infty$ polymeric chain.

Interestingly, the reaction of $\mathbf{L}^4$ with two molar equivalents of $AgPF_6$ in MeCN gives the 2D polymeric complex $\{[Ag_2(\mathbf{L}^4)][PF_6]_2\}_\infty$ [45]. In this compound the repeating unit is the binuclear complex cation $[Ag_2(\mathbf{L}^4)]^{2+}$ lying across a crystallographic inversion centre. Each [9]aneN$_3$ moiety in the ligand hosts one Ag^I centre, and each Ag^I ion in the repeating unit $[Ag_2(\mathbf{L}^4)]^{2+}$ is bound to five *N*-donors in a highly distorted trigonal bipyramical coordination geometry, with one axial position and two equatorial positions taken up by the three *N*-donors of a triaza ring moiety [Ag–N([9]aneN$_3$) 2.464(4)-2.483(3) Å], and the remaining sites occupied by the *N*-donors of nitrile groups from two different symmetry-related $[Ag_2(\mathbf{L}^4)]^{2+}$ units [Ag–N(C≡N) 2.262(4), 2.501(4) Å]. The two pentadentate compartments of $\mathbf{L}^4$ are arranged in an *anti* configuration and a 2D inorganic network is, therefore, formed in which each penta-coordinate Ag^I ion is connected to two other Ag^I centres through two NCH_2CH_2CN linkers of the pendant arms of $\mathbf{L}^4$, and to a third metal centre *via* an NCH_2CH_2N linker of ethylene bridge between the two [9]aneN$_3$ units of $\mathbf{L}^4$. As observed in the structure of the polymeric complex cation $\{[Ag_2(\mathbf{L}^2)_2]^{2+}\}_\infty$ (Fig. 2), ribbons of fused 12-membered rings connected at Ag^I spiro-centres can be envisaged within the 2D architecture of $\{[Ag_2(\mathbf{L}^4)]^{2+}\}_\infty$, each comprising two metal centres and two NCH_2CH_2CN linkers connecting them. These ribbons run along the [010] direction, are stacked along [100], and are connected at the metal centres *via* NCH_2CH_2N linkers. In terms of connectivity, the 2D network is perhaps best viewed as a distorted "brick wall" structure constructed using two different type of linkers (Scheme 2c). Thus, by simply changing the pendant arm length from CH_2CN in $\mathbf{L}^1$ to CH_2CH_2CN in $\mathbf{L}^2$, and the number of [9]aneN$_3$ moieties within the ligand from one in $\mathbf{L}^2$ to two in $\mathbf{L}^4$, different network motifs and dimensionalities for the resulting coordination polymers with Ag^I are obtained (Scheme 2).

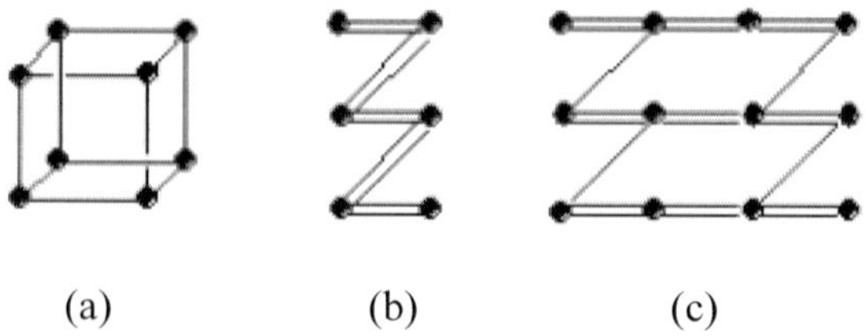

Scheme 2. Schematic representation of the connectivity in the structures of the polymeric complex cations: (a) $\{[Ag(\mathbf{L}^1)]^+\}_\infty$, (b) $\{[Ag_2(\mathbf{L}^2)_2]^{2+}\}_\infty$, and (c) $\{[Ag_2(\mathbf{L}^4)]^{2+}\}_\infty$.

2.2 REACTIVITY TOWARDS Cu^{II}

Further evidence for the inability of the nitrile groups in **L**[1]-**L**[3] to participate in efficient σ M←:N≡C–R *endo*-binding to a metal ion sitting within the [9]aneN_3 cavity is afforded by their reaction with one equivalent of $Cu(BF_4)_2{\cdot}4H_2O$ in MeOH at 65°C [47]. Methanolysis of two nitrile groups is observed in these reactions and blue crystalline complexes in which Cu^{II} is encapsulated by the resulting ligands **L**[5]-**L**[7] featuring imino-ether groups (Scheme 3) have been isolated and characterised.

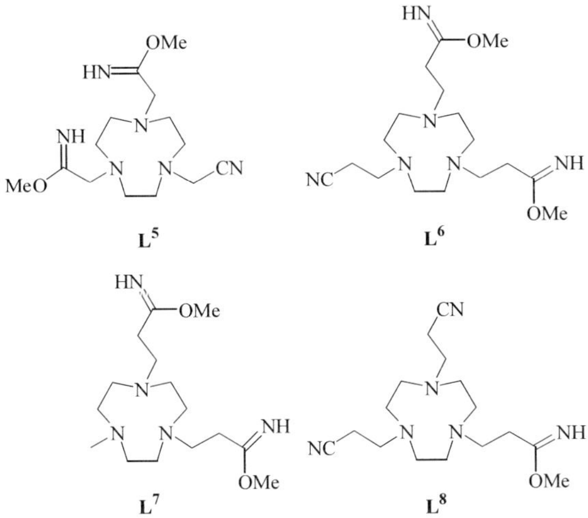

Scheme 3. Summary of the ligands obtainable starting from **L**[1]-**L**[3] by Cu^{II}-assisted methanolysis of cyano groups.

The stereochemistry at the Cu^{II} centres in these complexes is square-based pyramidal with two imine donors from the pendant arms and two tertiary amines from the macrocyclic framework occupying the basal positions. The apical position is occupied by the remaining N-donor of the [9]aneN_3 moiety bearing a nitrile-functionalised pendant arm (Fig. 3).

In $[Cu(\mathbf{L}^5)]^{2+}$ [Cu–N([9]aneN_3) 2.044(5)-2.249(5), Cu–N(C=N) 1.942(5), 1.947(5) Å] (Fig. 3a) three five-membered chelate rings are formed in the basal plane with very similar angles [82.8(2)-86.1(2)°] subtended at the metal centre. However, due to the high degree of planarity in the imino-ether moieties, the torsion angles about the central C–C bonds for the chelate rings involving the pendant arms [16.9(8) and 21.8(8)°] are significantly different from the corresponding torsion angle in the chelate ring involving the [9]aneN_3 framework [46.7(11)°].

In $[Cu(\mathbf{L}^6)]^{2+}$ [Cu–N([9]aneN_3) 2.050(3)-2.280(3), Cu–N(C=N) 1.979(3), 2.003(3) Å] (Fig. 3b), because of the longer pendant arms, one five- and two six-membered chelate rings are formed in the basal coordination plane, with angles subtended at the metal centre of 81.6(1)° for the former and of 90.7(1) and 92.3(1)° for the latter. While the five-membered ring adopts a *gauche* conformation with a torsion angle about the central C–C bond of 47.3(4)°, the two six-membered chelate rings show *pseudo*-boat conformations due to the planarity of the imino-ether fragments. Interestingly, for both complexes, $[Cu(\mathbf{L}^5)]^{2+}$ and $[Cu(\mathbf{L}^6)]^{2+}$, the nitrile group on the pendant arm attached to the apical *N*-donor does not undergo nucleophilic attack by MeOH, and this group is, therefore, oriented away from the square-based pyramidal sphere of the metal centre. To our knowledge, only one example of Cu^{II}-promoted nucleophilic attack on nitrile groups has been reported previously; this report concerns the alcoholysis of bis(2-cyanoguanidine) ligands with the formation of square-planar Cu^{II} complexes of the resulting (amidino-*O*-alkylurea) ligands [48]. This reaction takes place with either MeOH or EtOH and with different Cu^{II} salts (BF_4^-, Cl^-, Br^-, and SO_4^{2-}).

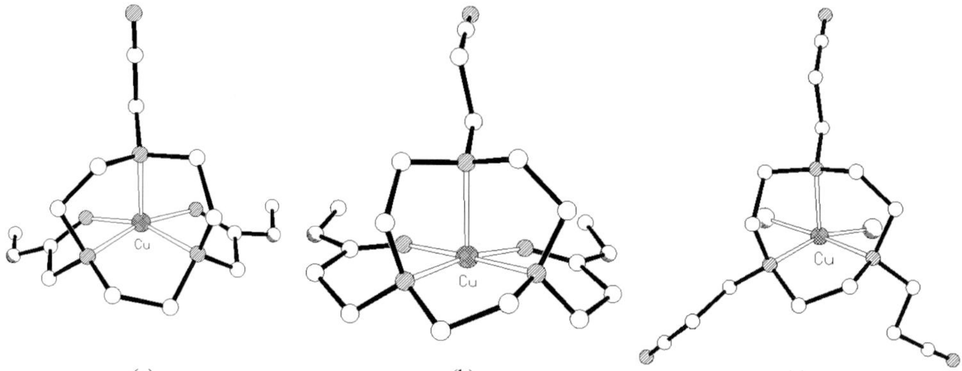

(a) (b) (c)

Figure 3. View of the coordination sphere in the complex cations: (a) $[Cu(\mathbf{L}^5)]^{2+}$; (b) $[Cu(\mathbf{L}^6)]^{2+}$; and (c) in the complex $[Cu(\mathbf{L}^2)Cl_2]$.

Treatment of $\mathbf{L}^2$ and $\mathbf{L}^3$ with one molar equivalent of $CuCl_2 \cdot 2H_2O$ in MeCN at room temperature (ie. under conditions where the the nitrile groups cannot undergo solvolysis) gives the mononuclear complexes $[Cu(\mathbf{L}^2)Cl_2]$ and$[Cu(\mathbf{L}^3)Cl_2]$ having two chloride anions coordinated to the metal centre [Fig. 3c for $[Cu(\mathbf{L}^2)Cl_2]$, Cu–N([9]aneN_3) 2.1203(17)-2.2368(16), Cu–Cl 2.2871(6), 2.2624(6) Å] [47]. In both structures, the basal positions of a distorted square-based pyramidal geometry are occupied by two tertiary *N*-donors from the [9]aneN_3 framework and by two chloride ligands, while the nitrile-functionalised pendant arms are oriented away from the metal centres.

Interestingly, the reaction of $\mathbf{L}^2$ with $CuCl_2 \cdot 2H_2O$ in refluxing MeOH gives upon slow diffusion of Et_2O vapour into the reaction mixture, first a small amount of $[Cu(\mathbf{L}^2)Cl_2]$, then a blue oil showing a band at *ca.* 1630 cm^{-1} typical for the ν(C=N) stretching vibration of an imino group [47]. On treating this blue oil with excess of Me_4NBF_4 in MeOH a very hygroscopic blue solid, $[Cu(\mathbf{L}^8)Cl]BF_4$, in which a nitrile group of $\mathbf{L}^2$ has undergone methanolysis, can be isolated by diffusion of Et_2O vapour into the resulting solution (Scheme 3) [47]. Scheme 4 summarises the different complexes that can be isolated from the reaction of $\mathbf{L}^2$ with $CuCl_2 \cdot 2H_2O$ and $Cu(BF_4)_2 \cdot 4H_2O$ as a function of solvent and reaction conditions.

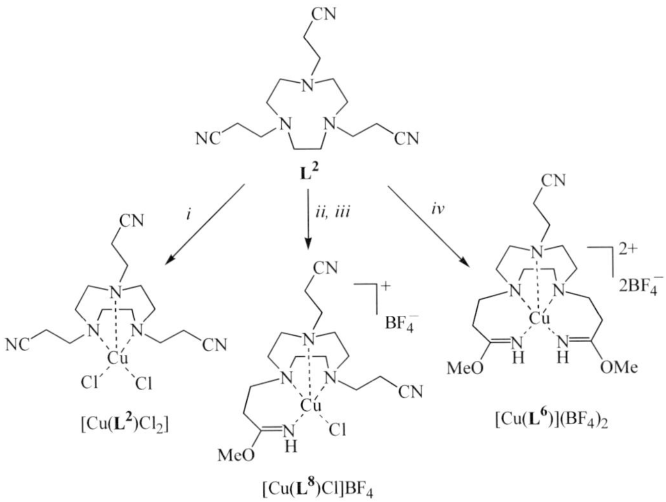

Scheme 4. Products obtained from the reaction of L^2 with $CuCl_2{\cdot}2H_2O$ and $Cu(BF_4)_2{\cdot}4H_2O$ under different experimental conditions. *i*: $CuCl_2{\cdot}2H_2O$ MeCN, r.t., 3 h; *ii*: $CuCl_2{\cdot}2H_2O$, MeOH, reflux, 5h; *iii*: Me_4NBF_4, MeOH; *iv*: $Cu(BF_4)_2{\cdot}4H_2O$, MeOH, reflux, 2 h.

These results support a reaction pathway for the Cu^{II}-assisted methanolysis of cyano groups in $\mathbf{L^1}$-$\mathbf{L^3}$ in which intermediate Cu^{II} complexes of $\mathbf{L^1}$-$\mathbf{L^3}$ having basal positions occupied by MeOH are initially formed [47]. The coordinated solvent molecules are then activated *via* metal-assisted deprotonation leading to nucleophilic attack on the nitrile group(s). The roles of the metal centre are, therefore, to template and activate the reaction of MeOH with nitrile groups of the ligands, and to bind and protect the resultant imino-ether groups in $\mathbf{L^5}$-$\mathbf{L^8}$, having different *N*-attached pendant-donors groups. Thus, selective methanolysis of one or two nitrile groups in $\mathbf{L^1}$-$\mathbf{L^3}$ can be readily achieved *via* control of choice of the coordinating properties of the anion in the starting Cu^{II} salt.

3. Amino-Functionalised Pendant arm Derivatives of [9]aneN_3

3.1 SYNTHESIS

The most successful synthetic procedure for the preparation of amino-functionalised pendant arm derivatives of [9]aneN_3 involves the reduction of pendant nitrile groups using 1M BH_3 solution in THF, followed by hydrolysis of the borane complexes in refluxing concentrated HCl solution. The free amine can be obtained from the resulting hydrochloride salt by passing an aqueous solution of this salt through a Dowex column. Using this route the ligands $\mathbf{L^{13}}$-$\mathbf{L^{16}}$ in Scheme 5, which feature tertiary *N*-donors within the [9]aneN_3 crown, can be prepared pure and in high yield [41-45]. Sargeson and Hammershøi have reported an alternative synthetic route for $\mathbf{L^{14}}$ which consists of a reductive alkylation of [9]aneN_3 with phthalimidoacetaldehyde in the presence of $NaBH_3CN$, followed by the acid hydrolysis of the recovered product to remove the protecting phthalol group [49]. The first reaction step has to be performed under anhydrous conditions in MeCN using phthalimidoacetaldehyde, which has previously been dehydrated by azeotropic removal of water with benzene, while the final product can only be partially purified by chromatography.

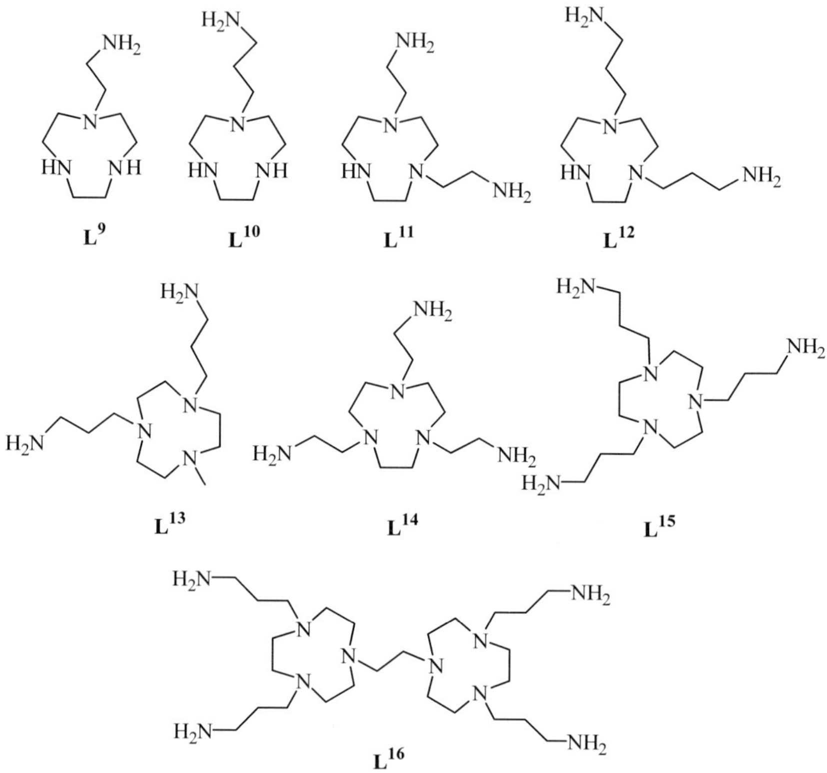

Scheme 5. Summary of amino-functionalised pendant arm derivatives of [9]aneN_3.

The synthesis of mono- and di-aminoalkyl pendant arm derivatives of [9]aneN_3 is less straightforward due to the necessity of preserving two or one secondary *N*-donors in the macrocyclic framework, respectively. McAuley and co-workers reported the Co^{III}, Cu^{II}, and Ni^{II} complexes of $\mathbf{L}^{10}$ and $\mathbf{L}^{12}$ and studied their properties both in solution and in the solid state [41, 42]. However, the two ligands were prepared as a mixture in low yield and their separation and purification were achieved only by column chromatography of the corresponding Ni^{II} complexes. An efficient synthetic method for the selective preparation of $\mathbf{L}^9$ and $\mathbf{L}^{11}$ has been reported recently [50, 51] (Scheme 6), and it illustrates the suitability of *p*-tolylsulfonyl derivatives of [9]aneN_3 as precursors to selectively functionalise this macrocyle.

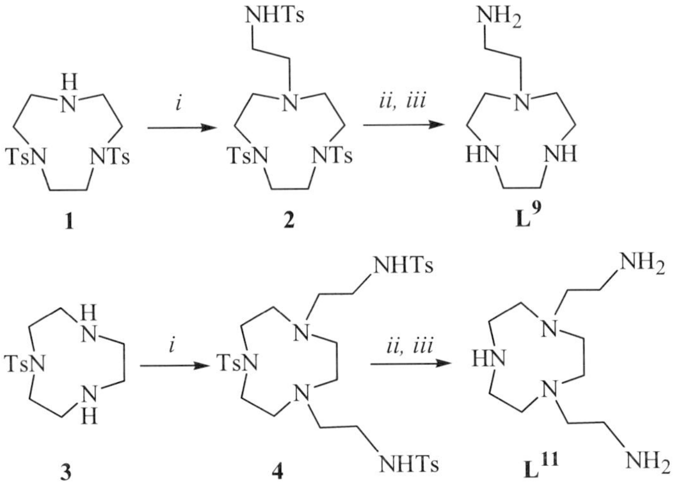

Scheme 6. Reagents and conditions for the synthesis of $\mathbf{L}^9$ and $\mathbf{L}^{11}$. *i*: Ts-aziridine, MeCN, reflux, 28 h for **2**, and 3 h for **4**; *ii*: H_2SO_4 conc. 110 °C, 72 h; *iii*: Amberlite IRA-416.

Reaction of 4,7-bis(*p*-tolylsulphonyl)-1,4,7-triazacyclononane (**1**) or 1-(*p*-tolylsulphonyl)-1,4,7-triazacyclononane (**3**) with a stoichiometric amount of *N*-tosyl-aziridine in MeCN under an atmosphere of N_2 gives high yields of **2** and **4**, respectively. These can be readily transformed into $\mathbf{L}^9$ and $\mathbf{L}^{11}$, respectively, by detosylation with concentrated sulphuric acid.

In principle, mono- and di-aminoalkyl pendant arm derivatives of [9]aneN_3 could be functionalised at the secondary macrocyclic *N*-centres with different pendant-donor groups. Any synthetic strategy designed for this purpose should provide for an efficient protection of the primary pendant-amine groups. Thus, direct [2+2] Schiff-base condensation of $\mathbf{L}^{11}$ with one molar equivalent of 2,6-diformyl-4-methylphenol in MeOH gives, after partial removal of the solvent and addition of light petroleum, the cofacial macropolycycle $\mathbf{H_2L}^{17}$ (Scheme 7) [50, 52]. The imine bonds in this new phenol-based compartmental system can be readily reduced with a mixture of $NaBH_4$/$NaBH_3CN$ in MeOH to give $\mathbf{H_2L}^{18}$ [53]. Having twelve potential coordination sites, both $\mathbf{H_2L}^{17}$ and $\mathbf{H_2L}^{18}$ can also be effectively used for the formation of binuclear complexes with various metal ions (see next paragraph) [52, 53]. Imines are not normally used as protective groups for amines because of their low stability and potential high reactivity, especially under acidic conditions. However, $\mathbf{H_2L}^{17}$ can be transformed to **5** and **6** *via* reaction with *tert*-butylbromoacetate and 2 bromoethanol, respectively (Scheme 7). **5** and **6** can be isolated as thick orange oils, but can also be used in *situ* for subsequent reactions. Infact, hydrolysis of the imino bonds in **5** and **6** with diluted aqueous HCl affords the hydrochloric salts of the two asymmetric derivatives of [9]aneN_3 $\mathbf{HL}^{19}$ and $\mathbf{L}^{20}$ having different *N*-attached pendant-donors groups [50]. In the case of **5**, the reaction with dilute aqueous HCl also hydrolyses the *tert*-butyl ester group to the free carboxylic acid function in $\mathbf{HL}^{19}$.

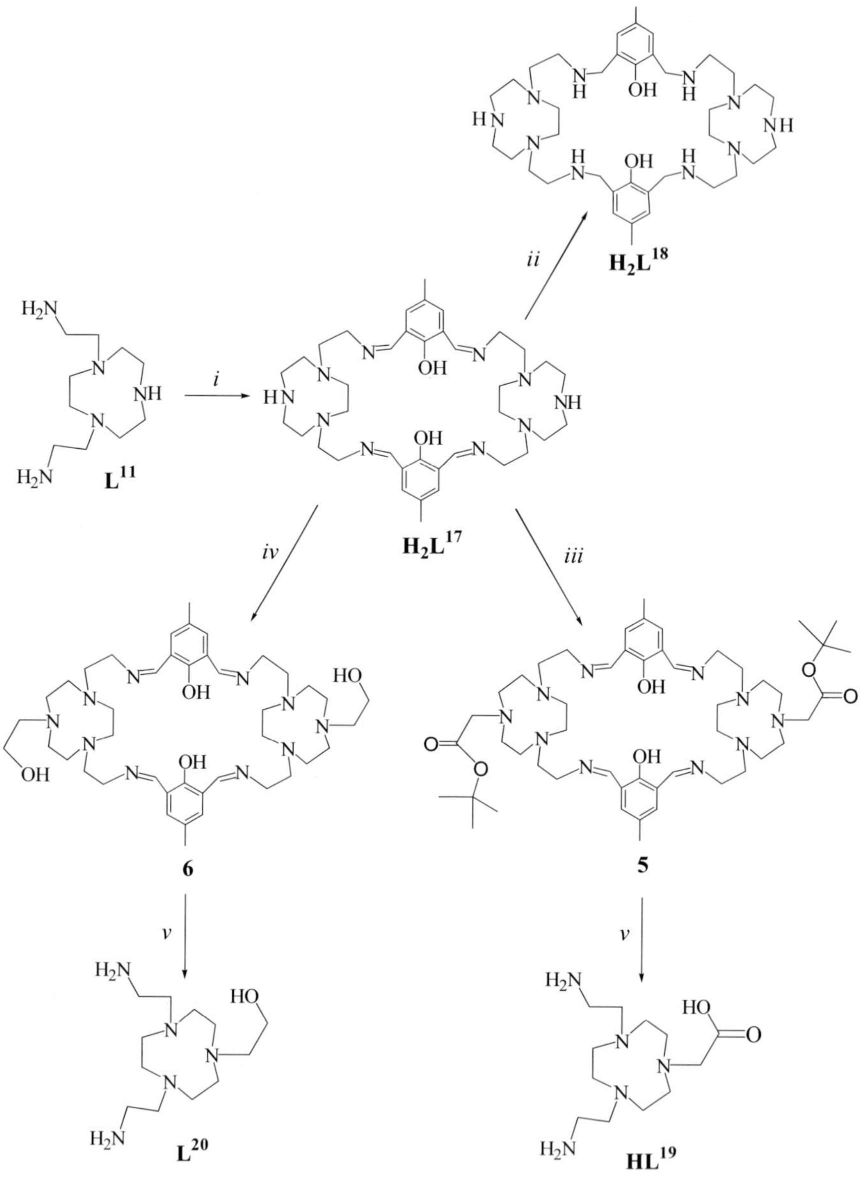

Scheme 7. Synthetic scheme for the asymmetric functionalisation of [9]aneN_3 with different pendant donor groups starting from $\mathbf{L^{11}}$. *i*: 2,6-diformyl-4-methylphenol, MeOH, reflux, 2 h; *ii*: $NaBH_4$/$NaBH_3CN$, MeOH, r.t., 30 h; *iii*: *tert*-butylbromoacetate, $CHCl_3$, $NEt(^iPr)_2$, r.t., 12 h; *iv*: 2-bromoethanol, EtOH, K_2CO_3, 50 °C, 16 h; *v*: HCl 0.01 M, r.t., 12 h.

Schiff-base condensation of aminoalkyl pendant arm derivatives of [9]aneN_3 with the appropriate carbonylic reagent can be used either to protect the primary pendant-amine groups, or to extend the pendant arms with additional coordinating groups. This latter approach affords ligands with an increased set of donor atoms capable of fully encapsulating metal ions of larger ionic radii. Thus, Schiff-base condensation in MeOH of $\mathbf{L}^{11}$ and $\mathbf{L}^{14}$ with two or three molar equivalents of sodium pyruvate in the presence of a lanthanide ion template, gives the complexes [Ln($\mathbf{L}^{21}$)X] (Ln = Y^{III}, Gd^{III}, Eu^{III}, Dy^{III} , X = Cl^- or CH_3COO^-) and [Ln($\mathbf{L}^{22}$)] (Ln = Y^{III}, La^{III}, Sm^{III}, Yb^{III}, Gd^{III}, Eu^{III}, Dy^{III}), respectively (Scheme 8) [43, 54].

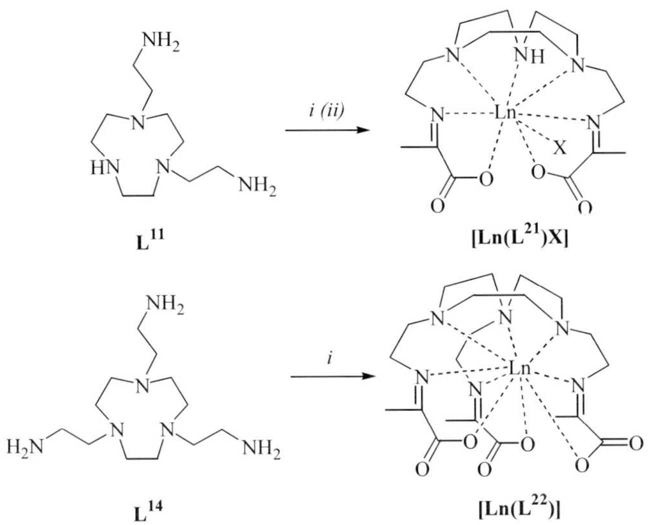

Scheme 8. Synthetic scheme for the preparation of complexes [Ln($\mathbf{L}^{21}$)X] (Ln = Y^{III}, Gd^{III}, Eu^{III}, Dy^{III} , X = Cl^- or CH_3COO^-) and [Ln($\mathbf{L}^{22}$)] (Ln = Y^{III}, La^{III}, Sm^{III}, Yb^{III}, Gd^{III}, Eu^{III}, Dy^{III}): *i*: MeCOCOONa, $LnCl_3$, MeOH, reflux, 2 h; *ii*: MeCOCOONa, $LnCl_3$, CH_3COONa, MeOH, reflux, 2 h.

$\mathbf{L}^{22}$ incorporates nine donor atoms which fulfil a slightly distorted tricapped trigonal prismatic stereochemistry at Ln^{III} centre affording thermodynamically and kinetically stable complexes [Ln($\mathbf{L}^{22}$)]. Acidic media hydrolyses the C=N bonds in [Ln($\mathbf{L}^{22}$)] within several hours [43, 54]. In contrast, $\mathbf{L}^{21}$ has only seven donor centres, providing an opportunity for coordination of anions or water molecules to the metal centre. Therefore, the complexes [Ln($\mathbf{L}^{21}$)X] can potentially act as MRI or ion recognition agents, while at the same time maintaining elements of stability of their [Ln($\mathbf{L}^{22}$)] analogues. Hydrolysis of the C=N bonds in [Ln($\mathbf{L}^{21}$)X] occurs within several hours at neutral pH. The number of water molecules bound to the metal centre estimated using NMR relaxivity measurements for Gd^{III} complexes, dysprosium induced shift (DIS) measurements for Dy^{III} complexes, and emission lifetime measurements for Eu^{III} complexes is close to one for complexes of $\mathbf{L}^{21}$ and zero for complexes of $\mathbf{L}^{22}$. Complexes structurally similar to [Ln($\mathbf{L}^{22}$)] have also been obtained from the Schiff-base condensation in MeOH of $\mathbf{L}^{14}$ with methyl sodium acetyl phosphonate and methyl sodium 4-methoxybenzoyl phosphonate in the presence of Y^{III}, La^{III}, Yb^{III}, and Gd^{III} [55].

3.2 COORDINATION CHEMISTRY

The protonation constants of the ligands $\mathbf{L}^9$, $\mathbf{L}^{11}$, $\mathbf{L}^{14}$, and $\mathbf{L}^{15}$ (Scheme 5), having one, two or three aminoethyl or three aminopropyl pendant arms, respectively, on the [9]aneN_3 framework, and the thermodynamic stabilities of their mononuclear complexes with Cu^{II} and Zn^{II} have been investigated by potentiometric measurements in aqueous solutions (Table 1) [51].

TABLE 1. Protonation constants of ligands $\mathbf{L}^9$, $\mathbf{L}^{11}$, $\mathbf{L}^{14}$, and $\mathbf{L}^{15}$ and formation constants of their Cu^{II} and Zn^{II} complexes in aqueous solution (0.1 mol dm^{-3} NMe_4Cl, 298.1 K).

Reaction	log *K*			
	$\mathbf{L}^9$	$\mathbf{L}^{11}$	$\mathbf{L}^{14}$	$\mathbf{L}^{15}$
$H^+ + L \rightleftharpoons [HL]^+$	10.7(1)	10.72(6)	10.77(6)	11.1(1)
$[HL]^+ + H^+ \rightleftharpoons [H_2L]^{2+}$	9.1(1)	9.32(5)	9.52(5)	10.1(1)
$[H_2L]^{2+} + H^+ \rightleftharpoons [H_3L]^{3+}$	5.3(1)	8.53(6)	8.72(6)	9.4(1)
$[H_3L]^{3+} + H^+ \rightleftharpoons [H_4L]^{4+}$	1.9(1)	2.2(8)	5.16(7)	8.5(1)
$[H_4L]^{4+} + H^+ \rightleftharpoons [H_5L]^{5+}$			2.43(5)	3.4(1)
$[H_5L]^{5+} + H^+ \rightleftharpoons [H_6L]^{6+}$				1.9(1)
$Cu^{2+} + L \rightleftharpoons [Cu(L)]^{2+}$	20.41(3)	23.78	22.0(1)	19.8(3)
$[Cu(L)]^{2+} + H^+ \rightleftharpoons [Cu(HL)]^{3+}$	2.92(1)	3.94(6)	9.5 (1)	10.1(3)
$[Cu(HL)]^{3+} + H^+ \rightleftharpoons [Cu(H_2L)]^{4+}$			3.4(1)	5.9(1)
$[Cu(H_2L)]^{4+} + H^+ \rightleftharpoons [Cu(H_3L)]^{5+}$				3.7(1)
$[Cu(L)]^{2+} + OH^- \rightleftharpoons [Cu(L)(OH)]^+$	4.57(3)	2.78		3.3(1)
$[Cu(L)]^{2+} + 2OH^- \rightleftharpoons [Cu(L)(OH)_2]$				3.0(1)
$Zn^{2+} + L \rightleftharpoons [ZnL]^{2+}$	13.81(3)	17.48(1)	21.28(6)	18.91(4)
$[Zn(L)]^{2+} + H^+ \rightleftharpoons [Zn(HL)]^{3+}$		3.2(1)	4.13(6)	5.97(5)
$[Zn(HL)]^{3+} + H^+ \rightleftharpoons [Zn(H_2L)]^{4+}$				5.44(5)
$[Zn(H_2L)]^{4+} + H^+ \rightleftharpoons [Zn(H_3L)]^{5+}$				4.9(1)
$[Zn(L)]^{2+} + OH^- \rightleftharpoons [Zn(L)(OH)]^+$	3.57(1)	3.74(3)	3.4(1)	3.55(6)
$[Zn(L)]^{2+} + 2OH^- \rightleftharpoons [Zn(L)(OH)_2]$	3.30(4)			3.47(5)

^1H NMR spectroscopic studies performed in D_2O as a function of pH allow determination of the protonation sites of these ligands. Generally, it is observed that the primary -NH_2 groups of the pendant arms are protonated first to minimize electrostatic repulsions with subsequent protonation of the secondary and tertiary macrocyclic *N*-centres. Interestingly, the crystal structure of $[H_2\mathbf{L}^9]Cl_2$ shows the two acidic protons localised on the secondary amine and stabilised by a hydrogen bonding network involving the neutral –NH_2 group of the side-arm and chloride counterions [51].

The data in Table 1 confirm that all ligands considered form very stable complexes in aqueous solutions with Cu^{II}. Comparing the binding abilities of $\mathbf{L}^9$, $\mathbf{L}^{11}$, $\mathbf{L}^{14}$, and $\mathbf{L}^{15}$, the stability constants of the complexes $[Cu(\mathbf{L})]^{2+}$ increase from $\mathbf{L}^9$ to $\mathbf{L}^{11}$ and then decrease from $\mathbf{L}^{11}$ to $\mathbf{L}^{14}$ and $\mathbf{L}^{15}$ (Table 1). However, $\mathbf{L}^9$ shows a binding ability toward Cu^{II} higher than [9]aneN_3 (log K = 20.41 for $\mathbf{L}^9$ *vs.* log K = 15.5 for [9]aneN_3) [51], indicating that in $[Cu(\mathbf{L}^9)]^{2+}$ the aminoethyl pendant is involved in metal coordination. The increased stability of $[Cu(\mathbf{L}^{11})]^{2+}$, therefore, can be ascribed to the coordination of both the $-NH_2$ groups of the side-arms to the metal centre, which is then penta-coordinated by the five *N*-donors of the ligand. The increased coordination number in $[Cu(\mathbf{L}^{11})]^{2+}$ compared with $[Cu(\mathbf{L}^9)]^{2+}$ is also confirmed by the lower constant for the addition of OH^- to the former (Table 1), which indicates an almost saturated coordination sphere for the metal in this complex.

$\mathbf{L}^{14}$ incorporates a further aminoethyl pendant arm as a potential donor group, but its Cu^{II} complex displays a somewhat lower stability than with $\mathbf{L}^{11}$, but at the same time, a much higher tendency to form a monoprotonated $[Cu(\mathbf{HL}^{14})]^{3+}$ complex (Table 1). This behaviour can be ascribed to the presence in $[Cu(\mathbf{L}^{14})]^{2+}$ of an uncoordinated aminoethyl pendant arm which can be readily protonated in aqueous solution.

The replacement of the aminoethyl pendant arms in $\mathbf{L}^{14}$ with aminopropyl ones in $\mathbf{L}^{15}$ leads to an overall weaker ligand-metal interaction, as shown by the lower formation constant for $[Cu(\mathbf{L}^{15})]^{2+}$ compared to $[Cu(\mathbf{L}^{14})]^{2+}$ (Table 1). This effect can be attributed to the larger N–Cu–N chelating bond angle arising from the larger bite angle of the $NCH_2CH_2CH_2N$ fragment, which reduces the stability of the complex $[Cu(\mathbf{L}^{15})]^{2+}$. However, like $[Cu(\mathbf{L}^{14})]^{2+}$, $[Cu(\mathbf{L}^{15})]^{2+}$ is readily protonated, and the particularly high value of the first protonation constant for this complex ($\log K = 10.1$, Table 1) strongly suggests that in this case too, at least one of the amino pendant-donor groups is not involved in binding to the meta ion.

As observed for the Cu^{II} complexes, the stability of the Zn^{II} complex with $\mathbf{L}^9$ is higher than that with [9]aneN_3 ($\log K = 13.81$ for $\mathbf{L}^9$ *vs.* $\log K = 11.62$ for [9]aneN_3) [51]; at the same time, $[Zn(\mathbf{L}^9)]^{2+}$ is considerably less stable than $[Zn(\mathbf{L}^{11})]^{2+}$ (Table 1). These observations suggest the involvement of all the *N*-donors of $\mathbf{L}^9$ and $\mathbf{L}^{11}$ in complex formation. A different behaviour is found on passing to $\mathbf{L}^{14}$ and $\mathbf{L}^{15}$, since the stability of the complexes $[Zn(\mathbf{L}^{14})]^{2+}$ and $[Zn(\mathbf{L}^{15})]^{2+}$ are higher than that found for $[Zn(\mathbf{L}^{11})]^{2+}$, the opposite trend to that observed for Cu^{II} complexation. This strongly suggests that in $[Zn(\mathbf{L}^{14})]^{2+}$ and $[Zn(\mathbf{L}^{15})]^{2+}$ all six *N*-donors are bound to the metal centre. Notably, these two Zn^{II} complexes also show a much lower tendency to protonate than the corresponding Cu^{II} complexes, which contain an uncoordinated aminoethyl group. The somewhat lower stability in solution for $[Zn(\mathbf{L}^{15})]^{2+}$ compared with $[Zn(\mathbf{L}^{14})]^{2+}$, can be ascribed, as in the case of Cu^{II}, to the replacement of ethylene chains in the side-arms with longer propylene ones.

The results of the potentiometric studies are confirmed by the crystal structure determinations on the complexes $[Cu(\mathbf{L}^9)(Br)]Br$, $[Zn(\mathbf{L}^9)(NO_3)]NO_3$, $[Cu(\mathbf{L}^{11})](ClO_4)_2$, $[Cu(\mathbf{L}^{14})](BF_4)_2 \cdot MeCN$, and $[Zn(\mathbf{L}^{15})](BF_4)_2 \cdot MeCN$ [51]. In both complex cations $[Cu(\mathbf{L}^9)(Br)]^+$ and $[Zn(\mathbf{L}^9)(NO_3)]^+$ all *N*-donors of $\mathbf{L}^9$ are involved in complex formation [Cu–N([9]aneN_3) 2.077(9)-2.119(10), Cu–N(NH_2) 1.955(9) Å; Zn–N([9]aneN_3) 2.083(3)-2.223(3), Zn–N(NH_2) 2.025(3) Å] and the two metal ions are five-coordinate (Fig. 4a,b). A Br^- ligand [Cu–Br 2.4089(14) Å] and a monodentate nitrate ion [Zn–O 2.128(2) Å] complete a distorted square-based pyramidal (SBP) and a distorted trigonal bipyramidal (TBP) environment in the Cu^{II} and Zn^{II} complexes, respectively. In $[Cu(\mathbf{L}^{11})]^{2+}$, the metal ion is still five coordinate, and according to the solution studies, all *N*-donors of $\mathbf{L}^{11}$ are involved in metal complexation [Cu–N([9]aneN_3) 2.033(2)-2.206(2), Cu–N(NH_2) 2.018(3), 2.019(3) Å]. They impose a distorted SBP geometry at the metal centre with the basal coordination sites occupied by the two primary *N*-donors from the pendant arms and two tertiary *N*-donors from the [9]aneN_3 framework. The apical position is occupied by the remaining secondary amino group of the pentadentate ligand $\mathbf{L}^{11}$ (Fig. 4c).

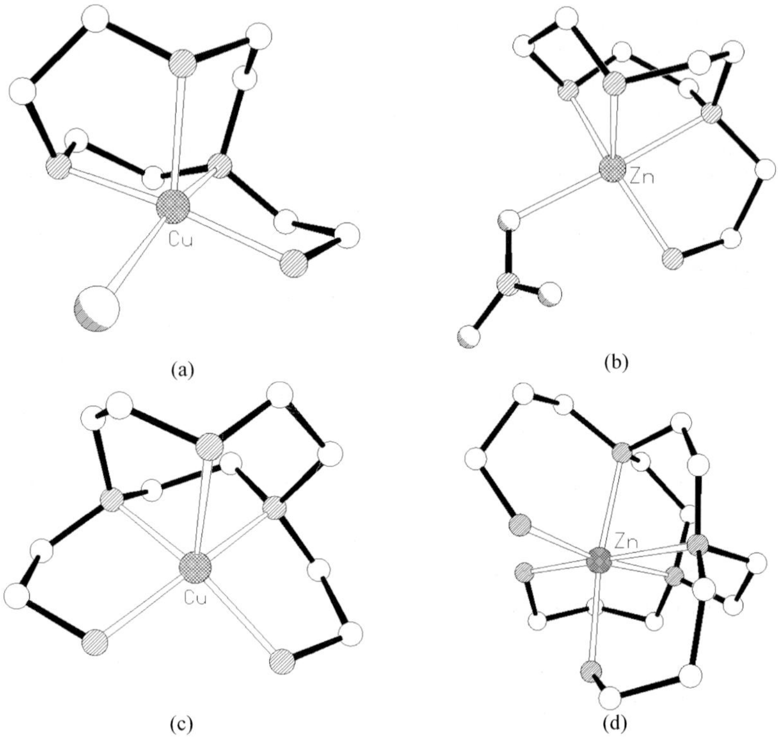

Figure 4. View of the coordination sphere in the complex cations: (a) $[Cu(\mathbf{L}^9)(Br)]^+$; (b) $[Zn(\mathbf{L}^9)(NO_3)]^+$; (c) $[Cu(\mathbf{L}^{11})]^{2+}$; (d) $[Zn(\mathbf{L}^{15})]^{2+}$.

The coordination sphere around Cu^{II} in $[Cu(\mathbf{L}^{11})]^{2+}$ is very similar to that observed in the complex cations $[Cu(\mathbf{L}^{12})]^{2+}$ [41], $[Cu(\mathbf{L}^{13})]^{2+}$ [44], and $[Cu_2(\mathbf{L}^{16})]^{4+}$ [44], (Scheme 5). Overall five coordination at Cu^{II} seems, therefore, to be the norm for complexes with these types of ligands. Indeed, the metal centre in $[Cu(\mathbf{L}^{14})]^{2+}$ is again five-coordinate via three *N*-donors of the [9]aneN_3 ring and two aminoethyl pendant arms [51]. The third pendant arm of $\mathbf{L}^{14}$ in $[Cu(\mathbf{L}^{14})]^{2+}$ is disordered over two sites with one of the two components being uncoordinated to the metal centre. This strongly support the results of the potentiomatric studies which indicate that in $[Cu(\mathbf{L}^{14})]^{2+}$ and $[Cu(\mathbf{L}^{15})]^{2+}$ one of the three amino-functionalised pendant arm is not involved in metal coordination and can be easily protonated in aqueous solution. Interestingly, the reaction of $\mathbf{L}^{15}$ with one molar equivalent of $Cu(BF_4)_2{\cdot}4H_2O$ in MeOH gives the complex $[Cu(\mathbf{L}^{11})](BF_4)_2$ *via* cleavage of one of the three aminopropyl pendant arms of $\mathbf{L}^{15}$. Thus, as already noted by McAuley on comparing the structure of $[Cu(\mathbf{L}^{12})]^{2+}$ with those of $[Ni(\mathbf{L}^{12})(H_2O)]^{2+}$ and $[Co(\mathbf{L}^{12})Cl]^{2+}$ [42], the very small crystal-field stabilisation energy difference between octahedral and SBP complexes for Cu^{II} with $\mathbf{L}^{15}$ does not compensate for the steric strain imposed by the addition of a sixth ligand to the metal centre in $[Cu(\mathbf{L}^{15})]^{2+}$. Unfortunately, the detailed mechanism of cleavage of one pendant arm in $\mathbf{L}^{15}$ upon reaction with Cu^{II} remains unclear.

According to the potentiometric studies, all six *N*-donors of $\mathbf{L}^{15}$ are bound to the metal centre in the complex cation $[Zn(\mathbf{L}^{15})]^{2+}$ [Zn–N([9]aneN_3) 2.219(4)-2.224(4), Zn–N(NH_2) 2.149(3)-2.167(3) Å] (Fig. 4d) with the coordination sphere around Zn^{II} showing a slight trigonal elongation from an ideal octahedral geometry [51]. Six coordination is also preferred by other first row transition metal ions such as Ni^{II}, Co^{III}, and Mn^{II} with amino-functionalised pendant arm derivatives of [9]aneN_3. In fact, the X-ray crystal structures of the complex cations $[Ni_2(\mathbf{L}^{10})_2(\mu\text{-}Cl)_2]^{2+}$ [42], $[Ni(\mathbf{L}^{11})(MeCN)]^{2+}$ [51], $[Ni(\mathbf{L}^{12})(H_2O)]^{2+}$ [42], $[Co(\mathbf{L}^{12})Cl]^{2+}$ [42], $[Co(\mathbf{L}^{15})]^{3+}$ [41], and $[Ni(\mathbf{L}^{15})]^{2+}$ [41], all show the metal ions in a distorted octahedral coordination geometry which is reached in same cases by coordination of exogenous ligands. In contrast, a distorted trigonal prismatic coordination geometry is observed for the complex cations $[Mn(\mathbf{L}^{11})(NO_3)]^{+}$ [50], $[Mn(\mathbf{L}^{15})]^{2+}$ [51], and $[Mn(\mathbf{L}^{19})]^{+}$ [50].

$\mathbf{H_2L}^{17}$ represents not only a useful intermediate in the synthesis of asymmetric derivatives of [9]aneN_3 having different *N*-attached pendant-donors groups (Scheme 7), but is also a new phenol-based compartmental system in which, for the first time, a preformed macrocycle is introduced as part of the lateral chains connecting the two *p*-cresol units to afford a large binucleating cofacial macropolycycle. In $\mathbf{H_2L}^{17}$, the particular binding properties of phenol-based compartmental macrocycles are combined with those of [9]aneN_3 to afford a ligand characterised by two large adjacent chambers, each with a potential N_5O_2-donor set capable of accommodating, in close proximity, two metal ions with large ionic radii. The ability of $\mathbf{H_2L}^{17}$ to form homo-binuclear complexes in which the two metal centres are bridged by the phenolate oxygen atoms depends on the nature of the metal ion.

(a) (b)

Figure 5. View of the coordination sphere in the complex cations (a) $[Cd_2(\mathbf{L}^{17})]^{2+}$; (b) $[Y_2(\mathbf{L}^{17})(OH)]^{3+}$.

In the binuclear Cd^{II} complex $[Cd_2(\mathbf{L}^{17})]^{2+}$ [Cd–N([9]aneN_3) 2.342(3)-2.511(3), Cd–N(C=N) 2.293(3), 2.313(3), Cd–O 2.283(2), Cd···Cd 7.017(1) Å] the two *O*-donors are not shared by the two metal centres, which are each bound within an independent N_5O-donating compartment (Fig. 5a) [52]. A face to face π π interaction between the two aromatic rings and additional hydrogen-bonding involving MeOH molecules force the *O*-donors of the anionic ligand $(\mathbf{L}^{17})^{2-}$ to assume a peculiar arrangement with the phenolate oxygens twisted in opposite direction and each pointing toward an individual metal ion. However, bridging phenolate *O*-donors are observed in the binuclear Y^{III} complex $[Y_2(\mathbf{L}^{17})(OH)]^{3+}$ in which the two eight-coordinated metal centres are also bridged by a hydroxo group [Y–N([9]aneN_3) 2.436(9)-2.655(7), Y–N(C=N) 2.418(8),

2.472(8), Y–O 2.277(7)-2.391(6), Y···Y 3.4295(19) Å] (Fig. 5b) [52]. It is important to note that the smaller Cd^{II} ion seldom adopts coordination numbers higher than six in these and related systems, while the larger Y^{III} ion normally shows higher coordination numbers of 8 or 9 in its complexes.

Reduction of the imine bonds in $\mathbf{H_2L^{17}}$ affords $\mathbf{H_2L^{18}}$ which is conformationally more flexible. Potentiometric and Uv spectrophotometric measurements in aqueous solutions reveals that $\mathbf{H_2L^{18}}$ binds up to seven protons in the pH range 2.5-10.5. Two further equilibria, relative to the ligand deprotonation [$\mathbf{H_2L^{18}} \rightleftharpoons (\mathbf{HL^{18}})^- + H^+$ and $(\mathbf{HL^{18}})^- \rightleftharpoons (\mathbf{L^{18}})^{2-} + H^+$], take place at pH>10.5 [53]. The calculated protonation constants suggest that successive protonation steps of $\mathbf{H_2L^{18}}$ (log K = 10.9-4.5) occur on the two separated pentamine compartments of the ligands, alternatively. Thus $(H_8\mathbf{L^{18}})^{6+}$ would contain six $-NH_2^+$ functions separated by a tertiary unprotonated *N*-donor to reduce the electrostatic repulsion between the charged ammonium functions. This hypothesis is corroborated by the crystal structure of $[H_8\mathbf{L^{18}}](ClO_4)_6{\cdot}8H_2O$, which shows protonation of all six secondary amines in the ligand $\mathbf{H_2L^{18}}$ [53]. The further protonation step (log K = 1.8) occurs on amine groups adjacent to already protonated nitrogens, thus decreasing the stability of the resulting species $(H_9\mathbf{L^{18}})^{7+}$.

In aqueous solution, $\mathbf{H_2L^{18}}$ shows a similar coordination behaviour toward Cu^{II}, Zn^{II}, Cd^{II}, and Pb^{II} giving rise to stable mono- and binuclear metal complexes in the presence of one or two molar equivalents of metal ion [53]. The mononuclear species $[M(\mathbf{L^{18}})]$ can bind up to five protons with high protonation constants, the first three being only 1-2 log units lower than the corresponding protonation constants of $\mathbf{H_2L^{18}}$ alone. This behaviour is typical of ditopic polyamine macrocycles which display two well-separated binding sites. In other words, in $[M(\mathbf{L^{18}})]$ (M = Cu^{II}, Zn^{II}, Cd^{II}, and Pb^{II}) the metal resides in one of the two pentamine compartments, while the other, which is not involved in binding to the metal, can be readily protonated. The mononuclear complexes $[M(\mathbf{L^{18}})]$ are much more stable than the corresponding complexes with $\mathbf{L^{11}}$ (log K = 30.1 and 23.8 for the Cu^{II} complexes, and 20.2 and 17.48 for the Zn^{II} complexes with $(\mathbf{L^{18}})^{2-}$ and $\mathbf{L^{11}}$, respectively) [53]; this suggests the involvement of deprotonated phenolate functions in metal coordination.

The mononuclear complexes show a marked tendency to encapsulate a second metal ion in aqueous solutions to give stable binuclear complexes. The formation of the binuclear species $[M_2(\mathbf{H_2L^{18}})]^{4+}$ with the neutral ligand $\mathbf{H_2L^{18}}$ at slightly acidic pH is followed by deprotonation of the Ar–OH functions in neutral or weakly basic solutions for Cu^{II} and Zn^{II}, and at basic solution for Cd^{II} and Pb^{II}, to give the species $[M_2(\mathbf{HL^{18}})]^{3+}$ and $[M_2(\mathbf{L^{18}})]^{2+}$, respectively [53]. The facile deprotonation of the Ar–OH functions, especially in the case of Cu^{II} and Zn^{II} binuclear complexes $[M_2(\mathbf{H_2L^{18}})]^{4+}$, which occurs, in the absence of metal cations, only at strongly alkaline pH, indicates the participation of the metal ions in this process. This feature is confirmed by the X-ray crystal structure of the complexes $[Cu_2(\mathbf{L^{18}})](BF_4)_2{\cdot}\frac{1}{2}MeCN$, $[Zn_2(\mathbf{HL^{18}})](ClO_4)_3{\cdot}\frac{1}{2}MeCN$, and $[Pb_2(\mathbf{L^{18}})](ClO_4)_2{\cdot}2MeCN$ (Fig. 6) [53].

In the complex cation $[Cu_2(\mathbf{L^{18}})]^{2+}$, each Cu^{II} ion is bound to an N_4O-donor set within a slightly distorted square-based pyramidal environment (Fig. 6a). The basal positions of the pyramidal coordination sphere are occupied by two tertiary *N*-donors from the [9]aneN_3 moiety [Cu–N 2.016(12)-2.102(12) Å], one secondary *N*-donor from the aliphatic chain of the anionic ligand $(\mathbf{L^{18}})^{2-}$ [Cu–N 2.008(12)-2.008(13) Å] and by one phenolate oxygen [Cu–O 1.909(9), 1.910(9) Å]. The apical position is occupied by the secondary *N*-donor from the [9]aneN_3 moiety [Cu–N 2.283(10), 2.312(11) Å].

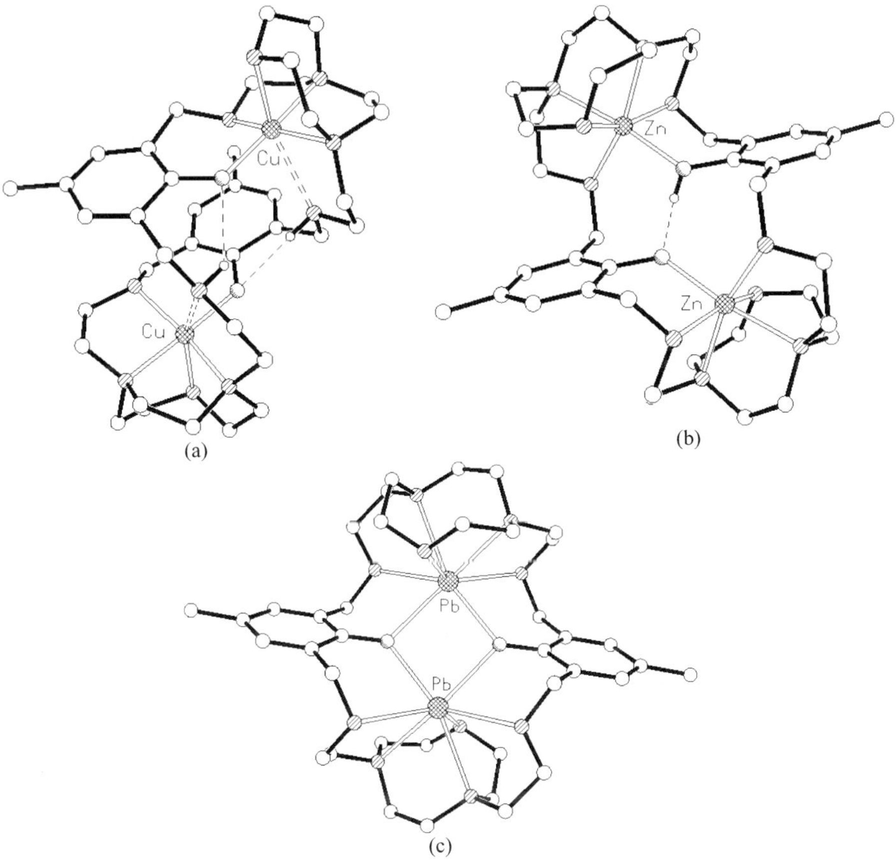

Figure 6. View of the coordination sphere in the complex cations (a) $[Cu_2(\mathbf{L^{18}})]^{2+}$; (b) $[Zn_2(H\mathbf{L^{18}})]^{3+}$; (c) $[Pb_2(\mathbf{L^{18}})]^{2+}$.

A long interaction with the remaining secondary *N*-donor of the anionic ligand [Cu–N 2.597(10), 2.593(11) Å] complete an overall N_5O-coordination around each metal centre. The two N_5O-donating compartments of $(\mathbf{L^{18}})^{2-}$ act, therefore, as isolated donor sets with the phenolate *O*-donors not bridging the metal ions [Cu···Cu 5.954(2) Å] [53].

Each Zn^{II} ion in the complex cation $[Zn_2(\mathbf{HL^{18}})]^{3+}$ is six-coordinate and bound by an N_5O-donor set within a slightly distorted octahedral environment [Zn–N 2.177(4)-2.204(4), Zn–O 2.058(3), Zn···Zn 5.572(1) Å] (Fig. 6b) [53]. Only one of the two Ar–OH groups of the starting ligand is deprotonated and is involved in a strong intramolecular hydrogen bond with the remaining –OH function. Thus, the two N_5O donating compartments of the macropolycyclic anion $(\mathbf{HL^{18}})^{-}$, although acting as isolated donor sets, are connected by an unusual Zn–O···H–O–Zn bridge.

Finally, the two Pb^{II} centres in the cation $[Pb_2(\mathbf{L^{18}})]^{2+}$ lie within the two N_5O-donating compartments of $(\mathbf{L^{18}})^{2-}$ [Pb–N 2.586(5)-2.744(5), Pb–O 2.487(4), 2.630(4) Å] and are bridged by the phenolate *O*-donors with a short interatomic Pb···Pb distance of 3.9427 (5) Å comparable to that observed between the two metal centres in $[Y_2(\mathbf{L^{17}})(OH)]^{3+}$ (Fig. 6c) [53]. On going from $[Cu_2(\mathbf{L^{18}})]^{2+}$ to $[Zn_2(\mathbf{HL^{18}})]^{3+}$ and to

$[Pb_2(\mathbf{L^{18}})]^{2+}$, we observe the ligand gradually adopting a more extended conformation (Fig. 6) allowing the *O*-donors to be shared between the two metal centres. Interestingly, and similar to the observations with the more rigid $\mathbf{H_2L^{17}}$ (Fig. 5), a bridging coordination mode for the phenolate *O*-donors is achieved with metal ions having a large ionic radius and high coordination number.

4. Acknowledgments

We thank EPSRC for support, and all the collaborators listed in the references for their invaluable contributions to this work.

5. References

[1] Chaudhuri, P., and Wieghardt, K.: The chemistry of 1,4,7-triazacyclononane and related tridentate macrocyclic compounds, *Prog. Inorg. Chem.* **35** (1987), 329-436.

[2] de Bruin, B., Donners, J.J.J.M., de Gelder, R., Smits, J.M.M., and Gal, A.W. : Mono-and dinuclear carbonyl complexes of (1,4,7-trimethyl-1,4,7-triazacyclononane)rhodium(I): facile migration of a C(O)OMe ligand at a dinuclear Rh(μ-CO)$_2$Rh core, *Eur. J. Inorg. Chem.* **3** (1998), 401-406.

[3] Diebold, A., Elbouadili, A., and Hagen, K.S.: Crystal structures and solution behaviour of paramagnetic divalent transition metal complexes (Fe, Co) of the sterically encumbered tridentate macrocycles 1,4,7-R$_3$-1,4,7-triazacyclononane: coordination numbers 5 (R = *i*-Pr) and 6 (R = *i*-Bu), *Inorg. Chem.* **39** (2000), 3915-3923.

[4] Wang, R., Eberspacker, T.A., Hasegawa, T., Day, V., Ware, D.C., and Taube, H.: tmtacn, tacn, and triamine complexes of (η^6-arene)Os^{II}: synthesis, characterizations, and photosubstitution reactions (tmtacn = 1,4,7-trimethyl-1,4,7-triazacyclononane; tacn = 1,4,7-triazacyclononane), *Inorg. Chem.* **40** (2001), 593-600.

[5] Graham, B., Spiccia, L., Bond, A.M., Hearn, M.T.W., and Kepert, C.M.: Structural, sepctroscopic and electrochemical studies of nickel(II) “sandwich” complexes with ligands featuring tethered 1,4,7-triazacyclononane, *Dalton Trans.* (2001), 2232-2238.

[6] Arnold, J., Knapp, V., Schmidt, J.A.R., and Shafir, A.: Reaction of *N,N',N''*-trimethyl-1,4,7-triazacyclononane with butyllithium reagents, *Dalton Trans.* (2002), 3273-3274.

[7] Cheng, P., Yan, S.-P., Xie, C.-Z., Zhao, B., Chen, X.-Y., Liu, X.-W., Li, C.-H., Liao, D.-Z., Jiang, Z.-H., and Wang, G.-L.: Ferromagnetic and antiferromagnetic polymeric complexes with the macrocyclic ligand 1,4,7-triazacyclononane, *Eur. J. Inorg. Chem.* **11** (2004), 2369-2378.

[8] Bernhardt, P.V., and Lawrence, G.A.: Complexes of polyazamacrocycles bearing pendent coordinating groups, *Coord. Chem. Rev.* **104** (1990), 297-343.

[9] Wainwright, K.P.: Synthetic and structural aspects of the chemistry of saturated polyaza macrocyclic ligands bearing pendant coordinating groups attached to nitrogen, *Coord. Chem. Rev.* **166** (1997), 35-90.

[10] Haines, R.I.: Pendant arm tri- and tetraazamacrocycles: synthesis, structure and properties of their first-row transition metal complexes, *Rev. Inorg. Chem.* **21** (2001), 165-205.

[11] Pavlishchuk, V., Birkelbach, F., Weyhermüller, T., Wieghardt, K., and Chaudhuri, P.: Polynuclear complexes of the pendent-arm ligand 1,4,7-tris(acetophenoneoxime)-1,4,7-triazacyclononane, *Inorg. Chem.* **41** (2002), 4405-4416.

[12] Weeks, J.M., Buntine, M.A., Lincoln, S.F., Tiekink, E.R.T., and Wainwright, K.P.: Diasteromeric discrimination in 1,4,7-tris((S)-2-hydroxypropyl)-1,4,7-triazacyclononane and its lithium(I), sodium(I) and Zn(II) complexes, *Dalton Trans.* (2001), 2157-2163.

[13] Gateau, C., Mazzanti, M., Pécaut, J., Dunand, F.A., and Helm, L. : Solid-state and solution properties of the lanthanide complexes of a new nonadentate tripodal ligand derived from 1,4,7-triazacyclononane, *Dalton Trans.* (2003), 2428-2433.

[14] Plush, S.E., Lincoln, S.F., and Wainwright, K.P.: Aminoacid *N*-substituted 1,4,7-triazacyclononane and 1,4,7,10-tetraazacyclododecane Zn^{2+}, Cd^{2+} and Cu^{2+} complexes. A preparative, potentiometric titration and NMR spectroscopy study, *Dalton Trans.* (2004), 1410-1417.

[15] Castro-Rodriguez, I., Olsen, K., Gantzel, P. and Meyer, K.: Uranium complexes supported by an aryloxide functionalised triazacyclononane macrocycle: synthesis and characterisation of a six-coordinate U(III) species and insight into its reactivity, *Chem. Commun.* (2002), 2764-2765.

[16] Kovacs, Z., and Sherry, A.D.: A general synthesis of mono- and disubstituted 1,4,7-triazacyclononanes, *Tetrahedron Lett.* **36** (1995), 9269-9272.

[17] Bambirra, S., van Leusen, D., Meetsma, A., Hessen, B., and Teuben, J.H.: Neutral and cationic yttrium alkyl complexes with linked 1,4,7-triazacyclononane-amide monoanionic ancillary ligands: synthesis and catalytic ethene polymerisation, *Chem. Comm.* (2001), 637-638.

[18] McGowan, P.C., Podesta, T.J., and Thornton-Pett, M.: *N*-Monofunctionalized 1,4,7-triazacyclononane macrocycles as building blocks in inorganic crystal engineering, *Inorg. Chem.* **40** (2001), 1445-1453.

[19] Penkert, F.N., Weyhermüller, T., Bill, E., Hildebrandt, P., Lecomte, S., and Wieghardt, K.: Anilino radical complexes of cobalt(III) and manganese(IV) and comparison with their phenoxyl analogues, *J. Am. Chem. Soc.* **122** (2000), 9663-9673.

[20] Schmidt, J.A.R., Giesbrecht, G.R., Cui, C., and Arnold, J.: Anionic triazacyclononanes: new supporting ligands in main group and transition metal organometallic chemistry, *Chem. Commun.* (2003), 1025-1033.

[21] Mondal, A., Klein, E.L. Kan, M.A., and Houser, R.P.: TACN-amino acid conjugates and thir copper(II) complexes, *Inorg. Chem.* **42** (2003), 5462-5464.

[22] Warden, A., Graham, B., Hearn, M.T.W., and Spiccia, L.: Synthesis of novel derivatives of 1,4,7-triazacyclononane, *Org. Lett.* **3** (2001), 2855-2858.

[23] André, J.P., Maecke, H.R., Zehnder, M., Macko, L., and Akyel, K.G.: 1,4,7-triazacyclononane-1-succinic acid-4,7-diacetic acid (NODASA): a new bifunctional chelator for radio gallium-labelling of biomolecules, *Chem. Commun.* (1998), 1301-1302.

[24] Grenz, A., Ceccarelli, S., and Bolm, C.: Synthesis and application of novel catalytically active polymers containing 1,4,7-triazacyclononane, *Chem. Commun.* (2001), 1726-1727.

[25] Bodsgard, B.R., and Burstyn, J.N.: Silica-bound copper(II) triazacyclononane: a robust material for the heterogeneous hydrolysis of a phosphodiester, *Chem. Commun.* (2001), 647-648.

[26] Rossi, P., Felluga, F., Tecilla, P., Formaggio, F., Crisma, M., Toniolo, C., and Scrimin, P.: A bimetallic helical heptapeptide as a transphosphorylation catalyst in water, *J. Am. Chem. Soc.* **121** (1999), 6948-6949.

[27] Reichenbach-Klinke, R., and Knöig, B.: Metal complexes of azacrown ethers in molecular recognition and catalysis, *Dalton Trans.* (2002), 121-130.

[28] Deck, K.M., Tseng, A.T., Burstyn, J.N.: Triisopropyltriazacyclononane copper(II): and efficient phosphodiester hydrolysis catalyst and DNA cleavage agent, *Inorg. Chem.* **41** (2002), 669-677.

[29] Lawrence, S.C., Ward, B.D., Dubberley, S.R., Kozak, C.M., and Mountford, P.: Highly efficient ethylene polymerisation by scandium alkyls supported by neutral *fac*-k^3 coordinated N_3 donor ligands, *Chem. Commun.* (2003), 2880-2881. Adams, N., Arts, H.J., Bolton, P.D., Cowell, D., Dubberley, S.R., Friederichs, N., Grant, C.M., Kranenburg, M., Sealey, A.J.,. Wang, B., Wilson, P.J., Cowley, A.R., Mountford P., and Schröder, M.: Discovery and evaluation of highly active imidotitanium ethylene polymerisation catalysts using high throughput catalyst screening, *Chem Comm.* (2004), 434-435.

[30] Lo, H.C., Chen, H., and Fish, R.H.: Synthesis of a [{mono-*N*-(4-vinylbenzyl)-1,4,7-triazacyclononane}$_2$Hg](OTf)$_2$ sandwich complex, polymerisation of this monomer with divinylbenzene, and Hg^{2+} ion selectivity studies with the demetallated resin, *Eur. J. Inorg. Chem.* **9** (2001), 2217-2220.

[31] Akkaya, E.U., Huston, M.E., and Czarnik, A.W.: Chelation-enhanced fluorescence of anthrylazamacrocycle conjugate probes in aqueous solution, *J. Am. Chem. Soc.* **112** (1990), 3590-3593.

[32] Koike, T., Abe, T., Takahashi, M., Ohtani, K., Kimura, E., and Shiro, M.: Synthesis and characterisation of the zinc(II)-fluorophore, 5-dimethylaminonaphthalene-1-sulfonic acid [2-(1,5,9-triazacyclododec-1-yl)ethyl]amide and its zinc(II) complex, *Dalton Trans.* (2002), 1764-1768.

[33] Pidwell, A.D., Collison, S.R., Coles, S.J., Hursthouse, M.B., Schröder, M., and Bruce, W.D.: The synthesis and properties of surfactants aza macrocycles, *Chem. Commun.* (2000), 955-956.

[34] Griffiths, P.C., Fallis, I.A., Willock, D.J., Paul, A., Barrie, C.L., Griffiths, P.M., Williams, G.M., Kinkg, S.M., Heenan, R.K., and Görgl, R.: The structure of metallomicelles, *Chem.-A Eur. Journal* **10** (2004), 2022-2028.

[35] Halfen, J.A., Jazdewski, B.A., Mahapatra, S., Berreau, L.M. Wilkinson, E.C., Que, L., Jr., and Tolman, W.B.: Synthetic models of inactive copper(II)-tyrosinate and active copper(II)-tyrosyl radical forms of galactose and glyoxal oxidase, *J. Am. Chem. Soc.* **119** (1997), 8217-8227.

[36] Sissi, C., Rossi, P., Felluga, F., Formaggio, F., Palombo, M., Tecilla, P., Toniolo, C., and Scrimin, P: Dinuclear Zn^{2+} complexes of synthetic heptapeptides as artificial nucleases, *J. Am. Chem. Soc.* **123** (2001), 3169-3170.

[37] Lin, G., Reid, G., and Bugg, T.D.H.: Extradiol oxidative cleavage of catechols by ferrous and ferric complexes of 1,4,7-triazacyclononane: insight into the mechanism of the extradiol catechol dioxygenases, *J. Am. Chem. Soc.* **123** (2001), 5030-5039.

[38] Tanase, T., Inukai, H., Onaka, T., Kato, M., Yano, S., and Lippard, S.J., Trinuclear Zn(II) and Cu(II) homo and heterotrimetallic complexes involving D-glucopyranosyl and biscarboxylate bridging ligands. A substrate binding model of Xylose isomerases, *Inorg. Chem.* **40** (2001) 3943-3953.

[39] Enomoto, M., and Aida, T.: Iron complexes capped with dendrimer-appended triazacyclononanes as the novel spatially encumbered models of non-eme iron proteins, *J. Am. Chem. Soc.* **124** (2002), 6099-6108.

[40] Wainwright, K.P.: Application for polyaza macrocycles with nitrogen-attached pendant arms, *Adv. Inorg. Chem.* **52** (2001), 293-334.

[41] Bushnell, G.W., Fortier, D.G., and McAuley, A.: Synthesis of the hexaamine ligand 1,4,7-tris(3-aminopropyl)-1,4,7-triazacyclononane: reactivity and X-ray crystal structures of the nickel(II) and Co(III) complexes, *Inorg. Chem.* **27** (1988), 2626-2634.

[42] Fortier, D.G., and McAuley, A.: Synthesis and crystallographic characterisation of pendant arm macrocyclic complexes derived from 1,4,7-triazacyclononane: mono- and di-(3-aminopropyl)-1,4,7-triazacyclononane as ligands, *Dalton Trans.* (1991), 101-109.

[43] Tei, L., Baum, G., Blake, A.J., Fenske, D., and Schrööder, M.: Lanthanide complexes of a new nonadentate ligand derived from 1,4,7-triazacyclononane: synthesis, structural characterisation and NMR spectroscopic studies, *J. Chem. Soc., Dalton Trans.* (2000), 2793-2799.

[44] Blake, A.J., Danks, J.P., Li, W.-S., Lippolis, V., and Schröder, M.: Synthesis and characterisation of pendant arm amino derivatives of 1,4,7-triazacyclononane and alkyl-bridged bis(1,4,7-triazacyclononane) macrocycles and complexation to Cu(II), *Dalton Trans.* (2000), 3034-3040.

[45] Tei, L., Blake, A.J., Cooke, P.A., Caltagirone, C., Demartin, F., Lippolis, V., Morale, F., Wilson, C., and Schröder, M.: Nitrile functionalised pendant arm derivatives of aza- and mixed donor macrocyclic ligands as new building blocks for inorganic crystal engineering, *Dalton Trans.* (2002), 1662-1670.

[46] Tei, L., Lippolis, V., Blake, A.J., Cooke, A.P., and Schröder M.: Nitrile functionalised pendant arm derivatives of [9]aneN_3 as new multidentate ligands for inorganic crystal engineering ([9]aneN_3 = 1,4,7-triazacyclononane), *Chem. Commun.* (1998), 2633-2634.

[47] Tei, L., Blake, A.J., Lippolis, V., Wilson, C., and Schröder, M.: Methanolysis of nitrile-functionalised pendant arm derivatives of 1,4,7-triazacyclononane upon coordination to Cu^{II}, *Dalton Trans.* (2003), 304-310.

[48] Blake, A.J., Hubberstey, P., Suksangpanya, U., and Wilson, C.: Anion recognition by conservation of hydrogen-bonding patterns in salts of copper(II) co-ordinated by tetradentate bis(amidino-*O*-alkylurea) ligands, *Dalton Trans.* (2000), 3873-3880.

[49] Hammershøi, A., and Sargeson, A.M.: Macrotricyclic hexaamine cage complexes of cobalt(III): synthesis, characterization, and properties, *Inorg. Chem.* **22** (1983), 3554-3561.

[50] Tei, L., Blake, A.J., Wilson, C., and Schröder, M.: Synthesis of asymmetric derivatives of 1,4,7-triazacyclonoane and trigonal prismatic Mn(II) complexes, *Dalton Trans.* (2002), 1247-1249.

[51] Tei, L., Bencini, A., Blake, A.J., Lippolis, V., Perra, A., Valtancoli, B., Wilson, C., and Schröder, M.: Co-ordination chemistry of amino pendant arm derivatives of 1,4,7-triazacyclononane, *Dalton Trans.* (2004), 1934-1944.

[52] Tei, L., Blake, A.J., Devillanova, F.A., Garau, A., Lippolis, V., Wilson, C., and Schröder, M.: A new cofacial binucleating macropolycycle: segregated *versus* encapsulated complexation, *Chem. Commun.* (2001), 2582-2583.

[53] Tei, L., Arca, M., Aragoni, M.C., Bencini, A., Blake, A.J., Caltagirone, C., Devillanova, F.A., Fornasari, P., Garau, A., Isaia, F., Lippolis, V., Schröder, M., Teat, S.J., and Valtancoli, B.: Coordination chemistry of a new cofacial binucleating macropolycycle derived from 1,4,7-triazacyclononane, *Inorg. Chem.* **42** (2003), 8690-8701.

[54] Tei, L., Blake, A.J., George, M.W., Weinstein, J.A., Wilson, C., and Schröder, M.: Lanthanide complexes of iminocarboxylate ligands derived from 1,4,7-triazacyclononane: structural characterisation and relaxivity of the Gd^{III} and luminescence of the Eu^{III} complexes, *Dalton Trans.* (2003), 1693-1700.

[55] Tei, L., Blake, A.J., Wilson, C., and Schröder, M.: Lanthanide complexes of new nonadentate imino-phosphonate ligands from 1,4,7-triazacyclononane: synthesis, structural characterisation and NMR studies, *Dalton Trans.* (2004), 1945-1952.

6.
AZAMACROCYCLIC SYSTEMS WITH DIFFERENT SUPRAMOLECULAR FUNCTIONS

BURKHARD KÖNIG AND JIŘÍ SVOBODA
Department of Chemistry, Universität Regensburg, 93040 Regensburg, Germany

1. Introduction

The creation of function and properties on the molecular level is one of the challenges of modern chemistry. A molecular function can be as simple as a selective recognition of a guest molecule or as complex as a molecular motor. Always reversible intermolecular non-covalent interactions are key elements in the design of molecular functionality. We summarise in this review recent examples of functional supramolecular compounds. All examples have in common an azamacrocycle as their core element, whereby we have to focus on cyclen and cyclam. We start with simple functions, such as the recognition or sensing of anions and cations, leading to examples where neutral guest molecules – as complex as a HIV-1 regulatory protein – are bound. In model compounds that mimic biological processes, such as hydrolysis or redox reactions, the ability to act as a functional model very much depends on their molecular structure. This allows to derive structure – property relations, from which the most significant structural parameters can be identified. From the numerous examples of azamacrocycle-based lanthanides chemosensors we can only present a few illustrative examples due to the limited size available for this review article. We apologise to all authors whose excellent contributions could not be included.

2. Recognition of Anions

Selective recognition of anions is important for environmental assays and analysis of biological samples. Due to their more complicated shape, charge distribution and smaller charge-to-radius ratios, recognition of anions is more difficult than the recognition of spherical cations. High selectivities can be achieved in neutral receptor molecules where the distance dependent Coloumb forces can be combined with the spatially more specific hydrogen bonding.

A neutral macrocyclic receptor of nitrate was designed by a rational approach.[1] A hexagonal grid was constructed around the trigonal nitrate anion and the molecular frame was cut from the chicken wire pattern (Scheme 1).

Karsten Gloe (ed.), Macrocyclic Chemistry: Current Trends and Future Perspectives, 87–103.
© 2005 *Springer. Printed in the Netherlands.*

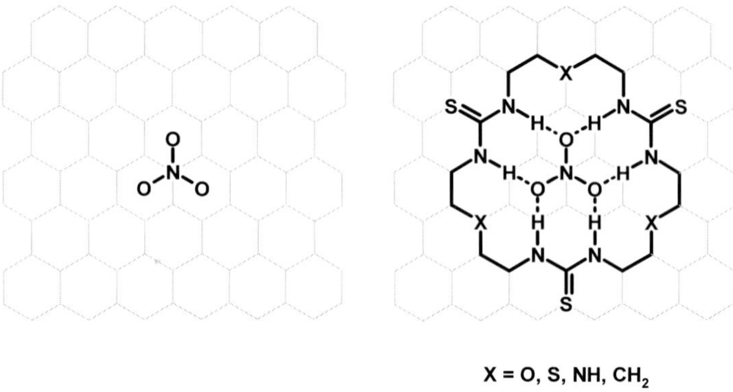

Scheme 1. Design of a neutral macrocyclic receptor using a chicken wire pattern

Thiourea units were chosen as the hydrogen bond donors, because this motif is often found in crystal structures of nitrate salts. Although this approach was crude, there were three positions (X) which could be further modified. Size and shape of the cavity should slightly change with different groups (S, O, NH, CH_2) in this position. To find the optimum arrangement around the nitrate anion, a density functional theory calculation was performed. The results revealed that molecules with oxygen and sulphur bridges are good candidates for the given task. Both molecules were synthesised and their binding properties studied by NMR titrations in dimethylsulphoxide. As the theoretical considerations indicated, the binding of nitrate anion to the receptor with oxygen bridges was among the strongest for neutral molecules with defined binding motifs. Small structural changes (sulphur bridges instead of oxygen bridges or more sterically demanding linkers between the thiourea units) resulted in a complete loss of the binding affinity. Surprisingly, the receptor also exhibited binding selectivity for bromide anions which are even poorer acceptors of hydrogen bonds than nitrates.

An important class of naturally occurring anions are the nucleoside phosphates (*i.e.* nucleotides) present in nucleic acids, sugar nucleotides for glycosylation of oligo-saccharides or proteins, activated forms of proteins and chemical mediators which play a central role in intracellular signals. Artificial phosphate receptors allow for detection and separation of biologically important compounds. However, most chemical receptors are soluble in solution and cannot be therefore separated easily from the solution binding the desired compound.

This problem was overcome by anchoring the recognition unit – zinc cyclene – onto a polymer.[2] Cyclene bearing a vinylbenzyl group was co-polymerised with an ethylene glycol dimethacrylate cross-linker, yielding a solid material (Scheme 2).

The amount of zinc(II) ions bound to the polymer was quite elegantly determined by the use of a zinc(II) selective fluorophore developed by the same group.[3] Adsorption of phosphates such as 5'-dAMP and 4-NPP, adenosine 3',5'-cyclic-monophosphate (3',5'-cAMP) and the corresponding dephosphorylated compounds, deoxyadenosine (dA) and 4-nitrophenol (4-NP) was studied. The polymer was stirred in a solution containing the studied guest in a buffered aqueous solution. The adsorption efficiency was determined by the decrease of the guest molecule concentrations by UV measurement.

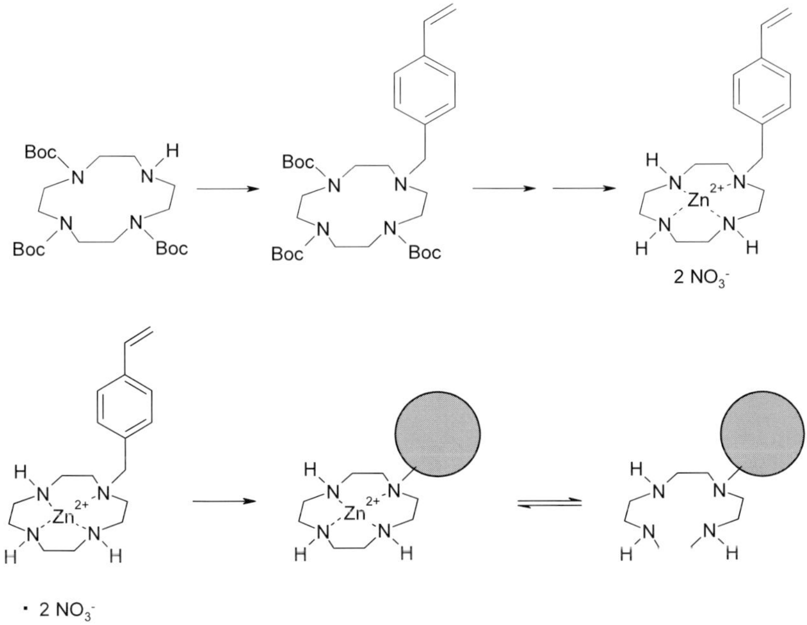

Scheme 2. Synthesis of a zinc(II) complex-conjugated polymer

It was found that the polymer exhibited selectivity towards phosphomonoester dianions. Less polar compounds were found to bind non-specifically to the polymer. The polymer was then used as a stationary phase for a HPLC column. A mixture containing dA, 5'-dAMP and 3',5'-cAMP was thus separated. As expected, the retention time of 5'-AMP was larger than those for dA and 3',5'-cAMP. The same was true for other nucleotides compared to the corresponding nucleosides. When the Zn^{2+}-free control polymer was used, all compounds were immediately eluted. The possibility to use polymer-anchored recognition units to separate biologically important phosphates was thus proved.

3. Recognition of Cations

Cyclenes are a useful component of metal sensor molecules. The common structure of such chemosensor is fluorophore – linker – sensor. Various fluorophores were connected to the cyclene which ensured the detection of the desired cation. These molecules were often water soluble and worked under physiological conditions what made them interesting from the biomedical point of view.

Recently, a selective receptor for copper(II) ion consisting of a ruthenium chromophore, lysine spacer and a cyclene unit was prepared (Scheme 3).[4]

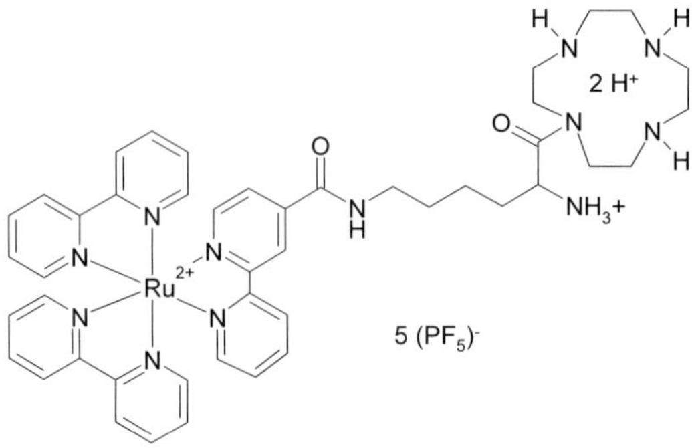

Scheme 3. Receptor for the selective recognition of copper(II)

The luminescence properties of the compounds were studied in solution alone and in the presence of various metal cations. The emission spectra of the free ligand were almost identical over the whole pH range studied (2 to 12). In the presence of nickel(II), a minimum at pH 9.5 was observed and the emission was quenched by ca. 20 %. Manganese(II), iron(II), cobalt(II) and zinc(II) did not result in pH-dependent emission quenching. However, the presence of copper(II) ions efficiently quenched the ruthenium-based emission at pH 6.8. Binding of copper(II) to the ligand was reversible upon protonation. When going from low pH value upwards, the emission intensity decreased to reach the minimum at pH 6.8. Back titration from pH 7 to 2 recovered the emission intensity. The cycle was almost reversible as shown by repetition of the experiment. Titrations of the ligand with nickel(II), copper(II) and zinc(II) ions at neutral pH (buffered aqueous solution) showed that the receptor was selective for copper(II) ions. Addition of other metal ions did not result in significant emission quenching. The copper(II) ion formed a 1:1 complex with the ligand and the quenching could be observed in low metal concentration (10^{-5} M). A competitive experiment was also performed and copper(II) ions were added to a solution which – apart from the ligand – already contained a 10-fold excess of zinc(II) and nickel(II) perchlorate. The emission was even in this case quenched equally well. Hence, a copper(II) selective ligand, undisturbed by the presence of other metal cations was prepared. The ligand worked in aqueous solutions at neutral pH and the binding of copper(II) ions was reversible.

On the other hand, Koike *et al.* [3] described a fluorescent ligand which was selective for zinc(II) ions. The photoactive part of the molecule was in this case the dansyl moiety, connected to a cyclene unit by a diethylamino linker. In order to shift the analytical range to higher concentrations, a similar structure containing [12]aneN_3 instead of cyclene was also prepared (Scheme 4).

The binding of metal cations to these ligands was studied by potentiometric pH titrations which showed i) a 1:1 stoichiometry of the complexes and ii) the [12]aneN_3 containing ligand may be used at intracellular zinc(II) concentration range. The effect of the binding event on absorbance and fluorescence was studied in an aqueous solution at pH 7.8. Upon addition of zinc(II) ions, the fluorescence intensity at 538 nm (λ_{ex} = 320 nm) increased 5.2 times. However, the fluorescence emission was almost unaffected by the presence of sodium(I), potassium(I), calcium(II), magnesium(II) or iron(III) ions under the same conditions.

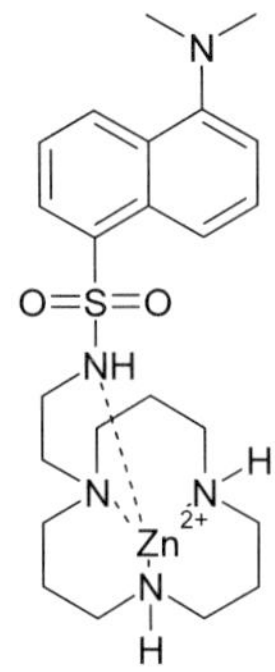

Scheme 4. Receptor for the selective recognition of zinc(II) cation

Only the presence of copper(II) quenched the fluorescence emission. Nevertheless, in ordinary biological systems, copper(II) ions are strongly bound to amino acids, peptides or proteins, and the ligands described may be therefore used for the dynamic analysis of the biologically important zinc(II) ions.

Later on, Aoki *et al.* [5] described a similar ligand, bearing an anthrylmethylamino group instead of the dansyl group (Scheme 5).

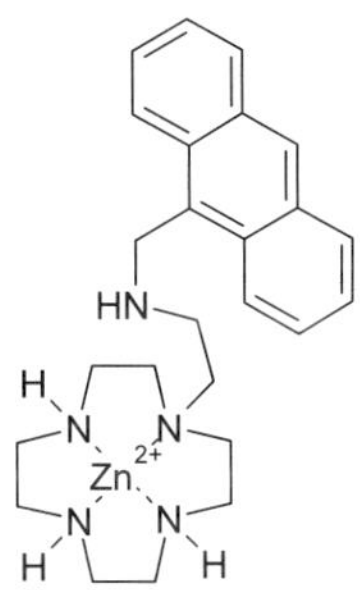

Scheme 5. A zinc(II) receptor with an anthryl group

Zinc(II) complexation properties were studied by potentiometric pH titrations and it was found that zinc(II) bound even more strongly to the anthryl-bearing ligand than to the ligand described previously. Fluorometric titrations at pH 7.4 showed that upon complexation of zinc(II) ion, the fluorescence emission from the anthryl group increased 3.1 times. The detection limit was 10^{-7} M of zinc(II) ions. The ligand signalled more strongly zinc(II) than cadmium(II). However, similar to the previous ligand described, the fluorescent emission was quenched upon the addition of copper(II). The effect of other cations [magnesium(II), calcium(II) and potassium(I)] or counterions (nitrate, perchlorate, sulphate, fluoride, chloride, bromide, iodide, hydrogenphosphate, thymidyl and barbiturate anion) on the zinc(II) signalling was negligible.

The same group also originated a selective fluorescent probe for yttrium(III) and lanthanum(III) ions based on the cyclene moiety.[6] Again, the cyclene ring bore a fluorescent dansyl group, but in this case, the other nitrogen atoms were substituted with carbamoyl groups (Scheme 6). These were believed to create a microenvironment around the dansyl group to increase its fluorescence, once the cyclene moiety was deprotonated.

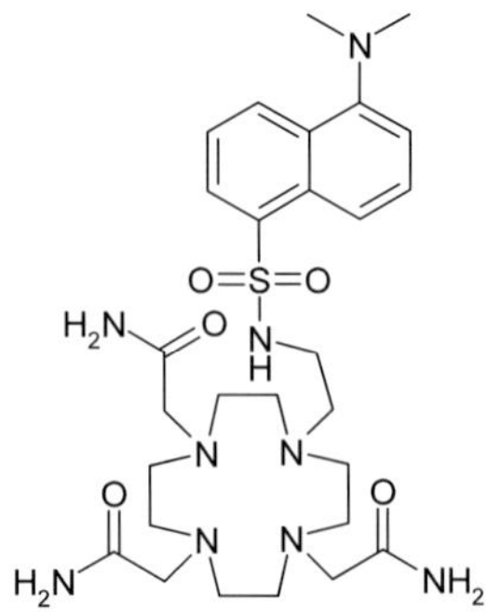

Scheme 6. A doubly-substituted cyclen as a receptor for Y^{3+} and La^{3+} ions

Upon pH-dependent fluorescence titration, the fluorescence of the dansyl group increased 8.5 times as the pH value increased. The quantum yield at pH 11.0 was 6 times higher than at pH 7.4. The fluorescence emission was greatly enhanced by the presence of yttrium(III) and lanthanum(III) ion (8.6 times and 3.8 times, respectively). On the other hand, the fluorescence was not affected by the presence of a great many similar cations. This means the ligand was a selective probe for Y^{3+} and La^{3+} working in aqueous solutions.

Lowe *et al.* [7] and Frias *et al.* [8] described complexes of cyclene-based molecules with lanthanoids. A gadolinium complex which would exhibit pH-dependent relaxivity thanks to a switch in hydration state was prepared.[7] Cyclene bore a sulphonamide substituent in order to achieve a variation of the coordination environment of the lanthanide centre as a function of pH (Scheme 7).

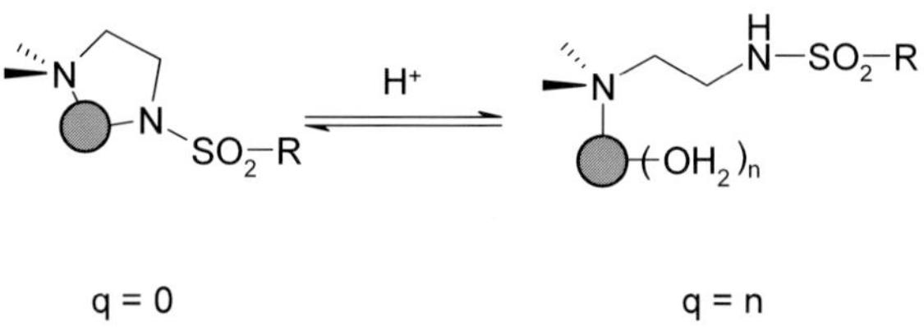

Scheme 7. pH-dependent microenvironment around the metal centre (q = number of coordinated water molecules)

Sensitivity of the lanthanide complex was considered likely to be determined by nature of the substituent R (R') (Scheme 8).

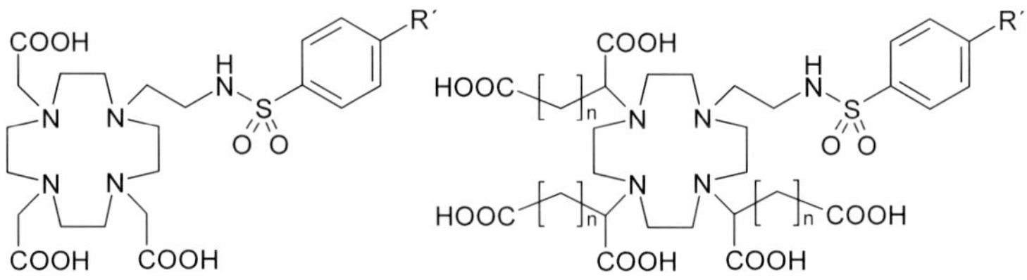

Scheme 8. Structure of the prepared ligands

The pH dependence of luminescence and relaxivity revealed that the complexes with europium, gadolinium and lanthanum might be used for MRI imaging. The concept was later verified by Frias *et al.* [8] by the use of structurally similar lanthanide complexes (Scheme 9).

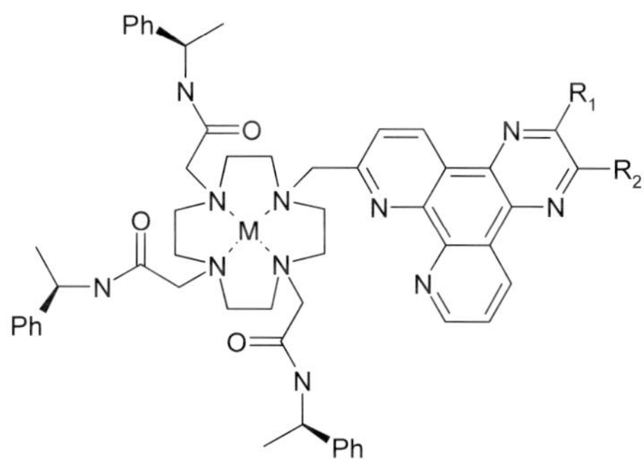

Scheme 9. Structure of the ligands for MRI imaging

The tetraazatriphenylene chromophore attached to the cyclene ring acted as an efficient sensitiser for Eu^{3+} and Tb^{2+} emission but also intercalated between the base pairs of DNA. The complexes were tested as cellular imaging and reactive probes using the mouse fibroblast cell line. The complexes were quickly taken up by the fibroblast cells and localised in nucleus and around the cell membrane. The process was visualised by fluorescence microscopy. Photolysis at 340 nm and 350 nm damaged plasmid supercoiled DNA producing nicked (form II) and linear (form III) DNA. DNA damage is known to induce apoptotic cell death and these complexes may be therefore considered for the development as therapeutic probes, for example in the treatment of accessible tumours, such as skin melanoma.

4. Recognition of Neutral Molecules

The technique of molecular imprinting was successfully used to create a polymer with specific recognition sites.[9] A template was used to organise monomers during the polymerisation process. After the polymerisation, it was washed away from the insoluble network, leaving behind domains of complementary size and shape (Scheme 10).

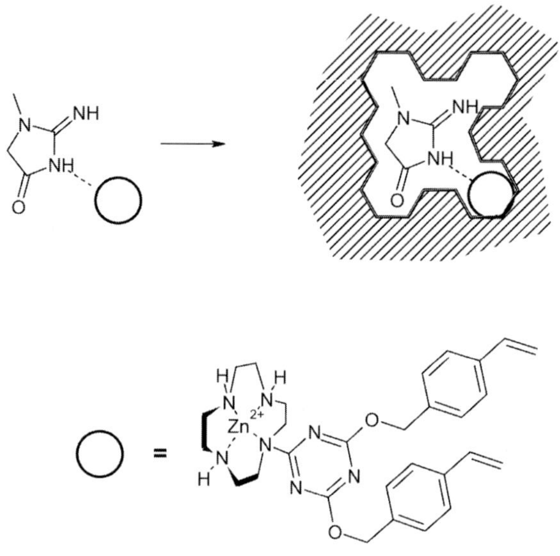

Scheme 10. Preparation of a molecular imprinted polymer

Potential candidates for the templates were e.g. riboflavin and thymine as they exhibit similar imide pK_a values but differ in sterical demands (Scheme 11).
A more challenging template was creatinine which is – due to its weaker electron acceptor properties and lower acidity – bound 34 times weaker to the zinc(II)-cyclene than thymine.

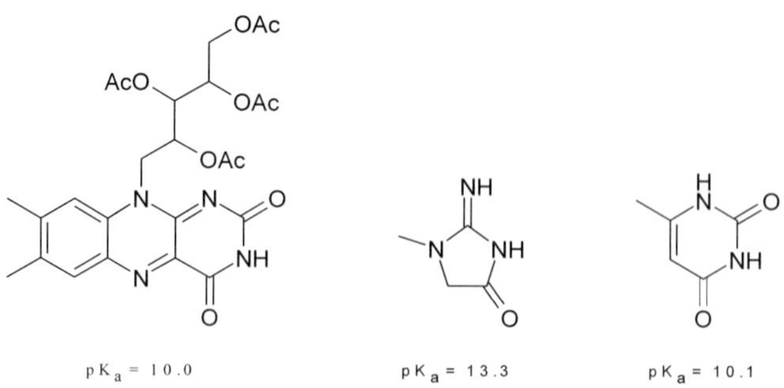

Scheme 11. Structures and pKa values of riboflavin tetraacetate, creatinine and thymine

Stoichiometric complexes of the zinc(II)-cyclene monomer with tetraacetylriboflavin and creatinine were prepared and co-polymerised with ethylene glycol dimethyl acrylate. After exhaustive extraction of the material which removed impurities and the templates, a functional polymer was formed. Control polymers without template and with cobalt(II) instead of zinc(II) were also prepared.

The studied polymer contained all accessible cyclene units as zinc(II) complexes. The recognition properties of the polymer were than studied by shaking the polymer with an aqueous buffered solution of thymin, creatinine, tetraacetylriboflavin or a mixture of thymin and creatinine. It was found that the cobalt(II)-containing control polymer did not bind thymin or creatinine, but irreversibly bound flavine. The control polymer without metal ions did not bind any guest, only weak non-specific adsorption to the polymer surface occurred. The binding of creatinine was found to be 97 % reversible and ca. 60 % of all the sites were accessible. For creatinine, a uniform binding situation was found, while thymine distinguished two types of sites: one with unrestricted access and the other with some sterical hindrance. The competition experiment revealed preferred binding for creatinine over thymine in a ratio of 3.5:1. In a polymer prepared without the imprint, a selectivity of 1:77 for thymine was found. This proved the fact that the imprinting process creates binding selectivity. Hence, binding of biologically important guests to molecular imprinted polymers containing metal binding sites under physiological conditions is possible.

Similar to the aforementioned polymer containing binding sites for nucleotides, a polymer for the selective binding and reversible release of riboflavin was prepared.[10] Cyclene was immobilised as a binding site on a polymer surface by reaction with Fractogel® EMD epoxy (Scheme 12).

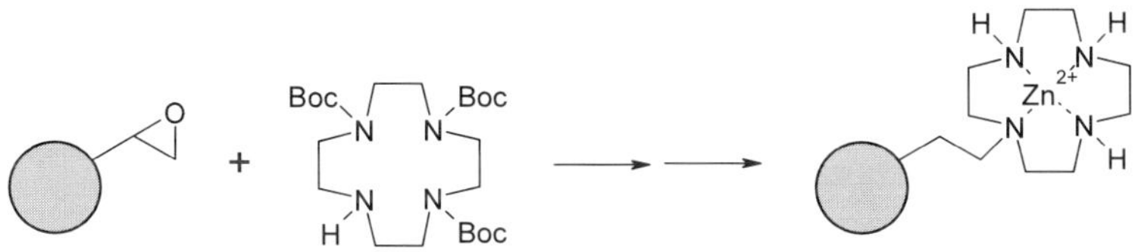

Scheme 12. Synthesis of a riboflavin-binding polymer

Remaining epoxide groups of the tentagel were then deactivated by reaction with glycine, Boc protecting groups of the cyclene were removed and a complex with zinc(II) was formed. Because of the large amount of the binding sites present on the polymer surface, the polymer was capable of riboflavin binding in aqueous solutions. Buffered solution of riboflavin was passed through a column filled with the polymer. Riboflavin was retained as proved by both naked eye observation and UV measurement. However, riboflavin was quantitatively released by using a more acidic buffer solution. The polymer exhibited an apparent binding constant to riboflavin in the millimolar range at pH 7.4 and selectively recognised imide bonds. To prove the ability of the polymer to selectively remove riboflavin from a mixture of compounds, the polymer was treated with a commercial vitamin dietary supplement, containing a mixture of vitamins B_1, B_2, B_6, B_{12}, C, E, folic acid, biotin, panthothenic acid and nicotinamide in amounts corresponding to recommended daily amounts for adults. Passage through the column filled with the polymer selectively removed riboflavin, whose typical absorption bands disappeared from the UV spectrum. Elution of the column with an acidic buffer solution yielded a solution of pure riboflavin whose amount corresponded to the amount present in the supplement tablet. Hence, a polymer which facilitates the isolation of riboflavin from natural source and analysis of riboflavin containing mixtures was found. Kikuta *et al.* [11] described a potent zinc(II)-cyclene based inhibitor of the HIV-1 TAR RNA-Tat peptide binding (Scheme 13).

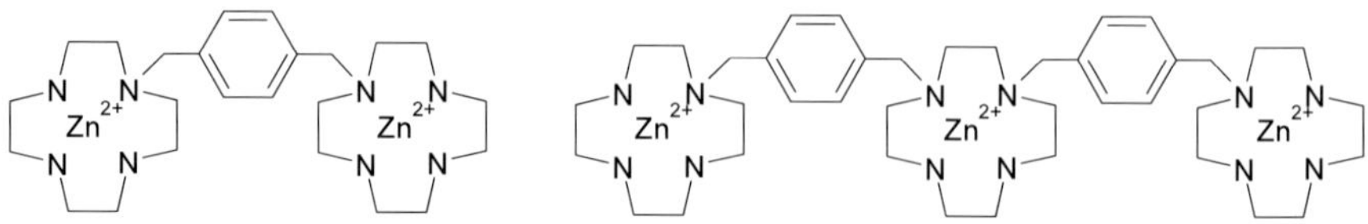

Scheme 13. Inhibitors of the HIV-1 TAR RNA-Tat binding

Single zinc(II)-cyclenes had been known to bind to uridine and thymidine nucleotides by the specific zinc(II)-imide N^- bonding. The repeated binding motif was found to selectively bind to a dinucleotide dTpdT or a trinucleotide dTpdTpdT, according to the number of zinc(II)-cyclene units.

Transcription of HIV-1 genome is facilitated by a HIV-1 regulatory protein Tat which activates the full-length synthesis of HIV-1 mRNA by binding to a *trans*-activation responsive (TAR) element RNA. The TAR element consists of the first 59 nucleotides of the HIV-1 primary transcript and adopts a hairpin structure with an uracil-rich bulge (UUU or UCU). The bulge is the Tat binding site. The tris-zinc(II)-cyclene complex was found to bind selectively to the TAR element and hence to prevent the TAR – Tat binding thanks to the extremely strong binding to the UUU bulge. Efficiency of the inhibition was studied by a footprinting analysis using micrococcal nuclease and a TAR_{33} model of the binding site. The footprinting analysis showed the tris-zinc(II)-cyclene complex was one of the most potent inhibitors of the binding ever described and may be therefore considered as a new “small molecule” targeting HIV-1 RNA.

Another group of cyclene-based molecules interacting with nucleic acids was described by us.[12] In this case, cyclene nitrogen atoms were substituted with carbamoyl groups which irreversibly reduced the basicity and binding abilities but were useful for the synthesis of more complex molecules (Scheme 14). Simple carbamoyl-substituted cyclenes were obtained by standard protection – acylation – deprotection procedures.

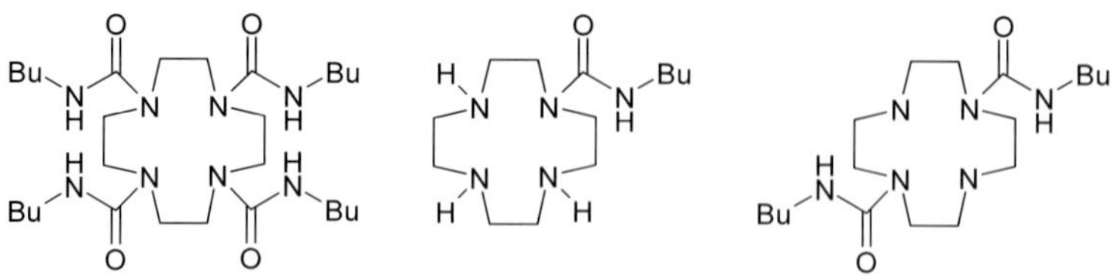

Scheme 14. Carbamoyl-substituted cyclenes

The use of diisocyanates (hexyl-1,6-diisocyanate and *m*-tolyl-diisocyanate) yielded bridged bis-cyclenes (Scheme 15). Complexes of the ligands with copper(II) and zinc(II) perchlorate were prepared.

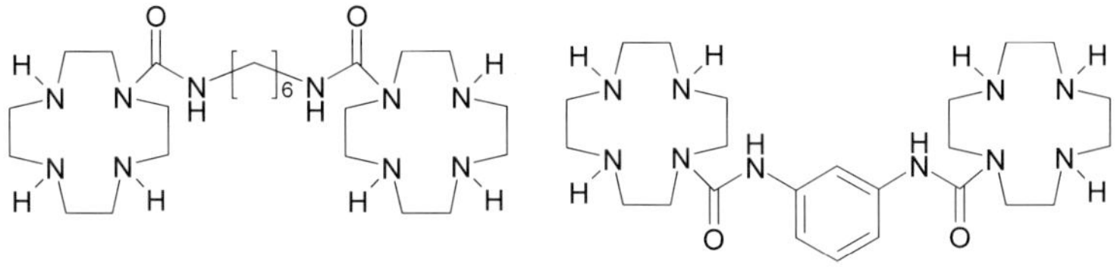

Scheme 15. Bridged bis-cyclenes

The ability to bind to DNA was studied by ethidium bromide displacement titrations. The phosphodiester backbone of DNA was fully stable in the presence of the coordination compounds. The apparent binding constants were dependent on the charge of the macrocycles. Apparent binding constants of the metal complexes were even stronger and co-operative effect could be observed for the dinuclear ligands.

Bridger *et al.* studied bis-tetraazamacrocycles which differed in the size of the azamacrocyclic ring.[13] These macrocycles had high potency and selectivity against HIV. The systematic study revealed that the most promising structure was the *para*-phenylenemethylenebis-[14]aneN_4 (Scheme 16).

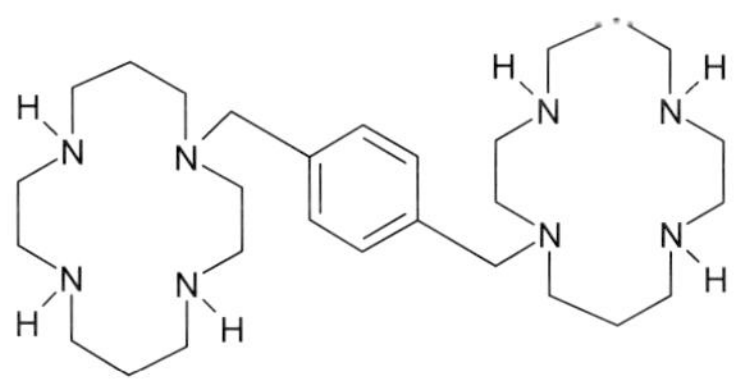

Scheme 16. Lead structure for the HIV inhibition

The anti-HIV activity of corresponding zinc(II), copper(II) and palladium(II) complexes was also evaluated. Activity inversely correlated with stability of the complexes: a kinetically labile bis-zinc(II) complex exhibited similar EC_{50} value against HIV-1 as the free ligand, but was 2-fold more potent against HIV-2. Derivatives with substituted aromatic linker were also studied. However, activity against HIV seemed independent of substituent electron properties. Bulky groups reduced the activity. In subsequent work [14], compounds with heteroaromatic linkers were evaluated, but none of the compounds was better than the original lead structures.

5. Biological System Models

A model of a flavin-based redox enzyme was prepared.[15] Redox enzymes are often flavoproteins containing flavin cofactors flavin adenine dinucleotide (FAD) or flavin mononucleotide (FMN). They mediate one- or two-electron redox processes at potentials which vary in a range of more than 500 mV. The redox properties of the flavin part must be therefore tuned by the apoenzyme to ensure the specific function of the enzyme. Influence by hydrogen bonding, aromatic stacking, dipole interactions and steric effects have been so far observed in biological systems, but coordination to metal site has never been found before. Nevertheless, the importance of such interactions for functions and structure of other biological molecules make this a conceivable scenario.

In order to predict feasibility of such a motif, changes in redox properties of 10-butylflavin and riboflavintetraacetate upon reversible coordination to Lewis-acidic zinc(II) complexes were studied, posing very simple models for metalloprotein binding (Scheme 17).

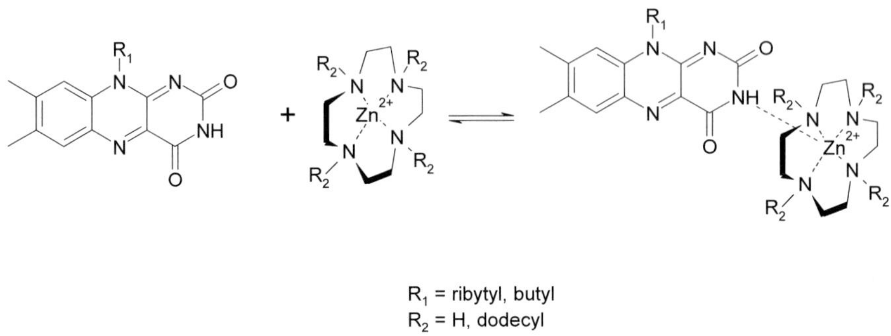

Scheme 17. A model system for flavoenzyme activity

The effect of complexation on redox properties was studied by cyclic voltammetry. Unbound flavin, dissolved in an aprotic solvent (dichloromethane), undergoes a two electron reduction perfectly explained by the ECE mechanism. Upon addition of cyclene ligand and coordination of flavin to the zinc ion complex, the flavohydroquinone redox state was stabilised.

Aprotic solvents mimic the hydrophobic protein interior. However, a functional artificial receptor for flavin binding under physiological conditions must be able to interact with the guest even in competitive solvents. As found by spectroscopic measurements with phenothiazene-labeled cyclene, the coordinative bond between flavin and Lewis-acidic macrocyclic zinc in methanol was strong enough for this function. Stiochiometry of the complex was proved by Job's plot analysis. Redox properties of the assemblies in methanol were studied by cyclic voltammetry which showed that the binding motif allowed interception of the ECE reduction mechanism and stabilisation of a flavosemiquinone radical anion in a polar solvent. As a consequence, the flavin chromophore switched from a two-electron-one-step process to a two-step-one-electron-each by coordination.

A different redox system model – the model for NADH – was also described by our group.[16] As electron transfer mediators, FMN and FAD accept two electrons from NAD(P)H and transfer one electron to metal centres in heme-containing proteins, nonheme iron, or molybdenum sites. However, the transfer of electrons between reduced pyridine – dinucleotide cofactors is slow under physiological conditions and must be catalysed by enzymes. Function of these enzymes was mimicked by a modification of the cofactor by a recognition site for its counterpart and, thus, efficient electron transfer was enabled directly. Functionalised 1,4-dihydronicotinamides bearing a recognition unit for flavins were synthesised (Scheme 18).

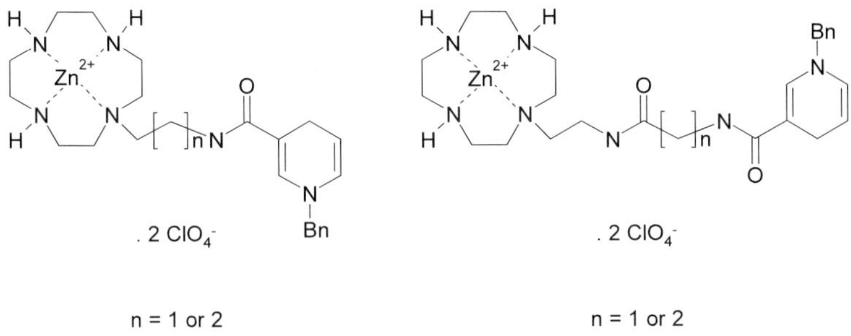

Scheme 18. NADH model systems

The prepared compounds systematically differed in the distance of the dihydropyridine and the flavin recognition part. Binding between flavin and the NADH model systems was proved by potentiometric pH titrations. Redox reaction between the NADH model systems and flavin was monitored by UV – VIS spectroscopy. The intensity of the long-wave absorption of flavin at 456 nm significantly decreased during the reaction and the decrease was attributed to the reduction of flavin to the fully reduced flavohydroquinone. At the same time, the intensity of the peak around 360 nm decreased as well, because of the reduction of flavin and the concerted oxidation of the 1,4-dihydronicotinamide to the corresponding pyridinium species. Kinetics of the electron transfer was studied and two reasonable kinetic models were proposed.

To compare these two mechanisms, an NADH model without the recognition site was synthesised. The contribution of the flavin binding to the rate constant was thus evaluated and it was shown that the proximity of flavin and NADH model influenced the electron transfer rate. Mechanistic computations helped to show that with the appropriate NADH model system, both components were optimally arranged for the electron transfer. Although the exact mechanism of the reaction is still under debate, the kinetic isotope effect experiment indicated that in this case, the hydrogen at 4-position was transferred in the rate determining step which supported the hydride mechanism.

By the use of a model system, Kimura *et al.* [17] tried to mimic the function of the two mechanistically most typical zinc(II) enzymes. Carbonic anhydrase (CA, EC 4.2.1.1) catalyses the reversible hydration of carbon dioxide to bicarbonate ion and its zinc(II) active site is bound to three histidine residues and a water molecule. Carboxypeptidase A (CPA, EC 3.4.17.1) catalyses the hydrolysis of the hydrophobic C-terminal amino acids from polypeptides, and its active-site zinc(II) is bound to two histidine residues, a glutamine residue and a water molecule which is hydrogen bound to a glutamine residue (Scheme 19).

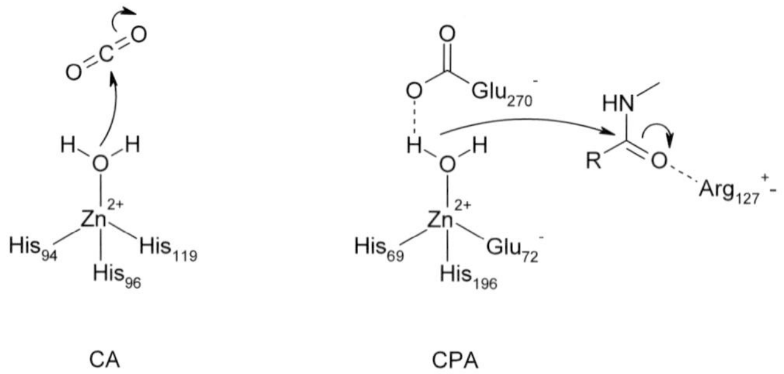

Scheme 19. Structure of CA and CPA active sites

The widely accepted Zn^{2+}-hydroxide mechanism of CA and CPA says the zinc(II)-bound waters that are generated at neutral pH are activated to attack polarised carbonyl substrates. It is noteworthy that carboxamides are substrates for CPA but inhibitors for CA. The visible spectral study of binding of iodoacetamide to colbalt(II)-substituted CA indicates coordination of the amidate N^- ion to the metal. In order to answer the question how do zinc(II) ions work differently toward the carboxamides in CA and CPA, the different modes of recognition were mimicked by the use of carboxamide-appended cyclenes (Scheme 20).

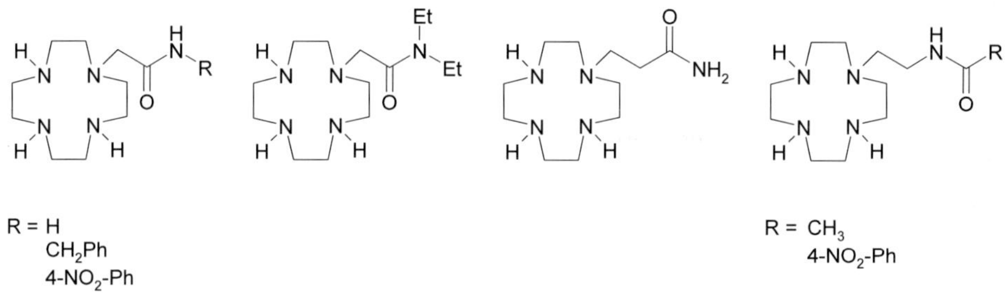

Scheme 20. Carboxamide-appended cyclenes

The protonation constants of the new ligands were determined by potentiometric pH titrations. Deprotonation of the amide hydrogen atoms was not observed. The 1:1 zinc(II) complexation equilibria were studied by potentiometric pH titrations with the protonated ligands and it was found that the complexes with the shorter pendant formed more stable complexes as the amide oxygen coordinated to the metal ion to form five-membered chelates. On the contrary, six membered chelates from the ligands with longer pendant were less likely. Under basic conditions, the coordination of the amidate nitrogen atom to the zinc(II) centre was proved by IR, UV and NMR measurements and finally by X-ray crystal analysis. Hence, the authors concluded, the reactive sites of the Zn^{2+}-cyclenes were occupied by the carboxamide oxygens in acidic pH and carboxamidato nitrogens in basic pH.

A biomimetic artificial photylase model was described by Wiest *et al.*[18] The model recognised pyrimidine dimers in both organic solvents and water. Excitation of the complex with visible light led to cycloreversion of the pyrimidine dimer through photoinduced electron-transfer catalysis (Scheme 21).

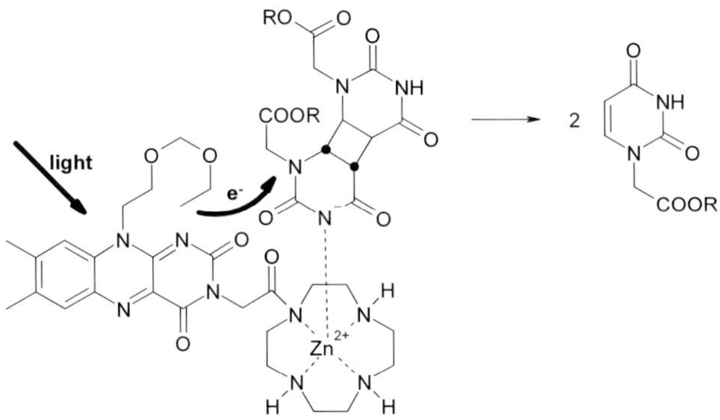

Scheme 21. An artificial photolyase model

6. Electron-transfer Dyads

The feasibility of intramolecular electron- and energy-transfer depends on distance and is usually studied in covalently linked systems. However, donor-acceptor dyads can be also arranged by self-assembly what resembles the situation of electron transfer in biological systems. Artificial dyads tethered by a small number of hydrogen bonds immediately dissociate in methanol or water. To improve the binding while keeping the reversibility, a photoinducible electron donor-acceptor dyad linked by a kinetically labile bond was designed.[19]

Zinc(II) cyclene was linked to phenothiazene group which served as electron-donor (Scheme 22). To the complexed zinc(II), a riboflavin tetraacetate molecule coordinated. Upon irradiation, the flavin became a strong oxidant and the transfer of electrons could be easily observed by emission quenching.

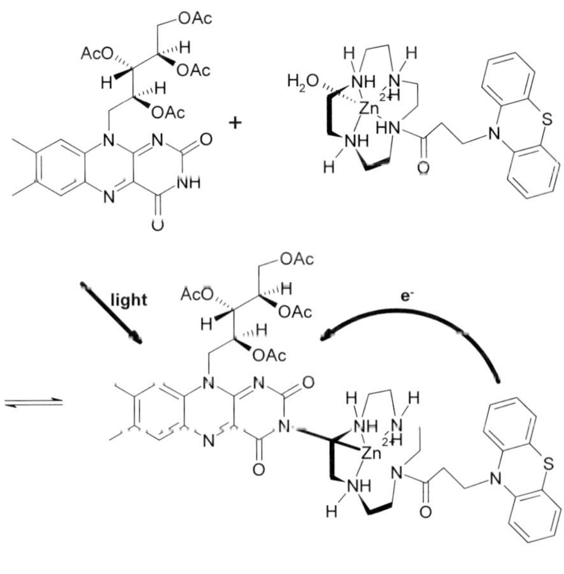

Scheme 22. A phenothiazine – riboflavin dyad

Binding of the components in a neutral aqueous solution was confirmed by potentiometric titrations. The feasibility of electron transfer between the components was predicted by cyclic voltammetry and an efficient outer-sphere fast electron transfer was foreseen. Fluorescence spectroscopy measurements showed that the formation of a defined donor-acceptor complex worked even in water at neutral pH. Electron transfer as the quenching mechanism was proved by laser flash photolysis.

A similar dyad was described by Cibulka *et al.*[20] The flavin part was in this case covalently bound to the zinc(II)-cyclene. 4-Methoxybenzyl alcohol was bound to the metal centre and was upon irradiation oxidised to the corresponding benzaldehyde (Scheme 23).

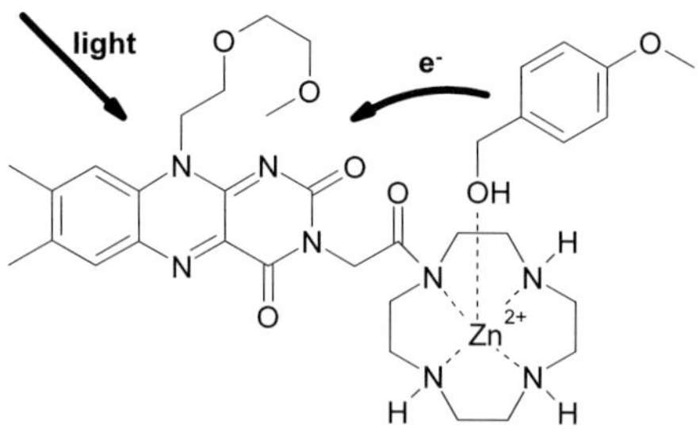

Scheme 23. A flavin – benzylalcohol dyad

7. Summary

The discussed examples clearly demonstrate the importance of azamacrocycles as structural element to create supramolecular function. Their rigid structure, the basicity and transition metal-ion coordinating ability make them suitable as scaffolds and binding sites. Lanthanide chemosensors containing azamacrocyclic ligands have already reached applications in medical diagnostics. Other applications of azamacrocyclic systems with supramolecular functions, particularly in biochemistry, will follow.

8. References

1. Herges, R., Dikmans, A., Umasish, J., Köhler, F., Jones, P. G., Dix, I., Fricke, T., König, B.: Design of a neutral macrocyclic ionophore: synthesis and binding properties for nitrate and bromide anions, *Eur. J. Org. Chem.* (2002), 3004 – 3014.
2. Aoki, S., Jikiba, A., Takeda, K., Kimura, E.: A zinc(II) complex-conjugated polymer for selective recognition and separation of phosphates, *J. Phys. Org. Chem.* **17** (2004), 489 – 497.
3. Koike, T., Abe, T., Takahashi, M., Ohtani, K., Kimura, E., Shiro, M.: Synthesis and characterization of the zinc(II)-fluorophore, 5-dimethylaminonaphthalene-1-sulfonic acid [2-(1,5,9-triazacyclododec-1-yl)-ethyl]amide and its zinc(II) complex, *J. Chem. Soc., Dalton Trans.* (2002), 1764 – 1768.
4. Geißer, B., König, B., Alsfasser, R.: Selective recognition of copper(II) by a water-soluble, emitter-receptor conjugate containing a ruthenium chromophore, a lysine bridge, and a cyclene unit, *Eur. J. Inorg. Chem.* (2001), 1543 – 1549.
5. Aoki, S., Kaido, S., Fujioka, H., Kimura, E.: A new zinc(II) fluorophore 2-(9-anthrylmethylamino)ethyl-appended 1,4,7,10-tetraazacyclododecane, *Inorg. Chem.* **42** (2003), 1023 – 1030.

6. Aoki, S., Kawatani, H., Goto, T., Kimura, E., Shiro, M.: A double-functionalized cyclen with carbamoyl and dansyl groups (cyclen = 1,4,7,10-tetraazacyclododecane): A selective fluorescent probe for Y^{3+} and La^{3+}, *J. Am. Chem. Soc.* **123** (2001), 1123 – 1132.
7. Lowe, M. P., Parker, D., Reany, O., Aime, S., Botta, M., Castellano, G., Gianolio, E., Pagliarin, R.: pH-dependent modulation of relaxivity and luminescence in macrocyclic gadolinium and europium complexes based on reversible intramolecular sulfonamide ligation, *J. Am. Chem. Soc.* **123** (2001), 7601 – 7609.
8. Frias, J. C., Bobba, G., Cann, M. J., Hutchison, C. J., Parker, D.: Luminescent nonacoordinate cations lanthanide complexes as potential cellular imaging and reactive probes, *Org. Biomol. Chem.* (2003), 905 – 907.
9. Subat, M., Borovik, A. S., König, B.: Synthetic creatinine receptor: imprinting of a Lewis acidic zinc(II) cyclene binding site to shape its molecular recognition selectivity, *J. Am. Chem. Soc.* **126** (2004), 3185 – 3190.
10. König, B., Gallmeier, H.-C., Reichenbach-Klinke, R.: Selective-binding and reversible release of riboflavin by polymer-bound zinc(II) azamacrocycles, *Chem. Comm.* (2001), 2390 – 2391.
11. Kikuta, E., Aoki, S., Kimura, E.: A new type of potent inhibitors of HIV-1 TAR RNA-Tat peptide binding by zinc(II)-macrocyclic tetraamine complexes, *J. Am. Chem. Soc.* **123** (2001), 7911 – 7912 and references therein.
12. König, B., Pelka, M., Subat, M., Dix, I., Jones, P. G.: Urea derivatives of 1,4,7,10-tetraazacyclododecane – synthesis and binding properties, *Eur. J. Org. Chem.* (2001), 1943 – 1949.
13. Bridger, G. J., Skerlj, R. T., Thornton, D., Padmanabhan, S., Martelluci, S. A., Henson, G. W., Abrams, M. J., Yamamoto, N., De Vreese, K., Pauwels, R., De Clercq, E.: Synthesis and structure-activity relationships of phenylenebis(methylene)-linked bis-tetraazamacrocycles that inhibit HIV replication. Effects of macrocyclic ring size and substituents on the aromatic linker, *J. Med. Chem.* **38** (1995), 366 – 378.
14. Bridger, G. J., Skerlj, R. T., Padmanabhan, S., Martelluci, S. A., Henson, G. W., Abrams, M. J., Joao, H. C., Witvrouw, M., De Vreese, K., Pauwels, R.: *J. Med. Chem.* **39** (1996), 109 – 119.
15. König, B., Pelka, M., Reichenbach-Klinke, R., Schelter, J., Daub, J.: A model system for flavoenzyme activity – binding of flavin and modulation of its redox potentials through coordination to a Lewis-acidic azamacrocyclic zinc(II) complex, *Eur. J. Org. Chem.* (2001), 2297 – 2303.
16. Reichenbach-Klinke, R., Kruppa, M., König, B.: NADH model systems functionalized with Zn(II)-cyclen as flavin binding site – Structure dependence of the redox reaction within reversible aggregates, *J. Am. Chem. Soc.* **124** (2002), 12999 – 13007.
17. Kimura, E., Gotoh, T., Aoki, S., Shiro, M.: Study of pH-dependent zinc(II)-carboxamide interactions by zinc(II)-carboxamide-appended cyclene complexes (cyclen = 1,4,7,10-tetraazacyclododecane), *Inorg. Chem.* **41** (2002), 3239 – 3248.
18. Wiest, O., Harrison, C. B., Saettel, N. J., Cibulka, R., Sax, M., König, B.: Design, synthesis and evaluation of a biomimetic artificial photolyase model, *J. Org. Chem.*, ASAP article DOI 10.1021/jo0494329.
19. König, B., Pelka, M., Zieg, H., Ritter, T., Bouas-Laurent, H., Bonneau, R., Desvergne, J.-P.: Photoinduced electron transfer in a phenothiazine – riboflavin dyad assembled by zinc – imide coordination in water, *J. Am. Chem. Soc.* **121** (1999), 1681 – 1687.
20. Cibulka, R., Vasold, R., König, B.: Catalytic photooxidation of 4-methoxybenzyl alcohol with a flavin zinc(II) cyclen complex, *Chem. Eur. J.*, in press.

7. SENSING, TEMPLATION AND SELF-ASSEMBLY BY MACROCYCLIC LIGAND SYSTEMS

PAUL D. BEER AND WALLACE W. H. WONG
Department of Chemistry, Inorganic Chemistry Laboratory, University of Oxford, South Parks Road, Oxford, OX1 3QR, U.K.

1. Introduction

For a number of years the research interests of the Beer group have covered many areas of macrocyclic coordination and supramolecular chemistry. This article reviews our latest results of current research by focusing on three major sections: sensing of cations and anions; anion templated assembly of pseudorotaxanes and rotaxanes, and metal-directed self-assembly using the dithiocarbamate ligand.

2. Sensing

2.1 INTRODUCTION

With the aim of advancing chemical sensor technology, the development of host molecules that contain signalling or responsive functional groups as an integral part of a host macrocyclic framework is a highly topical research field.[1] The incorporation of a redox- and/or photo-active signalling moiety in close proximity to a host binding site enables the host-guest recognition event to be sensed via a macroscopic electrochemical and/or photophysical response. An ongoing research programme in our group is concerned with the design and construction of innovative electrochemical and spectral sensory reagents for target cation and anion guest species of biological and environmental importance. The following sections review recent progress made with (i) a new family of redox-active ionophores[2] which exhibit remarkable selectivity preferences and substantial electrochemical recognition effects toward cesium and rubidium cations and (ii) anion recognition and sensing by self-assembled monolayers of amidoferrocene derivatives[3] and zinc metalloporphyrin-functionalised gold nanoparticles[4].

Karsten Gloe (ed.), Macrocyclic Chemistry: Current Trends and Future Perspectives, 105–119.
© 2005 *Springer. Printed in the Netherlands.*

2.2 CESIUM- AND RUBIDIUM-SELECTIVE REDOX-ACTIVE IONOPHORES

A number of research groups have incorporated redox-active transition metal and organic centres into a variety of frameworks based on crown ethers, cryptands and calixarenes, and have shown some of these systems to be selective and electrochemically responsive to the binding of metal cations, in particular lithium[5], sodium[6] and potassium[7]. However, the construction of redox-active ionophores for the selective recognition of the larger rubidium and cesium metal cations has not been described, a fact which is surprising in view of the environmental concern for monitoring radioactive cesium in nuclear waste streams[8] and the potential use of rubidium isotopes in radiopharmaceutical reagents[9].

Taking into account our calix[4]tube ligand design[10], the novel bis-calix[4]diquinone receptors were prepared via the initial synthesis of various bis-calix[4]arene derivatives connected by alkylene chain units and subsequent oxidation of the remaining phenolic groups to quinone moieties (*Scheme 1*).[2]

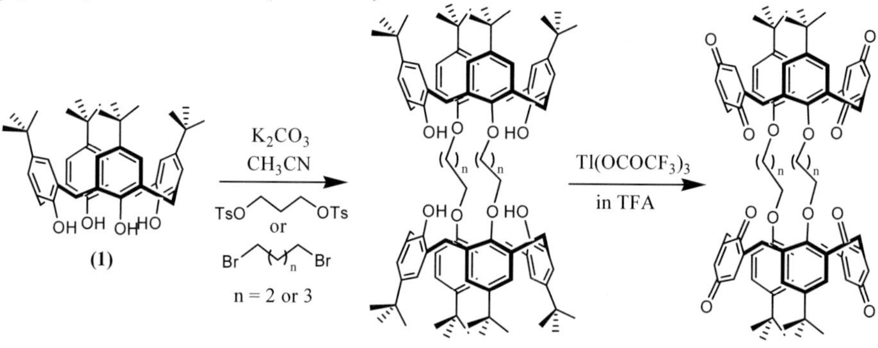

Scheme 1.

Stability constant determinations by UV-visible titration experiments with Group I metal cations in a highly competitive solvent mixture of 99:1 DMSO:water revealed the propylene spaced receptor **(4)** forms a strong selective complex with rubidium cations and exhibits the selectivity trend $Rb^+ > Cs^+ > K^+ >> Na^+$ (TABLE 1). In contrast, receptor **(5)** containing longer butylene spacers exhibits a remarkably high selectivity for cesium ions; in fact in this competitive aqueous DMSO solvent medium, **(5)** does not form complexes with either sodium or potassium cations (TABLE 1). This very high Cs^+/Na^+ selectivity preference is of real significance for the separation of radioactive cesium from nuclear waste as such solutions typically contain sodium ions at high concentrations.

TABLE 1. Stability Constant Data[a] for Group I Metal Complexes of **(4)** and **(5)**.

	K / M^{-1}	
	(4)	**(5)**
Na^+	b	b
K^+	7.9×10^3	b
Rb^+	6.3×10^4	40
Cs^+	1.3×10^4	1.6×10^3

[a] At 298K. Maximum error estimated to be ±10%. Experiments conducted in 99:1 DMSO:H_2O. [b] No evidence of complexation was observed.

An X-ray crystal structure of the cesium complex of butylene-bridged bis-calix[4]diquinone **(5)** illustrates that the cation resides within the cavity bonded to all eight oxygen donor atoms (*Figure 1*). Molecular modelling shows that the cavity size of **(5)** is highly complementary to the cesium ion. Molecular dynamics studies reveal effective strong metal cation binding is only observed for systems where the metal cation is of sufficient size to simultaneously bind to all eight oxygen donor atoms of the alkyl-bridged bis-calix[4]diquinone structural framework. In addition to metal cation size, the stability constant values reported in TABLE 1 show the length of the alkyl bridging chain between the two calix[4]diquinone moieties crucially dictates the selectivity and strength of Group I metal cation binding.

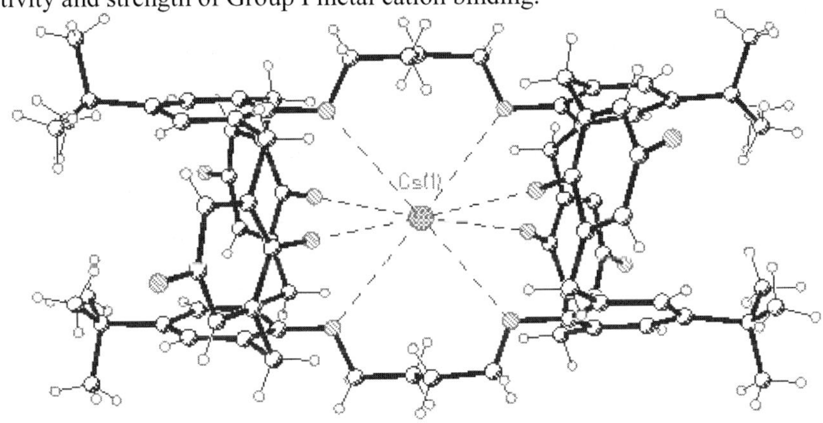

Figure 1. The structure of the cesium complex of bis-calix[4]arene **(5)**.

Cyclic and square wave voltammetric investigations in 4:1 CH_2Cl_2:CH_3CN solutions demonstrated that both **(4)** and **(5)** can electrochemically sense the binding of Group I metal cations via substantial anodic perturbations of the respective quinone reduction redox couples (TABLE 2). Electrochemical competition experiments showed that **(5)** is capable of selectively sensing cesium ions in the presence of large excess of sodium cations, confirming this ionophore's potential use as a novel prototype redox-responsive cesium ion sensor.

TABLE 2. Electrochemical data[a] for the binding of Group I metal cations by bis-calix[4]arenes **(4)** and **(5)**.

	(4)	**(5)**
$E_{½}$ (V)	-1.04, -1.16	-1.06, -1.15
E_{pc} (V)	-1.70	-1.75
ΔE (Na^+, mV)[b]	350	330
ΔE (K^+, mV)[b]	250	250
ΔE (Rb^+, mV)[b]	250	210
ΔE (Cs^+, mV)[b]	240	190

[a] Obtained in 4:1 CH_2Cl_2:CH_3CN solution containing 0.1M Bu_4NBF_4 as supporting electrolyte. Receptor concentration = 1 x 10^{-3}M. ΔE recorded upon addition of 5 equivalents of Group I metal cations. [b] Anodic shift of new redox couple relative to the first process of the original couple.

2.2 ANION RECOGNITION AND SENSING BY AMIDOFERROCENE SELF-ASSEMBLED MONOLAYERS ON GOLD ELECTRODES AND ZINC METALLOPORPHYRIN-FUNCTIONALISED GOLD NANOPARTICLES

Anions play numerous indispensable roles in biological and chemical processes, as well as contributing significantly to environmental pollution.[11] Since the beginning of anion co-ordination chemistry in the late 1960s, the main strategies in the design of synthetic anion complexing reagents have focused on cationic polyammonium[12], quaternary ammonium[13], guanidinium[14], expanded porphyrin host systems[15] and a variety of Lewis acid-containing receptors such as tin[16], boron[17], silicon[18], mercury[19] and uranyl[20]. Neutral organic receptors containing amide[21], urea[22], thiourea[11a] and pyrrole[23] groups which bind anions solely via hydrogen bonding interactions have also been exploited. Reviews on various aspects of anion coordination chemistry can be found in the references.[11,24]

In more recent years, attention has turned towards anion sensory reagents.[11,24,25] For example, we[1,25] and others[26-28] have exploited the redox-active ferrocene moiety in the selective electrochemical sensing of anions in organic and aqueous media. In particular, acyclic, macrocyclic and calixarene amide functionalised ferrocene derivatives have all been shown to undergo substantial perturbations of the respective metallocene redox couple in the presence of a variety of anions. In an effort to fabricate a robust prototype anion sensory device, we prepared the new 1,1'-bis(alkyl-*N*-amido)ferrocene derivative **(7)** which contains two disulfide functional groups for attachment to planar gold electrode surfaces (*Scheme 2.*)[3] Stable self-assembled monolayers (SAMs) of **(7)** were prepared on gold macroelectrodes which exhibited a reversible ferrocene/ferrocenium surface confined electrochemical response in both organic and aqueous electrolyte (*Figure 2*).

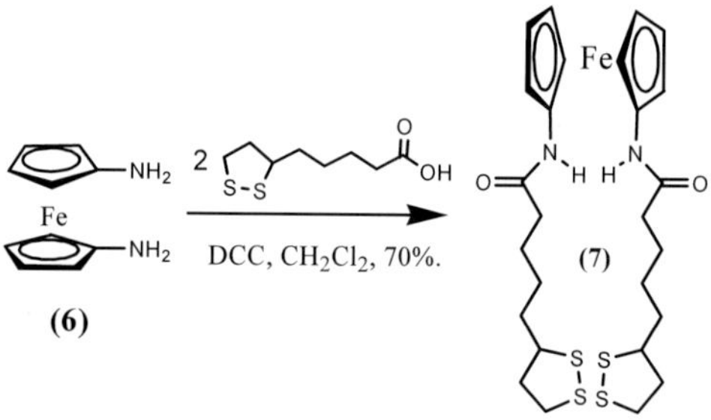

Scheme 2.

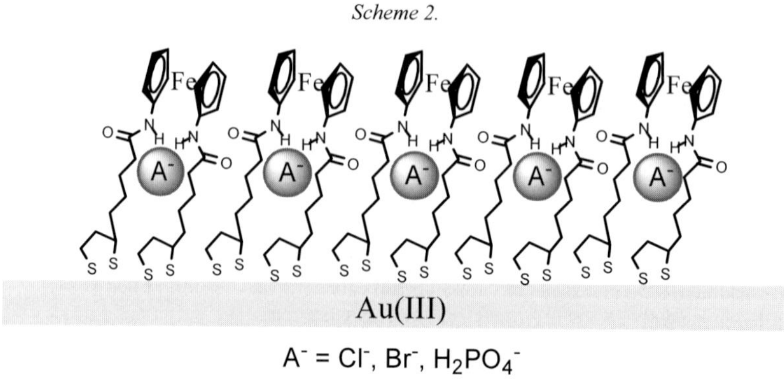

Figure 2. An illustration of a SAM of **(7)**.

TABLE 3 compares the diffusive electrochemical anion recognition behaviour of **(7)** with that of the SAM of **(7)**. The most striking aspect of data obtained on confining **(7)** to a monolayer is the consistent observation of "surface sensing amplification", that is, the magnitudes of anion-induced cathodic shifts on the SAM are consistently greater (*Figure 3*) than that observed diffusively, highlighting significant advantages associated with preorganisation of the host pseudo-macrocycle on a surface (*Figure 2*). Specifically, redox responses are more cathodically shifted by ca. 90 and 60 mV for phosphate and chloride respectively. Competition experiments carried out on the addition of $H_2PO_4^-$ to electrolytic solutions of **(7)**, 100 fold excess in halide, confirmed the ability of $H_2PO_4^-$ to compete effectively for the amide binding site; only the resultant $H_2PO_4^-$ shifted ferrocene couple is observed. This is the case both when **(7)** is analysed diffusively and surface confined.

TABLE 3. Perturbation of the ferrocene/ferrocenium couple of **(7)** and the SAM of **(7)** on addition of anions (8 mM).

Cathodic Shift ΔE (mV)	Cl^-	Br^-	$H_2PO_4^-$
(7)	40	20	210
SAM of **(7)**	100	30	300

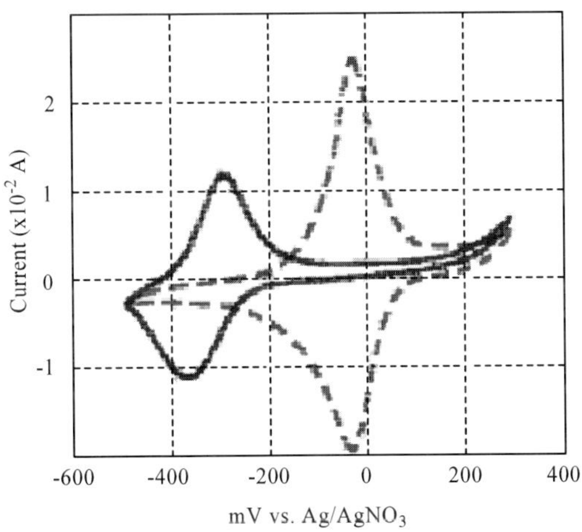

Figure 3. The voltammetric response of a monolayer of **(7)** *in the absence (dashed line) and presence of* $H_2PO_4^-$ *(8 mM)*

Coordination mediated anion detection in an aqueous medium is challenging due to the accompanying unfavourable energetic requirements of dehydration and the low binding constants of anions, particularly to neutral anionophores. Monolayers of **(7)** in water, however, give reproducible cathodic shifts (15-20 mV) in the presence of perrhenate, a comparatively poorly hydrated anion.

Metal nanoparticles have extraordinary size-dependent optical properties, not present in the bulk metal and have, consequently, been the subject of intense research during the past decade or so.[27] Attention has recently focused on functionalising colloidal nanoparticles with molecular recognition components for potential sensing applications.[28,29,30] We have prepared a new amido-disulfide functionalised zinc metalloporphyrin **(8)** which was self-assembled on to gold nanoparticles to produce a novel anion-selective optical sensing system **(9)** (*Scheme 3*).[4]

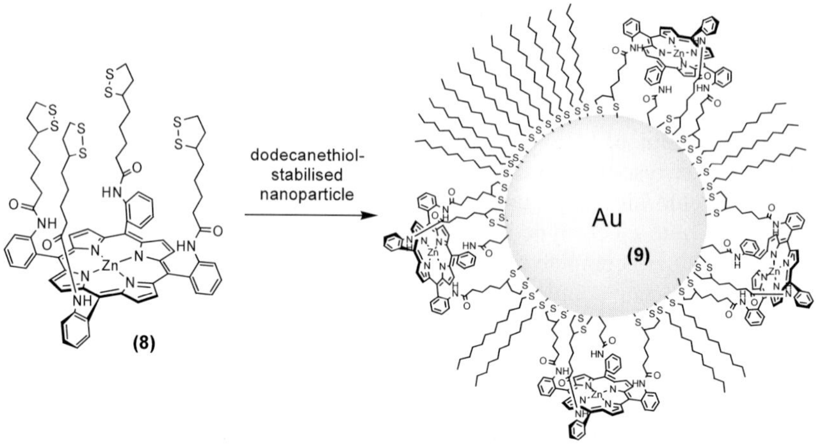

Scheme 3.

Comparative UV-visible titrations of the zinc metalloporphyrin **(8)** and the metalloporphyrin functionalised nanoparticle **(9)** with a variety of anions in CH_2Cl_2 and DMSO solutions gave association constant values shown in TABLE 4 and TABLE 5. This data highlights the striking result that the porphyrin functionalised nanoparticle **(9)** exhibits significantly larger magnitudes of anion association constant in comparison to the free zinc metalloporphyin **(8)**. In particular, in DMSO solution, the strength of binding of Cl^- by nanoparticle **(9)** is more than two orders of magnitude greater than that of the free metalloporphyrin **(8)** (TABLE 5).

TABLE 4. Association constant (log K) data in dichloromethane determined at 293 K, errors ±0.1.

Anion	(8)	(9)[a]
Cl^-	> 6	> 6
Br^-	4.1	5.0
I^-	3.2	4.0
NO_3^-	2.4	3.2
$H_2PO_4^-$	> 6	> 6
ClO_4^-	0	0

[a] Association constant values for the 1:1 porphyrin-anion complex on the nanoparticle surface.

TABLE 5. Association constants (log K) data in DMSO determined at 293 K, error ±0.1.

Anion	(8)	(9)[a]
Cl^-	<2	4.3
$H_2PO_4^-$	2.5	4.1

[a] Association constant values for the 1:1 porphyrin-anion complex on the nanoparticle surface.

This remarkable enhancement of anion complexed stability observed on confining the receptor to a surface is significant. Host-guest binding affinity is determined by factors controlling enthalpy and entropy. By preorganising receptors on surfaces, and thereby reducing their conformational flexibility, entropic contributions would be expected to be more favourable. Solvation effects associated with close packing of receptors in a dominantly hydrophobic self-assembled monolayer environment are also likely to be significant. The surface preorganisation of optical and/or electrochemical group functionalised host systems offers the opportunity to fabricate highly sensitive sensory devices.

3. Anion Templated Assembly of Pseudorotaxanes and Rotaxanes

Imaginative template methodologies are increasingly being used in the construction of complex supramolecular architectures.[31] ranging from molecular cage-like assemblies and helicates to mechanically interlocked supramolecules such as rotaxanes, catenanes and knots.[32] Cationic and neutral species have dominated the templated synthetic strategies reported to date by employing metal-ligand coordination[33], π–π stacking interactions[34], hydrogen bonding[35] and solvophobic effects[36] to effect assembly between two or more components. The challenge of exploiting anions to direct supramolecular assembly formation remains largely under-developed, which may be attributed to their diffuse nature, pH-dependence and relative high solvation energy as compared to cations.[11a] Although various serendipitous discoveries of where anions have templated the formation of, in particular, inorganic-based polymetallic cage complexes[37] and circular double helicates[38] have now appeared, strategic anion-templated syntheses and assemblies are rare.[39] This is especially the case when employing anions in the construction of interpenetrated compounds.

We have recently demonstrated the use of a rational design procedure to develop a general method of using anions to template the formation of a wide range of pseudorotaxanes based on the coupling of anion recognition with ion-pairing (*Figure 4*).[40,41,42] This templation strategy is designed to operate in non-competitive solvent media where the anion of the ion-pair is strongly associated with the potential cationic threading component and importantly remains coordinatively unsaturated. Subsequent anion recognition by the macrocyclic ligand results in pseudorotaxane formation as the cationic thread strongly associates with the complexed anion within the macrocyclic cavity.

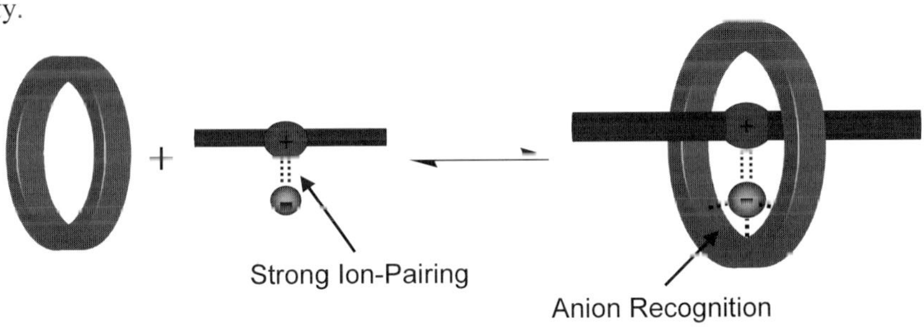

Figure 4. Anion-templated self-assembly of pseudorotaxanes. Recognition of the anion by macrocycle results in the formation of an interpenetrated structure.

Inspired by Crabtree and co-workers[43] who showed that simple isophthalamide molecules are receptors for anions in chloroform (*Figure 5*), the first success using this general anion template procedure was illustrated by the halide directed assembly of pyridinium pseudorotaxanes (*Figure 6* and *Figure 7*).[40] The efficacy of the pseudorotaxane formation was shown to depend critically on the nature of the halide anion, with association constant evaluations showing the pyridinium-halide ion-pair threading component's strength of binding order to be Cl^- > Br^- > I^-. No evidence of pseudorotaxane assembly was noted with the pyridinium PF_6^- salt.

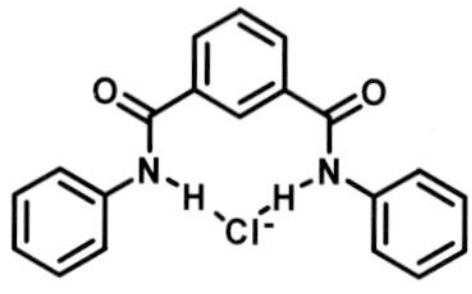

Figure 5. Crabtree's isophthalamide anion receptor.

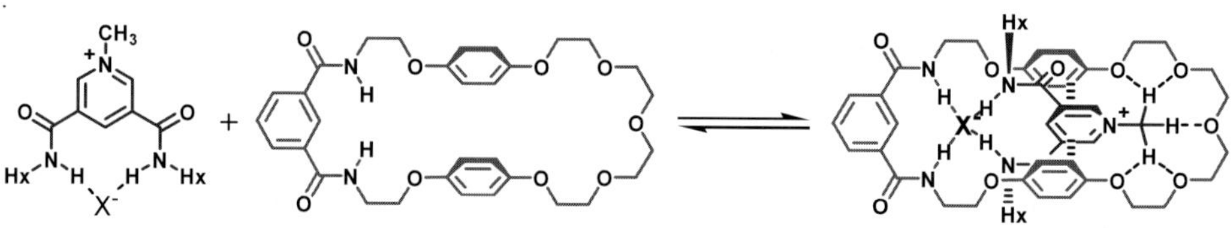

X^- = Cl^-, Br^-, I^-.

Figure 6. Assembly of pseudorotaxane from the ion-paired thread and the macrocycle.

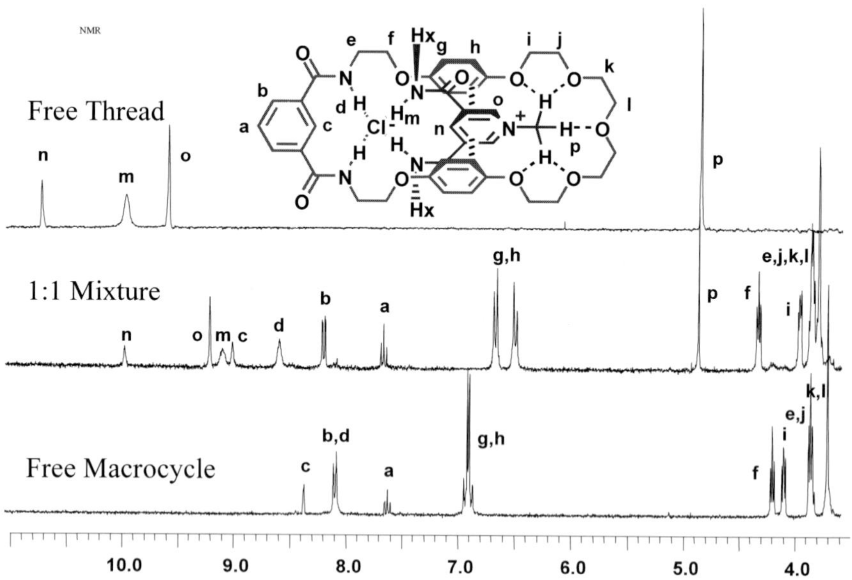

Figure 7. ¹H NMR spectra of the free thread, macrocycle and the 1:1 mixture of the two components.

The solid state structure of the pyridinium-chloride pseudorotaxane (*Figure 8*) reveals the interpenetrative nature of the components and provides evidence for anion complexation by the macrocycle's isophthalamide motif, π–π donor-acceptor interactions between the electron rich hydroquinone units of the macrocycle and electron deficient pyridinium group, together with hydrogen bonding between the pyridinium methyl group and the macrocycle polyether linkage.

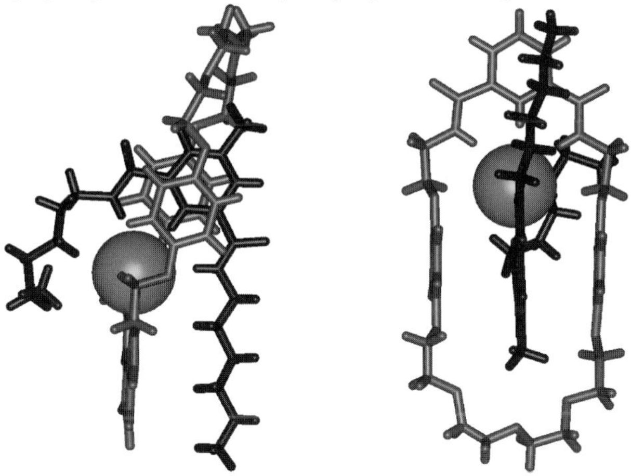

Figure 8. Two views of the structure of the pyridinium-chloride pseudorotaxane.

The versatility of this pseudorotaxane anion templation methodology was further illustrated with the assembly of a series of pseudorotaxanes containing pyridinium nicotinamide (*Figure 9*), imidazolium, benzimidazolium and guanidinium threading components.[41] Extensive ^{1}H NMR titration investigations revealed the thermodynamic stability of the pseudorotaxane assembly depends critically on the nature of the halide anion template, with chloride proving to be the optimum template anion, which correlates with this halide forming the strongest complexes with the isophthalamide macrocycles.

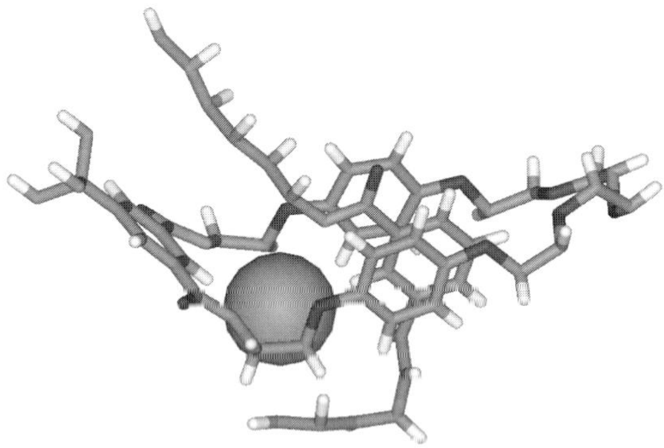

Figure 9. Structure of the nicotinamide-chloride pseudorotaxane.

Exploiting this work further led to the anion templated assembly of a [2]rotaxane.[12] An acyclic chloride anion template **(10)** based on a "stoppered" pyridinium thread was designed to act as the "axle" of the target rotaxane. The second component was a neutral acyclic molecule **(11)** incorporating an isophthaloyl anion binding cleft functionalised with two long hydroquinone containing side chains terminating with alkyl groups capable of ring closing metathesis (RCM). The two components associate strongly in non-competitive solvents and RCM reaction with Grubbs' catalyst facilitated rotaxane formation **(12a)** in yields of up to 47% (*Scheme 4*).

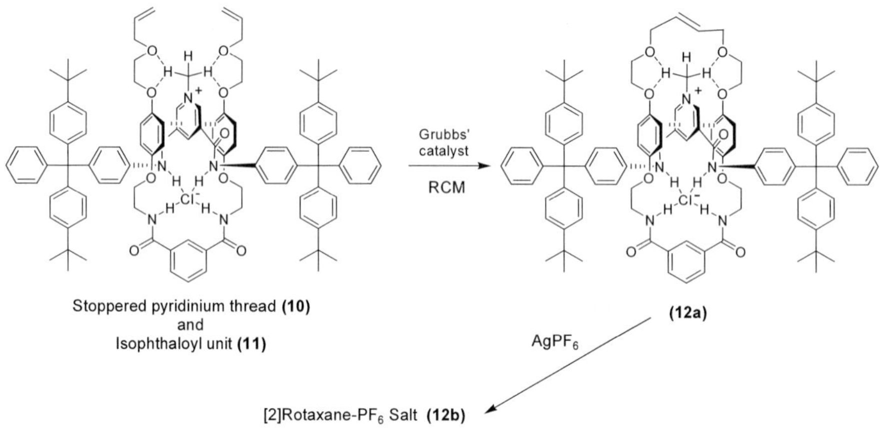

Scheme 4.

No rotaxane formation occurred with analogous stoppered pyridinium bromide or hexafluorophosphate salts, indicating the crucial templating role of the chloride anion. An impressive feature of this method of rotaxane assembly is that the resultant product retains a degree of functionality based on the anion template itself. Anion exchange of the templating chloride anion for the non-competitive hexafluorophosphate anion leaves a highly selective anion binding site within the rotaxane **(12b)** (*Scheme 4*). Complexation studies have shown that although the pyridinium PF_6^- thread alone binds anions with a selectivity trend $AcO^- >> H_2PO_4^- > Cl^-$, the rotaxane PF_6^- receptor exhibits a complete selectivity reversal with a high selectivity for chloride, the templating anion. The solid state molecular structure of the chloride complexed rotaxane (*Figure 10*) reveals this halide anion selectivity is a result of a unique hydrogen bond donating pocket within the [2]rotaxane superstructure which is of a complementary topology to the chloride guest anion. Lager anions cannot penetrate this diamide cleft and bind at the periphery of the rotaxane, or via a large displacement of the pyridinium axle thread from the macrocyclic cavity.

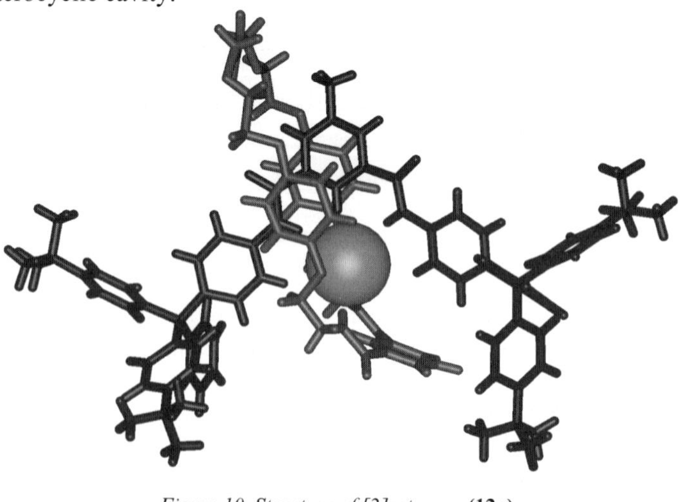

Figure 10. Structure of [2]rotaxane **(12a)**.

4. Metal-directed Self-assembly of Cryptands and Catenanes Using the Dithiocarbamate Ligand

Metal-directed self-assembly of three-dimensional macrocycles and cage-like structures including supramolecular architectures of nanoscale dimensions, is currently an area of intense research activity.[44] By judicious choice of metal and multidentate ligand components the resulting polymetallic assemblies may exhibit unique redox, magnetic and photochemical properties, and have the potential to bind guest substrates. With labile metals the assembly process is often thermodynamically controlled and yields of self-assembled constructs commonly exceed those of traditional covalent synthesis. To date the field has been dominated by multidentate pyridine ligands, followed by catecholates and others.[44] We recently reported for the first time the use of dithiocarbamate (dtc) ligand as a self-assembling construction motif in the synthesis of a range of structures including nano-sized resorcinarene-based assemblies[45], assorted macrocycles[46] and trinuclear cages[47].

The application of new bis-substituted dtc ligands containing various polyether and amide spacer groups has lead to the construction of a wide range of metal-directed assembled bimetallic macrocycles and cryptands (*Scheme 5*).[48] Cryptands can be assembled through careful choice of a suitable octahedral stereochemical directing metal centre such as iron(III) and cobalt(III).

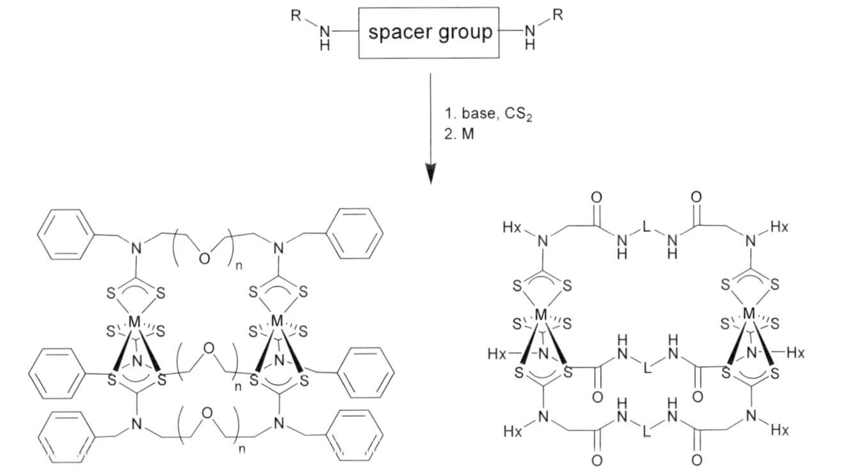

n = 1 (13a) M = Co(III) (13b) M = Fe(III)
n = 2 (14a) M = Co(III) (14b) M = Fe(III)
n = 3 (15a) M = Co(III) (15b) M = Fe(III)
L = ethyl (16a) M = Co(III) (16b) M = Fe(III)
L = *o*-phenyl (17a) M = Co(III) (17b) M = Fe(III)
L = *m*-phenyl (18a) M = Co(III) (18b) M = Fe(III)

Scheme 5.

The redox-active polyether containing cryptands were shown to complex Group I metal cations, initially by ESMS. The bis cobalt(III) cryptands also recognised these metal cations electrochemically via significant anodic shifts of the Co(IV)/Co(III) oxidation couple (TABLE 6). It is noteworthy that a correlation between magnitude of anodic shift and complementary metal cation:cryptand size was observed, and not with the metal cation polarising character. For example, with the large cryptand **(15a)**, Cs^+ causes the greatest perturbation of 45 mV whereas with the smaller cryptands **(14a)** and **(13a)**, K^+ and Na^+ respectively induce the largest anodic shifts.

Electrochemical recognition experiments with $Co(Et_2NCS_2)_3$ revealed no evidence of interactions with Group I metal cations.

TABLE 6. Electrochemical Group I metal cation recognition data[a]

	ΔE_p (mV)[b]			
	Li^+	Na^+	K^+	Cs^+
(13a)	20	25	15	5
(14a)	0	25	35	10
(15a)	0	20	30	45

[a] Square wave voltammogram recorded in 1:1 CH_2Cl_2/MeCN solutions containing 0.1 mol dm^{-3} NBu_4BF_4 as supporting electrolyte. [b] Anodic shift of the Co(IV)/Co(III) oxidation potential produced by presence of Group I metal cations (up to 5 equivalents) added as their perchlorate, hexafluorophosphate and triiodide salts.

The bimetallic amide containing macrocycles and cryptands recognise anionic guest species.[48] Although labile iron(III) complexes proved unsuitable for anion binding studies, the cobalt(III) complexes **(16a)** – **(18a)** were capable of electrochemically sensing anions. Cathodic perturbations of up to ΔE = 125 mV with $H_2PO_4^-$ of the respective Co(IV)/Co(III) redox couple were noted in dichloromethane solutions, compared to no anion-induced shifts with $Co(Et_2CS_2)_3$.

In a serendipitous fashion, a novel mixed valence tetranuclear copper(II)/copper(III) dithiocarbamate [2]catenane was prepared in near quantitative yield by partial chemical oxidation of a preformed dinuclear copper(II) naphthyl dtc macrocycle (*Scheme 6*).[49] X-ray structure, magnetic susceptibility, ESMS and electrochemical studies all support the tetranuclear catenane dication formulation. The combination of the lability of copper(II) dtc coordinate bonds and favourable copper(II) dtc-copper(III) dtc charge transfer stabilisation effects are responsible for the high yielding formation of the interlocked structure.[49]

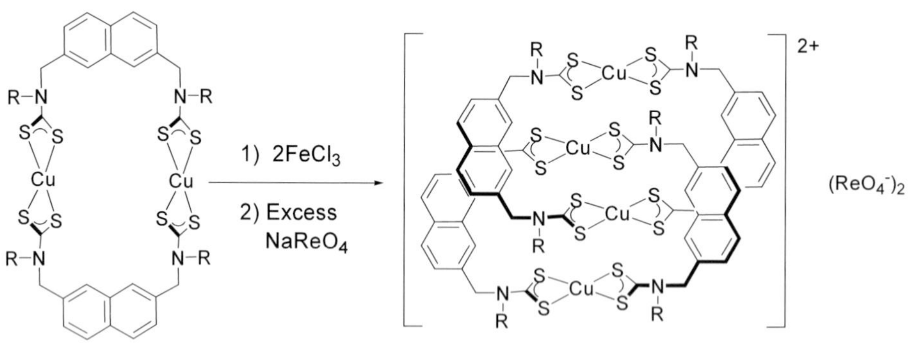

(19) R = butyl or hexyl. **(20)**

Scheme 6.

Exploiting the labile copper(II) dtc coordinate bond, a heteropolymetallic copper(II)/gold(III) catenane **(22)** has also been recently prepared in a 'magic ring' high yielding reaction simply by mixing the dinuclear copper(II) naphthyl dtc macrocycle **(19)** with the dicationic gold(III) cyclic analogue **(21)** (*Scheme 7*).[50]

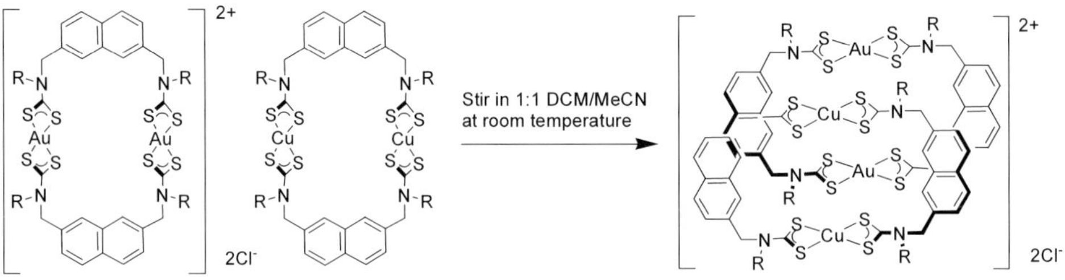

(21) **(19)** **(22)**

Scheme 7.

Applying this new exciting transition metal dtc-based catenane high yielding synthetic procedure to the construction of novel redox-controlled molecular machines and switches is the subject of ongoing research within the group.

5. Acknowledgements

P. D. B. thanks his postgraduate and postdoctoral co-workers for all their hard work and inspiration; their names are cited in the references. A special thank you goes to Professor M. G. B. Drew for his numerous structure determinations and molecular modelling expertise and to my colleague Dr. Jason Davis for collaborative research on anion sensing SAMs and nanoparticles. P. D. B. gratefully acknowledges the financial support of EPSRC, AWE, Johnson Matthey and NSERC (Canada).

6. References

1. (a). P. D. Beer, P. A. Gale, G. Z. Chen, *Adv. Phys. Org. Chem.* **1998**, *31*, 1; (b). P. D. Beer, P. A. Gale, G. Z. Chen, *Dalton Trans.* **1999**, 1897; (c). *Coord. Chem. Rev.* **2000**, *205, (1),* 1- 232.
2. P. R. A. Webber, P. D. Beer, G. Z. Chen, V. Felix, M. G. B. Drew, *J. Am. Chem. Soc.* **2003**, *125*, 5774.
3. P. D. Beer, J. J. Davis, D. A. Drillsma-Milgrom, F. Szemes, *Chem. Commun.* **2002**, 1716.
4. P. D. Beer, D. P. Cormode, J. J. Davis, *Chem. Commun.* **2004**, 414.
5. (a). H. Plenio, R. Diodone, *Inorg. Chem.*, **1995**, *34*, 3964; (b). H. Plenio, C. Aberle, *Organometallics* **1997**, *16*, 590.
6. (a). D. S. Shephard, B. F. G. Johnson, J. Matters, S. Parsons, *Dalton Trans.* **1998**, 2289; (b). J. C. Medina, T. T. Goodnow, M. T. Rojas, J. L. Atwood, B. C. Lynn, A. E. Kaifer, G. W. Gokel, *J. Am. Chem. Soc.* **1992**, *114*, 10583.
7. (a). P. D. Beer, P. A. Gale, Z. Chen, M. G. B. Drew, J. A. Heath, M. I. Ogden, H. R. Powell, *Inorg. Chem.* **1997**, *36*, 5880; (b). J.-P. Bourgeois, L. Echegoyen, M. Fibbioli, E. Pretsch, F. Diederich, *Angew. Chem. Int. Ed.* **1998**, *37*, 2118.
8. *Radioactive Waste Management and Disposal*, L. Cecille, Ed., Elsevier, New York, **1991**
9. C. J. Anderson, M. J. Welch, *Chem. Rev.* **1999**, *99*, 2219.
10. S. E. Matthews, P. Schmitt, V. Felix, M. G. B. Drew, P. D. Beer, *J. Am. Chem. Soc.* **2002**, *124*, 1341.
11. (a). P. D. Beer, P. A. Gale, *Angew. Chem. Int. Ed. Engl.* **2001**, *40*, 486; P. A. Gale, *Coord. Chem. Rev.* **2003**, *240*, 191; F. P. Schmidtchen, M. Berger, *Chem. Rev.* **1997**, *97*, 1609; P. D. Beer, D. K. Smith, *Progr. Inorg. Chem.* **1997**, *46*, 1; J. L. Atwood, K. T. Holman, J. W. Steed, *Chem. Commun.* **1996**, 1401; A. Bianchi, K. Bowman-James, E. García-Espãna (eds.), *Supramolecular Chemistry of Anions*, New York, Chichester, Wiley-VCH, **1997**; (b) P. D. Beer, *Acc. Chem. Res.* **1998**, *31*, 71; A. P. de Silva, H. Q. N. Guarantee, T. Gunnlaugsson, A. J. Huxley, C. P. McCoy, J. T. Rademacher, T. R. Rice, *Chem. Rev.* **1997**, *97*, 1515.
12. E. Graf, J.-M. Lehn, *J. Am. Chem. Soc.* **1976**, *98*, 6403; M. W. Hosseini, J.-M. Lehn, *J. Am. Chem. Soc.* **1982**, *102*, 3525.

13. F. B. Schmidtchen, *Angew. Chem. Int. Ed. Engl.* **1977**, *16*, 720.
14. D. W. Christianson, W. N. Lipscomb, *Acc. Chem. Res.* **1989**, *22*, 62.
15. J. L. Sessler, M. Cyr, H. Furuta, V. Kral, T. Mody, T. Morishima, M. Shionoya, S. Weghorn, *Pure and Appl. Chem.* **1993**, *65*, 393; J. L. Sessler, M. J. Davis, *Acc. Chem. Res.* **2001**, *34,* 989.
16. Y. Azuma, M. Newcomb, *Organometallics* **1984**, *3*, 9.
17. D. F. Shriver, M. J. Biallas, *J. Am. Chem. Soc.* **1967**, *89*, 1078.
18. K. Tamao, T. Hayashi, Y. Ito, *J. Organomet. Chem.* **1996**, *506*, 85.
19. X. Yang, C. B. Knobler, M. F. Hawthorne, *Angew. Chem. Int. Ed. Engl.* **1991**, *30*, 1507.
20. V. van Castelli, A. Dalla Cort, L. Mandolini, D. N. Reinhoudt, L. Schiaffino, *E. J. Org. Chem.* **2003**, 627.
21. R. A. Pascal, J. Spergel, D. V. Engbersen, *Tetrahedron Lett.* **1986**, *27*, 4099.
22. J. Scheerder, M. Fochi, J. F. J. Engbersen, A. Casnati, R. Ungaro, D. N. Reinhoudt, *J. Org. Chem.* **1995**, *60*, 6448; C. R. Bondy, P. A. Gale, S. J. Loeb, *J. Am. Chem. Soc.* **2004**, *126*, 5030.
23. J. L. Sessler, S. Camiolo, P. A. Gale, *Coordin. Chem. Rev.* **2003**, *240*, 17.
24. R. Martínez-Máñez, F. Sancenón, *Chem. Rev.* **2003**, *103*, 4419.
25. P. D. Beer, *Acc. Chem. Res.* **1998**, *31*, 71.
26. P. D. Beer, J. A. Cadman, *Coord. Chem. Rev.* **2000**, *205*, 131 and references cited therein.
27. O. Reynes, J. Moutet, J. Pecaut, G. Royal, E. Saint-Aman, *New. J. Chem.* **2002**, *26*, 9.
28. A. Labande, J. Ruiz, D. Astruc, *J. Am. Chem. Soc.* **2002**, *124*, 1782.
29. A. N. Shipway, E. Katz, I. Willner, *Chem. Phys. Chem.*, **2000**, *1*, 18; C. M. Neimeyer, *Angew. Chem. Int. Ed.* **2001**, *40,* 4123; A. C. Templeton, W. P. Wuelfing, R. W. Murray, *Acc. Chem. Res.* **2000**, *33*, 27; S. Chen, R. S. Ingram, M. J. Hostetler, J. J. Pietron, R. W. Murray, T. G. Schaaff, J. T. Khoury, M. M. Alvarez, R. L. Whetten, *Science* **1998**, *280*, 2098.
30. M. Daniel, J. Ruiz, S. Nlate, J. Blais, D. Astruc, *J. Am. Chem. Soc.* **2003**, *125,* 2617.
31. (a). F. M. Raymo, J. F. Stoddart, *Templated Organic Synthesis*; Wiley: Weinheim, Germany, **2000**; (b). *Comprehensive Supramolecular Chemistry*; J.-P. Sauvage, M. Hosseini, (eds.) Elsevier, New York, **1996**; Vol. 9; (c). A. V. Davis, R. M. Yeh, K. N. Raymond, *Proc. Natl. Acad. Sci.* **2002**, *99*, 4793; (d). R. L. E. Furlan, S. Otto, J. K. M. Sanders, *Proc. Natl. Acad. Sci.* **2002**, *99*, 4801; (e). M. J. Krische, J.-M. Lehn, *Structure and Bonding* **2000**, *96*, 3; (f). L. M. Greig, D. Philp, *Chem. Soc. Rev.* **2001**, *30*, 287; (f). P. A. Gale, *Annu. Rep. Prog. Chem. Sect. B* **2002**, *98*, 581; (g) M. D. Ward, *Annu. Rep. Prog. Chem. Sect A* **2002**, *98*, 285; (h). J. S. Lindsey, *New. J. Chem.* **1991**, *15*, 153.
32. (a). T. J. Hubin, D. H. Busch, *Coord. Chem. Rev.* **2000**, *200*, 5; (b). G. A. Breault, C. A. Hunter, P. C. Mayers, *Tetrahedron* **1999**, *55*, 5265; (c). I. G. Panova, I. N. Topchieva, *Russ. Chem. Rev.* **2001**, *70*, 23; (d). J.-P. Sauvage, *Acc. Chem. Res.* **1990**, *23*, 319.
33. (a). C. O. Dietrich-Buchecker, J.-P. Sauvage, *Tetrahedron Lett.* **1983**, *24*, 5095; (b). C. O. Dietrich-Buchecker, J.-P. Sauvage, J.-M. Kern, *J. Am. Chem. Soc.* **1984**, *106 (10)*, 3043; (c). M. Weck, B. Mohr, J.-P. Sauvage, R. H. Grubbs, *J. Org. Chem.* **1999**, *64 (15)*, 5463; (d). C. O. Dietrich-Buchecker, J.-P. Sauvage, *Angew. Chem. Int. Ed. Engl.* **1989**, *28*, 189-192; (e). M. Fujita, F. Ibukuro, H. Hagihara, K. Ogura, *Nature* **1994**, *367*, 720. (f). M. Fujita, F. Ibukuro, K. Yamaguchi, K. Ogura, *J. Am. Chem. Soc.* **1995**, *117*, 4175; (g) M. E. Padilla-Tosta, O. D. Fox, M. G. B. Drew, P. D. Beer, *Angew. Chem. Int. Ed. Engl.* **2001**, *40 (22)*, 4235; (h). C. Wu, P. R. Lecavalier, Y. X. Shen, H. W. Gibson, *Chem. Mater.* **1991**, *3 (4)*, 569.
34. (a). P. R. Ashton, R. Ballardini, V. Balzani, M. Bělohradsky, M. T. Gandolfi, D. Philp, L. Prodi, F. M. Raymo, M. V. Reddington, N. Spencer, J. F. Stoddart, M. Venturi, D. J. Williams, *J. Am. Chem. Soc.* **1996**, *118 (21)*, 4931; (b). J. F. Stoddart, H.-R. Tseng, *Proc. Natl. Acad. Sci.* **2002**, *99 (8)*, 4797; (c). F. M. Raymo, K. N. Houk, J. F. Stoddart, *J. Am. Chem. Soc.* **1998**, *120 (36)*, 9318; (d). P. R. Ashton, I. Baxter, M. C. T. Fyfe, F. M. Raymo, N. Spencer, J. F. Stoddart, A. J. P. White, D. J. Williams, *J. Am. Chem. Soc.* **1998**, *120 (10)*, 2297.
35. (a). C. A. Hunter, *J. Am. Chem. Soc.* **1992**, *114 (13)*, 5303; (b). H. Adams, F. J. Carver, C. A. Hunter, *Chem. Commun.* **1995**, 809; (c). F. Vögtle, S. Meier, R. Hoss, *Angew. Chem. Int. Ed. Engl.* **1992**, *31*, 1619; (d). S. Ottens-Hildebrandt, M. Nieger, K. Rissanen, J. Rouvinen, S. Meier, G. Harder, F. J. Vögtle, *Chem. Commun.* **1995**, 777; (e). A. G. Johnston, D. A. Leigh, R. J. Pritchard, M. D. Deegan, *Angew. Chem. Int. Ed. Engl.* **1995**, *34*, 1209. (f). G. Bottari, F. Dehez, D. A. Leigh, P. J. Nash, E. M. Perez, J. K. Y. Wong, F. Zerbetto, *Angew. Chem. Int. Ed.*, **2003**, *42*, 5886; (g). D. A. Leigh, J. K. Y. Wong, F. Dehez, F. Zerbetto, *Nature* **2003**, *424*, 174.
36. S. Anderson, H. L. Anderson, *Angew. Chem. Int. Ed. Engl.* **1996**, *35*, 1956.
37. J. S. Fleming, K. L.V. Mann, C.-A. Carraz, E. Psillakis, J. C. Jeffery, J. A. McCleverty, M. D. Ward, *Angew. Chem. Int. Ed. Engl.* **1998**, *37 (9)*, 1279.
38. (a). B. Hasenknopf, J.-M. Lehn, B. O. Kniesel, G. Baum, D. Fenske, *Angew. Chem. Int. Ed. Engl.* **1996**, *35(16)*, 1838. (b). B. Nasenknopf, J.-M. Lehn, N. Boumediene, A. Dupont-Gervais, A. Van Dorsselaer, B. Kneisel, D. Fenske, *J. Am. Chem. Soc.* **1997**, *119 (45)*, 10956.

39. (a). R. Vilar, *Angew. Chem. Int. Ed. Engl.* **2003**, *42 (13)*, 1460. (b). P. D. Beer, M. R. Sambrook, in *Encylopedia of Nanoscience and Nanotechnology*, Marcel Dekker Inc., New York, **2004**, 69.
40. J. A. Wisner, P. D. Beer, M. G. B. Drew, *Angew. Chem. Int. Ed.*, **2001**, *40*, 3606.
41. J. A. Wisner, P. D. Beer, N. G. Berry, B. Tomapatanaget, *Proc. Nat. Acad. Sci. USA*, **2002**, *99*, 4983.
42. J. A. Wisner, P. D. Beer, M. G. B. Drew, M. R. Sambrook, *J. Am. Chem. Soc.*, **2002**, *124,* 12469.
43. (a). K. Kavallieratos, C. M. Bertao, R. H. Crabtree, *J. Org. Chem.* **1999**, *64*, 1675. (b). K. Kavallieratos, S. R. de Gala, D. J. Austin, R. H. Crabtree, *J. Am. Chem. Soc.* **1997**, *119*, 2325.
44. (a). P. J. Stang, S. R. Seidel, *Acc. Chem. Res.* **2002**, *35*, 972; (b). R. W. Saalfrank, E. Uller, B. Demleitner, I. Bernt, *Struct. Bonding* **2000**, *96*, 149.
45. O. D. Fox, M. G. B. Drew and P. D. Beer, *Angew. Chem. Int. Ed. Engl.* **2000**, *39*, 136.
46. P. D. Beer, N. Berry, M. G. B. Drew, O. D. Fox, M. E. Padilla-Tosta, S. Patell, *Chem. Commun.* **2001**, 199; N. G. Berry, M. D. Pratt, O. D. Fox, P. D. Beer, *Supramolecular Chemistry* **2001**, *13*, 677.
47. P. D. Beer, A. G. Cheetham, M. G. B. Drew, O. D. Fox, E. J. Hayes, T. D. Rolls, *Dalton Trans.* **2003**, *4,* 603.
48. P. D. Beer, N. G. Berry, A. R. Cowley, E. J. Hayes, E. C. Oates, W. W. H. Wong, *Chem. Commun.* **2003**, 2408.
49. M. E. Padilla-Tosta, O. D. Fox, M. G. B. Drew, P. D. Beer, *Angew. Chem. Int. Ed.* **2001**, *40,* 4235.
50. J. Cookson, E. Evans, O. D. Fox, W. W. H. Wong, P. D. Beer, *Chem. Commun.* **2004**, in preparation.

8.
SIGNALLING REVERSIBLE ANION BINDING IN AQUEOUS MEDIA

RACHEL S. DICKINS AND DAVID PARKER
Department of Chemistry, University of Durham
South Road, Durham, DH1 3LE, U.K.

1. Introduction and Background

The study of selective anion recognition by synthetic receptors has been a key feature in the development of supramolecular chemistry [1]. Over the past 30 years, several groups have sought to devise systems that are able to bind certain anions selectively in an aqueous medium. Several of these, for example the protonated aza-cryptands, exhibit a marked size selectivity and hence are able to bind certain spherical anions with relatively high affinity, (e.g. halides) [2]. Examples of 'shape selectivity' in anion binding are also most strikingly exemplified by macropolycyclic receptors, and often rely on the appropriate disposition of hydrogen-bond donors [1]. Some of these systems have been inspired by our growing appreciation of the importance of anion recognition in biological chemistry. The sulfate and phosphate-binding proteins, for example, may be considered as refined examples of receptors relying upon shape and size selectivity determined by directed hydrogen-bonding [3,4]. Moreover, there are many metalloenzymes, whose function requires marked selectivity in anion binding at the metal centre. The zinc-containing enzymes, carbonic anhydrase and alkaline phosphatase are the classical examples of such systems [5], and the anion-binding step usually involves competitive displacement of a metal-bound water molecule.

2. Anion Binding in Aqueous Media

The majority of work published on anion-recognition has been devoted to the study of systems operating in non-aqueous media. This is not surprising, of course, because anion desolvation is often a key step in determining the free energy of binding to a given receptor.

Karsten Gloe (ed.), Macrocyclic Chemistry: Current Trends and Future Perspectives, 121–136.
© 2005 *Springer. Printed in the Netherlands.*

2.1 ENERGETICS AND SPECIATION

An analysis of the free energies of hydration of the more common anions [1,6] (Table 1) reveals some clear trends. Hydration free energy is generally larger for the smaller anions. It is also much greater for doubly/triply charged species and is therefore markedly sensitive to the degree of protonation (i.e. charge) of the anion. Thus, in the binding of key oxyanions hydrogencarbonate/carbonate and hydrogen phosphate/phosphate, the pH of the solution needs to be assessed and controlled in order to determine the relative proportion of these species. At pH 7.23, 50% of HPO_4^{2-} and $H_2PO_4^-$ exist in solution. Given that for both a charge neutral receptor possessing stabilising H-bonding donor sites and a positively charged receptor (e.g. a metal centre), it is electrostatic attraction which dominates the enthalpic contribution, the more highly charged anionic species is likely to be bound at equilibrium, notwithstanding the energetic cost of desolvation. The enhanced free energy of hydration of the more charge dense and of the doubly charged anionic species in Table 1 also reflects the enhanced electrostatic ordering of the water molecules in the second hydration sphere.

TABLE 1. Size, shape and hydration free energies[b] for common anions [1,6].

Shape	Thermochemical[a] radius (nm)	ΔG^0_{hyd} ($kJmol^{-1}$)
Spherical		
F^-	0.126	–465
Cl^-	0.172	–340
Br^-	0.188	–315
I^-	0.210	–275
S^{2-}	0.191	–1315
Linear		
SH^-	0.207	–295
OH^-	0.133	–430
N_3^-	0.195	–295
CN^-	0.191	–295
Trigonal planar		
NO_3^-	0.196	–300
HCO_3^-	0.156	–335
CO_3^{2-}	0.178	–1315
OAc^-	0.162	–365
Tetrahedral		
$H_2PO_4^-$	0.200	–465
PO_4^{3-}	0.238	–2765
SO_4^{2-}	0.230	–1080
ClO_4^-	0.240	–205

[a] Derived from analysis of lattice energies and typically within 5% of the crystallographic ionic radii.
[b] Refers to transfer of the gas-phase ion to aqueous solution at 298K, (ΔG^0_{hyd} [H^+] –1056 $kJmol^{-1}$).

Given that water is one of the best solvents at promoting the favourable interaction between receptors containing significant structural apolar regions and anions that possess some apolar region or surface in their structure, it is also appropriate to consider the local hydrophobicity of the anion binding pocket in the receptor. The contribution of this "hydrophobic effect" to the overall free energy of binding can therefore be enhanced by local structural modification of the receptor. An intrinsic component of this term is receptor desolvation and the less the degree of local hydration around the receptor's anion binding site, the greater the overall free energy change accompanying anion binding. A simple example of such an effect is provided by the enhanced binding affinity of the N-alkylated derivatives, [Ln.**1b**]$^{3+}$ versus the derivatives bearing an N-H group, [Ln.**1a**]$^{3+}$. Free energies of binding (298 K), in the former case are typically 7 to 10 kJ mol^{-1} greater, for anions such as HCO_3^-, acetate and lactate [7,8].

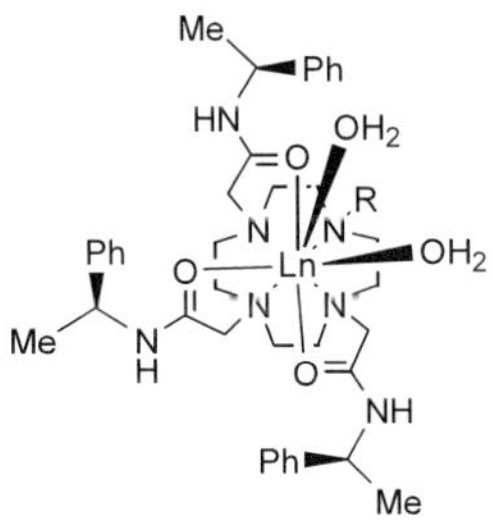

[Ln.1a]$^{3+}$ R = H
[Ln.1b]$^{3+}$ R = Me

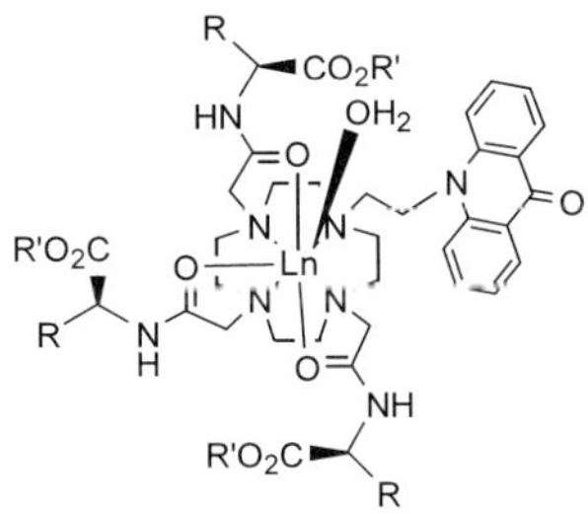

[Ln.2a]$^{3+}$ R = Me, R' = Et
[Ln.2b]$^{3-}$ R = $CH_2CH_2CO_2^-$, R' = H
[Ln.2c] R = Me, R' = H

2.2 STEREOELECTRONICS AND ELECTROSTATICS

The differing size, shape and charge density of the common anions are issues requiring careful consideration in the design of appropriate receptor systems. Spherical anions obviously exhibit no directional preference and can only function as a monodentate ligand to a single metal ion centre. Trigonal oxyanions frequently chelate to octahedral (90°, $O\hat{M}O$) and even more commonly to square antiprismatic (70°, $O\hat{M}O$) metal centres. On the other hand, phosphate tends to act uniformly as a monodentate ligand to a single metal centre. Indeed, in CCD database surveys of phosphate binding to Na^+, K^+, Mg^{2+}, Ca^{2+} and Zn^{2+}, no examples of phosphate chelation were revealed [9].

For a given anion, modulation of the overall charge on the receptor will also define the strength of the binding interaction. Thus, in the series of complexes [Eu.**2**], variation of the peripheral substituents allowed macrocyclic complexes with overall charges at ambient pH of +3, 0 and −3 to be defined. The anionic complex bound most weakly to HCO_3^- in a high salt aqueous environment, allowing the effective association constant to be tuned to the desired local analyte concentration [10]. In this example the dissociation constant characterising hydrogen carbonate binding varied from 11 mM (cationic) and 32 mM (neutral zwitterion) to 100 mM for the anionic system.

3. Signalling Methods

Signal transduction is a key element in any recognition process. Ideally, the anion binding event should be signalled by the modulation of a measurable parameter that is independent of local receptor concentration and is easily calibrated to the desired anion concentration range. The response time and spatial resolution of the analytical measurement are other salient features to be considered. Magnetic resonance measurements afford very limited spatial resolution, are rather insensitive, require expensive dedicated instrumentation but give a significant amount of structural information through analysis of chemical shift or relaxation rate data. Electrochemical methods rely upon the measurement of a change in current (amperometry) or emf (potentiometry) that is proportional to the concentration of the anion present. Potentiometric ion selective electrodes have been devised based on the incorporation of neutral anion receptors (e.g. alkyltin or metalloporphyrins) into plasticised lipophilic polymeric membranes [11,12]. Such methods are intrinsically invasive, require careful calibration and possess rather slow response times. In contrast, optical methods of signalling anion binding are sensitive and fast. They rely upon a change in absorption or emission intensity (preferably as a ratio of two absorption/emission bands), although variations in emission lifetime or polarisation may also be used.

The use of luminescent probes dominates biological and clinical applications for the detection and monitoring of a wide range of chemical species, from simple cations to complex bioactive molecules. Many fluorescent organic molecules are available [13], either incorporating or attached to suitable binding sites for the target species. This area of research is dominated by purely organic molecules, but the number of systems capable of detecting anions is very small. Most success has come in the development of chloride sensors operating in the 10 to 100 mM range, in which the emission of a zwitterionic quinolinium dye is quenched by electron transfer from the halide analyte. The method is limited by the absence of a ratiometric analogue, being dependent purely on an emission intensity change that is proportional to chloride concentration.

A considerable effort is being directed to the development of new emissive sensory systems using luminescent metal complexes, including those of transition metal ions [14] and of the lanthanide (III) ions [15]. Metal complexes may possess an extensive excited state chemistry, owing to the presence of metal-centred (e.g. *dd* and *ff*) and charge-transfer states primarily involving the metal, in addition to the singlet, triplet and charge transfer excited states of the ligand. Thus they not only offer additional opportunities for perturbing the emission characteristics but also may give rise to longer-lived luminescence. The lowest energy, metal-centred excited states of several lanthanide (III) ions possess long radiative lifetimes in the millisecond range. Such long-lived emission is an attractive feature in a luminescent probe or sensor, as it allows the use of time-resolved detection methods, affording very good discrimination between probe and background emission. A time delay may be introduced between the pulsed excitation of the sample and the measurement of the probe's luminesence. During this period the shorter-lived background fluorescence characteristic of biological and clinical samples decays to negligible levels. Such procedures obviate experimental problems associated with light scattering. The large Stokes' shifts of emissive lanthanide complexes also negates the problem of autofluorescence, that can restrict the use of sensors based on organic fluorophores.

4. Reversible Anion Binding at Lanthanide Centres

The ability of the lanthanide ions to exhibit high coordination numbers (generally 8/9) allows scope for the design of specific anion receptors wherein the affinity and selectivity may be modulated by suitable choice of ligand and lanthanide ion. Lanthanide complexes with heptadentate macrocyclic ligands are particularly effective. They form well-defined, relatively stable 1:1 ML complexes and the bound water molecules may be displaced by a variety of monodentate (e.g. F^-, HPO_4^-) or chelating (e.g. HCO_3^-, lactate) anions [7,8]. Furthermore, the rich optical and magnetic properties exhibited by the lanthanide ions allows anion binding to be signalled and structural information to be gleaned through NMR (Eu, Yb), luminescence (Eu, Tb, Yb) and chiroptical techniques (Eu, Yb).

4.1 NMR PROPERTIES

The paramagnetic properties of the lanthanides have been exploited in NMR spectroscopy for many years [16,17]. The paramagnetic lanthanide induced shift is generally considered as the sum of the contact and pseudocontact terms (Equation 1) [16-18]

$$\delta_{para} = \langle S_z \rangle_j F_i + C_j B_0{}^2 G_i = \delta_{ij}^{c} + \delta_{ij}^{pc} \tag{1}$$

where $\langle S_z \rangle_j$ is the reduced value of the average spin polarisation, F_i is the contact term proportional to the hyperfine Fermi constant (A_i), C_j is the Bleaney constant dependent on the given $4f^n$ electronic configuration, $B_0{}^2$ is the second order crystal field coefficient and G_i is a geometrical term containing structural information ($G_i \propto (3\cos^2\theta - 1) / r^3$). The pseudocontact contribution usually predominates for the lanthanides as the unpaired electron spin density largely resides on the Ln^{3+} ion. In the case of axially symmetric systems Equation 2 provides a good approximation for the pseudocontact NMR shift and is frequently used to estimate the lanthanide induced shift in structural analyses, even for complexes lacking an axis of symmetry [8,18].

$$\delta_{ij}^{pc} = -\frac{2C_j\beta^2}{(kT)^2}\frac{[3\cos^2\theta - 1]}{r^3}B_0^2 \tag{2}$$

The angular and distance information provided by the lanthanide induced shift has found widespread application from the determination of solution structures of Ln chelates [18,19] to gaining structural information on proteins, nucleotides and amino acids [19]. More recently anion binding to coordinatively unsaturated lanthanide complexes has been effectively signalled as the observed lanthanide induced shift has been directly correlated to the nature of the donor atom in the axial position [8,20,21]. It is the polarisability of the axial donor that ranks the second order crystal field coefficient, $B_0{}^2$, and hence determines the magnitude of the observed shift. Values of the mean shift of the four most-shifted axial protons of the 12-N_4 ring for $[Yb.\mathbf{1a}]^{3+}$ are collated in Table 2.

TABLE 2. Effect of the axial donor on the ^{1}H NMR spectral range observed for $[Yb.1a]^{3+}$ in D_2O (295 K, 200 MHz).

Anion	Axial donor	Shift range (ppm)	Mean shift H_{ax} [a] (ppm)
Phosphate	H_2O	+124 to –98	100
Water	H_2O	+110 to –82	75
Lactate	OH	+96 to –74	71
Acetate	$CO^{\delta-}$	+96 to –100	68
Succinate[b]	$CO^{\delta-}$	+98 to –108	73
Gly-Ala	NH_2	+65 to –59	50
Glycine	NH_2	+68 to –56	43
Malonate	CO^-	+57 to –55	39
Oxalate	CO^-	+55 to –41	37
Carbonate	CO^-	+45 to –40	27

[a] Mean chemical shift (ppm) of the four most-shifted axial ring protons of the 12-N_4 ring.
[b] ^{1}H NMR titration and ESMS analysis give no evidence for a succinate bridged 2:1 complex.

Hard, uncharged donors such as water give rise to the largest shifts whereas more polarisable donors (e.g. carbonate, serine) give much smaller shifts. This simple correlation has proved to be invaluable in determining the binding modes for a variety of chelating anions. For example, NMR data for each of the α-amino acid adducts with $[Yb.\mathbf{1a}]^{3+}$ is consistent with a common binding mode [21]. The observed chemical shift range (av. +68 to -56 ppm) is indicative of a chelated structure involving the more polarisable amine N in the axial position and a carboxylate O in the equatorial position, forming a favourable 5-ring chelate. The observed chemical shift range rules out bidentate chelation through the C-terminal carboxylate oxygens (c.f. acetate : +96 to –100 ppm) or through side-chain functionality (e.g. serine, threonine, aspartate, glutamate). Similarly, an NMR analysis on simple peptide adducts ruled out five of the six possible binding modes, revealing the preferred chelate to involve the terminal amine (axial) and proximate amide oxygen (equatorial) [21].

The use of aqueous chiral lanthanide complexes in the determination of the enantiomeric purity of chiral α-hydroxy acids has also been assessed by ^{1}H NMR [21]. Large lanthanide induced shifts, chemical shift non-equivalence and an apparent absence of kinetic resolution in complex formation is observed upon addition of racemic lactate to $[Yb.\mathbf{3a}]^{3+}$ (Figure 1). The lactate CH_3 resonances are clearly resolved for the R and S diastereomers ($\Delta\Delta\delta = 10$ ppm) and experience a large lanthanide induced shift with the CH_3 resonating at +21 and +31 ppm respectively for (*R*)- and (*S*)-lactate respectively (c.f. +1.3 ppm in the free form). The large shifts exhibited by this unique chiral derivatising agent may be compared to the much smaller $\Delta\Delta\delta$ values reported for shift reagents which are generally of the order 1 ppm, and often < 0.1 ppm.

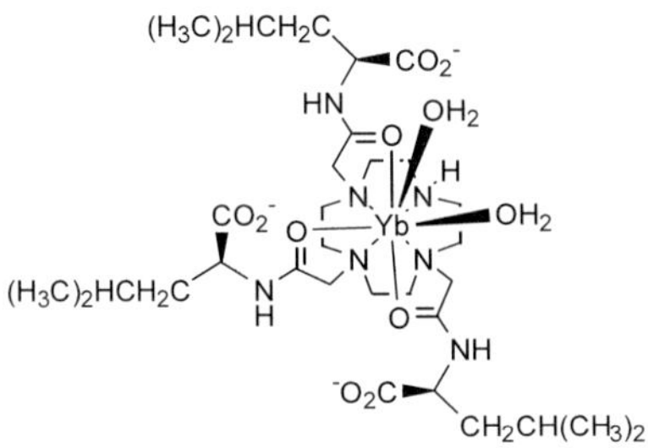

[Yb.3a]

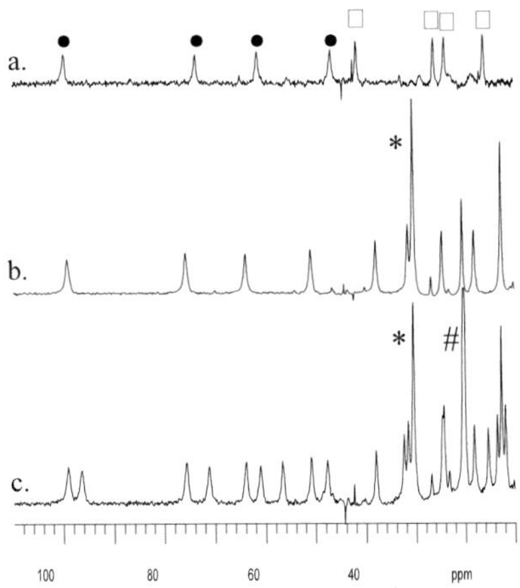

Figure 1. Partial ^{1}H NMR (200MHz, D_2O) of (a) **[Yb.3a]** (diaqua species) showing four axial (•) and four equatorial (□) ring protons; (b) in the presence of 5 equivalents of (*S*) lactate [(*) (*S*)-lactate CH_3]; (c) 5 equivalents of racemic lactate [(#) (*R*)-lactate CH_3].

4.2 LUMINESCENCE ASPECTS

Both the excited state lifetime and intensity of lanthanide emission are sensitive to changes in the local coordination environment about the Ln ion, providing an effective probe to signal anion binding. Analysis of the rate constants for radiative decay of the Eu ${}^{5}D_0$, Tb ${}^{5}D_4$ and Yb ${}^{2}F_{5/2}$ excited states in H_2O and D_2O allows the solution hydration state of emissive Ln complexes to be estimated, using established methodology [23]. As a consequence, information concerning the mode of binding of a variety of anions may be assessed. Hydration states (q) are tabulated for ternary complexes of $[Tb.\mathbf{1a}]^{3+}$ (Table 3) [7].

TABLE 3. Effect of added anion on the rate constants (± 10%) for depopulation of the excited states of $[Tb.1a]^{3+}$ (295 K, 1 mM complex, 10 mM anion) and derived hydration numbers, q (± 20%).[a]

Anion	k_{H2O} $(ms)^{-1}$	k_{D2O} $(ms)^{-1}$	Δk_{corr} $(ms)^{-1}$	q
Triflate	0.84	0.39	0.39	2.0
Cl^-, Br^-, I^-	0.87	0.39	0.42	2.1
F^-	0.61	0.36	0.19	1.0
HPO_4^{2-}	0.57	0.37	0.14	0.7
Lactate	0.56	0.40	0.10	0.5
Citrate	0.55	0.41	0.08	0.4
Acetate	0.58	0.45	0.07	0.3
Malonate	0.49	0.43	0.0	0
CO_3^{2-}/HCO_3^-	0.53	0.45	0.02	0.1

[a] Apparent values of q were derived using $q = 5\ \Delta k_{corr}$ after correcting for the estimated effect of unbound water molecules ($0.06 ms^{-1}$).

Corrections were applied to account for quenching contributions of closely diffusing (second sphere) water molecules [23]. The [Tb.**1a**]$^{3+}$ triflate complex possesses two bound water molecules and a q = 2 hydration state was also observed with Cl^-, Br^- and I^-. Addition of acetate, malonate and carbonate gave a q = 0 complex, consistent with bidentate chelation involving displacement of the two bound water molecules (Figure 2). Phosphate, however, did not chelate, preferring a monodentate binding mode with one water molecule remaining (q = 1), consistent with water as the axial donor as observed in the ^{1}H NMR shift for [Yb.**1a**]$^{3+}$ discussed above. The α-hydroxy acids, lactate and citrate also appeared to chelate, as expected, but gave a residual q value (0.4-0.5) due to the presence of an (exchangeable) OH oscillator close to the Ln ion which may independently quench the excited state with approximately half the efficiency of a coordinated water molecule.

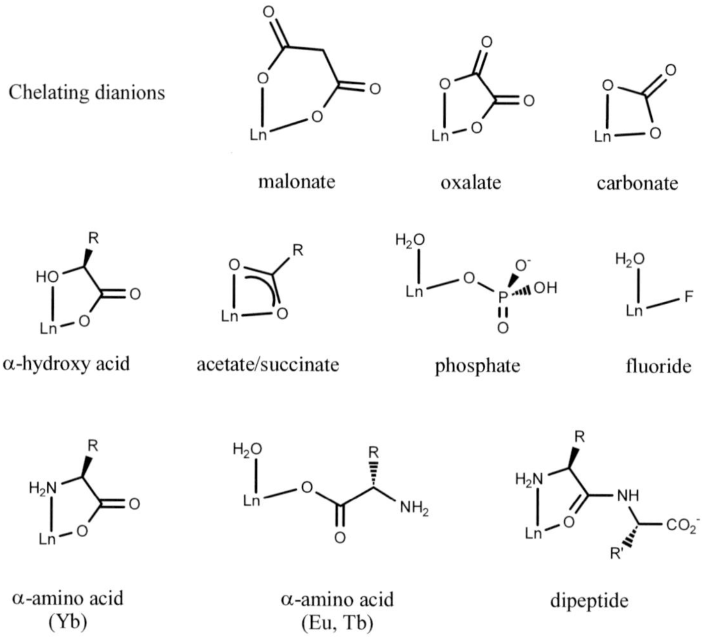

Figure 2. Structures of Anion Adducts Established by ^{1}H NMR, Ln Emission and X-Ray.

The emissive $^5D_0$ excited state for Eu is non-degenerate and Eu^{3+} species give rise to relatively simple emission spectra that may be readily analysed [17]. For the highest energy $^5D_0 - {}^7F_0$ ($\Delta J = 0$) emission band at 580 nm one component arises for every chemically distinct species that is not undergoing chemical exchange on the millisecond scale. The transition is sensitive to the ligand field and symmetry [23]. The intensity of the magnetic-dipole allowed $^5D_0 - {}^7F_1$ transition ($\Delta J = 1$, 591 nm) is relatively independent of the coordination environment. Two transitions are allowed if the complex possess a C_3 or C_4 axis, the separation of which is a direct measure [24] of B_0^2, and three transitions in complexes of lower symmetry. Both the hypersensitive $^5D_0 - {}^7F_2$ ($\Delta J = 2$, 619 nm) and the $^5D_0 - {}^7F_4$ transitions ($\Delta J = 4$, 690 nm) are predominantly electric dipole in character and the intensity is highly sensitive to ligand field, especially the nature and polarisability of the axial donor, if present [24,26]. Furthermore, the ratio of ΔJ=2/ΔJ=1 band intensities is a useful parameter in evaluating changes in the Eu coordination environment [7,27]. In establishing an order of Ln ion donor preferences, it was found that the bigger this ratio, the more polarisable the axial donor and the greater the affinity for the lanthanide ion [20].

Displacement of water molecules by a coordinating anion is signalled by an increase in the emission intensity (as coordinated and closely diffusing OH and NH oscillators quench the $^5D_0$ excited state of Eu) [23] as depicted in Figure 3.

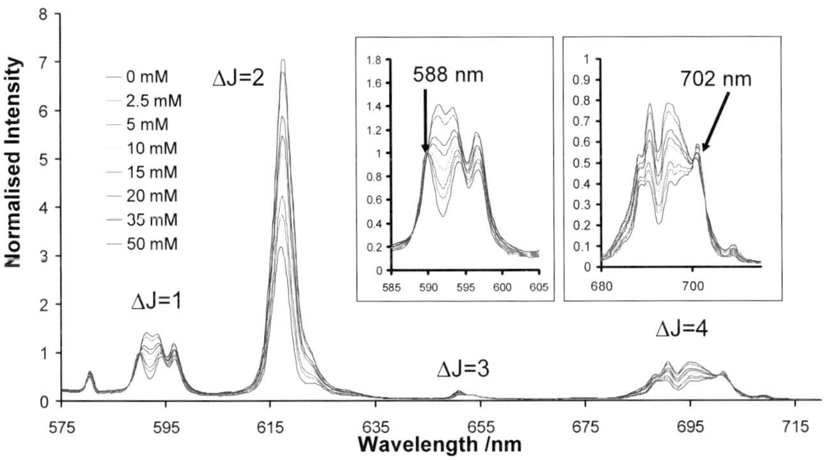

Figure 3. Variation in the Eu emission spectrum for [Eu.**2c**] (pH 7.4, 0.1M MOPS) following incremental addition of sodium hydrogencarbonate; spectra in the insets show formation of isoemissive points at 588 and 702 nm.

The increase in the relative intensity of the ΔJ = 2 band at 616 nm is particularly apparent, following incremental addition of HCO_3^- to [Eu.**2c**] (pH 7.4, 0.1 M MOPS buffer). The process is further characterised by the formation of isoemissive points at 588 and 702 nm, allowing ratiometric analyses to be used to characterise the binding of HCO_3^-.

4.3 CHIROPTICAL SPECTROSCOPY

Chiral lanthanide complexes are amenable to study by circular dichroism, CD, and circularly polarised luminescence, CPL. CD is the differential absorption of right and left handed circularly polarised light and reflects ground state structure. As most f-f transitions are parity forbidden and have low molar absorptivities, the CD is rather weak and can be difficult to observe. CPL is the emission analogue of CD and probes the chirality of the excited state. CPL is very sensitive, combining the specificity of natural optical activity with the measurement sensitivity of optical emission techniques. Chiroptical properties of the lanthanides are particularly sensitive to ligand coordination geometry and stereochemistry, electronic state structure and the local coordination environment [28]. With the exception of Eu^{3+} in CPL, spectral analysis is not straightforward due to the large number of transitions possible between the different J states. Therefore full exploitation of structural studies awaits further spectral-structure correlation. However, CD and CPL have both been used to signal anion binding at coordinatively unsaturated lanthanide complexes [7,21]. The form of the CD and CPL spectra is remarkably sensitive to the nature of the axial donor ligand.

Yb absorbs in the near-IR region and is the most CD-sensitive of all the lanthanides, exhibiting bands centred around 980 nm due to magnetic-dipole allowed $^2F_{7/2}$ - $^2F_{5/2}$ transitions. Although spectra are complicated by the large number of sub-levels of the $^2F_{7/2}$ state, a distinct trend is apparent. The band centred around 990 nm moves to shorter wavelength as the polarisability of the axial donor increases (phosphate > diaqua > lactate > acetate > serine > carbonate) as depicted in Figure 4 [7].

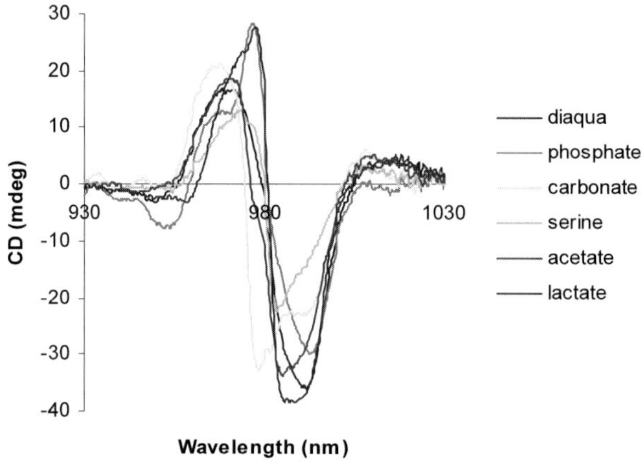

Figure 4. Near-IR CD spectra of **[Yb.1a]**$^{3+}$ in the presence of added anions (295 K, 10 mmol, H_2O).

CPL spectroscopy may also be used to signal the nature of the ternary anion adduct. In the series of phospho-anion complexes of [Eu.**1a**]$^{3+}$, Figure 5, complexes of HPO_4^{2-} and glucose-6-phosphate give identical CPL spectra – consistent with their very similar ^1H NMR spectral profile. Adducts with phospho-tyrosine, N-acetyl phospho-Tyr and a short peptide (phosphorylated at the Pyr residue) are also near-identical, consistent with chemoselective binding of the Tyr-OP phospho-anion.

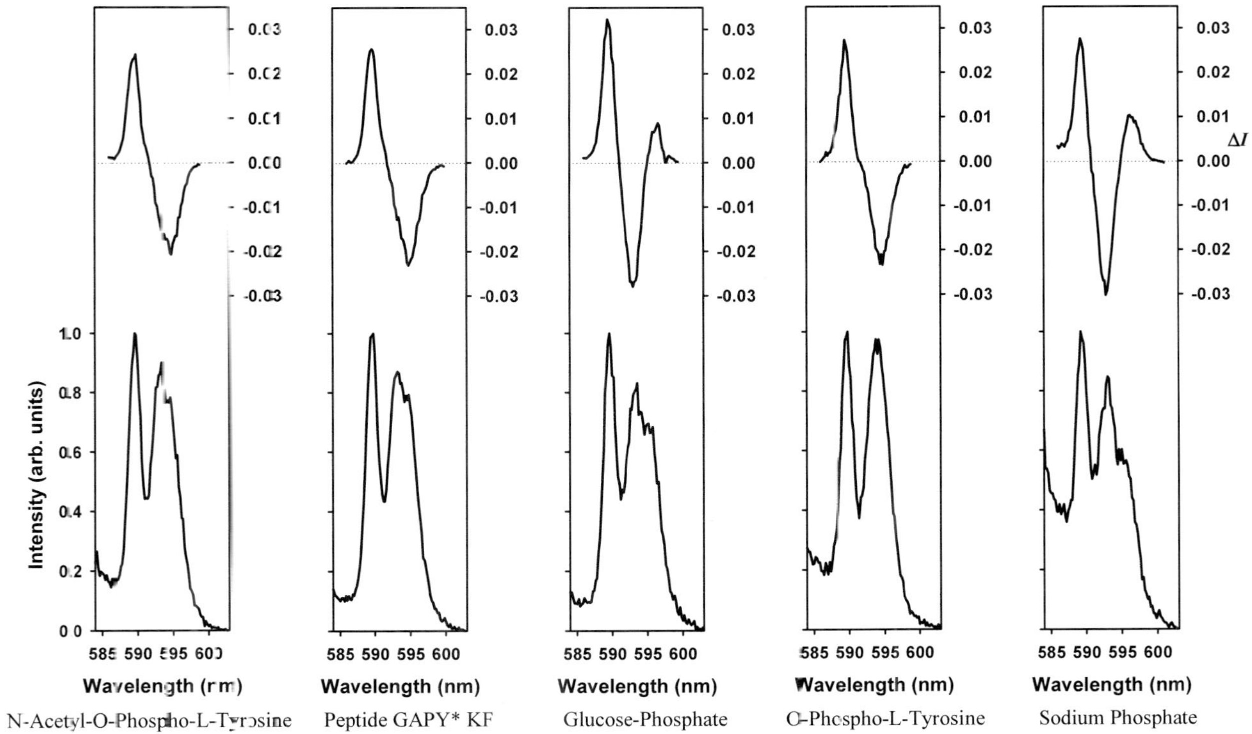

Figure 5. CPL and emission spectra showing the ΔJ=1 transition for $[Eu.\mathbf{1a}]^{3+}$ (1mM) in the presence of 10 equivalents of the added anions.

4.4 STRUCTURAL ANALYSIS

X-Ray crystallographic analysis used alongside solution phase techniques such as NMR, is invaluable in the design of selective anion receptors as it can be used to define the solid-state structure of the anion-bound complex. Several X-ray structures of anion-bound complexes of [Ln.**1a**]$^{3+}$ have been defined recently (acetate, lactate, citrate, alanine, glycine, methionine, serine and threonine) [8,21,22]. The complexes adopt a monocapped square antiprismatic structure with one base comprising of four N atoms of the macrocycle and the other base containing three O atoms of the pendent arms. The latter base is completed by a carboxylate O donor of the anion, which binds in a bidentate manner and simultaneously caps this base by O^- (acetate), OH (α-hydroxy acids) or NH_2 (α-amino acids). The nature of the donor atom in the capping (or axial) position correlates well with the observed NMR shift in the solution phase.

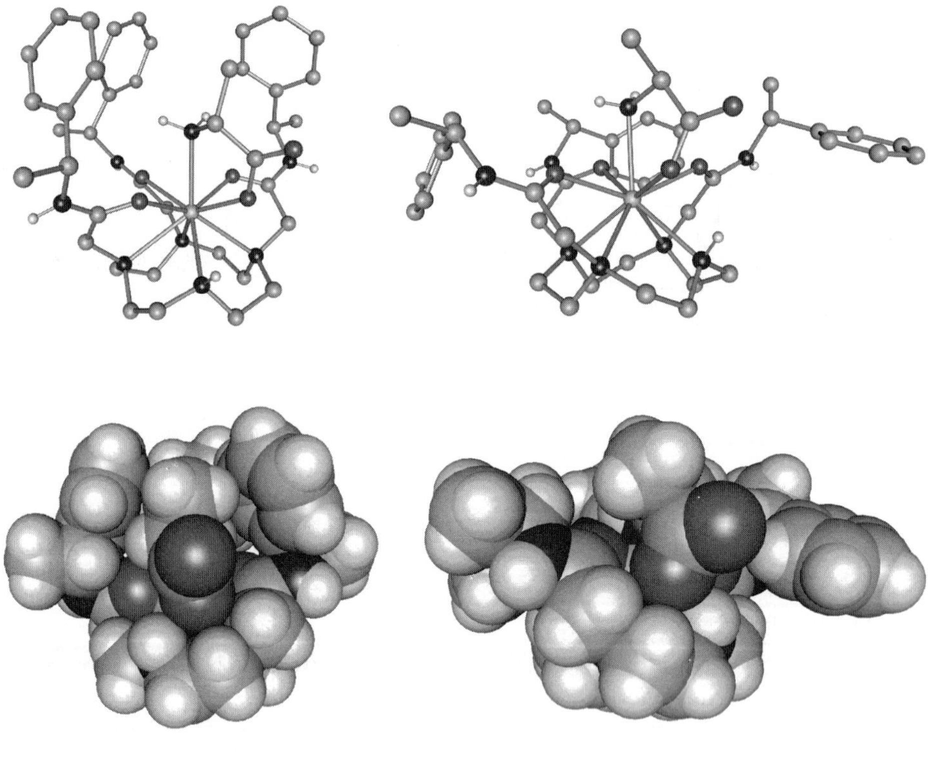

Figure 6. Crystal structures of the ternary complexes of [Yb.**1a**]$^{3+}$ with (*R*)-alanine (left) and (*S*)-alanine (right). Note the axial and equatorial conformations of the pendent side chains, respectively, and the opposite chirality of the macrocyclic ring and pendent arm configuration.

Depicted in Figure 6 are the X-ray structures for the ternary complexes of (RRR)-[Yb.**1a**]$^{3+}$ with (*R*)- or (*S*)-alanine respectively. ^{1}H NMR of the two ternary complexes show the presence of two isomers in solution, Λ(δδδδ) and Δ(λλλλ) [21,29]. With an (*R*)-configuration at the remote chiral centre of the pendent amide arm of [Yb.**1a**]$^{3+}$ the Λ(δδδδ) isomer adopts a more 'open' structure, with the pendent arms arranged in an equatorial position. However, the Δ(λλλλ) isomer adopts a 'pocket-like' structure with the pendent arms axially disposed [30]. The relative proportion of the 'open' Λ(δδδδ) and more 'pocket-like' Δ(λλλλ) diastereomeric complexes with all the amino acids is independent of amino acid configuration and is a function of the nature of the α-substituent [21]. Substituents with sterically demanding side-chains or the potential to undergo stabilising H bond interactions with solvent or counter-ions (e.g. threonine, Figure 7) favour the open Λ(δδδδ) structure. Substituents that offer opportunities for hydrophobic interactions or closer van der Waals contact with the ligand, stabilise the Δ(λλλλ) complex (e.g. methionine, Figure 7).

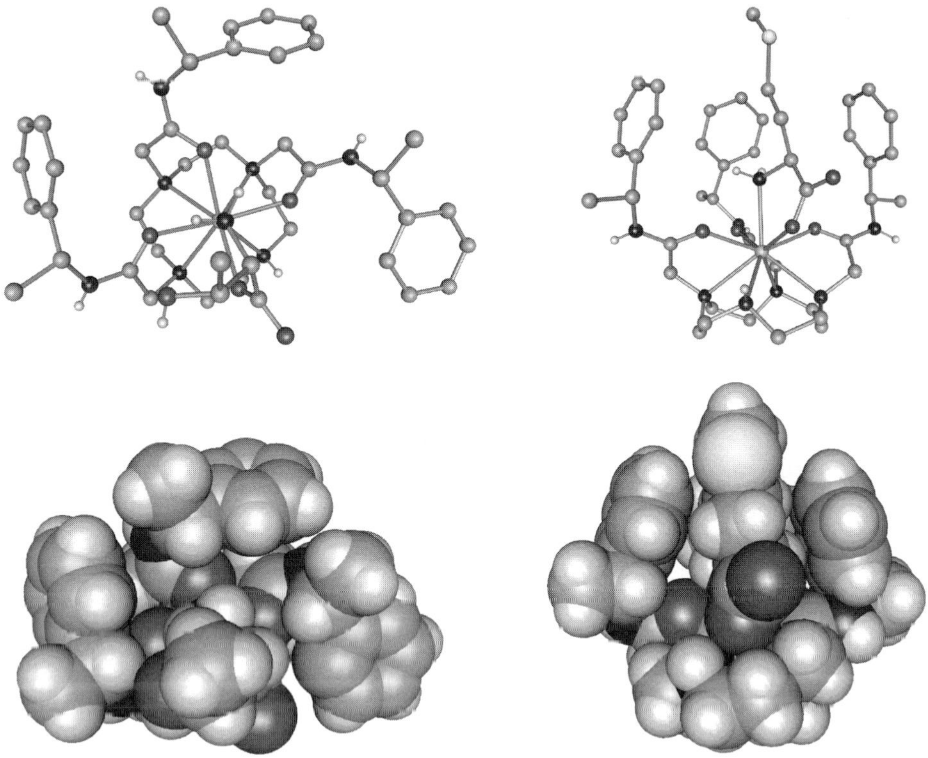

Figure 7. Crystal structures of the ternary complexes of [Yb.**1a**]$^{3+}$ with (*S*)-threonine (left) and (*S*)-methionine (right). The methinione side chain is enclosed in a tight hydrophobic pocket formed by the axial arrangement of the pendent arms.

4.5 SELECTIVITY IN ANION BINDING

Although lanthanide complexes are highly effective at signalling anion binding, they must show selectivity in a competitive anion background to be useful as anion receptors. The affinity of an anion for the lanthanide centre may be modulated by variation of the lanthanide ion: the higher the charge density, the greater the affinity (Yb > Tm > Tb > Eu), or by modification of the ligand structure. Tb adducts of a variety of bound anions (e.g. hydrogencarbonate, phosphate, lactate, acetate) are found to be an order of magnitude more stable than their Eu analogues, reflecting the differential complex hydration and enhanced Lewis acidity [7]. Similarly, the increased charge density of the Yb ion favours a chelated structure with the common α-amino acids, whereas Eu and Tb analogues exist as monoaqua species with an apical bound water molecule (Figure 2) [8]. Variation in the overall charge of the lanthanide complex modulates binding affinity, the order being dictated by the degree of electrostatic repulsion (e.g. cationic > neutral > anionic) [10]. The actual ligand structure also plays an important role, for example, binding affinities for N-methylated analogues [Ln.**1b**]$^{3+}$ are an order of magnitude greater [7] than their NH analogues, [Ln.**1a**]$^{3+}$. NMR and emission studies reveal that selectivity in binding is also observed with these complexes. Lanthanide complexes may bind to the phosphorylated amino acids OPSer, OPThr and OPTyr either via phosphate coordination or chelation to the amino acid, the degree of which is determined by the lanthanide ion and ligand structure. Moving along the Ln series favours amino acid chelation: [Eu.**1b**]$^{3+}$ binds through the phosphate group (>95%) whereas the increased charge density of ytterbium and thulium analogues favour the amino acid chelate (>95%) [31]. Amino acid chelation is also more competitive for the N-methylated analogues [Ln.**1b**]$^{3+}$ compared to the parent complexes [Ln.**1a**]$^{3+}$: [Yb.**1a**]$^{3+}$ binds to OPTyr via phosphate and amino acid (~50:50) whereas [Yb.**1b**]$^{3+}$ favours the amino acid chelate (>95%). Furthermore, chemoselective ligation of phosphorylated tyrosine has been observed in competitive media, both for the free Tyr-OP and in a doubly phosphorylated model hexapeptide (Gly-Ser-Pro-Tyr-Lys-Phe) with [Ln.**1a**]$^{3+}$ and [Ln.**2**], probably due to the lesser degree of hydration around the anion and a subsequent lower desolvation energy contribution to the overall free energy of binding [31].

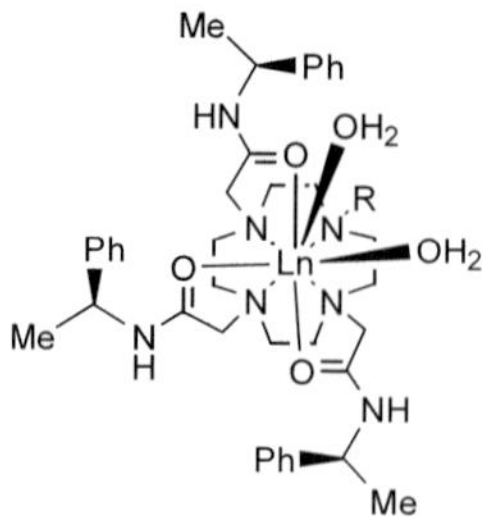

[Ln.1a]$^{3+}$ R = H
[Ln.1b]$^{3+}$ R = Me

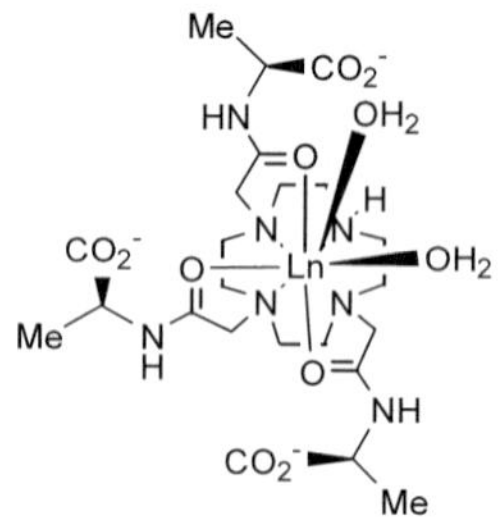

[Ln.3b]

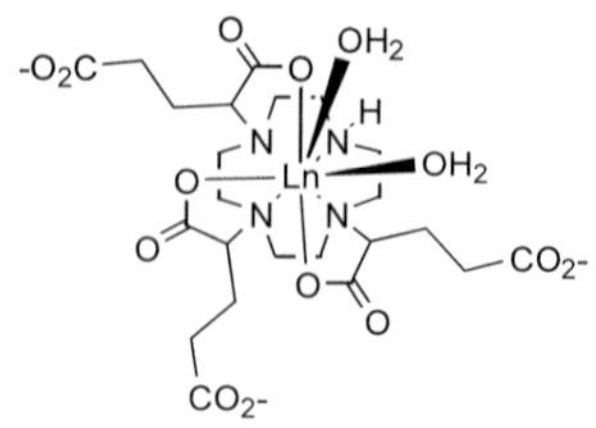

[Ln.4]$^{3-}$

Selectivity for carbonate in a competitive anion background (30 mM carbonate, 100 mM NaCl, 0.9 mM phosphate, 2.3 mM lactate, 0.13 mM citrate) simulating an extracellular anionic environment has been assessed [7]. The solution pH determines the effective concentration of HCO_3^-, falling from ~90 % of all dissolved C_1 species at pH = 8.9 to <2 % at pH = 4.9 [32]. The emission spectra for the cationic [Eu.**1a**]$^{3+}$, zwitterionic [Eu.**3b**] and anionic [Eu.**4**]$^{3-}$ complexes at pH = 10 in the stated anion mixture resembles that of the carbonate adducts. On reducing the pH, the effective concentration of first carbonate then hydrogencarbonate decreases, reaching a limiting value below pH = 6.5 in all cases. The emission spectra at pH = 5.5 reflects bound lactate for the cationic complexes [Eu.**1a**]$^{3+}$ whereas the diaqua complex is evident for the anionic complex [Eu.**4**]$^{3-}$.

At pH 7.4, in buffered media containing the common endogenous anions, incremental addition of HCO_3^- to a solution containing these Eu complexes is signalled by an increase in the ratio of the $\Delta J=2/\Delta J=1$ transitions. A ratiometric analysis method has now been devised [10] in which the apparent dissociation constant, K_d, is tuned to the required, physiologically relevant concentration range by permutation of the peripheral ligand substitution pattern. Moreover, such complexes, e.g. [Ln.**2**] are not toxic in cells such as NIH-3T3 (mouse-skin) and HEK-293 (human kidney) and show some encouraging uptake behaviour that will stimulate further work. Key issues to address will include pinpointing the structural factors that determine the efficiency of complex uptake, their localisation profile (compartmentalisation, speed of uptake and egress) and thereby define their potential to act as genuine practicable probes for the intracellular environment or related biologically oriented applications.

5. References

1. A. Bianchi, K. Bowman-James, E. García-Espana (eds.), *Supramolecular Chemistry of Anions*, Wiley-VCH, New York, 1997.
2. E. Graf and J.-M. Lehn, *Helv. Chim. Acta*, **64** (1981) 1040.
3. J.W. Pflugrath, F.A. Quiocho, *Nature*, **314** (1985) 257; J.W. Pflugrath and F.A. Quiocho, *J. Mol. Biol.*, **200** (1988) 163; Z.F. Kanyo and D.W. Christianson, *J. Biol. Chem.*, **266** (1991) 4264.
4. I. Sekler, R.S. Lo, R.R. Kopito, *J. Biol. Chem.*, **270** (1995) 28751.
5. D.W. Christianson and C.A. Fierke, *Acc. Chem. Res.*, **29** (1996) 331; L. Ma, T.T. Tibbitts, E.R. Kantrowity, *Protein Sci.*, **4** (1995) 1498.
6. Y. Marcus, *Faraday Trans.*, **87** (1991) 2995.
7. J.I. Bruce, R.S. Dickins, L.J. Govenlock, T. Gunnlaugsson, S. Lopinski, M.P. Lowe, D. Parker, R.D. Peacock, J.J.B. Perry, S. Aime, M. Botta, *J. Am. Chem. Soc.*, **122** (2000) 9674.
8. R.S. Dickins, S. Aime, A.S. Batsanov, A. Beeby, M. Botta, J.I. Bruce, J.A.K. Howard, C.S. Love, D. Parker, R.D. Peacock, H. Puschmann, *J. Am. Chem. Soc.*, **124** (2002) 12697.
9. B. Schneider, M. Kabelac, *J. Am. Chem. Soc.*, **120** (1998) 161; R.J. Alexander, Y.F. Kanyo, L.E. Chirlian, D.W. Christianson, *J. Am. Chem. Soc.*, **112** (1990) 933.
10. Y. Bretonniere, M.J. Cann, D. Parker, R. Slater, *Org. Biomol. Chem.*, (2004) 1624.
11. S.C. Glazier, M.A. Armold, *Anal. Chem.*, **63** (1991) 754.
12. P. Schultess, D. Ammann, B. Krautler, C. Caderas, R. Stepanek, W. Simon, *Anal. Chem.*, **57** (1985) 1397.
13. R.P. Haugland: *Handbook of Fluorescent Probes and Research Chemicals*, 9th Edn, Molecular Probes, Eugene, Oregan, 2002.
14. J.N. Demas, B.A. DeGraff, *Coord. Chem. Rev.*, **211** (2001) 317.
15. D. Parker, J.A.G. Williams: *Metal Ions in Biological Systems: the Lanthanides and their Interrelutions with Biosystems*, eds. H. Sigel, A. Sigel, Marcel Dekker, New York (2003), ch. 7, pp. 233-280.
16. I. Bertini, C.H. Luchinat, *Coord. Chem. Rev.*, **150** (1996) 1; T.G. Wenzel, *NMR Shift Reagents;* CRC Press, Boca Raton, 1987.
17. D. Parker, R.S. Dickins, H. Puschmann, C. Crossland, J.A.K. Howard, *Chem. Rev.*, **102** (2002) 1977.
18. J.A. Peters, J. Huskens, D.J. Raber, *Prog. Nucl. Magn. Reson. Spectrosc.*, **28** (1996) 283.
19. C.F.G.C. Geraldes: *NMR in Supramolecular Chemistry*, M. Pans, Ed., Kluwer, Netherlands, (1999), pp.133-154.

20. R.S. Dickins, D. Parker, J.I. Bruce, D. Tozer, *Dalton Trans.*, (2003) 1264.
21. R.S. Dickins, A.S. Batsanov, D. Parker, H. Puschmann, S. Salamano, J.A.K. Howard, *Dalton Trans.*, (2004) 70.
22. R.S. Dickins, C.S. Love, H. Puschmann, *Chem. Commun.*, (2001), 2308.
23. A. Beeby, I.M. Clarkson, R.S. Dickins, S. Faulkner, D. Parker, L. Royle, A.S. de Sousa, J.A.G. Williams, M. Woods, *Perkin Trans. 2*, (1999) 493.
24. R.D. Peacock, *Struct. Bonding (Berlin)*, **22** (1975) 83.
25. B.R. McGarvey, *J. Magn. Reson.,* **33** (1979) 445; V.S. Mironov, Y.G. Galyametdinov, A. Ceulemans, C. Gorlier-Walrand and K. Binnemans, *J. Chem. Phys.*, **116** (2002) 4673; J.M. Ren, S.R. Zhang, A.D. Sherry, C.F.G.C. Geraldes, *Inorg. Chem. Acta*, **339** (2002) 273.
26. J.I. Bruce, D. Parker, D.J. Tozer, *Chem. Commun.*, (2001) 2250.
27. D.R. Forster, F.S. Richardson, *Inorg. Chem.*, **22** (1983) 3996; M.F. Reid, F.S. Richarson, *J. Phys. Chem.*, **88** (1984) 3579.
28. H.G. Brittain, *Chirality*, **8** (1996) 357; J.P. Riehl, F.S. Richardson, *Methods Enzymol.*, **226** (1993) 539; K Nakanishi, N. Berova, R.W. Wood, *Circular Dichroism, Principles and Applications*, VCH, New York, 1994.
29. For a detailed explanation of the isomerism exhibited by such macrocyclic complexes see reference 17.
30. For the enantiometric [SSS-Yb.**1a**]$^{3+}$ ternary adducts, Δ(λλλλ) gives the open structure whereas λ(δδδδ) gives the more pocket-like structure.
31. P. Atkinson, Y. Bretonniere, D. Parker, *Chem. Commun.*, (2004), 438.
32. A.K. Covington, *Chem. Soc. Rev.*, **14** (1985) 265; C.T.G. Flear, A.V. Covington, J.C. Stoddart, *Ann. Intern. Med.*, **144** (1984) 2285.

9.

ION-PAIR RECOGNITION BY DITOPIC MACROCYCLIC RECEPTORS

BRADLEY D. SMITH
Department of Chemistry and Biochemistry, University of Notre Dame Notre Dame IN 46556, U.S.A.

1. Introduction

Over the past thirty years, a large number of macrocyclic receptors have been synthesized and evaluated for their abilities to bind cations. More recently, increased attention has been directed towards receptors for anions. Many of these synthetic receptors are uncharged molecules and operate in organic solvents. Under these conditions the target salts exist as associated ion-pairs which can hinder the single-ion recognition process. [1] A strategy to circumvent this problem is to design a single receptor with specific cation and anion binding sites. In other words, a heteroditopic receptor that can simultaneously bind both of the salt ions. The chronological development of salt receptors has been reviewed a number of times in the past few years, [2-6] and so the exercise will not be repeated here. Instead, the chapter will focus on the more specific topic of ion-pair recognition.

2. Ditopic Receptors for Separated Ions

Most salt receptors in the literature bind the cation and the anion as spatially separated ions. Early examples are receptors **1** and **2** that were developed by groups lead by Reinhoudt, [7] and Beer, [8] respectively. In both cases, the metal cation binds to the crown ether rings and the anion binds simultaneously to the Lewis acidic uranyl center in the case of **1**, and to the amide NH residues in the case of **2**. A more recent example is Kilburn's ditopic receptor **3** which coordinates the metal cation via the calixarene oxygens, and binds the anion in a hydrogen bonding pocket formed by the two thiourea groups. [9] A conceptually different receptor for separated ions is Gellman's macrocyclic phosphine oxide **3**. [10] Hydrogen bonding with a monoalkylammonium cation on one face of the macrocycle induces receptor polarization and preorganization which promotes association with a Cl^- counter-ion on the reverse face. In other words, the receptor is inserted between the two ions.

Karsten Gloe (ed.), Macrocyclic Chemistry: Current Trends and Future Perspectives, 137–151.
© 2005 *Springer. Printed in the Netherlands.*

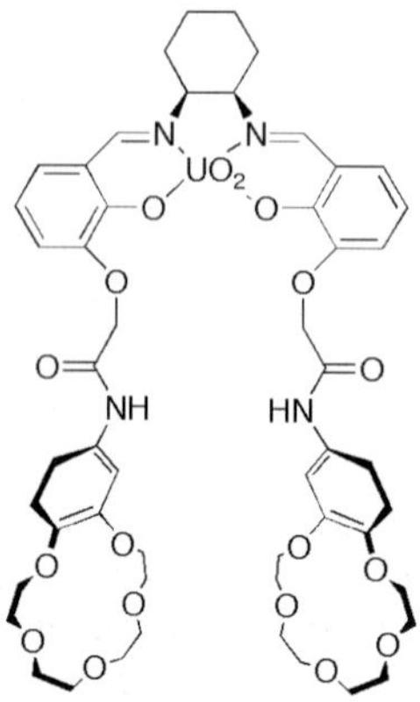

1

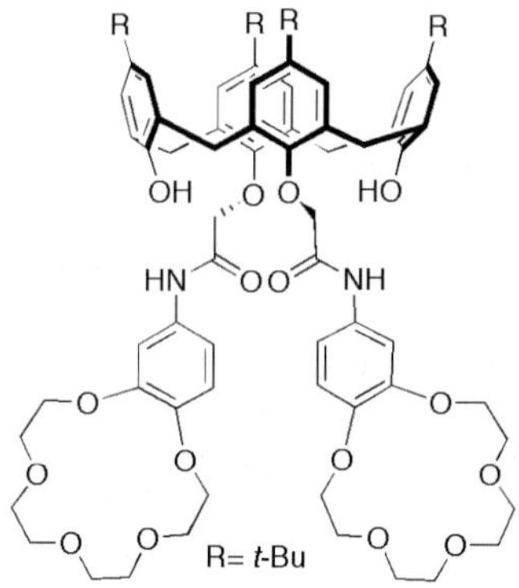

2

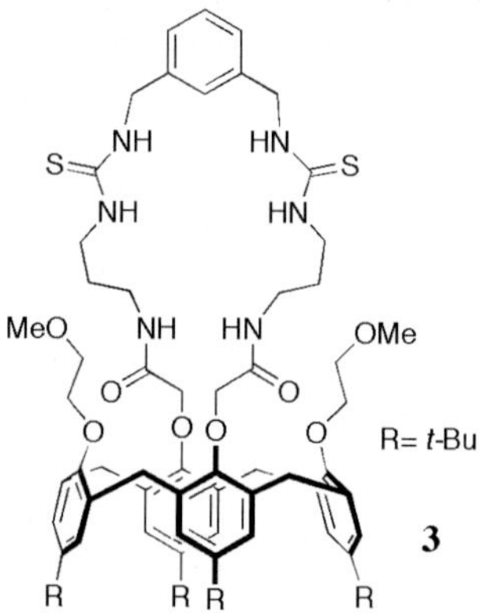

3

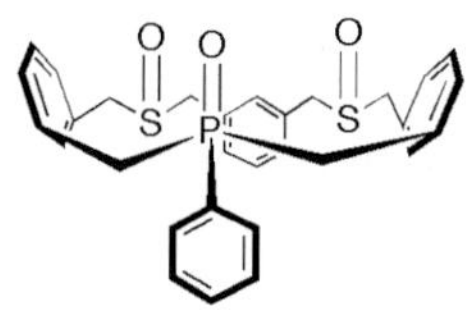

4

A fascinating property with some of these ditopic salt receptors is the feature of complexation by induced fit; that is, the binding of one ion induces a major conformational change in the receptor such that affinity for the counter-ion is improved. The mechanism is conceptually related to the allosteric action of enzymes and biological receptors. Two examples are illustrated here. The first is the Lockhart system **5** where the central polyamine linker wraps around a Cl^- anion which brings the two benzo-5-crown-15 ether rings together and allows them to bind a K^+ cation. [11] The second example is Kubo's receptor **6** which acts in reverse, that is, the dibenzo-30-crown-10 wraps around a K^+ cation and forms a preorganized binding pocket for a phosphate dianion. [12]

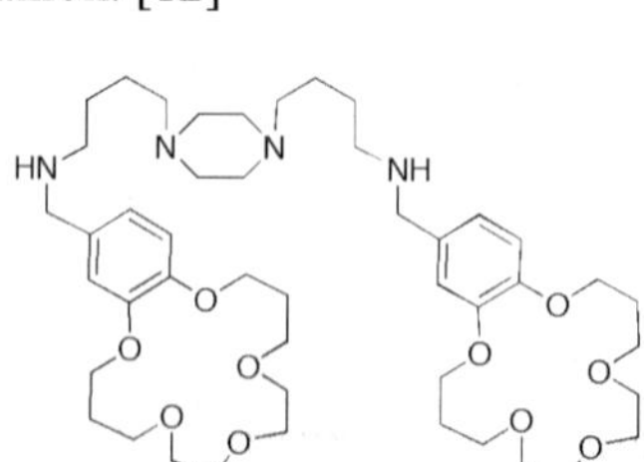

5

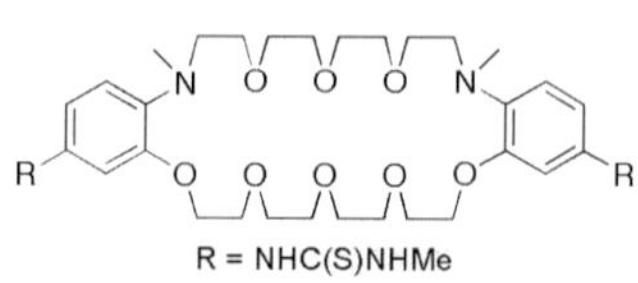

6

3. Ditopic Receptors for Associated Ion-Pairs

A potential drawback with ditopic receptors that bind salts as separated ions is the Coulombic penalty that must be paid to enforce charge separation. This problem is circumvented if the receptor binds the salt as an associated ion-pair; thus, ditopic receptors for associated ion-pairs are expected to have generally superior affinities. However, the design of convergent heteroditopic receptors is quite a challenge because the ion binding sites have to be incorporated into a suitably preorganized scaffold that holds them in close proximity, but not so close that the sites interact. One of the first successful examples of a ditopic salt receptor for associated ion-pairs is compound **7** reported by Reetz. [13] The Lewis acidic boron atom can form a reversible dative bond with a F^- anion which promotes simultaneous coordination of a K^+ cation by the oxygen atoms in the surrounding crown ether. Other early examples are the Kilburn macrobicycle **8**, which appears to bind the mono-potassium salts of dicarboxylic acid acids as contact ion-pairs, [14] and our receptor **9** which was shown by X-ray diffraction to complex NaCl as a solvent separated ion-pair. [15] More recently, the Beer group has reported that macrocycle **10** is able bind an organic ion-pair in solution and form the pseudo rotaxane **11**. [16]

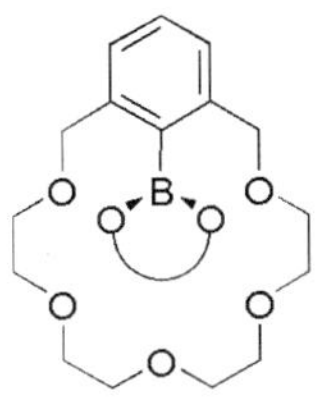

7

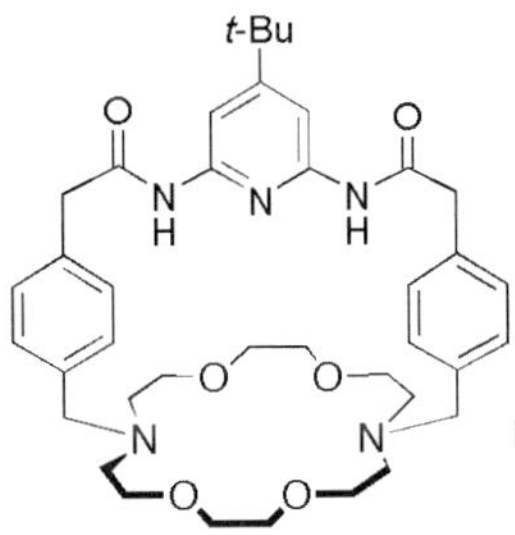

8

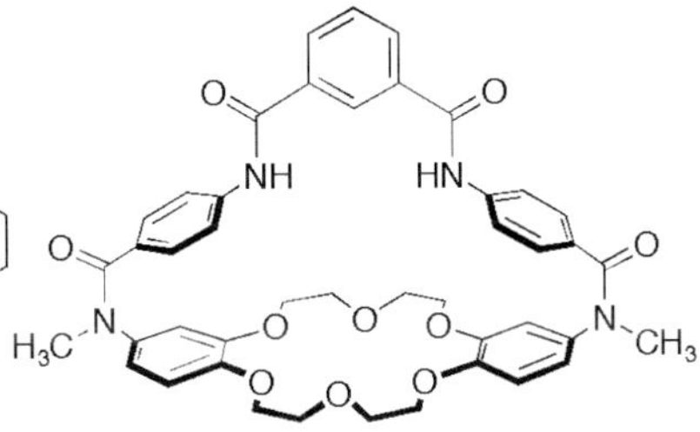

9

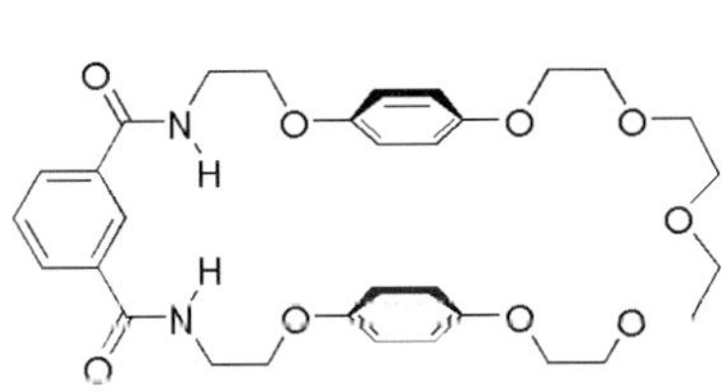

10

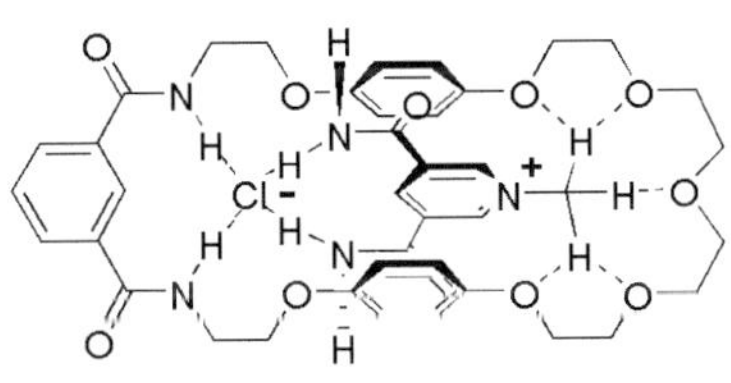

11

A growing area of supramolecular research is the topic of controlled self-assembly, and a number of groups have shown how salts can be used to template the dimerization of receptors with water often acting as a stabilizing agent. [17,18] In some cases, the assembled aggregate is large enough to act as a capsule and completely encapsulate both ions of the salt. [19] A related design is the so-called "venus fly trap" capsule.

An example of this design was recently reported by Atwood and coworkers, who showed that an extended, deep-cavity resorcinarene derivative can completely encapsulate a NMe_4^+ cation with a Cl^- anion positioned by hydrogen bonding at the capsule entrance (see schematic receptor **12**). [20]

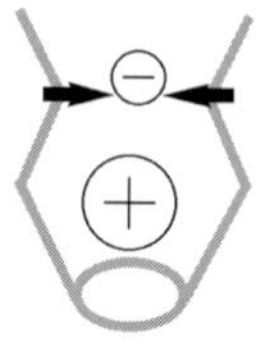

12

4. Ion-Pair Recognition Using Ditopic Receptor 13

As stated above, the most effective way to enhance salt binding by electrostatic effects is to use a receptor that binds the salt as a contact ion-pair. In 2001, our group prepared the simple ditopic receptor **13** and evaluated its ability to bind salts as associated ion-pairs (Scheme 1). We discovered that the receptor is able to extract a wide range of monovalent salts into weakly polar solvents. Furthermore, the resulting receptor/salt complexes are very stable. For example, most can survive column chromatography using silica gel and weakly polar solvents. Shown in Figure 1 is a series of ^1H NMR spectra that monitor the receptor mediated extraction of KCl into $CDCl_3$. Note that the exchange of salt between occupied and unoccupied receptor is slow on the NMR time scale; thus, the complexation system provides a unique opportunity to investigate the structure and dynamics of an isolated, associated ion-pair. After each extraction is completed, the uncomplexed salt is removed by filtration, and the filtrate evaporated to leave the **13**/salt complex as a solid residue which most times can be readily recrystallized. The resulting X-ray structures provide high-resolution pictures of the complexes, and the following sections illustrate the mechanisms that receptor **13** employs to recognize a wide range of salts with different shapes and coordination properties.

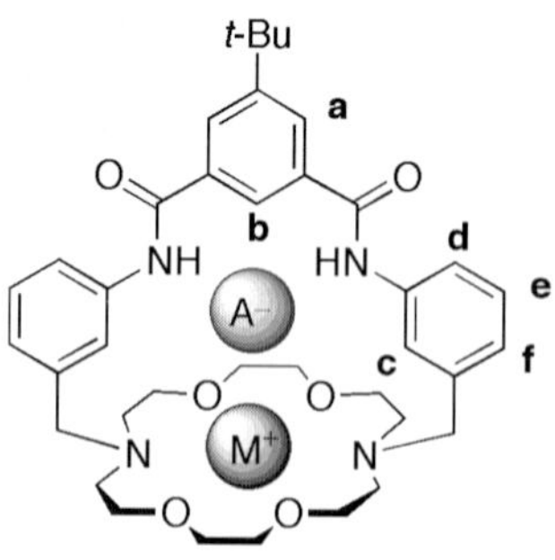

Scheme 1. Receptor **13** with bound salt.

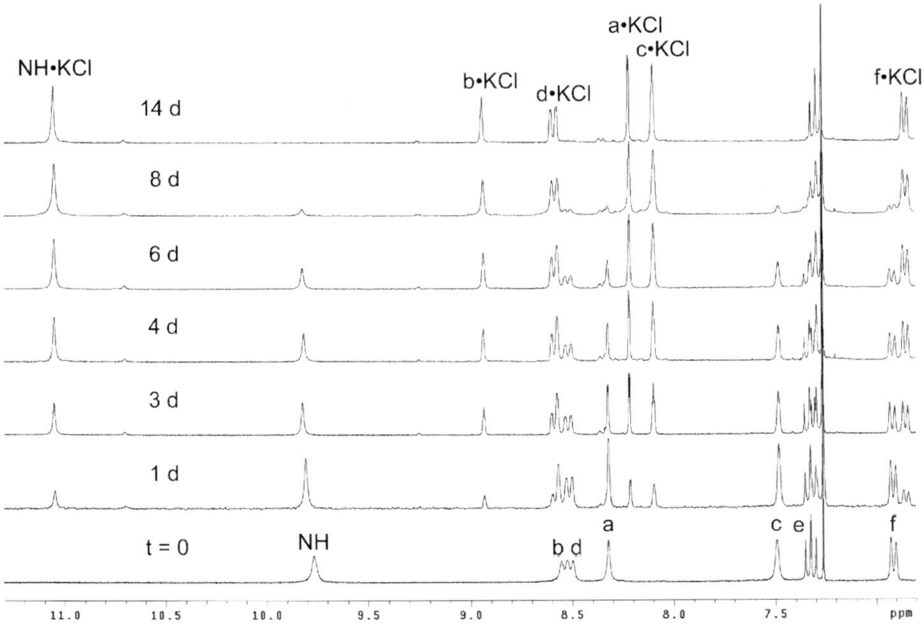

Figure 1. Partial ^{1}H NMR spectra of receptor **13** in $CDCl_3$ at 295K after addition of solid KCl. See structure of **13** for proton labeling. (Reprinted with permission from ref. 21. Copyright 2001 American Chemical Society.)

4.1 COMPLEXATION OF ALKALI HALIDES

NMR titration experiments showed that the presence of an alkali metal cation greatly enhances halide-binding constants. [21] For example, the **13**/Cl^- association constant in $CDCl_3$:DMSO-d_6 (85:15) was increased from 80 M^{-1} to 2.5 x 10^4 M^{-1} by the presence of one molar equivalent of potassium tetraphenylborate. Additional NMR titration experiments showed that receptor **13** binds KCl better than NaCl. X-ray analysis of the **13**•KCl complex uncovered two independent but structurally similar complexes in the unit cell. One of the structures is shown in Figure 2. The salt is clearly a contact ion–pair whose K-Cl distance of 2.989 Å is slightly shorter than that observed in solid KCl. We also obtained the X-ray structures of receptor **13** complexed with LiCl, LiBr, NaCl, NaBr, NaI, KCl, KBr, and KI. [22,23] As expected, the larger anions do not fit perfectly inside the macrocyclic cavity, and so salt affinities are decreased.

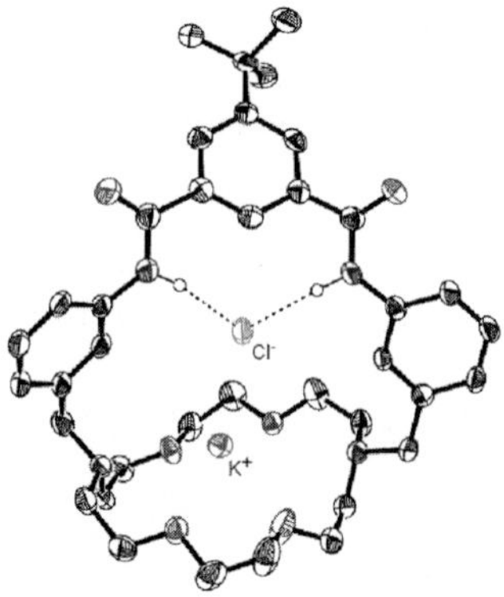

Figure 2. X-ray crystal structure of [**13**•K^+•Cl^-].
(Reprinted with permission from ref. 21. Copyright 2001 American Chemical Society.)

We evaluated the ability of **13** to transport salts through a liquid organic membrane. [22] Transport experiments using a supported liquid membrane, and high salt concentration in the source phase, showed that **13** can transport alkali halides up to ten times faster than a monotopic cation receptor or a monotopic anion receptor. All transport systems exhibited the same qualitative order of ion selectivity, that is, for a constant anion, the cation selectivity order is $K^+ > Na^+ > Li^+$, and for a constant cation, the anion transport selectivity order is $I^- > Br^- > Cl^-$. These trends are in general agreement with the Hofmeister series, a solvation-based selectivity bias that is typically observed for liquid/liquid partitioning processes. [24] Transport fluxes decrease with the smaller, more charge-dense ions because they have a more unfavorable Gibbs free energy for aqueous to organic transfer. [25] It appears that the Hofmeister bias overwhelms any difference in receptor/salt binding affinities. Receptor **13** can also transport NaCl or KCl across vesicle membranes. [26] Chloride efflux from unilamellar vesicles was monitored using a chloride selective electrode and significant transport was observed even when the transporter/phospholipid ratio was as low as 1:2500. Mechanistic studies indicate that the facilitated efflux is due to the uncomplexed transporter diffusing into the vesicle and the transporter/salt complex diffusing out.

Receptor **13** was also evaluated in competitive solid/liquid extraction experiments. Remarkably, the cation selectivity order is strongly reversed when the receptor extracts solid alkali chlorides and bromides into organic solution; that is, the process is highly lithium selective. For a three-component mixture of solid LiCl, NaCl and KCl, the ratio of salts extracted and complexed to the receptor in $CDCl_3$ was 94:4:2, respectively. The same strong lithium selectivity was also observed in the case of a three-component mixture of solid LiBr, NaBr and KBr where the ratio of extracted salts was 92:5:3. These contrasting results can be rationalized in terms of the equilibria that govern aqueous/organic extraction and solid/organic extraction (Scheme 2). Extraction of a salt, M^+A^-, into an organic phase mediated by a salt receptor, R, can be considered as a two step process. The first equilibrium, K_p, involves partitioning of the salt into the organic phase and the second equilibrium, K_a, concerns association of the partitioned salt with the receptor. In the case of aqueous/organic extraction (eq 1), the equilibrium step controlling liquid membrane transport and determining cation selectivity is K_p (which follows the Hofmeister series). The selectivity of the solid/liquid extraction mediated by **13** (eq 2) is also due to large differences in the partitioning equilibrium K_p, however, the order of K_p for solid/liquid extraction is opposite to that for aqueous/organic extraction.

Solid LiCl and solid LiBr are significantly more soluble in non-polar solvents than the corresponding sodium or potassium salts. The bonding in alkali halides, including lithium halides, is predominantly ionic, however, many lithium salts are known to have unusually low melting points and good solubilities in organic solvents. [27] This is due to the small size of the lithium cation and the molecular nature of its associated ion-pairs. The mechanism for salt transfer from aqueous to organic is not the same as the mechanism for solid/organic partitioning. Aqueous/organic partitioning involves the transfer of individual, hydrated ions that subsequently associate in the organic phase; whereas, solid/organic partitioning more likely involves the transfer of associated ion-pairs from solid to organic phase. In this latter case, it appears that receptor **13** binds the solubilized ion-pairs, and retains them in organic solution, which converts the large differences in solid/organic K_p into a potentially useful, lithium-selective extraction process. These results suggest that solid/liquid extraction may be a purification strategy that is applicable to other salts. The aim would be to design and construct multitopic receptors with an ability to extract the solid salts as associated ion-pairs.

$$M^{+}_{aq} + A^{-}_{aq} \underset{}{\overset{K_p}{\rightleftharpoons}} (M^{+}A^{-})_{org} \xrightarrow[]{\overset{K_a}{R}} R\bullet(M^{+}A^{-})_{org} \quad \textbf{(eq 1)}$$

$$(M^{+}A^{-})_{solid} \overset{K_p}{\rightleftharpoons} (M^{+}A^{-})_{org} \overset{\overset{K_a}{R}}{\rightleftharpoons} R\bullet(M^{+}A^{-})_{org} \quad \textbf{(eq 2)}$$

Scheme 2. Two-step process for extraction of salt, $M^{+}A^{-}$, using a salt receptor R. Aqueous/organic extraction (eq 1) and solid/organic extraction (eq 2). (Reprinted with permission from ref. 23. Copyright 2004 American Chemical Society.)

In an effort to increase salt binding affinities we collaborated with the Gale research group and prepared the second-generation macrobicyclic receptor **14**, which contains a bridging 2,5-diamidopyrrole group. [28] NMR titration experiments indicated that **14** has a three-fold higher affinity for Cl^{-} than **13**. The **14**/Cl^{-} association constant is hardly changed by the presence of one molar equivalent of Na^{+} ions but it is increased substantially by the presence of K^{+} ions. As expected, an X-ray structure of the [**14**•NaCl] complex confirmed that the receptor binds NaCl as a contact ion-pair (Figure 3). The crystal structure shows clearly why **14** exhibits enhanced Cl^{-} affinity relative to **13**. Not only does the Cl^{-} form hydrogen bonds with both amide but also with the pyrrole NH. The Na-Cl distance of 2.65 Å is shorter than the Na-Cl distance when sodium chloride is bound in **13** (2.70 Å).

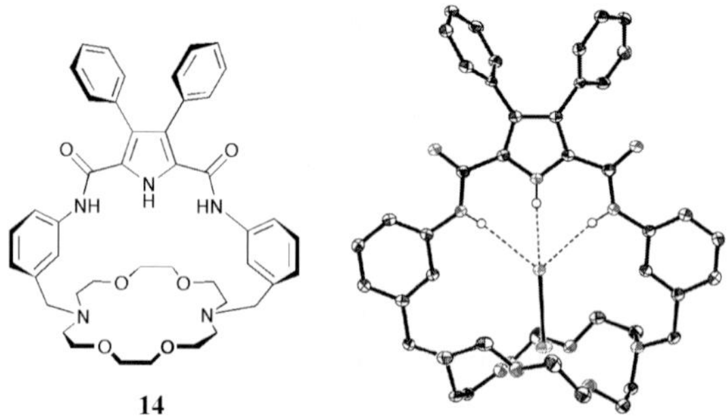

Figure 3. Chemical structure of **14** and X-ray structure of [**14**•Na^+•Cl^-].

4.2 COMPLEXATION OF ALKYLAMMONIUM SALTS

Receptor **13** can also bind monoalkylammonium salts as contact ion-pairs. [29] The X-ray structure of the **13**•$MeNH_3Cl$ complex is shown in Figure 4. The methylammonium cation fits deeply into the binding pocket of the receptor and forms three hydrogen bonds; one to a crown oxygen, one to a crown nitrogen, and one to the chloride which is in turn hydrogen bonded to the two receptor NH residues. The X-ray structure suggests that the macrocyclic cavity can only accommodate alkylammonium cations with small or narrow alkyl groups. This hypothesis was tested by measuring the ability of receptor **13** to bind various alkylammonium chloride salts in 85:15 $CDCl_3$:DMSO-d_6, a solvent system where host/guest exchange is rapid on the NMR time scale. As shown in Table 1, the association constant for NBu_4•Cl is 50 M^{-1}. A control experiment with NBu_4PF_6 confirmed that the tetrabutylammonium cation does not bind to the receptor; thus, the association constant is a measure of Cl^- affinity for **13**. The association constant for Et_2NH_2•Cl is 10 M^{-1} which indicates that the diethylammonium cation lowers the Cl^- affinity by sequestering the Cl^- away from receptor **13**. In the case of *i*-$PrNH_3$•Cl and *n*-$PrNH_3$•Cl the association constants are 2.0 x 10^2 and 2.0 x 10^4 M^{-1}, respectively. In the case of *n*-$PrNH_3$•Cl binding, a Job plot indicated that the complex stoichiometry is 1:1. The one hundred-fold selectivity for *n*-$PrNH_3$•Cl over *i*-$PrNH_3$•Cl was confirmed by a competitive binding experiment where ^1H NMR showed that one molar equivalent of *n*-$PrNH_3$•Cl can completely displace *i*-$PrNH_3$•Cl from a complex of [**13**•*i*-$PrNH_3$•Cl] in $CDCl_3$. In addition, receptor **13** has an affinity for *n*-$PrNH_3$•Cl that is two hundred times stronger than for *n*-$PrNH_3$•AcO and *n*-$PrNH_3$•*p*-TsO (Table 1). The relatively large changes in chemical shift for several diagnostic receptor hydrogens upon salt binding provides good evidence that the mode of binding in solution is very similar to that observed in the solid state.

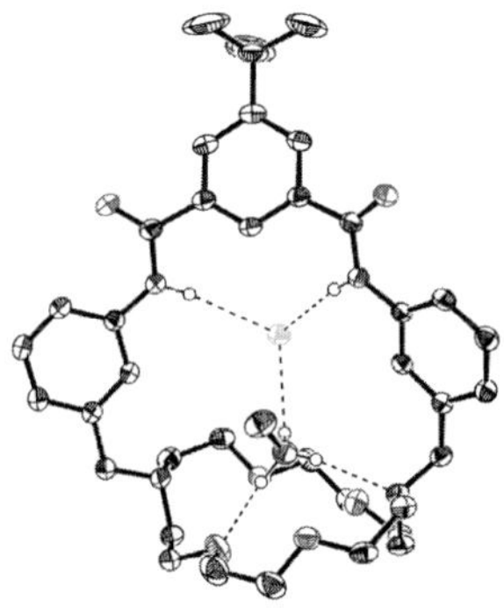

Figure 4. X-ray crystal structure of [**13**•$MeNH_3^+$•Cl^-].
(Reprinted with permission from ref. 29. Copyright 2003 American Chemical Society.)

TABLE 1. Association constants (K_a) and chemical shift changes (Δδ) for receptor **13**.

Guest	K_a (M^{-1})[a]	NH Δδ[b]	H_b Δδ[b]	H_c Δδ[b]
Bu_4N•Cl	50	+0.90	+0.44	+0.44
n-$PrNH_3$•Cl	2.0 x 10^4	+1.00	+0.11	+0.39
i-$PrNH_3$•Cl	2.0 x 10^2	+1.02	+0.28	+0.42
Et_2NH_2•Cl	10	+0.78	+0.19	+0.31
n-$PrNH_3$•*p*-TsO	1.0 x 10^2	+0.30	-0.31	+0.22
Bu_4N• *p*-TsO	4	+0.08	+0.01	+0.08
n-$PrNH_3$•AcO	1.2 x 10^2	+1.10	-0.11	+0.25
Bu_4N•AcO	20	+1.82	+0.44	+0.45

[a]In $CDCl_3$:DMSO-d_6 85:15, T = 295 K, initial [**13**] = 10 mM. Uncertainty ±40%. [b]Change in receptor chemical shift (ppm) after addition of 200 mM guest salt. See structure of **13** for proton labeling.

4.3 COMPLEXATION OF SALTS WITH TRIGONAL OXYANIONS

Most recently we have evaluated the ability of **13** to recognize alkali AcO^- and NO_3^- salts. We were particularly interested in seeing how these trigonal oxyanions, with lone pair and π-electron density, simultaneously interact with the NH residues on **13** and the bound alkali metal cation. We were motivated by a report by Hay and coworkers who found crystallographic evidence indicating that trigonal oxyanions (like NO_3^- and AcO^-) prefer to form hydrogen bonds with R-H acceptors that have H···O-A angles near 120° and R-H···O-N dihedral angles near 0° (Scheme 3). [30] In other words, the donor hydrogen atom lies within the plane of the trigonal anion. Hay calculates that the preference for hydrogen bonding to the oxygen lone pairs over the π electron density is about 2 kcal/mol. [31]

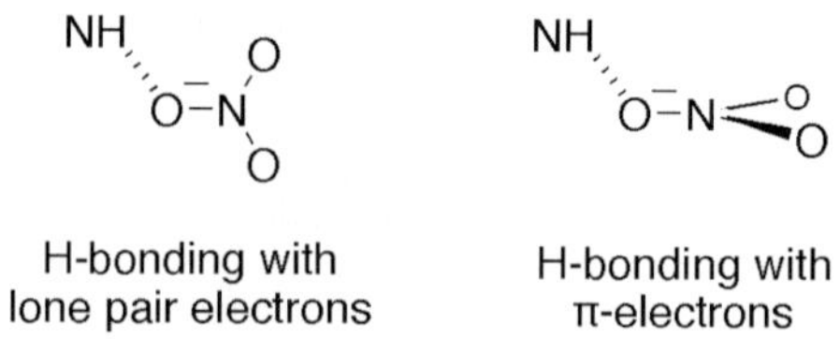

Scheme 3. Hydrogen bonding with nitrate lone pair electrons (H-O-N-O dihedral angle 0°) is favored over hydrogen bonding with π-electrons (H-O-N-O dihedral angle 90°).

Shown in Figures 5-7 are three X-ray structures of receptor **13** complexed with $NaNO_3$, KNO_3, or $LiNO_3$. [32] The X-ray structure of **13**•$NaNO_3$ (Figure 5) shows that the NO_3^- is located deep inside the macrocyclic cavity and chelates the bound Na^+. The receptor NH residues form hydrogen bonds with the non-chelating NO_3^- and the two H-O-N-O dihedral angles are 34° and 78°. In the case of solid-state **13**•KNO_3, the larger K^+ cation forces the chelating NO_3^- to sit further out of the receptor cavity, which gives the NO_3^- more freedom to move (Figure 6). Indeed, the NO_3^- flips between two unequally occupied positions between the two receptor NH residues. The 70 % occupancy structure is shown and has two quite different intermolecular N···O distances to the non-chelating NO_3^- oxygen and two quite different H-O-N-O dihedral angles of 15° and 70°. This asymmetrical positioning of the NO_3^- allows the receptor to better align one of its two NH residues with the more basic lone pair electrons, thus forming a stronger hydrogen bond. The X-ray structure of the **13**•$LiNO_3$ complex (Figure 7) differs from the sodium and potassium analogs in a number of ways. The Li^+ is coordinated by five heteroatoms, one nitrogen and two oxygens from the crown and two water oxygens (derived adventitiously from the atmosphere during the crystallization process). One of the water molecules bridges the cation and anion. The receptor/NO_3^- orientation is rotated almost 90°, compared to the sodium and potassium structures, which means that the receptor NH residues are directed primarily towards the two lone pairs on one of the NO_3^- oxygens (the two H-O-N-O dihedral angles are 28° and 37°), and the bridging water OH is directed towards the NO_3^- oxygen's π-electrons (H-O-N-O dihedral angle of 90°). Thus, with these three receptor/NO_3^- salt structures, the directionality of the hydrogen bonding between the complexed NO_3^- and the receptor NH is strongly influenced by the identity of the counter cation.

It appears that NO_3^- orientation is controlled by a complex interplay of steric factors, coordination bonding to the metal cation, and hydrogen bonding with the receptor NH residues. This latter factor includes a modest preference to direct the two receptor NH residues towards the more basic oxyanion lone pairs. The directionality can be overwhelmed by stronger bonding effects such as ion-pairing. For example, in the case of the sodium and potassium structures, the NO_3^- is directly chelated to the counter cation, but this is only achieved by twisting the receptor/NO_3^- orientation such that one or both of the receptor NH residues are pointing substantially at the NO_3^- π-surface. This weaker hydrogen bonding arrangement is more than offset by the formation of strong coordination bonds.

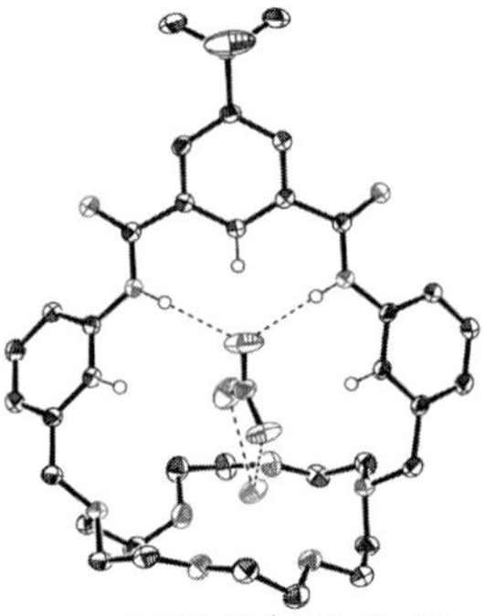

Figure 5. X-ray crystal structure of [**13**•Na^+•NO_3^-]. (Reprinted with permission from ref. 32. Copyright 2004 American Chemical Society.)

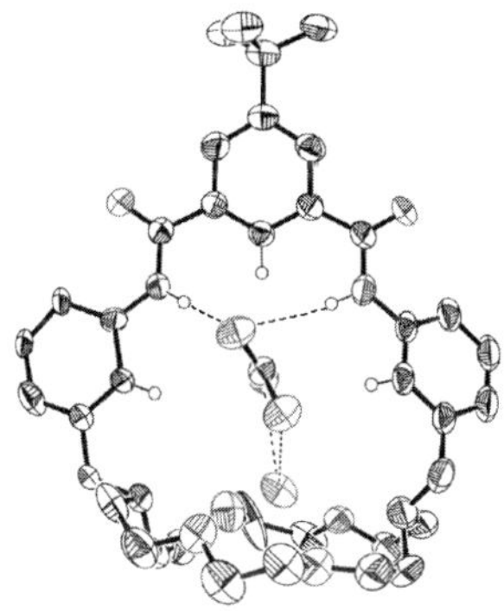

Figure 6. X-ray crystal structure of [**13**•K^+•NO_3^-]. The 70% occupancy orientation is shown for nitrate. (Reprinted with permission from ref. 32. Copyright 2004 American Chemical Society.)

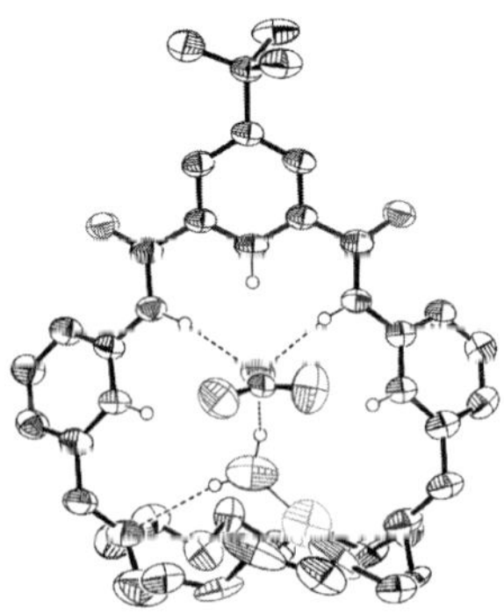

Figure 7. X-ray crystal structure of [**13**•Li^+•$2H_2O$•NO_3^-]. Absent is a second water molecule that is located underneath the crown and coordinated to the Li^+. (Reprinted with permission from ref. 32. Copyright 2004 American Chemical Society.)

The preceding X-ray data provides a structural basis for interpreting the following unusual NMR data. Listed in Table 2 are the changes in ^1H NMR chemical shifts for the receptor signals upon complexation with a variety of salts. In the case of NaCl, NaAcO and KAcO, the NH signals moves downfield 0.7–0.9 ppm as expected. [33] On the other hand, complexation with the nitrate salts induces quite different changes in chemical shift. In particular, the NH signal moves upfield by 0.05–0.22 ppm (Table 2 and Figure 8). To the best of our knowledge, an upfield shift of a neutral receptor amide NH signal upon anion complexation is unprecedented. Another unusual change is the large upfield shift of equivalent protons **c** (see Table 2 for a comparison of complexed-induced-shifts). These unusual changes in chemical shift upon complexation indicate that the magnetic shielding environment around the NO_3^- is anisotropic. Indeed, we used Density Functional Theory to calculate the shielding surface around the NO_3^- anion and found that it is deshielding around the peripheral plane of the molecule and shielding in a region above the central nitrogen (Scheme 4).

TABLE 2. Change in ^1H NMR chemical shift (Δδ) of selected protons upon saturation of **13** with salt.[a]

	Δδ (ppm)[b]			
Salt	a	b	NH	c
NaCl	+0.02	+0.66	+0.94	+0.87
NaAcO	0.00	+0.45	+0.72	+0.11
KAcO	-0.03	+0.48	+0.87	+0.05
$NaNO_3$	0.00	+0.21	-0.22	-0.18
KNO_3	0.00	+0.32	-0.05	-0.21
$LiNO_3$	-0.01	+0.20	-0.22	-0.09
Bu_4NNO_3[c]	-0.03	+0.04	+0.51	+0.27

[a]Solid salt extracted into solution of **13** (10 mM) in $CDCl_3$ at T = 295 K. [b]See structure of **13** for proton labeling. Δδ = δ **13**•salt - δ **13**. [c]Data obtained after mixing 150 mM Bu_4NNO_3 and 10 mM **13** in $CDCl_3$.

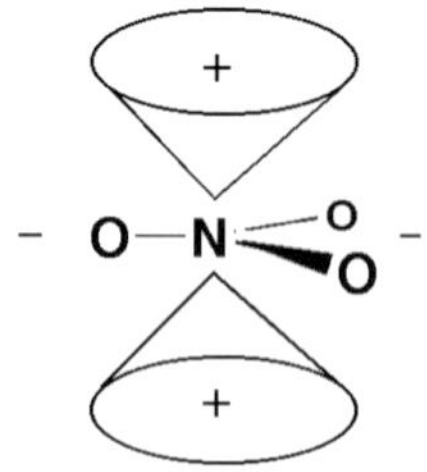

Scheme 4. Anisotropic shielding surface surrounding the nitrate anion.

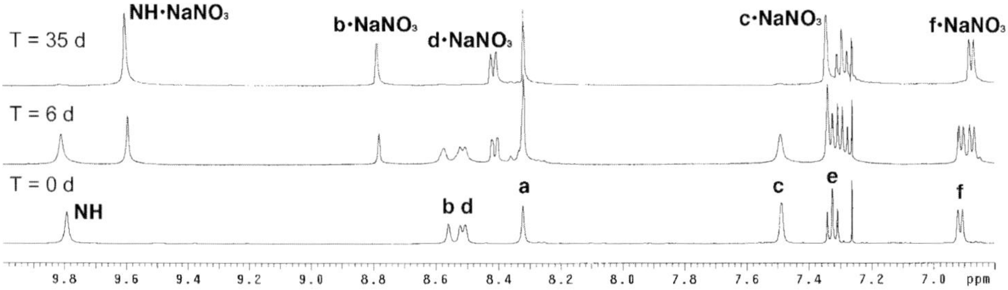

Figure 8. Partial ^{1}H NMR spectra of receptor **13** in $CDCl_3$ at 295K after addition of solid $NaNO_3$. See structure of **13** for proton labeling. (Reprinted with permission from ref. 32. Copyright 2004 American Chemical Society.)

Having established that NO_3^- has an anisotropic shielding surface, the next question was whether the complexation-induced changes in chemical shifts that are listed in Table 2 can be used to elucidate the structure of the receptor/salt complexes in solution. In some cases, the solution state NMR data seems to match with the solid state structure. For example, the signals for the NH and **c** protons in **13** move upfield by -0.22 and -0.18 ppm, respectively, when **13** is saturated with $NaNO_3$. The X-ray structure of **13**•$NaNO_3$ (Figure 5) suggests that this is because the NH and **c** protons are located in shielding zones above and below the plane of the encapsulated NO_3^-. However, when **13** is saturated with $LiNO_3$, the solution state NMR data does not appear to match with the solid state structure. For example, the receptor NH and **c** protons move upfield by -0.22 and -0.09 ppm, respectively, but the X-ray structure of **13**•$2H_2O$•$LiNO_3$ (Figure 7) shows that the NH and **c** protons are located in the peripheral plane around the encapsulated NO_3^- (deshielding zones). This suggests that the X-ray structure is not the predominant structure in solution. Evidence in favor of this hypothesis was gained from a variable temperature ^{1}H NMR study of the complex. The system undergoes dynamic exchange because at low temperature the spectrum splits into three sets of signals. Shown in Figure 9 are the signals for the NH residues at 213 K. The NH peak at 10.09 ppm corresponds to free receptor; whereas, the upfield peak at around 9.8 ppm is attributed to a dehydrated **13**•$LiNO_3$ complex with a structure that is analogous to **13**•$NaNO_3$ in Figure 5 (*i.e.*, the NO_3^- is chelated to the Li^+ inside the cavity of the receptor). The downfield NH peak at 10.15 ppm is attributed to a hydrated complex with a structure that is very similar to **13**•$2H_2O$•$LiNO_3$ in Figure 7. The relative ratio of these three signals depends on the amount of water in the sample. As depicted in Figure 9, the peak at 9.8 ppm (corresponding to dehydrated salt complex with chelated $LiNO_3$) is diminished when water is added to the sample.

The large complexation-induced changes in ^{1}H chemical shift for the protons that line the cavity of receptor **13** indicate how the oxyanion salts bind to the receptor cavity in solution. In principle, the direction and magnitude of the shieldings can be used to elucidate the relative orientation of the encapsulated anisotropic anion, however, this requires quantitative mapping of the shielding surface around the anion (which we have done in the case of NO_3^-) and knowledge of the receptor/salt dynamics. In certain cases (*e.g.*, the **13**•$LiNO_3$ system above), the signal averaging due to dynamic exchange can be eliminated by acquiring the NMR spectrum at low temperature. Overall, the use of ditopic salt receptors, such as **13**, to solubilize salts as discrete, slowly exchanging, associated ion-pairs, is a new and effective way to characterize the structure of ion-pairs.

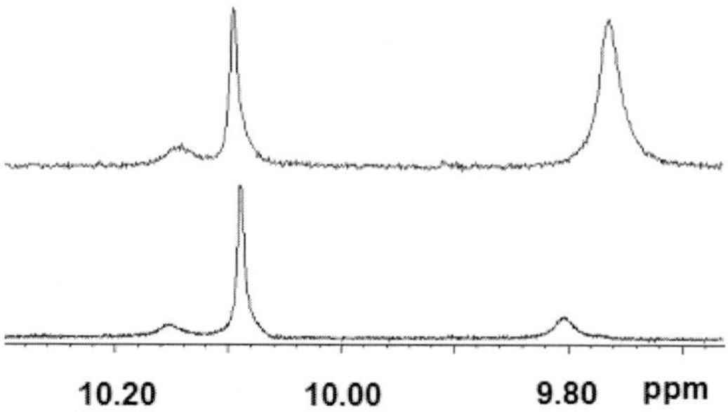

Figure 9. Partial ^{1}H NMR spectrum of **13**•$LiNO_3$ at 213 K. Top: sample prepared by solid-liquid extraction using freshly opened $CDCl_3$. Bottom: sample prepared with water-saturated $CDCl_3$. (Reprinted with permission from ref. 32. Copyright 2004 American Chemical Society.)

5. Summary

In the future, salt-binding receptors will be employed in various separation and sensing applications. The work described in this chapter demonstrates that ditopic receptors, with an ability to bind the salts as contact ion-pairs, have particularly attractive properties as extraction and transport agents. Another future direction is the utilization of salts as "molecular glue" to assemble complex supramolecular structures that have dynamic properties and the capability to behave as molecular machines.

6. Acknowledgements

I warmly thank the students and postdoctoral associates in my research group who made the excellent experimental contributions described in this chapter. In particular, I acknowledge the work of Dr. Joseph M. Mahoney. This study was supported by the University of Notre Dame and the National Science Foundation.

7. References

1. Shukla, R., Kida, T., and Smith, B.D., *Org. Lett.* **2**, (2000) 3039-3248.
2. Gale, P.A., *Coord. Chem. Rev.* **240**, (2003) 191-221.
3. Smith, B.D. and Mahoney, J.M., *Encyclopedia of Supramolecular Chemistry*, Marcell Dekker: New York, 2004, 1291-1294.
4. Mahoney, J.M., Davis, J.P., and Smith, B.D., In, *Fundamentals and Applications of Anion Separations*, Moyer, B.A., Singh, R., Eds.; ACS Symposium Series, ACS: Washington, 2004, 115-124.
5. Beer, P.D., and Gale, P.A., *Angew. Chem. Int. Ed.* **40**, (2001) 486-516.
6. Kirkovits, G., J., Shriver, J.A., Gale, P.A., and Sessler, J.L., *J. Incl. Phenom. Macrocycl. Chem.* **40**, (2001) 69-75.
7. Rudkevich, D.M., Brzozka, Z., Palys, M., Visser, H.C., Verboom, W.; and Reinhoudt, D.N., *Angew. Chem. Int. Ed. Engl.* **33**, (1994) 467-469.
8. Beer, P.D., Drew, M.G., Knubley, R.J., and Ogden, M.I., *Dalton Trans.* (1995) 3117-3123.
9. Tumcharern, G., Tuntulani, T., Coles, S.J., Hursthouse, M.B., and Kilburn, J.D., *Org. Lett.* **5**, (2003) 4971-4974.
10. Savage, P.B., Holmgren, S.K., and Gellman, S.H., *J. Am. Chem. Soc.* **116**, (1994) 4069-4070.

11. Arafa, E.A., Kinnear, K.I., and Lockhart, J.C., *Chem. Commun.* (1992) 61-64.
12. Tozawa, T., Misawa, Y., Tokita, S., and Kubo, Y., *Tetrahedron Lett.* **41**, (2000) 5219-5223.
13. Reetz, M.T., Niemeyer, C.M., and Harms, K., *Angew. Chem. Int. Ed. Engl.* **30**, (1991) 1474-1476.
14. Flack, S.S., Chaumette, J-L., Kilburn, J.D., Langley G.J., and Webster, M., *Chem. Commun.* (1993) 399-400.
15. Deetz, M.J., Shang, M., and Smith, B.D., *J. Am. Chem. Soc.* **122**, (2000) 6201-6207.
16. Wisner, J.A., Beer, P.D., Berry, N.G., and Tomapatanaget, B., *Proc. Natl. Acad. Sci. USA* **99**, (2002) 4983-4986.
17. Shi, X., Mullaugh, K.M., Fettinger, J.C., Jiang, Y., Hofstadler, S.A., and Davis, J.T., *J. Am. Chem. Soc.* **125**, (2003) 10830-10841.
18. Manisikkamäki, H., Nissinen, M., Schalley, C.A., and Rissanen, K., *New. J. Chem.* **27**, (2003) 88-97.
19. Bourgeois, J., Fujita,M., Kawano, M., Sakamoto, S., and Yamaguchi, K., *J. Am. Chem. Soc.* **125**, (2003) 9260-9261.
20. Atwood, J.L., and Szumna, A., *Chem. Comm.* (2003) 940-941.
21. Mahoney, J.M., Beatty, A.M., and Smith, B.D., *J. Am. Chem. Soc.* **123**, (2001) 5847-5848.
22. Mahoney, J.M., Nawaratna, G.U., Beatty, A.M., Duggan, P.J., and Smith, B. D., *Inorg. Chem.* **43**, (2004) 5902-5907.
23. Mahoney, J.M., Beatty, A.M., and Smith, B.D., *Inorg. Chem.* **43**, (2004) ASAP.
24. Levitskaia, T., Marquez, M., Sessler, J.L., Shriver, J.A., Vercouter, T., and Moyer, B.A., *Chem. Comm* (2003) 2248-2249.
25. Olsher, U., Kankins, M., Kim, Y.D., and Bartsch, R.A., *J. Am. Chem. Soc.* **115**, (1993) 3370-3371.
26. Koulov, A.V., Mahoney, J.M., Beatty, A.M., and Smith, B.D., *Org. Biomol. Chem.* **1**, (2003) 27-29.
27. Snaith, R. and Wright, D.S., In *Lithium Chemistry – A Theoretical and Experimental Overview*, Sapse, A.–M., Von Rogué Schleyer, P., Eds.; Wiley: New York, 1995, Ch. 8.
28. Mahoney, J.M., R. Marshall, R.A., Beatty, A.M., Smith, B.D., Camiolo, S., and Gale, P.A., *J. Supramol. Chem.* **1**, (2001) 289-292.
29. Mahoney, J.M., Davis, J.P., and Smith, B.D., *J. Org. Chem.* **68**, (2003) 9819-6820.
30. Hay, B.P., Dixon, D.A., Bryan, J.C., and Moyer, B.A., *J. Am. Chem. Soc.* **124**, (2002) 182-183.
31. Hay, B.P., Gutowski, M., Dixon, D.A., Garza, J., Rubicelia, R., and Moyer, B.A., *J. Am. Chem. Soc.* **126**, (2004) 7925-7934.
32. Mahoney, J.M., Stucker, K.A., Jiang, H., Carmichael, I., Brinkmann, N.R., Beatty, A.M., and Smith, B.D., manuscript submitted.
33. Konrat, R., Tollinger, M., Kontaxis, G., and Kräutler, B., *Monatsh. Chem.* **130**, (1999) 961-982.

10. CYCLIC AND ACYCLIC AMIDOPYRROLE CONTAINING ANION RECEPTORS

S. J. BROOKS AND P. A. GALE
School of Chemistry, University of Southampton,
Southampton, SO17 1BJ, U.K.

1. Introduction

Intense recent interest in anion complexation has resulted in significant advances in the design of potent anion binding agents. The development and implementation of the amidopyrrole motif for the complexation of anions is a relatively new and exciting area of anion coordination chemistry. The motif is of particular interest due to the convergent hydrogen bond donor characteristics that distinguish it from other dual hydrogen bond donor systems such as urea (*Figure 1*). The advances in this area are outlined in this short review.

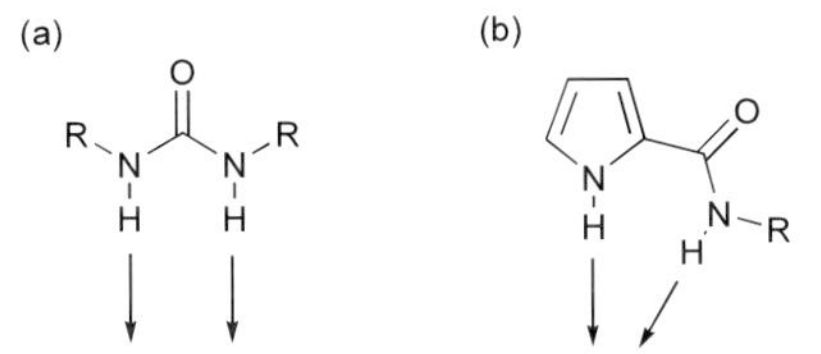

Figure 1. Urea and 2-amidopyrrole provide parallel and convergent hydrogen bond donation respectively.

2. 2-Amido and 2,5-Diamidopyrroles

In the latter half of the 1990's, Crabtree and co-workers[1,2] demonstrated that isophthalamides are excellent receptors for fluoride and chloride in organic solution. (*Figure 2*).

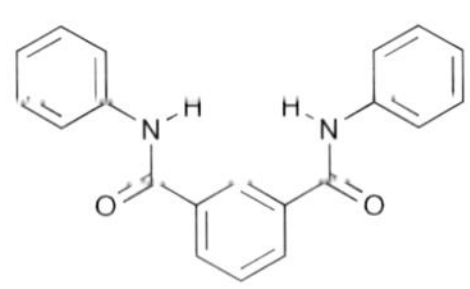

Figure 2. The isophthalamide motif has been employed in a variety of receptors possessing high affinities for anions.

Karsten Gloe (ed.), Macrocyclic Chemistry: Current Trends and Future Perspectives, 153–172.
© 2005 *Springer. Printed in the Netherlands.*

2,5-Diamidopyrroles may be regarded as analogues of this type of receptor. In addition to the possession a third hydrogen-bond donor group which may be involved in anion complexation, replacement of a six-membered aryl ring with a five membered pyrrole causes a relative increase in the internal angles within the cleft relative to isopthalamide derived clefts. Compounds **1** and **2** were synthesized from 3,4-diphenyl-1H-pyrrole-2,5-bisacid chloride which was condensed with aniline and *n*-butylamine respectively to afford the products in 18 and 47% yields (*Figure 3*).[3] Analysis of the stability constants of these compounds with a variety of anions was determined by ^{1}H NMR titration techniques fitted to a 1:1 binding model (performed in CD_3CN and DMSO-d_6/H_2O 0.5% respectively due to the relative solubility of the two compounds), revealing a preference for *oxo*-anion complexation in both cases.

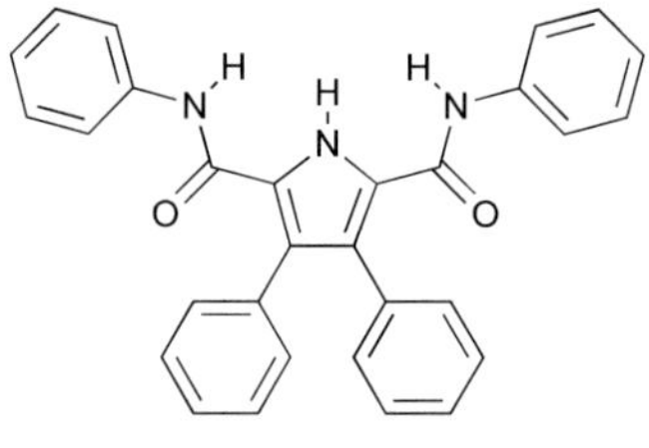

1

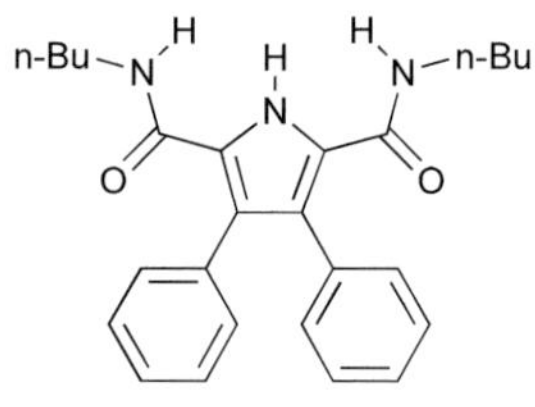

2

Figure 3. First generation 2,5-diamido-1H-pyrrolic amide clefts.

Amongst the putative anionic guests studied, receptor **1** was found to bind benzoate most strongly with a stability constant of 2.5×10^3 M^{-1} whilst compound **2** bound dihydrogen phosphate most strongly in DMSO-d_6/H_2O 0.5% (used for solubility reasons) with a stability constant of 1.5×10^3 M^{-1}.

In an attempt to more fully understand the interactions between these novel pyrrolic clefts with anions, in particularly the observed high *oxo*-anion affinity, 2-amido-5-methylpyrrole analogues **3** and **4** were synthesized (*Figure 4*).[4]

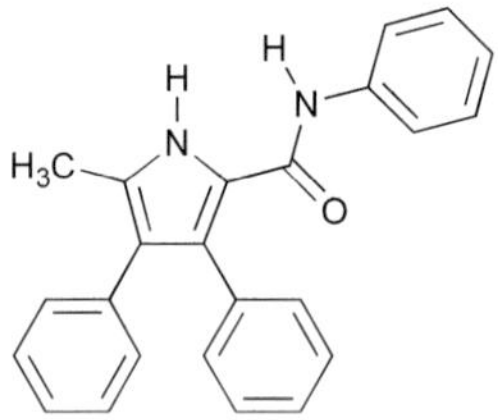

3

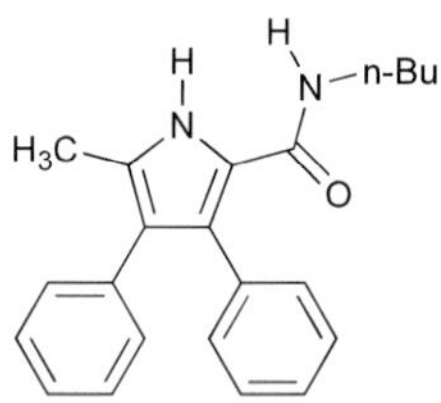

4

Figure 4. 2-Amido-5-methyl-1H-pyrrole based analogues.

Problems resulting from precipitation when solutions of compound **4** underwent addition of anions prevented an accurate titration data set from being collected, however a direct comparison was possible between compounds **1** and **3**, for which ^{1}H NMR titrations which could be titrated under the same solvent (CD_3CN). The comparison of the binding data revealed a significant reduction in the stability constants for *oxo*-anions of the mono-amide relative to the bis-amide (stability constants for dihydrogen phosphate and benzoate for **1** [3.6×10^2 M^{-1} and 2.5×10^3 M^{-1}] whilst association constants for **3** with the same anions [8.9×10^1 M^{-1} and 2.0×10^2 M^{-1}]). This important result gives a strong indication that during *oxo*-anion complexation, all three hydrogen bond donor groups form significant interactions with the guest.

A subsequent development was the synthesis of a class of 2,5-diamidopyrroles in which differently functionalized pendent arms were appended. Bowman-James[6] had previously shown that amine containing macrocycles exhibited a strong selectivity for protonated *oxo*-anions partially due to a proton transfer from the anion to the receptor resulting in a more highly charged anion receptor pair. The resulting structure contained protonated quaternary amines that introduced a significant additional electrostatic attractive force between receptor and substrate.

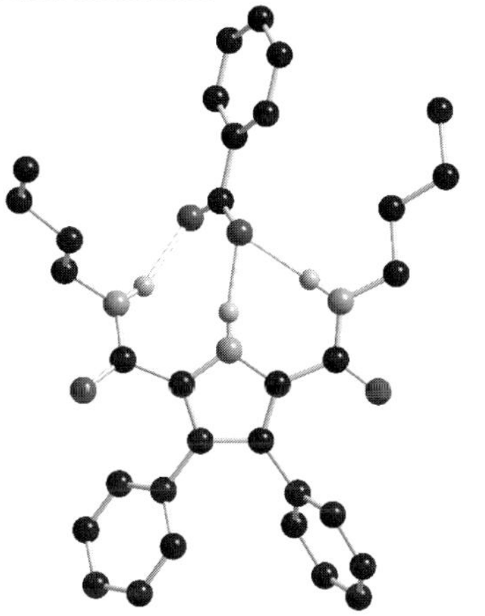

Figure 5. The X-ray crystal structure of the benzoate complex of receptor **1** (tetrabutylammonium counter cation and non-acidic hydrogen atoms have been omitted for clarity).

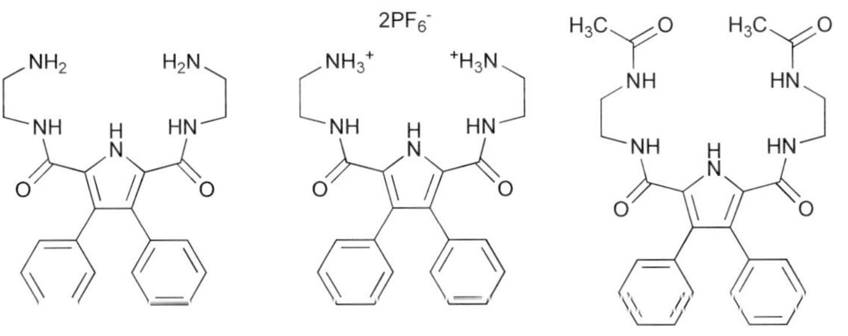

5 **6** **7**

Figure 6. Pendent arm 2,5-diamidopyrrole receptors.

By introducing basic groups tethered to the receptor through an amide linkage it was hoped that the oxo-affinity that had been displayed by the diamidopyrrole motif could be enhanced in the presence of the appropriate anions. A range of compounds **5-7** were synthesized in order to determine and quantify any possible additional binding interactions (*Figure 6*).[7] Compound **5** could undergo protonation at the amine site upon addition of a protonated anion, such as hydrogen sulfate, thus facilitating the additional electrostatic interaction whilst compound **6**, which was a pre-protonated form, could not remove a proton from the anion under the investigated binding conditions, thus preventing SO_4^{2-} formation. Compound **7**, an amide derivative would provide a neutral analogue for comparison.

^1H NMR titration studies performed in DMSO-d_6/0.5% H_2O provided strong evidence that pendent amine groups could indeed undergo a proton transfer from appropriate *oxo*-anions. For instance, hydrogen sulphate was found to have a stability constant of greater than 10^4 M^{-1} with **5** and less than 20 M^{-1} when titrated against **6** whilst no binding with hydrogen sulphate was observed with receptor **7**. As all three receptors have the same hydrogen bonding motif in terms of the pyrrolic centre the additional binding effect of the protonatable amine derived pendant arms appears highly significant.

Ferrocene appended 2,5-diamidopyrole receptors were subsequently developed in order to produce electrochemical sensors for anions (*Figure 7*).[8] The anion complexation of these compounds was reported through the results of both ^1H NMR titration and cyclic voltammetry techniques.

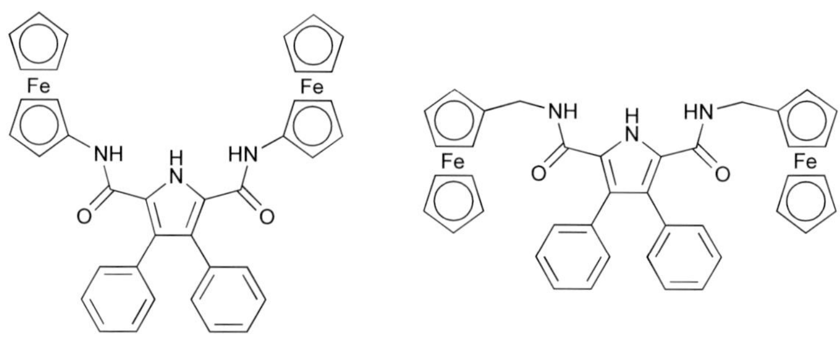

8 **9**

Figure 7. Ferrocene appended amidopyrrole electrochemical molecular sensors.

^1H NMR titrations with a variety of anions were performed in DMSO-d_6/0.5% H_2O and revealed that compound **8** showed significantly higher stability constants for all anions investigated by approximately an order of magnitude as compared to receptor **9** possibly due to the higher rigidity of receptor **8**. The presence of the methylene spacer group between the ferrocene and anion binding site is postulated to provide extra flexibility within the receptor and providing a mechanism in which the reporter group can occupy a position directly adjacent to the cleft effectively reducing the binding affinity of the receptor. Association constants calculated for fluoride reveal values of 705 and 170 M^{-1} for **8** and **9**.

The electrochemical behaviour of these two compounds was investigated using cyclic voltammetry, performed in dichloromethane with tetrabutylammonium tetrafluoroborate as the base electrolyte using a platinum disk microelectrode. Unfortunately both for hydrogen sulphate and dihydrogen phosphate passivation of the electrode resulted in a seriously distorted voltammetric wave, and in the presence of all other anions, with the exception of bromide which did not significantly interact with the receptors, the observed results were anodic shifts of the ferrocene/ferrocinium redox couple. Addition of both fluoride and benzoate resulted in significant shifts in these potentials for both of the investigated sensors with the observed shifts for **8** higher than those of **9** (shifts in redox couple for fluoride –255 and–130 mV and benzoate –120 and –60mV for **8** and **9** respectively). Presumably, a through bond mechanism[9] is operating upon addition of anions to receptor **8** which is not observed in receptor **9** due to the methylene carbons breaking conjugation between the anion binding site and the redox-active ferrocene groups.

The introduction of electron-withdrawing substitutents to the 3- and 4-positions of the pyrrole ring of a 2,5-diamidopyrrole resulted in the formation of a unique hydrogen bonded dimer.[10] Compounds **10** and **11** were synthesized and their stability constants measured with a variety of anions (*Figure 8*). For chloride a stability constant of 2×10^3 M^{-1} was obtained for **11** in acetonitrile, whilst the analogous 3,4-diphenyl receptor **2** binds chloride with a stability constant of 1.4×10^2 M^{-1}. The increased affinity for chloride of the 3,4-dichloro derived compound presumably results from the increase in electron withdrawing nature of the chloride group compared to the phenyl substituents in the first generation compounds.

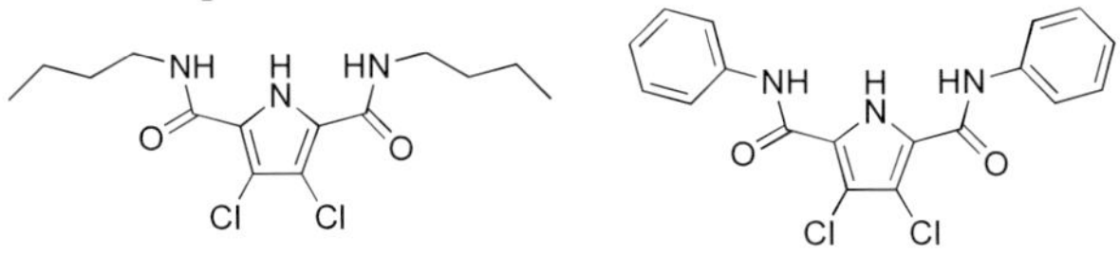

10 **11**

Figure 8. 2,5-Diamido-3,4-dichloropyrrole anion receptors.

Upon addition of tetrabutylammonium fluoride in CD_2Cl_2, it was found that the ^1H NMR titration curve produced when following the amide NH resonance chemical shift for **11**, shifted downfield until one equivalent of fluoride had been added (*Figure 9*). Following this an upfield shift in the resonance was observed to reach a plateau after two equivalents and then remaining constant at approximately 9.3 ppm. This behaviour appeared indicative of a initially a binding event followed by a deprotonation process. Support for this hypothesis was provided by treatment of receptor **11** with one equivalent of tetrabutylammonium hydroxide to yield the tetrabutylammounium salt of **11**. The chemical shift of the amide NH of the salt was found to be the same chemical shift as that observed when the free receptor was treated with two equivalents of fluoride.

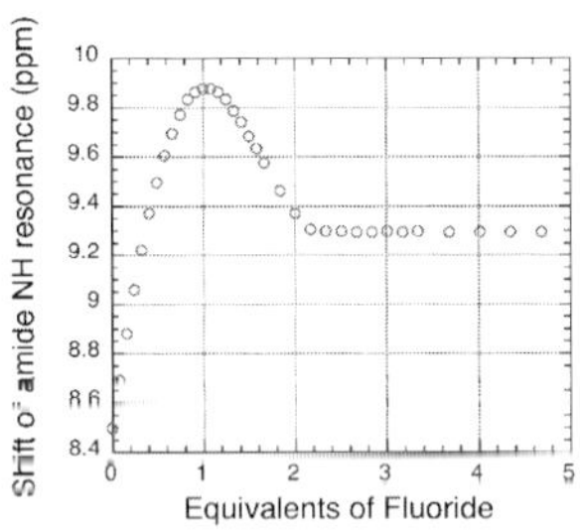

Figure 9. ^1H NMR titration curve for **11** vs. TBA.F in CD_2Cl_2.

Crystals of the tetrabutylammonium salt of (**11**-H^+) were obtained by slow evaporation of a dichloromethane solution of the receptor in the presence of excess tetrabutylammonium fluoride. X ray structure analysis revealed that the deprotonated pyrrole was stabilized via the formation of a ‘narcissistic’ dimer, in which the amide groups of a second deprotonated receptor molecule formed hydrogen bond interactions with the deprotonated pyrrolic nitrogen (*Figure 10*). In order to accommodate this arrangement, the two amide groups adopt the *syn-syn* arrangement, whilst the two receptor molecules arrange themselves in an orthogonal arrangement in space relative to one another, so minimizing the steric interactions between phenyl groups.

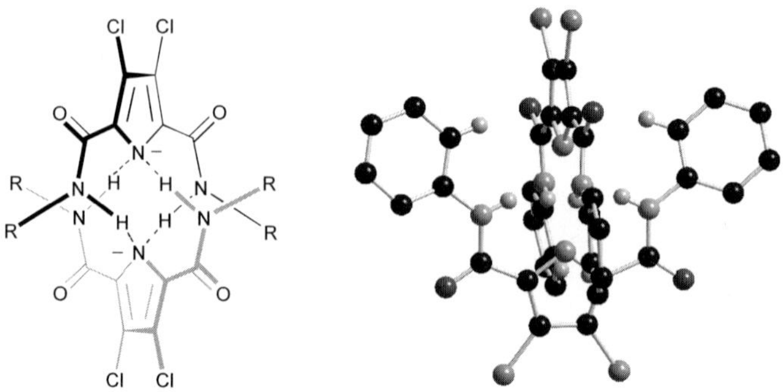

Figure 10. Schematic representation and X-ray crystal structure of the Narcissistic interlocking deprotonated dimer of **11**. (Non-acidic hydrogen atoms and counter cations are omitted for clarity).

By combining two bis-amido-3,4-dichloropyrroles into one molecule, bridged by an inflexible linker unit a new class of interlocking anionic polymer has been produced (*Figure 11*).[11] Compounds **12** and **13** were synthesized and exhibited limited solubility in a variety of solvents although addition of tetrabutylammonium fluoride resulted in an increase solubility to a point where full characterization could be determined.

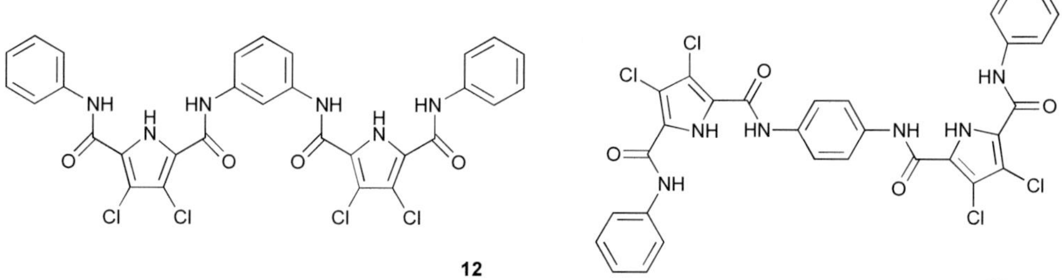

12

13

Figure 11. Compounds **12** and **13**.

Crystals of both compounds were obtained by slow evaporation of acetonitrile solutions in the presence of excess tetrabutylammonium fluoride. It was determined that both compounds crystallized as the tetrabutylammonium salts of the doubly deprotonated pyrrole anions (*Figure 12*). Each individual dianion essentially possesses a planar conformation and forms the interlocking arrangement previously seen (*Figure 10*).

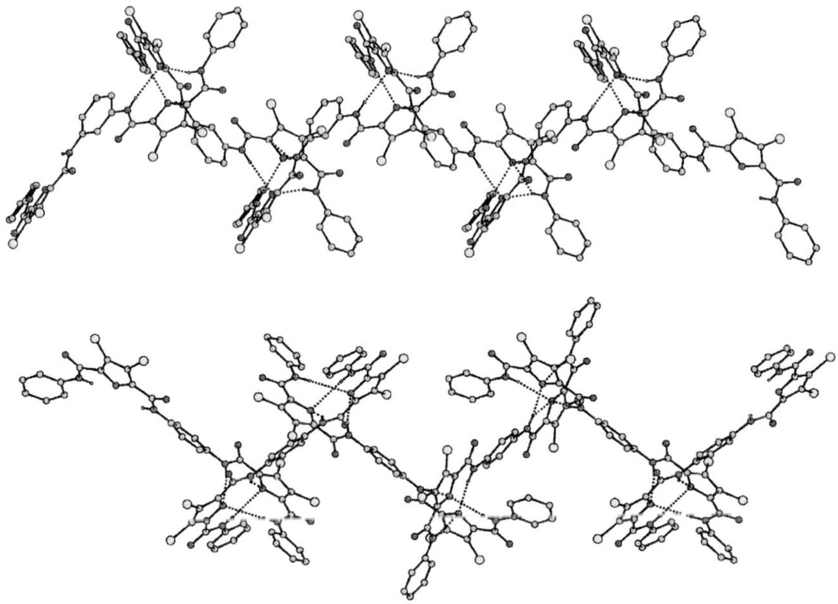

Figure 12. Crystallographic representations of polymeric structure of **12** (top) and **13** (bottom). Reproduced with permission from *J. Am. Chem. Soc.* (2002) **124**, 11228. Copyright American Chemical Society 2002.

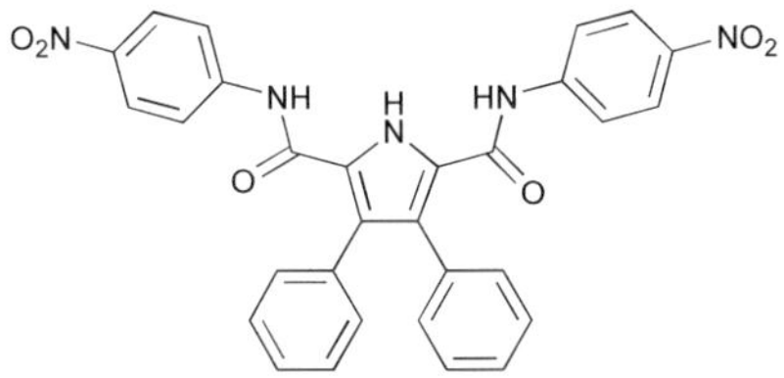

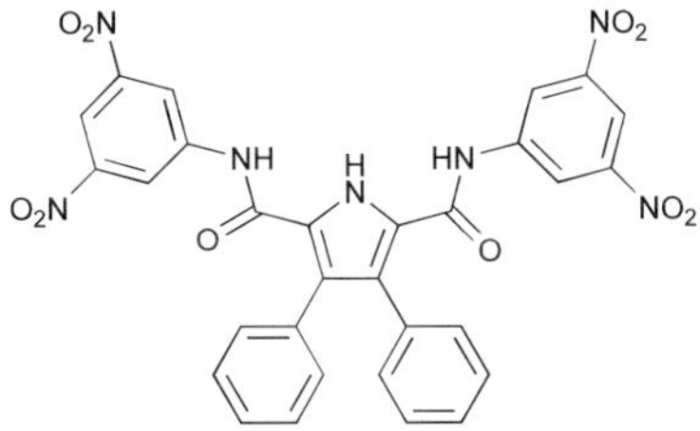

14 **15**

Figure 13. Mono- and dinitro-derived pyrrolic receptors.

After the discovery of the observed deprotonation of **12** and **13**, the effect of the introduction of electron-withdrawing amide substituents was investigated.[12] Synthesis of compounds **14** and **15** where achieved by reaction of 3,4-diphenyl-1*H*-2,5-dicarbonyl chloride with 4-nitro- and 3,5-dinitroaniline respectively, in dichloromethane in the presence of triethylamine and DMAP to yield the receptors in 43 and 11% respective yields (*Figure 13*). Comparison of the chemical shifts of the amide resonances of compounds **1**, **14** and **15** (9.36, 10.19 and 11.29 ppm) illustrate a trend indicative of the increasing degree of de-shielding of this proton as the number of nitro groups increases. Upon the addition of 10 equivalents of tetrabutylammonium fluoride to receptor **15** in acetonitrile, an intense colour change was observed resulting in the formation of a deep blue solution.

^{1}H NMR titrations performed on compound **14** in DMSO-d_6/0.5% H_2O for a variety of anions, show a general increase in association constants when compared to the phenyl derivative (fluoride is bound with a association constants of 1245 M^{-1} by **14** and 74 M^{-1} by **1**). When fluoride was titrated against **15** an unusual three-step titration curve was observed (*Figure 14*). This may be due to initial binding of the anion by the amide and pyrrolic NH, a secondary deprotonation step of the pyrrole NH, followed by binding in the cavity formed between the two amides and two of the ortho phenyl CH, which are acidic due to the presence of the electron withdrawing nitro- groups. X-ray crystallographic data appears to support this proposal. Crystals obtained via slow evaporation of an acetonitrile solution of receptor **15** with excess tetrabutylammonium fluoride. The species which crystallized was again a tetrabutylammonium salt of the deprotonated pyrrole however, in this case, an adventitious chloride anion was bound to the deprotonated pyrrole (*Figure 15*). The anion was observed to form hydrogen bonds with both amides and ortho positioned CH groups. Combining strongly electron withdrawing groups on an aryl amide, with an open cleft generated by the presence of five-membered core ring system can be clearly be seen to effectively produce a secondary anion binding site within receptors of this nature.

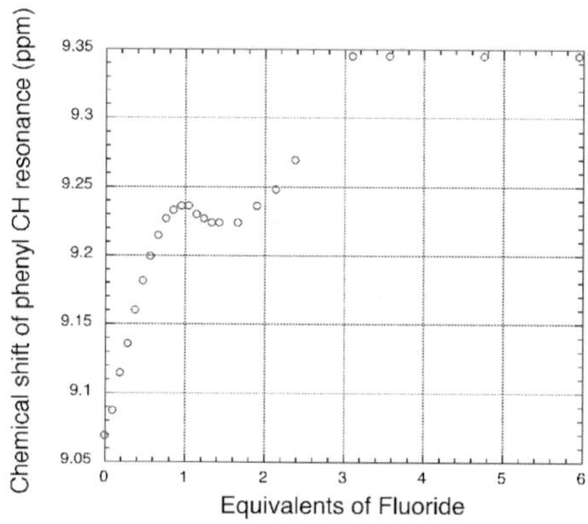

Figure 14. ^{1}H NMR titration curve of **15** vs. TBA.F. performed in DMSO-d_6/0.5% H_2O.

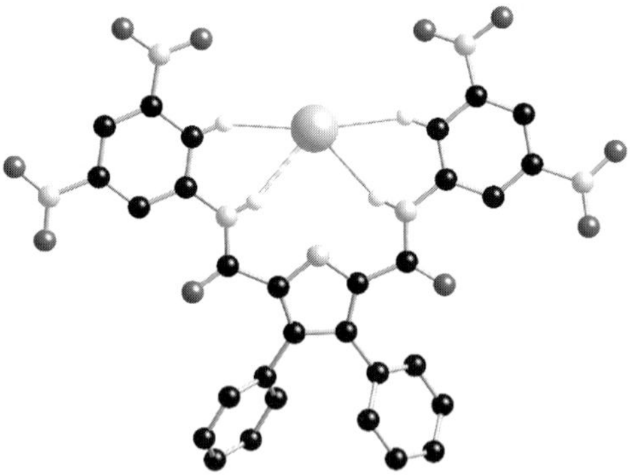

Figure 15. The X-ray crystal structure of the chloride complex of **15**-H^+ (tetrabutylammonium counter cations and non-acidic hydrogens are omitted for clarity).

By appending crown ether moieties to the amidopyrrole skeleton it had been shown that it is possible to both enhance the binding of anions to the cleft by the introduction of caesium cations but only with a limited degree of success (*Figure 16*).[13] Compounds **16** and **17** were synthesized by reaction of the 4′-aminobenzo-15-crown-5 with the 2-methyl-5-mono-acid chloride-3,4-diphenylpyrrole and the analogous 2,5-bis-acid chloride in 28 and 33% respective yields.

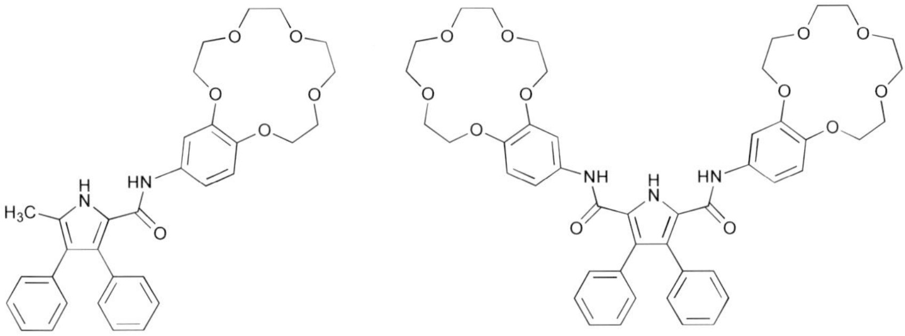

16 **17**

Figure 16. Mono and di-crown appended amido pyrrole salt receptors.

Unexpectedly introduction of sodium and potassium did not result in a significant increase the binding potential of these receptors to bind anions a result that might be expected if the receptor has effectively gained a positive charge. What appears to occur is that the ion-pairing effect between free ion in solution is greater a joint association through the receptor of both anion and cation. The only exception where the anion binding affinity of these receptors was observed when fluoride was titrated with receptor **17** in the presence of caesium ions in which the association constants increased from 67 to 308 M^{-1}, an increase by a factor of four. This may be due to the combined facts that caesium due to its large size and diffuse charge, is inefficient in forming ion pairs.

A more successful development of a salt receptor based upon the 2,5-diamidopyrrole motif was synthesized via a collaboration with Bradley Smith at The University of Notre Dame (*Figure 17*).[14] In the case of compound **18** the crown used to coordinate the cation was appended as a bridge between the amide groups. This configuration allows the anion and to bind as a contact ion pair. Stability constants of the receptor with chloride in DMSO-d_6 were determined in the absence of metal cations (109 M^{-1}), and in the presence of one equivalent of sodium (128 M^{-1}) and with potassium (540 M^{-1}). Interestingly, the addition of cations appears to substantially improve subsequent anion complexation and the highest association in the presence of potassium reflects the association constants of cations with the appropriate crown size. It was also found that the presence of the bridging group improved the association relative to the corresponding acyclic diphenyl receptor **1** by an order of magnitude, reflecting the increase in the pre-organized structure of the receptor.

Crystallographic evidence of the contact ion pairing of sodium and chloride in the presence of the receptor **18** (*Figure 18*) clearly shows the two ions within contact distance (2.65 Å) and interacting with the crown and the cleft respectively.

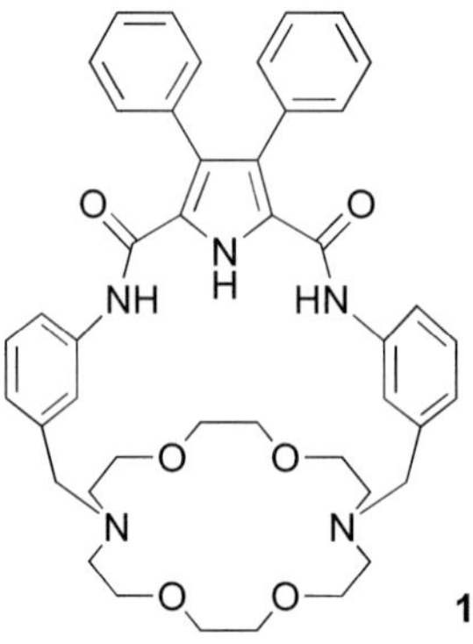

18

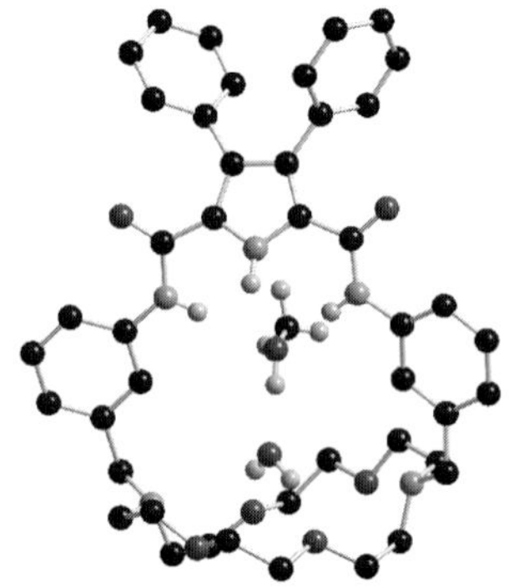

Figure 17. Contact ion-pair salt receptor schematic and crystallographic representation (X-ray structure shows receptor with methanol and water bound. Non-acidic receptor hydrogen atoms removed for clarity).

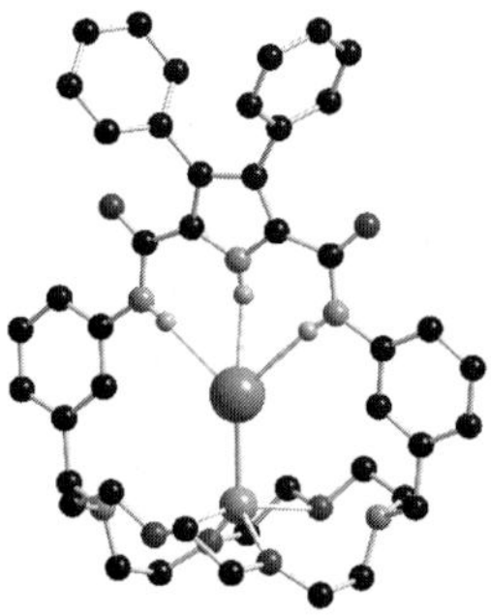

Figure 18. Crystallographic representation of **18** with bound NaCl (non-acidic receptor hydrogens removed for clarity).

A more recent development has been the synthesis of bisamidopyrrolylmethane based anion receptor systems (*Figure 19*).[15] These receptors might be regarded as containing 'half a calix[4]pyrrole' combined with the 2-amido appendages in common with the pyrrolic amide cleft. Compounds **19** and **20** were synthesized by reaction of diethyl-5,5′-methylenebis(4-ethyl-3-methyl-2-pyrrole) carboxylate with aniline or n-butylamine in the presence of trimethylaluminium in dry dichloromethane at 35°C in 40 and 43% respective yields.

Initial ^{1}H NMR titration data performed in DMSO-d_6/5%H_2O revealed that whilst all other anions (chloride, bromide, hydrogen sulphate and benzoate) could be fitted to a 1:1 binding model, association constants could not be reliably calculated for fluoride and dihydrogen phosphate due to the sharp curve obtained. Consequently for these anions, titrations were repeated in a more competitive solvent mix of DMSO/25%H_2O, providing stability constants with **19** of 114 and 234 M^{-1} and with **20** 11 and 20 M^{-1} respectively.

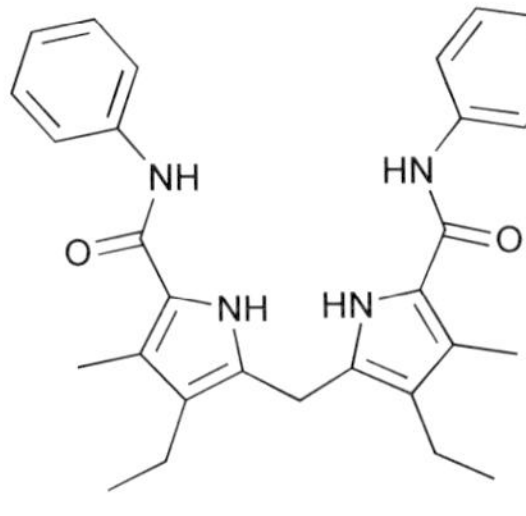

19

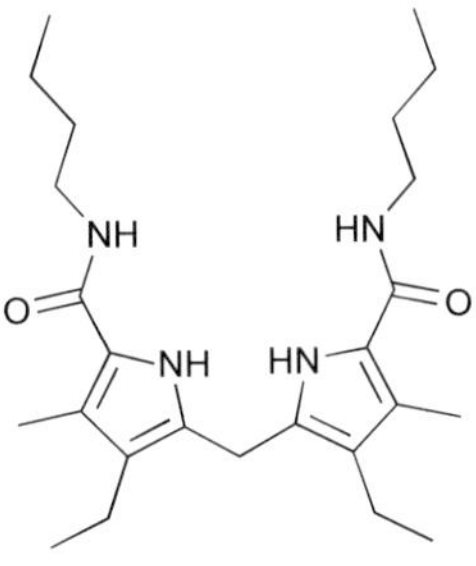

20

Figure 19. Bis-pyrrolylmethane based anion receptors.

Unfortunately it was discovered that after a few days in solution, a red discolouration was observed (particularly with compound **19**). Mass spectrometry revealed the loss of two mass units, consistent with oxidation of the product to the dipyrromethene.

In order to overcome this oxidation problem the synthesis of **21** and **22** was achieved in which two methyl groups occupy the positions bridging the two pyrrolic groups (*Figure 20*).[16] Although compounds **21** and **22** were found to be stable, less stable complexes with anions were formed than with the first generation dipyrrolylmethane systems. However, despite this reduction in affinity, the stability constants for these receptors could only be calculated in DMSO-d_6/5% H_2O for fluoride (124 and 89 M^{-1}), dihydrogen phosphate (1092 and 81 M^{-1}), benzoate (1092 and 81 M^{-1}) calculated for receptors **21** and **22** respectively.

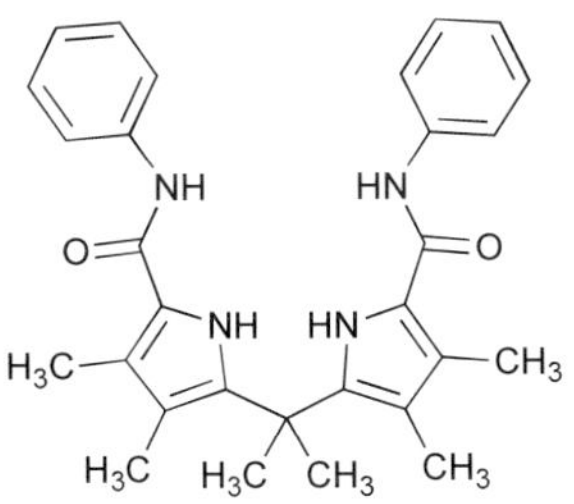

21

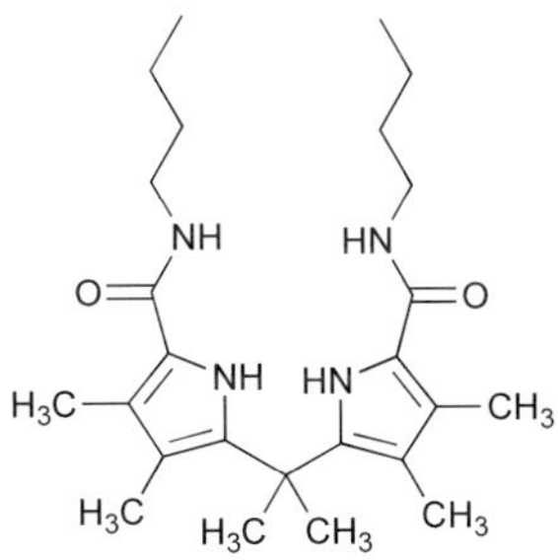

22

Figure 20. Second generation diamidodipyrrolylmethanes.

3. Guanidiniocarbonylpyrroles

Schmuck and coworkers have developed a range of guanidiniocarbonylpyrroles that have been shown exhibit exceptionally high stability constants, in competitive solvents, for binding carboxylate groups.

In early work, Schmuck synthesised the simple 2-guanidiniocarbonylpyrrole **23** (*Figure 21*) by the reaction of ethyl-1*H*-pyrrole-2-carboxylate with guanidinium hydrochloride in the presence of sodium methoxide affording the receptor in 52% yield.[17]

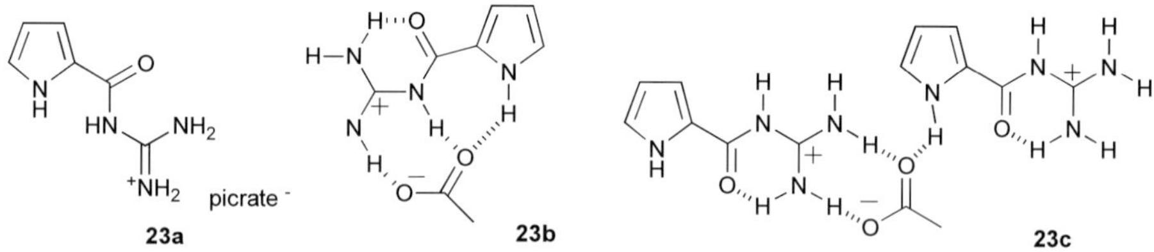

Figure 21. 2-Guanidiniocarbonylpyrrole with proposed solution and solid-state acetate binding modes.

^{1}H NMR titrations revealed that association of **23** with acetate was so strong that accurate determination of the association constants became difficult (in DMSO-d_6/50% H_2O association constant of the order of 10^3 M^{-1}). A simple guanidinium cation titrated in the same conditions showed no signs of association, whilst an acetyl derived guanidinium cation produced association constants of around three times lower.

This effect shows that the pyrrole NH is indeed involved in complexation of carboxylates and comparison of all of the NH signals in **23** upon addition of anion reveal downfield shifts, indicative of the formation of a tridentate binding mode (figure 23b).

Solid state analysis of crystals obtained of receptor with acetate from water/methanol solutions reveal that once again the guanidiniocarbonylpyrrole cation exists in an extended conformation however in contrast to the proposed solution state model, the pyrrole NH points away from the guanidium NHs with the acetate anions bound by two separate receptor groups with solvent molecules helping to create a series of 2D arrays (figure 23c).

The binding of anionic substrates has been further investigated with examples detailing the complexation of short chain peptides[18,19] and asymmetric carboxylates.[20]

Many of Schmuck's guanidiniocarbonylpyrroles show an exceptional ability to self-associate in solution through a combination of ion pairing and mutual hydrogen bonding due to the complementary fit of 2,5- substituted heteroditopic charged species for dimerization or the 2,4- substituted systems for the formation of chain like oligomers.

By the synthesis of **24** the advantageous introduction of charged groups onto the pyrrole skeleton causes the main interaction between discrete molecules to now be ion pairing (*Figure 22*).[21] As a result of this the self-association process can now occur in more highly competitive solvent mixtures.

The result of the positioning of the groups in the 2 and 5 positions allows dimerization of discrete molecules to occur through the complementary geometrical arrangement of the zwitterionic guanidiniocarbonyl and carboxylate groups.

Whilst the fully protonated version of the compound **24a** was readily soluble in methanol and water, solubility problems of the neutral zwitterionic form **24b** meant that analysis could only be performed in DMSO-d_6 and then only up to concentrations of around 5mM. Further deprotonation of the guanidine residue by treatment of a second equivalent of base allows the formation of the anion **24c** that is once again freely soluble in methanol and water.

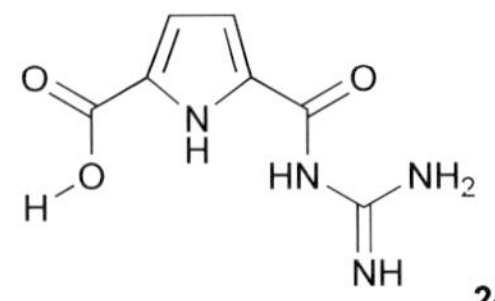

24

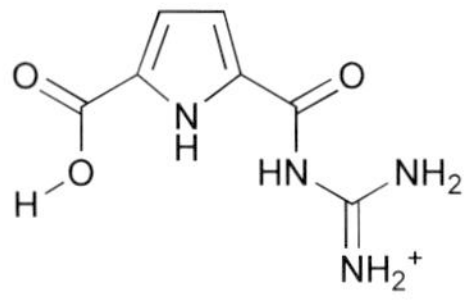

24a

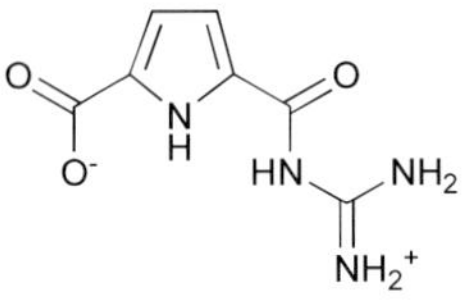

24b

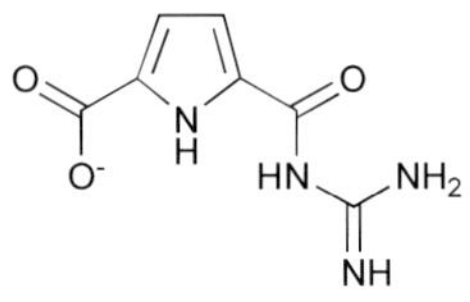

24c

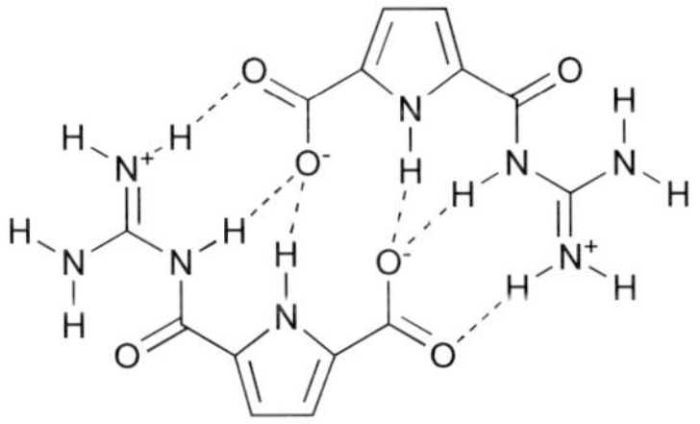

Figure 22. 5-(Guanidiniocarbonyl)-1*H*-pyrrole-2-carboxylate, its various ionic forms and proposed dimeric structure adopted in DMSO solution.

Analysis of the ^{1}H NMR of the protonated cation (in the form of the picrate salt) in DMSO-d_6 revealed the expected signals, guanidinium (8.2 ppm), carboxylic acid (13.2 ppm), amide (11.1) and pyrrole (12.7) signals were all broad resonances. As no splitting was observed due to the four guanidinium NH protons shows that in DMSO there is no interactions between these protons and the adjacent carbonyl group. Upon deprotonation of the acid group a markedly downfield shift can be observed. The guanidinium signal splits into two distinct signals at chemical shifts of (8.1 and 9.8 ppm) correspond to two NH protons each. This corresponds to the adoption of a more rigid structure. The drastic downfield shift of only two of the guanidinium protons indicates that only these protons are involved in hydrogen bonding, whilst the remaining two are not, consistent with dimer formation. Downfield shifts also observed of the amide and pyrrole protons (14.8 and 13.1 ppm respectively) indicative of hydrogen bond formation also supporting the formation of dimers in solution.

Due to the apparent very high stability constant of the dimer and the lack of solubility in more competitive media, estimation of the association constant of the dimer formation was made by comparing the association constants of various smaller components in which the strength of the individual interactions could be calculated and used to give an approximate value of the dimer association constant in DMSO-d_6 of 10^{12} M^{-1}.

Control over the structure adopted in solution by selective introduction of appropriate anions has been detailed by Schmuck.[22] Compound **25** was synthesized and it was observed that in DMSO discrete dimeric units could be observed by ^{1}H NMR and ESI mass spectrometry analysis. In the presence of halide anions, no discernable change was noted, implying that there was no effect upon the relative stability of the dimer. Introduction of picrate anions that have the ability to complete with the π-stacking interaction, results in a break up of the dimer and the formation of a 1:1 host:guest system (*Figure 23*).

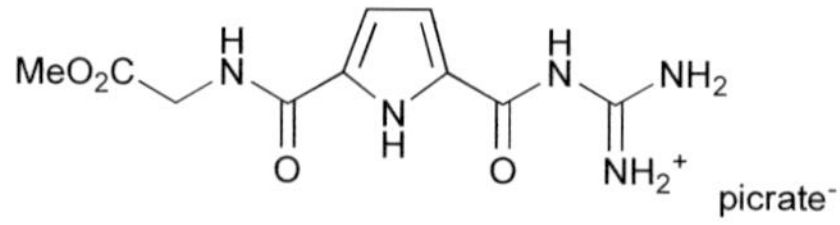

25

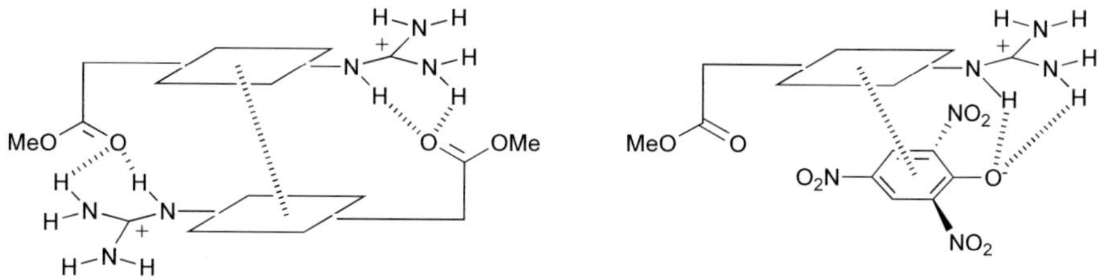

Figure 23. Picrate disrupts the dimerization of **25** in DMSO-d_6

By studying the concentration dependence of the ^{1}H NMR spectra performed in DMSO-d_6 the association constant of the dimer could be calculated at 673 M^{-1} at 298 K. This value is surprisingly high for a cationic species in a polar solvent such as DMSO and is indicates that the binding within the dimer is not the result purely of hydrogen bonding, with π-stacking and hydrophobic interactions potentially important.

The structure of the dimer was elucidated using ROESY NMR measurements performed in DMSO-d_6 with the guanidioniocarbonyl group appearing to be in an orientation in which the adjacent carbonyl group is in plane with the pyrrolic NH, as only NOE interactions observed with the pyrrolic CH. The orientation of the amino acid arm could not be determined due to the presence of various NOE interactions along the chain.

Using data obtained from the ROESY experiments confirming the conformation adopted by the guanidinum residue molecular modelling experiments were performed which lead to structural elucidation of the dimer. According to this model the planar aromatic parts of the molecules form π-stacks and although the amino acid amide is not involved in forming the dimer, is probably extremely likely to participate in anion coordination. The importance of the formation of π-stacking as a stabilising force of the dimer in the presence of anions can be seen when the ^{1}H NMR spectrum of the picrate salt is examined, revealing no signs of dimerization has occurred, the result of anion binding of the picrate anion in solution.

Positioning of a carboxylate group in the 4-position (*Figure 24*) as opposed the 5-position as seen with **25**, the preference for dimerization formation can be overcome and allow oligomerization to become the dominant structure in solution.[23]

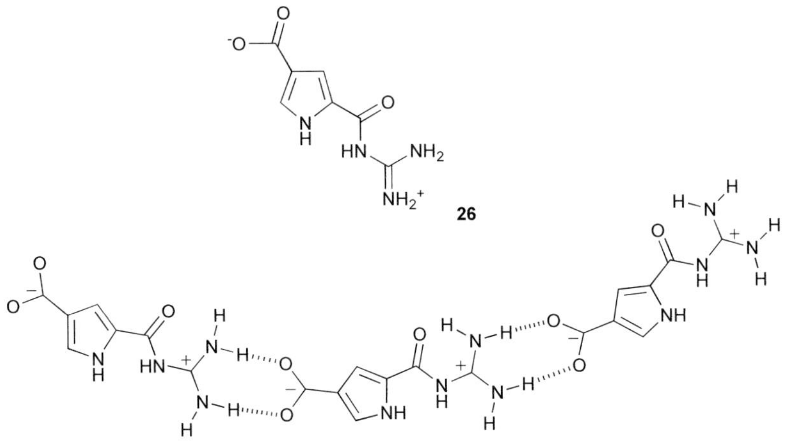

Figure 24. 2,4-Substituted systems form linear coordination polymers.

In contrast with previous compound, the ^1H NMR spectrum of **26** does not show any signs of any intermolecular interactions at sub millimolar concentrations in DMSO-d_6, however there is indications of concentration dependent shifts that are consistent with the formation of linear aggregates in solution with interactions between the carboxylate and guanidinium moieties of neighbouring molecules.

The binding constant of this process was determined by analysis of the concentration dependence observed in the ^1H NMR spectrum in DMSO-d_6 through the concentration range of 1-100 mM with the guanidinum NH resonance followed. A plot of the observed chemical shift against concentration gives an isothermic binding curve that indicates a concentration dependent intermolecular association is occurring. As the pyrrole NH is not involved in the carboxylate binding mode due to geometric reasons this compound binds the carboxylate in a bidentate fashion, as opposed to the tridentate fashion seen previously and in this respect exhibits association constants that are constant to those displayed by simple guanidinium cation binding. The association constant of 22.2 M^{-1} is consistently smaller with the bidentate verses tridentate binding mode argument and is in agreement with similar values that are quoted in the literature.

Examination of the association process over a range of temperatures from 303 to 363 K using an NMR dilution experiment in which once again the guanidinium resonances were followed gave association constants that varied from 22.2 M^{-1} at 303 K to 70.5 M^{-1} at 363 K. Using a van't Hoff plot of the calculated binding constants, association in the temperature range studied was revealed to be endothermic and as a result the process driving oligomerization must therefore be entropic, the process presumablely be driven by the release of solvent molecules from the binding site increasing the entropy of the system.

Compounds **25** and **26** both demonstrated the use of intermolecular interactions for the formation of solution phase structures, however by combining carboxylate and guanidinium residues into the same molecule which sufficient flexibility in which the two groups can come into contact with one another has allowed Schmuck and co-workers to form self folding, well-defined stable loops in DMSO.[24] As such these molecules might present a way to access macrocyclic structures that in which the structure, and therefore the resulting properties, may be controlled by affecting the polarity of the solvent (*Figure 25*).

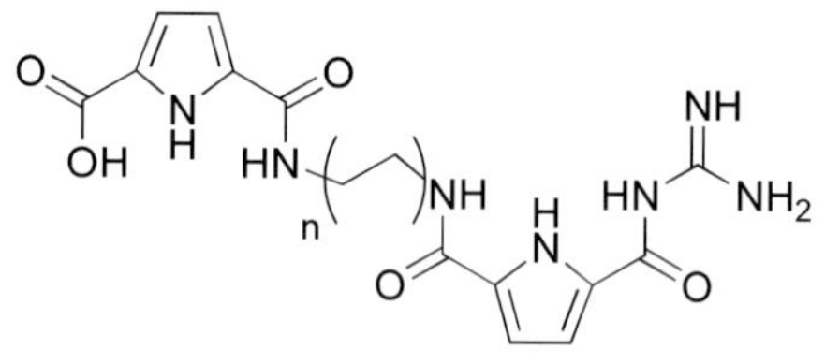

27 n = 1
28 n = 2

Figure 25. Carboxylate-guanidinium containing loop forming compounds.

Compounds **27** and **28** were synthesized in order to demonstrate possible self-association between carboxylate and guanidinium residues and thus the formation of folded structures in solution. Determining the length of the linker between the two pyrrolic centres allows control over the creation of loops in solution to be exerted upon the system, the ethylene linker in **27** being too short to facilitate intramolecular folding of the molecule whilst the butylene linked **28** possesses extra flexibility and thus should allow the complementary groups to come into contact with one another.

^{1}H NMR spectra performed in *DMSO-d6* of the associated zwitterions of the two compounds are markedly different, with the **27** displaying the typical guanidinocarbonyl pyrrole spectra previously seen when no intermolecular/intramolecular interactions are present, i.e. all signals are broad and relatively upfield with only one signals corresponding to the guanidinium protons. The ^{1}H NMR spectrum of **28** displays completely different behaviour in the same conditions, with the guanidinium protons split into two separate signals each now comprising of two protons, one of which displays a significant downfield shift relative to the corresponding chemical shift observed with **27**. Also both the pyrrole and amide display significant downfield shifts, with the amide shifting 3 ppm downfield. In theory this behaviour could be the result of either an intra- or intermolecular complexation process, however as this behaviour is not seen with **27** it is more likely that an intramolecular association is taking place.

The intramolecular self-association of **28** has been conformed by ROESY NMR experiment that conform that the amide NH at the carboxylate terminus shows NOE signals to all four CH_2 groups of the linker and the neighbouring pyrrole NH, indicating that folding of the molecule must occur in order to bring these groups close enough for the observed interactions to occur.

4. Macrocyclic Systems

Sessler has combined amidopyrroles with the ferrocene moiety has resulted in new macrocyclic anion receptors in which the anion-binding cavity could be adjusted in size by variation of the bridging alkyl chain to allow selectivity of this class of receptor to be fine-tuned (*Figure 26*).[25]

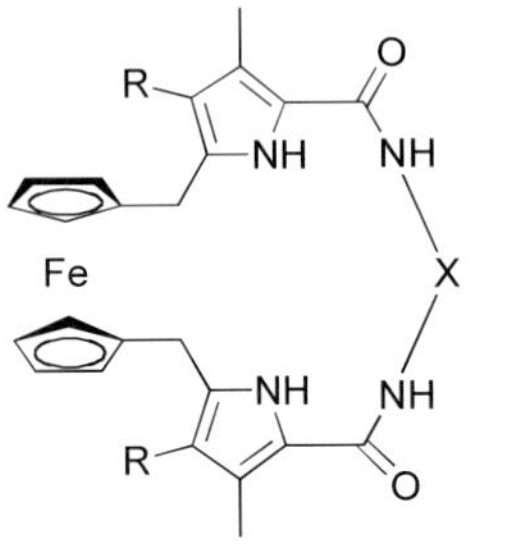

29 R = CH_2CH_3, X = $CH_2(CH_2OCH_2)_2CH_2$
30 R = CH_3, X = $(CH_2)_5$
31 R = CH_3, X = $CH_2CH_2OCH_2CH_2$
32 R = CH_3, X = $CH_2(CH_2OCH_2)_2CH_2$

Figure 26. Pyrrole containing *ansa*-ferrocenes.

A crystal structure was obtained of **29** in which a water molecule was observed to be contained in the cavity, with a second water molecule associated to it revealed hydrogen-bonding interactions between the water oxygen and the pyrrole and amide NH groups, and also between the water OH and the ethylene oxide oxygen atom. This demonstrated a possible binding mode with dihydrogen phosphate in which alternate hydrogen bond donor/acceptor groups in both the host and guest could result in a high degree of association.

Subsequently three derivatives were synthesized in which the length of the bridging arm was varied to determine the relative effect of chain length upon binding of dihydrogen phosphate.

Association constants for **30-32** were determined by ^1H NMR titrations performed in dichloromethane-d_2/2% DMSO-d_6 and revealed a general increase in the binding affinity upon increasing numbers of oxygen atoms in the ethylene oxide based linker. Compound **30** that possesses no hydrogen bond acceptor sites had a stability constant of 4050 M^{-1}, whilst compound **31** which contains one oxygen atom had a stability constant of 13200 M^{-1}, and compound **32** which contains two oxygen atoms has a stability constant of 81400 M^{-1}. This substantial increase relative to the number of oxygen atoms in the chain suggests that the oxygen atoms present in the introduced chains are actively involved in the complexation.

Square wave electrochemical analyses confirmed that the affinity of the receptors for dihydrogen phosphate increase with increasing numbers of oxygen atoms in the bridging chain, however the fact that there is no observed increase in the electrochemical response between one and two oxygen atoms indicates that in these two receptors, complexed dihydrogen phosphate occupies the same location within the cavity, relative to the ferrocene moiety.

Whilst although not technically an amidopyrrole in the sense that the previous examples, macrocycle **33** synthesised by Sessler, Ustynuk and co-workers displays a high sulfate-to-nitrate selectivity (*Figure 27*).[26]

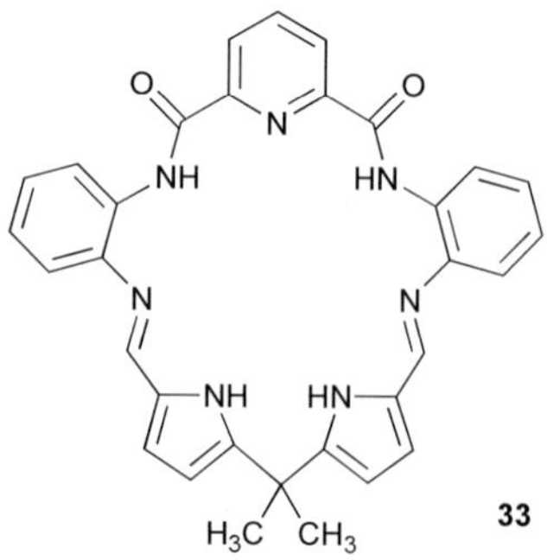

33

Figure 27. Diamidodipyrromethane macrocycle that displays high sulphate to nitrate association.

The pyrrole units in this macrocycle are arranged in a manner identical to examples **21** and **22**, however instead of appended amides, imides are present. An amide cleft is present in this receptor in the form of the pyridine 2,5-diamide component in which the favourable geometry of the amides is reinforced through the presence of the pyridine nitrogen atom. The anion complexation of this receptor was measured by UV/vis titrations performed in acetonitrile and it was discovered that **33** bound sulfate strongly in a 1:1 fashion with an association constant of 64,000 M^{-1} whilst there was no observed affinity for nitrate. Strong association constants were also observed for dihydrogen phosphate (342,000 M^{-1}), acetate (38,000 M^{-1}) and cyanide (12,000 M^{-1}). Dihydrogen phosphate bound in a 2:1 anion receptor stoichiometry with subsequent DFT calculations revealing the presence of a secondary binding site that can be formed upon complexation of one equivalent of anion as a result of the conformational shape adopted by the receptor upon binding.

Catenanes containing bipyrroleamides have successfully demonstrated anion-binding characteristics with selectivity observed for dihydrogen phosphate **34** (*Figure 28*).[27]

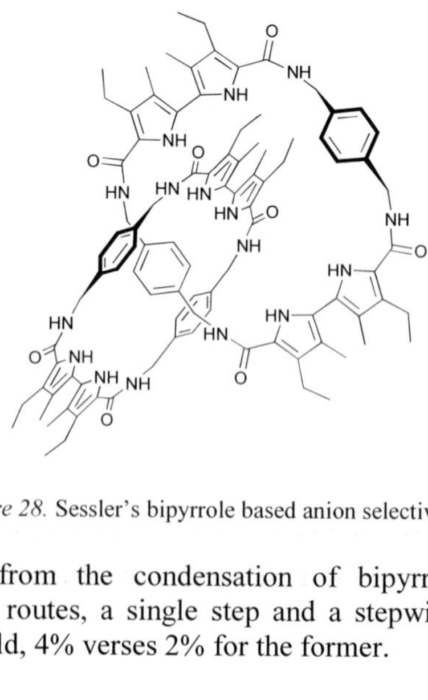

34

Figure 28. Sessler's bipyrrole based anion selective catenane.

Formed primarily from the condensation of bipyrrole-diacid chloride and p-xylenediamine via two routes, a single step and a stepwise approach, with the latter resulting in a higher yield, 4% verses 2% for the former.

The conformational properties of catenane **34** were investigated by ^{1}H NMR spectroscopy in order to more fully understand the dynamics of the system. In a range of investigated solvents, at ambient temperature, it was observed that broad peaks were observed which indicates both the effects of intramolecular hydrogen bonding as well as the dynamics associated with various circumrotation processes. Varying the temperature of the solution could also change the behaviour of the catenane. At higher temperatures it was determined that signals generally broadened and shifted and in the case of the xylene protons coalesced.

Titration of fluoride into a 1,1,2,2-tetrachloroethane-d_2 solution of the catenane revealed through NMR peak sharpening that the circumrotation process could in effect be "frozen out" after the addition of less than two equivalents of fluoride as the effect of anion binding contributions from each of the catenane macrocycles. It has been theorised that the binding of the fluoride could be occurring through the formation of a tetragonal anion-binding pocket that is comprised of a pyrrole and amide functional group from both of the macrocycles.

Titrations performed with a range of anions in 1,1,2,2-tetrachloroethane revealed that the association constants calculated for dihydrogen phosphate ($>1 \times 10^7$ M^{-1}), chloride (3.55×10^6 M^{-1}) and acetate (9.63×10^5 M^{-1}) proved to be higher than that calculated for fluoride (1.48×10^5 M^{-1}).

5. Conclusions

The amidopyrrole motif is proving to have wide applicability in the formation of new anion receptor species. It is clear that the use of this motif in macrocyclic systems is yet to be fully exploited. We are currently working on a variety of macrocyclic amidopyrrole containing species. The result of these studies will be reported in due course.

6. Acknowledgements

We would like to thank the EPSRC for a DTA studentship (to S.J.B) and for use of the crystallographic facilities at the University of Southampton. P.A.G. would like to thank the Royal Society for a University Research Fellowship.

7. References

1. Kavallieratos, K., de Gala, S. R., Austin, D. J., Crabtree, R. J. (1997) A readily available non-preorganized neutral acyclic halide receptor with an unusual nonplanar binding conformation, *J. Am. Chem. Soc.* **119**, 2325-2326.
2. Kavallieratos, K., Hwang, S., Crabtree, R. J. (1999) Aminoferrocene derivatives in chloride recognition and electrochemical sensing, *Inorg. Chem.* **38**, 5184-5186.
3. Gale, P. A., Camiolo, S., Chapman, C. P., Light, M. E., Hursthouse, M. B. (2001) Hydrogen-bonding pyrrolic amide cleft anion receptors, *Tetrahedron Lett.* **42**, 5095-5097.
4. Gale, P. A., Camiolo, S., Tizzard, G. J., Chapman, C. P., Light, M. E., Coles, S. J., Hursthouse, M. B. (2001) 2-Amidopyrroles and 2,5-diamidopyrroles as simple anion binding agents, *J. Org. Chem.* **66**, 7849-7853.
5. Camiolo, S., Gale, P. A., Hursthouse, M. B., Light, M. E. (2002) Confirmation of the 'cleft-mode' of binding in a 2,5-diamidopyrrole anion receptor in the solid state, *Tetrahedron Lett.* **43**, 6995-6996.

6. Hossain, M. A., Llinares, J. M., Powell, D., Bowman-James, K. (2001) Multiple hydrogen bond stabilisation of a sandwich complex of sulfate between two macrocyclic tetraamides, *Inorg. Chem.* **40**, 2936-2937.
7. Navakhun, K., Gale, P. A., Camiolo, S., Light, M. E., Hursthouse, M. B. (2002) Pendant arm pyrrolic amide cleft anion receptors, *Chem. Commun.* **18**, 2084-2085.
8. Denuault, G., Gale, P. A., Hursthouse, M. B., Light, M. E., Warriner, C. N. (2002) Anion complexation and elctrochemical behaviour of ferrocene-appended amido-pyrrole clefts, *New. J. Chem.* **26**, 811-813.
9. Beer, P. D., Gale, P. A., Chen, G.Z. (1999) Electrochemical molecular recognition: Pathways between complexation and signalling, *Dalton. Trans.* **12**, 1897-1909.
10. Camiolo, S., Gale, P. A., Hursthouse, M. B., Light, M. E., Shi, A. J. (2002) Solution and solid-state studies of 3,4-dichloro-2,5-diamidopyrroles: Formation of an unusual anionic narcisstic dimer, *Chem. Commun.* **7**, 758-759.
11. Gale, P. A., Navakhun, K., Camiolo, S., Light, M. E., Hursthouse, M. B. (2002) Anion-anion assembly: A new class of anionic supramolecular polymer containing 3,4-dichloro-2,5-diamido-substituted pyrrole anion dimers, *J. Am. Chem. Soc.* **124**, 11228-11229.
12. Camiolo, S., Gale, P. A., Hursthouse, M. B., Light, M. E. (2003) Nitrophenyl derivatives of pyrrole 2,5-diamides: structural behaviour, anion binding and colour change signalled deprotonation, *Org. Biomol. Chem. 1.* **4**, 741-744.
13. Camiolo, S., Coles, S. J., Gale, P. A., Hursthouse, M. B., Tizzard, G. J. (2003) Crown ether appended amidopyrrole clefts, *Supramol. Chem.* **15**, 231-234.
14. Mahoney, J. M., Andrew Marshall, R., Beatty, A. M., Smith, B. D., Camiolo, S., Gale, P. A. (2001) Complexation of alkali chloride contact ion-pairs using a 2,5-diamidopyrrole crown macrobicycle, *J. Supramol. Chem.* **1**, 289-292.
15. El Drubi Vega, I., Camiolo, S., Gale, P. A., Hursthouse, M. B., Light, M. E. (2003) Anion complexation properties of 2,2-bisamidopyrrolylmethanes, *Chem. Commun.* **14**, 1686-1687.
16. El Drubi Vega, I., Gale, P. A., Hursthouse, M. B., Light, M. E. (2004) Anion binding properties of 5,5-dicarboxamido-dipyrrolylmethanes, *Org. Biomol. Chem.* (2004) DOI: 10.1039/B409115A.
17. Schmuck, C., Lex, J. (1999) Acetate binding within a supramolecular network formed by a guanidiniocarbonyl pyrrole cation in the solid state, *Org. Lett.* **3**, 1253-1256.
18. Schmuck, C., Heil, M. (2003) Peptide binding by one-armed receptors in water: Screening of a combinatorial library for the binding of Val-Val-Ile-Ala, *Chembiochem.* **4**, 1232-1238.
19. Schmuck, C., Bickert, V. (2003) N-alkylated guanidiniocarbonylpyrroles: New receptors for amino acid derivatives, *Org. Lett.* **5**, 4579-4581.
20. Schmuck, C. (2000) Carboxylate binding by 2-(guanidiniocarbonyl)pyrrole receptors in aqueous solvents: Improving the binding properties of guanidinium cations through additional hydrogen bonds, *Chem. Eur. J.* **6**, 709-718.
21. Schmuck, C. (1999) Highly stable self association of 5-(guanidiniocarbonyl)-1H-pyrrole-2-carboxylate in DMSO - The importance of electrostatic interactions, *Eur. J. Org. Chem.* **9**, 2397-2403.
22. Schmuck, C., Heil, M. (2001) Anion-dependent dimerisation of a guanidiniocarbonyl pyrrole cation in DMSO, *Org. Lett.* **3**, 1253-1256.
23. Schmuck, C. (2001) Self-assembly of 2-(guanidiniocarbonyl)-pyrrole-4-carboxylate in dimethyl sulfoxide: an entropy driven oligomerisation, *Tetrahedron* **57**, 3063-3067.
24. Schmuck, C. (2000) Self-folding molecules: A well defined, stable loop formed by a carboxylate-guanidinium zwitterion in DMSO, *J. Org. Chem.* **65**, 2432-2437.
25. Sessler, J. L., Zimmerman, R. S., Kirkovits, G. J., Gebauer, A., Scherer, M. (2001) Synthesis and dihydrogen phosphate binding properties of pyrrole containing ansa-ferrocenes, *J. Organomet. Chem.* **637**, 343-348.
26. Sessler, J.L., Katayev, E., Pantos, G. D., Ustynuk, Y. A. (2004) Synthesis and study of a new diamidodipyrromethane macrocycle, An anion receptor with a high sulfate-to-nitrate binding selectivity. *Chem. Commun.* **11**, 1276-1277.
27. Andrievsky, A., Ahuis, F., Sessler, J. L., Vögtle, F., Gudat, D., Moini, M. (1998) Bipyrrole-based [2]catenane: A new type of anion receptor, *J. Am. Chem. Soc.* **120**, 9712-9713.

11.
STRUCTURAL ASPECTS OF HALIDES WITH CRYPTANDS

MD. ALAMGIR HOSSAIN, SUNG OK KANG AND
KRISTIN BOWMAN-JAMES
Department of Chemistry, University of Kansas
Lawrence, Kansas 66045 U.S.A.

1. Introduction

Anion coordination chemistry is a growing field within the inorganic and organic communities [1,2]. Findings compiled over the last several decades show that anions form complexes with receptors (ligands) *via* hydrogen bonding interactions as opposed to the coordinate covalent bonds operable in transition metal coordination. As binding and structural data accumulate, more informed strategies are being devised to achieve selective binding of anions of different topologies. Nonetheless, in terms of corollaries with transition metal ions, the spherical anions, namely the halides, present the simplest of topologies and the closest structural analogies. In fact, the first synthetic anion receptors were the halide binding bicyclic receptors, **1**, known as *katapinands,* and reported by Park and Simmons in 1968 [3] (*Figure 1*). Several years later, crystallographic studies confirmed the encapsulation of chloride in $H_2\mathbf{1}^{2+}$ (n = 9) [4].

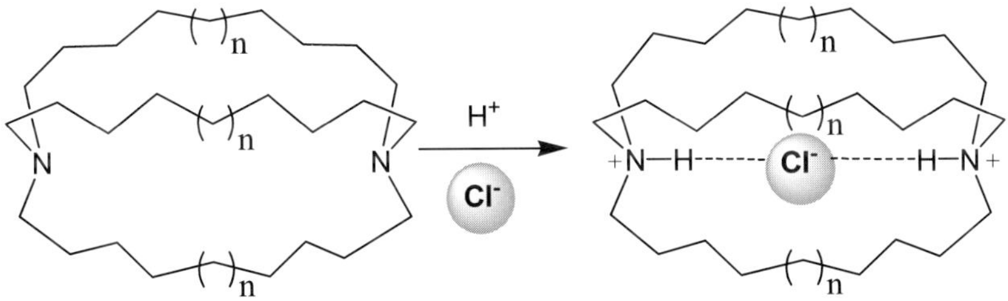

1 (n = 1, 2)

Figure 1. Encapsulation of chloride by diprotonated *katapinands*.

Receptors for anions vary widely, and are in many instances either the same or modified versions of the ligands that bind transition metal ions [5-20]. For example, poly*amine* macrocycles can bind transition metal ions, while poly*ammonium* macrocycles bind anions. In many instances the same ligand can bind either a transition metal ion [21-24] or an anion [25-29].

Karsten Gloe (ed.), Macrocyclic Chemistry: Current Trends and Future Perspectives, 173–188.
© 2005 *Springer. Printed in the Netherlands.*

That being the case, it is appealing to refer to anion receptors as ligands. However, traditionally the commonly understood definition of ligand has been a Lewis base capable of forming a coordinate-covalent bond with a metal ion. Since for anions, the "ligand" actually behaves as a Lewis acid, the definition needs to be expanded to include both Lewis acid and Lewis base behavior.

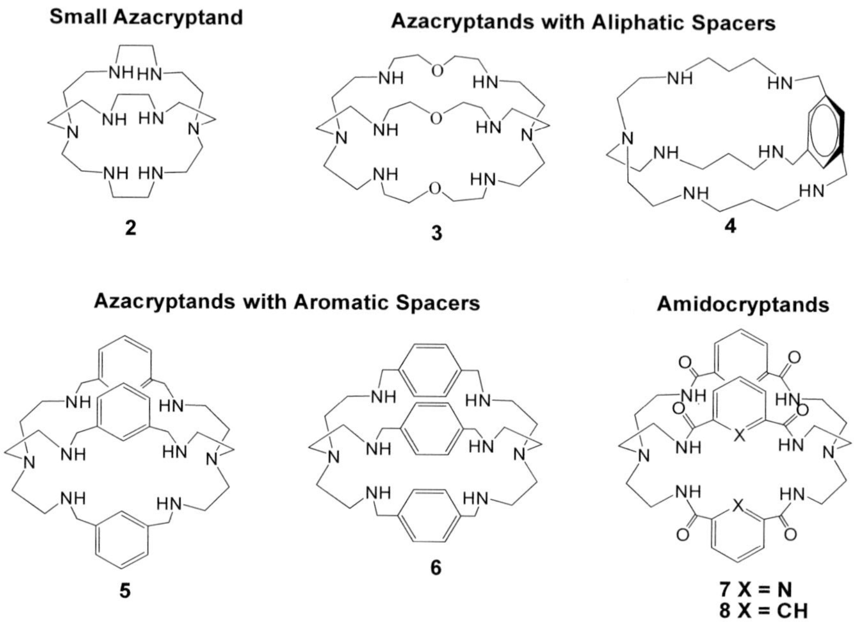

Figure 2. Azacryptands and amidocryptands of varying size and flexibility.

Over the last several years we have explored anion binding by systematically varying ligands with similar frameworks with respect to hydrogen bond donor, charge, and dimensionality [30-42]. We have examined polyammonium [30-38], polyamide [39-41], and polythioamide ligands [42]; monocycles and bicycles; and neutral and charged ligands. Crystallographic findings show that bicyclic ligands or cryptands, as coined first by Lehn [43], more frequently bind anions *via* encapsulation, while monocycles often do not. This chapter will compare structural effects of a series of cryptands, including some reported by us and some by others, ranging in size from very small and capable of encapsulating just the smallest of halides, to a size large enough to hold multiple species within the cavity. These ligands include the tiny "octaazacryptand," **2**; larger flexible azacryptands with aliphatic spacers, **3** and **4**; more "rigid" azacryptands with aromatic spacers, **5** and **6**; and as corollaries to the rigid azacryptands, the amidocryptands, **7** and **8** (*Figure 2*). Since this chapter focuses on encapsulated anions, most of the structures shown will include only the ligand and encapsulated anion.

2. Azacryptands

2.1 CHARACTERISTICS

Azacryptands tend to be highly water soluble, which makes them ideal for studying biological systems. However, affinities for anions vary depending on the degree of protonation. Binding usually commences only after several protons are added. For the systems described in this chapter, the onset of protonation normally occurs somewhere around pH 9-10, and the ligands are multiply protonated by pH 7. This results in significant anion binding capability at physiological pH, an essential criterion for biomimetic studies.

2.2 SMALL OCTAAZACRYPTAND, **2**

The small octaazacryptand, **2**, is selective for fluoride [44], with extremely high affinity, log K_a = 10-11 in aqueous solution [44-46]. As anticipated, the crystal structure of **2** with fluoride showed the halide to be encapsulated (*Figure 3*) [44]. Later theoretical and modeling studies indicated that the small size of the cavity could preclude encapsulation of larger anions [45].

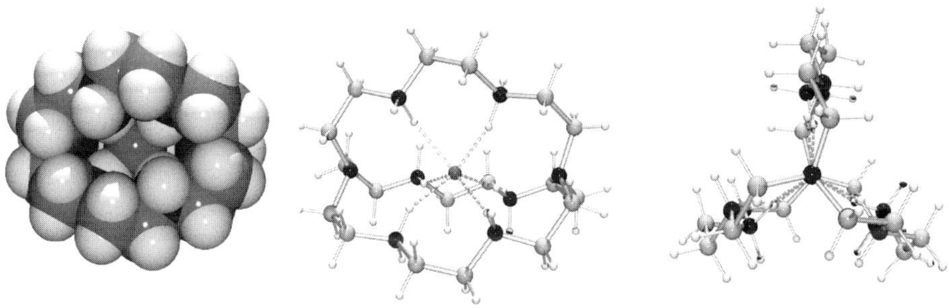

A **B** **C**

Figure 3. Crystal structure of the fluoride complex of **2** showing the encapsulated fluoride in the space filling model (**A**), in the perspective side view (**B**), and in the view down the pseudo-threefold axis (**C**).

More recent crystallographic studies on crystals isolated at pH < 2, indicated that **2** can also bind chloride internally (*Figure 4*) [44]. The internal chloride is hydrogen bonded with the six protons on the secondary nitrogens, with an average N-H···Cl distance = 3.10 Å [32,36]. The distance between the two bridgehead nitrogens is 6.59 Å, slightly shorter than that observed for the fluoride complex (6.65 Å) [44].

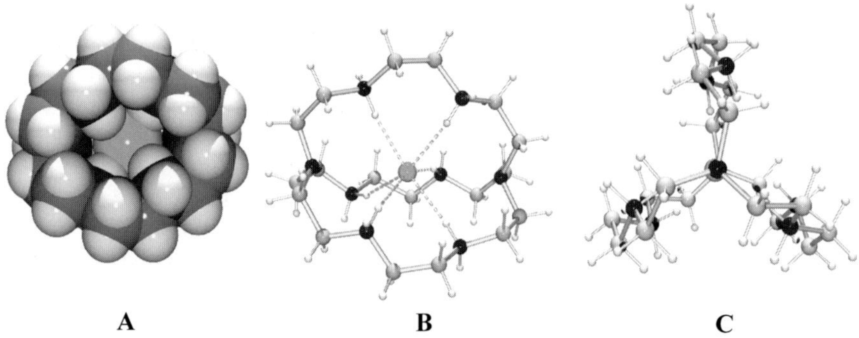

Figure 4. Crystal structure of the chloride complex of **2** showing the chloride in the space filling model (**A**), in the perspective side view (**B**), and in the view down the pseudo-threefold axis (**C**).

We also observed interactions of **2** with chloride during a routine NMR study. In a simple titration of a series of anions with **2**, large chemical shift changes were observed at pHs below 2.5 for chloride compared to the uncomplexed ligand [32,36], similar to what was observed for fluoride (*Figure 5A*). Other halides did not show these shifts. By examining the entire range of pHs, it can be seen that chloride resonances are not shifted from the signals for the uncomplexed **2** until pD 2.5, at which point a sharp downfield shift occurs (*Figure 5B*).

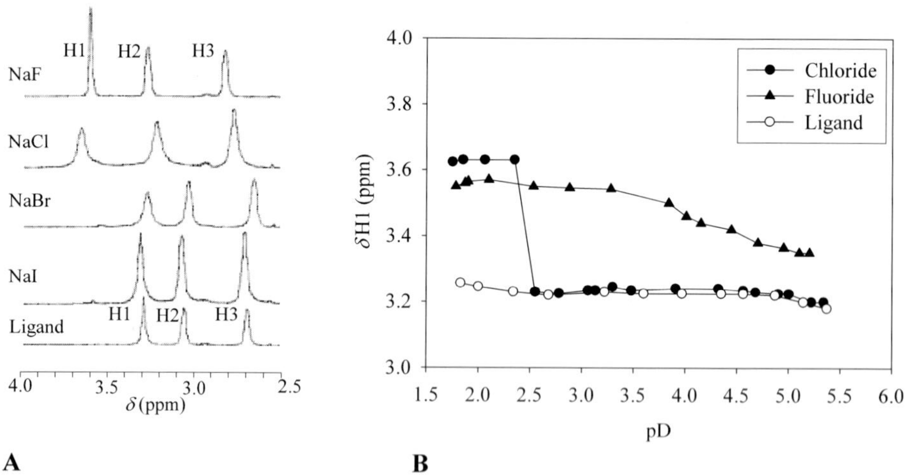

Figure 5. (**A**) ^1H NMR spectra of **2** with one equivalent of halide salt (NaX) at pD = 2.0 ± 0.1 in D_2O. (**B**) Titrations of **2** showing the change in chemical shift of the aliphatic protons (H1) for free and 1:1 adducts of chloride and fluoride with $[H_6\mathbf{2}][TsO]_6$. H1 = $HNCH_2$, H2 = NCH_2CH_2, H3 = NCH_2CH_2. The pD was adjusted with TsOH and NaOD.

Crystals grown from the iodide salt of **2**, indicate a molecule of water, not iodide, in the cavity. In this structure, the tetraprotonated ligand is surrounded by four iodides and four water molecules with an additional molecule of water inside the cavity [36]. The two lone pairs of the internal water molecule are hydrogen bonded to two of the protonated cryptand hydrogens, and the two hydrogens on the water are associated with the lone electron pairs of the two neutral secondary amines. The iodide ions sit outside the cavity and form hydrogen bonds with the remaining water molecules and amines of the ligand (*Figure 6*). The average hydrogen bond distance between the internal water and the amines is both NH···O and OH···N = 2.77 Å, and the distance between the bridgehead amines is 6.63 Å.

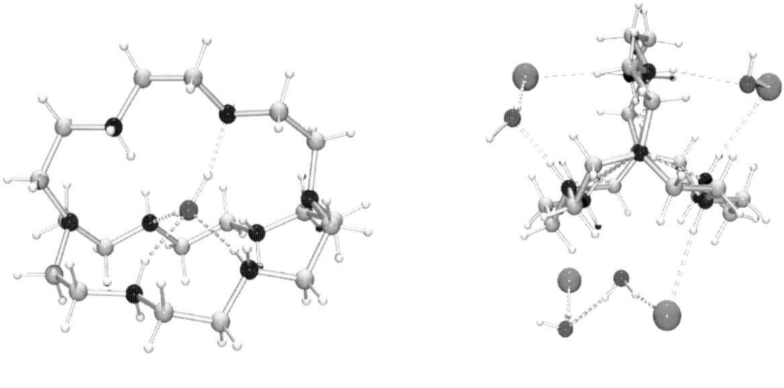

A B

Figure 6. Crystal structure of the iodide complex of **2** showing encapsulated water as viewed from the side (**A**), and additionally the external iodides and waters as viewed down the pseudo-threefold axis (**B**).

2.3 AZACRYPTANDS WITH ALIPHATIC SPACERS, 3 AND 4

Lehn and coworkers were the first to evaluate the binding of anions in the larger azacryptands [47]. After structurally characterizing the pseudo-halide azide complex encapsulated in the cryptand "bis-tren," **3**, they proceeded to examine the binding of other halides, including fluoride, chloride and bromide [45,48]. The structure of the fluoride complex revealed a single encapsulated fluoride located closer to one tren unit than the other, and a very distorted cryptand (*Figure 7A*). Disappointingly perhaps, there was no triatomic bifluoride, F-H-F$^-$, in the cavity that would resemble the azide structure. Instead, the single fluoride exhibits pseudo-tetrahedral coordination with four of the hexaprotonated cryptand amines. The cryptand distortion was attributed to the mismatch of size between the small fluoride and the relatively large cryptand [48]. The distance between the bridgehead amines is 7.66 Å.

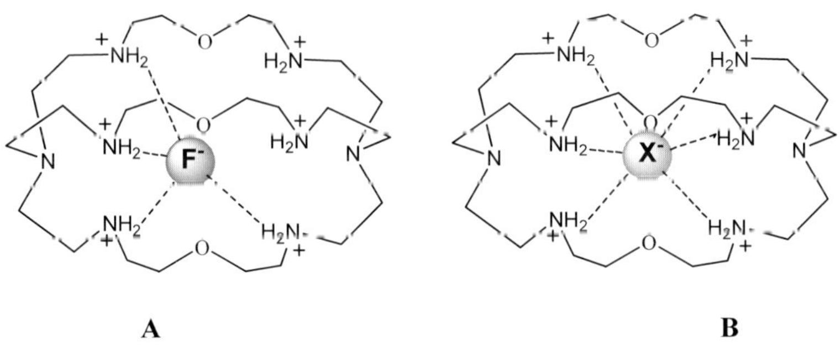

A B

Figure 7. Schematic representation of bis-tren with fluoride (**A**), and with chloride and bromide (**B**).

In both the chloride and bromide structures, the halide was also found to be encapsulated, but was positioned more centrally within the cavity, as explained by a better size match between the ligand and halide (*Figure 7B*) [48]. In both cases the cryptand is hexaprotonated and all six of the protonated amines exhibit hydrogen bonding with the internal halide, with average NH···X distances of 3.30 Å for chloride and 3.39 Å for bromide. The distance between the bridgehead amines is 7.40 Å and 7.50 Å for the chloride and bromide complexes, respectively.

Recently, Steed and coworkers reported a cryptand, **4** (*Figure* 2), linking a tren unit with an aliphatic spacer to a rigid tripodal aromatic phenyl group (*Figure 8*) [49]. The new ligand forms 1:1 inclusion complexes with fluoride, chloride, bromide, and iodide in the solid state, with all four structures displaying pseudo-octahedral coordination geometries. While it might be anticipated that structural characteristics of **4** would parallel those of bis-tren, the binding more closely approximated that observed in the small octaazacryptand, **2** [32,36,44-46]. The halide in the cavity forms hydrogen bonds with the three secondary amines of the tren group and three methylene hydrogens of the aliphatic spacers instead of amines closer to the tripodal aromatic group. This hydrogen bonding pattern results in six-membered chelate rings between the halides and the ligand as seen for **2** both with fluoride and chloride. Potentiometric studies indicated exceptionally high binding of $H_6[\mathbf{4}]^{5+}$ for fluoride (log K_a = 9.64 in water), almost as high as the binding constants observed for fluoride with **2**.

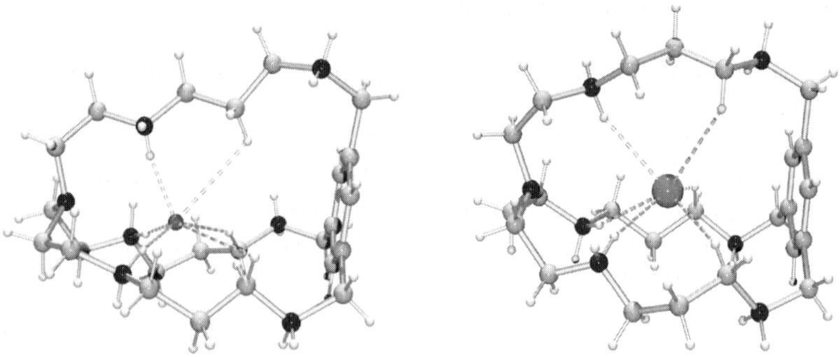

A B

Figure 8. Crystal structures of **4** with fluoride (**A**) and iodide (**B**) showing the single anion encapsulated in the cavity.

2.4 AZACRYPTANDS WITH AROMATIC SPACERS, 5 AND 6

A drawback to cryptands with aliphatic spacers such as **3** is that the synthetic routes are long and tedious, requiring time-consuming protection and deprotection techniques in order to accomplish the condensation reactions with reasonable yields. In the late 1980s however, a couple of groups realized that azacryptands could readily be obtained via Schiff base condensations between aromatic or heterocyclic dialdehydes and tren, followed by borohydride reductions [50,51]. This simple pathway opened the door for a number of new anion ligands, examples of which include **5** and **6**.

In the smaller cryptand, **5**, with the *m*-xylyl spacers, the structure of crystals obtained from the reaction of the cryptand with HF indicated a single fluoride in the cavity with a water molecule companion (*Figure 9*) [31,33]. The structure contained two bifluorides, but outside the cavity, in addition to seven waters and 1.5 SiF_6^{2-} ions. The internal fluoride is pseudo-tetrahedrally hydrogen bonded to the internal molecule of water plus to three of the protonated amines of one tren unit, with hydrogen bonds of 2.84 Å to the former and averaging 2.68 Å for the latter.

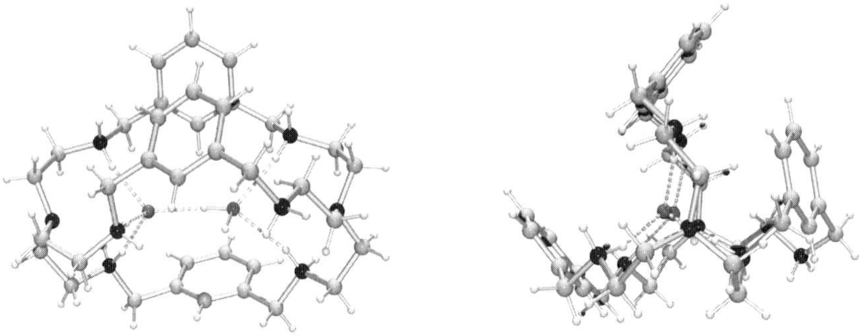

A B

Figure 9. Crystal structure of **5** with fluoride showing the side view (**A**) and view down the pseudo-threefold axis (**B**).

In an effort to examine the influence of cavity expansion, the fluoride complex of **6** with *p*-xylyl spacers was prepared. The resulting tritopic complex contained two fluoride ions, bridged by a molecule of water [35] (*Figure 10*). Such a complex is reminiscent of the widely studied cascade metal complexes, referring to the term coined by Lehn and coworkers in the late 1970s [52]. The two fluorides lie on the axis joining the bridgehead nitrogens, and the water molecule is slightly displaced from the axis with its hydrogen atoms linking the two fluorides. Each fluoride exhibits tetrahedral coordination with the three NH groups and the bridging water molecule, with N-H···F distances ranging from 2.60 – 2.72 Å. The distance between the bridgehead nitrogens of the cascade complex is 10.72 Å, which is larger than the corresponding distance (9.22 Å) in the fluoride complex of **5**.

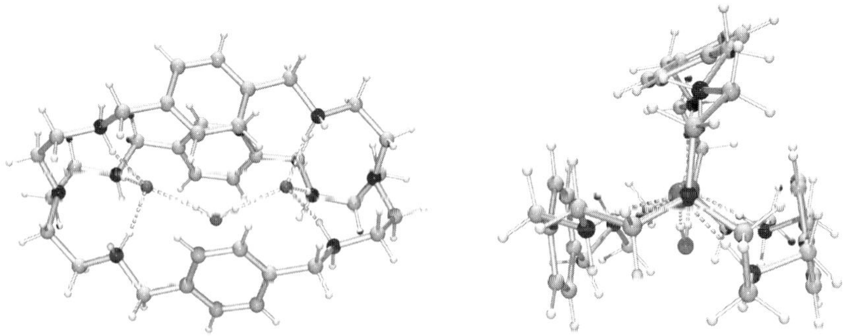

A B

Figure 10. Crystal structure of the fluoride cascade complex of **6** showing the side view (**A**) and view down the pseudo-threefold axis (**B**).

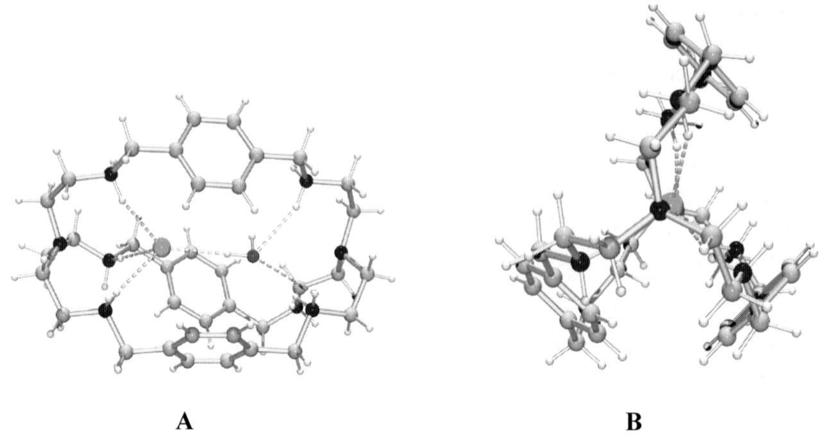

Figure 11. Crystal structures of the complex of **6** with chloride showing the side view (**A**) and view down the pseudo-threefold axis (**B**).

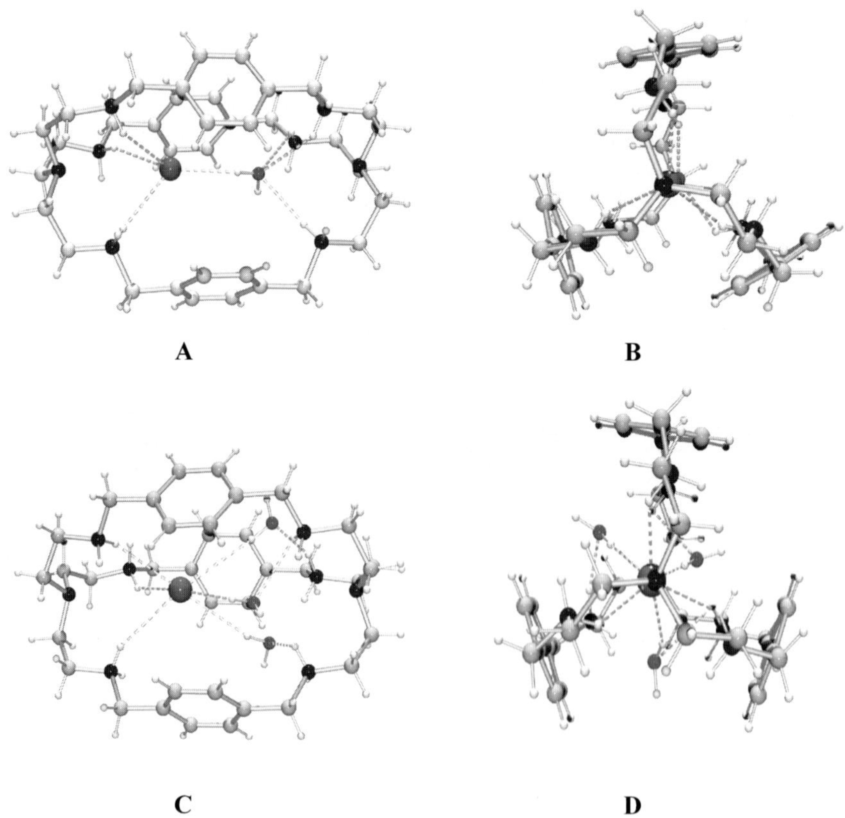

Figure 12. Crystal structure of the bromide complex with **6** showing the side (**A**) and end-on (**B**) views for the bromide and one water molecule, and the side (**C**) and end-on (**D**) views for the bromide and three water molecules in the two crystallographically independent cryptates.

The chloride structure with **6** indicated ditopic binding, with a chloride and water in the cavity (*Figure 11*), as in the fluoride structure with **5** (*Figure 9*). The bromide structure was slightly different from the chloride, exhibiting two crystallographically independent cryptate units within the unit cell. In one, the coordination is analogous to that observed for the chloride structure, with the same pseudo-tetrahedral coordination (*Figure 12A* and *B*). The other cryptate also contained a single bromide on one side of the cavity, but three water molecules sitting between the three arms of the cryptand at the other (*Figure 12C* and *D*) [53]. In the first unit, the bromide exhibits the commonly observed pseudo-tetrahedral geometry, while in the second, the hydrogen bond network forms a pseudo-octahedral coordination environment. The distance between the bridgehead nitrogens is 10.09 Å in the chloride complex, and for the two bromide units distances are 10.38 and 10.15 Å, for the tetrahedral and octahedral forms, respectively.

A comparison of the binding geometries of the three halide complexes with **6** is shown in *Figure 13*. Here the two most common geometries that are repeated in other cryptand complexes are observed, tetrahedral (*Figure 13A – C*) and octahedral (*Figure 13D*).

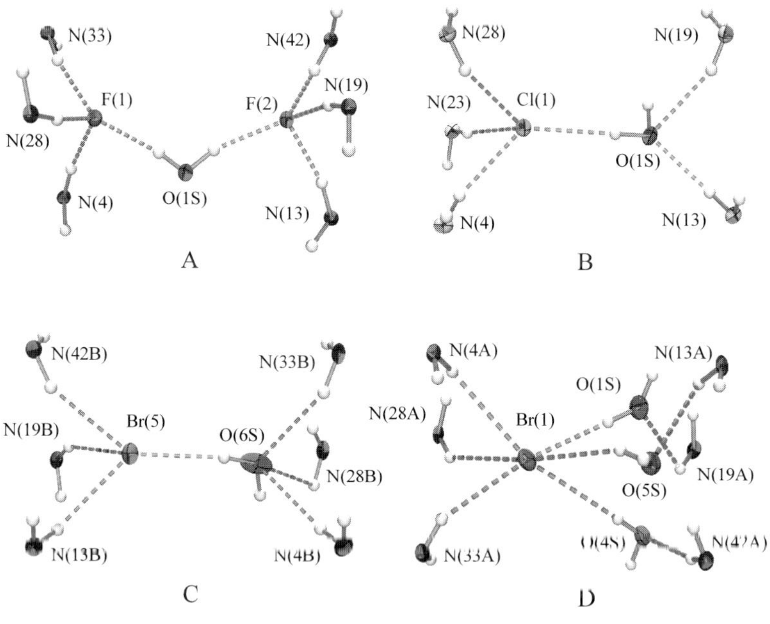

Figure 13. ORTEP views showing the hydrogen bonding interactions in the complexes of **6** with guests, F-H_2O-F (**A**), Cl-H_2O (**B**), Br-H_2O (**C**) and Br-$(H_2O)_3$ (**D**).

3. Amidocryptands

3.1 CHARACTERISTICS

Amidocryptands were until recently unknown in anion circles [39-41,54-56]. However, Raymond and coworkers previously synthesized catechol-based amidocryptands as models for siderophores [57]. The appeal of polyamide cryptands over polyamines is related to their solution properties, namely the former class of ligands is less susceptible to variations of pH. Furthermore, they are usually not very soluble in aqueous solutions, which leads to more flexibility in analytical applications.

3.2 AMIDOCRYPTANDS WITH FLUORIDE, 7 AND 8

Two crystal structures of fluoride complexes have been obtained with the new amidocryptands, **7** and **8** [41,58] (*Figure 14*). The structures are very similar in that a single fluoride is encapsulated in the center of the cavity with hydrogen bonds to all six amide hydrogens. The difference between the two structures in the solid state, however, is that in **8** the fluoride also exhibits hydrogen bonding with the phenyl hydrogens that are directed inside the cavity. The result is a nine-coordinate, pseudo-tricapped trigonal prism geometry for **8**, compared to a twisted trigonal prism for **7**. The hydrogen bond distances are shorter in **7**, ranging from 2.84-2.89 Å, while in **8** they vary over a wider range and are longer, 2.95-3.11 Å, averaging 3.05 Å for both the NH···F and CH···F bonds. In both structures the trigonal prism portion, consisting of the six amide groups, is twisted slightly from ideal averaging about 37°.

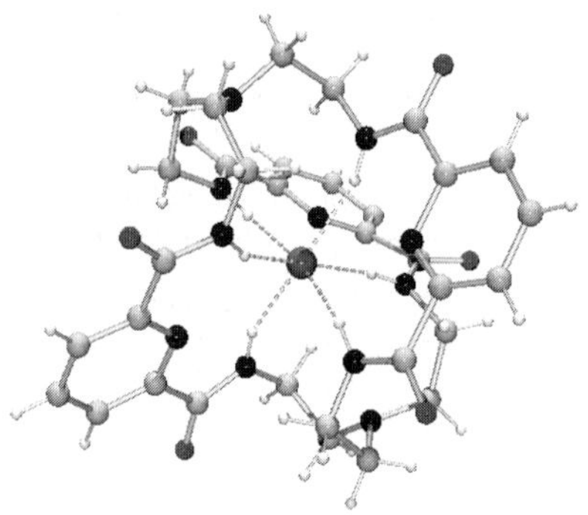

A

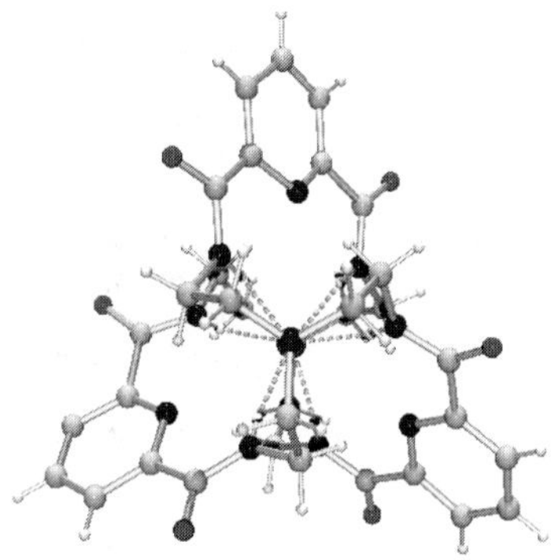

B

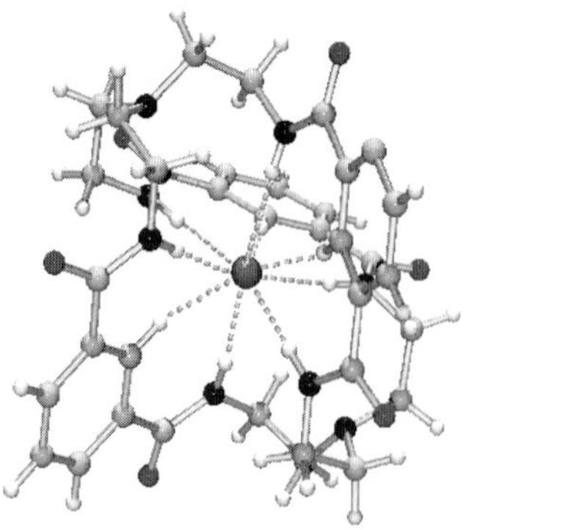

C

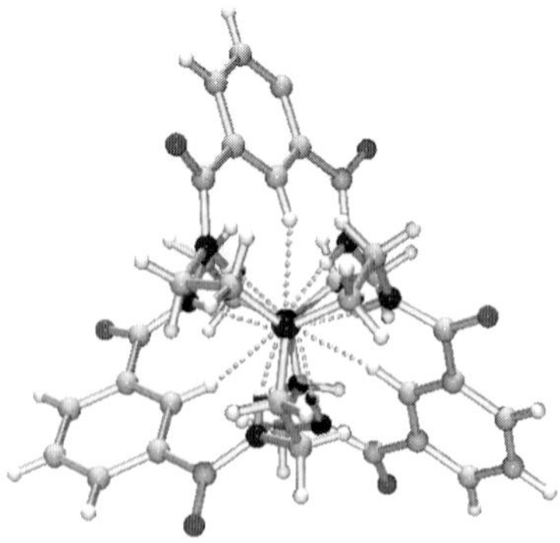

D

Figure 14. The crystal structures of [**7**(F)]$^-$ and [**8**(F)]$^-$ showing the side (**A**) and end-on (**B**) views for [**7**(F)]$^-$ and the side (**C**) and end-on (**D**) views for [**8**(F)]$^-$.

^{19}F NMR spectroscopy is especially informative for probing solution structure in fluoride receptors, and can provide an answer the question of whether or not fluoride is encapsulated in solution. In both **7** and **8** the ^{19}F NMR spectra indicate encapsulation in solution. For **7** a clearly defined septet is observed at –111.6 ppm, resulting from coupling between the encapsulated fluoride and all six amide hydrogens (*Figure 15*). A similar definitive coupling pattern is seen for **8**, but with an unresolved multiplet assigned to a 10-line signal [58].

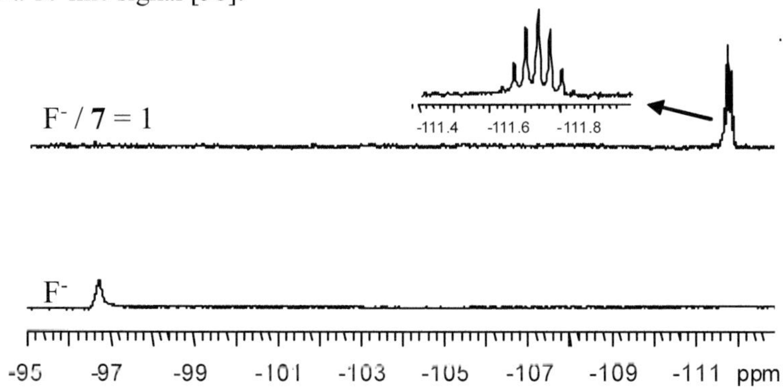

Figure 15. ^{19}F NMR spectra of free [n-Bu_4N][F] and [n-Bu_4N][F] with 7 in DMSO-d_6.

In addition to determining the solution structure of the fluoride complex, ^{19}F NMR is also useful in following chemical reactions in solution. Fluoride is a very strong base in DMSO, known to deprotonate weak acids [59]. We have observed this phenomenon for both **7** and **8**, where the weak acids are the ligands and DMSO-d_6. In the presence of the fluoride complexes, hydrogens on the ligand exchange with deuteriums on the DMSO-d_6. Using a 2:1 ratio of F^-:**7** a series of seven signals are observed, due to sequential deuterium exchange between the amide hydrogens and the DMSO-d_6 (*Figure 16*). The series culminates in a singlet indicative of no ^1H-^{19}F coupling and fully deuterated ligand as shown schematically in *Figure 17* [58]. A similar process is seen for **8**; however, in this case the series of signals culminates in a quartet, the result of fluoride coupling with the three non-exchangeable phenyl hydrogens.

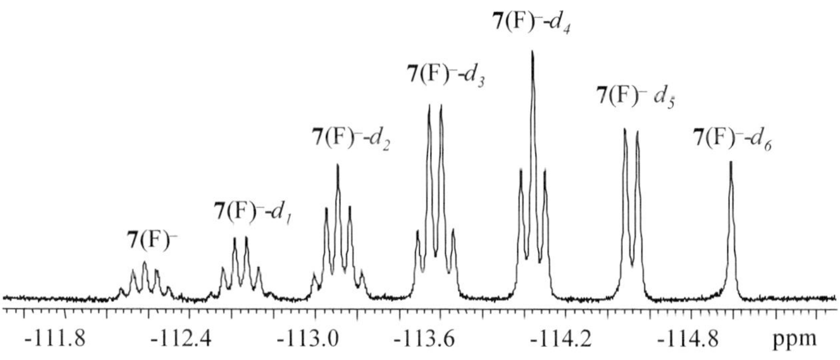

Figure 16. ^{19}F NMR spectrum of F^-:**7** = 2:1 after 10 days.

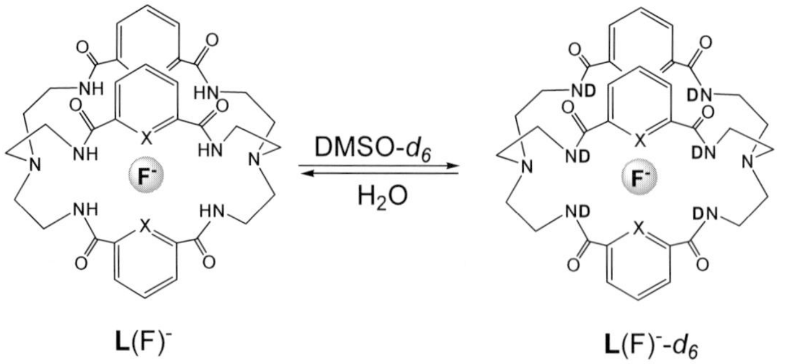

Figure 17. Deuterium exchange reaction for [**7**(F)]$^-$ and [**8**(F)]$^-$, X = N for **7** and CH for **8.**

3.3 AMIDOCRYPTAND WITH CHLORIDE, 7

The third structure obtained to date in this new class of amidocryptands is the monohydrochloride salt of the cryptand **7** [41]. In this structure the bridgehead amine of the cryptand is protonated and a single chloride is sitting in the cavity (*Figure 18*). The chloride is six-coordinate, with hydrogen bond distances ranging from 3.20-3.60 Å. The structure helped to explain puzzling solution phenomena observed in NMR studies in $CDCl_3$. While spectral data obtained in DMSO-d_6 showed the expected five-line spectrum, the spectrum in $CDCl_3$ indicated what appeared to be contamination and additional signals (*Figure 19A*). These signals increased in intensity over several days, while the signals due to the uncomplexed cryptand decreased (*Figure 19B*). This process has now been identified as the uptake of residual HCl in the chloroform solution. With time, however, even these new signals began to decrease in intensity, being replaced by new resonances, until finally a very simple, six-line spectrum was obtained (*Figure 19C)*. The second chemical process is attributed to uptake of another HCl, with the cryptand becoming symmetrical with both bridgehead amines protonated (and possibly two symmetrically-placed chlorides in the cavity). Additional signals due to protonation of the bridgehead amines are observed downfield at about 11 ppm in both the mono- and di-hydrochloride spectra. A schematic representation of the proposed reaction sequence is shown in *Figure 20.*

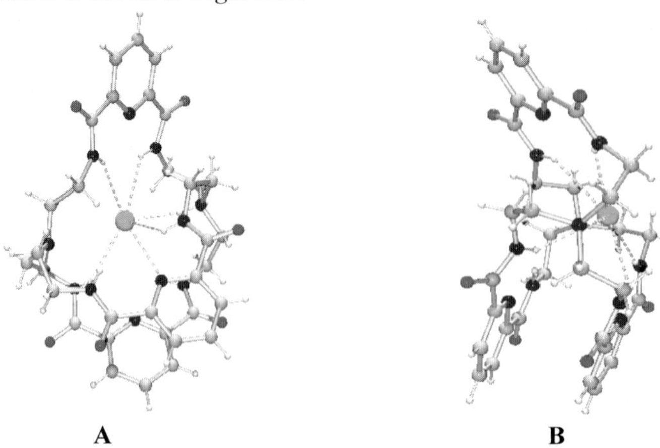

Figure 18. The crystal structures of [H7(Cl)] showing the side (**A**) and end-on (**B**)

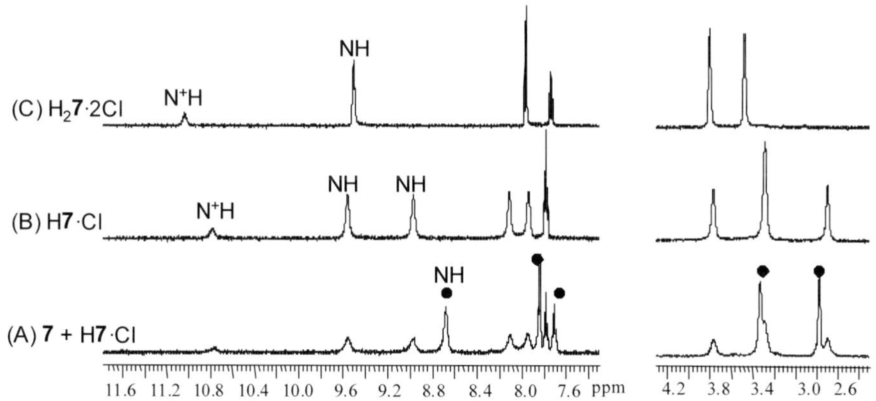

Figure 19. ^{1}H NMR spectrum of **7** in $CDCl_3$ (**A**) immediately, (**B**) after 2 days, and (**C**) after 7 days.

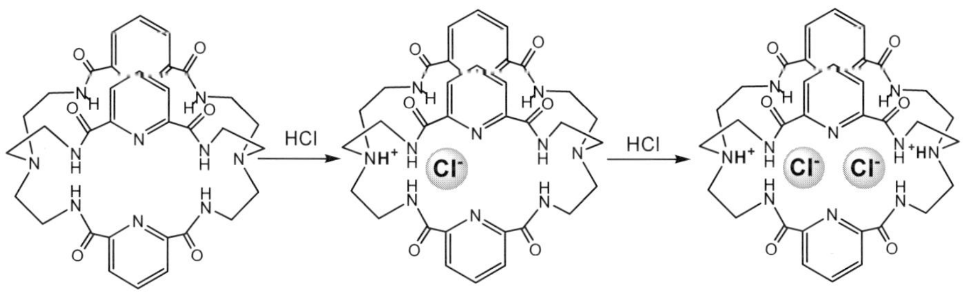

7 H7(Cl) $H_27(Cl)_2$

Figure 20. Proposed reaction of **7** with trace HCl.

4. Summary

Now that an entire series of cryptand structures with halides has been obtained, in which mono-, di-, and tri-topic encapsulation are observed, some simple conclusions can be drawn concerning "anion capacity," i.e., the number of anions that can be incorporated in the cavity. As might be anticipated, there appears to be a correlation between the distances between the bridgehead amines and the number of the encapsulated species as shown in *Figure 21*. For the tiny octaazacryptand, the distance between the bridgehead amines is short and limited by the ethylene spacers between the two tren units. While all of the cryptands with longer spacers might be anticipated to show multitopic behavior, none of the structures with **3**, containing the ether spacers, shows multitopic binding of halides (except for the triatomic encapsulation of the pseudohalide azide). Likewise, structural findings for the amidocryptands, **7** and **8** show only monotopic behavior (at least to date), which is curious since the structurally similar azacryptands, **5** and **6** both show extensive ditopic and even tritopic inclusion species.

Distances between the bridgehead amines vary according to the "topicity" of the receptors. For monotopic complexes, the bridgehead distances are between 6.5 and 7.9 Å. In terms of incorporating multiple species, however, the azacryptands appear to be the most flexible, as seen additionally in the dinitrate structure with **5** [30]. When multitopic binding occurs, the bridgehead distances increase dramatically to between 9 and 11 Å, the largest being the 10.72 Å observed for the tritopic fluoride cascade complex with **6**.

In conclusion, just within this series of tren-based cryptands, a breadth of coordination chemistry exists. Coordination geometries of the encapsulated species vary, with four- and six-coordination being more commonly encountered, although even a nine-coordinate structure has been isolated (**8** with fluoride). The field will certainly continue to elicit excitement as more structural and binding data are forthcoming, aiding in the design of elegant and highly selective ligands for anions.

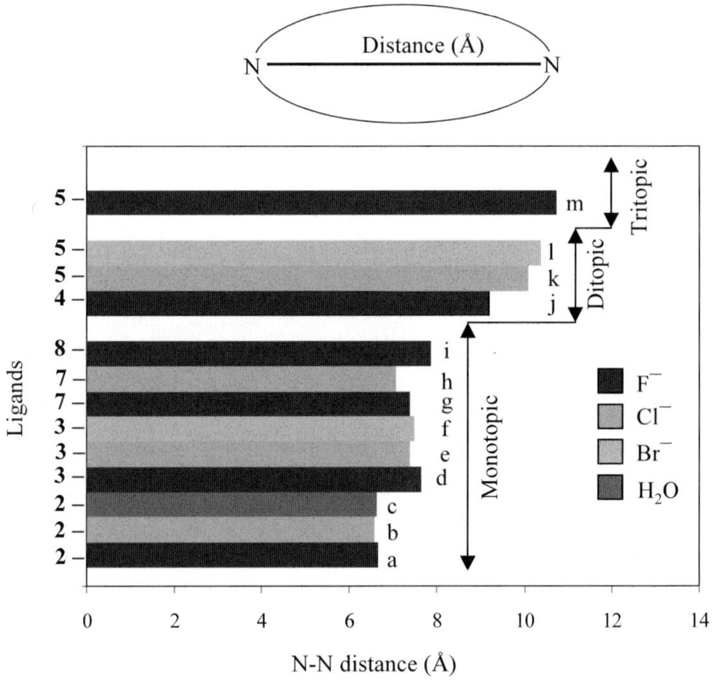

Figure 21. Encapsulated species in the cryptand as a function of N-N distances: a =6.65 Å [44], b= 6.59 Å [32,36], c= 6.63 Å [36], d = 7.66 Å [45,48], e = 7.40 Å [45,48], f = 7.50 Å [45,48], g = 7.39 [41], h = 7.07 [41], i = 7.87 [58], j = 9.21 [31], k = 10.09 [53], l = 10.38 [53], m = 10.72 (Å) [35,53].

5. References

1. *Supramolecular Chemistry of Anions*, Bianchi, A.; Bowman-James, K.; García-España, E., Eds.; Wiley-VCH: New York; **1997**.
2. Special Issue: 35 Years of Synthetic Anion Receptor Chemistry, *Coord. Chem. Rev.* **2003**, *240(1-2)*.
3. Park C. H.; Simmons, H. E.: Macrobicyclic amines, III. Encapsulation of halide ions by *in,in*-1,(*k*+2)-diazabicyclo [*k.l.m.*] alkane ammonium ions, *J. Am. Chem. Soc.* **1968**, *90*, 2431-2452.
4. Bell, R. A.; Christoph, G. G.; Fronczek, F. R.; Marsh, R. E.: Cation $H_{13}O_6^+$. Short, symmetric hydrogen bond, *Science* **1975**, *190*, 151-152.
5. Vögtle, F.; Sieger, H.; Müller, W. M.: Complexation of uncharged molecules and anions by crown-type host molecules, *Top. Curr. Chem.* **1981**, *98*, 107-161.
6. Dietrich, B.; Guilhem, J.; Lehn, J-M.; Pascard, C.; Sonveaux, E.: Molecular recognition in anion coordination chemistry, *Helv. Chim. Acta* **1984**, *67*, 91 -104.
7. Kimura, E.: Macrocyclic polyamines as biological cation and anion complexones - an application to calculi dissolution, *Top. Curr. Chem.* **1985**, *128*, 113-141.

8. Izatt, R. M.; Pawlak, K.; Bradshaw, J. S.; Bruening, R. L.: Thermodynamic and kinetic data for macrocycle interaction with cations and anions, *Chem. Rev.* **1991**, *91*, 1721-2085.
9. Katz, H. E.: Recent advances in multidentate anion complexation in *Inclusion Compounds*, Atwood, J. L.; Davies, J. E. D.; MacNichol, D. D., Eds.; Oxford University Press: Oxford; **1991**, 391-405.
10. Beer, P. D.; Wheeler, J. W.; Moore, C.: New approaches to anion coordination chemistry. Acyclic quaternary polybipyridinium receptors designed to bind anionic guest species, in *Supramolecular Chemistry*, Balzani, V.; De Cola, L., Eds.; Kluwer Academic Publishers: Netherlands; **1992**, 105-118.
11. Dietrich, B.: Design of anion receptors: applications, *Pure & Appl. Chem.* **1993**, *65*, 1457-1464.
12. Sessler, J. L.; Cyr, M.; Furuta, H.; Kral, V.; Mody, T.; Morishma, T.; Shionaya, M.; Weghorn, S.: Anion binding: A new direction in porphyrin-related research, *Pure & Appl. Chem.* **1993**, *65*, 393-398.
13. Beer, P. D.; Smith, D. K.: Anion binding and recognition by inorganic based receptors, *Prog. Inorg. Chem.* **1997**, *46*, 1-96.
14. Schmidtchen, F. P.; Berger, M.: Artificial organic host molecules for anions, *Chem. Rev.* **1997**, *97*, 1609-1646.
15. Antonisse, M. M. G.; Reinhoudt, D. N.: Neutral anion receptors, *Chem. Commun.* **1998**, 443-448.
16. Beer, P. D.; Gale, P. A.; Chen. G. Z.: Mechanisms of electrochemical recognition of cations, anions and neutral guest species by redox-active receptor molecules, *Coord. Chem. Rev.* **1999**, *185-186*, 3-36.
17. Gale, P. A.: Anion coordination and anion-directed assembly: Highlights from 1997 and 1998, *Coord. Chem. Rev.* **2000**, 199, 181-233.
18. Beer P. D.; Gale, P. A.: Anion recognition and sensing: The state of the art and future perspectives, *Angew. Chem. Int. Ed.* **2001**, *40*, 486-516.
19. Ilioudis, C. A.; Georganopoulou, D. G.; Steed, J. W.: Insights into supramolecular design 2: Analysis of anion coordination geometry of oxoanions in a protonated polyamine matrix, *CrysEngComm.* **2002**, 26-36.
20. Fitzmaurice, R. J.; Kyne, G. M.; Douheret, D.; Kilburn, J. D.: Synthetic receptors for carboxylic acids and carboxylates, *J. Chem. Soc., Perk. Trans. 1* **2002**, 841-864.
21. Menif, R.; Reibenspies, J.; Martell, A. E.: Synthesis, protonation constants, and copper(II) and cobalt(II) binding constants of a new octaaza macrobicylic cryptand: $(MX)_3(TREN)_2$. Hydroxide and carbonate binding of the dicopper(II) cryptate and crystal structures of the cryptand and of the carbonato-bridged dinuclear copper(II) cryptate, *Inorg. Chem.* **1991**, *30*, 3446-3454.
22. Marrs, D. J.; McKee, V.; Nelson, J.; Lu, Q.; Harding, C. J.: Pyridinophane cryptand hosts with dicopper(I) or dicopper(II) preferences, *Inorg. Chim. Acta* **1993**, *211*, 195-202.
23. Lu, Q.; Reibenspies, J. H.; Carroll, R. I.; Martell, A. E.; Clearfield, A.: Interaction of mono- and triphosphate anions with a protonated polyaza macrocycle and its Cu(II) complexes, *Inorg. Chim. Acta* **1998**, *270*, 207-215.
24. Dussart, Y.; Harding, C.; Dalgaard, P.; McKenzie, C.; Kadirvelraj, R.; McKee, V.; Nelson, J.: Cascade chemistry in azacryptand cages: bridging carbonates and methylcarbonates, *J. Chem. Soc., Dalton Trans.* **2002**, 1704-1713.
25. Dietrich, B.; Hosseini, M. W.; Lehn, J.-M.; Sessions, R. B.: Anion receptor molecules. Synthesis and anion-binding properties of polyammonium macrocycles, *J. Am. Chem. Soc.* **1981**, *103*, 1282-1283.
26. Boudon, S.; Decian, A.; Fischer, J.; Hosseini, M. W.; Lehn, J.-M.; Wipff, G.: Structural and anion coordination features of macrocyclic polyammonium cations in the solid, solution, and computational phases, *J. Coord. Chem.* **1991**, *23*, 113-135.
27. Lu, Q.; Motekaitis, R. J.; Reibenspies, J. J.; Martell, A. E.: Molecular recognition by the protonated hexaaza macrocyclic ligand 3,6,9,16,19,22-hexaaza-27,28-dioxa-tricyclo[22.2.1.1^{11,14}]octacosa-1(26),11,13,24-tetraene, *Inorg. Chem.* **1995**, *34*, 4958-4964.
28. Morgan, G.; McKee, V.; Nelson, J.: Caged anions: Perchlorate and perfluoroanion cryptates, *J. Chem. Soc., Chem. Commun.* **1995**, 1649-1652.
29. Nation, D. A.; Reibenspies, J.; Martell A.: Anion binding of inorganic phosphates by the hexaaza macrocyclic ligand 3,6,9,17,20,23-hexaaza-tricyclic[23.3.1.1^{11,15}]triaconta-1(29),11(30),12,14,25,27-hexaene, *Inorg. Chem.* **1996**, *35*, 4597-4603.
30. Mason, S.; Seib, L.; Bowman-James, K.: Unusual encapsulation of two nitrates in a single bicyclic cage, *J. Am. Chem. Soc.* **1998**, *120*, 8899-8900.
31. Mason, S.; Llinares, J. M.; Morton, M.; Clifford, T.; Bowman-James, K.: Snapshots of fluoride binding in an azacryptand, *J. Am. Chem. Soc.* **2000**, *122*, 1814-1815.
32. Hossain, M. A.; Llinares, J. M.; Miller, C.; Seib, L.; Bowman-James, K.: Further insight to selectivity issues in halide binding in the tiny octaaza cryptand, *Chem. Commun.* **2000**, 2269-2270.
33. Aguilar, J. A.; Clifford, T.; Danby, A.; Llinares, J. M.; Mason, S.; García-España, E.; Bowman-James, K.: Fluoride ion receptors: A comparison of a polyammonium monocycle versus its bicyclic corollary, *Supramol. Chem.* **2001**, *13*, 405-417.

34. Clifford, T.; Danby, A.; Llinares, J. M.; Mason, S.; Alcock, N. W.; Powell, D.; Aguilar, J. A.; García-España, E.; K. Bowman-James, K.: Anion binding with two polyammonium macrocycles of different dimensionality, *Inorg. Chem.* **2001**, *40*, 4710-4720.
35. Hossain, M. A.; Llinares, J. M.; Mason, S.; Morehouse, P.; Powell, D.; Bowman-James, K.: Parallels in cation and anion coordination: a new class of cascade complexes, *Angew. Chem., Int. Ed.* **2002**, *41*, 2335-2338.
36. Hossain, M. A.; Llinares, J. M.; Powell, D.; Bowman-James, K.: Structure and binding properties of the tiny octaaza cryptand, *Supramol. Chem.* **2002**, *2,* 143-151.
37. Morehouse, P; Hossain, M. A.; Llinares, M. J.; Powell, D.; Bowman-James, K.: Ditopic azacryptate proton cage, *Inorg. Chem.* **2003**, *42,* 8131-8133.
38. Hossain, M. A.; Liljegren, J. A.; Powell, D.; Bowman-James, K.: Anion binding with a tripodal amine, *Inorg. Chem.* **2004**, *43*, 3751-3755.
39. Hossain, M. A.; Llinares, J. M.; Powell, D.; Bowman-James, K.: Multiple hydrogen bond stabilization of a sandwich complex of sulfate between two macrocyclic tetraamides, *Inorg. Chem.* **2001**, *40*, 2936-2937.
40. Hossain, M. A.; Kang, S. O.; Powell, D.; Bowman-James, K.: A new class of amide/quaternized amine macrocycles and the chelate effect, *Inorg. Chem.* **2003**, *42*, 1397-1399.
41. Kang, S. O.; Llinares, J. M.; Powell, D.; VanderVelde, D.; Bowman-James, K.: New polyamide cryptand for anion binding, *J. Am. Chem. Soc.* **2003**, *125,* 10152-10153.
42. Hossain, M. A.; Kang, S. O.; Llinares, J. M.; Powell. D.; Bowman-James, K.: Elite new anion ligands: polythioamide macrocycles, *Inorg. Chem.* **2003**, *42*, 5043-5045.
43. Dietrich, B.; Lehn, J.-M.; Sauvage, J. P.:Les cryptates, *Tetrahedron Lett.* **1969**, *10*, 2889-2892.
44. Dietrich, B.; Lehn, J.-M.; Guilhem, J.; Pascard, C.: Anion receptor molecules: synthesis of an octaaza-cryptand and structure of its fluoride cryptate, *Tetrahedron Lett.* **1989**, *30*, 4125–4128.
45. Dietrich, B.; Dilworth, B.; Lehn, J.-M.; Souchez, J.-P.; Cesario, M.; Guilhem, J.; Pascard, C.: 52. Anion cryptates: synthesis, crystal structures, and complexation constants of fluoride and chloride inclusion complexes of polyammonium macrobicyclic ligands, *Helv. Chim. Acta* **1996**, *79*, 569-587.
46. Reilly, S. D.; Khalsa, G. R. K.; Ford, D. K.; Brainard, J. R.; Hay, B. P.; Smith, P. H.: Octaazacryptand complexation of the fluoride ion, *Inorg. Chem.* **1995**, *34*, 569–575.
47. Lehn, J. M.; Sonveaux, E.; Willard, A. K. Molecular recognition. Anion cryptates of a macrobicyclic receptor molecule for linear triatomic species, *J. Am. Chem. Soc.* **1978**, *100,* 4914-4916.
48. Dietrich, B.; Guilhem, J.; Lehn, J.-M.; Pascard, C.; Sonveaux, E.: Molecular recognition in anion coordination chemistry, *Helv. Chim. Acta* **1984**, *67,* 91-104.
49. Ilioudis, C. A.; Tocher, D. A.; Steed, J. W.: A highly efficient, preorganized macrobicyclic receptor for halides base on CH··· and NH··· anion interactions, *J. Am. Chem. Soc.* **2004,** *126*, 1239-2402.
50. MacDowell, D.; Nelson, J.: Facile synthesis of a new family of cage molecules, *Tetrahedron Lett.* **1988,** *29*, 385-386.
51. Chen, D.; Martell, A. E.: The synthesis of new binuclear polyaza macrocyclic and macrobicyclic molecules, *Tetrahedron*, **1991**, *47*, 6895-6902.
52. Lehn, J.-M.; Pine, S. H.; Watanabe, E. I.; Willard, A. K.: Binuclear cryptates. Synthesis and binuclear cation inclusion complexes of bis-tren macrobicyclic ligands, *J. Am. Chem. Soc.* **1977**, *99*, 6766-6768.
53. Hossain, M. A.; Morehouse, P.; Powell, D.; Bowman-James, K.: Tritopic (cascade) and ditopic complexes of halides with an azacryptand, *Inorg. Chem.* in press.
54. Bisson, A. P.; Lynch, V. M.; Monahan, M. K. C.; Anslyn, E. V.: Recognition of anions through NH-π hydrogen bonds in a bicyclic cyclophane-selectivity for nitrate, *Angew. Chem. Int. Ed.* **1997**, *36*, 2340-2342.
55. Pieters, R. J.; Diederich, F.: Enantioselective complexation of excitatory amino acid derivatives by chiral, cage-like C3-symmetrical receptors, *Chem. Commun.* **1996**, 2255-2256.
56. Lecollinet, G.; Dominey, A. P.; Velasco, T.; Davis, A. P.: Highly selective disaccharide recognition by a tricyclic octaamide cage, *Angew. Chem. Int. Ed.* **2002**, *41*, 4093-4096.
57. Karpishin, T. B.; Stack, T. D. P.; Raymond, K. N.: Octahedral versus trigonal prismatic geometry in a series of catechol macrobicyclic ligand-metal complexes, *J. Am. Chem. Soc.* **1993**, *115*, 182-192.
58. Kang, S. O.; VanderVelde, D.; Powell, D.; Bowman-James, K.: Fluoride-facilitated deuterium exchange from DMSO-d_6 to polyamide-based cryptands, *J. Am. Chem. Soc.* **2004**, *126*, 12272-12273.
59. Clark, J. H.: Fluoride ion as a base in organic synthesis, *Chem. Rev.* **1980**, *80*, 429-452.

12.
OXOANION SELECTIVITY WITH PROTONATED AZACRYPTATE HOSTS: THE INFLUENCE OF HYDRATION ON STRUCTURE AND STABILITY

J. NELSON[a], V. McKEE[b] AND R. M. TOWN[c]
[a] *Biomedical Science, University of Ulster, Coleraine, BT52 1SA, U.K.*
[b] *Chemistry Department, Loughborough University, Leicestershire, LE11 3TU, U.K.*
[c] *Department of Chemistry, University of Southern Denmark, Campusvej 55, 5230 Odense M, Denmark*

1. Introduction

The pioneering work of Lehn established the fundamental strategy of using preformed three dimensional cage ligands, dubbed cryptands, to ensure enhanced thermodynamic and kinetic stability for cation complexation [1]. Positively charged cryptands, typically the N-donor azacryptands, can also serve to encapsulate anions.

Several anions currently cause environmental concern: The adverse environmental effect of escalating concentrations of nitrate and phosphate in surface waters has been recognised for some years. Other anions are of biological interest: perchlorate is present in solid rocket fuel, and there may be links between its contamination of drinking water and abnormal human thyroid activity [2,3]; use of nitrate-contaminated drinking water to prepare infant formula is a well-known risk factor for so-called blue baby syndrome (infant methemoglobinemia) [4]; chromate is a widespread contaminant of groundwaters [5] in the form of hydrogen chromate and chromate (both of which are toxic to humans [6]); carboxylates have many important biochemical roles, so that analysis of e.g., oxalate in body fluids is important in some clinical situations, such as primary hyperoxaluria and chronic renal failure. There is thus a need to realise strong and selective anion complexation for applications in monitoring and clean-up.

Rational design of efficient anion binding cryptands and characterisation of their properties are challenging tasks. Anions are larger and more variable in shape than cations, and any potential complexant must overcome the strong competition presented by solvation or hydration equilibria. Earlier work has shown that for oxoanions, stability constants show a charge-dependent increase from mononegative anions such as perchlorate to di- and tri-negative anions such as sulfate and phosphate, as well as a dependence on the match between host and guest dimensions [7-9].

Karsten Gloe (ed.), Macrocyclic Chemistry: Current Trends and Future Perspectives, 189–201.
© 2005 *Springer. Printed in the Netherlands.*

For greatest binding efficiency, the size and shape of the host should be complementary to that of the anionic guest. The literature abounds with reports of carefully designed receptors with demonstrated efficiency for isolated anionic targets [10,11]. However, the field is now at a stage where a generic understanding of the principles involved in host-guest association is required. Progress towards this objective requires systematic quantitative analysis of solution equilibrium parameters (both enthalpy- and entropy- related) and of structural features of the host-guest associates. Such information should, through elucidation of the significance of factors such as host-guest structural complementarity and the relative acidities/basicities of these partners, together with consideration of changes of solvation upon aggregation, lead to improved design of efficient, selective systems.

As a step towards this goal, we present a systematic structural and solution phase study of complexation of a suite of oxoanionic guests by a series of protonated azacryptand hosts. Over a series of anions, the relative values of free energy of hydration of the different anions constitute a significant, sometimes dominant, contribution to the free energy of complexation, while comparisons of this parameter for the same anion with different cryptate hosts are affected by the linked properties of host basicity and solvation. The consequences of steric matching of host and guest also contribute in some degree to the observed complexation constants, so we first compare structural data on anion cryptates. This body of data provides a database for development and validation of models. Although pH-metric data had been obtained for earlier macrobicyclic hosts, such as O-bistren, there was little corresponding structural information. It must be kept in mind that X-ray crystallographic evidence cannot be uncritically extrapolated to the solution phase: solubility effects can easily lead to crystallisation of minor constituents of an equilibrium mixtures. Nevertheless, the links between structural features and solution equilibria can be strengthened if a comprehensive set of data, for a range of hosts and guests, shows consistent trends.

2. Structural Features

The series of azacryptands used in this study are relatively easily made via 2+3 Schiff-base condensation of the triamine, tren, with the appropriate dialdehyde, followed by borohydride reduction of the imine functions (Scheme1). Adjustment of pH in the presence of the appropriate anion followed by slow evaporation of solvent yields the anion cryptate, often in a form suitable for X-ray diffraction analysis.

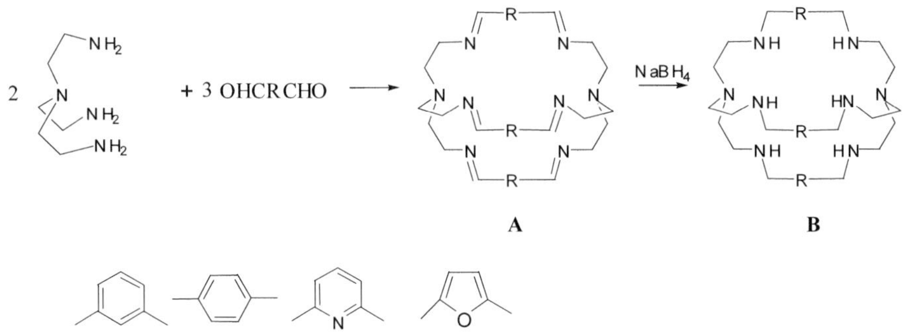

A:	3Bm	3Bp	3P	3F
B:	R3Bm	R3Bp	R3P	R3F

Scheme 1

2.1 STRUCTURES OF THE MORE HIGHLY PROTONATED CRYPTATES

Earlier discussion [12] of the structures of protonated cryptand/oxoanion assemblies was based on consideration of H-bonds between the encapsulated anion and the NH^+ donors of the cryptand. These interactions are assumed to be responsible for retention of the guest anion in the host cryptand cavity, both in the solid state and in solution. We have shown that, in all cases, anion cryptates exhibit at least three, and often more, direct H-bond NH^+-O_{anion} contacts tethering the included oxoanion within the crypt.

These are often supported by indirect (water-mediated) NH^+-O_W-O_{anion} H-bonds, both components shorter than the direct bonds because absence of geometric constraint allows the water bridge to position itself to best advantage. Fig. 1 illustrates the binding in a typical mononegative anion complex: the ClO_4 cryptate of hexaprotonated R3Bm.

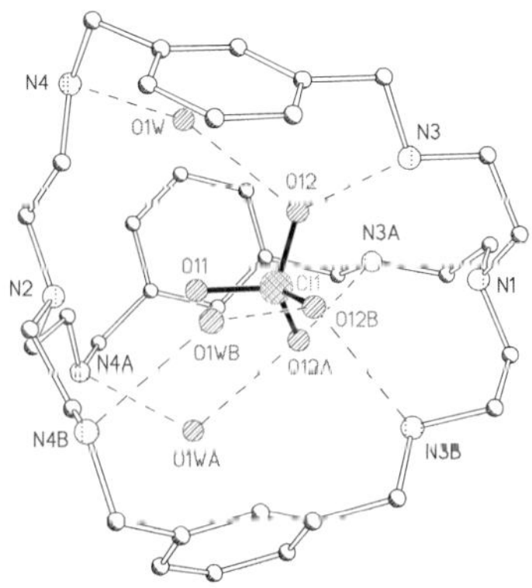

Figure 1. Perchlorate cryptate of $[H_6R3Bm]^{6+}$, N..O 2.790(5), N..O_w 2.712(6), O_w..O2.782(5) Å.

In the analogous tetrahedral dinegative oxoanion cryptates a similar mix of direct and indirect water-bridged H-bonds links the encapsulated oxoanion with the NH^+ donors. The direct NH^+--O^- contacts are (Table 1) on average, slightly shorter than those made by mononegatively charged oxoanionic guests. The cryptand hosts $[H_6R3Bm]^{6+}$ and $[H_6R3F]^{6+}$ behave similarly in furnishing a cavity of adequate size for oxoanion encapsulation, although there are minor differences in host dimensions leading to better fit of dinegative anions within the $[H_6R3F]^{6+}$ cavity, which behaves as though smaller and more spherical than $[H_6R3Bm]^{6+}$. In consequence of a slight degree of mismatch, one NH^+ donor of $[H_6R3Bm]^{6+}$ is sometimes left unexploited in these dinegative anion cryptates (see colour pages Figs. 2-4) which leads to disorder in the corresponding cryptand strand [13].

Where there is good complementarity of fit, the trigonal symmetry of host and guest may be reflected in the appearance of a three-pronged crown motif, whereby NH^+ functions which are individually chelated (via bifurcated H-bonds) by a pair of adjacent oxoanion O-acceptors, in turn chelate each oxoanion O-acceptor (Figs. 4-6) [12-14]. This motif is prominent in binding of a pair of nitrate ions within $[H_6R3Bm]^{6+}$. This unusual dinuclear binding situation, however, does not extend to aqueous solution [15].

Anion encapsulation by these hosts is not constrained by a requirement for trigonal symmetry in the guest. The dicarboxylate ion, oxalate, which is strongly complexed by the protonated cryptand hosts, binds all six NH^+ donors via their triple chelation to each of the two carboxylate O^--acceptors. Differentiation of the originally identical pair of carboxylate oxygens has occurred, via polarisation of the $-CO_2^-$ functions into one relatively long and one relatively short C-O bond; the longer C-O bond acting as H-bond acceptor for the triply chelating set of NH^+ donors. This polarisation has also resulted in twisting of oxalate ion away from its normal planar geometry [16] (Fig. 7).

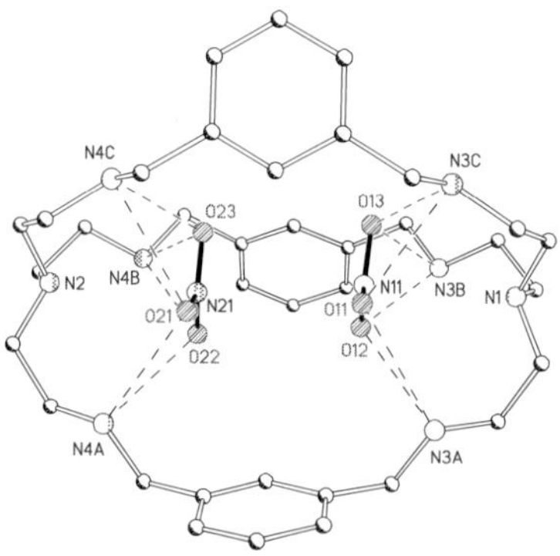

Figure 6. Dinitrate cryptate of $[H_6R3Bm]^{6+}$ N..O 2.84-3.04, N-N3.391(1) Å

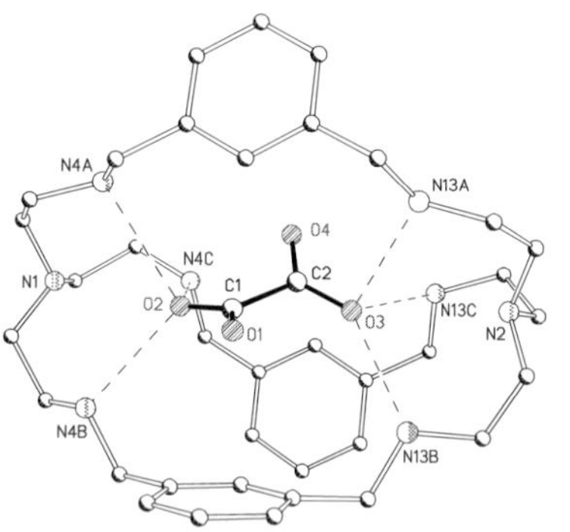

Figure 7. Oxalate crypt of $[H_6R3Bm]^{6+}$ N..O distances in range 2.76-3.04 Å.

The third cryptand studied, $[H_6R3P]^{n+}$, behaves quite differently from the other two in its reluctance to present a cavity for anion encapsulation. Instead it prefers to adopt cleft-binding conformations, where the associated oxoanion may be hydrogen bonded via two of its O-acceptors to some or all of the NH^+ donors incorporated in the cleft (Figs. 8, 9). With oxalate, this conformation generates a binding situation quite different from that of the cavity-encapsulating hosts. The oxalate guest slots as far as possible into the relatively shallow groove presented by the cleft, leaving the other end protruding. Under aqueous synthetic conditions protonation occurs, the consequence of which is the generation of a short, possibly symmetric oxalate/semioxalate hydrogen bond, bringing the pair of cryptates together into a dimer [17] (Fig. 10).

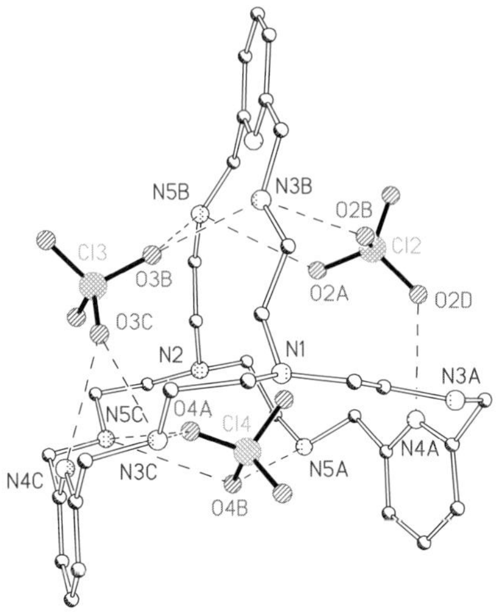

Figure 8. Perchlorate cryptate of $[H_8R3P]^{8+}$.

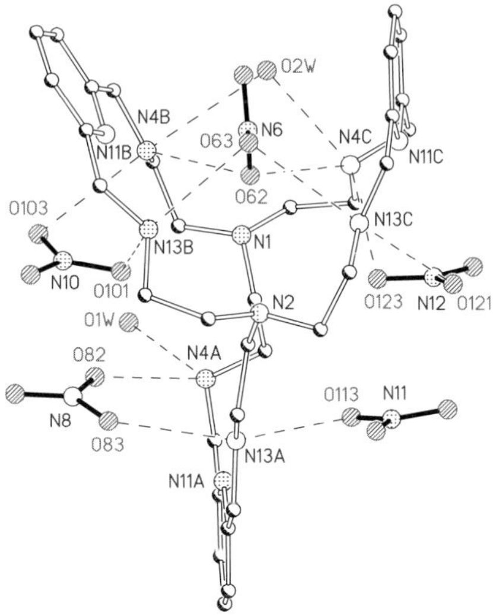

Figure 9. Nitrate cryptate of $[H_6R3P]^{6+}$.

To summarise: for the tren-derived azacryptands, hydrogen bonding derives from protonation of all six secondary amines, but not the tertiary bridgeheads. Consequently each $RR'NH_2^+$ unit can act as donor for two hydrogen bonds so that the protonated cryptand can form 12 hydrogen bonds, which is in many cases the observed pattern. Geometric considerations mean that no more than six of the N-H bonds can be oriented convergently toward an anionic guest in the cavity. The others diverge and can lead to extensive hydrogen-bonded networks running through the lattice, typically involving the cryptate, unencapsulated anions and water molecules. Examples involving six direct NH..O hydrogen bonds are relatively uncommon (Figs. 4-6) and this pattern appears to indicate a good fit. Where the fit is less good, the hydrogen bond capacity of anion and cryptand can be better satisfied by "indirect" N-O_w-O links in which the protonated amine is hydrogen-bonded to a water molecule which is in turn hydrogen-bonded to the encapsulated anion (Figs. 1-3). In other examples, [21] one oxygen atom of the bound oxoanion is not involved (Fig. 1) in any hydrogen bonding.

At first sight it may appear surprising that the anion-cation contacts are not more dramatically shortened on doubling the charge of the encapsulated anion. However, as structural data accumulates it becomes clear that the "direct" charge-assisted hydrogen bonds are not, in general, the shortest H-bond contacts in the lattice. This is presumably for geometric reasons: the siting of the oxoanion H-bond acceptor is not so well optimised for effective H-bonding with the relatively fixed NH^+ donors as with the relatively unrestrained μ^1 O_w bridge. Indeed it appears that the single-O_w bridge atoms are still subject to minor restraint, as the shortest H-bond contacts tend to manifest themselves in connection with the pair of water molecules in the bridging $O_w..O_w'$ linking unit, which is a common motif in dinegative anion structures (see colour pages Figs. 2-5).

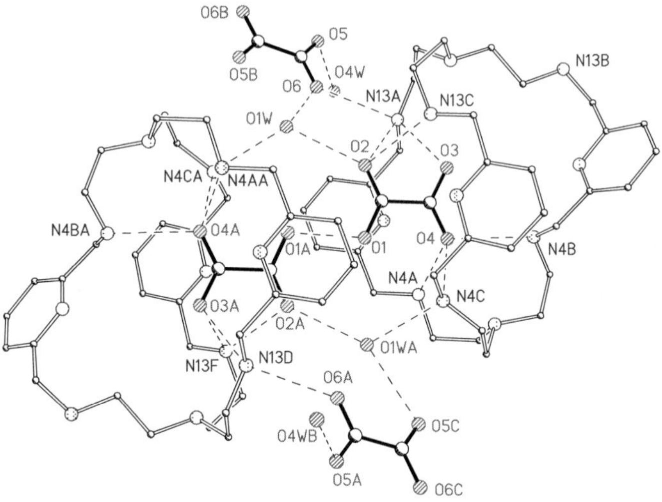

Figure 10. Oxalate cryptate of $[H_6R3P]^{6+}$ showing H-bond dimer linked by O1-O1A 2.474 Å.

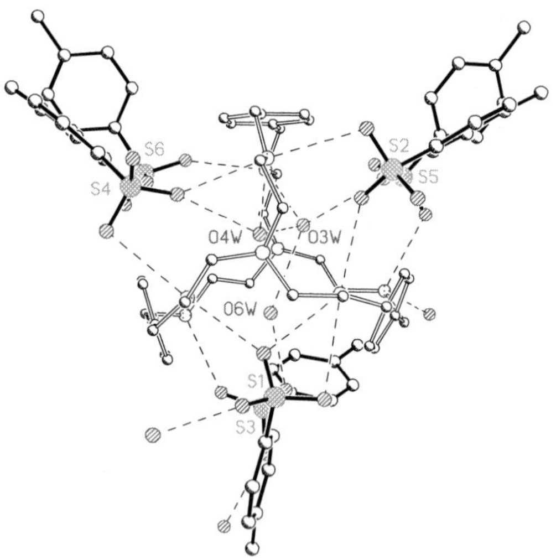

Figure 11. Tosylate crypt of $[H_6R3Bm]^{6+}$, N..O distances in range 2.76-3.04 Å showing 3 water molecules encapsulated in preference to OTs^-, N..O(OTs) 2.73-3.03 Å.

In **1** (Table 1a) the direct $NH^+..O_{oxoanion}$ contacts (Fig. 5) lie in the range 3.21 – 2.85 Å while the $O_{oxoanion} – O_w$ contacts average to 2.72 Å and the $O_w..O_w$' contacts to 2.73 Å.

TABLE 1.

(a) $[H_6R3Bm]^{6+}$ H-bonds to encapsulated anions; **3** = CrO_4^{2-}; **1**= $S_2O_3^{2-}$; **4**= SeO_4^{2-}

3	**4**	**1**
2.766(8) (N3B-O1)	2.820(6) (N3A-O1)	3.059(4) (N3B-O1)
2.841(9) (N3A-O1)	2.924(6) (N3B-O1)	3.214(5) (N3A-O1)
2.950(9) (N4B-O2)	2.927(6) (N4A-O2)	3.298(3) (N4B-S2)
3.214(10) (N4A-O2)	3.116(6) (N4B-O2)	3.455(3) (N4A-S2)
		3.491(3) (N4C-S2)
		2.852(4) (N3C-O3)
		2.957(4) (N3B-O3)
2.75(3) (N3C-O4)	2.653(12) (N3C-O4)	2.861(4) (N3A-O4)
2.75(4) (N3C′-O4)	2.913(13) (N3C′-O4)	3.087(4) (N3C-O4)

(b) $[H_6R3F]^{6+}$ H-bonds to encapsulated anions; **3a** = CrO_4^{2-}; **1a**= $S_2O_3^{2-}$; **2a**= SeO_4^{2-}

3a	**2a**	**1a**
2.970 (N4A-O11)	3.009 (N11B-O11)	2.861 (N12A-O11)
2.972 (N4C-O11)		3.080 (N12C-O11)
2.964 (N4A-O12)	2.830 (N11A-O12)	3.256 (N4A-S2)
2.806 (N4B-O12)	3.057 (N11C-O12)	3.237 (N4B-S2)
3.059 (N4B-O13)	2.959 (N11B-O14)	3.043 (N4B-O13)
2.960 (N4C-O13)	2.850 (N11C-O14)	3.185 (N4C-O13)
2.792 (N12A-O14)	2.925 (N3A-O13)	2.841 (N4A-O12)
2.862 (N12B-O14)	2.898 (N3B-O13)	2.846 (N4C-O12)
2.783 (N12C-O14)	3.018 (N3C-O13)	

In **3** (Fig. 3) a combination of O_w O_w' and single O_w bridging strategies is used. Here the direct $NH^+..O_{oxoanion}$ contacts cover the range 3.21 – 2.75 Å, the singly bridging O_w-$O_{oxoanion}$ contacts average to 2.85 Å, while that to the $O_w..O_w$' strand is shorter at 2.81Å, and the $O_w..O_w$' contact shortest of all at 2.70 Å. Once again in **4**, (Colour plate Fig. 2) we see that the direct $NH^+..O_{oxoanion}$ contacts are relatively long, in the range 3.11 – 2.78 Å (apart from the apparently shorter contact (2.65 Å) presenting as one component of disorder at N3C). The single-water bridge contacts $O_{oxoanion}$-O_w average to 2.82 Å, and those to and within the double-water O_w-O_w' bridged strand are at 2.85 and 2.71 Å respectively. Finally (see colour pages Fig. 4) in **2a** where only single-O_w bridges are used, direct $NH^+..O_{oxoanion}$ contacts are in the range 2.83 – 3.02 Å, and the average contact from the oxoanion to the O_w bridge is 2.73 Å.

From data such as this we conclude that polarisation of bridging water via its linkage to the doubly charged oxoanion enhances its H-bonding properties. This polarisation is presumed responsible for the extensive water-mediated H-bonding interaction running through these lattices, interconnecting O_w and O_w' molecules with lattice oxoanions and NH^+ donors from neighbouring protonated hosts, which is particularly noticeable in dinegative structures. Although we do not suggest that this lattice H-bonding will persist in aqueous solution, its importance suggests that the solvation which replaces it can be expected to impinge significantly on the energetics of complexation.

In the cleft-bound complexes, there is less concentration of charge on the anionic guest, but more opportunity for intermolecular interaction, given the incomplete protection of the anion from the lattice environment. One example is the perchlorate cryptate of $[H_8R3P]^{8+}$ [18], the only octaprotonated cryptate we have structurally characterised, where the cleft bound anions connect both intramolecularly and intermoleculary *via* O_w chains containing distances as short as 2.66 Å, to make an extensive lattice network. The direct NH^+ - $O_{oxoanion}$ contacts are, in contrast, close to 2.9 Å on average.

Finally the structure of the tosylate cryptate of R3Bp demonstrates the size exclusion expected of the large cation (Fig. 11). Even so, it is not devoid of interaction with the cationic host, in the lattice at least, as each anion exhibits one moderately short H-bond contact to one of the NH^+ functions of the cryptate. These direct H-bond contacts are often supported by indirect water-mediated links of shorter dimensions, acting as part of branched hydrate chains which run through the less hydrophobic section of the lattice.

3. Solution Equilibria

Complexation constants for binding of mono- and di-negative oxoanions by the cryptands can be determined by pH-potentiometry. For the more weakly-binding mononegative systems, NMR shifts can provide complementary information. For these studies we use the large size-excluded tosylate anion as the supporting electrolyte and all log *K* values are thus relative to this notional standard [19].

3.1 MONONEGATIVE OXOANIONS

NMR titrations (of anion into ligand at fixed pH) and pH-potentiometric titrations (of pH at fixed anion:ligand ratios) provide comparable values of the stability constants for binding of mononegative oxoanions by protonated R3Bm, R3F, and R3P hosts [15,20,21]; Table 2. The weak complexation at hexaprotonated levels for tetrahedral monoanionic oxoanions makes it difficult to obtain reliable data for protonation levels below 5. This has however been achieved for nitrate with the cleft binding host R3P as well as for ReO_4^- with the most basic cryptand R3Bm.

Although these binding constants for mononegative oxoanions are relatively low, the value for H_6R3Bm-nitrate is approximately an order of magnitude greater than that reported for O-bistrenH_6^{6+} (in tosylate supporting electrolyte) [22], and the detectable binding of perchlorate is noteworthy. Previous studies have reported either no perchlorate complexation [23] or tentative values of the order of 1 [7,24], and some workers use perchlorate as a supposedly inert electrolyte for potentiometric studies of other anions [25-27]. In general, we observe that nitrate and perrhenate are more efficiently complexed than perchlorate, and that complexation by the encapsulating cryptands, R3Bm and R3F, at protonation levels above 5 at least, is stronger than by the cleft-binding R3P; Table 2. The lower binding constants for R3F, as compared to R3Bm is attributed to the lower basicity of the otherwise similar ligand, R3F: overall $\Sigma \log K_i$ = 43.46 at the hexaprotonated level *cf.* 46.76 for R3Bm. Both of these cryptands have lower basicity than O-bistren (for which the $\Sigma \log K_i$ at this level is 47.90 [22]), C3-bistrpn (52.35 [28]), or the hepta-methylene-linked bistrien macrocyle (54.35). However, stability constants recorded [19] for oxoanion complexation by O-bistrenH_6^{6+} are in general one or two orders of magnitude smaller than those with $[H_6R3Bm]^{6+}$ or $[H_6R3F]^{6+}$ apart from the log *K* for nitrate binding by O-bistrenH_6^{6+}, 2.80 [22], which is practically the same as that determined by us for R3F. The lack of correlation with basicity in the comparison of O-bistren with our azacryptand hosts may derive from an overriding desolvation difference between the hosts. For a specific anion the efficacy of a particular host will be influenced by its desolvation cost as well as its basicity: the more hydrophilic the host, the larger this cost. O-bistren can be expected to have substantially greater hydration in its uncomplexed state than the more hydrophobic aromatic-linked azacryptand analogues, and thus to incur a larger enthalpic cost from desolvation.

TABLE 2. Stepwise formation for binding of mononegative oxoanions

Equilibria	R3Bm	R3F	R3P[f]
Nitrate			
$H_6L^{6+} + NO_3^- \rightleftharpoons H_6LNO_3^{5+}$	3.41±0.03[a] 3.74±0.09[d]	3.00±0.03[e]	2.67±0.04[b] 2.77±0.08[c]
$H_5L^{5+} + NO_3^- \rightleftharpoons H_5LNO_3^{4+}$	2.53±0.06[b]	2.22±0.08[e]	2.29±0.06[e]
Perchlorate			
$H_6L^{6+} + ClO_4^- \rightleftharpoons H_6LClO_4^{5+}$	3.24±0.04[b] 3.53±0.04[d]	2.34±0.03[e] 2.66±0.13[d]	nd[e] 2.56±0.05[d]
Perrhenate[g]			
$H_6L^{6+} + ReO_4^- \rightleftharpoons H_6LReO_4^{5+}$	3.71±0.10[b] 3.76±0.10[c]	nd	3.20±0.10[d]
$H_5L^{5+} + ReO_4^- \rightleftharpoons H_5LReO_4^{4+}$	3.45±0.09[b] 3.66±0.06[c]	nd	nd

nd = not detectable; [a] pH-metry [15]; [b] pH-metry [21]; [c] NMR [21]; [d] NMR [15]; [e] pH-metry [20];
[f] log *K* for $[H_4LNO_3]^{3+}$, $[H_3LNO_3]^{2+}$, $[H_2LNO_3]^{+}$: 2.18, 1.93, 1.70;
[g] log *K* for $[H_4LReO_4]^{3+}$, $[H_3LReO_4]^{2+}$, $[H_2ReO_4]^{+}$: 3.06, 2.81, 2.72.

The relative stabilities of cryptand associations with various oxoanions in solution are often consistent with structural features. The structure of the H_6R3Bm-ReO_4 cation [21] is not dramatically different from that of the ClO_4 analogue [15]: there is a similar mix of direct $NH^+–O_{oxoanion}$ and indirect (water-mediated) H-bonds of similar length, and one of the oxoanion O atoms in each case points in the direction of the bridgehead N uninvolved in H-bonding. However, in contrast to perchlorate, the perrhenate is involved in direct H-bonding with NH^+ donors from both ends of the cryptand. Furthermore, the larger size, and hence tighter cavity fit of the perrhenate anion (thermochemical radius [29] 260 pm for ReO_4^- vs 240 pm for ClO_4^-) may partly explain the higher stability of the perrhenate complexes, as could the lower competition expected from hydration in this larger and thus less polar oxoanion [29]. In the case of $[H_6R3F]^{6+}$, the larger size of the oxoanion perrhenate may preclude encapsulation.

3.2 TETRAHEDRAL DINEGATIVE OXOANIONS

Electrostatic interactions are accepted as a major determinant of stability in anion complexation [30]. We have data for the dinegative tetrahedral oxoanions, thiosulfate, sulfate, chromate, and selenate, for which charge is the major or only differentiator from the mononegative analogues. As discussed previously [12], the greater stability of the dianion-cyptand complexes means that ^1H NMR titrations are not an appropriate means for determining the formation constants. In contrast to the mononegative analogues, we have been able via pH-potentiometric titrations to determine significant binding of dinegative oxoanions down to the diprotonated host level for all cryptands studied (Table 3). As expected, for all cryptands, the stability of the complexes formed with the dinegative oxoanions decreases as the protonation level of the host decreases. At the hexaprotonated level, the cavity-binding hosts, R3Bm and R3F, form complexes of the greatest stability. However, R3P becomes the strongest complexant at lower protonation levels, i.e. levels of 2 for sulfate, and 3 or 4 for the other two oxoanions. It is probable that this greater stability arises from the greater conformational flexibility of the cleft-binding R3P host, which allows for more energetically advantageous positioning of the guest in lower protonation states than is possible in the cavity-binding hosts. The cleft conformation allows a smaller number of NH^+ donors to bind any particular anion [21] and is more flexible in response to repulsive interactions. However, extraction studies demonstrate lower lipophilicity of R3P as compared to the other cryptands [21] which will also impact on the size of the anion complexation constants via larger desolvation costs. We do not yet have supporting calorimetric data to establish the relative importance of enthalpic versus entropic contributions to the formation constants, but consider calorimetric experiments to be of the highest priority for future work.

Considering the dinegative anions alone, free energies of hydration are in all cases large. Literature data show the value for sulfate to be significantly larger than that of selenate [29,31] and while no measurements of the hydration energy of thiosulfate in aqueous solution have been reported, a gas phase study [31] shows that in that medium at least, its hydration energy, as measured by the loss of one water molecule at a given hydration level, is significantly less than that of selenate. Hydration is by no means entirely lost on complexation, as the structural discussion illustrates, but it is certainly reduced, and so the relative magnitude of an oxoanions hydration energy is expected to be influential in determining its complexation energetics.

TABLE 3. Stepwise formation for binding of dinegative oxoanions†

Equilibria	R3Bm	R3F	R3P
Sulfate			
$H_6L^{6+} + SO_4^{2-} \rightleftharpoons H_6LSO_4^{4+}$	6.57 ± 0.04	7.21 ± 0.03	6.08 ± 0.03
$H_5L^{5+} + SO_4^{2-} \rightleftharpoons H_6LSO_4^{3+}$	4.72 ± 0.07	5.21 ± 0.06	5.55 ± 0.03
$H_5L^{4+} + SO_4^{2-} \rightleftharpoons H_6LSO_4^{2+}$	3.70 ± 0.12	4.32 ± 0.06	5.19 ± 0.03
$H_5L^{3+} + SO_4^{2-} \rightleftharpoons H_6LSO_4^{+}$	3.47 ± 0.05	4.02 ± 0.03	4.84 ± 0.02
$H_5L^{2+} + SO_4^{2-} \rightleftharpoons H_6LSO_4$	3.06 ± 0.10	3.37 ± 0.06	4.12 ± 0.04
Selenate			
$H_6L^{6+} + SeO_4^{2-} \rightleftharpoons H_6LSeO_4^{4+}$	7.24 ± 0.05	7.27 ± 0.06	5.38 ± 0.06
$H_6L^{5+} + SeO_4^{2-} \rightleftharpoons H_6LSeO_4^{3+}$	5.39 ± 0.04	5.38 ± 0.08	4.93 ± 0.06
$H_6L^{4+} + SeO_4^{2-} \rightleftharpoons H_6LSeO_4^{2+}$	4.77 ± 0.06	4.53 ± 0.08	4.66 ± 0.08
$H_6L^{3+} + SeO_4^{2-} \rightleftharpoons H_6LSeO_4^{+}$	4.18 ± 0.04	4.15 + 0.06	4.34 ± 0.06
$H_6L^{2+} + SeO_4^{2-} \rightleftharpoons H_6LSeO_4$	3.64 + 0.08	3.52 + 0.08	3.87 ± 0.07
Thiosulfate			
$H_6L^{6+} + S_2O_3^{2-} \rightleftharpoons H_6LS_2O_3^{4+}$	8.51 ± 0.05	7.65 ± 0.07	6.00 ± 0.06
$H_6L^{5+} + S_2O_3^{2-} \rightleftharpoons H_6LS_2O_3^{3+}$	6.40 ± 0.09	5.11± 0.10	5.63± 0.05
$H_6L^{4+} + S_2O_3^{2-} \rightleftharpoons H_6LS_2O_3^{2+}$	5.49 ± 0.08	5.09 ± 0.10	5.30 ± 0.06
$H_6L^{3+} + S_2O_3^{2-} \rightleftharpoons H_6LS_2O_3^{+}$	4.74 ± 0.05	4.56± 0.06	4.95 ± 0.05
$H_6L^{3+} + S_2O_3^{2-} \rightleftharpoons H_6LS_2O_3$	4.12 ± 0.10	3.83 ± 0.08	4.25 ± 0.06

† Analogous data for carboxylate anions is presented and discussed in references 12 and 14

Less efficient competition from solvation equilibria may explain the generally larger free energies of complexation of the thiosulfate ion by these protonated cryptand hosts and vice versa in the case of sulfate. Other things being equal, selenate might then be expected to be more efficiently complexed than sulfate by all three hosts. The fact that this expectation is fulfilled only for R3Bm implicates geometric factors in modifying the complexation energetics across the series of cryptates. Indeed, we observe that the influence of the basicity of the host can be overridden by geometric factors in binding of dinegative oxoanions, as illustrated by the complementarity of fit for sulfate with $[H_6R3F]^{6+}$ (which behaves as though smaller/more spherical than the $[H_6R3Bm]^{6+}$ analogue), and may explain the corresponding larger log K values in this system, Table 3 [14]. However, where comparisons are possible, the overall H-bond distances do not consistently reflect the difference in strength of interaction between host and guest as measured in solution, at least within the series of dinegative anion cryptates. This is despite the demonstration by quantitative solution measurements (Table 3) of much higher (by several orders of magnitude) formation constants in specific cases.

The lack of correlation with enthalpic (i.e. bond length-related) data highlights the complexity of the system, and the need to consider solvation/desolvation-related effects as contributing significantly to the free energies of complexation.

4. Conclusions

All structures presented here show remnant hydration of the encryptated oxoanions. The direct charge-assisted NH^+-$O_{oxoanion}$ hydrogen bonds are not generally the shortest H-bond contacts. Oxoanion contacts to bridging water are, in general, noticeably shorter, and particularly those involving the O_w-O_w' links. This arises from lack of geometric restraint affecting these hydrate water molecules together with their polarisation deriving from the influence of dinegatively charged anions.

The tenacious retention of molecules of hydrated water by the oxoanions adds to the difficulty of designing receptors, for example, using computational methods. In most circumstances the anion must be considered as including some level of remnant hydration involving bound water molecules which alter its shape, polarity and hydrogen bonding properties. On the basis of present structural evidence we can see no way to predict the disposition of such hydration even in the solid state.

Formation constants of dinegatively charged anion cryptates are much larger than the analogous values for the mononegative analogues perchlorate, perrhenate, or nitrate, which are not large to enough to generate interference in the determination of dinegatively charged ions.

The effects of solvation/desolvation equilibria on the thermodynamics of complexation can be discerned in the pattern of formation constant values across a series of azacryptand hosts or anionic guests. In the dianionic series thiosulfate, selenate, sulfate, formation constants, to a first approximation, decrease as the hydration energy of the anion increases, reflecting the enthalpic cost of desolvation.

Solvation/desolvation effects in the cryptand also complicate the expected simple dependence of stability constant on host basicity. For example the aliphatic cryptand O-bistren shows lower formation constants than the less basic aromatic analogues such as R3F, which we attribute to the greater desolvation cost of complexation with the former, more hydrophilic host.

The effects of host conformation cannot be ignored and these are particularly evident with the cleft-binding pyridine-spaced host R3P. In consequence of relative access to the appropriate number of NH^+ donors, this host is less effective than cavity binding analogues R3Bm and R3F at hexaprotonated levels, but more effective at protonation levels less than four.

The effectiveness of the cleft-binding strategy is well demonstrated for oxalate, which in addition to moderately strong H-bonds retaining one end of the carboxylate in the cleft, also exhibits a short, possibly symmetric, H-bond linking the protruding carboxylate functions of the other end.

5. Acknowledgements

We thank the trustees of the Analytical Trust Fund of the Royal Society of Chemistry UK for a SAC research studentship, and are grateful to EPSRC for access to the mass-spectrometry service at Swansea.

6. References

1. Lehn, J.-M. (1978) *Pure Appl. Chem.* **50**, 871-892.
2. Christen, K. (2000) *Environ. Sci. Technol.* **34**, 374A-375A.
3. Greer, M.A., Goodman, G., Pleus, R.C., Greer, S.E. (2002) *Environ. Health Perspect.* **110**, 927-937.
4. Knobeloch, L., Salna, B., Hogan, A., Postle, J., Anderson, H. (2000) *Environ. Health Perspect.* **108**, 675-678.
5. Fruchter, J. (2002) *Environ. Sci. Technol.* **36**, 464A-472A.
6. Nriagu, J.O., Niebor, E. (eds.) (1988) *Chromium in the Natural and Human Environment.* Wiley, New York.
7. Lehn, J.-M., Sonveaux, E., Willard, A.K. (1978) *J. Am. Chem. Soc.* **100**, 4914-4916.
8. Dietrich, B., Fyles, D.L., Fyles, T.M., Lehn, J.-M. (1979) *Helv. Chim. Acta* **62**, 2763-2787.
9. Bazzicalupi, C., Bencini, A., Bianchi, A., Cecchi, M., Escuder, B., Fusi, V., Garcia-España, E., Giorgi, C., Luis, S.V., Maccagni, G., Marcelino, V., Paoletti, P., Valtancoli, B. (1999) *J. Am. Chem. Soc.* **121**, 6807-6815.
10. Scheerder, J., Engerberson, J.F.J., Reinboudt, D.N. (1996) *Rec. Trav. Chim. Pays-Bas* **115**, 307-320.
11. Schmidtchen, F.P., Berger, M. (1997) *Chem. Rev.* **97**, 1609-1646.
12. McKee, V., Nelson, J., Town, R.M. (2003) *Chem. Soc. Rev.* **32**, 309-325.
13. Nelson, J., McKee, V., Maubert, B., Pál, I., Town, R.M. (2001) *Dalton Trans.* 1395-1397.
14. Nelson, J., Nieuwenhuyzen, M., Pál, I., Town, R.M. (2004) *Dalton Trans.* 2303-2308.
15. Hynes, M.J., Maubert, B., McKee,V., Town, R.M., Nelson, J. (2000) *Dalton Trans.* 2853-2859.
16. Nelson, J., Nieuwenhuyzen, M., Pál, I., Town, R.M. (2002) *Chem. Commun.* 2266-2267.
17. Nelson, J., Nieuwenhuyzen, M., Pál, I., Town, R.M. (2004) *Dalton Trans.* 229-235.
18. McKee, V., Morgan, G.G. (2003) *Acta Cryst.* **C59**, 150-154.
19. Dietrich, B. (1993) *Pure Appl. Chem.* **65**, 1457-1464.
20. Pál, I. Ph.D. Thesis, Queen's University Belfast, 2002.
21. Farrell, D., Gloe, K., Gloe, K., Goretzki, G., McKee, V., Nieuewenhuyzen, M., Nelson, J., Pál, I., Stephan, H., Town, R.M., Wichmann, K. (2003) *Dalton Trans.* 1961-1968.
22. Dietrich, B., Guilheim, J., Lehn, J.-M., Pascard, C., Sonveaux, E. (1984) *Helv. Chim. Acta* **67**, 91-104.
23. Heyer, D., Lehn, J.-M. (1989) *Tetrahedron Lett.* **27**, 5869-5872.
24. Cullinane, J., Gelb, R.I., Maregulis, T.N., Zompa, L.J. (1982) *J. Am. Chem. Soc.* **104**, 3048-3053.
25. Motekaitis, R.J., Martell, A.E., Lehn, J.-M., Watanabe, E. (1982) *Inorg. Chem.* **21**, 4253-4257.
26. Motekaitis, R.J., Martell, A.E., Murase, I., Lehn, J.-M., Hosseini, M.W. (1988) *Inorg. Chem.* **27**, 3630-3636.
27. Bazzicalupi, C., Bencini, A., Bianchi, A., Fusi, V., Paoletti, P., Vantancole, B. (1994) *Perkin Trans. 2,* 815-820.
28. Hosseini, M.W., Lehn, J.-M. (1988) *Helv. Chim. Acta* **71**, 749-756.
29. Moyer, B.A., Bonnesen, P.V. (1997) in *Supramolecular Chemistry of Anions*. Bianchi, A., Bowman-James, K., Garcia-España, E. (eds.), Wiley-VCH, New York, pp. 1-41.
30. Arranz, P, Bencini, A., Bianchi, A., Diaz, P., Garcia-España, E., Giorgi, C., Luis, S.V., Queroi, M., Valtancoli, B. (2001) *Perkin Trans.* 1765-1770.
31. Blades, A.T., Klassen, J.S., Kebarle, P. (1995) *J. Am. Chem. Soc.*,**117**, 10563-10571.

13. CYCLODEXTRIN COMBINATIONS WITH AZOCOMPOUNDS

E. LUBOCH, Z. POLESKA-MUCHLADO,
M. JAMRÓGIEWICZ AND J. F. BIERNAT
Gdańsk University of Technology,
Narutowicza St 11-12, 80-952 Gdańsk, Poland

1. Introduction

Cyclodextrins (CDs) are cyclic oligosaccharides consisting of various numbers of α-1,4-linked glucose units (α-CD, β-CD and γ-CD for hexa-, hepta- and octamers, respectively). Larger homologues are recently described, cf., [1,2]. CDs form inclusion complexes with a variety of organic compounds [3]. Complexation influences solubility of organic guests in water, and stabilizes the guest molecules towards oxidation and other unwanted reactions. Also CDs catalyze some reactions playing role of microreactors or serve as building blocks for nanotubes. For latest papers or books dealing with cyclodextrin chemistry see [4-7] .

In this review are considered combinations of CDs and their derivatives, in which compounds bearing azo unit are chemically bonded or mechanically associated with the host. In this case, the binding properties are modified by light when the guest molecule undergoes *trans* $\rightleftharpoons$ *cis* isomerization [6a].

2. Combinations of Cyclodextrins with Guests

2.1 CYCLODEXTRINS AND DIAZONIUM SALTS

Cyclodextrins form complexes with diazonium salts. Rates and product distribution for spontaneous dediazoniation of isomeric methylbenzenediazonium salts in the presence of α-, β-, and γ-CD in aqueous acidic solution were found to be independent on acidity and CDs concentration [8], cf. [9].

Exhaustive studies of β-CD inclusion complex with model *p*-nitrobenzenediazonium (PNBD) tetrafluoroborate in acidic aqueous solutions were performed by Gonzalez-Romero *et al.* [10]. Spectrophotometric and polarographic experiments suggest that β-CD promotes homolytic dediazoniation of the guest because large amount of the reduction product nitrobenzene is formed at the expense of *p*-nitrophenol. Quantitative conversion to products is achieved at any concentration of β-CD when pH < 5.

Karsten Gloe (ed.), Macrocyclic Chemistry: Current Trends and Future Perspectives, 203–218.
© 2005 *Springer. Printed in the Netherlands.*

β-CD added to the diazonium salt solution hinders the reduction of the nitro group since apparent peak potential E_{app} ($-NO_2$) is shifted toward more negative values, in contrast with the E_{app} shift observed for the one-electron reduction of the $-N_2^+$ group to form the aryldiazenyl radical, $ArN_2^{\cdot}$, which is shifted toward more positive values. Quantitative analysis of the E_{app} and peak current i_p values allowed estimations of the binding constant of ArN_2^+ and, for the first time, of the binding constant of $ArN_2^{\cdot}$ (or $Ar^{\cdot}$) with β-CD, which is much higher than that of the parent ArN_2^+. Either the decrease in peak currents or E_{app} shifts are interpreted in terms of the formation of a 1:1 inclusion complex between β-CD and PNBD cation, with the $-NO_2$ group inserted into the CD cavity. The specific spatial orientation of complexed PNBD ions allows solvation of the positive charge of the $-N_2^+$ group mainly by the secondary hydroxyl groups of β-CD, favoring further reaction to yield a highly unstable transient intermediate, a *Z*-diazo ether, which was detected experimentally.

Gonzalez-Romero *et al.* [9] also found that cyclodextrin does not affect coupling reactions of 2-, 3-, and 4-methyl benzenediazonium salt with typical electrophiles. Recent study on coupling of *o*-ethoxybenzenediazonium salt with pyrrole revealed important influence of β-cyclodextrin on the product distribution and the yield of particular products [11,12]. Figure 1 shows changes of the ratio of mono- and disubstituted pyrrole that are formed as main products. Similarly, cyclodextrin affects coupling of *o*-nitrobenzenediazonium salt with pyrrole and imidazole. Cyclodextrin also significantly increases the yield of macrocyclic products obtained by coupling pyrrole and imidazole with bisdiazonium salts, cf. [13,14].

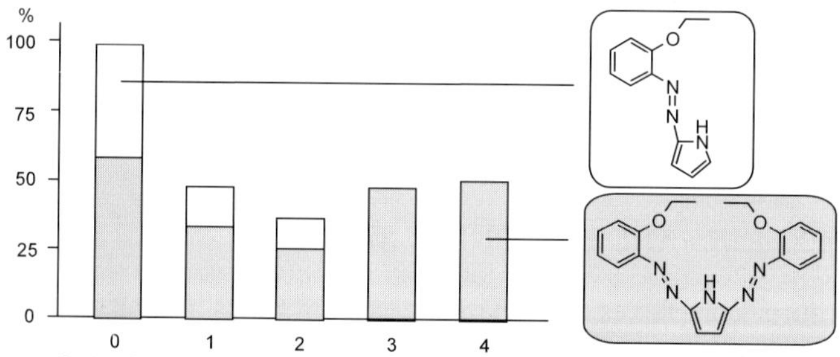

Figure 1. Total yield [%] and ratio of 2-substituted to 2,5-disubstituted pyrrole coupling products with *o*-ethoxybenzenediazonium salt in the presence of β-CD.

2.2 CYCLODEXTRINS AND AZOALKANES

Some papers are devoted to analysis of co-conformation of CD complexes with azo-compounds. The structures of β-cyclodextrin complexes of 2,3-diazabicyclo[2.2.2]oct-2-ene **1** and its 1-isopropyl-4-methyl derivative **2** (Figure 2) in solution have been investigated by means of induced circular dichroism (ICD) and MM3-92 force-field calculations [15,16]. The experimental and theoretical ICD as well as the calculated low-energy complex geometries suggest solution co-conformations in which **1** adapts a lateral arrangement with the ethano bridge of the guest penetrating deepest into the cavity and the azo group aligning parallel to the plane of the upper rim. In contrast, the alkyl derivative **2** prefers a frontal co-conformation with the *iso*-propyl group penetrating deepest into the cavity and the azo group aligning perpendicular to the plane of the upper rim. With respect to the co-conformational variability of the complexes of the two azoalkanes, it was observed that the nearly spherical guest **1** forms a geometrically better defined complex than the sterically biased, alkyl-substituted derivative **2**.

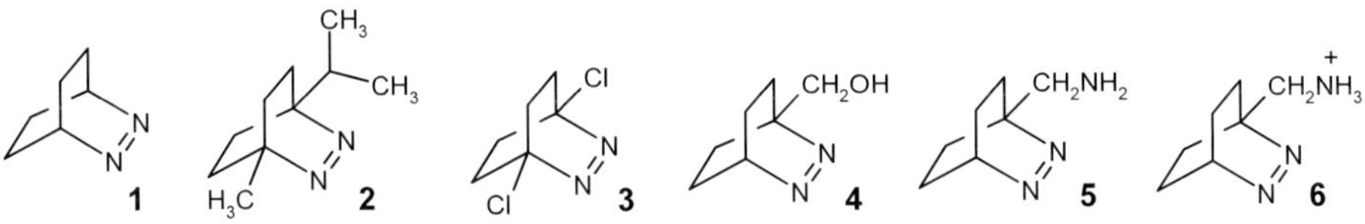

Figure 2

Consecutively, the structural, kinetic and thermodynamic investigation of substituent effect on host-guest complexation with *β*-CD for compounds **1** – **6** (Figure 2) was also performed by time-resolved and steady-state fluorimetry, UV spectrophotometry, ICD measurements, and ^{1}H NMR spectroscopy [17]. The binding constants for **1** (1100 M^{-1}), **2** (900 M^{-1}), **3** (1900 M^{-1}), **4** (180 M^{-1}), **5** (250 M^{-1}), and **6** (ca. 20 M^{-1}) were obtained by UV, NMR, and ICD titrations. The ICD was employed for the assignment of the solution structures of the complexes, in particular the relative orientation of the guest in the host.

2.3 COMPLEXES OF AZO DYES WITH CYCLODEXTRINS

Numerous papers on CDs complexation of azo dyes concern stability, thermodynamics and kinetics of complexation and co-conformation. The studies are followed frequently by theoretical calculations.

Hirai *et al.* [18] found that *β*-CD forms 1:1 complexes with Methyl Orange and Congo Red in water, whereas *γ*-CD forms 1:2 complex with Methyl Orange and 1:1 complex with Congo Red. Induced circular dichroism of the dyes was studied allowing determination of stability constants.

Yoshida and Hayashi [19] determined formation constants for *α*-CD inclusion complexes with azobenzene derivatives **1** listed in Figure 3 that are in the range of 5.0 x 10^3 to 1.75 x 10^4 $mol^{-1}\cdot dm^3$. The rate constants for inclusion reaction and its mechanism are dependent on the substituent group Y. For Y = CO_2^- the reaction is very fast (milliseconds) and proceeds as one step process. When Y = AsO_3H^- the reaction is slow. When Y = SO_2NH_2 or SO_3^-, the inclusion proceeds as a two-step process.

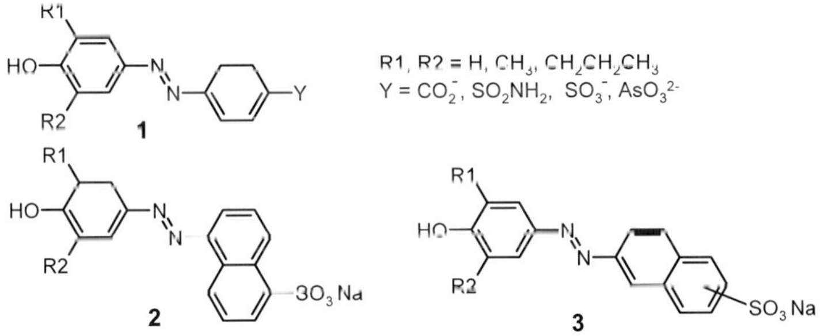

Figure 3

The naphthalene derivatives **2** are included into CD from phenolic residue site. For compounds **3** with different location of sulfonic group in naphthalene residue the inclusion direction depends on the degree of steric hindrance, cf. [20].

The mechanism of guest complexation by *α*-cyclodextrin was studied by high-pressure investigations. The first volume profiles obtained for azo dyes **1** – **4** (Figure 4) are presented as a new approach in understanding inclusion phenomena.

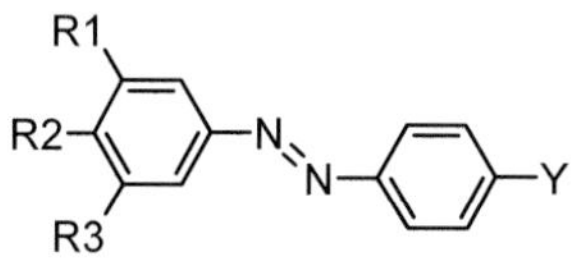

	R1	R2	R3	Y
1.	H	$N(Et)_2$	H	SO_3^-
2.	CH_3	OH	COO^-	SO_3^-
3.	CH_3	OH	CH_3	SO_3^-
4.	CH_3	OH	COO^-	SO_2NH_2

Figure 4

The behavior of the uncomplexed dyes was first studied in aqueous solutions to rule out any competition reaction [21]. At pH ca. 6.5 the azo guests **1**, **3** and **4** exist as a monovalent anion and **2** as a divalent anion. Under the experimental conditions used for the stopped-flow kinetic studies, only monomeric dyes are present indicating no aggregation by π-π stacking interactions. Two-step kinetics of complexation was evidenced. NMR experiments and kinetic evidences have shown that only directional binding of the dye *via* the sulfonate/sulfonamide group through the wide rim of the *α*-CD was possible. 2D NMR experiments served for a molecular dynamics calculation leading to a structural representation of the intermediate and final complexes.

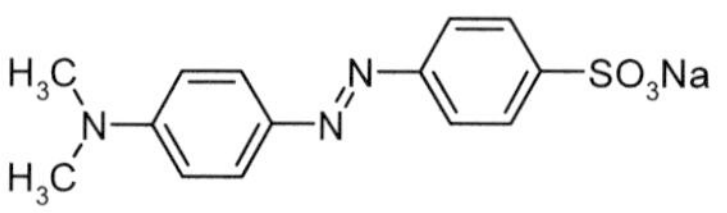

Methyl Orange

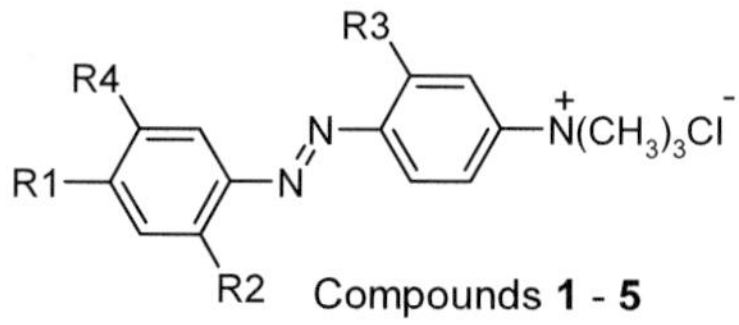

Compounds **1** - **5**

Compound	**R1**	**R2**	**R3**	**R4**	**Association constants in Water** K **(mol^{-1} dm^3)**
Methyl Orange					3600
1	H	H	H	H	860
2	CH_3	H	H	H	1300
3	CH_3	CH_3	CH_3	H	550
4	Cl	CH_3	CH_3	H	460
5	H	CH_3	CH_3	NO_2	290

Figure 5

The equilibrium constants (K) for the inclusion complex formation of *β*–cyclodextrin with Methyl Orange (MO) and substituted azoanilinium chlorides (Figure 5) was determined spectrophotometrically [22]. Based on the results, the substituent effect on the inclusion complexation of *β*-CD was discussed. Further, the solvent effect on the inclusion complexation of Methyl Orange with *β*-CD and heptakis(2,6-di-O-methyl)-*β*-cyclodextrin were examined in aqueous organic mixtures. The authors suggest orientation of compounds **3** and **5** inside the CD as given in Figure 6.

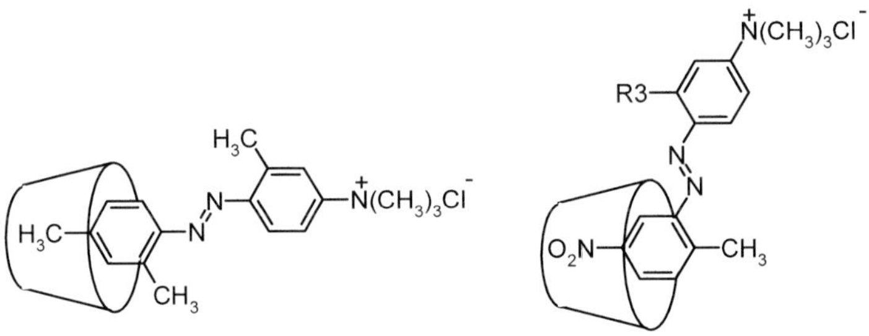

Figure 6

The anionic azo dye Orange II complexation with α-, and β-CD follow one-step reaction mechanism [23]. Data obtained by the stopped-flow method and UV as well as NMR are rationalized in terms of inclusion direction of the naphthalene side of the guest azo molecule along the short axis into the cavity of both α- and β-CD.

A method for the determination of Orange II has been established and the supramolecular system of Orange II with cyclodextrins has been studied by polarography. A sensitive and stable linear-sweep voltametric peak was obtained at 364 mV (*vs.* SCE) in HAc – NaAc (pH 3.62) buffer solution [24]. Polarography has demonstrated that Orange II forms 1:1 inclusion complex with α-, β-, γ-CDs and three modified CDs. Modified β-CDs such as hydroxypropyl-β-CD, dimethyl-β-CD and trimethyl-β-CD exhibit stronger inclusive ability. The inclusive ability of γ-CD with Orange II is the strongest. The inclusive ability of α-CD is very weak. Suggested structures of complexes are shown in Figure 7.

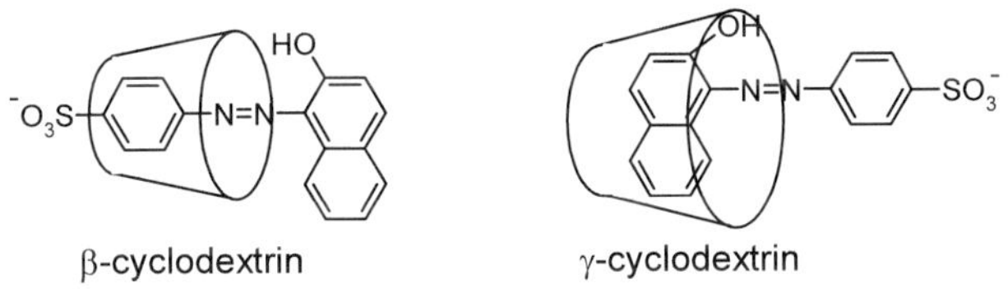

Figure 7

NMR studies of aqueous solutions (pH 7) of γ-CD and the azo dye Congo Red show distinct, concentration-independent ^{1}H NMR signals. A very stable 1:1 pseudorotaxane (K_{11} = 38,000 M^{-1}) is formed. The structure of the 1:1 pseudorotaxane involves fast motion of the γ-CD ring along the Congo Red backbone, leaving the outer naphthalene rings free. This entity undergoes structural reorganization and dimerizes to form the 2:2 adduct (K_{22} = 13 M^{-1}). Variable-temperature spectra did not lead to coalescence and allowed for the calculation of the corresponding thermodynamic parameters. Formation of the 1:1 complex is favorable and exothermic, whereas formation of the 2:2 entity is also favorable but endothermic. The corresponding values for entropy change are both positive. Isothermal titration calorimetry studies confirm the NMR findings [25].

Effect of β-cyclodextrin on *cis* $\rightleftarrows$ *trans* isomerization of azobenzenes was studied by Sanchez and de Rossi [26]. It was found that the *cis-trans* thermal isomerization of *p*-Methyl Red, *o*-Methyl Red and Methyl Orange is inhibited by β-CD at constant pH. The isomerization rate decreases 4, 8, and 1.67 times, respectively, in a solution containing 0.01 M β-CD. This effect was attributed to the formation of inclusion complexes hindering rotation of the –N=N– bond. Isomerization of Methyl Yellow and naphthalene-1-azo-[4'-(dimethylamino)benzene] requiring mixed organic-aqueous solvent to increase solubility is not affected by the presence of CD as the solvent replaces azobenzene derivatives from cyclodextrin cavity. More, in the case of the two last dyes the isomerization proceeds more probably by inversion of nitrogen atom of –N=N– group. The inversion requires less volume changes than rotation, so it is less hindered by the complexation with *β*-CD. The arrangement of each of the above dyes in *β*-cyclodextrin inclusion complex, the strain energy involved and the interaction forces driving towards the different kinds of stable structures were theoretically assessed [27]. The same authors studied theoretically inclusion complexes of 1:2 stoichiometry between azobenzenes and cyclodextrins [28].

Organoselenium bridged *β*-cyclodextrin dimers as specific host molecules were investigated [29]. The authors showed that the "dimeric" molecules (Figure 8) not only significantly enhance the binding affinity of the parent *β*-cyclodextrin but also remarkably extend its molecular recognition abilities towards different host molecules, including Methyl Orange and Methyl Red.

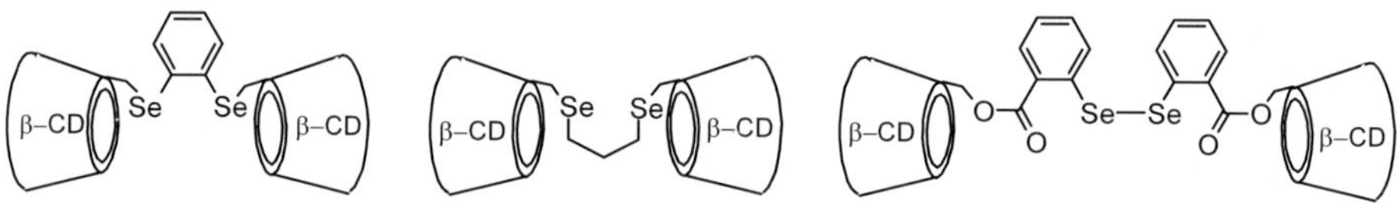

Figure 8

2.4 MOLECULAR MACHINES

Due to different complexation ability of *cis*- and *trans*-azobenzenes their assemblies with CDs show signal-triggering functions leading to molecular machines. Tailoring of such machines remains challenging research area. New materials based on cyclodextrin–polyrotaxane have been investigated and resumed by Nepogodiev and Stoddart [30], cf., [31]. Murakami *et al.* [32] reported synthesis of the first light driven molecular shuttle based on rotaxane consisting of CD and a compound bearing azobenzene with two paraquat units. In this rotaxane the *α*-cyclodextrin moves forth and back between two "stations" in response to external stimuli, i.e., by alternating illumination of UV and visible light. The cyclodextrin forms inclusion complex with *trans*-azobenzene moiety (Figure 9, cf. [30]).

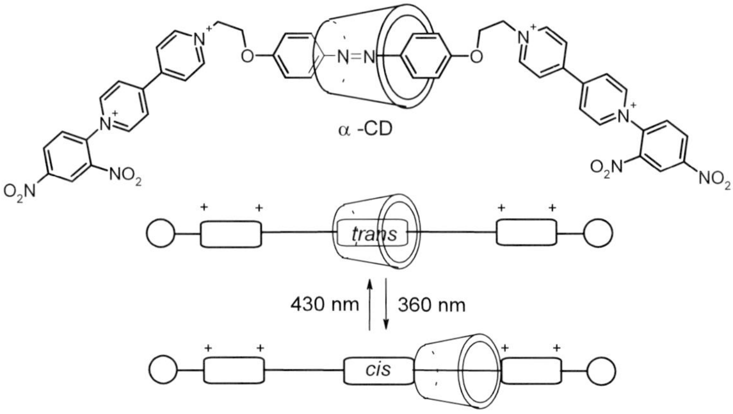

Figure 9

Intermolecular interaction between the pseudorotaxane molecules (Figure 10) caused by N....H-O hydrogen bonds with *cis*-azobenzene-group was found to retard the photoinduced *cis* ⇌ *trans* isomerization [33]. The *cis*-azobenzene is sterically less crowded and has a higher proton affinity then the *trans*-azobenzene groups. By forming the hydrogen bond, the *cis* form of azobenzene seems to be stabilized. Addition of large excess of γ-cyclodextrin to a DMSO solution of **3** (*cis*) retards the photoinduced *cis*- to *trans*-isomerization, but γ-cyclodextrin contained in **2** blocks the isomerization much more effectively.

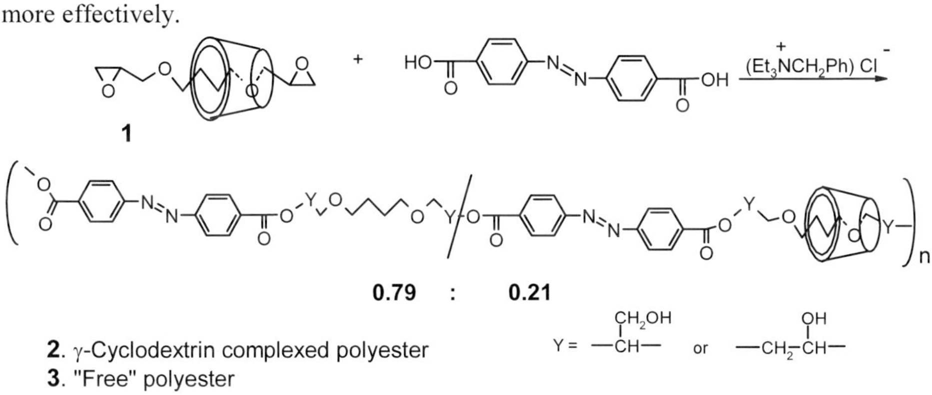

Figure 10

Anderson *et al.* [34] described rotaxanes of cyclodextrins with azo dyes that were synthesized by coupling of bis-diazonium salt **1** or **2** with water-soluble 2-naphthol derivative in the presence of α- or β-cyclodextrin **3** (Figure 11). The products were isolated by paper chromatography. The yield for rotaxane **4** varied from 12 to 15 % for α- and β-CD, respectively

If the 2-naphthol derivative was replaced by 2,6-dimethylphenol, the main unexpected product is the [3]rotaxane (12%) among [2]rotaxane (9%) and the unencapsulated dye **5** (Figure 11) [35]. The reaction mixture was separated by ultrafiltration. In [3]rotaxane there are three possible relative orientations of the CD units, but the compound is produced as a single stereoisomer. In DMSO [2]rotaxane exists preferentially in the conformation shown in Figure 11.

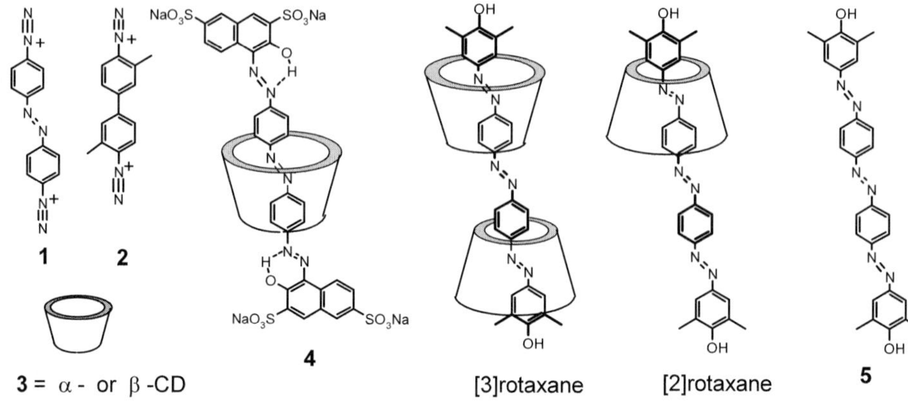

Figure 11

A review on molecular machines useful for the design of chemosensors emphasizing, among others, modified CDs as molecular sensors, and on rotaxanes and catenanes is published [36].

3. Cyclodextrins and Polymers

Electrostatic self-assembly was combined with supramolecular chemistry to obtain inclusion complexes of a polymeric nonlinear optical (NLO) active dye and modified β-cyclodextrin with induced chromophore orientation [37]. The polyanion is a N,N-diallyl-aniline and sodium-2-acrylamido-2-methylpropanesulfonate copolymer functionalized with pendant azo group. The modified β-cyclodextrin oligo-cation was obtained by treatment of heptakis(6-deoxy-6-iodo-β-cyclodextrin) with excess pyridine. A linear polyamine, chitosan, was also combined with the polyanion, for comparison. Films were deposited on glass slides by dipping them alternatively in aqueous solutions of the cation and the polyanion. UV-visible spectra indicate dye aggregation and suggest the formation of an inclusion complex of the dye with the cyclodextrin, thus isolating the chromophores.

The adsorption isotherms of acid azo dyes onto water soluble and insoluble polymers containing cyclodextrin were measured in aqueous solution. The adsorption of dyes on both types of polymers increased with increase in the ratio of hydrophobic components in the dyes [38]. Dyes derivative of dialkylaminobenzene were used for the dyeing of nylon 6 and 6,6 in the presence of interacting β-CD [39]. β-CD showed good levelling properties in the dyeing of polyamide fibers. The observed effect can be due to the formation of complexes between β-cyclodextrin and dyes.

The copolymerization of a methylated-β-cyclodextrin 1:1 host-guest compound of styrene with various molar ratios of sodium 4-(acrylamido)-phenyldiazosulfonate carried out in water with free radical initiator is described [40]. Depending on the amount of sodium 4-(acrylamido)-phenyldiazosulfonate incorporated in the copolymer, water- or DMF-soluble copolymers of high molar mass were obtained. Irradiation of the copolymers with UV light in solution resulted in rapid decomposition of the azo chromophore. Irradiation of the polymers as films led to crosslinking and thus to insolubility.

A reactive dye inclusion compound **1** (Figure 12) was prepared by condensing disodium salt of 7-[4-(methylamino)phenylazo]-2,4-naphthalenedisulfonic acid with cyanuric chloride and sodium 3-aminobenzenesulfonate (1:1:1) in the presence of hexakis(2,3,6-tri-O-methyl)-α-cyclodextrin [41]. Solutions of the bright yellow rotaxane **1** are more resistant to chemical bleaching than the uncomplexed dye. This compound anchored to mercerized cotton **2** also showed more photofading resistance [42].

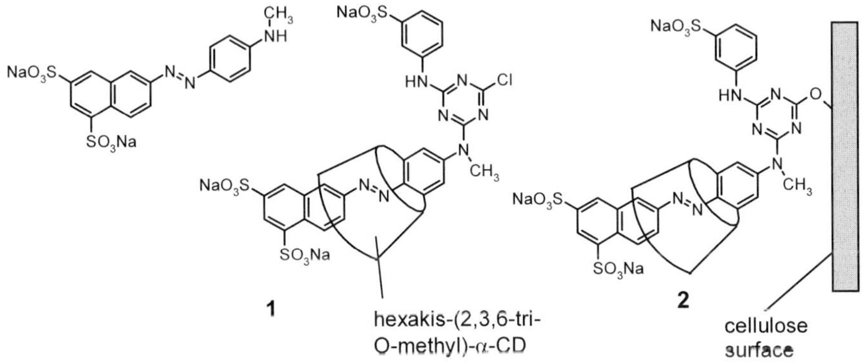

Figure 12

4. Azobenzenes Chemically Bonded to Cyclodextrins

The hetero-dimerization behavior of dye-modified β-cyclodextrins with native CDs was investigated by means of absorption and induced circular dichroism spectroscopy in aqueous solution [43]. Three types of azo dye-modified β-CDs show different association behavior, depending on the positional difference and the electronic character of substituent connected to the CD unit in the dye moiety. *p*-Methyl Red-modified β-CD (**1**), which has a 4-(dimethylamino)azobenzene moiety connected to the CD unit at the 4' position by an amido linkage, forms an intramolecular self-complex, inserting the dye moiety in its β-CD cavity (Figure 13). **1** also associates with native α-CD by inserting the dye residue into the α-CD cavity. The association constants for such hetero-dimerization are 198 M^{-1} at pH 1.00 and 305 M^{-1} at pH 6.59, which are larger than the association constants of **1** for β-CD (43 M^{-1} at pH 1.00).

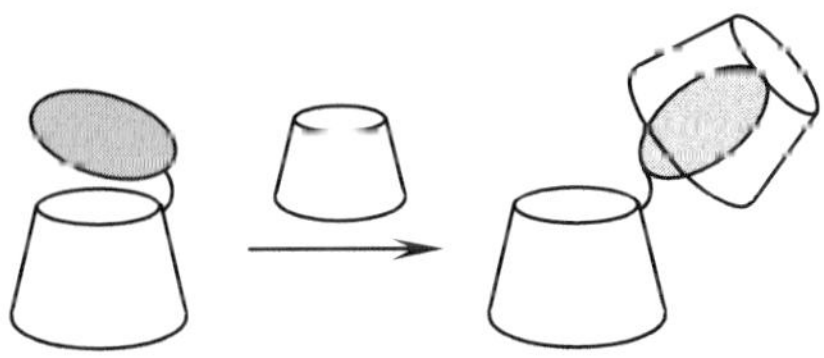

Figure 13

Permethylated tosylated α-cyclodextrin was coupled with aminohydroxyazobenzene. The product was dimerized to Janus [2]pseudorotaxane, which in turn was bis-azo coupled with 2-naphthol-3,6-disulfonic acid. The product undergoes self-association as shown in Figure 14 [44]. Note that the final product has *o*-quinone hydrazone structure.

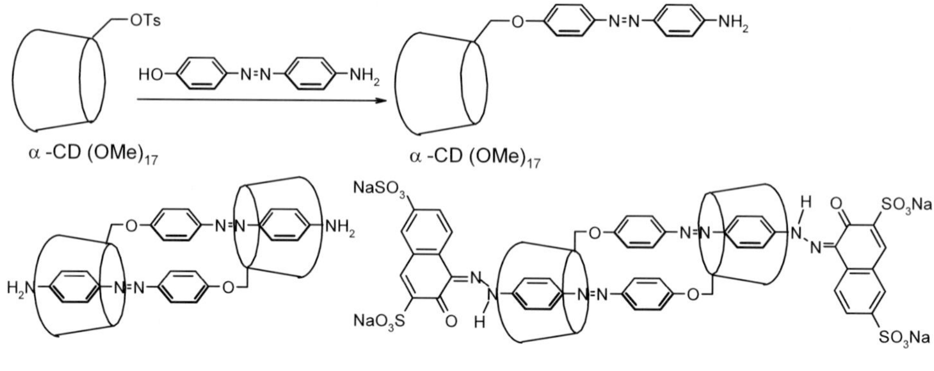

Figure 14

Ionically self-assembled monolayers possessing exceptional temporal and thermal stability of χ relative to poled polymers were prepared [45]. Molecules of non-polymeric azo dyes (Mordant Orange 10) form pseudorotaxanes with β-cyclodextrin in monolayers and polylayers.

The helical structure of peptides is stabilized by intramolecular host(CD) – guest(azobenzene) bridges involving side amino-acid's chains. The amide of N-acetyl-heptadecapeptide incorporating one aminoacid bearing azobenzene unit exhibit 54% helix content [46]. The same peptide incorporating both *trans*-azobenzene and α-cyclodextrin derivatives of two different amino acid residues (Figure 15) shows 82% helix content. The helix content changed by *trans* → *cis* photoisomerization indicating exclusion of *cis*-azobenzene unit from α-cyclodextrin cavity. Replacement of α-CD by β-CD in this peptide creates 94% helix content when azobenzene unit is in *cis* form, whereas for *trans* isomer the helix content decreases to 87%. The photoregulation of catalytic activity toward ester hydrolysis by prepared azobenzene tagged cyclodextrine-peptide hybrid with histidine unit was examined.

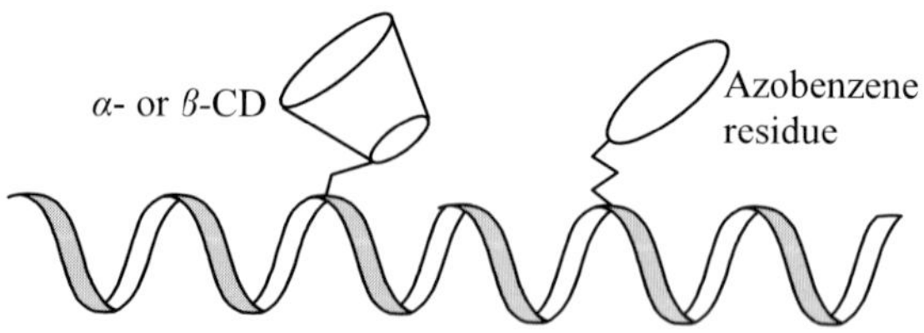

Figure 15

Inclusion properties of molecular nanotubes composed of crosslinked α-cyclodextrin was investigated [47]. Induced circular dichroism was used to probe the formation and dissociation of complexes between the nanotubes and azobenzene modified poly(ethylene glycol), either with or without a hydrophobic alkyl chain. The inclusion complex between the nanotubes and polymers formed at room temperature, and the polymers dissociated from the nanotubes with increasing temperature.

Flexible compound consisting of two aminoethylamino-β-cyclodextrin residues linked by azobenzene dicarbonyl moiety, and bearing two dansylglycyl groups (Figure 16) was synthesized. It was applied to detect chenodeoxycholic acid, ursodeoxycholic acid and hyodeoxycholic acid with high selectivity [48].

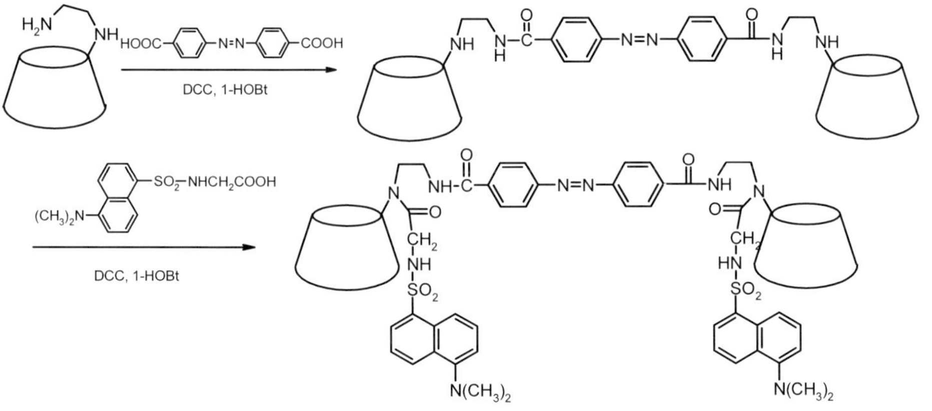

Figure 16

The series of photoresponsive, azobenzene modified CDs were prepared by Ueno *et al.* [49-51]. These compounds form complexes, in which either the *E*- and *Z*-forms of the azobenzene moiety is intramolecularly included [52]. In concentrated solutions, the parent compounds reversibly form association dimers (Figure 17). The intramolecularly complexed residues undergo expulsion upon purposely-added guest molecules to form new 1:1 complexes. The binding constants of the complexes with adamantanol and adamantane carboxylic acid are approximately five times stronger for *E*- than for *Z*-form leading to photocontrol of complexation. The authors suggest usefulness of such systems for construction of molecular devices with which reactions and catalytic properties can be regulated by light in an on-off fashion.

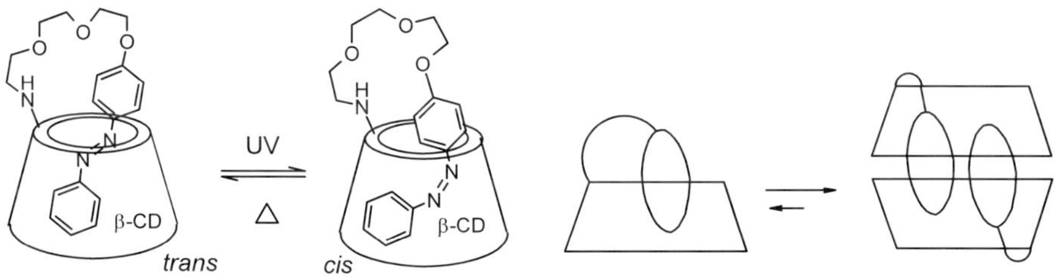

Figure 17

Fukushima *et al.* reported analogously modified β-CD [50] and γ-CD [51] with shorter spacer. Association of regioisomer *E* takes place in concentrated solutions. Upon UV irradiation the β-CD dimer dissociate and returns to its original form by visible light illumination. Isomer *Z* of this compound does not form dimers. In aqueous solution, the γ-CD derivative upon UV induced photoisomerization afforded 79% *cis* form at the photostationary state. The half-life of this isomer is 34 h in dark.

Ueno et al. found that amino-β-cyclodextrin acylated with Methyl Red (Figure 18) undergoes selfcomplexation [52]. Although Methyl Red itself changes color from yellow to red at pH 4-5; the cyclodextrin derivative is still yellow even at pH 1.6 suggesting prohibiting protonation of the dye moiety included in CD. However, aqueous solution of this cyclodextrin derivative at pH 1.6 changes color from yellow to red when the azobenzene moiety is thrown out from the cavity by purposely-added competing guest molecules. Camphor, fenchone, menthol, geraniol, nerol, cyclohexanol, and cyclooctanol compete with azobenzene residue to a lesser extent than *l*-borneol, 1-adamantanol, or 1-adamantanecarboxylic acid [53,54].

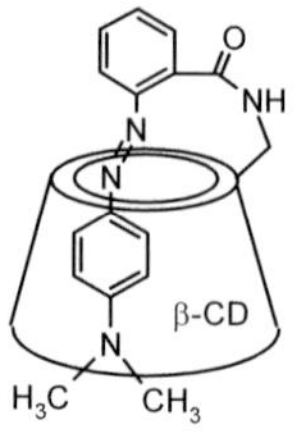

Figure 18

Ueno *et al.* [55,56] showed that *β*-cyclodextrin arched with azobenzene-4,4'-dicarbonyl- and azobenzene-4,4'-disulfonyl- residues (Figure 19) have enlarged hydrophobic cavity, particularly in the case of the *cis* isomers.

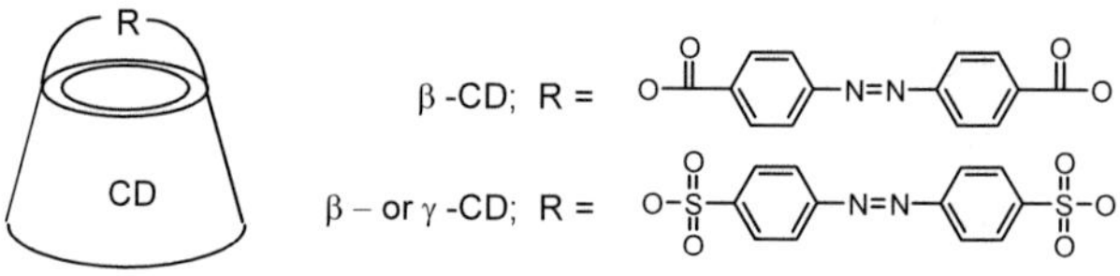

Figure 19

The respective AD and AE regioisomers of azobenzene-4,4'-disulfonyl derivative of *γ*-cyclodextrin were obtained by Hamada *et al.* [57] (Figure 19). The guest binding ability of *β*-CD is higher for *cis* compared to *trans*-azobenzene shallow cap. In contrast, complexation of *l*-borneol, 1-adamantane-carboxylic acid, cyclododecanol, nerol and *l*-menthol by *γ*-CD derivative before and after illumination suggest that the *cis* form is stabilized by including the *cis*-azobenzene moiety into the cavity (c.f. Figure 20), and that the *trans*-azobenzene moiety acts as an effective hydrophobic cap increasing the hydrophobicity of the environment around the cavity. Complexation process is photocontrolled; the guest releases from the cavity by UV illumination and binds back upon visible light irradiation or after spontaneous isomerization. At the photostationary state, the *cis*-isomer content of the AE regioisomer of *γ*-CD equals 47 %. The *cis* form converts back to *trans* form in dark with a half-life of 262 h at 25°C.

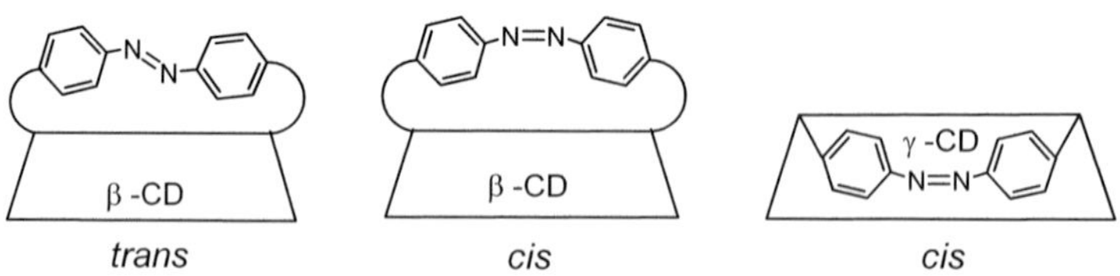

Figure 20

Jung *et al.* [58] obtained permethylated *α*-CD bridged with (nitrophenylazo)-phenol dye (Figure 21). This compound exhibits time-dependend UV-Vis and NMR spectra in aqueous solutions at room temperature in the dark. This phenomenon was interpreted in terms of self-aggregation involving intermolecular inclusion of the *p*-nitrophenylazophenol group with one of the two permethylated cyclodextrin residues of the original compound.

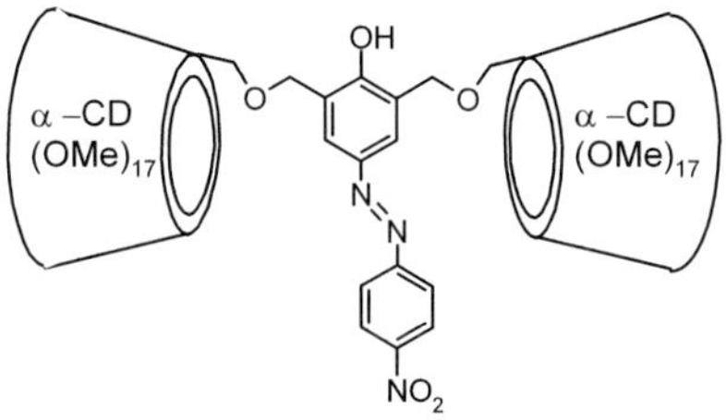

Figure 21

4,4'-Dihydroxyazobenzene derivative bearing two β-cyclodextrin units connected by ether bonding at both ends was synthesized by Aoyagi *et al.* [59]. This compound undergoes photoisomerization. The formed *cis* isomer returns to the stable form with half life of 54.8 h at 25°.

The first [5]super-cyclodextrin whose nano-sized cyclo-pentameric array is held only by a mechanical bond was synthesized by the pentakis-azo coupling of a new hermaphrodite monomer with 2-naphthol as a stopper, isolated by chromatography, and characterized by MS, 2D NMR, and visible spectral methods with the help of computer simulation [60]. Cyclic pentamer (as red film, 15% yield) accompanied with the corresponding monomer (20%) (Figure 22) was obtained.

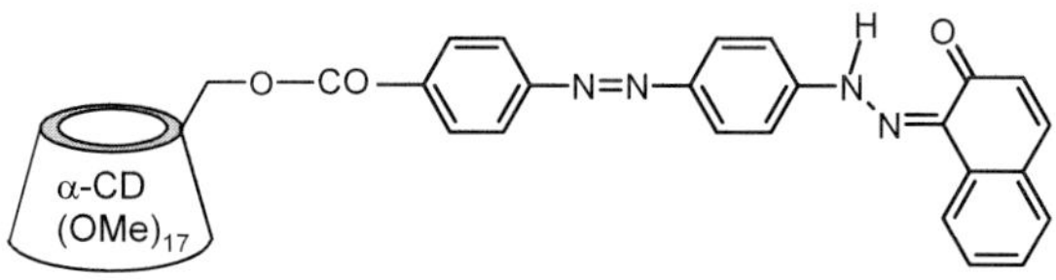

Figure 22

5. Chromatographic Application of Cyclodextrins

Adsorption chromatography on cellulose was examined for azo dyes that form complexes with CDs using aqueous solutions of α-CD as eluents. The cyclodextrin substantially increases the R_F values [61]. Commercial α-, β-, and γ-CD polymers were used for the same purpose, however changes of R_F were observed only for some azo dyes [62]. 2-Hydroxypropyl-β-cyclodextrin used in 10-20% concentration causes complete elution of many compounds [63].

Chirality of derivatized cyclodextrin was used for recognition of stereoisomers. Phenylazobenzoyl modified γ-cyclodextrin was anchored onto silica gel used as stationary phase in HPLC and photoresponsive chromatographic behavior of dansyl amino acid enantiomers was studied [64].

6. Acknowledgements

Financial support from Gdańsk University of Technology, Grant No. DS 014668/003, and from State Committee for Scientific Research, Grants No. 3 T09A 151 27 and 3 T09A 128 27, are kindly acknowledged.

7. References

1. Ueda, H. (2002) Physicochemical Properties and Complex Formation Abilities of Large-Ring Cyclodextrins, *J. Incl. Phenom.* **44**, 53-56.
2. Ueda, H. Wakisaka, M. Nagase, H. Takaha, T. and Okada, S. (2002) Physicochemical Properties of Large-Ring Cyclodectrins ($CD_{18} - CD_{21}$), *J. Incl. Phenom.* **44**, 403-405.
3. Szejtli, J. (1988) *Cyclodextrin Technology*, in: J.E.D. Davies (ed.), *Topics in Inclusion Science*, Kluwer Academic Publishers, Dordrecht/Boston/London.
4. Szejtli J. (1998) Introduction and General Overview of Cyclodextrin Chemistry, *Chem. Rev.* **98**, 1743-1753.
5. Frőmming, K.-H. and Szejtli, J. (1994) *Cyclodextrins in Pharmacy*, in J.E.D. Davies (ed.) *Topics in Inclusion Science*, Kluwer Academic Publishers, Dordrecht/Boston/London.
6. Lehn, J.-M. Atwood, J.L. Davies, J.E.D. McNicol, D.D. and Vögtle F. (eds.) (1995) *Comprehensive Supramolecular Chemistry, Pergamon, New York,* 1996; 6a) Shinkai, S. Switchable Guest-binding Receptor Molecules, Vol. 1; G.W. Gokel (volume ed.), pp. 671-700.
7. Proceedings of the 11th International Cyclodextrin Symposium, (2002) *J. Incl. Phenom.* Special issue, Vol. **44**.
8. Bravo-Diaz, C. Sarabia-Rodriguez, M.J. Barreiro-Sio, P. and Gonzalez-Romero, E. (1999) Heterolytic Dediazoniation of 2-, 3-, and 4-Methylbenzenediazonium Tetrafluoroborate in the presence of α-, β-, and γ-Cyclodextrins, *Langmuir* **15**, 2823-2828.
9. Bravo-Diaz, C. Romero-Nieto, M.E. and Gonzalez-Romero, E. (2000) Micellar-Promoted Homolytic Dediazoniation of *p*-Nitrobenzenediazonium Tetrafluoroborate, *Langmuir* **16**, 42-48.
10. Gonzalez-Romero, E. Malvido-Hermelo, B. and Bravo-Diaz, C. (2002) Effects of β-cyclodextrin on the electrochemical behavior of a model arenediazonium ion. Kinetics and mechanism of the reaction, *Langmuir* **18**, 46-55.
11. Jamrógiewicz, M. Wagner-Wysiecka, E. Luboch, E. and Biernat, J.F. (2004) Effect of β-cyclodextrin on coupling of diazonium salts, *XXIX International Symposium on Macrocyclic Chemistry*, Cairns, Australia, Abstracts, KL – ISMC – 9.
12. Jamrógiewicz, M. Luboch, E. and Biernat, J.F. (2004) Wpływ β-cyklodekstryny na reakcje sprzęgania soli diazoniowych, *XLVII Meeting of Polish Chemical Society*, Wrocław, Poland, Abstracts, Vol. II, K001, p. 780.
13. Wagner-Wysiecka, E. Luboch, E. Kowalczyk, M. and Biernat, J.F. (2003) Chromogenic macrocyclic derivatives of azoles – synthesis and properties, *Tetrahedron* **59**, 4415 – 4420.
14. Jamrógiewicz, M. Wagner-Wysiecka, E. and Biernat, J.F. Cyclodextrin mediated synthesis of macrocyclic imidazole derivatives with two azo units, *manuscript in preparation.*
15. Zhang, X. and Nau, W.M. (2000) Chromophore Alignment in a Chiral Host Provides a Sensitive Test for the Orientation – Intensity Rule of Induced Circular Dichroism, *Angew. Chem., Int. Ed. Engl.* **39**, 544-547.
16. Mayer, B. Zhang, X. Nau, W.M. and Marconi, G. (2001) Co-conformational Variability of Cyclodextrin Complexes Studied by Induced Circular Dichroism of Azoalkanes, *J. Am. Chem. Soc.* **123**, 5240-5248.
17. Zhang, X. Gramlich, G. Wang, X. and Nau, W.M. (2002) A Joint Structural, Kinetic, and Thermodynamic Investigation of Substituent Effects on Host-Guest Complexation of Bicyclic Azoalkanes by β-Cyclodextrin, *J. Am. Chem. Soc.* **124**, 254-263.
18. Hirai, H., Toshima, N., and Uenoyama, S. (1985) Inclusion Complex Formation of γ-Cyclodextrin. One Host-Two Guest Complexation with Water-soluble Dyes in Ground State, *Bull. Chem. Soc. Jpn.* **58**, 1156-1164.
19. Yoshida, N., and Hayashi, K. (1994) Dynamic Aspects in Host-Guest Interactions. Part 2. Directional Inclusion Reactions of Some Azo Guest Molecules with α-Cyclodextrin, *Perkin Trans. 2* 1285-1290.
20. Yoshida. N. (1995) Dynamic acpects in host-guest interactions. Part 4. Kinetic and ^{1}H NMR evidence for multi-step directional binding in the molecular recognition of some 2-naphthylazophenol guests with α-cyclodextrin, *Perkin Trans. 2* 2249-2256.
21. Abou-Hamdan, A. Bugnon, P. Saudan, Ch. Lye, P.G. and Merbach, A.E. (2000) High-Pressure Studies as a Novel Approach in Determining Inclusion Mechanisms: Thermodynamics and Kinetics of the Host-Guest Interactions for α-Cyclodextrin Complexes, *J. Am. Chem. Soc.* **122**, 592-602.
22. Sueishi, Y. Kasahara, M. Inoue, M. and Matsueda, K. (2003) Effect of substituent and solvent on inclusion complexation of β–cyclodextrins with azobenzene derivatives, *J. Incl. Phenom.* **46**, 71-75.
23. Zheng, P. Li, Z. Tong, L. and Lu, R. (2002) Study of Inclusion Complexes of Cyclodextrins with Orange II, *J. Incl. Phenom.* **43**, 183-186.
24. Guo, Y.-J. Pan, J.-H. and Jing, W.-J. (2004) Determination of Orange II and the Supramolecular System of Orange II with Cyclodextrins by Polarography, *Dyes and Pigments* **63**, 65-70.

25. Mourtzis, N. Cordoyiannis, G. Nounesis, G. and Yannakopoulou, K. (2003) Single and Double Threading of Congo Red into γ-Cyclodextrin. Solution Structures and Thermodynamic Parameters of 1:1 and 2:2 Adducts, as Obtained from NMR Spectroscopy and Microcalorimetry, *Supramol. Chem.* **15**, 639-649.
26. Sanchez, A.M., and de Rossi, R.H. (1996) Effect of β-cyclodextrin on the thermal *cis-trans* isomerization of azobenzenes, *J. Org. Chem.* **61**, 3446-3451.
27. Barbiric, D.J. Castro, E.A. and De Rossi, R.H. (2000) A molecular mechanics study of 1:1 complexes between azobenzene derivatives and β-cyclodextrin, *J. Mol. Struct. (Theochem.)* **532**, 171-181.
28. Barbiric, D.J. De Rossi, R.H. and Castro, E.A. (2001) Inclusion complex of 1:2 stoichiometry between azobenzenes and cyclodextrins: a molecular mechanics study, *J. Mol. Struct.* (*Theochem.)* **537**, 235-243.
29. Liu, Y. Li, L. and Chen, Y. (2002) Molecular Recognition Studies on Supramolecular Systems. Part 38. Inclusion Complexation of Organic Dyes by Organoselenium Bridged Bis-(β-cyclodextrin)s with a Short Linker, *J. Incl. Phenom.* **42**, 151-155.
30. Nepogodiev, S.A., and Stoddart, J.F. (1998) Cyclodextrin-Based Catenanes and Rotaxanes, *Chem. Rev.* **98**, 1959-1976.
31. Harada, A. (1998) Polyrotaxanes, *Acta Polymer.* **49**, 3-17.
32. Murakami, H., Kawabuchi, A., Kotoo, K., Kunitake, M., and Nakashima, N. (1997) A light-Driven Molecular Shuttle Based on a Rotaxane, *J. Am. Chem. Soc.* **119**, 7605-7606.
33. Yamaguchi, I. Osakada, K. and Yamamoto, T. (2000) Pseudopolyrotaxane composed of an azobenzene polymer and γ-cyclodextrin. Reversible and irreversible photoisomerization of the azobenzene groups in the polymer chain, *Chem. Commun.* 1335-1336.
34. Anderson, S., Claridge, T.D.W., and Anderson, H.L. (1997) Azo-Dye Rotaxanes, *Angew. Chem. Int. Ed. Engl.* **36**, 1310-1312.
35. Craig, M.R., Claridge, T.D.W., Hutchings, M.G., and Anderson, H.L. (1999) Synthesis of cyclodextrinazo dye [3]rotaxane as a single isomer, *Chem. Commun.* 1537-1338.
36. Shinkai, S. Takeuch, M. and Ikeda, A. (2000) in Osada Y. De Rossi D.E. (eds.) *Polym. Sens. Actuators*, Springer-Verlag, Berlin, Germany,183-206; Harada, A. (2001) Cyclodextrin Based Molecular Machines, *Acc. Chem. Res.* **34**, 456 – 464.
37. Fischer, P. Koetse, M. Laschewsky, A. Wischerhoff, E. Jullien, L. Persoons, A. and Verbiest, T. (2000) Orientation of Nonlinear Optical Active Dyes in Electrostatically Self-Assembled Polymer Films Containing Cyclodextrins, *Macromolecules* **33**, 9471-9473.
38. Ihara, Y. (2000) Adsorption of acid dyes from aqueous solution onto polymers containing cyclodextrin, *Yamaguchi-kenritsu Daigaku Seikatsu Kagakubu Kenkyu Hokoku* **25**, 23-28; C.A. (2001) **134**, 257290.
39. Savarino, P. Parlati, S. Buscaino, R. Piccinini, P. Degani, I. and Barni, E. (2004) Effect of additives on the dyeing of polyamide fibers. Part I. β-Cyclodextrin, *Dyes and Pigments* **60**, 223-232 and papers cited therein.
40. Storsberg, J. Glockner, P. Eigner, M. Schnoller, U. Ritter, H. Voit, B. and Nuyken, O. (2001) Cyclodextrins in polymer synthesis: photocrosslinkable films via free radical copolymerization of methylated β-cyclodextrin-complexed styrene with sodium 4-(acrylamido)-phenyldiazosulfonate in aqueous medium, *Designed Monomers and Polymers* **4**, 9-17.
41. Craig, M.R. Hutchings, M.G. Claridge, T.D.W. and Anderson, H.L. (2001) Rotaxane-encapsulation enhances the stability of an azo dye, in solution and when bonded to cellulose, *Angew. Chem., Int. Ed.* **40**, 1071-1074.
42. Anderson, H.L. Craig, M.R. and Hutchings, M.G. (2002) PCT Int. Appl. WO 2002024816 A1 28 Mar 2002, 28 pp.; C.A. (2002) **136**, 280778.
43. Kuwabara, T. Aoyagi, T. Takamura, M. Matsushita, A. Nakamura, A. and Ueno, A. (2002) Hetero-Dimerization of Dye-Modified Cyclodextrins with Native Cyclodextrins; *J. Org. Chem.* **67**, 720-725.
44. Fujimoto, T. Sakata, Y. and Kaneda, T. (2000) The first Janus [2]rotaxane, *Chem. Commun.* 2143-2144.
45. Gibson, H.W. Guzy, M.T. Shah, S. Wang, H. Neyman, P.J. Van Cott, K.E. Davis, R.M. Figura, C. Brands, C. and Heflin, J.R. (2001) Second-order nonlinear optical properties of ionically self-assembled films, *Polymeric Materials: Science & Engineering of Composite Materials* **84**, 251-252.
46. Ueno, A. Shimizu, T. Mihara, H. Hamasaki, K. and Pitchumani, K. (2002) Supramolecular Chemistry of Cyclodextrin-Peptide Hybrids: Azobenzene-Tagged Peptides, *J. Incl. Phenom.* **44**, 49-52.
47. Shimomura, T. Funaki, T. Ito, K. Choi, B.-K. and Hashizume, T. (2002) Circular dichroism study of the inclusion-dissociation behavior of complexes between a molecular nanotube and azobenzene substituted linear polymers, *J. Incl. Phenom.* **44**, 275-278.
48. Kikuchi, T. Narita, M. and Hamada, F. (2002) Synthesis of bis dansyl-modified β-cyclodextrin dimer linked with azobenzene and its fluorescent molecular recognition, *J. Incl. Phenom.* **44**, 329-333.
49. Ueno, A. Fukushima, M., and Osa, T. (1990) Inclusion Complexes and *Z-E* Photoisomerization of β-Cyclodextrin Bearing an Azobenzene Pendant, *Perkin Trans. 2* 1067-1072.

50. Fukushima, M., Osa, T., and Ueno, A. (1990) Photocontrol of Molecular Association Attained by Azobenzene-modified Cyclodextrin, *Chem. Commun.* 15-17.
51. Fukushima, M. Osa, T. and Ueno, A. (1991) Photoswitchable Multi-response Sensor of Azobenzene-modified γ-Cyclodextrin for Detecting Organic Compounds, *Chem. Lett.* 709-712.
52. Ueno, A. Kuwabara, T. Nakamura, A. and Toda, F. (1992) A modified cyclodextrin as a guest responsive color-change indicator, *Nature* **356**, 136-137.
53. Ueno, A. (1993) Fluorescent Sensors and Color-Change Indicators for Molecules, *Adv. Mater. (Weinheim)* **5**, 132-134.
54. Ueno, A. (1993) Modified cyclodextrin as supramolecular sensors of molecular recognition, *New Funct. Mater.* C. 521-526.
55. Ueno, A. Yoshimura, H. Saka, R. and Osa, T. (1979) Photocontrol of Binding Ability of Capped Cyclodextrin, *J. Am. Chem. Soc.* **101**, 2779-2780.
56. Moriwaki, F. Kaneko, H. Ueno, A. Osa, T. Hamada, F. and Murai, K. (1987) Excimer Formation and Induced-Fit Type of Complexation of β-Cyclodextrin Capped by Two Naphthyl Moieties, *Bull. Chem. Soc. Jpn.* **60**, 3619-3623.
57. Hamada, F. Fukushima, M. Osa, T. Ikeda, H. Toda, F. and Ueno, A. (1993) Azobenzene-capped γ-cyclodextrin as photo-switchable host, *Makromol. Chem., Rapid Commun.* **14**, 287-291.
58. Jung, J.H., Takehisa, C., Sakata, Y., and Kaneda, T. (1996) *p*-(4-Nitrophenylazo)phenol Dye-bridged Permethylated α-Cyclodextrin Dimer: Synthesis and Self-aggregation in Dilute Aqueous Solution, *Chem. Lett.* 147-148.
59. Aoyagi, T. Ueno, A. Fukushima, M. and Osa, T. (1998) Synthesis and photoisomerization of an azobenzene derivative bearing two β-cyclodextrin units at both ends, *Macromol. Rapid Commun.* **19**, 103-105.
60. Kaneda, T. Yamada, T. Fujimoto, T. and Sakata, Y. (2001) The first [5]supercyclodextrin whose cyclopentameric array is held only by a mechanical bond, *Chem. Lett.* 1264-1265.
61. Huynh, T.K.X. Lederer, M. and Leipzig-Pagani, E. (1995) Adsorption chromatography on cellulose XII. General effects of aqueous solutions of α-cyclodextrin as eluent, *J. Chromatogr.* **695**, 155-159.
62. Lederer, M. and Nguyen, H.K.H. (1996) Adsorption chromatography on cellulose XIV. Some results using aqueous solutions of soluble cyclodextrin polymers as eluents, *J. Chromatogr.* **723**, 405-409.
63. Hostettmann, K. Lederer, M. and Marston, A. (2000) A study of the cyclodextrin complexes of flavonoids and azodyes by thin layer chromatography. Part II. Hydroxypropyl-cyclodextrins, *Phytochemical Analysis* **11**, 380-382.
64. Yoshida, K. Nakagama, T. Uchiyama, K. and Hobo, T. (1998) Photo responsive chromatographic behavior of dansyl amino acid enantiomers using azobenzene-modified cyclodextrin stationary phases in HPLC, *Chromatography* **19**, 310-311.

14. Ru(II) AND Os(II) COMPLEXES OF A SHAPE-PERSISTENT MACROCYCLIC LIGAND: SYNTHESIS, PHOTOPHYSICAL PROPERTIES, AND ELECTROCHEMICAL CHARACTERIZATION

MARGHERITA VENTURI[a], VINCENZO BALZANI[a], FILIPPO MARCHIONI[a], DORINA M. OPRIS[b], PETER FRANKE[b] AND A.-DIETER SCHLÜTER [b, c]
[a] *Dipartimento di Chimica "G. Ciamician", Università di Bologna, via Selmi 2, 40126 Bologna, Italy*
[b] *Freie Universität Berlin, Institut für Chemie, Takustrasse 3,14195 Berlin, Germany*
[c] *Present Address: Department of Materials, Institute of Polymers ETH-Hönggerberg, HCI J 541, 8093 Zürich, Switzerland*

1. Introduction

Much attention is currently devoted to the synthesis and properties of shape-persistent macrocycles[1]. Such compounds are interesting for a variety of reasons including formation of columnar stacks potentially capable of performing as nanopores for incorporation into membranes or for the generation of nanowires[2]. Furthermore, in shape-persistent macrocycles incorporating coordination units, *endo*-cyclic metal-ion coordination may be exploited to generate nanowires[3], whereas *exo*-cyclic coordination can be used to construct large arrays of polynuclear metal complexes[4]. Shape-persistent macrocycles with reactive substituents may also be linked to other units to yield multicomponent, hierarchical structures.

In the past few years it has been shown that suitably designed molecular and supramolecular species can perform as nanoscale devices and machines[5]. An important role in this regard could be played by shape-persistent macrocycles exhibiting properties modifiable by photochemical, electrochemical, or acid/base inputs. Particularly interesting are shape-persistent macrocyclic ligands containing 2,2'-bipyridine (bpy) units and their Ru(II) and/or Os(II) complexes. It is, indeed, since long known that Ru(II) and Os(II) polypyrindine-type complexes[6] exhibit suitable excited-state and redox properties to be used as building blocks for the construction of photo- and redox-active multicomponent systems[7].

Karsten Gloe (ed.), Macrocyclic Chemistry: Current Trends and Future Perspectives, 219–234.
© 2005 *Springer. Printed in the Netherlands.*

As an extension of our previous studies[4,8], we report here the syntheses, the photophysical properties, and the electrochemical chracterization of the shape-persistent macrocyclic ligand **1**, containing one bpy unit, and its $[Ru(bpy)_2(\mathbf{1})]^{2+}$ and $[Os(bpy)_2(\mathbf{1})]^{2+}$ complexes (Scheme 1). We also report the syntheses of simpler Ru(II) and Os(II) complexes that together with other previously synthesized species (Scheme 2) have been used as model compounds for interpreting the behavior of the new macrocyclic ligand and its complexes.

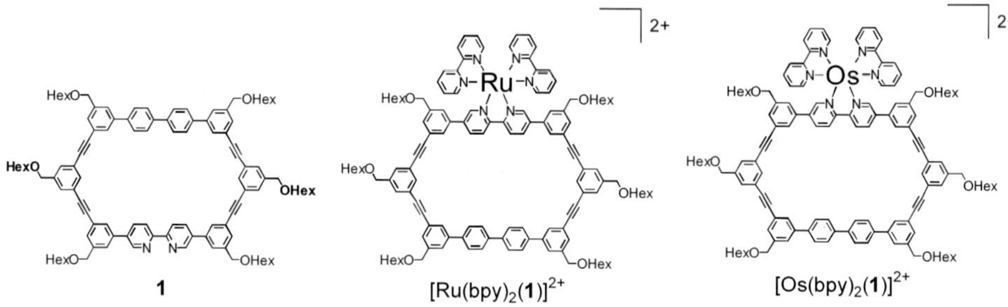

Scheme 1. Structures of the macrocyclic ligand **1** and its Ru and Os complexes.

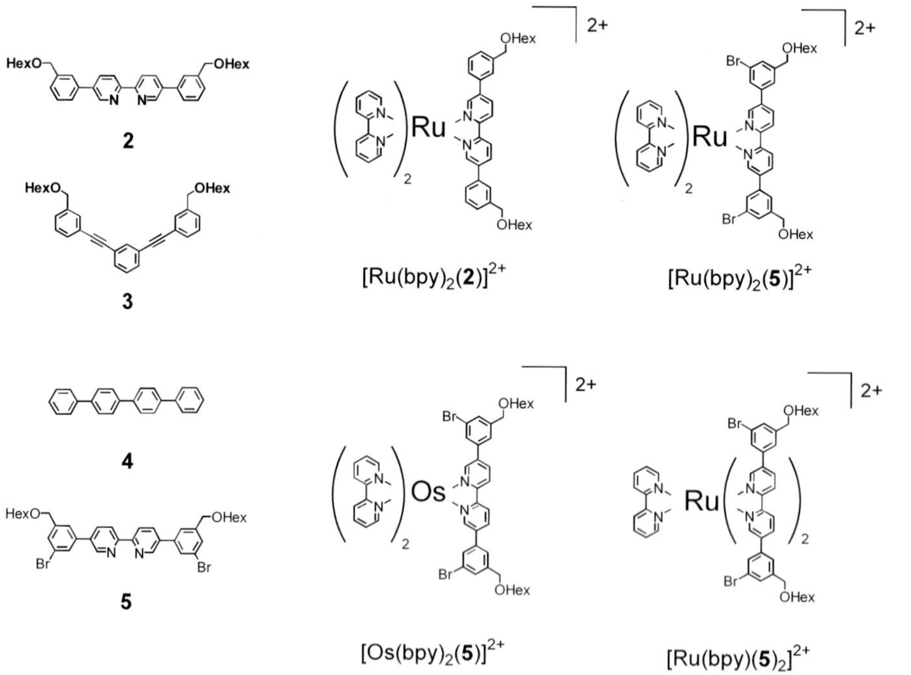

Scheme 2. Structures of model compounds used for the component units of macrocyclic ligand **1** and of its Ru and Os complexes.

2. Synthesis

The synthetic route to macrocycle **1** is shown in Scheme 3. It was constructed from the building blocks **8b** and **9**. The synthesis of **8b** starts from the boronic acid ester **6**[9] which was coupled with the diiodobiphenyl **7** to give **8a**. Its silyl groups were virtually quantitatively converted into the diiodo analogue **8b** with iodochloride[9]. The other building block, **9**, was prepared as described[10]. Both cycle precursors were obtained as analytically pure materials on the gram scale (**8b**: 1.7 g, **9**: 2.3 g). The ring closure was done according to a literature procedure[1a] and gave cycle **1**, a presumably cyclic compound with double molar mass ([**1**]$_2$), and linear oligomers. Cycles **1** and [**1**]$_2$ (not described here[11]) were obtained on the 300 mg (Yield: 20%) and 100 mg scale, respectively, after preparative gel permeation chromatography. The oligomeric material, though it may be of interest for its capacity to form helical structures[12] was not further investigated[13]. The ^{1}H and ^{13}C NMR spectra prove the cyclic nature of **1**. Specifically signals of end groups are completely absent. The FAB mass spectrum shows a signal at *m/z* = 1544 Da which corresponds to the molecular ion.

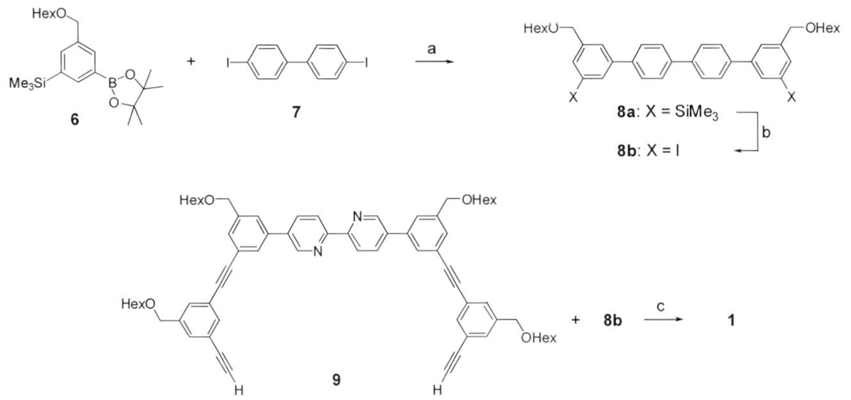

Scheme 3. a) Pd(PPh$_3$)$_4$, 1M Na$_2$CO$_3$/water, toluene, r.f. 3 d (57%); b) ICl, DCM, -15°C, 30 min. (97%); c) Pd(PPh$_3$)$_4$, CuI, toluene, triethylamine, 60°C for 4 d, 95°C for 1 d (20%).

Bipyridines have been intensely used as ligands in transition metal complexes[6]. Within a broader investigation of the redox and photophysical behavior of Ru and Os complexed macrocycles[4,8], the Ru(II) and Os(II) complexes shown in Schemes 4 and 5 were prepared.

The complexations were done with either [Os(bpy)$_2$Cl$_2$], [Ru(bpy)$_2$Cl$_2$], or [Ru(bpy)Cl$_3$]$_x$. Refluxing of the respective bipyridine (**5**, **10**, Scheme 4, and **1**, Scheme 5) in either ethanol (or mixtures of it with dioxane and water) or butanol afforded the corresponding complexes [Os(bpy)$_2$(**5**)]$^{2+}$, [Ru(bpy)(**5**)$_2$]$^{2+}$, and [Ru(bpy)(**11**)$_2$]$^{2+}$ (Scheme 4), and [Ru(bpy)$_2$(**1**)]$^{2+}$ and [Os(bpy)$_2$(**1**)]$^{2+}$ (Scheme 5), in yields of 38, 84, 84, 56, and 50%, respectively. During complexation the colors changed from purple to green for the Os and from red to orange for the Ru complexes. Purification was done by column chromatography through neutral aluminium oxide or silica gel. **10** carries the rather sensitive THP protecting group. This was cleaved off during complexation to give the diol [Ru(bpy)(**11**)$_2$]$^{2+}$, instead of the expected THP-protected analogue.

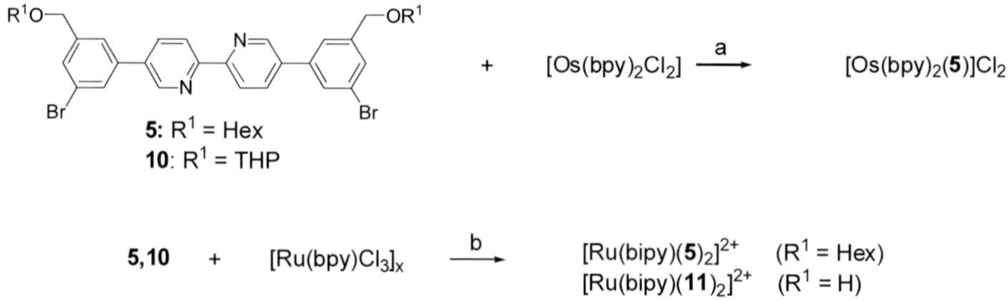

Scheme 4. a) ethanol, r.f., 3 d, (38%); b) ethanol, dioxane, water, r.f., 24 h, (84%).

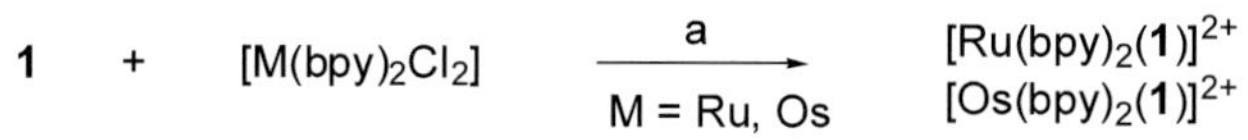

Scheme 5. a) for $[Ru(bipy)_2(\mathbf{1})]^{2+}$: dioxane, propylene glycol, ethanol, r.f., 24 h (56%); for $[Os(bipy)_2(\mathbf{1})]^{2+}$: buthanol, r.f., 5 d, (50%).

The complexes were characterized by FAB ($[Os(bpy)_2(\mathbf{5})]^{2+}$, $[Ru(bpy)(\mathbf{5})_2]^{2+}$, and $[Ru(bpy)(\mathbf{11})_2]^{2+}$) and MALDI-TOF mass spectrometry ($[Ru(bpy)_2(\mathbf{1})]^{2+}$ and $[Os(bpy)_2(\mathbf{1})]^{2+}$) as well as high field ^{1}H and ^{13}C NMR spectroscopy. They have two enantiomeric forms (Λ and Δ). Figure 1 shows the ^{1}H NMR spectrum of complex $[Ru(bpy)_2(\mathbf{1})]^{2+}$ with a signal assignment which was possible by means of HMQC and COSY pulse sequences. Two sets of signals were observed for each of the two pyridines of the bpy ligands and one set for those of the cycle's bpy. The MALDI-TOF mass spectra of $[Ru(bpy)_2(\mathbf{1})]^{2+}$ and $[Os(bpy)_2(\mathbf{1})]^{2+}$ were recorded in dithranol matrix and show the signals at m/z = 2102, 1957 for $[Ru(bpy)_2(\mathbf{1})]^{2+}$ and m/z = 2192, 2047 for $[Os(bpy)_2(\mathbf{1})]^{2+}$ which are characteristic for $M^+(-PF_6^-)$ and $M^+(-2PF_6^-)$, respectively (Figure 2).

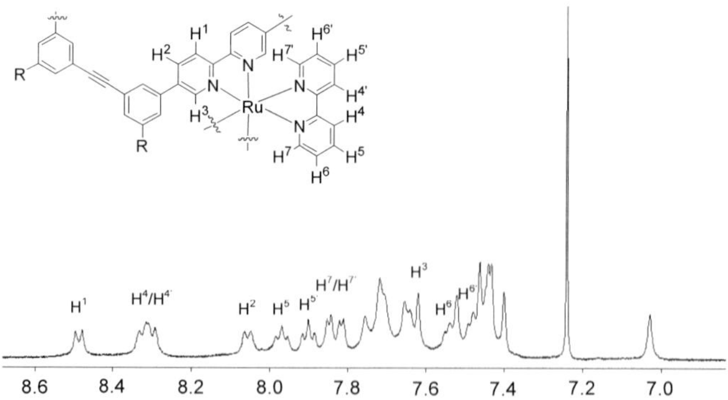

Figure 1. ^{1}H NMR spectrum (500 MHz) of $[Ru(bipy)_2(\mathbf{1})]^{2+}$ complex in $CDCl_3$ at 20 °C with partial signal assignment based on COSY pulse sequences.

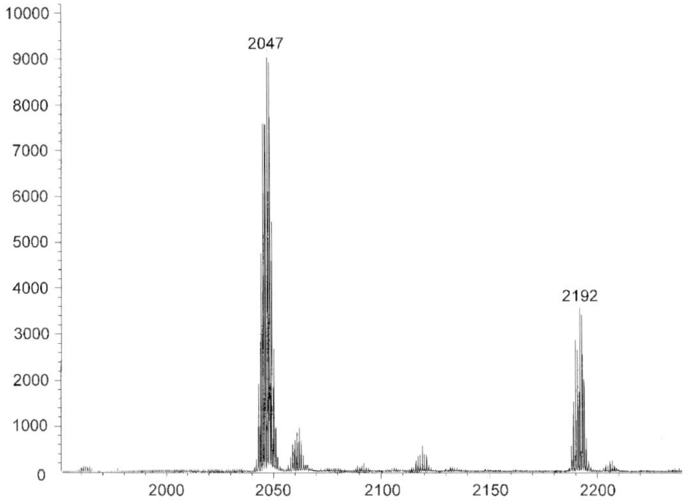

Figure 2. MALDI-TOF mass spectrum (dithranol matrix) of $[Os(bipy)_2(\mathbf{1})]^{2+}$ complex with partial assignment.

The syntheses of compounds **2**[8], **3**[14], $[Ru(bpy)_2(\mathbf{2})]^{2+}$[8], and $[Ru(bpy)_2(\mathbf{5})]^{2+}$[4b] have been previously reported.

3. Photophysical Properties

Macrocyclic ligand **1** (Scheme 1) is soluble in dichloromethane and tetrahydrofuran. It shows in dichloromethane solution (Figure 3) a structured and intense absorption in the UV region, that is practically the sum of the absorption spectra of compounds **2**, **3**, and **4** (Scheme 2, Figure 3), taken as models of the macrocycle's component units. This result indicates that in **1** the component units behave independently, i.e., they do not interact in the ground state. A different behavior, however, is observed as far as the luminescence properties are concerned. A comparison with the emission spectra in dichloromethane solution of compounds **2**, **3**, and **4** (Table 1, Figure 4) shows that the strong and structured emission band of **1** is very similar in shape and energy (Table 1, Figure 4) to that of compound **2**. Therefore it can be concluded that light excitation of **1** is followed by energy transfer to the component unit **2**, whose fluorescent excited state lies at lowest energy than the fluorescent excited states of compounds **3** and **4**. The fact that **1** shows a quantum yield higher than that of compound **2** is most probably ascribed to its rigid structure which slows down radiationless deactivation.

	298 K[a]			77 K[b]	
	λ_{max}, nm	Φ	τ, ns	λ_{max}, nm	τ, ns
1	380	0.7	0.7	378[c]	0.8
2	372	0.4	0.5	372[c]	0.8
3	340	0.3	2.7	340[c]	4.4
4	365	0.5	0.8	358[c]	0.5
$[Ru(bpy)_2(\mathbf{1})]^{2+}$	619	0.05	733	593	5500
$[Os(bpy)_2(\mathbf{1})]^{2+}$	744	0.002	80	710	1400
$[Ru(bpy)_3]^{2+}$	600	0.03	338	570	5800
$[Ru(bpy)_2(\mathbf{2})]^{2+}$	607	0.07	630	591	5700
$[Ru(bpy)_2(\mathbf{5})]^{2+}$	626	0.05	775	601	5600
$[Ru(bpy)(\mathbf{5})_2]^{2+}$	618	0.05	679	598	5700
$[Os(bpy)_3]^{2+}$	730	0.004	68	715	1300
$[Os(bpy)_2(\mathbf{5})]^{2+}$	754	0.003	62	727	1100

TABLE 1. Luminescence properties

[a]Air-equilibrated acetonitrile. [b]Dichloromethane/methanol (1:1 v/v) rigid matrix, unless otherwise noted. [c]Dichloromethane/chloroform (1:1 v/v) rigid matrix.

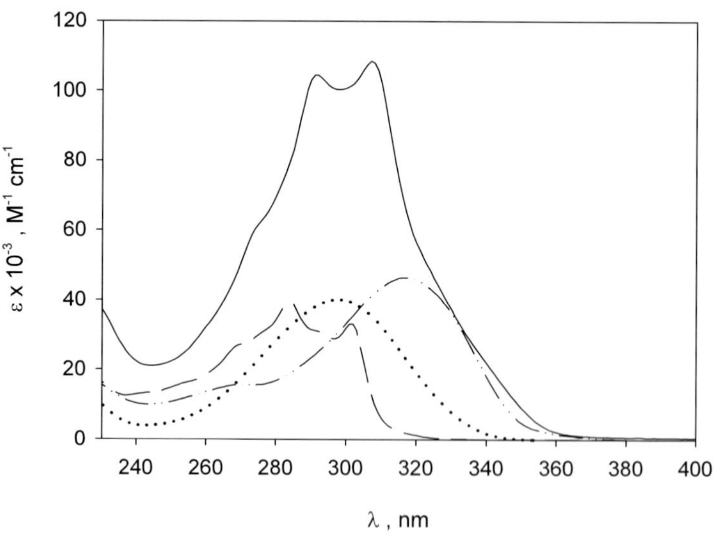

Figure 3. Absorption spectra in dichloromethane solution of macrocyclic ligand **1** (full line), and its model compounds **2** (dashed and dotted line), **3** (dashed line), and **4** (dotted line).

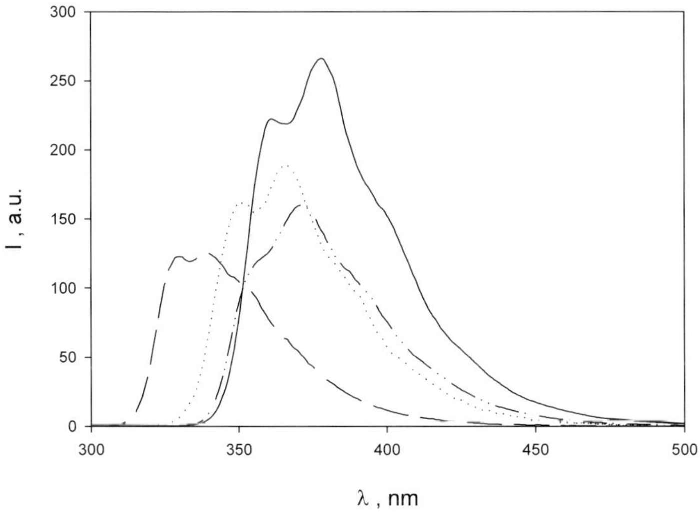

Figure 4. Emission spectra in dichloromethane solution of macrocyclic ligand **1** (full line), and its model compounds **2** (dashed and dotted line), **3** (dashed line), and **4** (dotted line).

In dichloromethane solution, the $[Ru(bpy)_2(\mathbf{1})]^{2+}$ complex (Scheme 1) exhibits an absorption band at 455 nm (ε_{max} = 10400 $M^{-1}cm^{-1}$, Figure 5) and an emission band at 619 nm (τ = 733 ns, ϕ = 0.05, Figure 5, Table 1). These bands can straightforwardly be assigned to spin-allowed and, respectively, spin-forbidden metal-to-ligand-charge-transfer (MLCT) excited states, characteristic of Ru(II) polypyridine complexes[6a,c,e].

In a rigid dichloromethane/methanol (1:1 v/v) matrix at 77 K (Table 1), $[Ru(bpy)_2(\mathbf{1})]^{2+}$ shows a structured emission band with λ_{max} = 593 nm (Figure 5) and τ = 5.5 μs, again typical of Ru(II) polypyridine complexes[6a,c,e]. As far as the absorption properties in dichloromethane solution are concerned, the complexes $[Ru(bpy)_3]^{2+}$ (λ_{max} = 450 nm, ε_{max} = 9900 $M^{-1}cm^{-1}$), $[Ru(bpy)_2(\mathbf{2})]^{2+}$($\lambda_{max}$ = 455 nm, ε_{max} = 10000 $M^{-1}cm^{-1}$), and $[Ru(bpy)_2(\mathbf{5})]^{2+}$ (λ_{max} = 455 nm, ε_{max} = 11000 $M^{-1}cm^{-1}$) can be considered good model compounds (Scheme 2) of $[Ru(bpy)_2(\mathbf{1})]^{2+}$. A slightly different behavior is, however, observed in emission (Table 1, Figure 6) showing that the different nature of the third ligand affects the emitting excited state. At room temperature the maximum of the emission band is shifted to lower energy along the series $[Ru(bpy)_3]^{2+}$ > $[Ru(bpy)_2(\mathbf{2})]^{2+}$ > $[Ru(bpy)_2(\mathbf{1})]^{2+}$ > $[Ru(bpy)_2(\mathbf{5})]^{2+}$. This shift can be attributed to the different electron acceptor abilities of the ligands, as shown by the reduction potential values (see electrochemical characterization).

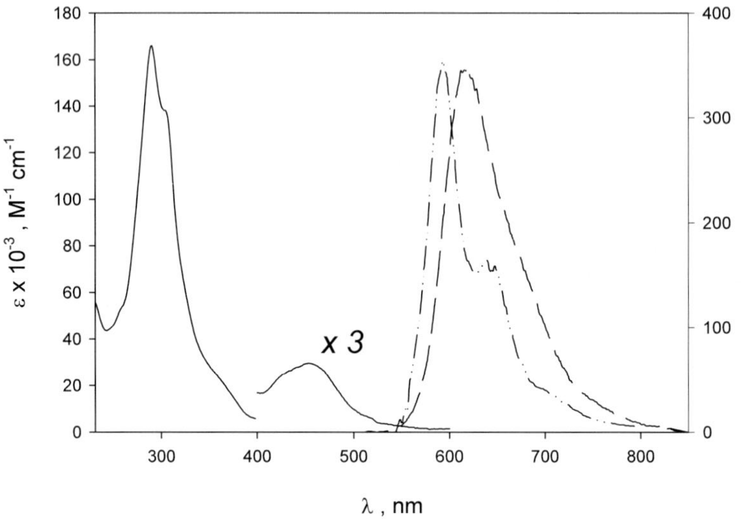

Figure 5. Absorption (dichloromethane solution, full line) and emission (dichloromethane solution at 298 K, dashed line; dichloromethane/methanol rigid matrix at 77 K, dashed and dotted line) spectra of ($[Ru(bpy)_2(\mathbf{1})]^{2+}$ complex.

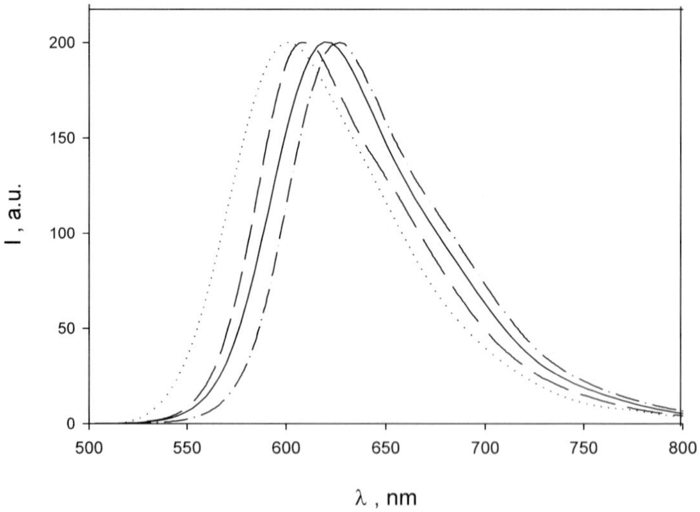

Figure 6. Comparison of the emission bands in dichloromethane solution at 298 K of $[Ru(bpy)_2(\mathbf{1})]^{2+}$ (full line) and its model compounds $[Ru(bpy)_3]^{2+}$ (dotted line), $[Ru(bpy)_2(\mathbf{2})]^{2+}$ (dashed line), and $[Ru(bpy)_2(\mathbf{5})]^{2+}$ (dashed and dotted line).

Concerning $[Ru(bpy)(\mathbf{5})_2]^{2+}$ (Scheme 2), it is interesting to note that its absorption band in the visible region (λ_{max} = 462 nm, ε_{max} = 11700 $M^{-1}cm^{-1}$) is similar to that of $[Ru(bpy)_2(\mathbf{5})]^{2+}$, while its emission band (Table 1), contrary to the expectations, is blue shifted compared to $[Ru(bpy)_2(\mathbf{5})]^{2+}$. This behavior can be attributed to the fact that the presence of two ligands easy to reduce withdraws negative charge from the metal, thus moving the metal-to-ligand-charge-transfer excited state at higher energy[6a]. All the analyzed Ru-based model compounds show lifetimes at room temperature and in rigid matrix at 77 K (Table 1) typical of Ru(II) polypyridine complexes[6a,c,e].

In dichloromethane solution, the $[Os(bpy)_2(\mathbf{1})]^{2+}$ complex (Scheme 1) exhibits absorption (λ_{max}= 490 nm, ε_{max} = 8000 $M^{-1}cm^{-1}$; λ_{max}= 650 nm, ε_{max} = 2000 $M^{-1}cm^{-1}$) and emission bands (λ_{max} = 745 nm, τ = 80 ns, ϕ = 0.002, Table 1) in the visible region. These bands, straightforwardly assigned to spin-allowed and, respectively, spin-forbidden MLCT excited states, are characteristic of the Os(II) polypyridine complexes[6c,e], as also shown (Table 1) by the model compounds $[Os(bpy)_3]^{2+}$ and $[Os(bpy)_2(\mathbf{5})]^{2+}$ (Scheme 2). In a rigid dichloromethane/methanol (1:1 v/v) matrix at 77 K (Table 1), $[Os(bpy)_2(\mathbf{1})]^{2+}$ shows a structured emission band with λ_{max} = 710 nm and τ = 1.4 μs, again typical of Os(II) polypyridine complexes[6c,e].

4. Electrochemical Characterization

The electrochemical properties of macrocyclic ligand **1** were investigated in purified tetrahydrofuran under vacuum conditions. In the accessible potential window (+1/–3 V vs SCE), compound **1** undergoes two reversible and monoelectronic reduction processes at –1.78 and –2.02 V *vs* SCE, and no oxidation process (Table 2). A comparison with the data obtained under the same conditions for ligand **2**, which can be considered as a model compound of the bpy unit of **1**, and bpy (Table 2) shows that the macrocyclic ligand **1** is easier to reduce than the other two ligands. Ligand **5** shows only weak and overlapping irreversible processes, difficult to attribute, a behavior that is most probably caused by the presence of the Br substituents. Although it is impossible to obtain quantitative information from the electrochemical analysis, it is expected that ligand **5**, because of the electron-withdrawing effect of the two Br substituents, is easier to reduce than the ligand **2** and also the macrocyclic ligand **1**, in agreement with the photophysical results and the electrochemical data obtained for $[Ru(bpy)_2(\mathbf{5})]^{2+}$ (see below).

In the electrochemical experiments performed in argon-purged dichloromethane solution (accessible potential window: +1.7/–1.7 V vs SCE), $[Ru(bpy)_2(\mathbf{1})]^{2+}$ undergoes a reversible monoelectronic oxidation at +1.45 V *vs* SCE (Table 2), that can be assigned to the oxidation of the metal-center[6a,e]. The potential value at which the Ru ion is oxidized is similar to that observed for the metal oxidation in the model compounds $[Ru(bpy)_2(\mathbf{2})]^{2+}$, $[Ru(bpy)_2(\mathbf{5})]^{2+}$, and $[Ru(bpy)_3]^{2+}$ (Table 2). On reduction, two reversible and monoelectronic processes are observed in the accessible potential window, which are attributed to the reduction of the ligands (Table 2)[6a,e]. By comparison with the data obtained for $[Ru(bpy)_2(\mathbf{2})]^{2+}$, $[Ru(bpy)_2(\mathbf{5})]^{2+}$, and $[Ru(bpy)_3]^{2+}$ (Table 2), the first process at –1.12 V *vs* SCE is assigned to the reduction of the bpy unit of the macrocycle **1**, considering that such a ligand is easier to reduce than bpy, whereas the second process at –1.47 V *vs* SCE involves one of the bpy ligands. This process occurs at a potential value which is (a) more negative than that of the first reduction of $[Ru(bpy)_3]^{2+}$, because in $[Ru(bpy)_2(\mathbf{1})]^{2+}$ the bpy reduction is preceded by the reduction of the bpy unit of **1**, and (b) less negative than that of the second reduction of $[Ru(bpy)_3]^{2+}$ because of the greater delocalization of the charge introduced by replacement of one bpy with the ligand **1**.

TABLE 2. Redox potentials.[a]

	Ligand centered reduction $E_{1/2}$, V vs SCE			Metal centered oxidation Os	Metal centered oxidation Ru
1 [b]		−2.02	−1.78		
2 [b]		−2.13	−1.85		
bpy [b]		−2.69 [c]	−2.18		
$[Ru(bpy)_2(\mathbf{1})]^{2+}$		−1.47	−1.12		+1.45
$[Os(bpy)_2(\mathbf{1})]^{2+}$		−1.36	−1.03	+1.03	
$[Ru(bpy)_3]^{2+}$		−1.50	−1.24		+1.43
$[Ru(bpy)_2(\mathbf{2})]^{2+}$		−1.43	−1.12		+1.45
$[Ru(bpy)_2(\mathbf{5})]^{2+}$		−1.44	−1.07		+1.47
$[Ru(bpy)(\mathbf{5})_2]^{2+}$	−1.65	−1.24	−0.99		+1.54
$[Os(bpy)_3]^{2+}$		−1.35	−1.06	+1.03	
$[Os(bpy)_2(\mathbf{5})]^{2+}$		−1.36	−1.03	+1.04	

[a]Room temperature argon-purged dichloromethane solution and reversible and monoelectronic processes, unless otherwise noted; tetrabutylammonium hexafluorophosphate as supporting electrolyte, glassy carbon as working electrode. [b]Purified tetrahydrofurane under vacuum conditions. [c]Irreversible process; potential value estimated by the DPV peak.

The same considerations can be used to compare the behavior of $[Ru(bpy)_2(\mathbf{2})]^{2+}$, $[Ru(bpy)_2(\mathbf{5})]^{2+}$, and $[Ru(bpy)_3]^{2+}$. A closer inspection of the data obtained for $[Ru(bpy)_2(\mathbf{1})]^{2+}$ and $[Ru(bpy)_2(\mathbf{5})]^{2+}$ (Table 2) shows that in the complex containing ligand **5** the metal oxidation and the first reduction process occur at more positive and less negative potential values, respectively, than in the complex containing ligand **1**. This result indicates, in agreement with the expectation and the photophysical results, that ligand **5** is easier to reduce than macrocyclic ligand **1**.

Concerning the $[Ru(bpy)(\mathbf{5})_2]^{2+}$ complex (Table 2), it is interesting to note that the metal-centered oxidation is shifted to a more positive potential value in comparison to the other complexes, as expected because of the presence of two ligands easier to reduce than bpy. On reduction, this complex shows three monoelectronic and reversible processes (Table 2) that, on the basis of the electron-acceptor abilities of the two different ligands present in the complex, can be assigned, the first two, to the reduction of the two ligands **5**, and, the third one to the reduction of the bpy ligand.

In the electrochemical experiments performed in argon-purged dichloromethane solution, $[Os(bpy)_2(\mathbf{1})]^{2+}$ undergoes a reversible monoelectronic oxidation at +1.03 V *vs* SCE (Figure 7, Table 2), that can be assigned to the oxidation of Os(II)[6c,e]. The potential value at which the metal ion is oxidized is very close to those obtained for the metal oxidation in the model compounds $[Os(bpy)_2(\mathbf{5})]^{2+}$ and $[Os(bpy)_3]^{2+}$ (Table 2). On reduction, two reversible and monoelectronic processes are observed in the accessible potential window (Figure 7, Table 2).

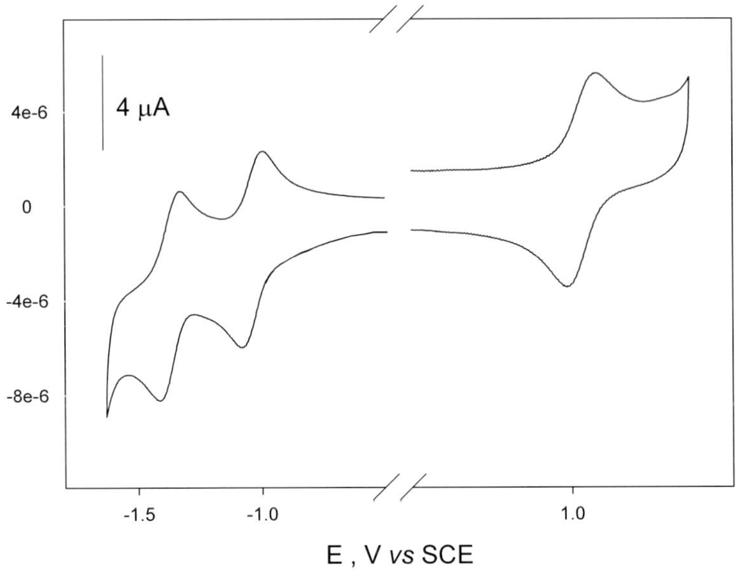

Figure 7. Cyclic voltammogram obtained for $[Os(bpy)_2(\mathbf{1})]^{2+}$ complex (argon-purged dichloromethane; complex concentration 4×10^{-4} M; glassy carbon as working electrode, scan rate of 100 mV/s).

In agreement with the above discussed Ru-based complex and by the comparison with the results obtained for its model compounds (Table 2), the first process at –1.03 V *vs* SCE is assigned to the reduction of ligand **1**, while the second process at –1.36 V *vs* SCE concerns the reduction of one of the two bpy ligands. The fact that in $[Os(bpy)_2(\mathbf{1})]^{2+}$ and $[Os(bpy)_2(\mathbf{5})]^{2+}$ these processes occur at potential values very similar to those observed for $[Os(bpy)_3]^{2+}$ can be explained by considering that the Os ion is capable of delocalizing the introduced charges better than the Ru ion[15].

5. Conclusions

The shape-persistent macrocycle **1** exhibits interesting photophysical properties, including a fluorescence band with very high quantum yield. Because of the presence of one bpy coordination unit, macrocycle **1** reacts with $Ru(bpy)_2Cl_2$ and $Os(bpy)_2Cl_2$ to give the $[Ru(bpy)_2(\mathbf{1})]^{2+}$ and $[Os(bpy)_2(\mathbf{1})]^{2+}$ complexes, respectively. These complexes do not show any ligand-centered emission and exhibit absorption and emission bands in the visible region, characteristic of the Ru(II) and Os(II) polypyridine-type complexes. The electrochemical investigation has evidenced metal-centered oxidation and ligand-centered reduction processes, again in agreement with the behavior of the Ru(II) and Os(II) polypyridine-type complexes. For a conclusive interpretation of the results, model compounds of the component units of macrocycle **1** and of the Ru and Os complexes have been synthesized and investigated. Among the Ru and Os compounds, $[Ru(bpy)(\mathbf{5})_2]^{2+}$ and $[Ru(bpy)(\mathbf{11})_2]^{2+}$ are of additional interest as models for future nanoconstructions with shape-persistent macrocycles containing two bipyridine units.

6. Experimental Section

6.1 SYNTHESIS

All the substances were purchased from Aldrich or Acros and were used without further purification. Compounds $[Ru(bpy)_3]^{2+}$, and $[Os(bpy)_3]^{2+}$, $[Ru(bpy)_2(\mathbf{2})]^{2+}$, and $[Ru(bpy)_2(\mathbf{5})]^{2+}$, and ligand **2** were available from previous studies. Compounds **6**[9], **9**[10], **5**[16], **10**[16], $[(bpy)RuCl_3]_x$[17], and $[Os(bpy)_2Cl_2]$[18] were prepared according to the literature. Toluene and triethylamine were distilled from sodium/benzophenone ketyl and CH_2Cl_2 from CaH_2. All reactions sensible to oxygen were carried out under oxygen-free conditions. Column chromatography: Merck flash silica gel 60, 0.040 - 0.063 nm (230 - 400 mesh) or Fluka aluminium oxide neutral 507 C, 0.05-0.15 mm. NMR: Bruker AC 250, AM 270, AMX 500 (^{1}H: $CDCl_3$ at δ = 7.24, 13C: $CDCl_3$ at δ = 77.00 as internal standards, 20°C). MS: Perkin - Elmer Varian MAT 711, electron-impact (EI) mode. Elemental analyses: Perkin - Elmer EA 240. The experimental values for $[Ru(bpy)_2(\mathbf{1})]^{2+}$ and $[Os(bpy)_2(\mathbf{1})]^{2+}$ differ from the calculated ones by more than what is normally acceptable. The reason is seen in their hygroscopy.

*5,5'''-Bis-hexyloxymethyl-3,3'''-bis-trimethylsilanyl-[1,1';4',1'';4'',1''']-quaterphenyl (**8a**)*: to a degassed mixture of **6** (3.62 g, 9.27 mmol) and 4,4'-diiododiphenyl **7** (1.79 g, 4.40 mmol) in toluene (60 ml) and aq. 1M Na_2CO_3 (60 ml), $Pd(PPh_3)_4$ (280 mg, 0.24 mmol) was added and the mixture refluxed for 72 h. The layers were separated and the aqueous one was extracted with toluene (3 x 50 ml). The combined organic phases were dried over $MgSO_4$ and the solvent was removed. The yellow oil was purified by column chromatography through silica gel (hexane/ethyl acetate 30:1) to get 1.7 g (57%) of **8a** as a white solid. R_f(hexane/ethyl acetate 30:1) = 0.29; m.p. 73.5-75°C. ^{1}H NMR ($CDCl_3$, 250 MHz): δ = 0.33 (s, 18 H, $Si(CH_3)_3$), 0.93 (t, 6 H, CH_3), 1.24-1.45 (m, 12 H, γ-, δ-, ε-CH_2) 1.65 (m, 4 H, β-CH_2), 3.53 (t, 4 H, α-CH_2), 4.60 (s, 4 H, benzyl-CH_2), 7.47 (s, 2 H, phenyl-H), 7.61 (s, 2 H, phenyl-H), 7.71 (s, 10 H, phenyl-H); ^{13}C NMR ($CDCl_3$, 63 MHz): δ = -1.10, 14.02, 22.59, 25.91, 29.73, 31.66, 70.59, 72.92, 126.98, 127.25, 127.64, 131.16, 131.52, 138.40, 139.50, 140.08, 140.41, 141.08; MS (EI): *m/z* (%) = 678 (100), 663 (5.47), 5.77 (6.00), 91 (27.32); Elem. anal. calcd (%) for: $C_{44}H_{62}O_2Si_2$ (679.13): C 77.82, H 9.20; found: C 77.45, H 8.83.

*5,5'''Bis-hexyloxymethyl-3,3'-diiodo-[1,1';4',1'';4'',1''']quaterphenyl (**8b**)*: to a stirred solution of **8a** (1.52 g, 2.24 mmol) in dry CH_2Cl_2 (60 ml) a solution of ICl (1.10 g, 6.77 mmol) in CH_2Cl_2 (15 ml) was added dropwise over a period of 30 min at -15°C. The resulting mixture was stirred at this temperature for 30 min. Then a saturated solution of $Na_2S_2O_5$ (30 ml) was added. The layers were separated, the aqueous one was extracted with CH_2Cl_2 (2 x 30 ml). The combined organic phases were washed with water (20 ml), dried over $MgSO_4$ and the solvent removed to give 1.72 g (97%) of **8b** as a pure colorless powder. R_f (hexane/ethyl acetate 10:1) = 0.32; m.p. 58-59°C. ^{1}H NMR ($CDCl_3$, 250 MHz): δ = 0.89 (t, 6 H, CH_3), 1.27-1.38 (m, 12 H, γ-, δ-, ε-CH_2) 1.62 (m, 4 H, β-CH_2), 3.50 (t, 4 H, α-CH_2), 4.50 (s, 4 H, benzyl-CH_2), 7.55 (s, 2 H, phenyl-H), 7.64 (d, 4 H, phenyl-H), 7.67 (d, 4 H, phenyl-H), 7.67 (s, 2 H, phenyl-H), 7.89 (s, 2 H, phenyl-H); ^{13}C NMR ($CDCl_3$, 63 MHz): δ = 14.04, 22.62, 25.87, 29.68, 31.66, 70.89, 71.90, 94.86, 125.41, 127.40, 127.52, 135.01, 135.25, 138.60, 139.90, 141.53, 142.74; MS (EI): *m/z* (%) = 786 (100), 685 (7.5), 559 (9.73); Elem. anal. calcd. (%) for: $C_{38}H_{44}O_2I_2$ (786.568): C 58.03, H 5.64; found: C 58.02, H 5.41.

*Macrocycle **1***: a solution of **9** (993 mg, 0.98 mmol) and **8b** (771 mg, 0.98 mmol) in a mixture of triethylamine (320 ml) and toluene (320 ml) was carefully degassed. Then $[Pd(PPh_3)_4]$ (85 mg, 0.04 equiv) and CuI (14 mg, 0.04 equiv) were added and the reaction was stirred at 60°C for 4 d and then at 95°C for 1 d. The solvent was removed and purification of the residue by GPC gave 302 mg (20%) of **1** as a white powder. ^{1}H NMR ($CDCl_3$, 250 MHz): δ = 0.93 (m, 18 H, CH_3), 1.38 (m, 36 H, γ-, δ-, ε-CH_2), 1.69 (m, 12 H, β-CH_2), 3.55 (m, 12 H, α-CH_2), 4.45 (s, 4 H, benzyl-H), 4.48 (s, 4 H, benzyl-H), 4.53 (s, 4 H, benzyl-H), 7.40 (s, 6 H, phenyl-H), 7.48 (s, 2 H, phenyl-H), 7.56 (s, 4 H, phenyl-H), 7.66 (m, 14 H, benzyl-H), 8.03 (d, ^3J = 6.4 Hz, pyridyl-H), 8.51 (d, ^3J = 6.6 Hz, 2 H, pyridyl-H), 8.91 (s, 2 H, pyridyl-H); ^{13}C-NMR ($CDCl_3$, 63 MHz): δ = 14.02, 22.63, 25.91, 29.75, 31.71, 70.84, 70.91, 71.02, 72.00, 72.19, 72.39, 88.72, 89.36, 89.53, 90.15, 121.57, 123.41, 123.51, 123.70, 124.07, 125.56, 125.76, 127.17, 127.27, 129.12, 129.49, 129.97, 130.07, 131.84, 134.35, 135.49, 137.05, 138.72, 139.44, 139.53, 139.64, 140.20, 140.46, 146.74; MS (FAB) m/z (%) = 1544 (100), 1459 (59.21); elemental analysis calcd. (%) for $C_{108}H_{122}N_2O_6$ (1544.13): C 84.01, H 7.96, N 1.81; found: C 82.26, H 7.54, N 1.50.

[Os(bpy)$_2$(5)](PF$_6$)$_2$: [Os(bpy)$_2$Cl$_2$] (41 mg, 0.072 mmol) and **5** (50 mg, 0.072 mmol) were dissolved in ethanol (20 ml) and the mixture was stirred under nitrogen at reflux for 3 d. The solvent was removed to give a dark green residue which was purified by column chromatography through neutral aluminium oxide (CH_2Cl_2/methanol 95:5). The green fraction was collected and the solvent removed. The solid was dissolved in methanol (1 ml) and added to a solution of NH_4PF_6 (240 mg) in water (2 ml). The precipitated was separated by filtration, washed with water (5 x 2 ml) and dried in vacuum to give 41 mg (38%) of [Os(bpy)$_2$(**5**)](PF$_6$)2 as a green solid. ^{1}H NMR ($CDCl_3$, 500 MHz): δ = 0.86 (t, 6 H, CH_3), 1.27-1.36 (m, 12 H, γ-, δ-, ε-CH_2), 1.59 (m, 4 H, β-CH_2), 3.47 (t, 4 H, α-CH_2), 4.45 (s, 4 H, benzyl-H), 7.24 (s, 4 H, phenyl-H), 7.44 (t, 4 H, pyridyl-H), 7.53 (s, 2 H, phenyl-H), 7.56 (s, 2 H, pyridyl-H), 7.78 (d, ^3J = 5.5 Hz, 2H, pyridyl-H), 7.77-7.82 (m, 6 H, pyridyl-H), 7.93 (d, ^3J = 7.5 Hz, 2 H, pyridyl-H) 8.36 (d, ^3J = 8 Hz, 4 H, pyridyl H), 8.48 (d, ^3J – 9 Hz, 2 H, pyridyl H). ^{13}C NMR ($CDCl_3$, 63 MHz): δ = 14.06, 22.61, 25.80, 29.66, 31.65, 71.17, 71.32, 123.43, 124.04, 124.39, 124.85, 124.98, 128.86, 129.36, 131.57, 135.68, 136.19, 137.24, 137.69, 139.90, 142.91, 147.42, 151.11, 151.32, 157.45, 157.87, 158.55; MS (FAB): *m/z* (%): 1343 (2.13) [M+H-PF$_6$]$^+$, 1198 (2.38) [M+H-2PF$_6$]$^+$; Elem. anal. calcd. (%) for $C_{56}H_{58}Br_2F_{12}N_6O_2OsP_2$ (1487.07): C 45.23, H 3.93, N 5.65; found: C 44.43, H 4.07, N 5.53.

[Ru(bpy)(5)$_2$](PF$_6$)$_2$: A solution of **5** (76 mg, 0.11 mmol) and [Ru(bpy)Cl$_3$]$_x$ (20 mg, 0.055 mmol) in dioxane (2 ml), ethanol (3.5 ml), water (1.5 ml) was refluxed for 24 h. The solvent was removed and the residual orange material purified by column chromatography through silica gel (methanol/2M NH_4Cl/nitromethane 7:2:1). The combined orange fractions were diluted with chloroform and the organic phase was separated, and the solvent removed. The orange precipitate was then dissolved in the minimal amount of methanol (2 ml) and added to a solution of NH_4PF_6 (200 mg) in water (3 ml). The precipitated was separated by filtration, washed with water (6 ml), and dried in vacuum to give 88 mg (83%) of [Ru(bpy)(**5**)$_2$](PF$_6$)2 as an orange solid. ^{1}H NMR ($CDCl_3$, 500 MHz): δ = 0.79 (t, 12 H, CH_3), 1.14-1.3 (m, 24 H, γ-, δ-, ε-CH_2) 1.45 (m, 4 H, β-CH_2), 1.5 (m, 4 H, β-CH_2), 3.33 (t, 4 H, α-CH_2), 3.42 (t, 4 H, α-CH_2), 4.33 (s, 4 H, benzyl-CH_2), 4.4 (s, 4 H, benzyl-CH_2), 7.13 (s, 4 H, phenyl-H), 7.21(s, 4 H, phenyl-H), 7.44 (s, 2 H, phenyl-H), 7.5 (s, 2 H, phenyl-H), 7.58 (s, 2 H, pyridyl-H), 7.67 (d, 4 H, pyridyl-H), 7.9 (s, 2 H, pyridyl-H), 7.98 (s, 2 H, pyridyl-H), 8.13 (s, 4 H, pyridyl-H), 8.45 (s, 2 H, pyridyl-H), 8.64 (s, 4 H, pyridyl-H); ^{13}C NMR ($CDCl_3$, 63 MHz): δ = 13.91, 22.47, 25.65, 25.68, 29.48, 31.52, 71.18, 123.35, 123.59, 124.40, 124.84, 125.13, 125.50, 128.66, 131.46, 135.94, 136.22, 136.33, 136.72, 138.34, 139.36, 142.69, 143.05, 147.98, 152.15, 155.11, 155.81, 156.35; MS (FAB): *m/z* (%): 1791 [M-PF$_6$]$^+$ (100), 1646 [M-2 x PF$_6$]$^{2+}$ (90.29); Elem. anal. calcd. (%) for $C_{82}H_{92}Br_4N_6O_4P_2F_{12}Ru$ (1936.26): C 50.86, H 4.79, N 4.34; found: C 51.04, 4.73, N 4.11.

[Ru(bpy)(11)$_2$](PF$_6$)$_2$: a stirred solution of **10** (150 mg, 0.21 mmol) and [Ru(bpy)Cl$_3$]$_x$ (39 mg, 0.10 mmol) in dioxane (4 ml), ethanol (4 ml), water (2 ml) was refluxed for 24 h. The solvent was removed and the residual orange material purified by column chromatography through silica gel (methanol/2M NH_4Cl/nitromethane 7:2:1). The combined orange fractions were diluted with chloroform, the organic phase separated, and the solvent removed. The orange precipitate was then dissolved in methanol (20 ml) and added to a solution of NH_4PF_6 (200 mg) in water (3 ml). The precipitated was separated by filtration, washed with water (6 ml), and dried in vacuum to give 144 mg (84%) of [Ru(bpy)(**11**)$_2$](PF$_6$)2 as an orange solid. ^{1}H NMR (DMSO, 500 MHz): δ = 4.45 (s, 4 H, benzyl-CH_2), 4.54 (s, 4 H, benzyl-CH_2), 5.39 (s, 4 H, -OH), 7.43 (s, 4 H, phenyl-H), 7.56 (s, 2 H, phenyl-H), 7.61 (s, 2 H, pyridyl-H, 4 H, phenyl-H), 7.65 (s, 2 H, phenyl-H), 7.86 (s, 2 H, pyridyl-H), 7.92 (s, 2 H, pyridyl-H), 8.11 (d, 2 H, pyridyl-H), 8.27 (t, 2 H, pyridyl-H), 8.62

(d, 2 H, pyridyl-H), 8.69 (d, 2 H, pyridyl-H), 8.88 (d, 2 H, pyridyl-H), 8.95 (d, 2 H, pyridyl-H), 9.02 (d, 2 H, pyridyl-H); ^{13}C NMR (DMSO, 126 MHz): δ = 62.26, 122.85, 124.35, 125.14, 128.13, 130.29, 135.82, 136.13, 136.73, 137.06, 137.33, 137.86, 138.24, 146.84, 149.31, 149.66, 153.03, 155.77, 156.16, 157.30; MS (FAB): *m/z* (%): 1451 (2.21) $[M-PF_6]^+$, 1306 (3.28) $[M-PF_6]^{2+}$; Elem. anal. calcd. (%) for $C_{58}H_{40}F_{12}Br_4N_6O_4P_2Ru$ (1595.59): C 43.66, H 2.53, N 5.27; found: C 44.06, H 2.84, N 5.15.

$[Ru(bpy)_2(1)](PF_6)_2$: a stirred solution of **1** (50 mg, 0.032 mmol) and $[Ru(bpy)_2]Cl_2$•$2H_2O$ (24 mg, 0.066 mmol) in dioxane (5 ml), propylene glycol (3 ml) and ethanol (3 ml) was refluxed for 24 h under nitrogen. The solvent was removed and the brown-orange solid purified by column chromatography through silica gel (CH_2Cl_2/methanol 95:5). The solvent was removed to give an orange solid which was dissolved in 1 ml methanol. This solution was added to a solution of NH_4PF_6 (50 mg) in H_2O (2 ml). The precipitated was separated by filtration, washed with H_2O (4 x 2 ml) and dried in vacuum to give 40 mg (56%) of $[Ru(bpy)_2(\mathbf{1})](PF_6)_2$ as an orange solid. R_f = 0.54 (dichloromethane/methanol = 95:5). ^{1}H NMR ($CDCl_3$, 500 MHz): δ = 0.89 (m, 18 H, CH_3), 1.25-1.42 (m, 36 H, γ-, δ-, ε-CH_2), 1.53-1.70 (m, 12 H, β-CH_2), 3.47 (t, 4 H, α-CH_2), 3.52 (m, 8 H, α-CH_2), 4.40 (s, 4 H, benzyl-H), 4.49 (s, 4 H, benzyl-H), 4.54 (s, 2 H, benzyl-H), 7.05 (s, 2 H, phenyl-H), 7.42 (s, 2 H, phenyl-H), 7.45 (s, 4 H, phenyl-H), 7.46 (s, 2 H, phenyl-H) 7.48 (s, 2 H, phenyl-H), 7.90 (t, ^3J = 6.5 Hz, 2 H, pyridyl-H), 7.54 (s, 2 H, phenyl-H), 7.56 (t, ^3J = 6.5 Hz, 2 H, pyridyl-H), 7.62 (s, 2 H, pyridyl-H), 7.63-7.69 (mc, 4 H, phenyl-H), 7.71-7.74 (m, 6 H, phenyl-H), 7.77 (s, 2 H, phenyl-H), 7.84 (d, ^3J = 4.5 Hz, 2 H, pyridyl-H), 7.87 (d, ^3J = 4.5 Hz, 2 H, pyridyl-H), 7.92 (t, ^3J = 7 Hz, 2 H, pyridyl-H), 7.99 (t, ^3J = 7 Hz, 2 H, pyridyl-H), 8.07 (d, ^3J = 8 Hz, 2 H, pyridyl-H), 8.33 (2d overlapped, ^3J = 8 Hz, 4 H, pyridyl-H), 8.51 (d, ^3J = 8.5 Hz, 2 H, pyridyl-H); ^{13}C NMR ($CDCl_3$, 500 MHz): δ = 14.19, 22.76, 25.98, 29.86, 31.82, 71.00, 71.08, 71.19, 71.76, 72.05, 72.48, 88.82, 89.17, 90.15, 90.40, 123.61, 123.85, 124.62, 124.82, 125.39, 127.35, 127.51, 128.39, 128.95, 129.58, 129.84, 130.11, 130.22, 130.26, 131.16, 134.69, 134.94, 136.52, 137.80, 138.25, 138.81, 139.50, 139.74, 139.90, 140.60, 141.07, 148.03, 151.99, 152.32, 155.30, 156.15, 156.75; MALDI-TOF *m/z*: 2103 $[M-PF_6]^+$, 2025 $[M+Na-PF_6-OC_6H_{13}]^+$, 1958 $[M-2PF_6]^+$; Elem. anal. calcd. for $C_{128}H_{138}N_6O_6RuP_2F_{12}$ (2247.50): C 68.40, H 6.19, N 3.74; found: C 67.13, H 6.19, N 3.41.

$[Os(bpy)_2(1)](PF_6)_2$: a stirred solution of **1** (40 mg, 0.026 mmol) and $[Os(bpy)_2]Cl_2$ (17 mg, 0.030 mmol) in butanol (15 ml) was refluxed for 5 d. The solvent was removed and the brown-green solid purified by column chromatography on neutral aluminium oxide (CH_2Cl_2/methanol 95:5) to remove unreacted macrocycle. Then the complex was eluted off the column using methanol. The green fraction was collected and the solvent removed to give a green solid which was solved in methanol (2 ml). This solution was added to a solution of NH_4PF_6 (86 mg) in H_2O (2 ml). The precipitated was separated by filtration, washed with H_2O (4 x 3 ml), and dried in vacuum to give 30 mg (50%) of $[Os(bpy)_2(\mathbf{1})](PF_6)_2$ as a green solid. ^{1}H NMR ($CDCl_3$ 500 MHz): δ = 0.88 (m, 18 H, CH_3), 1.24-1.44 (m, 36 H, γ-, δ-, ε-CH_2), 1.56-1.72 (m, 12 H, β-CH_2), 3.47 (t, 4 H, α-CH_2), 3.52 (m, 8 H, α-CH_2), 4.42 (s, 4 H, benzyl-H), 4.48 (s, 4 H, benzyl-H), 4.54 (s, 4 H, benzyl-H), 7.05 (s, 2 H, phenyl-H), 7.42 (s, 4 H, phenyl-H), 7.45-7.53 (m, 10 H, 6 phenyl-H, 4 pyridyl-H), 7.56 (s, 4 H, 2 pyridyl-H, 2 phenyl-H), 7.64-7.84 (m, 20 H, 12 phenyl-H, 8 pyridyl-H), 7.91 (d, ^3J = 8 Hz, 2 H, pyridyl-H), 8.32 (t, ^3J = 9 Hz, 4 H, pyridyl-H), 8.5 (d, ^3J = 8 Hz, 2 H, pyridyl-H); ^{13}C NMR ($CDCl_3$, 500 MHz): δ = 14.16, 22.74, 25.97, 29.83, 31.81, 71.00, 71.06, 71.17, 71.75, 72.00, 72.48, 88.8, 88.84, 90.16, 90.40, 99.57, 102.14, 116.68, 123.28, 123.6, 123.85, 123.97, 124.39, 124.57, 125.00, 125.47, 126.05, 127.36, 127.50, 128.87, 129.19, 129.35, 129.57, 129.80, 130.22, 131.16, 132.54, 134.56, 134.89, 135.92, 136.51, 137.11, 137.63, 138.81, 138.99, 139.53, 139.74, 139.89, 140.26, 140.62, 141.06; MS (MALDI-TOF): 2192 $[M-PF_6]^+$, 2047 $[M-2PF_6]^+$; elemental analysis calcd for $C_{128}H_{138}N_6O_6OsP_2F_{12}$ (2336.66): C 65.79, H 5.95, N 3.60; found: C 64.23, H 5.75, N 3.29.

6.2 PHOTOPHYSICS AND ELECTROCHEMISTRY

The equipments used for investigation of absorption spectra and luminescence properties (in fluid solution and in rigid matrix) have been previously described [19]. Anthracene in air-equilibrated ethanol ($\phi = 0.21$)[20] and $[Ru(bpy)_3]^{2+}$ in air-equilibrated aqueous solution ($\phi = 0.028$)[21] were used as standards for evaluating the luminescence quantum yield of **1** and the Ru and Os complexes, respectively. Air equilibrated solutions were used.

The electrochemical investigation of ligands **1**, **2**, **5** and bpy was carried out by using the equipment and procedure previously described [8]. The equipment[19] and procedure [8] used for investigation of the electrochemical behavior of the complexes in argon-purged Hi-dry dichloromethane have been previously described.

Experimental errors: molar absorption coefficients, +/–5%; luminescence quantum yield, +/–10%, luminescence lifetime, +/–5%; redox potentials, 10 mV and 20 mV for reversible and irreversible processes, respectively.

7. Acknowledgements

This research was supported by MIUR (Supramolecular Devices Project), the University of Bologna (Funds for Selected Topics), and Deutsche Forschungsgemeinschaft (Sfb 448, TP A1) which are gratefully acknowledged. The authors thank Ms. E. Franzus and Dr. G. Holzmann for their competent help with EI and FAB mass spectrometric analyses and Ms. C. Zimmermann for her help with GPC analyses.

8. References

[1] (a) Henze, O., Lentz, D., and Schlüter, A. D. (2000) Synthesis and an X-ray structure of soluble phenylacetylene macrocycles with two opposing bipyridine donor sites, *Chem. Eur. J.* **6**, 2362-2367. (b) Grave, C. and Schlüter, A. D. (2002) Shape-persistent, nano-sized macrocyles, *Eur. J. Org. Chem.* 3075-3098. (c) Zhao, D. and Moore, J. S. (2003) Shape-persistent arylene ethynylene macrocycles: syntheses ans supramolecular chemistry, *Chem. Commun.* 807-818. (d) Baxter, P. N. W. (2003) Synthesis of a hexagonal nanosized macrocyclic fluorophore with integrated endotopic terpyridine metal-chelation sites, *Chem. Eur. J.* **9**, 5011-5022. (e) Grave, C., Lentz, D., Schäfer, A., Samorì, P., Rabe, J. P., Franke, P., and Schlüter, A. D. (2003) Shape-persistent macrocycles with terpyridine units: synthesis, characterization, and structure in the crystal, *J. Am. Chem. Soc.* **125**, 6907-6918. (f) Höger, S. (2004) Shape-persistent macrocycles: from molecules to materials, *Chem. Eur. J.* **10**, 1320-1329. (g) Fischer, M., Lieser, G., Rapp, A., Schnell, I., Mamdouh, W., De Feyter, S., De Schryver, F. C., and Höger, S. (2004) Shape-persistent macrocycles with intraanular polar groups: synthesis, liquid crystallinity, and 2D organization, *J. Am. Chem. Soc.* **126**, 214-222.

[2] Cobden, D. H. (2001) Nanowires begin to shine, *Nature*, **409**, 32-33.

[3] Scott, L. T., DeCicco, G. J., and Reinhardt, G. (1985) Pericyclynes of order [5], [6], [7], and [8]. Simple convergent syntheses and chemical reactions of the first homoconjucated cyclic polyacetylenes, *J. Am. Chem. Soc.* **107**, 6546-6555.

[4] Henze, O., Lentz, D., Schäfer, A., Franke, P., and Schlüter, A. D. (2002) Phenylacetylene macrocycles with two opposing bipyridine donor sites: synthesis, X-ray structure determinations, and Ru complexation, *Chem. Eur. J.* **8**, 357-365 and references therein. (b) Henze, O., Ph.D. Thesis, FU Berlin, Germany, **2000**, http://www.diss.fu-berlin.de/2000/47/.

[5] (a) Balzani, V., Credi, A., Raymo, F. M., and Stoddart, J. F. (2000) Artificial molecular machines, *Angew. Chem. Int. Ed.* **39**, 3348-3391. (b) Balzani, V., Credi, A., and Venturi, M. *Molecular Devices and Machines – A Journey into the Nanoworld*, Wiley-VCH, Weinheim, **2003**.

[6] (a) Juris, A., Balzani, V., Bangelletti, F., Campagna, S., Belser, P., and von Zelewsky, A. (1988) Ru(II) polypyridine complexes: photophysics, photochemistry, electrochemistry, and chemiluminescence, *Coord. Chem. Rev.* **84**, 85-277. (b) Kober, E. M., Caspar, J. V., Sullivan, B. P., and Meyer, T. J. (1988) Synthetic routes to new polypyridyl complexes of osmium (II), *Inorg. Chem.* **27**,4587-4598. (c) Kalyanasundaram, K. *Photochemistry of Polypyridine and Porphyrin Complexes*, Academic Press, London, **1992**. (d) Reitsma, D. A., and Keene, F. R., (1993) Diastereoisomeric forms of ligand-bridged dimetallic diruthenium (II) and ruthenium (II)-osmium (II) species containing bidentate polypyridyl ligands, *J. Chem. Soc. Dalton. Trans.* 2859-2860. (e) Balzani, V., Juris, A., Venturi, M., Campagna, S., and Serroni, S. (1996) Luminescent and redox-active polynuclear transition metal complexes, *Chem. Rev.* **96**, 759-834.

[7] (a) Balzani, V. and Scandola, F. *Supramolecular Photochemistry*; Horwood, Chichester, **1991**. (b) Balzani, V., Ceroni, P., Juris, A., Venturi, M., Campagna, S., Puntoriero, F., Serroni, S. (2001) Dendrimers based on photoactive metal complexes. Recent advances, *Coord. Chem. Rev.* **219**, 545-572. (c) Juris, A., Venturi, M., Ceroni, P., Balzani, V., Campagna, S., and Serroni, S. (2001) Dendrimers based on electroactive metal complexes. A review of recent advances, *Collect. Czech. Chem. Commun.* **66**, 1-32.

[8] Venturi, M., Marchioni, F., Balzani, V., Opris, D. M., Henze, O., and Schlüter, A. D. (2003) A photophysical and electrochemical investigation on a phenylacetylene macrocycle containing two 2,2'-bipyridine units, its protonated forms, and Ru^{II} and Os^{II} complexes, *Eur. J. Org. Chem.* 4227-4233.

[9] Manickam, G. and Schluter, A. D. (2000) New parts for a construction set of bifunctional oligo(het)arylene building blocks for modular chemistry, *Synthesis*, **3**, 442-446.

[10] Opris, D. M. and Schlüter, A. D. manuscript in preparation.

[11] The MALDI TOF mass spectrum of $[\mathbf{1}]_2$ showed the molecular ion peak at m/z = 3089

[12] (a) Hill, D. J., Mio, M. J., Prince, R. B., Hughes, T. S., and Moore, J.S. (2001) A field guide to foldamers, *Chem Rev.* **101**, 3893-4011. (b) Balbo Block, M. A., Kaiser, C., Khan, A., and Hecht S. (in press) Discrete organic nanotubes based on a combination of covalent and non-covalent approaches, *Top. Curr. Chem.*

[13] The GPC molar mass versus polystyrene standard is: Mn = 13100 g/mol, Mw = 14300 g/mol, PDI: 1.09 (in THF).

[14] Opris, D. M., Ph.D. Thesis, 2004, FU Berlin, Germany, in preparation.

[15] Giuffrida, G. and Campagna, S. (1994) Influence of peripheral ligands on the metal-metal interaction in dinuclear metal complexes with *N*-heterocyclic bridging ligands, *Coord. Chem. Rev.* **135/136**, 517-531.

[16] Henze, O., Lehmann, U., Schlüter, A. D. (1999) Synthesis of 5,5'-disubstituted 2,2'-bipyridines for modular chemistry, *Synthesis,* **4**, 683-687.

[17] Kelch, S. and Rehahn, M. (1997) High-molecular-weight ruthenium(II) coordination polymers: synthesis and solution properties, *Macromolecules*, **30**, 6185-6193.

[18] (a) Buckingham, D. A., Dwyer, L. F. P., Goodwin, H. A., and Sargeson, A. M. (1964) Mono- and bis-(2,2'-bipyridine)and (1,10-phenantroline) chelates of ruthenium and osmium, *Aust. J. Chem.*, **17**, 325-336. (b) Lay, P. A., Sargeson, A. M., and Taube, H. (1986) cis-Bis(2,2'-bipyridine-N,N') complexes of ruthenium (III)/(II) and osmium (III)/(II), *Inorg. Synth.* **24**, 291-298.

[19] Amabilino, D. B., Asakawa, M., Ashton, P. R., Ballardini, R., Balzani, V., Belohradsky, M., Credi, A., Higuchi, M., Raymo, F. M., Shimizu, T., Stoddart, J. F., Venturi, M., and Yase, K. (1998) Aggregation of self-assembling branched [*n*]rotaxanes, *New J. Chem.* 959-972.

[20] Melhuish, W. H. (1961) Quantum efficiencies of fluorescence of organic substances: effect of solvent and concentration of the fluorescent solute, *J. Phys. Chem.* **65**, 229-235.

[21] Nakamaru, K. (1982) Synthesis, luminescence quantum yields, and lifetimes of trischelated ruthenium(II) mixed-ligand complexes including 3,3'-dimethyl-2,2'-bipyridyl, *Bull. Chem. Soc. Jpn.* **55**, 2697-2705.

15.
MACROCYCLIC SYSTEMS WITH PHOTOSWITCHABLE FUNCTIONS

O. A. FEDOROVA[a], E. N. USHAKOV[b], Y. V. FEDOROV[a],
Y. P. STROKACH[a] AND S. P. GROMOV[a]
*[a]Photochemistry Center of Russian Academy of Sciences,
Novatorov St., 7a, 119421 Moscow, Russia*
*[b]Institute of Problems of Chemical Physics, Russian Academy of Sciences,
142432 Chernogolovka, Moscow Region, Russia*

1. Introduction

Photochromism is a reversible transformation of a single chemical species between two states, the absorption spectra of which are clearly different, the transition in at least one direction being induced by electromagnetic radiation [1]. The widest and most important group of the photochromic system is based on electrocyclic reactions [2,3]; a few have been commercially successful (polymer-based photochromic eyewear, novelty items and security printing inks). Several other photochromic systems based on *E,Z-isomerization*, cycloaddition reaction, electron or proton transfer have potential industrial applications [4].

Photochromic macrocyclic compounds constitute a new class of artificial receptors in which the recognition of metal cations induces a conformational change in the receptor framework accompanied by signalling (coloration). In addition, since binding of cations is sensitive to the ligand environment, the binding constant can be controlled effectively by employing photochromic systems which change in the light [5,6].

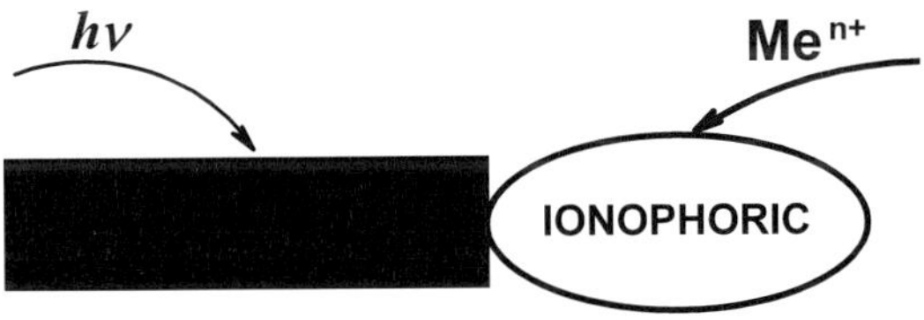

This review surveys our studies devoted to the photoswitchable molecular receptors based on photochromic crown ethers. Photochromic systems described in the review may be classified into three groups according to the reaction types: *E,Z-isomerization*, [2+2]-photocycloaddition reactions and electrocyclic reaction. It has proved the groups to be an especially suitable basis for photochromic systems, and promising for the industrial applications.

Karsten Gloe (ed.), Macrocyclic Chemistry: Current Trends and Future Perspectives, 235–252.
© 2005 *Springer. Printed in the Netherlands.*

It is known from literature that several reversible photochemical reactions, such as geometric isomerism of azobenzene [7], electrocyclic reaction of dihydroindolizines, fulgides and diarylethylenes with heterocyclic groups [8-10], dimerization of anthracene [11], and photochromic reaction of spirocompounds [12] have been also employed to provide photocontrol over metal-ion binding ability of crown ethers.

2. Macrocyclic Systems with Ability for *E,Z*-isomerization

It was found that the linking of styryl dye fragment to benzocrown ether results in novel photochromic compounds CESD (Crown Ether Styryl Dyes) possessing interesting physico-chemical properties (Scheme 1) [13]. The dyes are intensively colored and show significant hypsochromic shifts upon complexation with alkaline earth metal cations in acetonitrile solution. Reversible photochemical reaction *E,Z-isomerization* is observed for both dyes and their complexes.

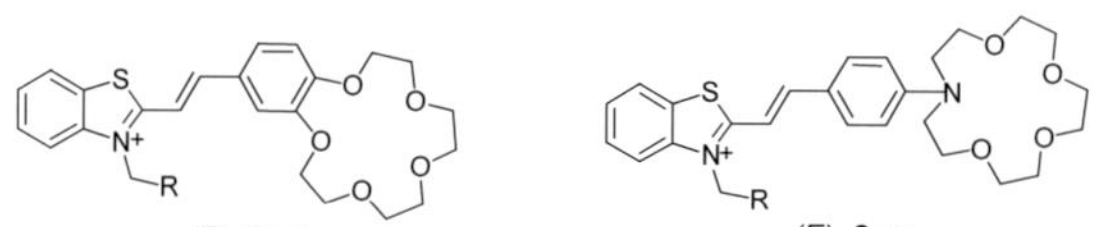

(*E*)- **1a–f** (*E*)- **2a,c**

a: R = Me (anion ClO_4^-), **b:** R = $(CH_2)_2SO_3^-$, **c:** R = $(CH_2)_3SO_3^-$, **d:** R = $(CH_2)_4SO_3^-$,
e: R = *o*-$CH_2C_6H_4SO_3^-$, **f:** R = *p*-$CH_2C_6H_4SO_3^-$

Scheme 1

The CESD **1c-f** with terminal sulfo group are able to form anion-“capped” complexes (Scheme 2) upon *E,Z*-photoisomerization owing to the interaction of the sulfo group of the *N*-substituent (spacer) with a metal cation in the crown-ether cavity. However, the dates obtained that the *Z*-isomer of **1a,b** is unable to form an anion-“capped” complex [14].

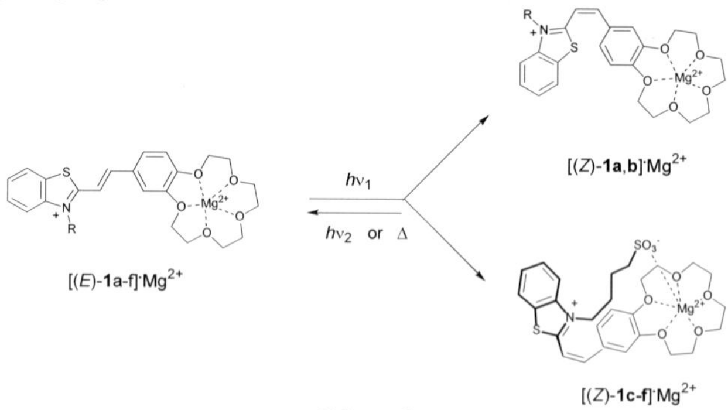

Scheme 2

It should be noted that the influence of the spacer structure on the stability of the complexes is much more pronounced in the case of *Z*-isomers. This refers most of all to CESD **1e,f** with conformationally rigid spacers. The ratio of the stability constants of the anion-“capped” complexes [(*Z*)-**1f**]·Mg^{2+} and [(*Z*)-**1e**]·Mg^{2+} is about 500 (Scheme 3) [15,24].

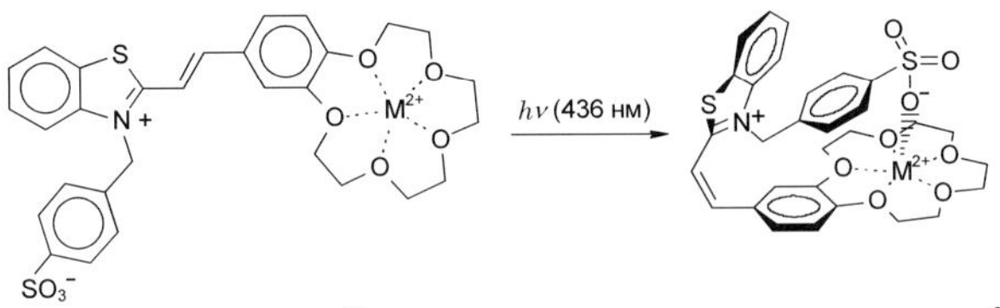

[(*E*)-**1f**]·Mg^{2+} [(*Z*)-**1f**]·Mg^{2+}

Scheme 3

The *E,Z-isomerization* of the styryl dyes **2c** containing aza-15-crown-5 ether moiety is characterized by great hypsochromic shifts, equal to 170 nm, of the spectrum of [(*Z*)-**2c**]·Ca^{2+} with respect to the spectrum of [(*E*)-**2c**]·Ca^{2+} (see Figure 1). Apparently, it can be explained by the fact that in anion-"capped" complex the molecule of dye acquires a twisted conformation with marked disruption of the conjugation in the chromophoric system. When (*E*)-**2a** is converted into the *Z*-form, the stability of complex increases approximately 2.5 fold. On going from the cationic dye **2a** to the betaine **2c**, stability of the complexes formed by *Z*-isomer increases by more than three orders of magnitude [16].

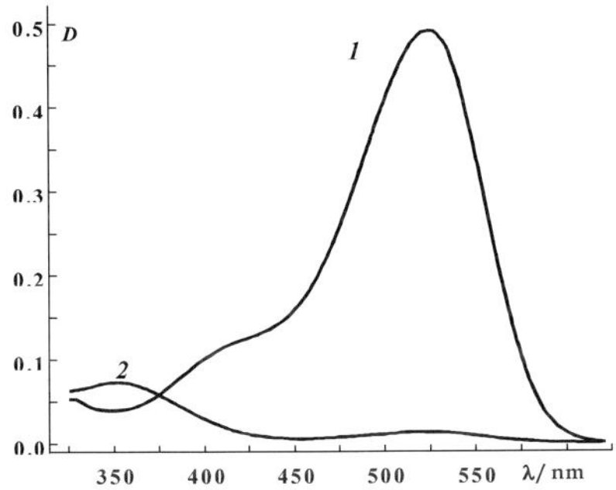

Figure 1. Absorption spectra of CESD **2c** in MeCN (C_L = 1.0×10^{-5}mol L^{-1}) in the presence of $Ca(ClO_4)_2$ at C_M/C_L = 100; E-isomer (1); E-Z-photosteady (2) state formed on exposure to light with a wavelength of 546 nm and intensity of ~ $1\cdot10^{16}cm^{-1}s^{-1}$.

Styryl dyes CESDs **3a-e, 4a-e** incorporating thiacrown ether moieties exhibit a strong preference for the formation of complexes with heavy metal salts (Hg^{2+}, Pb^{2+}, Cd^{2+}, Ni^{2+} or Zn^{2+} perchlorates) [17,18].

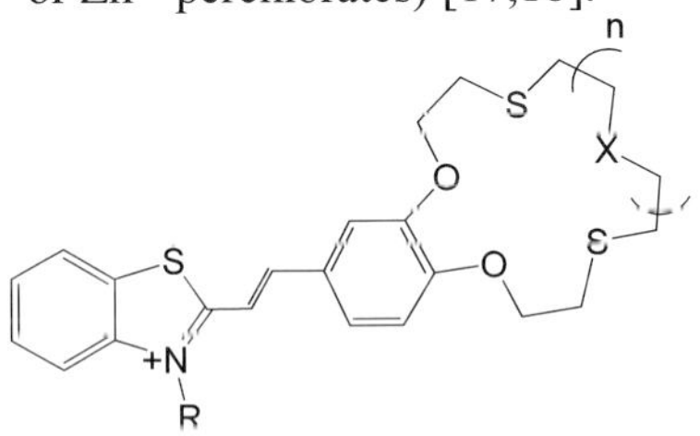

3a-e	**4a-e**
R = Me (ClO_4^-);	R = $(CH_2)_4SO_3^-$;
3a: n= **0**;	**4a**: n= **0**;
3b: n= **1**, **X** = O;	**4b**: n= **1**, **X** = O;
3c: n= **2**, **X** = O;	**4c**: n= **2**, **X** = O;
3d: n= **3**, **X** = O;	**4d**: n= **3**, **X** = O;
3e: n= **2**, **X** = S	**4e**: n= **2**, **X** = S

Scheme 4

Exposure of (*E*)-**3a-e, 4a-c** in MeCN solutions in the presence of Hg^{2+} or Pb^{2+} to visible light, leads to UV-Vis spectral changes typical of *E,Z*-isomerization [18]. The behaviour of complexes [(*E*)-**4a-d**]·Hg^{2+} is analogous to that previously described for the [(*E*)-**1c**]·Mg^{2+} complex, whose *Z*-isomer is anion-"capped". Observed long absorption band (LAB) shifts of 50-70 nm on complexation of (*Z*)-**4a-d** with Hg^{2+} are consistent with diminished conjugation between the three moieties comprising the

CESD chromophore, due to the pronounced departure from planarity that accompanies formation of anion-"capped" complexes. Enhanced stability due to intramolecular coordination in the anion-"capped" [(*Z*)-**4c**]·Hg^{2+} complexes are reflected in the stability constants in Table 1 (K_c/K_t equals 10 for **4c**·Hg^{2+} and 15 for **4c**·Pb^{2+}) and causes sharp decelerations of their dark $Z \rightarrow E$ isomerizations. The saturated coordination capacity of Hg^{2+} in complex [(*E*)-**4e**]·Hg^{2+} explains why its *Z* isomer does not form an anion-"capped" complex. For the *Z* isomers of CESDs **4a-d** no spectral evidence of anion-"capped" complex formation could be found when Cd^{2+}, Ni^{2+} or Zn^{2+} was substituted for Pb^{2+} or Hg^{2+}. As expected, based on the theory of "hard" and "soft" acids and bases, the strength of the coordination bond between Cd^{2+}, Ni^{2+} or Zn^{2+} and the SO_3^- group appears to be to small to stabilize the Z isomer [19,25].

TABLE 1: Equilibrium Constants for Pb^{2+} and Hg^{2+}Complex Formation with CESDs

Complex	log *K*	Complex	log *K*
(*E*)-**3b**·Hg^{2+}	15.9(1)[a]; 15.7(1)[b]	(*E*)-**4b**·Hg^{2+}	18.0(1)[b]
(*E*)-**3c**·Hg^{2+}	18.2(1)[a]; 18.0(1)[b]	(*E*)-**4c**·Hg^{2+}	19.8(1)[b,c]
(*E*)-**3c**·Pb^{2+}	5.61(2)[b]	(*E*)-**4c**·Pb^{2+}	7.57(3)[b]
(*E*)-**3e**·Hg^{2+}	21.0(1)[a]	(*E*)-**4e**·Hg^{2+}	20.7(1)[b]
BDT12C4[e]·Hg^{2+}	22.7(2)[a,d]	(*Z*)-**4c**·Hg^{2+}	20.8(1)[b,c]
BDT15C5[e]·Hg^{2+}	18.0(1)[a]	(*Z*)-**4c**·Pb^{2+}	8.75(3)[b]
BDT18C6[e]·Hg^{2+}	19.5(1)[a,c]		
BDT18C6[e]·Pb^{2+}	7.26(2)[b]		

[a] These values were determined polarographically for 0.01 M solution of Et_4NClO_4 in MeCN at 20 °C; values in parentheses are uncertainties in the last significant figure shown. [b] These values were determined spectrophotometrically in MeCN at 20 °C; values in parentheses are uncertainties in the last significant figure shown. [c] These values were taken from ref 19. [d] For complex 2 ligand : 1 metal. [e] BDT12C4 –benzodithia-12-crown-4; BDT15C5 – benzodithia-15-crown-5; BDT18C6 – benzodithia-18-crown-6.

Substantial changes are observed in the ^1H NMR spectra when a solution of [(*E*)-**4c**]·Pb^{2+} is exposed to visible light (Figure 2). In the novel photoproduct, the spin-spin coupling constants for the olefinic proton signals at 6.70 and 7.52 ppm were 12.2 Hz, which implied the formation of [(*Z*)-**4c**]·Pb^{2+}. In the *Z*-isomer, the signals of aromatic protons of the benzocrown ether moiety and of the olefinic protons shift upfield, while the benzothiazole proton signals shift downfield relative to those of [(*E*)-**4c**]·Pb^{2+}. The pronounced spectral difference between the two photoisomers is due to the substantial conformational rearrangement induced by the formation of the anion-"capped" complex. The formation of this complex enforces a twisted conformation on the chromophore, resulting in a distortion of the conjugated system and concentration of the positive charge in the benzothiazole fragment of the molecule. The greatest changes were found for the chemical shifts of the H(a) and H(2') protons, which is apparently due to the fact that the protons fall into the areas of shielding of the benzothiazole fragment and the double C=C bond [20].

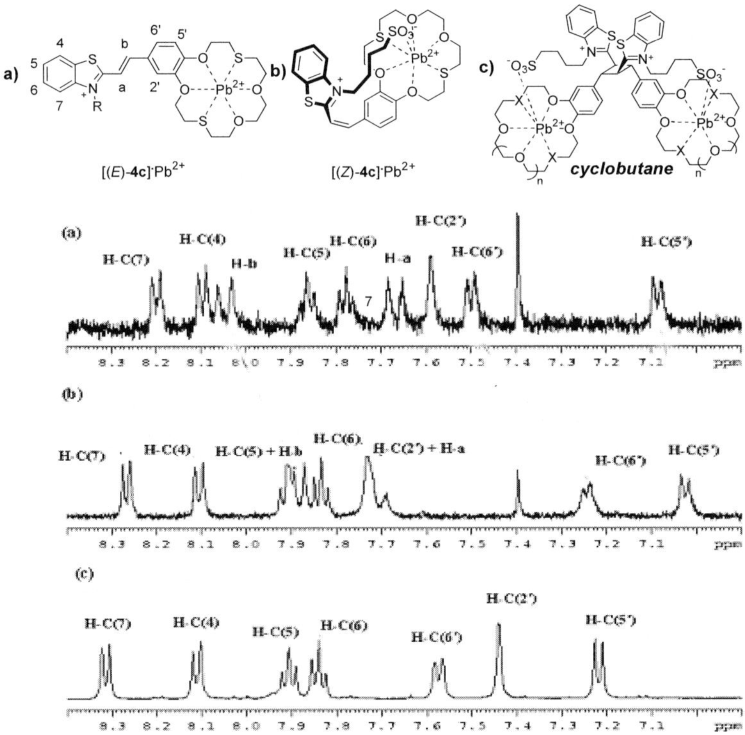

Figure 2. ^{1}H NMR spectra of the complexes [(E)-**4c**]$\cdot Pb^{2+}$(a), [(Z)-**4c**]$\cdot Pb^{2+}$(b) and cyclobutane (c) ($C_L = 1\times10^{-3}$ M; $C_{Pb}{}^{2+} = 5\times10^{-3}$ M) in CD_3CN at 50°C.

Thus, the CESD with anionic terminal substituent are able of anionic "capping" complex formation during the irradiation and thus to achieve photochemical control over binding of metal cations.

3. Macrocyclic Systems with Ability for [2+2]-Photocycloaddition Reaction

A promising tool for controlling the regio- and stereoselectivity of photocycloaddition reaction (PCA) as well as the efficiency may be provided by assembling alkenes into a supramolecular structure with pre-organization of reactants such that the spatial arrangement of molecules would be favorable for the formation of only one cyclobutane isomer in a high yield.

This idea was realized using crown ether styryl dyes (CESD) **1c,d, 4c** (Scheme 1,4). The compounds **1c,d, 4c** having betaine structures form supramolecular dimers with a crossed arrangement of molecules (*anti*-head-to-tail) in the presence of ions, due to the intermolecular interaction between the sulfo group of one of the molecules and a ion located in the crown-ether cavity of the other molecule [20,21]. It was shown that photoirradiation of solutions of dimer results in stereospecific PCA giving only one of the 11 possible derivatives of cyclobutane, which is expected in conformity with the concerted superficial (s,s) addition of the reactants (Scheme 5) [22,23]. It is noteworthy that neither (*E*)-**1c,d, 4c** without metal cations nor complexes of (*E*)-**1a,b** with Mg^{2+}, Ca^{2+}, Ba^{2+} undergo PCA even in saturated solutions.

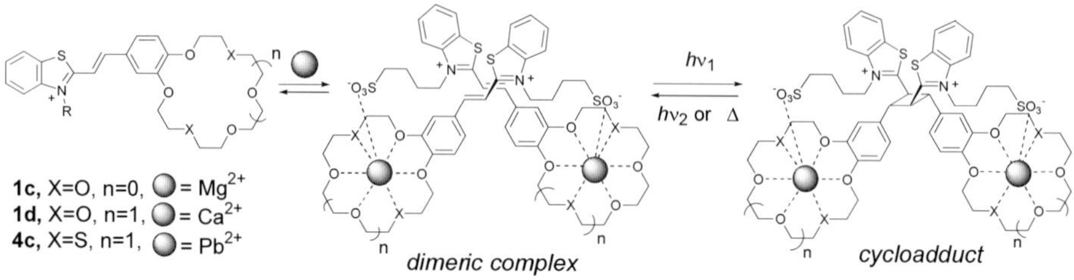

Scheme 5

Transition from spacers with flexible polymethine chains to *N*-substituents in which the sulfo group is rigidly arranged in space (CESD **1e,f**), makes it possible to influence the efficiency of these photochemical reactions and also to change the route of transformation of CESD [15,24]. Thus, upon irradiation with UV light reversible [2+2]-photocycloaddition reaction takes place only in case of styryl dye **1e** with *ortho*-sulfobenzyl substituent.

The single [2+2]-photocycloadduct obtained by irradiating thiacrown-containing dye **4a-e** in the presence of either Pb^{2+} or Hg^{2+} was assigned the cyclobutane structure (see figure 2c). The quantum yield of photodimer formation is substantially more efficient in the presence of Pb^{2+} than in the presence of Hg^{2+} (Table 2) [20,25]. The efficiency of cycloadduct formation depends strongly on the size and S/O ratio of the crown ether moiety. It is not known at present whether the quantum yield difference is due to a) a more favorable orientation of the dye units dimeric complex, b) a longer excited state lifetime dimeric complex and/or c) a higher dimeric complex concentration as a result of stronger Pb^{2+}/SO_3^- than Hg^{2+}/SO_3^- coordination bonds.

TABLE 2: Photodimerization Quantum yields of the Hg^{2+} and Pb^{2+} Complexes of CSDs (E)-4a–e[a]

Complex	$10^4 \varphi_{PCA}$
[*(E)*-**4a**]·Hg^{2+}	1.3
[*(E)*-**4b**]·Hg^{2+}	28
[*(E)*-**4c**]·Hg^{2+}	1.3
[*(E)*-**4c**]·Pb^{2+}	10
[*(E)*-**4d**]·Hg^{2+}	0.79
[*(E)*-**4e**]·Hg^{2+}	-

[a] Excitation intensity = 2.3 x 10^{-6} einstein/(s L); ligand concentrations were 4.0 x 10^{-5} M, and $Pb(ClO_4)_2$, and $Hg(ClO_4)_2$ concentrations were 7.0 x 10^{-5} and 4.1 x 10^{-5} M, respectively, in acetonitrile.

Isomeric chromogenic 15-crown-5-ethers of the quinoline series **5** and **6** were shown to undergo PCA in acetonitrile to give cyclobutane derivatives only in the presence of $Mg(ClO_4)_2$ and $Ca(ClO_4)_2$ [26,27]. The modification of the benzocrown-ether position from 2 to 4 appears to change in principle the PCA route (Scheme 6). The overall quantum yield in the PCA of **5**·Mg^{2+} was 0.0007. In the case of the PCA of **6**·Mg^{2+}, the high quantum yield (0.13) was found, which indicates that the degree of dimerization of the complexes of this dye remains rather high even in a very dilute solution and that the spatial structure of dimeric complexes is probably rather favourable for PCA.

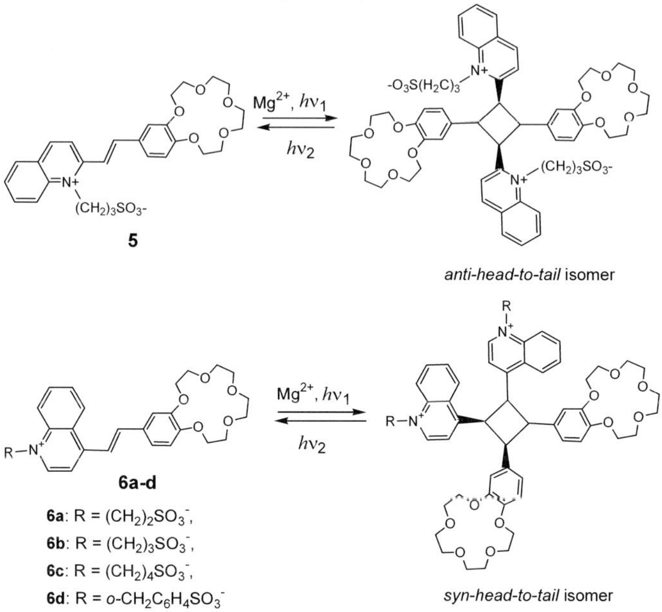

Scheme 6

The structure of the spacer has a significant effect on the efficiency of photocycloaddition reaction of **6a-d** [27]. The PCA quantum yield is of 0.20 (**6a**), 0.13 (**6b**), 0.073 (**6c**) and 0.024 (**6d**) (irradiation at 436 nm). The reaction is photochemically reversible. Upon irradiation of the cycloadducts with 313 nm light, the initial dyes are formed with a quantum yield ranging from 0.02 to 0.01.

Bis-crown-containing styryl dye (*E,E*)-**7** having the structure and properties of molecular pincers has been synthesized. In acetonitrile solution, dye (*E,E*)-**7** is able to form intramolecular sandwich complexes with Ca^{2+} and Ba^{2+} cations (Scheme 7) [28,29].

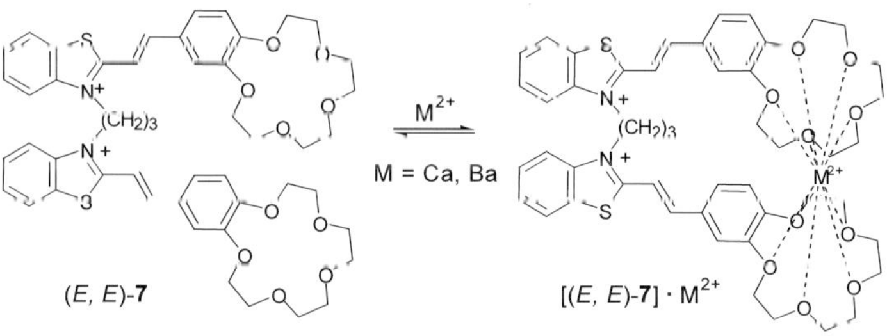

Scheme 7

The analysis of 2D COSY and NOESY NMR spectra of the PCA products showed the presence of two isomeric crown-containing cyclobutane derivatives **7a** and **7b** in 17 : 83 ratio (Scheme 8).

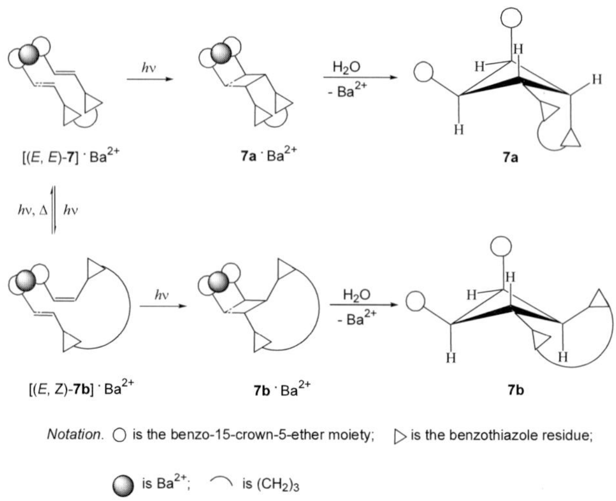

Scheme 8

Recently, synthesis and investigation of stilbene analogues containing a crown ether fragment were developed [30,31] (Scheme 9).

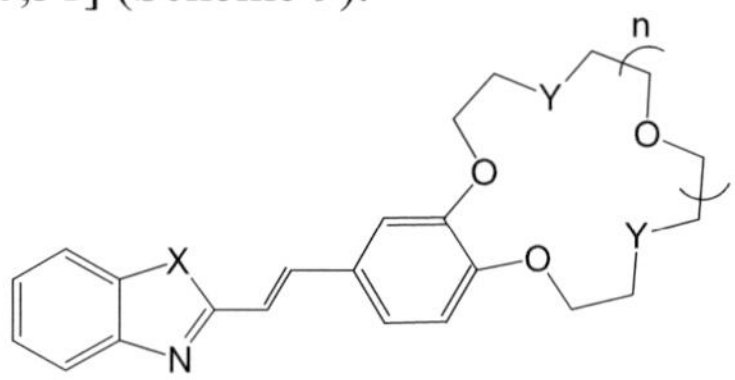

X = S, O, CH=CH, Y = O, S, n=0-3

Scheme 9

The alkaline earth metal cations with a large diameter (Sr^{2+}, Ba^{2+}) and benzo-15-crown-5 stilbene derivatives are characterized by the formation of the 2 : 1 sandwich-type complexes (Scheme 10). In the case of Ba^{2+} the sandwich complex possesses the highest stability. According to the NMR studies and X-ray analysis of the sandwich complex of **8**, molecules of dye are arranged in a "head-to-head" stacking.

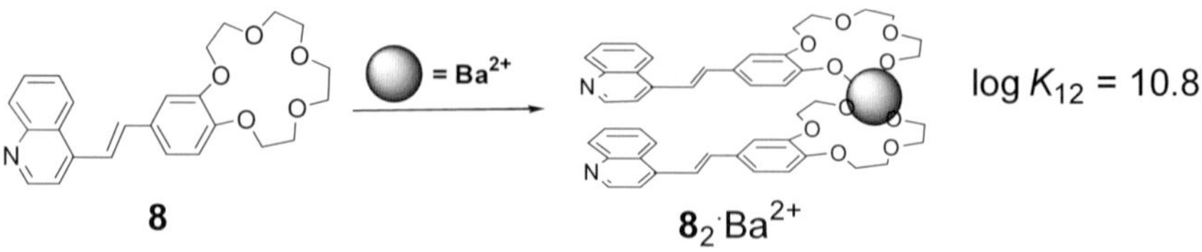

Scheme 10

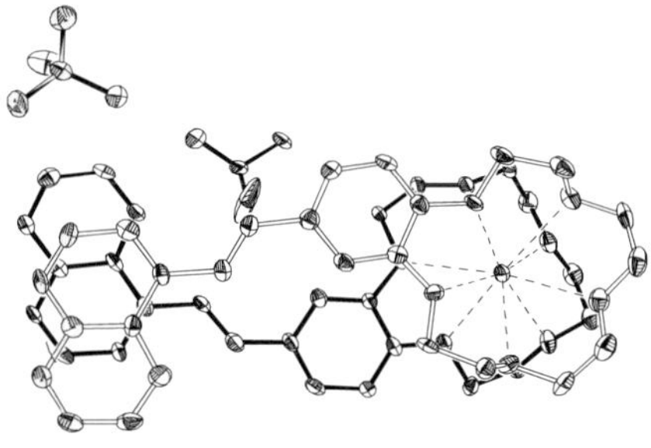

Figure 3. X-Ray structure of the [$\mathbf{8}_2$·Ba^{2+}](ClO_4^-)$_2$ complex

In the sandwich complex with Ba^{2+} cations, the mutual arrangements of 2-styrylbenzothiazole molecules **9** are favourable for the occurring of the reaction of the photocycloaddition. Upon irradiation with blue light, **9** undergoes stereoselective [2+2]-photocycloaddition (PCA) to afford isomeric cyclobutane derivatives "*E*+*E*" (78%) and "*E*+*Z*" (22%) isomers with relatively high combined quantum yield (0.13) (Scheme 11) [32].

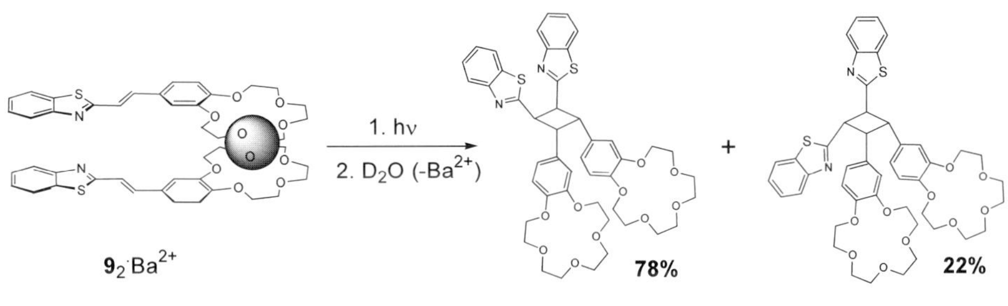

Scheme 11

The photochemical investigation of the bis-crown-containing benzobis(thiazole) **10** showed that the irradiation of the compound results in the proceeding of two photochromic processes: *E,Z*-photoisomerization and [2+2]-photocycloaddition (Scheme 12). The last process occurs due to organization of molecules in dimers through the π-stacking interaction of large chromophore systems of molecules. The complex formation of sandwich type between **10** and Ba^{2+} cations increases the effectivity of photocycloaddition reaction what fully suppresses the *E,Z*-photoisomerization around the double C=C bonds. The photochemical behaviour of the complex **10**·$(Mg^{2+})_2$ is similar to one of the free ligand **10** [33].

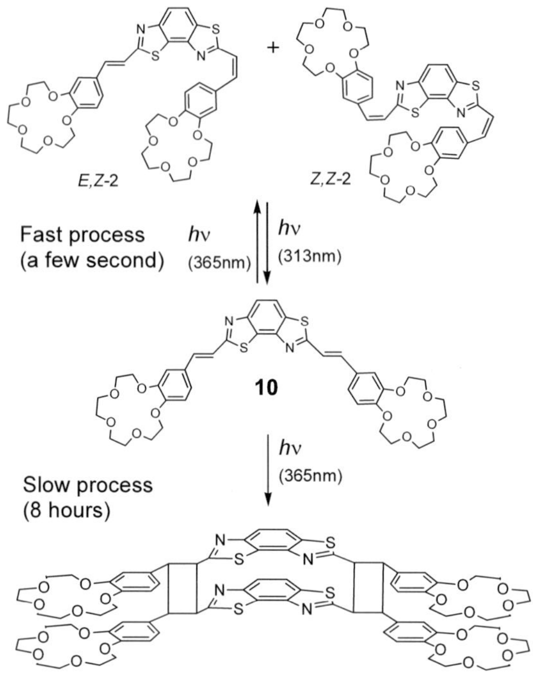

Scheme 12

Thus, the CESD are capable to form supramolecular complexes containing two dye molecules and metal cations. The self-assembly of the dimeric complexes was shown to be a unique tool for controlling regio- and stereoselectivity of [2+2]-photocycloaddition reaction. The variation of the structure of CESD would make it possible to change the supramolecular spatial structure of the dimer in a desired direction and thus to control the efficiency of interaction and stereochemistry of the final product of PCA.

4. Macrocyclic Systems whose Photochromism is Based on Electrocyclic Reaction

4.1. CROWN-CONTAINING SPIROOXAZINES

The structure and the position of the crown-containing fragments in spiro compound are important factors influencing the stability of the merocyanine form upon complex formation. Thus, investigations into spiroindolinonaphthoxazines bearing an aza-15-crown-5 moiety in 9′-positions (**11a-c**) of naphthalene ring showed insignificant influence of the metal cation presence on the spectral and kinetic characteristics of compounds (Scheme 13) [34]. Otherwise, spironaphthoxazine derivatives with a monoazacrown ether moiety at the 5′-position (**12a-c**) are discovered to be sensitive to the presence of metal cations [35].

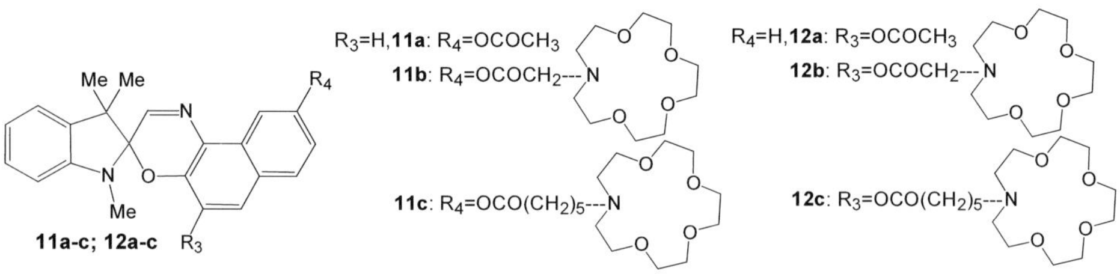

Scheme 13

Binding of alkaline earth metal ions (Mg^{2+}, Ca^{2+} and Ba^{2+}) by the crown moieties in **12a-c** leads to isomerization of the crowned spironaphthoxazines even in the dark. UV-Vis, NMR and Raman spectroscopy suggests that a metal ion complexed by the crown moiety in the merocyanine isomer is subject to intramolecular interaction with the phenoxide anion. In fact, the metal ion is bound more strongly than in the corresponding closed isomer, owing to the additional-binding-site effect. On exposure to visible light, the cation-bound merocyanine form readily reverts to the spiro form, releasing the metal ions to some extent. The alternating irradiation with UV and visible light or alternating switching-on and -off of the visible light causes isomerization of the crowned spironaphthoxazines, which, in turn, provides a tool for controlling their cation-complexing capacity (Scheme 14).

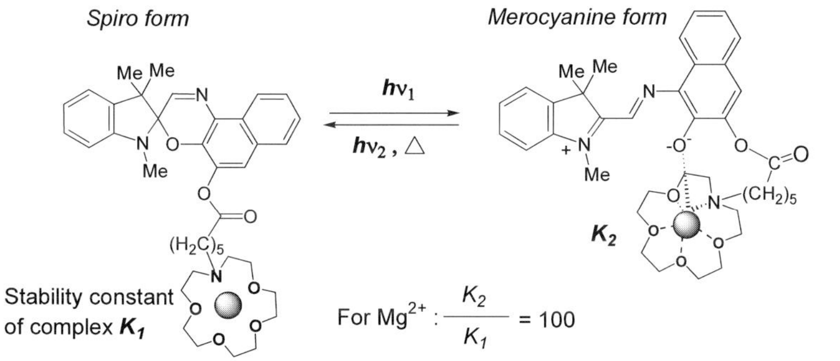

Scheme 14

For the spironaphthoxazines conjugated with aza-15(18)-crown-5(6)-ether moieties at 6'-position of naphthalene fragment (**13a,b**) it was found that the addition of Li^{+} and alkaline earth (Mg^{2+}, Ca^{2+}, Sr^{2+} and Ba^{2+}) metal cations to **13a,b** solutions results in a hypsochromic shift of the UV absorption band of the spiro form and a bathochromic shift of the absorption band of the merocyanine form in the visible region [36]. In addition, the equilibrium shifts to the merocyanine form, and the lifetime of the photoinduced merocyanine form increases (Scheme 15). The isomerization of crown-containing compound **13a,b** to the colored merocyanine form was promoted most strongly by the presence of metal ions, which are expected to be the best recognized by the crown ether ring (Scheme 15).

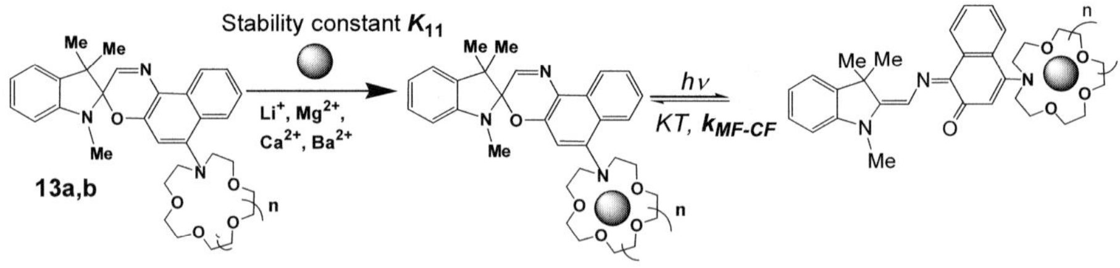

lg K_{11}	Li^+	Mg^{2+}	Ca^{2+}	Ba^{2+}
13a: n=1	3.3	3.2	3.6	3.1
13b: n=2		2.2	4.6	5.6

k_{MF-CF} (rate constant of ring-closure reaction)

	Li^+	Mg^{2+}	Ca^{2+}	Sr^{2+}	Ba^{2+}
n=1	7.7	1.2	1.2	3.6	5.6
n=2	8.7	0.9	2.2	0.8	0.8

Scheme 15

The UV-induced isomerisation of **13a,b** into the merocyanine form causes a decrease of the cation binding ability (Scheme 16).

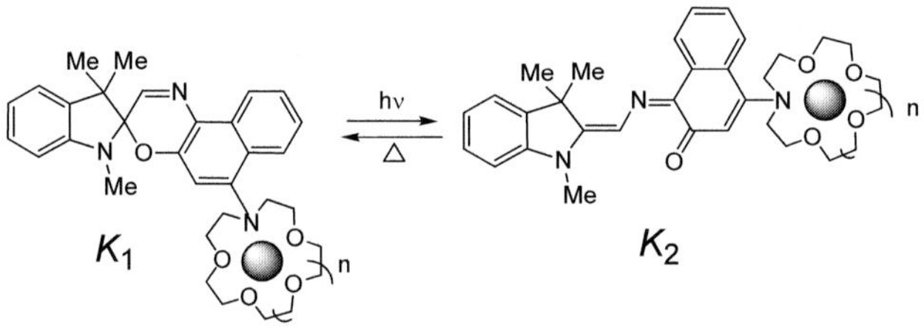

Ca^{2+}	lgK_1	lgK_2
n = 1	3.6	2.4
n = 2	4.6	2.3

Scheme 16

The method for synthesis of the spirobenzothiazolinonaphthoxazine **14** stable in merocyanine form and containing crown ether fragment in the heterocyclic part was developed [37]. The presence of alkali earth metal cations in the solution of crown ether containing merocyanine dye in MeCN results in coordination of metal cations with two binding centers: crown ether fragment and merocyanine oxygen atom (Scheme 17). These complexes are structurally differ from each other. The complex formation process causes the changes in spectral characteristics and influences on the photochromic behaviour of the prepared compound.

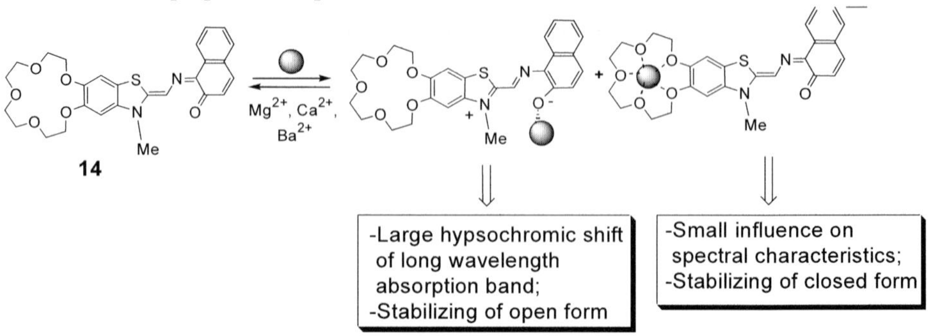

Scheme 17

4.2. CROWN-CONTAINING BENZOCHROMENES

The complex formation and photochromic behaviour of the azacrown-containing chromene **15b** and its crown-free analogue **15a** were studied in detail [38,39]. The addition of $Ca(ClO_4)_2$ to a solution of **15b** led to the shift of the band at 400 nm, indicating the binding of Ca^{2+} by the macrocyclic unit of **15b**. In contrast, the presence of $Ca(ClO_4)_2$ in a solution of **15a** even at a high concentration of 0.1 mol dm^{-3} had no effect on the absorption spectrum of **15a**.

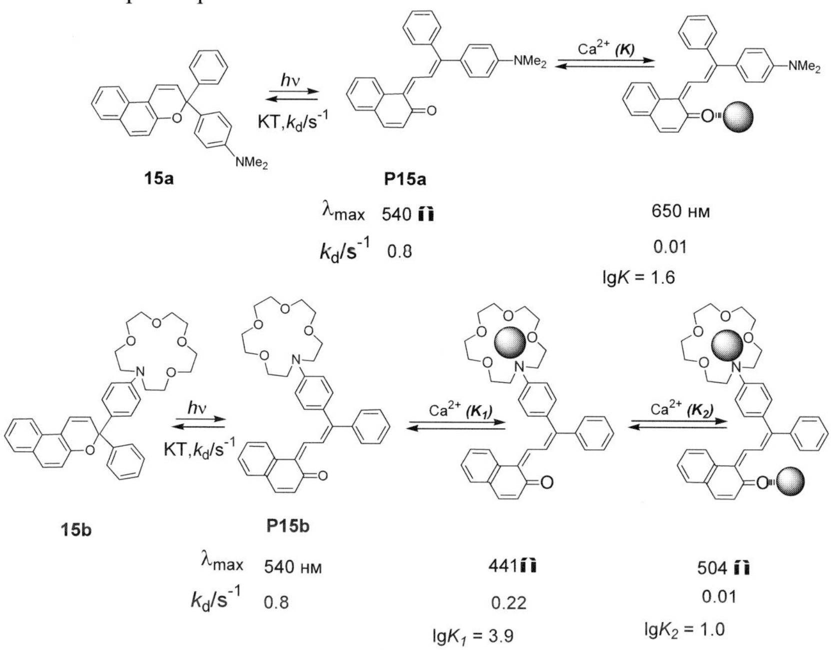

Scheme 18

The photochromic transformation of the closed form of **15b** to merocyanine form (**P15a**) leads to a marked decrease in the ability of **15b** to bind Ca^{2+}. Thus, the stability constant for the merocyanine complex **P15b**•Ca^{2+} is almost one order of magnitude lower than that for the corresponding complex of the closed form. The dimethylamino substituted chromene **15a** unable to bind Ca^{2+} in the dark shows small Ca^{2+}-binding ability upon UV irradiation. The formation of a week 1:1 complex between merocyanine form of **P15a** and Ca^{2+} leads to a large bathochromic effect and to a significant decrease in the rate constant of dark ring-closure reaction of **P15a**, indicating that the metal ion in the 1:1 complex **P15a**•Ca^{2+} coordinates to the oxygen atom of the merocyanine (Scheme 18). The formation of a 1:1 complex between merocyanine form of **P15b** and Ca^{2+} leads to a large hypsochromic effect. For **P15b**, the decrease in the rate constant of dark ring-closure reaction upon 1:1 complexation is less considerable than in the case of **P15a** (Scheme 18). This indicates that the Ca^{2+} ion in this complex coordinates to the crown ether moiety. In addition to the 1:1 complex, the photoinduced merocyanine **P15b** is able to form a very week 1:2 complex **P15b**•$(Ca^{2+})_2$ involving two Ca^{2+} ions. Judging from the spectrokinetic data, the 1:2 complex arises from the coordination of Ca^{2+} to the metal-free carbonyl oxygen in the **P15b**•Ca^{2+} complex (Scheme 18).

The results of the spectral and kinetic investigations of the benzochromenes **16a-c** and their complexes with Mg^{2+}, Ba^{2+} and Pb^{2+} were presented in [40] (Scheme 19).

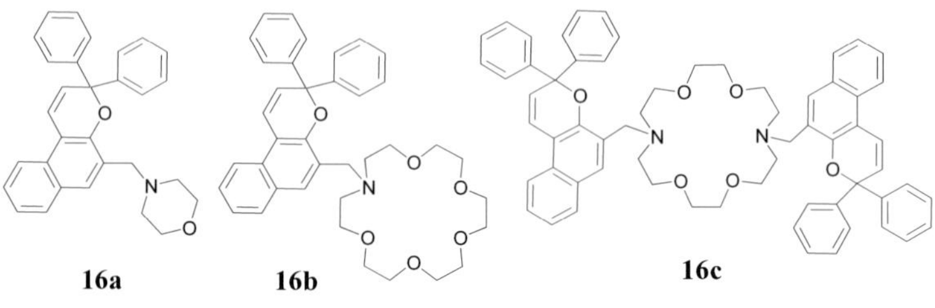

Scheme 19

Compounds **16a-c** in MeCN exhibited very similar UV-Vis absorption spectra. The addition of Mg^{2+}, $Ba^{2+}$and Pb^{2+} cations to a solution of **16a-c** led to the small changes in absorption spectra. In contrast, the presence of the metal cations in the solution of photomerocyane form of **16a-c** led to a significant change in the dark lifetime of photomerocyanines and caused strong shifts in their absorption spectra, indicating that these compounds were able to bind metal cations.

The formation of a 1:1 complex between open form of **16a-c** and Mg^{2+} leads to an increase in the rate constant for the dark ring-closure reaction of **16a-c** (Scheme 20, Table 3). The Mg^{2+} ion binds with carbonyl oxygen atom what leads to the increase of the polarity of molecule (Scheme 20). In the polar form the dark ring-closure reaction occurs more easily.

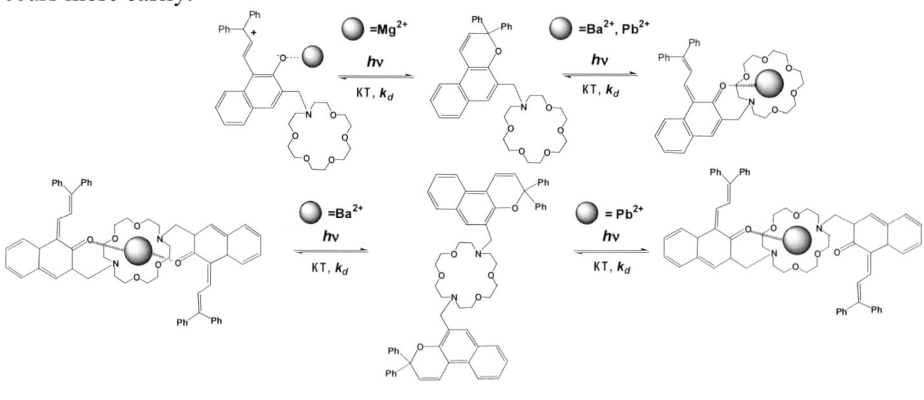

Scheme 20

The formation of a 1:1 complex between open form of **16b,c** and Ba^{2+}, Pb^{2+} leads to a large bathochromic effect and the decrease in the rate constant for dark ring-closure reaction. The effect is explained by the formation of the anion-"capped" complex of metal cation located in crown ether cavity with carbonyl oxygen atom (Scheme 20). Substantial difference in the changes of magnitude of k_d/s^{-1} was obtained for the complexes of mono- and bischromene **16b,c** with Ba^{2+} (Table 3). The difference is due to in the complex of **16c** with Ba^{2+} the carbonyl atoms of both chromene units participate in the formation of anion-"capped" complex, whereas, the anion-"capped" complex between **16b** and Ba^{2+} is formed by formation of one coordination bond (Scheme 20). For the complexes of mono- **16b** and bischromenes **16c** with Pb^{2+} the magnitudes of k_d/s^{-1} are closed to each other, what means the participation of only one chromene unit of **1c** in the formation of anion-"capped" complex with Pb^{2+} (Scheme 20).

TABLE 3. Effect of the Mg^{2+}, $Ba^{2+}$and Pb^{2+} presence on the rate constant k_d / $s^{-1}$of dark ring-closure reaction, relationship ligand:metal cations=1:1

Chromene	k_d / s^{-}			
	---	**Mg^{2+}**	**Ba^{2+}**	**Pb^{2+}**
16a	0.11	0.44	0.18	0.12
16b	0.11	0.49	0.0016	0.00035
16c	0.15	0.19	0.00034	0.00054

Thus, the study of the crown containing systems whose photochromism is based on the electrocyclic transformation demonstrates that the complex formation process can be successfully applied for controlling of its spectral and photochemical properties.

5. Conclusions

The intense research efforts made in photochromic crown ethers convincingly demonstrated that this type of systems represent a novel class of photochromic molecules suited for some possible applications. A substantial influence of complex formation on the spectral characteristics of the molecules and on the kinetics of phototransformation was found. Thus, incorporation of a crown ether moiety, which is able to bind metal ions into the photochromic skeleton, can help to tune the photochromic properties by using complex formation. The development of these photochromic systems is aimed at improving the photostability, increasing the sensitivity and obtaining a broader range of operating wavelengths.

In turn, the conformational transformations of molecules that accompany the photoreaction sharply influence the ability of molecules to bind metal cations. This implies that photocontrolled complex formation is possible in this type of system.

Examples of potential applications utilizing the physical or chemical changes that accompany the observed shift of the absorption maxima include:

- photoswitching extraction of metal cations;
- optoelectronic systems;
- photoswitching transport through membranes;
- optical information storage;
- photochemical switchable enzymatic systems;
- nonlinear optical devices.

The results obtained were extended to novel series of photochromic systems. The important objectives for the future research include: a) synthesis of new compounds; b) more extensive and more detailed investigation of their physico-chemical properties in order to find structure-property correlations; and c) modification of already known systems by incorporating them physically or chemically into liquid crystals or polymers in order to develop new effective materials based on the novel photochromic molecules.

6. Acknowledgements

This work was supported by financial assistance from the RFBR 02-03-32058, 03-03-32849, 03-03-32178, 03-03-32888, CRDF RC2-2344-MO-02, INTAS 03-51-4696 and 01-0267, and Ministry of Sciences and High Education (Program Integration).

7. References

[1] Brown, G.H. (ed.) (1971) *Photochromism*, Wiley-Interscience, New York.
[2] Durr, H., Bouas-Laurent, H. (eds.) (1990) *Photochromism: Molecules and Systems*, Elsevier, Amsterdam.
[3] Crano, J.C., Guglielmetti, R. (eds.) (1999) *Organic Photochromic and Thermochromic Compounds*, Plenum Press, New York, v. 2.
[4] McArdle, C.B. (ed.) (1992) *Applied Photochromic Polymer Systems*, Blackie, New York.
[5] Alfimov, M.V., Fedorova, O.A., Gromov, S.P. (2003) Photoswitchable molecular receptors, *Photochem. Photobiol. A*,. 183-198.
[6] Inoue, Y., Gokel, G. W. (eds.) (1990) *Cation Binding by Macrocycles. Complexation of Cationic Speacies by Crown Ethers*, Marcel Dekker Inc., New York and Basel.
[7] a) Shinkai, S., Minami, T., Kusano, Y., Manabe, O. (1982) Photoresponsive crown ethers. 5. Light-driven ion transport by crown ethers with a photoresponsive anion cap, *J. Am. Chem. Soc.*, ***104***, 1967-1972; b) Shinkai, S., Ishihara, M., Ueda, K., Monabe, O. (1985) Photoresponsive crown ether. Part 14. Photoregulated crown – metal complexation by competative intramolecular tail(ammonium)-biting, *J. Chem. Soc. Perkin Trans. I*, 511-518; c) Shinkai, S., Shigematsu, K., Sato, M., Manabe, O. (1982) Photoresponsive crown ether. Part 6. Ion transport mediated by photoinduced *cis-trans* interconversion of azobis(benzocrown ethers), *J. Chem. Soc., Perkin Trans. 1*, 2735-2739.
[8] a) Durr, H., Thome, A., Kranz, C., Bossman, S. (1992) Supramolecular effects on photochromism-properties of crown ether-modified dihydroindolizines, *J. Phys. Org. Chem.*, 111-149; b) Durr, H., Kranz, C., Kilburg, H. (1997) Supramolecular effects on photochromism-properties of crown ether-modified dihydroindolizines, *Mol. Cryst. Liq. Cryst.*, ***298***, 89-96.
[9] a) Guo, Z., Wang, G., Tang, Yi., Song, X. (1997) A crown ether bearing fulgides: the regulation of photochromism by supramolecular effect, *Liebigs Ann. /Recueil.*, 941-942; b) Kawai, S. H. (1998) A crown ether bearing fulgides: the regulation of photochromism by supramolecular effect, *Tetrahedron Lett.*, ***39***, 4445-4448.
[10] Takeshita, M., Irie, M. (1998) Photoresponsive tweezers for alkali metal ions. Photochromic diarylethanes having two crown ether moieties, *J. Org. Chem.*, ***63***, 6643-6649.
[11] a) Bouas-Laurent, H., Castellan, A., Desvergne, J.-P., Lapouyade, R. (2001) Photodimerization of anthracenes in fluid solutions. Part 2. Mechanistic aspects of the photocycloaddition and of the photochemical and thermal cleavage, *R. Chem. Soc. Rev.*, ***30***, 248-263; b) Desvergne J.-P., Perez-Inestrosa E., Bouas-Laurent H., Jonusauskas G., Oberle J., Rullier C., in Valeur, B., Brochon J.-C. (eds.) (2001) *New Trends in Fluorescence Spectroscopy. Application in Chemical and Life Sciences*, Springer-Verlag, Berlin, Heidelberg, chapter 8; c) Desvergne, J.-P., Bouas-Laurent, H., Perez-Inestrosa, E., Marsua, P., Contrait, M. (1999) Photoinduced control of cation binding ability of non-conjugated bichromophoric receptors, *Coord. Chem. Rev.* ***185-186***, 357-371.
[12] a) Inouye, M. (1996) Artificial-signaling receptors for biologically important chemical species, *Coord. Chem. Rev.* ***148***, 265-283; b) Inouye, M. (1998) «Purposive» molecular design for multifunctional artificial receptors, *J. Incus. Phenom. Molecul. Recogn. Chem.*, ***32***, 223-233; c) Kimura, K., Sakamoto, H., Nakamura, M. (2003) Molecular design and application of photochromic crown compounds – How can we manipulate metal ions photochemically, *Bull. Chem. Soc., Jpn.*, **76**, 225-245; d) Tanaka, M., Kamada, K. (2003) Ion-responsive photochromic materials. Crowned spiropyranes, *J. Synth. Org. Chem. Jpn.*, ***61***, 322-330.
[13] Gromov, S.P., Alfimov, M.V. (1992) Supramolecular photochemistry of the crown ether styryl dyes, *Russ. Chem. Bull.*, ***46***, 611-636.
[14] a) Gromov, S.P., Fedorova, O.A., Ushakov, E.N., Stanislavskii, O.B., Alfimov, M.V. (1991) Crown-containing styryl dyes. 2. Synthesis, photoinduced and dark complex formation of betainic chromogenic 15-crown-5-ether, *Dokl. Akad. Nauk*, ***319***, 1134-1139; b) Gromov, S.P., Fedorova, O.A., Ushakov, E.N., Stanislavskii, O.B., Alfimov, M.V. (1991) Crown-containing styryl dyes. The size of helate macrocycle and complex formation of betainic photochromic 15-crown-5-ether, *Dokl. Akad. Nauk*, ***321***, 104-107; c) Gromov, S.P., Golosov, A.A., Fedorova, O.A., Levin, D.E., Alfimov, M.V. (1995) Nature of the heterocyclic moiety, complexation, and electronic absorption and fluorescence spectra of *trans-* and *cis-* isomers of photochromic 15-crown-5-ethers, *Russ. Chem. Bull.*, ***44***, 124-130.

[15] Gromov, S.P., Fedorova, O.A., Ushakov, E.N., Baskin, I.I., Lindeman, A.V., Malysheva, E.V., Balashova, T.A., Arsen'ev, A., Alfimov, M.V. (1998) Crown ether styryl dyes. 24. Synthesis of multiphotochromic 15-crown-5 ethers with rigid spacers, their anion-"capped" complexes, and stereospecific [2+2]-autophotocycloaddition, *Russ. Chem. Bull.*, ***47***, 97-105.
[16] a) Ushakov, E.N., Gromov, S.P., Fedorova, O.A., Alfimov, M.V. (1997) Crown-containing styryl dyes. 19. Complexation and cation-induced aggregation of chromogenic aza-15-crown-5-ethers, *Russ. Chem. Bull.*, ***46***, 463-471; b) Gromov, S.P., Ushakov, E.N., Fedorova, O.A., Soldatenkova, V.A., Alfimov, M.V. (1997) Crown-containing styryl dyes. 22. Synthesis and complexation of photochromic aza-15-crown-5-ethers , *Russ. Chem. Bull.*, ***46***, 1143-1147.
[17] Alfimov, M.V., Fedorov, Yu.V., Fedorova, O.A., Gromov, S.P., Hester, R.E., Lednev, I.K., Moore, J.N., Oleshko, V.P., Vedernikov, A.I. (1996) Synthesis and spectroscopic studies of novel photochromic benzodithiacrown ethers and their complexes, *J. Chem. Soc., Perkin Trans. 2*, 1441-1447.
[18] Alfimov, M.V., Gromov, S.P., Fedorov, Yu.V., Fedorova, O.A., Vedernikov, A.I. (1997) Crown-containing styryl dyes. 23. Synthesis and complexation of cis-isomers of photochromic dithia-15(18)-crown-5(6)-ethers, *Russ. Chem. Bull.*, ***46***, 2099-2106.
[19] Alfimov, M.V., Vedernikov, A.I., Gromov, S.P., Fedorov, Yu.V., Fedorova, O.A., Churakov, A.V., Kuz'mina, L.G., Howard, J.A.K., Bossmann, S., Braun, A., Woerner, M., Sears, D.F., Saltiel, J. (1999) Synthesis, structure and ion selective complexation of *trans* and *cis* isomers of photochromic dithia-18-crown-6 ethers, *J. Am. Chem. Soc.*, ***121***, 4992-5000; b) Stanislavskii, O.B., Ushakov, E.N., Gromov, S.P., Fedorova, O.A., Alfimov, M.V. (1996) Crown-containing styryl dyes. 14. The influence of N-substitute length on the complex formation of betainic chromogenic 15-crown-5-ether with alkaline earth metal cations, *Russ. Chem. Bull.*, ***45***, 564-572.
[20] Fedorova, O.A., Fedorov, Y.V., Vedernikov, A.I., Yescheulova, O.V., Gromov, S.P., Alfimov, M. V., Kuz'mina, L.G., Churakov, A.V., Howard, J.A.K., Zaitsev, S.Yu., Sergeeva, T.I., Möbius, D. (2002) Supramolecular assemblies of photochromic benzodithia-18-crown-6 ethers in crystals, solutions, and monolayers, *New J. Chem.*, ***26***, 543-553.
[21] Barzykin, A.V., Fox, M.A., Ushakov, E.N., Stanislavsky, O.B., Gromov, S.P., Fedorova, O.A., Alfimov, M.V. (1992) Dependence of metal ion complexation and intermolecular aggregation on photoinduced geometric isomerism in a crown ether styryl dye, *J. Am. Chem. Soc.*, ***114***, 6381-6385.
[22] Alfimov, M.V., Gromov, S.P., Stanislavskii, O.B., Ushakov, E.N., Fedorova, O.A. (1993) Crown-containing styryl dyes. 8. Cation-dependent concerted [2+2]-autophotocycloaddition of photochromic 15-crown-5 ether betaines, *Russ. Chem. Bull.*, ***42***, 1385-1389 .
[23] Gromov, S.P., Ushakov, E.N., Fedorova, O.A., Buevich, A.V., Alfimov, M.V. (1996) Crown-containing styryl dyes. 18. Synthesis, anion-"capped" complexes and ion-selective stereospecific [2+2]-autophotocycloaddition of photochromic 18-crown-6 ether, *Russ. Chem. Bull.*, ***45***, 654-661 .
[24] Ushakov, E.N., Gromov, S.P., Buevich, A., Baskin, I.I., Fedorova, O.A., Vedernikov, A.I., Alfimov, M.V., Eliasson, B., Edlund, U. (1999) Crown-containing styryl dyes: cation-induced self-assembly of multiphotochromic 15-crown-5 ethers into photoswitchable molecular devices, *J. Chem. Soc., Perkin Trans. 2*, 601-607.
[25] Fedorova, O.A., Fedorov, Y.V., Vedernikov, A. I., Gromov, S. P., Yescheulova, O.V., Alfimov, M.V., Woerner, M., Bossmann, S., Braun, A., Saltiel, J. (2002) Thiacrown Ether Substituted Styryl Dyes: Synthesis, Complex Formation and Multiphotochromic Properties, *J. Phys. Chem. A*, ***106***, 6213-6222.
[26] Gromov, S.P., Fedorova, O.A., Ushakov, E.N., Buevich, A.V., Alfimov, M.V. (1995) Crown-containing styryl dyes. 15. Synthesis and two pathways of the regio- and stereospecific cation-dependent [2+2]-autophotocycloaddition of chromogenic 15-crown-5 ether betaines of quinoline series, *Russ. Chem. Bull.*, ***44***, 2131-2136 .
[27] Gromov, S.P., Ushakov, E.N., Fedorova, O.A., Baskin, I.I., Buevich, A.V., Andryukhina, E.N., Alfimov, M.V., Johnels, D., Edlund, U.G., Whitsel, J.K., Fox, M.A. (2003) Novel photoswitcable receptors: synthesis and cation-induced assembly into dimeric complexes leading to stereospecific [2+2]-photocycloaddition of styryl dyes containing a 15-crown-5 ether unit, *J. Org. Chem.*, ***68***, 6115-6125.
[28] Gromov, S.P., Fedorova, O.A., Ushakov, E.N., Buevich, A.V., Baskin, I.I., Pershina, Y.V., Eliasson, B., Edlund, U., Alfimov, M.V. (1999) Photoswitchable molecular pincers: synthesis, self-assembly into sandwich complexes and ion-selective intramolecular [2+2]-photocycloaddition of an unsaturated bis-15-crown-5 ether, *J. Chem. Soc., Perkin Trans. 2*, 1323-1329.
[29] Ushakov, E.N., Gromov, S.P., Fedorova, O.A., Pershina, Y.V., Alfimov, M.V., Barigelletti, F., Flamigni, L., Balzani, V. (1999) Sandwich-Type Complexes of Alkaline-Earth Metal Cations with a Bisstyryl Dye Containing Two Crown Ether Units, *J. Phys. Chem. A*, ***103***, 11188-11193.
[30] Fedorov, Y.V., Fedorova O.A., Gromov S.P., Bobrovsky M.B., Andrukhina E.N., Alfimov M.V. (2002) Simulation of optical responce to complex formation of crown-containing 2-styrylbenzothiazoles with alkaline-earth metals, *Russ. Chem. Bull.*, ***51***, 789-795.

[31] Fedorov, Yu.V., Fedorova, O.A., Andryukhina, E.N., Gromov, S.P., Alfimov, M.V., Kuzmina, L.G., Churakov, A.V., Howard, J.A.K., Aaron, J.-J. (2003) Ditopic complex formation of the crown-containing 2-styrylbenzothiazole, *New J. Chem.*, 280–288.
[32] Fedorova, O.A., Fedorov, Y.V., Andryukhina, E.N., Gromov, S.P., Alfimov, M.V. (2003) Cation-dependent photochromic properties of the novel ditopic receptors, *Pure&Applied Chem.*, 1077-1084.
[33] Fedorov, Yu., Fedorova, O., Schepel, N., Gromov, S., Alfimov, M., J. Saltiel (2004) Synthesis and multitopic complex formation of the photochromic bis-crown ether based on benzobis(thiazole), *J. Phys. Org. Chem.*, in press.
[34] Fedorova, O.A., Gromov, S.P., Pershina, Y.V., Sergeev, S.S., Strokach, Y.P., Barachevsky, V.A., Alfimov, M.V., Pèpe, G., Samat, A., Guglielmetti, R. (2000) Novel azacrown ether-containing spiro[indoline-2,3'-naphthoxazines]: design, synthesis and cation-dependent photochromism, *J. Chem. Soc., Perkin Trans. 2*, 563-571.
[35] Fedorova, O.A., Gromov, S.P., Strokach, Yu.P., Pershina, Yu.V., Sergeev, S.A., Barachevsky, V.A., Pepe, G., Samat, A., Guglielmetti, R., Alfimov, M.V. (1999) Crown-containing spironaphthoxazines and spiropyrans. 1. Synthesis and anion-"capped" complexes of photochromic aza-15-crown-5 ethers with flexible spacers, *Russ. Chem. Bull.*, ***48***, 1950-1959.
[36] Fedorova, O.A., Strokach, Y.P., Gromov, S.P., Koshkin, A.V., Valova, T.M., Alfimov, M.V., Feofanov, A.V., Alaverdian, I.S., Lokshin, V.A., Samat, A., Guglielmetti, R., Girling, R.B., Moore, J.N., Hester, R.E. (2002) Effect of metal cations on the photochromic properties of spironaphthoxazines conjugated with aza-15(18)-crown-5(6) ethers, *New J. Chem.*, ***26***, 1137-1145.
[37] Fedorova, O.A., Koshkin, A.V., Gromov, S.P., Avakyn, V.G., Nazarov, V.B., Brichkin, S.B., Vershinnikova, T.G., Nikolaeva, T.M., Chernih, L.A., Alfimov, M.V. (2002) Crown-containing spironaphthoxazines and spiropyrans. 3. Synthesis and investigation of the open merocyanine form of the crown-containing spirobenzothiazolylnaphthoxazine, *Russ. Chem. Bull.*, ***51***, 1441-1450.
[38] Ushakov, E.N., Nazarov, V.B., Fedorova, O.A., Gromov, S.P., Chebun'kova, A.V., Alfimov, M.V., Barigelletti, Fr. (2003) Photocontrol of Ca^{2+} complexation with an azacrown-containing chromene, *J. Phys. Org. Chem.*, ***16***, 306-309.
[39] Fedorova, O.A., Maurel, Fr., Ushakov, E.N., Nazarov, V.B., Gromov, S.P., Chebun`kova, A.V., Feofanov, A.V., Alaverdian, I.S., Alfimov, M.V., Barigelletti, Fr. (2003) Synthesis, photochromic behaviour and light-controlled complexation of 3,3-diphenyl-3*H*-benzo[*f*]chromenes containing a dimethylamino group or an aza-15-crown-5 ether unit, *New J. Chem.*, ***27***, 1155-1167.
[40] Lokshin, V., Samat, A., Chebun'kova, A.V., Fedorova, O.A., Gromov, S.P., Strokach, Yu. P., Valova, T.M., Alfimov, M.V. (2004) Investigation of the Azacrown-Ether Substituted Naphtopyranes, *Mol. Cryst. Liq. Cryst.*, in press.

16.
MODEL SYSTEMS FOR BIOLOGICAL PROCESSES

GEORGE W. GOKEL[a,b], W. MATTHEW LEEVY[b] AND
MICHELLE E. WEBER[b]
Departments of [a]Chemistry and [b]Molecular Biology & Pharmacology,
Washington University School of Medicine,
660 South Euclid Ave., Campus Box 8103, St. Louis, MO 63110 U.S.A.

1. Crown Ethers and Biological Activity

"Biological activity" refers to a compound's ability either to alter or to mimic a living system or one of its components. A living system could be an entire animal, or even a cell in a Petri dish. Components of a living system would include organelles, proteins, or DNA. Crown ether compounds have been studied to determine their effects at these various levels of life. Crown ethers interact directly with such molecules as DNA and enzymes. Numerous crowns have also been tested for their toxic effects to various mammalian and bacterial cell lines, as well as animals including mice.

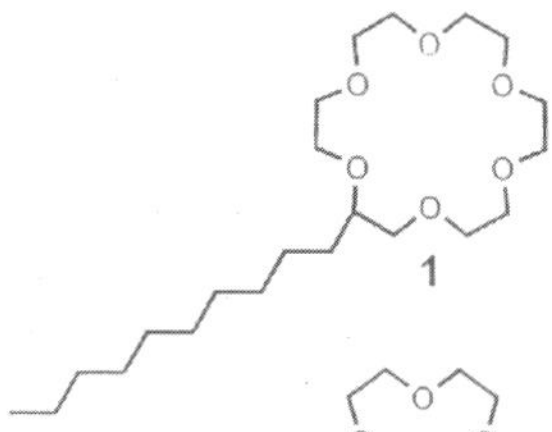
1

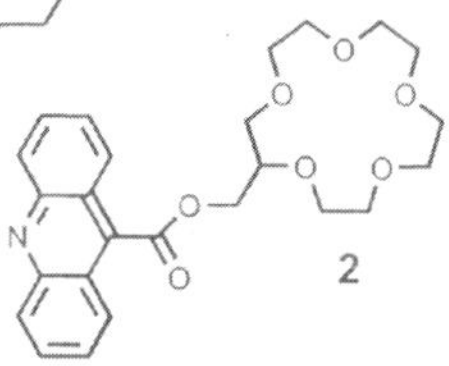
2

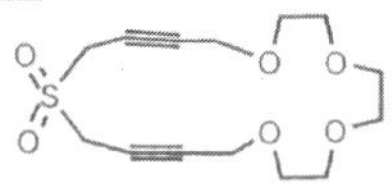
3

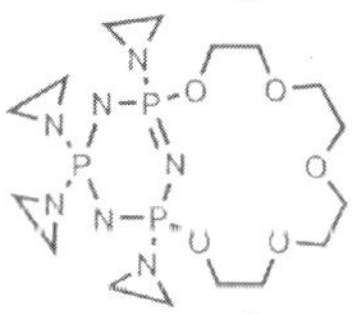
4

Pedersen's reports of the compounds he called crown ethers (Pedersen, 1967) began a world wide synthetic effort to prepare novel macrocycles, to define the limits of crown ether structure, and to assess the range of their biological and chemical properties. Among the latter, great effort was expended to define and understand cation complexation by these remarkable molecules. On the biological side, the toxic effects of crown ethers to cell lines and animals were assayed to understand their inherent safety or danger and the pharmacological value, if any, of crown compounds.

Early reports (Takayama *et al.*, 1978; Hendrixson *et al.*, 1978; Gad *et al.*, 1978, 1985) indicated that the macrocycles 18-crown-6, 15-crown-5 and 12-crown-4 were all well tolerated by dogs, rabbits and mice, with a toxicity profile generally similar to aspirin.

Karsten Gloe (ed.), Macrocyclic Chemistry: Current Trends and Future Perspectives, 253–265.
© 2005 *Springer. Printed in the Netherlands.*

However, alkyl-substituted crown ethers such as (**1**) were found to be toxic to Gram-positive bacteria and some species of yeast when administered at concentrations in the 5-10 μM range (Konup *et al.*, 1989; Kato *et al.*, 1980; Yagi *et al.*, 1984).

Recently, the crown ether based hydraphile channels of Gokel (see Section 5.3) were shown to be lethal to the Gram-negative *E. coli* when administered in the 2-10 μM range (Leevy *et al.*, 2002, 2004). While activity levels for these compounds surpass those of many antibiotics, including penicillin, selective killing between bacteria and mammalian cells remains the limiting factor in the use of macrocycles as antibiotics (Huang *et al.*, 2002).

Crown ethers have been tested for their interactions with a number of biological molecules, including proteins and DNA. Crowns bearing an acridine-terminated sidearm were found to intercalate into DNA and to protect the double helix from enzymatic cleavage (Basak and Dugas, 1986). The strength of the binding by (**2**) to DNA was enhanced by the presence of a Na^+ or K^+ cation present in the crown ring that could interact with the negatively charged phosphate backbone (Fukuda *et al.*, 1990). Crown ether compounds have been developed that cleave DNA. Kerwin reported the synthesis of a bis(propargylic) sulfone crown ether, (**3**), that could cleave DNA when co-incubated with it in buffer for 24 h (Kerwin, 1994). These compounds were shown to operate in the 90-200 μM range; the cleavage reaction was enhanced by the presence of Li^+, Na^+, or K^+ cation (Kerwin, 2000). These compounds were tested against over 50 cancer cell types, and found to inhibit their growth at concentrations ranging from 1-20 μM (McPhee and Kerwin, 2001). Another crown containing an integral aziridinylcyclophosphazene subunit (**4**) was developed that cleaves DNA and was highly active against AIDS-related lymphoma cell lines at a concentration of 0.6 μM (Brandt *et al.*, 2001). The latter compound is being evaluated *in vivo* at the National Cancer Institute.

Certain crown ethers have been found to interact with enzymes in organic solvents. It is known, for example, that simple macrocycles, such as 18-crown-6 and 15-crown-5, aid the solubilization of proteins in organic solvents (Odell and Earlam, 1985). Furthermore, crowns can enhance the catalytic activity of enzymes in certain instances. The reaction rates of lipase enzymes in methanol were enhanced by up to 87-fold by the use of crown ethers and thiol-crown ethers (Tsukube *et al.*, 2001). The presence of these crown compounds also enhanced the enantioselectivity of the enzymes (Itoh *et al.*, 1996). Similar effects were noted with the enzyme *subtilisin Carslberg* (van Unen *et al.*, 1998; Santos *et al.*, 2001) and also α-chymotrypsin (van Unen *et al.*, 2002). Crowns are thought to complex amine residues, such as lysine, that are present on the surface of the protein. This complexation is thought to help protect the enzymes during lyophilization (freeze-drying), and also to stabilize the protein once dissolved in non-aqueous solvents (van Unen *et al.*, 2002). The basis for crown-enhanced enzyme activity is currently unclear, but these systems remain under investigation.

Crown compounds have been explored in a number of biological contexts. They have shown activity in toxicity studies to bacteria and cancer cell lines. A very recent and striking example of the biological activity of crown compounds involves the tris(macrocyclic) hydraphile channels developed in the authors' lab. These purely synthetic, non-peptide compounds were recently shown to form channels in the bilayer of living human embryonic kidney (HEK 293) cells (Leevy *et al.*, 2004). Their activity was up to 4-fold higher than native cation channels, and these molecules may hold potential under the developing concept known as "channel replacement therapy" (Cook *et al.*, 2004). These current advances, and those noted above, ensure that crown ether chemistry has clearly found a niche in biology.

2. Biological Look-alikes

Interest in biological mimics probably started in the late 1800s with the discovery of cyclodextrin (CD), obtained from the starch digest of a strain of *Bacillus*. The realization that CDs could form host-guest complexes with a variety of small molecules, and the myriad of uses implied therein, led to extensive study in both academics and industry.

Early work in the area of crown ether biological mimics was reported by Vögtle and coworkers (Tummler, 1977). These early compounds were made in an effort to duplicate natural ion carriers in the sense that they could complex biologically important ions for transport. The early, two-dimensional crowns (such as **5**) and non-cyclic polyethers gave way to the three-dimensional cryptands (**6**), which generally complexed ions more tightly but which lack the dynamics of podands, crown ethers, or lariat ethers.

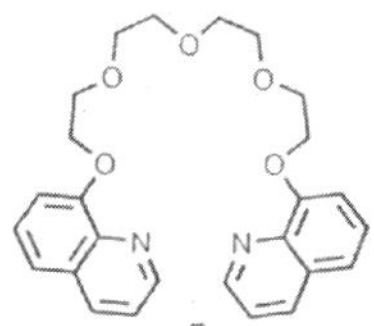
5

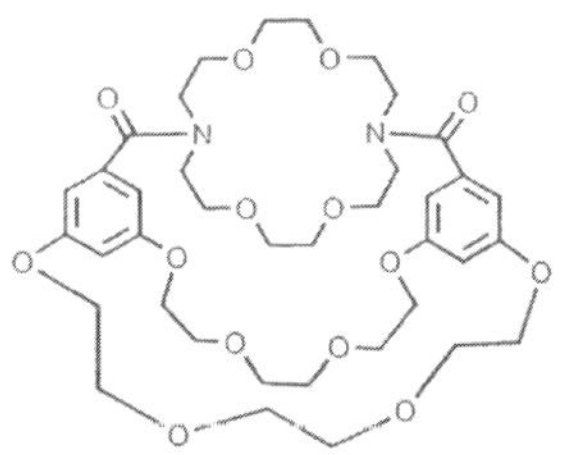
6

Mertes and Mertes used ammonium crowns and their protonated forms as enzyme mimics (Mertes and Mertes, 1990; Bianchi *et al.*, 1997). These receptors afforded a platform for enzyme mimetic chemistry due to their synthetic accessibility. An advantage of the approach to (**7**) is that a modular synthetic route provided a scaffold for changes in hydrophobicity, shape, electronic distribution, and redox potential, among other properties. Mertes and Mertes reported that these crowns demonstrated activity as ATPases, presumably owing to their ability to interact both with the ATP phosphate groups and with magnesium and calcium ions in solution. Formyl transferase activity by these enzyme mimics was also noted.

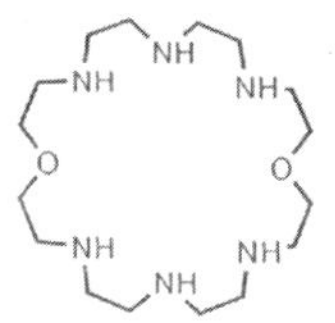
7

Compounds that combine crown ethers with porphyrins have been developed in order to mimic the binuclear metal binding sites of some metalloproteins. Compound (**8**) was designed to provide two proximate metal binding sites, similar to those that might be found in natural systems (Chang, 1977). This "crowned porphyrin" can bind a transition metal ion and an alkali metal ion simultaneously.

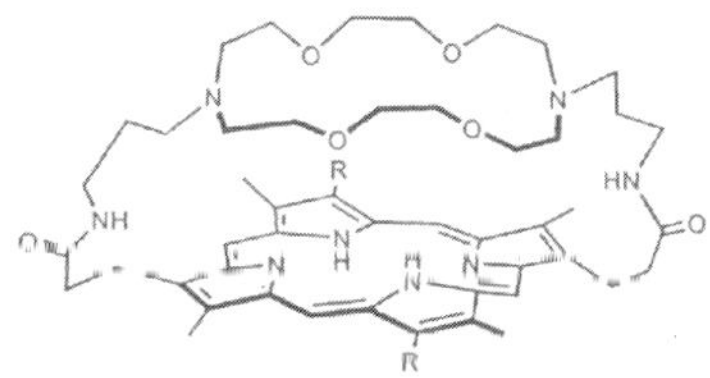
8

In other work, Dandliker *et al.* have reported the inclusion of iron porphyrins within dendrimers to serve as functional mimics of redox-based proteins (Dandliker *et al.*, 1994, 1995, 1997). These redox-switchable porphyrins show that the Fe^{3+}/Fe^{2+} redox couple can be altered by the polarity of the surrounding environment. By changing the polarity imposed by the tightly packed branches of the dendritic core, the authors have illustrated that electrochemical behavior can be controlled by slight and subtle through-space environmental factors. These mimics may potentially model a wide variety of redox-driven enzymes and possibly provide mechanistic insights into their function.

Attempts by chemists to develop enzyme mimics have ranged from the use of dendrimers ("dendrizymes," reviewed by Brunner, 1995) to metal-templated superoxide dismutatese (SOD) mimics (reviewed in Riley, 1999, 2000) to supramolecular cytochrome P450 mimics. The role of cytochrome P450 in biology is diverse. A known chemical role is to split molecular oxygen, incorporating one atom into its organic substrates and converting the other to water. The oxidative substrates include both endogenous molecules and xenobiotics; there are several P450 enzymes that do not have a known oxidative role. This surprising diversity presents both biologists and chemists with the challenge to determine the enzyme's mechanism of action.

Several cytochrome P450 mimics have been reported. Early model systems utilized iron porphyrins having sterically bulky groups arrayed about the periphery. Evolution of these systems has focused on functionalized porphyrins, as well as variation of the complexed metal. More recently, attachment of cyclodextrin (CD) or other residues onto the porphyrin has been reported. Breslow and coworkers have prepared several CD-based porphyrins that have demonstrated catalytic activity for a variety of hosts (reviewed in Breslow and Dong, 1998). Activity and specificity as an enzyme mimic were improved by increasing the number of attached CD units. One example, which contains four CD units, was shown to bind cholesterol and alkene derivatives in a site-specific manner and to permit stereoselective hydroxylation as known for the cyctochrome P450 enzyme itself (Breslow *et al.*, 1997; Yang *et al.*, 2002). Others, including Tabushi and Ogoshi, have reported related compounds and activities (Tabushi, 1982, 1986; Tabushi *et al.*, 1988; Kuroda *et al.*, 1991).

Iron transporters known as siderophores occur in various bacteria. They coordinate iron in a complex that involves three catechol residues. A natural host molecule called enterobactin is shown along with a cryptand-like molecule (**9**) that is one of several that were devised to mimic this complexation behavior (reviewed in Roosenberg, 2000; Raymond, 2003). The catechols deprotonate to the catecholate anions, which provide six oxygen donors for ferric ion. The host thus completely envelops the cation permitting transport as the complex.

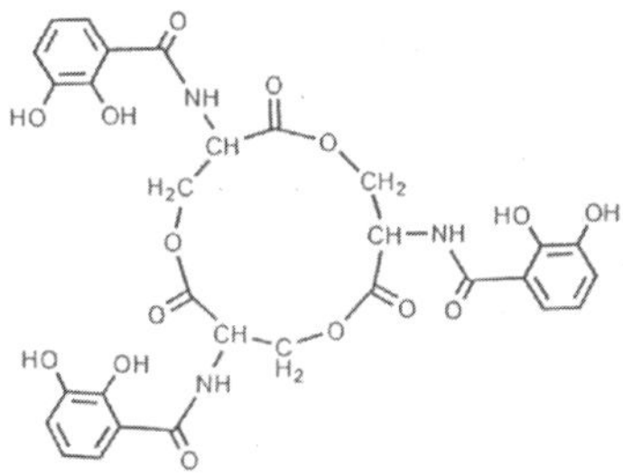

enterobactin

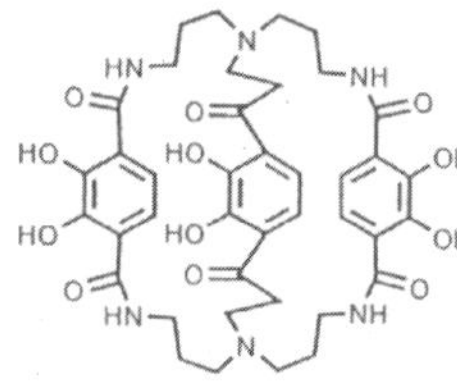

9

3. Development of Ionophore Mimetics

Lariat ethers, which were designed to add dimensionality to the essentially flat crown ether, have been used as synthetic ionophores for decades. The binding properties of crown ethers are now well documented and generally well understood. Most such studies have been conducted with metal ions, although complexation of ammonium species and some neutral species (Gokel, 1973; Kyba, 1977) have also been reported.

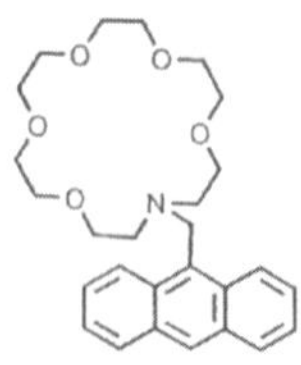

10

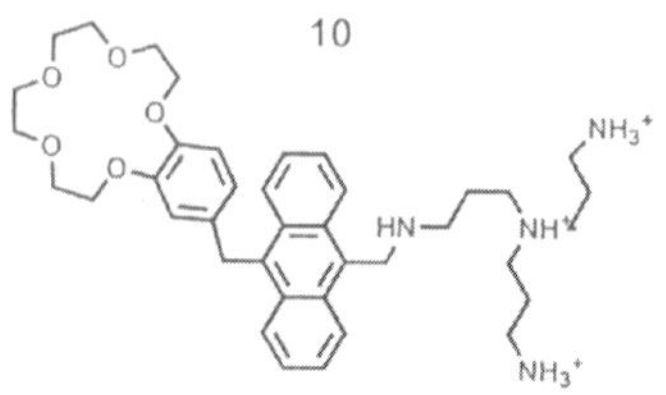

11

There are hundreds of reports in the literature of crown ethers that have been used in ion-selective electrodes (ISEs). Many such compounds were designed to differentiate the ion of interest from other similar metal ions. Among the many potential applications of these systems is the removal of ionic species from wastewaters.

Other crown ether ionophores have been developed as ion sensors. The first reports of crown ethers as dyes appeared in the 1970s. These derivatives of benzo-15-crown-5 undergo a color change when bound to specific ions and can be used as efficient extractors for ions from aqueous solution. Numerous other chromoionophores have also been prepared and were reviewed by Takagi (1984).

The anthracenylmethyl lariat ether shown as (**10**) was reported by de Silva and coworkers (1986) as a crown that is useful for fluorescence sensing. Binding of K^+ resulted in a detectable fluorescence emission signal. Recent advances in this area include the development of compounds that can sense ion pairs. The anthracenyl crown shown as (**11**) was designed to simultaneously sense sodium and phosphate ions (de Silva, 2003). Excitation of the complex with ultraviolet radiation results in a fluorescent output only when Na^+ is complexed in the crown and the ammonium groups are bound to the anion. The complex self-quenches in the case when either no or only one of the ions is present.

Extensive effort in our own laboratory has been devoted to documenting cation-π interactions. Lariat ethers have proved to be an excellent vehicle for these studies. Using a variety of sidearms, especially the sidechains of the aromatic amino acids (Meadows, 2001; Hu, 2002), double bonds (Hu, 2001a), and triple bonds (Hu, 2001b), we have clearly demonstrated cation-π interactions. A complex of diaza-18-crown-6 having phenethyl sidearms is shown in the accompanying illustration. It is the KI complex. The potassium cation is completely encapsulated by the macrocycle and the π-donor sidearms. The anion, which has an obvious affinity for the cation, is completely excluded from the solvation sphere in the solid state. Numerous solid-state structures of this type have been obtained for K^+ and Na^+ with a variety of counterions. Solution NMR studies (both cation-induced shifts and NOEs) confirm the presence of these complexes in solution (Meadows, 2001). Given that 1 in every 11 amino acids in nature has an aromatic side chain, coupled with the high concentrations of sodium and potassium *in vivo*, it is possible that nature has developed a purpose for such an interaction.

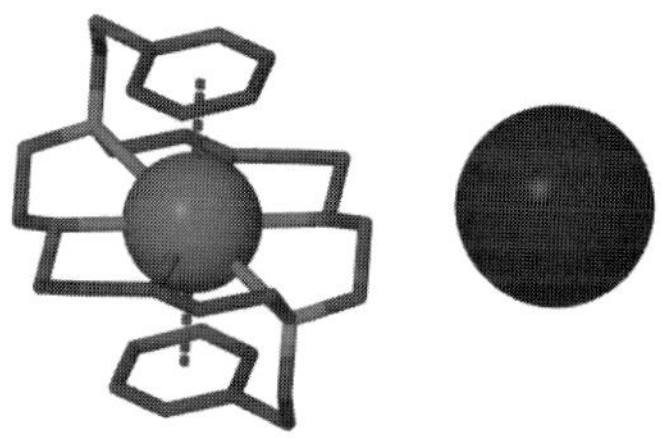

4. Crown Ethers as Membrane Amphiphiles

The presence of multiple heteroatoms in crown ethers makes them potential headgroups in the amphiphilic sense. Two groups in Japan recognized the possibility of making crown ether based amphiphiles in the 1980s. Okahara (Okahara *et al.*, 1980; Gu *et al.*, 1986; Ikeda *et al.*, 1986, 1988, 1990), Kuwamura (Kuwamura and Yoshida, 1980; Inokuma *et al.*, 1988, 1989; Furusawa *et al.*, 1990; Matsumura *et al.*, 1990) and coworkers prepared and studied several crown ether systems. Much of this work was published in Japanese journals and is not as widely known as it should be. Our own efforts in this area involved mostly azalariats (**14**) (Gokel *et al.*, 1987; Echegoyen *et al.*, 1988) that formed either micelles or bilayer vesicles, depending on their chain length and structure. In addition, twin-crown "bola-amphiphiles" (**12**) were prepared that were found to form stable vesicles (Muñoz *et al.,* 1992, 1993). The ability of crown ethers to serve as headgroups in amphiphiles was critical to the development of synthetic ion channels, as described in section 5.

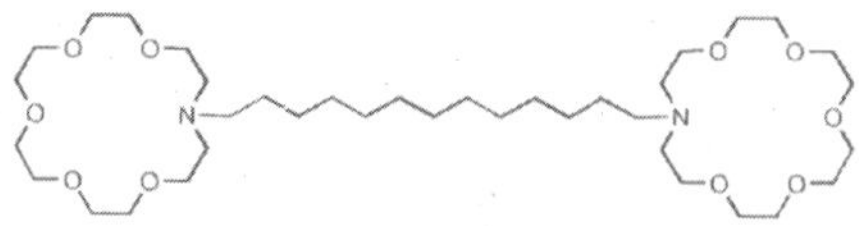

12

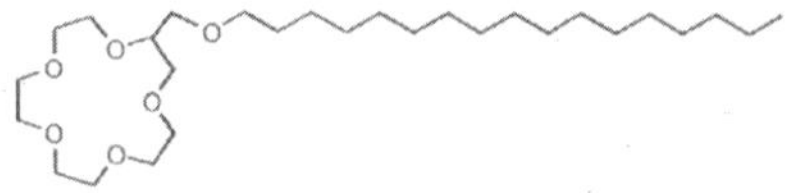

13

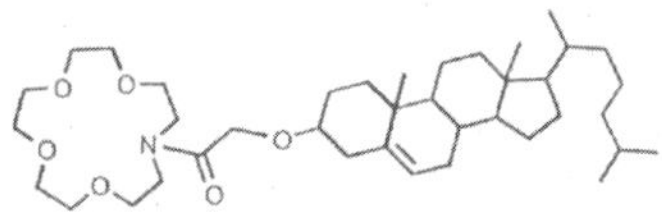

14

Crowns have also been used in concert with other amphiphiles to form mixed bilayers. An example is the mixture of amphiphilic crown and amphiphilic ammonium salt (**16**) blended to form a mixed bilayer. Circular dichroism (CD) was used to detect the presence of Na^+ ion (Nakashima *et al.,* 1987). Amphiphilic 12-crown-4 derivatives prepared by Katoh and coworkers inserted into lipid bilayers and showed significant transmembrane currents in aqueous NaCl, KCl, and RbCl solutions (Katoh *et al.,* 1996). X-Ray scattering of amphiphilic crowns has been conducted at the air-water interface. Increased surface pressure gave results consistent with two-layer packing (Larson *et al.,* 2001).

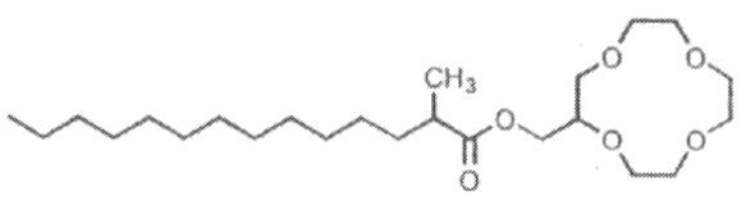

15

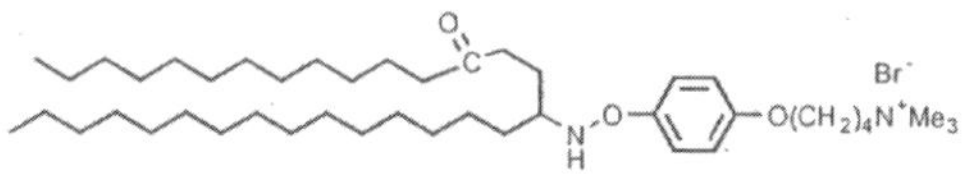

16

Shinkai and coworkers prepared numerous novel amphiphilic crowns (Shinkai, 1990) and incorporated them into membranes, formed membranes from them, or used them in liquid crystalline assemblies to control properties (He *et al.,* 1990). Interest in this area continues. Four chiral amphiphilic crown ethers were recently reported that recognize enantiomeric amino acids when examined as Langmuir films (Badis *et al.,* 2004). Finally, it is interesting to note that liposomes formed from amphiphiles (*e.g.,* crown ethers) having neutral headgroups (*i.e.,* niosomes) have been studied as drug delivery vehicles (Uchegbu and Vyas, 1998).

5. Ion Channel Models

From the biological perspective, an ion channel is a compound that transports ions through a phospholipid bilayer membrane. Crowns interact more or less selectively with a whole range of cations, but the ones most relevant to biology are H^+, Na^+, K^+, and Ca^{2+}. The transport of these ions is closely regulated *in vivo*. For example, typical cells have high (~150 mM) concentrations of K^+ within them and Na^+ without. Internal Na^+ concentrations are typically 5-10 mM as are external K^+ concentrations. The ideal synthetic ion channel would insert in a phospholipid bilayer and transport only one of these ions (*i.e.*, show selectivity) predominantly in a single direction (*i.e.*, exhibit rectification). At present, these goals have not been completely met, but progress has been made.

An important observation needs to be made about channel models and, indeed, model systems in general. Chemists can design molecules to have remarkable shapes and sizes. For a model system, however, it is the properties that determine whether the compound is relevant. A compound that looks like it should be a channel is, as Fitzmaurice has put it, only a "long thin thing" absent a demonstration of efficacy (Fitzmaurice, 2004). There is no rule that demands selectivity for the biologically relevant ions. Indeed, transport of divalent cobalt has been studied. A cobalt-transporting channel is not, however, a biological mimic so far as is currently known.

5.1 ADVANTAGES OF CROWN ETHERS IN SYNTHETIC CHANNEL MODELS

Simple crown ethers possess alternating ethylene units and oxygen atoms. As such, they are both hydrophobic and hydrophilic. When a long hydrocarbon chain is attached, they become excellent amphiphiles in which the polar oxygens comprise the polar end. The oxygen atoms also serve as donors for Lewis acids generally. Thus, amphiphilic crown ethers can anchor in the membrane and the macrocycle can serve as a headgroup. The crown can also serve as an entry portal that may confer ion selectivity upon the system. It is problematic to assume that a cation-selective crown will impart the same selectivity to the channel, however. After all, crown selectivity has to do with binding, and channel selectivity has to do with transport or passage of the ion.

5.2 EARLY CROWN-ETHER CONTAINING CHANNEL MODELS

Probably the best known early crown ether example is the "chundle" reported by Jullien and Lehn (Jullien and Lehn, 1988). Their strategy used a central crown ether unit with sidearms radiating from it. The stereochemistry of the sidearms was fixed by incorporation of tartaric acid units within the macrocycle. The name was given because the compound was a channel formed from a bundle of fibers. In this first report, no information about insertion or transport appeared, and the assertion that the compound was a channel apparently rested on the intent of the design. Later work from this group showed that related compounds, called "bouquet" molecules, did conduct cations, albeit rather slowly (Canceill *et al.*, 1992).

Drenth, Nolte and their coworkers developed polyisonitriles as channel mimics in the 1980s (Beijnen *et al.*, 1982). These compounds were characterized more fully later (Roks and Nolte, 1992), but this pioneering never provided true biomimetic activity.

5.3 SYSTEMATIC EFFORTS TO DEVELOP BIOLOGICALLY-RELEVANT CHANNEL MODELS

In the late 1980s, efforts initiated by Fyles in Canada (Carmichael *et al.*, 1989) and by our own group (Nakano *et al.*, 1990) resulted in two different channel models, both of which incorporated crown ethers. Both of these efforts resulted in channels that show significant transport of alkali metal cations through phospholipid bilayers. Both groups (Gokel and Murillo, 1996) have undertaken extensive structure-activity studies to characterize their function.

The Fyles family of channels is illustrated schematically in the accompanying figure. The tartaric-acid-based crown ether central unit defines the position of the sidearms and serves as a selectivity filter and central relay within the bilayer membrane. Four phospholipid molecules are also shown in cartoon form to illustrate the position in which the channel is thought to reside. The spacer chains or "walls" of the channel are based on substituted succinic acid groups, terminated at each membrane boundary by a polar residue, designated "H" for headgroup in the figure (Fyles *et al.*, 1993). A typical headgroup is ~SCH_2COOH. The channels differed in the number of spacer chains (4 or 6) and in the identities of the strands designated "A" and "B" in the figure. An interesting finding of these studies is that channel activity was higher when these chains were alkyl rather than ethyleneoxy. Presumably, in the latter case, cations were better bound and therefore transported more slowly.

The Fyles approach illustrates an important aspect of design strategy. The channels are highly modular and individual elements can be added, eliminated, or exchanged fairly readily. Once a working model is in hand, such a design strategy permits one to assess how each of the various structural elements affects channel function. A competition between metal cation and proton transport was used to assay function in these systems. In addition, the voltage clamp technique common to biology was applied to study transport and voltage gating (Fyles *et al.*, 1998).

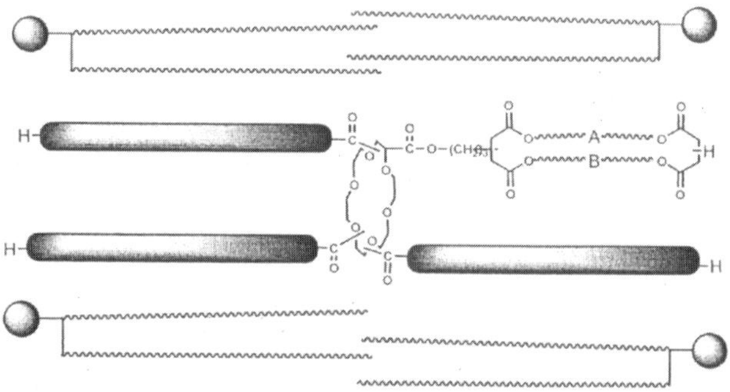

Our initial channel design incorporated three crown ethers rather than the single macrocycle used by Lehn or by Fyles. Our design required that two of the crowns would serve both as headgroups in the amphiphilic sense and as entry portals for cations. The third macrocycle would be positioned at the midplane of the bilayer. We felt that this would increase channel efficacy by reducing the distance that a cation must travel in the nonpolar regime of the membrane. We referred to the third proximal macrocycle as the "central relay." At the time of the design, no channel crystal structure was known and the presence of such a structure (a water and ion-filled capsule) in protein ion channels was only revealed in 1998 by the Nobel-prize winning structural efforts of the MacKinnon lab (Doyle *et al.*, 1998).

Simple hydrocarbon spacers connected the three macrocycles in our design. Two additional chains of comparable length were appended to each distal macrocycle. It was thought that these hydrocarbon chains would fold back into the bilayer and comprise structural counterparts to the spacers linked covalently to both distal and proximal crowns. The overall structure was made flexible so that it could adjust after insertion into the bilayer. Certainly at that time, and even today, many of the properties of bilayers remain uncertain (Tristram-Nagle and Nagle, 2004). The spacer chains and side chains were both dodecyl units in the original design. This gave the channel, in the presumed active conformation, a span between the two distal macrocycles of about 30-35 Å. This corresponded to the thickness of the bilayer's insulating regime ("hydrocarbon slab"). The first channel we prepared and reported is shown in the accompanying figure. It is shown in the top part of the figure in the conformation we assumed it would have in the bilayer. The actual conformation, determined by extensive structural, analytical, and biophysical studies, is shown below (Gokel, 2000).

Stylized phospholipid membrane monomers are illustrated in the figure. The conformation was established as follows. First, replacement of the central macrocycle either by diaza-15-crown-5 or diaza-12-crown-4, or open-chained units gave a channel that was functional, albeit diminished in transport activity (Murray *et al.*, 2000). If so, the smaller ring should have slowed transport or the channel should simply have failed to conduct cations. A relative of the channel shown above was prepared in which the dodecyl sidechains were replaced by fluorescent dansyl groups. The shift in dansyl fluorescence observed when this functional channel was inserted in the bilayer showed that this side chain experienced a dielectric of ε = 25. If the side chain was in contact with the aqueous phase, the dielectric should be 80. Within the hydrocarbon insulator regime, ε = 2-5. Further, fluorescence depth quenching showed that the dansyl groups were about 30 Å apart, as expected for the indicated conformation. Finally, fluorescence resonance energy transfer (FRET) experiments involving the dansyl channel and another fluorescent channel gave an aggregation number for the system of 1.1. This comports with the monomer structure shown in the figure above (Abel *et al.*, 1997).

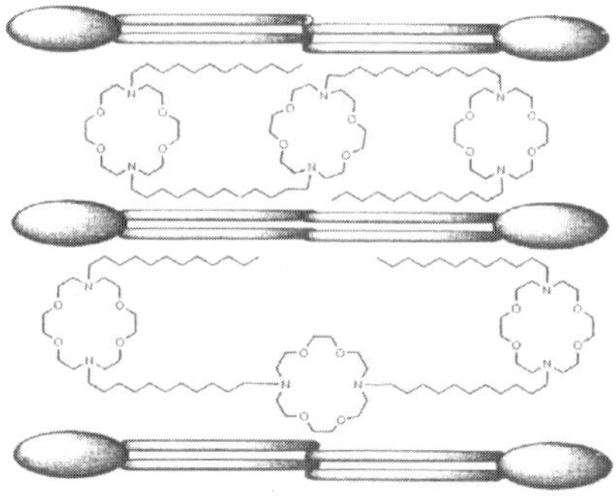

Additional studies have disclosed significant biological activity. We call the compound in which two benzyl groups replace the dodecyl sidearms of the channels shown above "benzyl channel." The benzyl channel is toxic to the Gram negative bacterium *E. coli*. A corresponding dibenzyl structure that has 8-carbon spacers and is too short to effectively function as a channel is at least 15-fold less toxic (Leevy *et al.*, 2002). In very recent work, the benzyl channel has been found to function in live cells. This has been determined by the electrophysiological technique known as whole cell patch clamping (Leevy *et al.*, 2004).

A number of other synthetic ion channels have been developed that incorporate crown ethers as critical elements. They cannot all be described and illustrated in this short chapter but it is important to note them. Voyer and coworkers developed channels that use an α-helical backbone to align a series of crowns into a channel that was both functional and biologically active (Voyer and Robataille, 1995; Voyer *et al.*, 1997). Pechulis and coworkers developed a channel in which a central crown used tartaric acid subunits to anchor sterolds, which formed the channel's walls (Pechulis *et al.*, 1997). Mendoza and coworkers prepared an active channel based on a calixarene central unit but that had crown ether headgroups (de Mendoza *et al.*, 1998). Hall and coworkers modified the tris(macrocycle) design originating in our lab to form a redox-switchable crown that was active in bilayers (Hall *et al.*, 1999, 2003).

6. Conclusions

Crown ether compounds now have a nearly 40-year history. Early studies sought to define their variety and properties. As the class became well defined, crowns became useful as scaffolds for the development of novel and complex systems that model natural structures and processes.

7. Acknowledgements

The authors gratefully acknowledge support of their work, some of which is described herein, by the NIH under GM-36262 (previously GM-29150) and by the NSF and PRF. WML (T32-GM04892-11) and MEW (T32-GM08785) thank the NIH for training grant support.

8. References

Abel, E.; Maguire, G. E. M.; Meadows, E. S.; Murillo, O.; Jin, T.; Gokel, G. W., (1997) "Planar bilayer conductance and fluorescent studies confirm the function and location of a synthetic sodium-ion-conducting channel in a phospholipid bilayer membrane"; *J. Am. Chem. Soc.* **119**, 9061-9062.

Badis, M.; Tomaszkiewicz, I.; Joly, J.-P.; Rogalska, E., (2004) "Enantiomeric recognition of amino acids by amphiphilic crown ethers in langmuir monolayers"; *Langmuir* **20**, 6259-6267.

Basak, A.; Dugas, H., (1986) "Design and synthesis of DNA intercalating crown ether molecules"; *Tetrahedron Letters*. **27**, 3-6.

Beijnen, A. J. M. v.; Nolte, R. J. M.; Zwikker, J. W.; Drenth, W., (1982) "A molecular cation channel"; *Recl. Trav. Pays-Bas* **101**, 409-410.

Bianchi, A.; Bowman-James, K.; Garcia-España, E. (1997) *Supramolecular Chemistry of Anions*; Wiley-VCH: New York, 461 pp.

Brandt, K.; Kruszynski, R.; Bartczak, T. J.; Porwolik-Czomperlik, I., (2001) "AIDS-related lymphoma screen results and molecular structure determination of a new crown ether bearing aziridinylcyclophosphazene, potentially capable of ion-regulated DNA cleavage action"; *Inorganica Chimica Acta*. **322**, 138-144.

Breslow, R.;Huang, Y.;Zhang, X., Yang, J., (1997) "An artificial cytochrome P450 that hydroxylates unactivated carbons with regio-and stereoselectivity and useful catalytic turnovers"; *Proc. Natl. Acad. Sci. U.S.A.* **94**, 11156-11158.

Breslow, R., Dong, S. D., (1998) "Biomimetic reactions catalyzed by cyclodextrins and their derivatives"; *Chem. Rev.* **98**, 1997-2011.

Brunner, H., (1995) "Dendrizymes: Expanded ligands for enantioselective catalysis"; *J. Organomet. Chem.* **500**, 39-46.

Canceill, J.; Jullien, L.; Lacombe, L.; Lehn, J.-M., (1992) "62. Channel-type molecular structures. Synthesis of bouquet-shaped molecules based on a B-cyclodextrin core"; *Helv. Chim. Acta* **75**, 791.

Carmichael, V. E.; Dutton, P. J.; Fyles, T. M.; James, T. D.; Swan, J. A.; Zojaji, M., (1989) "Biomimetic ion transport: a functional model of a unimolecular ion channel"; *J. Am. Chem. Soc.* **111**, 767-769.

Chang, C. K., (1977) "Stacked double-macrocyclic ligands. 1. Synthesis of a "crowned" porphyrin"; *J. Am. Chem. Soc.* **99**, 2819-2822.

Cook, G. A.; Prakash, O.; Zhang, K.; Shank, L. P.; Takeguchi, W. A.; Robbins, A.; Gong, Y.; Iwamoto, T.; Schultz, B.D.; Tomich, J.M. (2004) "Activity and structural comparisons of solution associating and monomeric channel-forming peptides derived from the glycine receptor M2 segment"; *Biophys. J.* **86,** *1424-35.*

Dandliker, P. J.; Diederich, F.; Gross, M.; Knobler, C. B.; Louati, A., Sanford, E., (1994) "Dendritic porphyrins: Modulating redox potentials of electroactive chromophores with pendant multifunctionality"; *Angew. Chem. Int. Ed. Engl.* **33**, 1739-1742.

Dandliker, P. J.;Diederich, F.;Gisselbrecht, J.-P.;Louati, A., Gross, M., (1995) "Water-soluble dendritic iron porphyrins: Synthetic models of globular heme proteins"; *Angew. Chem. Int. Ed. Engl.* **34**, 2725-2728.

Dandliker, P. J.; Diederich, F.;Zingg, A.; Gisselbrecht, J.-P.; Gross, M.; Louati, A., Sanford, E., (1997) "121. Dendrimers with porphyrin cores: Synthetic models for globular heme proteins"; *Helv. Chim. Acta* **80**, 1773-1801.

de Mendoza, J.; Cuevas, F.; Prados, P.; Meadows, E. S.; Gokel, G. W., (1998) A synthetic cation-transporting calix[4]arene derivative active in phospholipid bilayers; *Angew. Chem., Int. Ed. Engl.* **37**, 1534-1537.

de Silva, A. P., de Silva, S. A., (1986) "Fluorescent signaling crown ethers: Switching on of fluorescence by alkali metal ion recognition and binding in situ"; *Chem. Commun.* 1709-1710.

de Silva, A. P.;McClean, G. D., Pagliari, S., (2003) "Direct detection of ion pairs by fluorescence enhancement"; *Chem. Commun.* 2010-2011.

Doyle, D. A.; Cabral, J. M.; Pfuetzner, R. A.; Kuo, A.; Gulbis, J. M.; Cohen, S. L.; Chait, B. T.; MacKinnon, R., (1998) "The structure of the potassium channel: molecular basis of K^+ conduction and selectivity"; *Science* **280**, 69-77.

Echegoyen, L. E.; Hernandez, J. C.; Kaifer, A.; Gokel, G. W.; Echegoyen, L., (1988) "Aggregation of steroidal lariat ethers: The first example of nonionic liposomes (niosomes) formed from neutral crown ether compounds"; *Chem. Commun.* 836-837.

Fitzmaurice, D., (July, 2004) Lecture at XIII International Symposium on Supramolecular Chemistry, University of Notre Dame.

Fukuda, R.; Takenaka, S.; Takagi, M., (1990) "Metal ion assisted DNA-intercalation of crown ether-linked acridine derivatives"; *Chem. Commun.* 1028-1030.

Furusawa, K.; Obata, C.; Matsumura, H.; Takeuchi, M.; Kuwamura, T., (1990) "Bilayer membranes of amphiphilic crown ethers with amino acid residue"; *Chem. Lett.* 1047-1050.

Fyles, T. M.; James, T. D.; Kaye, K. C., (1993) "Activities and modes of action of artificial ion channel mimics"; *J. Am. Chem. Soc.* **115**, 12315-12321.

Fyles, T. M.; Loock, D.; Zhou, X., (1998) "A voltage-gated ion channel based on a bis-macrocyclic bolaamphiphile"; *J. Am. Chem. Soc.* **120**, 2997-3003.

Gad, S. C.; Conroy, W. J.; McKelvey, J. A.; Turney, R. A., (1978) "Behavioral and neuropharmacological toxicology of the macrocyclic ether 18-crown-6"; *Drug Chem. Toxicol.* **1**, 339-353.

Gad, S. C.; Reilly, C.; Siino, K.; Gavigan, F. A.; Witz, G., (1985) "Thirteen cationic ionophores: their acute toxicity, neurobehavioral and membrane effects"; *Drug Chem. Toxicol.* **8**, 451-468.

Gokel, G. W., Cram, D. J., (1973) "Molecular complexes of arenediazonium and benzoyl cations by macrocyclic polyethers"; *Chem. Commun.* 481-482.

Gokel, G. W.; J. C. Hernandez; A. M. Viscariello; K. A. Arnold; C. F. Campana; L. Echegoyen; F. R. Fronczek; R. D. Gandour; C. R. Morgan; J. E. Trafton; C. Minganti; D. Eiband; R. A. Schultz; M. Tamminen, (1987) "Steroidal lariat ethers: a new class of macrocycles and the crystal structure of *N*-(cholesteryloxycarbonyl)aza-15-crown-5"; *J. Org. Chem.* **52**, 2963-2968.

Gokel, G. W.; Murillo, O., (1996) "Synthetic organic chemical models for transmembrane channels"; *Acc. Chem. Res.* **29**, 425-432.

Gokel, G. W., (2000) "Hydraphiles: design, synthesis, and analysis of a family of synthetic, cation-conducting channels"; *Chem. Comm.* 1-9.

Gu, X.-P.; Ikeda, I.; Okahara, M.; Kim, J., (1986) "3,5-bis(2,5,8-trioxaeicosyl)-18-crown-6 and its formation of bilayer membranes."; *Chem. Lett.* 1715-1718.

Hall, A. C.; Suarez, C.; Hom-Choudhury, A.; Manu, A. N. A.; Hall, C. D.; Kirkovits, G. J.; Ghiriviga, I., (2003) "Cation transport by a redox-active synthetic ion channel"; *Org. Biomol. Chem.* **1**, 2973-2982.

Hall, C. D.; Kirkovits, G. J.; Hall, A. C., (1999) "Towards a redox-active artificial ion channel"; *Chem. Commun.* 1897-1898.

He, G.-X.; Wada, F.; Kikukawa, K.; Shinkai, S.; Matsuda, T., (1990) "Syntheses and thermal properties of new liquid crystals bearing a crown ether ring: cation binding in the nematic phase"; *J. Org. Chem.* **55**, 541-548.

Hendrixson, R. R.; Mack M. P.; Palmer, R. A.; Ottolenghi, A.; Ghirardelli, R.G., (1978) "Oral toxicity of the cyclic polyethers—12-crown-4, 15-crown-5, and 18-crown-6—in mice"; *Toxicol. Appl. Pharmacol.* **44**, 263-268.

Hu, J.; Barbour, L. J., Gokel, G. W., (2001a) "Solid state evidence for pi-complexation of sodium cation by carbon carbon double bonds"; *Chem. Commun.* 1858-1859.

Hu, J.; Barbour, L. J., Gokel, G. W., (2001b) "Solid-state evidence for pi-complexation of sodium and potassium cations by carbon-carbon triple bonds"; *J. Am. Chem. Soc.* **123**, 9486-9487.

Hu, J.; Barbour, L. J., Gokel, G. W., (2002) "Probing alkali metal-pi interactions with the side chain residue of tryptophan"; *Proc. Natl. Acad. Sci. U.S.A.* **99**, 5121-5126.

Huang, S.; Kuo, H.; Hsiao, C.; Lin, Y., (2002) "Efficient synthesis of 'redox-switched' naphthoquinone thio-crown ethers and their biological activity evaluation"; *Bioorg. Med. Chem.* **10**, 1947-1952.

Ikeda, I.; Iwaisako, K.; Nakatsuji, Y.; Okahara, M., (1986) "Surface active properties of long chain alkoxymethyl hydroxy crown ethers." *Yukagaku* **35**, 1001 (*Chem. Abstr.* 106:86714).

Ikeda, I.; Tsuji, T.; Okahara, M., (1988) "Synthesis of amphiphilic crown ethers and their ability to form molecular aggregation film." *Kenkyu Hokoku - Asahi Garasu Kogyo Gijutsu Shoreikai* **53**, 311-315 [*Chem. Abstr.* 112:98500].

Ikeda, I.; Takayuki, K.; Tanaka, K.; Takeda, T.; Kurosawa, H.; Okahara, M., (1990) "Synthesis and vesicle-forming ability of an amphiphilic compounds composed of a single lipophilic chain and a crown ether ring as a hydrophilic moiety." *Nippon Kagaku Kaishi* 1059-1064 [*Chem. Abstr.* 114:26193].

Inokuma, S.; Ito, D.; Kuwamura, T., (1988) "Surface active crown ethers. XII. Synthesis and properties of amphiphilic monoazacrown ethers possessing a hydroxyl group in the side chain"; *Yukugaku* **37**, 441-446 [*Chem. Abstr.* 109:131255]

Inokuma, S.; Okada, H.; Kuwamura, T., (1989) "Surface active crown ethers. XV. Syntheses and properties of amphiphilic aza crowns derivatives possessing cation-ligating double side chains"; *Yukugaku* **38**, 705-709 [*Chem. Abstr.* 112:58794].

Itoh, T.; Takagi, Y.; Murakami, T.; Hiyama, Y.; Tsukube. H, (1996) "Crown ethers as regulators of enzymatic reactions: enhanced reaction rate and enantioselectivity in lipase-catalyzed hydrolysis of 2-cyano-1-methylethyl acetate"; *J. Org. Chem.* **61**, 2158-2163.

Jullien, L.; Lehn, J.-M., (1988) "The 'chundle' approach to molecular channels. Synthesis of a macrocycle-based molecular bundle"; *Tetrahedron Lett.* 3803-3806.

Kato, N.; Ikeda, I.; Okahara, M.; Shibasaki, I., (1980) "Antimicrobial activity of alkyl crown ethers"; *Res. Soc. Antibac. Antifung. Agents.* **8**, 532-533.

Katoh, A.; Ishida, S.; Ohkanda, J.; Washizu, M., (1996) "Synthesis of novel amphiphilic compounds containing aza-12-crown-4 or D-glucosamine and their ion permeability"; *Heterocycles* **43**, 589-599.

Kerwin, S. M., (1994) "Synthesis of a DNA-cleaving bis(propargylic) sulfone crown ether"; *Tetrahedron Lett.* **35**, 1023-1026.

Kuroda, Y.;Sera, T., Ogoshi, H., (1991) "Regioselectivities and stereoselectivities of singlet oxygen generated by cyclodextrin sandwiched porphyrin sensitization. Lipooxygenase-like activity"; *J. Am. Chem. Soc.* **113**, 2793-2794.

Kuwamura, T.; Yoshida, S., (1980) "Surface-active crown ethers. 2. Synthesis, surfactant properties, and catalytic action of macrocyclic polyethers possessing a 4-alkyl-1,3,5-triazine subunit"; *Nippon Kagaku Kaishi* **427** [*Chem. Abstr.* 493:28168e].

Kyba, E. P.; Helgeson, R. C.; Khorshed, M.; Gokel, G. W.; Tarnowski, T. L.; Moore, S. S., Cram, D. J., (1977) "Host-guest complexation. 1. Concept and illustration"; *J. Am. Chem. Soc.* **99**, 2564-2571.

Larson, K.; Vaknin, D.; Villavicencio, O.; McGrath, D.; Stephenson, N.; Tsukruk, V. V., (2001) Synchrotron studies of amphiphiles with crown polar heads at air-water interface; *Polymeric Mater. Sci. Engin.* **85**, 229-230.

Leevy, W. M.; Donato, G. M.; Ferdani, R.; Goldman, W. E.; Schlesinger, P. H.; Gokel, G. W., (2002) "Synthetic hydraphile channels of appropriate length kill *Escherichia coli*"; *J. Am. Chem. Soc.* **124**, 9022-9023.

Leevy, W. M.; Huettner, J.E.; Pajewski, R.; Schlesinger, P. H.; Gokel, G. W. (2004) "Synthetic ion channel activity documented by electrophysiological methods in living cells"; *J. Am. Chem. Soc.* in press.

Matsumura, H.; Kuwamura, T.; Furusawa, K., (1990) "Coagulation behavior of vesicles composed of crown ethers and phospholipids caused by bolaform surfactants"; *J. Colloid Interface Sci.* **139**, 331-336.

McPhee, M. M.; Kern, J. T.; Hoster, B. C.; Kerwin, S. M, (2000) "Propargylic sulfone-armed lariat crown ethers: alkali metal ion-regulated DNA cleavage agents"; *Bioorg. Chem.* **8**, 98-118.

McPhee, M. M.; Kerwin, S. M, (2001) "Synthesis, DNA cleavage, and cytotoxicity of a series of bis(propargylic) sulfone crown ethers"; *Bioorg. Med. Chem.* **9**, 2809-2818.

Meadows, E. S.; DeWall, S. L.; Barbour, L. J., Gokel, G. W., (2001) "Alkali metal cation-pi interactions observed by using a lariat ether model system"; *J. Am. Chem. Soc.* **123**, 3092-3107.

Mertes, M. P., Mertes, K. B., (1990) "Polyammonium macrocycles as catalysts for phosphoryl transfer: The evolution of an enzyme mimic"; *Acc. Chem. Res.* **23**, 413-418.

Muñoz, S.; Mallén, J. V.; Nakano, A.; Chen, Z.; Echegoyen, L.; Gay, I.; Gokel, G. W., (1992) "Lariat ether bola-amphiphiles: formation of crown ether based bola-amphisomes"; *Chem. Commun.* 520-522.

Muñoz, S.; Mallén, J.; Nakano, A.; Chen, Z.; Gay, I.; Echegoyen, L.; Gokel, G. W., (1993) "Ultrathin monolayer lipid membranes from a new family of crown ether-based bola-amphiphiles"; *J. Am. Chem. Soc.* **115**, 1705-1711.

Murray, C. L.; Shabany, H.; Gokel, G. W., (2000) "The central 'relay' unit in hydraphile channels as a model for the water-and-ion 'capsule' of channel proteins"; *Chem. Commun.* 2371-2372.

Nakano, A.; Xie, Q.; Mallen, J. V.; Echegoyen, L.; Gokel, G. W., (1990) "Synthesis of a membrane-insertable, sodium cation conducting channel: kinetic analysis by dynamic sodium-23 NMR"; *J. Am. Chem. Soc.* **112**, 1287-1289.

Nakashima, N.; Moriguchi, I.; Nakano, K.; Takagi, M., (1987) "Design of a novel bilayer system responsive to chemical signals. Selective discrimination of sodium ions by a spectroscopic method"; *Chem. Commun.* 617-619.

Odell, B.; Earlam, G., (1985) "Dissolution of proteins in organic dolvents using macrocyclic polyethers: association constants of a cytochrome c-[1,2-$^{14}C_2$]-18-crown-6 complex in methanol"; *Chem. Commun.* 359-361.

Okahara, M.; Kuo, P. L.; Yamamura, S.; Ikeda, I., (1980) "Effect of metal salts on the cloud point of alkyl crown compounds." *Chem. Commun.* 586-587.

Pechulis, A. D.; Thompson, R. J.; Fojtik, J. P.; Schwartz, H. M.; Lisek, C. A.; Frye, L. L., (1997) The design, synthesis and transmembrane transport studies of a biomimetic sterol-based ion channel; *Bioorg. Med. Chem.* **5**, 1893-1901.

Pederson, C.J. (1967) "Cyclic polyethers and their complexes with metal salts"; *J. Am. Chem. Soc.* **89**, 7017-7036.

Raymond, K. N.;Dertz, E. A., Kim, S. S., (2003) "Enterobactin: An archetype for microbial iron transport"; *Proc. Natl. Acad. Sci. U.S.A.* **100**, 3584-3588.

Riley, D. P., (1999) "Functional mimics of superoxide dismutase enzymes as therapeutic agents"; *Chem. Rev.* **99**, 2573-2587.

Riley, D. P., (2000) "Rational design of synthetic enzymes and their potential utility as human pharmaceuticals: Development of manganese(II)-based superoxide dismutase mimics"; *Adv. Supramol. Chem.* **6**, 217-244.

Roks, M. F. M.; Nolte, R. J. M., (1992) "Biomimetic macromolecular chemistry: design and synthesis of an artificial ion channel based on a polymer containing cofacially stacked crown ether rings. Incorporation in dihexadecyl phosphate vesicles and study of cobalt ion transport"; *Macromolecules* **25**, 5398-5407.

Roosenberg, J. M. I.; Lin, Y.-M.; Lu, Y., Miller, M. J., (2000) "Studies and synthesis of siderophores, microbial iron chelators, and analogs as potential drug delivery agents"; *Curr. Med. Chem.* 7, 159-197.

Santos, A.M.; Vidal, M.; Pacheco, Y.; Frontera, J.; Baez, C.; Ornellas, O.; Barletta, G.; Griebenow, K. (2001) "Effect of crown ethers on structure, stability, activity, enantioselectivity of *Subtilisin Carlsberg* in organic solvents." *Biotech. Bioengin.* **74**, 296-308.

Shinkai, S., (1990) "Functionalization of crown ethers and calixarenes: new applications as ligands, carriers, and host molecules"; *Bioorg. Chem. Front.* **1**, 161-195.

Tabushi, I., (1982) "Cyclodextrin catalysis as a model for enzyme action"; *Acc. Chem. Res.* **15**, 66-72.

Tabushi, I., (1986) "Chiral selection and chiral induction by the use of regiospecifically di-(or poly)substituted cyclodextrins"; *Pure Appl. Chem.* **58**, 1529-1534.

Tabushi, I.; Kuroda, Y., Yamada, M., (1988) "Dynamic molecular motions of guest molecules in modified b-cyclodextrins"; *Tetrahedron Lett.* **29**, 1413-1426.

Takagi, M., Ueno, K., (1984) "Crown complexes as alkali and alkaline earth metal ion selective chromogenic reagents"; *Top. Curr. Chem.* **121**, 39-65.

Takayama, K.; Hasegawa, S.; Sasagawa, S.; Nambu, N.; Nagai, T., (1977) "Apparent oral toxicity of 18-crown-6 in dogs"; *Chem. Pharm. Bull.* **25**, 3125.

Tristram-Nagle, S.; Nagle, J. F., (2004) "Lipid bilayers: thermodynamics, structure, fluctuations, and interactions"; *Chem Phys Lipids* **127**, 3-14.

Tsukube, H.; Yamada, T.; Shinoda, S., (2001) "Crown ether strategy toward chemical activation of biological protein functions"; *J. Heterocyclic Chem.* **38**, 1401-1408.

Tummler, B.; Maass, G.; Weber, E.; Wehner, W., Voegtle, F., (1977) "Noncyclic crown-type polyethers, pyridinophane cryptands, and their alkali metal ion complexes: Synthesis, complex stability and kinetics"; *J. Am. Chem. Soc.* **99**, 4683-4689.

Uchegbu, I. F.; Vyas, S. P., (1998) Nonionic surfactant based vesicles (niosomes) in drug delivery; *Int. J. Pharmaceutics* **172**, 33-70.

van Unen, D-J; Sakodinskaya, I.K.; Engbersen, J. F. J.; Reinhoudt, D.N. (1998) "Crown ether activation of cross-linked *Subtilisin Carlsberg* crystals in organic solvents"; *Perkin Trans. 1*, 3341-3343.

van Unen, D-J; Engbersen, J. F. J.; Reinhoudt, D.N. (2002) "Why do crown ethers activate enzymes in organic solvents?"; *Biotech. Bioengin.* **77**, 248-255.

Voyer, N.; Robataille, M., (1995) A novel functional artificial ion channel; *J. Am. Chem. Soc.* **117**, 6599-6600.

Voyer, N.; Potvin, L.; Rousseau, E., (1997) Electrical activity of artificial ion channels incorporated into planar lipid bilayers; *Perkin Trans.* 2, 1469-1471.

Yagi, K.; Garcia, V.; Rivas, M. E.; Salas, J.; Camargo, A.; Tabata, T., (1984) "Antifungal activity of crown ethers"; *J Incl. Phenomena.* **2**, 179-184.

Yang, J.; Gabriele, B.; Belvedere, S.; Huang, Y., Breslow, R., (2002) "Catalytic oxidations of steroid substrates by artificial cytochrome P450 enzymes"; *J. Org. Chem.* **67**, 5057-5067.

17. RECOGNITION OF CYTOCHROME *c* BY TETRAPHENYLPORPHYRIN-BASED PROTEIN SURFACE RECEPTORS

RISHI K. JAIN, LUN K. TSOU AND ANDREW D. HAMILTON
Department of Chemistry, Yale University
P.O. Box 208107, New Haven, CT 06520-8107 U.S.A.

1. Introduction

Cytochrome *c* (cyt. *c*) has become a major protein for testing new approaches and techniques in protein science.[1] This is partly due to the venerable position that cytochrome *c* holds in the field of biochemistry since it was isolated and characterized more than 70 years ago. Cyt. *c* was one of the first proteins to be sequenced,[2] and to have its X-ray structure determined in 1967.[3] Cyt. *c* also has the advantage of stability and a spectroscopically distinct heme group. More than 23,000 articles mentioning cyt. *c* were published between 1945-2002 (ISI Web of Science). Here, we describe an approach to tetraphenylporphyrin-based protein surface receptors and the characterization of their interactions with the principal target cyt. *c*.

2. Interactions of Cytochrome *c* with Physiological and Non-physiological Binding Partners

The principal function of cyt. *c* is to form complexes through a defined interface with protein partners in our cells. This is most established for eukaryotic cytochrome *c* within the mitochondrial electron transport chain (ETC), a process required for carrying out the oxidative phosphorylation of ATP.[4] Formation of a complex with cyt. *c* reductase (an electron-donor protein from complex III) and cyt. *c* oxidase (an electron-acceptor protein from complex IV) leads to the transfer of electrons between otherwise separated proteins. More recently cyt. *c* has been found to play a critical role in the process of apoptosis or programmed cell death. This in turn has led to a resurgence of interest in all aspects of cyt. *c* research.[5] Again protein-protein interactions have been shown be essential with mitochondrial cyt. *c* binding to such proteins as APAF-1 to form the multi-protein species known as the apoptosome that is now thought to be a requirement for apoptosis.[6,7]

Karsten Gloe (ed.), Macrocyclic Chemistry: Current Trends and Future Perspectives, 267–275.
© 2005 *Springer. Printed in the Netherlands.*

At physiological pH, horse heart cytochrome *c* is strongly positively charged with approximately 8 protonated lysines and arginines, and a pI value of 10. The crystal structures of horse cyt. *c*, (Figure 1a)[8] and its complex with yeast cytochrome *c*-peroxidase (ccp) (Figure 1b)[9] show that a critical array of lysine and arginine residues are found near the exposed heme edge of cyt. *c*. In conjunction with these structural studies, mutation analysis has allowed the identification of the energetic contributions made by these cationic residues on the binding to CCP. Margoliash and co-workers have generated several cyt. *c* mutants with certain surface lysines modified by carboxy-dinitrophenyl groups.[10] Measurement of the interaction of these mutants with various protein partners, including cyt. *c* reductase, cyt. *c* oxidase, sulfite oxidase, and ccp allowed an assessment of which lysines were important for stabilizing the complex. The most important residues were found on the face of the protein (lysines 13, 27, 72, 73, 86, 87) surrounding the exposed heme-edge. In some cases, where mutations were distant from the heme edge, changes in activity could be ascribed to changes in the dipole moment of the protein. The critical interaction domain of cyt. *c* with its partners was thus clearly established to be close to the heme edge, defined by the surrounding lysines.

Negatively charged species other than proteins have been shown to interact, albeit weakly (mM-μM), with the positively-charged heme edge domain of cyt. *c*. These include small peptides,[11-12] lipid vesicles,[13] inorganic and organic anions[14] (phosphate, oxalate), and the egg yolk storage protein phosvitin.[15] An early strategy for the development of synthetic binding agents for cyt. *c* is exemplified by the work by Fine and coworkers.[16-18] Anionic sulfated sugar polymers were prepared and shown to bind cyt. *c*, through charge complementarity and hydrophobic forces leading to significant affinity. In 1985, Fisher showed that a functionalized tetracarboxyphenylporphyrin (TCPP) possessed the correct size and charge complementarity to match the hydrophobic and charged residues on the heme edge surface of cyt. *c* and established a moderate binding affinity for the protein (K_d ~5μM).[19]

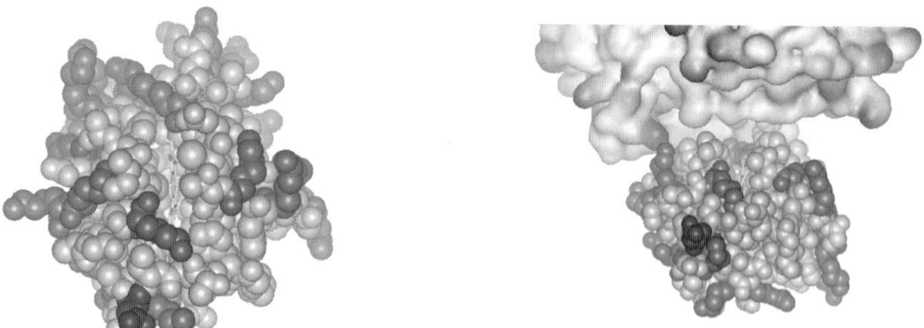

Figure 1. (a) X-ray crystal structure of horse-heart ferricytochrome *c*.[8] All protein atoms are shown in the C.-P.-K. form, while the heme group is shown in the stick form. All Arg and Lys residues are colored blue, while Glu and Asp are colored in red, to contrast the destribution of the most ionizable side chains. (b) The X-ray crystal structure of horse heart ferricytochrome *c* in complex with horse cytochrome *c* peroxidase (ccp).[9] The peroxidase is shown as a molecular surface model, with blue regions depicting positive and red representing negative electrostatic potential. Note the cluster of negative potential on ccp that surrounds the contact interface. A coloured version is given in the Appendix.

3. Designs Based on Macrocyclic Scaffolds Appended with Amino Acids and Peptides

Macrocylic structures, in general, are ideal molecular scaffolds for constructing synthetic protein surface receptors. A wide range of macrocyclic structures have been designed and exploited in all areas of host-guest chemistry.[20] Among the most attractive features of macrocycles are the potential for small molecule binding to the inner cavity and the possibility of derivatization around the perimeter to construct an "external" binding site. Importantly, most macrocycles are relatively restrained with respect to their conformation, primarily due to the loss of rotational degrees of freedom upon cyclisation, and by stabilization due to intramolecular forces.[21] Finally, many macrocycles have subnano-to nanoscale dimensions spanning a large area of space. This type of dimension in addition to their conformational rigidity would be ideal for projecting multiple binding groups towards a protein surface in a directed manner.

While cyclic peptides have proven to be problematic, we believe that amino acids are ideal candidates for derivatization of our macrocyclic scaffolds. Many natural and unnatural amino acids with appropriately protected side chains are commercially available or can be readily prepared, providing facile access.[22] The α-amino- and α-carboxy- groups common to all of these will provide constant sites for attachment to a macrocyclic scaffold core. Side chains varying in aromatic, aliphatic, polar and ionic characters should provide sufficient chemical diversity. Finally, amino acids may be combined in many ways to form short acyclic peptides, allowing access to more diverse chemical properties not found in individual amino acids.

4. Design and Synthesis of Tetraphenylporphyrin-based Protein Surface Receptors

Among the world of macrocyclic molecules, porphyrins[23] hold a particularly important place due to their extensive synthesis and applications literature and their potential compatibility with the biological milieu.[24] Both natural and unnatural porphyrins and porphyrinoids have been used in medical applications such as tumour photodynamic therapy (PDT)[25] and cancer cell imaging.[26] Furthermore, the very presence of metalloporphyrin derivatives in the heme and chlorin prosthetic groups of hemoglobin (and the cytochromes) and the photosynthetic apparatus confirm their central role in biochemistry. As a result porphyrin derivatives have been intensely investigated in terms of their structural characterization, their physicochemical properties and their synthetic accessibility.[27] The consequence of this detailed investigation is that many porphyrin derivatives have been used as key structural components in the construction of synthetic hosts, artificial enzymes and various model systems in bioorganic and bioinorganic chemistry.[28]

The most widely studied porphyrin core is 5,10,15,20-*meso*-tetrakis-phenyl porphine (TPP) and this derivative is particularly suited to the purpose of protein surface recognition. A wide range of TPP derivatives have been investigated and there are several methods of preparation available. The most frequently used approach for the synthesis of TPP involves the condensation of benzaldehyde and pyrrole under acidic conditions, followed by air oxidation of the intermediate porphyrinogen.[29] More recently, Lindsey[30] has developed a much improved method of TPP synthesis involving equilibrium control under high dilution and significantly milder conditions. As a result many chemically sensitive functional groups can be tolerated in the reaction. Various TPP derivatives have also been shown to dissolve in a wide range of aqueous and organic solvents.

Depending on the nature of the peripheral functionalization, aggregation can occur and this aspect of their chemistry has been extensively investigated. These studies are all made more straightforward by the highly chromogenic and flurogenic properties of most TPPs. A critical issue in the design of protein surface receptors is the large surface area defined by most protein interfaces. The size and shape of TPP offers the potential of complementing a large fraction of the protein surface in the design of receptors. Molecular modeling using the MM-2 force field shows (Figure 2) that the distance between two diagonally opposed phenyl groups (from their para positions) is approximately 19 Å.

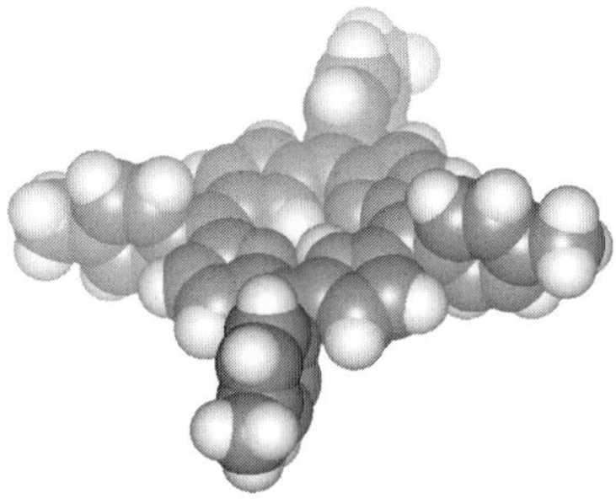

Figure 2. C.-P.-K. model of energy minimized (MM-2, chem 3D) meso-tetrakis (*p*-methyl) phenyl porphyrin.

As peptidic appendages were chosen as elements for receptor diversification, porphyrins that allow the attachment of peptides through their amino- or carboxy termini were desirable. Analogous to the calix[4]arene tetraacid, we chose a tetraacid derivative of TPP, 5,10,15,20-*m*-tetrakis-*p*-carboxyphenyl porphyrin (TCPP) **1** as the porphyrin scaffold. The carboxylic acid allows the linkage of peptides and amino acids through amide bond formation. The synthesis of our TPP-based protein surface receptors is outlined in Figure 3.

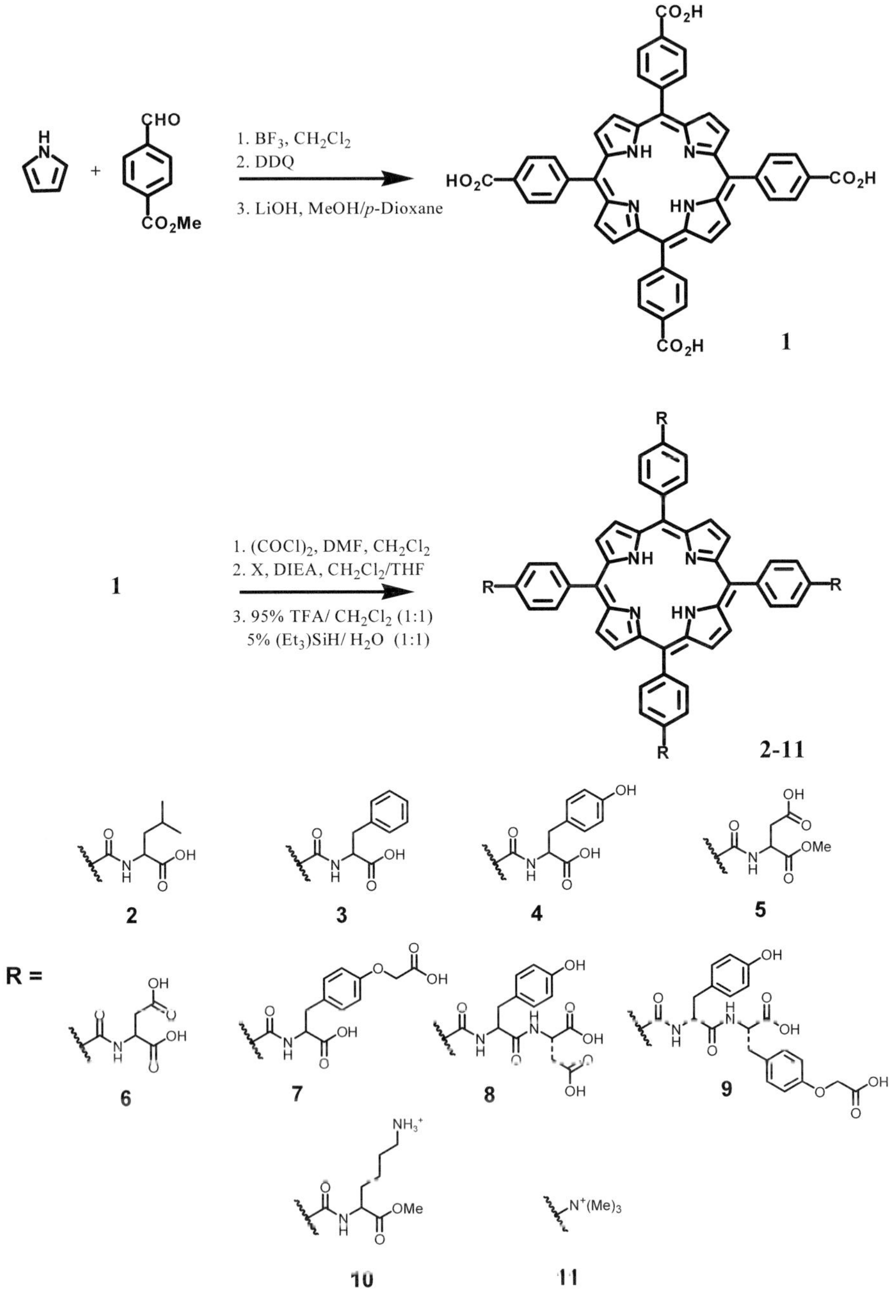

Figure 3. Synthetic scheme for the preparation of porphyrin-amino acid or porphyrin- peptide receptors **2 - 11** for targeting protein surfaces. **1** (TCPP) and other TPP variants are conveniently prepared using the Lindsey procedure (BF_3, methylene chloride, dilute). The carboxylic acids provide an attachment site for various amino acids and peptides, via amide bond formation. **X** corresponds to appropriately protected amino acids to make **2-11**.

5. Tetraphenylporphyrin-based Synthetic Receptors for Targeting the Cytochrome *c* Surface

Porphyrin-based receptors **2** – **12,** and **12** (coproporphyrin I), a naturally occurring porphyrin, were screened for binding against horse heart ferricytochrome c. Rodgers and co-workers have previously reported on the use of fluorescence quenching for convieniently determining the Kd's of porphyrin guests against cyt. *c*.[31] Quenching is believed to occur due to the enforced proximity of the Fe (III) heme to the porphyrin macrocycle (Figure 4). All porphyrins synthesized possessed similar fluorescence properties, thus allowing for direct comparison of K_d values obtained by quenching titrations.

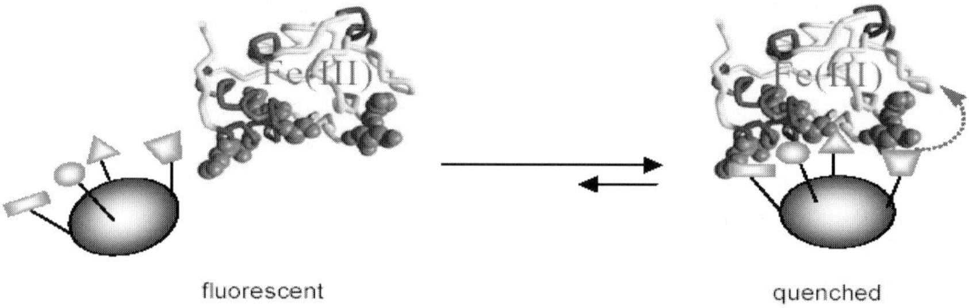

Figure 4. A schematic depiction of the fluorescence quenching-based binding assay. The synthetic receptor is fluorescent when unbound but when the porphyrin macrocycle is forced into proximity with the Fe(III) containing heme, fluorescence is quenched. The ratio between free and bound receptor can be equated to the change in fluorescence due to protein binding.

Addition of cytochrome c to solutions containing some of our TPP-based receptors resulted in quenching of porphyrin fluorescence emission (Ex = 420, Em = 650 nm). In contrast, titrations with tetracationic *m*-tetrakis-(4-trimethylaminophenyl) porphyrin (**11**,TTMAPP) and cationic TPP-based receptor **10** showed no quenching even at high concentrations. This demonstrates the clear preference for anionic species by the cationic cytochrome *c* recognition surface, in agreement with previously reported studies.

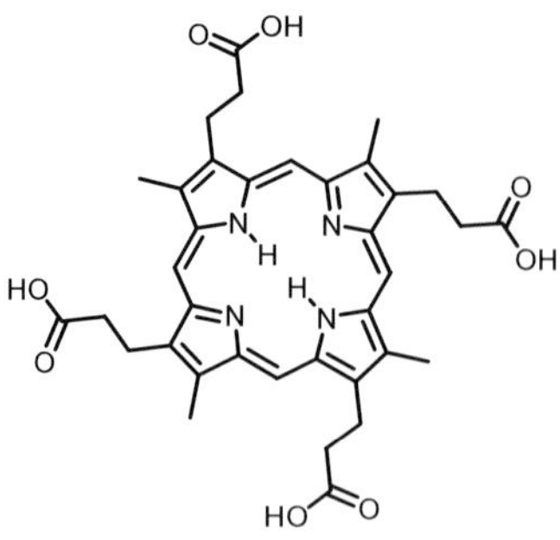

12
Coproporphyrin I

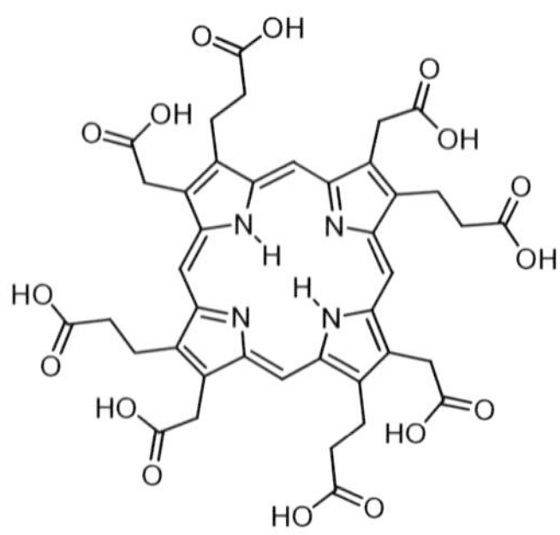

13
Uroporphyrin I

Titrations were performed by addition of cyt. *c*., and the change in fluorescence was plotted against the concentration of cyt. *c*. Dissociation constants were derived by curve fitting to a 1:1 binding equation. Significant changes in affinity for cyt. *c* were observed by altering the relative proportions of acidic and aromatic functionalities in the receptors. Receptors **5** and **6** differ only by the substitution of four methyl esters by carboxylic acids, respectively, providing controls for probing the charge requirements for cyt. *c* recognition. A ~5 fold increase in affinity was seen, on going from receptor **5** to receptor **6**, indicating a preference for octaanionic receptors over their tetraanionic counterparts. Receptor **5** and **TCPP-1**, with the same number and type of charged groups, show little difference in their binding affinities. Similar trends were observed when aromatic groups were incorporated into the receptor while keeping the number of charged groups constant. Rodgers had earlier shown that **13** (uroporphyrin I), containing eight carboxylate groups, binds cytochrome c with μM affinity. However, receptor **6** which has eight carboxylate groups and 4 aryl groups binds to cytochrome c approximately six fold stronger than **13** (uroporphyrin I). In a similar analysis, receptor **5** binds cytochrome c nine-fold stronger than tetraanionic **12** (coproporphyrin I). These results suggested that an appropriate combination of charged and hydrophobic groups on the porphyrin periphery would give a molecule with exceptionally high affinity for cytochrome c. To confirm this, we designed receptor **8** to contain 8 carboxylate groups and 8 aryl groups. The titration curve (Figure 5) shows a sharper achievement of saturation with **8** compared to **6**.

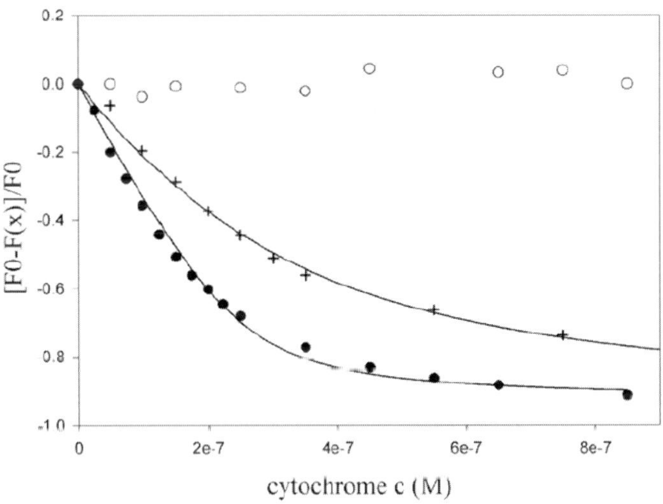

Figure 5. Fluorescence quenching of **8** (black dot), **6** (+), and **11** (O) upon addition of cytochrome c. Curve fit indicates a K_d of 20 nM ,160 nM respectively.

This corresponds to a K_d of 20 nM for **8** binding to cytochrome c and represents an eight fold increase in affinity compared to **6**. A receptor containing peptidomimetic sidechains of **8**, receptor **7**, was synthesized to determine if the observed trend applies to receptors of comparable charge and hydrophobicity but with different connectivity. We saw no significant affinity change by truncating the Tyr(OH)-Asp-OH dipeptide on **8** to Tyr(OCH2COOH)-OH on receptor **7**, a modified tyrosine with a carboxymethyl group on the side chain phenol. These results strongly indicated that the side chains from the porphyrin scaffold do not have to be stringently oriented as long as the ionizable and hydrophobic groups are in the general proximity of interacting protein side chains.

6. Acknowledgments

We thank the National Institutes of Health (GM35208) for financial support of this work.

7. References

(1) *Cytochrome c, a multidisciplinary approach*; Scott, R.A.; Mauk, A.G. Eds. University Science Books: Sausalito CA, 1996.
(2) "Complete amino-acid sequence," Margoliash, E.; Tuppy, H.; Smith, E.L.; Kreil, G. *Nature* **1961**, *192*, 1125.
(3) "Location of heme in horse heart ferricytochrome c by x-ray diffraction," Dickerson, R.E.; Kopka, M.L.; Weinzier, J.; Varnum, J.; Eisenberg, D.; Margoliash, E. *J. Biol. Chem.* **1967**, *242,* 3015-3018.
(4) *The History of Cell Respiration and Cytochrome*; Keilin, D.; University Press,: Cambridge, 1966.
(5) "The mitochondrial apoptosome: a killer unleashed by the cytochrome seas," Adrain, C.; Martin, S.J. Trends. Biochem. Sci. **2001**, *26*, 390-397.
(6) "The biochemistry of apoptosis," Hengartner, M.O. *Nature* **2000**, *407*, 770-776.
(7) "Stoichiometry, free energy, and kinetic aspects of cytochrome c: Apaf-1 binding in apoptosis," Purring C.; Zou H.; Wang X.D.; McLendon G. *J. Am. Chem. Soc.* **1999**, *121*, 7435-7436.
(8) "High-resolution 3-dimensional structure of horse heart cytochrome c," Bushnell, G.W.; Louie, G.V.; Brayer, G.D. *J. Mol. Biol.* **1990**, *214*, 585-595.
(9) "Crystal-structure of a complex between electron-transfer partners, cytochrome-c peroxidase and cytochrome c," Pelletier, H.; Kraut, J. *Science* **1992**, *258* , 1748-1755.
(10) "The asymmetric distribution of charges on the surface of horse heart cytochrome c-Functional implications," Koppenol, W.H.; Margoliash, E. *J. Biol. Chem.* **1982**, *257*, 4426-4437
(11) "Interactions of cytochrome c and cytochrome f with aspartic acid peptides," Hirota S., Endo M., Hayamizu K., Tsukazaki T., Takabe T., Kohzuma T., Yamauchi O. *J. Am. Chem. Soc.* **1999**, *121*, 849-855.
(12) "Effects of charged peptides on electron transfer from [Fe(CN)6]4- to cytochrome c or plastocyanin," Hirota, S.; Endo, M.; Tsukazaki, T.; Takabe, T.; Yamauchi O. *J. Biol. Inorg. Chem.* **1998**, *3*, 563-569.
(13) "Interactions of cytochrome c and [14C] carboxymethylated cytochrome c with monolayers of phophatidylcholine, phosphatidic acid and cardiolipin," Quinn, P.J.; Dawson, R.M.C. *Biochem J.* **1969**, *115*, 65.
(14) "Phosphate binding by cytochrome c-Specific binding-site involved in the formation and reactivity of a complex of ferricytochrome-c, ferrous ion, and phosphate," Taborsky, G.; McCollum, K. *J. Biol. Chem.* **1979**, *254*, 7069-7075.
(15) "Interaction of cytochrome-c and phophoprotein phosvitin-Formation of a complex with an intact 695-mmu band," Taborsky, G. *Biochemistry* **1970**, *9*, 3768.
(16) "Reversible inhibition of cytochrome system components by macromolecular polyions," Person, P.; Fine, A.S. *Arch. Biochem. Biophys.* **1961**, *94*, 392.
(17) "Macroion interactions involving components of cytochrome system 4. Polyion induced reversible oxidations and reductions of cytochrome c," Person, P.; Fine, A.S.; Marta, P.T.; Zipper, H. *J. Biol. Chem.* **1965**, *240*, 3159.
(18) "Macroion interactions involving components of cytochrome system 5. Reduction of cytochrome c by synthetic polysaccharides," Mora, P.T.; Creskoff, E.L.; Person, P. *J. Biol. Chem.* **1967**, *242*, 1280.

(19) "Topographical mimicry of the enzyme binding domain of cytochrome-c," Clark-Ferris, K.K.; Fisher, J. *J. Am. Chem. Soc.* **1985**, *107*, 5007-5008.
(20) "Macrocyclic Receptor Molecules," J.-M Lehn, *Pure and App. Chem.* **1979**, *51* 979-997.
(21) *Macrocyclic Chemistry*; Dietrich B.; Viout, P.; Lehn, J.-M.; VCH: Weinheim, 1993.
(22) "Contemporary methods for peptide and protein synthesis," Aimoto, S. *Curr. Org. Chem.* **2001**, *5*, 45-87.
(23) *The Porphyrins*; Dolphin, D., Ed.; Academic Press: New York, 1978.
(24) *The Colours of Life: An Introduction to the Chemistry of Porphyrins and Related Compounds*; Milgrom, L.R.; Oxford University Press: New York, 1997.
(25) "Photodynamic therapy," Dougherty, T.J.; Gomer, C.J.; Henderson, B.W.; Jori, G.; Kessel, D.; Korbelik, M.; Moan, J.; Peng, Q. *J. Natl.Cancer Inst.* **1998**, *90*, 889-905.
(26) "*In vivo* fluorescence spectroscopy and imaging for oncological applications," Wagnieres, G.A.; Star, W.M.; Wilson, B.C. *Photochem. Photobiol.* **1998**, *68*, 603-632.
(27) *The Porphyrin Handbook*; Kadish, K.; Smith, K.; Guilard, R., ed.; Academic Press: New York, 2000.
(28) "Porphyrin building-blocks for modular construction of bioorganic model systems," Lindsey, J.S.; Prathapan, S.; Johnston, T.E.; Wagner, R.W. *Tetrahedron* **1994**, *50*, 8941-8968.
(29) "A simplified synthesis for meso-tetraphenylporphyrin," Adler, A.D.; Longo, F.R.; Finarell, J.D.; Goldmach, J.; Assour, J.; Korsakof, L. *J. Org. Chem.* **1967**, *32*, 476.
(30) "Rothemund and Adler-Longo reactions revisited - Synthesis of tetraphenylporphyrins under equilibrium conditions," Lindsey, J.S.; Schreiman, I.C.; Hsu, H.C.; Kearney, P.C.; Marguerettaz, A.M. *J. Org. Chem.* **1987**, *52*, 827-836.
(31) "Photoinduced electron-transfer in self-associated complexes of several uroporphyrins and cytochrome c," Zhou, J.S.; Granada, E.S.V.; Leontis, N.B.; Rodgers, M.A.J. *J. Am. Chem. Soc.* **1990**, *112*, 5074 5080.

18.
SUPRAMOLECULAR COMPLEXES WITH MACROCYCLES: SURPRISES AND INSIGHTS

HANS-JÖRG SCHNEIDER
FR Organische Chemie der Universität des Saarlandes,
66041 Saarbrücken, Germany

"All we can and need do is create theories and eliminate error."
(Karl Popper)

1. Introduction

Progress in fundamental science lives on falsification of existing theories or hypotheses.[1] This is a major reason why we need experiments, and why we must not only be prepared for surprises, but should actually seek for those. Chemists all too often find satisfaction only if their results confirm their anticipation. Progress in supramolecular chemistry as a relatively new field of science especially relies on conclusions from unexpected results. Although there are many reviews and already several monographs in the field,[2] authors and even more students understandably prefer rules and orders, and dislike complications. This could lead to the impression that as e.g. in classical mechanics only the intelligent application of recognized laws is what will in future count in our field. However, even if one only would be interested in making e.g. perfect new receptor molecules a better understanding of the relevant interaction mechanisms will be of significant help in a rational approach to such goals. Synthetic supramolecular chemistry is particularly suited for experiments which challenge existing hypotheses, as entirely new experimental systems such as new host-guest complexes can be designed in order to answer specific questions. In the present chapter an attempt is made to illustrate results in macrocyclic chemistry, which at least at first sight are unexpected, but can lead to new insights in theory. Most of the illustrations will be taken from work of the author's laboratory; these and the selected other examples are meant to be an incentive for both experimentalists and theoreticians for the future development of one of the most promising fields in chemistry.

2. Structure of Macrocyclic Complexes - Solution vs. Solid State

The examples given in Figure 1 illustrate not only the persisting problems to identify the structure of supramolecular complexes, but at the same time represent an interesting piece of history. The complex of a benzidino-cyclophane with a benzene guest molecule inside the macrocycle is likely the oldest published picture of such a supramolecular complex.[3]

Karsten Gloe (ed.), Macrocyclic Chemistry: Current Trends and Future Perspectives, 277–289.
© 2005 *Springer. Printed in the Netherlands.*

It was based on the finding from e.g. elemental analysis that crystals showed a 1:1 composition of host and guest molecule. Some 27 years later X-ray crystallography revealed, that in these crystals the benzene was in fact not inside the macrocyle cavity, but outside.[4] Again, some decades later, work in our lab has now shown, that in aqueous solution the benzene indeed is complexed inside the cavity.[5] This is evident not only from the observed complex stability itself -which exceeds the immeasurably low affinity to a corresponding monomer-, but mainly from the NMR shielding effects of the macrocyclic host on the benzene protons. The NMR shifts are small as a result of averaging between six protons, which only partially are exposed to the in addition rather asymmetric shielding cones of the host phenyl rings. With p-toluenesulfonic acid as guest the shielding effects on the macrocyclic host protons illustrate the intracavity inclusion also.

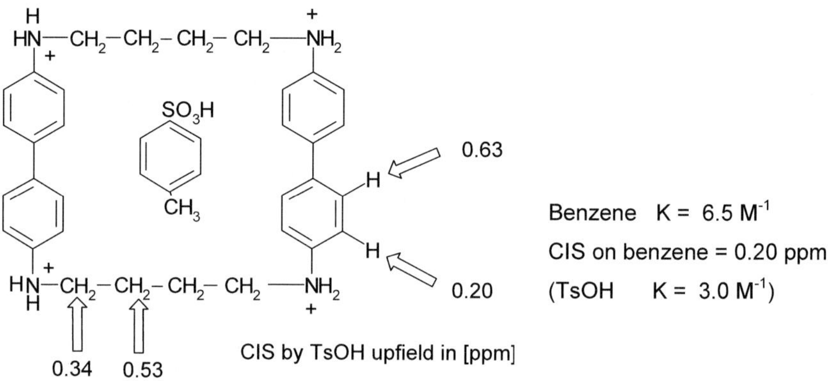

Figure 1. Stetter's cyclophane (1955) , and evidence for complexation in aq. solution (unpublished).

An opposite example of intracavity guest inclusion in the solid state, but not in solution is documented in Figure 2. X-ray crystallography shows as expected innersphere coordination of the lanthanide ion with the [2.2.2] cryptand.[6] However, if one dissolves the solid Eu complex in methanol, one observes only at the beginning NMR signals which correspond to the D_{3h} symmetry depicted as structure **I**, and characterize the innersphere complex also by large upfield shifts and linewidth increases of the CH_2 protons. After short time the metal ion moves out and forms a complex **II** with C_{2v} symmetry , as obvious from the number of now nine CH_2 signals.[7] After longer time the free diprotonated ligand **III** appears; in water **III** is formed more rapidly, with a rate constant of 8.7×10^{-4} s^{-1}. Particularly in the solid state it is also not unusual to find not only the expected guest inside the cavity, as shown e.g. with a macrocyclic octalactam host encapsulating besides two Cl^- ions also two water molecules.[8]

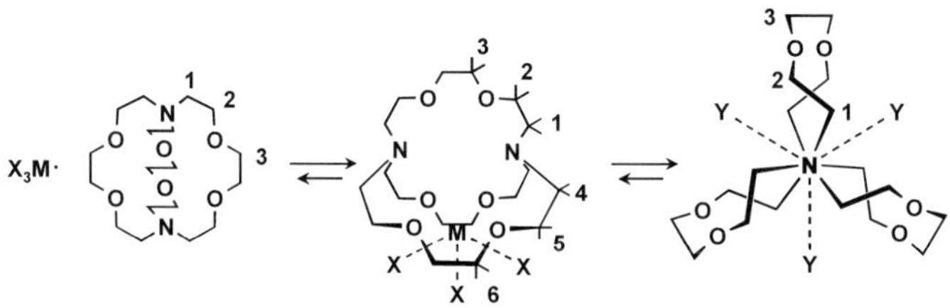

Figure 2. Complexation of the [2.2.2] cryptand with $EuCl_3$ (=MX_3); symmetries derived from the number of NMR signals (the metal ion M is covered in the front view of structure **I**).

3. Some Unusual Properties of Crown Ether and Calixarene Complexes

Crown ethers continue to be one of the most useful parts of supramolecular chemistry.[9] From the beginning computations of metal ions complexes with synthetic ionophores,[10] which have been aptly reviewed,[11] emphasized the importance of including explicitly solvation in free energy calculations, also with ab initio calculations on calixarene complexes.[12] Molecular dynamics simulations of 18-crown-6 ether complexes in aqueous solutions predict too low affinities, but at least correctly reproduce the sequence trend $K^+ > Rb^+ > Cs^+ > Na^+$. However, only the selection of K^+ over Rb^+ and Cs^+ is ascribed to the cation size relative to that of the crown cavity, whereas K^+ appears in these calculations to be selected over Na^+ as consequence of the greater free energy penalty involved in displacing water molecules from Na^+.[13]

One of the early puzzles in host-guest chemistry was the difference of complex stabilities with the crown ether 18-C-5 in comparison to 18-C-6.[14] Although 18-C-5 has only 20% less donor atoms than 18-C-6, its stability constant with e.g. K^+ is more than 4000 times smaller than with 18-C-6 (Figure 3). Molecular mechanics calculations have shown, that the exchange of only one out of six oxygen atoms by a methylene group leads to the almost identical basic conformation (see Figure 3), but to the presence of a single C-H pointing inwards the 18-C-5 cavity, thus generating a steric hindrance of insertion and preventing full contact of the ion with the other oxygen atoms.[15] Such distortions suggest to be careful with the interpretation of experiments with e.g. site-directed mutagenesis, where usually one interacting function is replaced by another one under the assumption that no other significant changes at the binding sites will occur.[16]

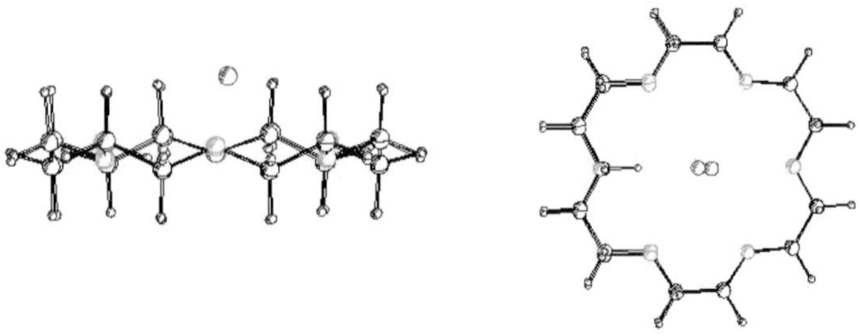

KCl	with 18-C-5:	0.26	with 18-C-6	1 120	K_6/K_5= 4310
$PhNH_3Cl$		0.025		12	480

Figure 3. Stability constants K (in 10^3 M^{-1} units) for 18-C-5 in comparison to 18-C-6 and superposition of K^+ complexes with 18-crown-6 and 18-crown-5, showing the same conformation of both (in side and top view), but the expulsion of the metal ion out of the cavity (after energy minimization in gas phase with MM2, Ph.D. Dissertation of V. Rüdiger, Universität des Saarlandes, Saarbrücken).

Another significant deviation from known rules has been observed in the rather low affinity of K^+ with the 1,10-diaza crown 18-C-6 with only ΔG=10 kJ/mol (in methanol). It has been shown that the free energy of binding ΔG in crown ether and cryptand complexes usually is an additive function of number and electron donicity of the host donor atoms which are in contact with the metal ion.[17] Molecular mechanics calculations suggest the reduced affinity with the diazacrown to be due to the N-lone pairs in pseudoaxial position, pointing away from the metal ion (Figure 4). This has led to experiments with the N-methyl substituted crown: here the N-alkyl substituents would clash which each other inside the macrocycle, therefore a pseudoequatorial lone pair orientation towards the cation is enforced, and the stability of the complex indeed returns to the normal scale with an increase to 29.5 kJ/mol.[18]

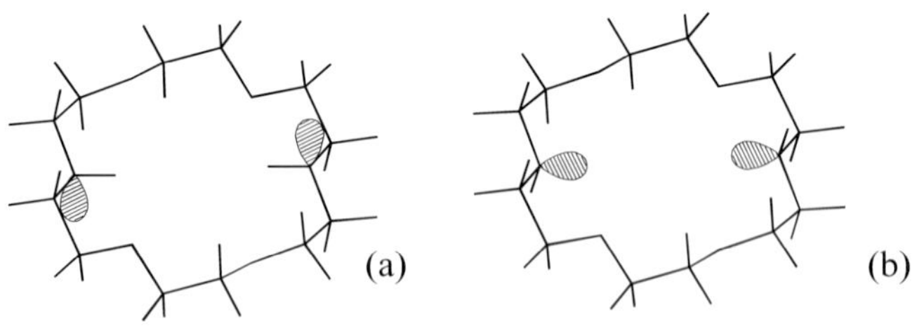

Figure 4. Conformations of 1,10-diaza-18-crown-6 with either diaxial orientation of the N-lone pairs (a), or pseudo-equatorial orientation of the lone pairs (b) (conformation (a) has lower energy if substituent at N is R = H, but higher energy with R = Me, favoring then conformer (b); models after energy minimization in gas phase with MM2, Ph.D. Dissertation of V. Rüdiger, Saarbrücken), Exp. observed (in MeOH): ΔG(complex): for R = H 10.0 kJ/mol; for R = Me : 29.5 kJ/mol.

Of the many other experimental studies revealing at first sight unexpected structures of crown ether complexes we can only cite some recent ones. Unusually weak cation affinities of diaza-18-crown-6 ethers having aromatic sidearms was found to due to interference between the crown's macroring and both sidearms, involving intramolecular C-H···O contacts.[19] In 11-membered ring crown trithioethers the lowest energy conformation are not suitable for tridentate coordination with first row transition metals , so that the lack of ideal preorganization dominate the complexation behavior of the ligands.[20] In the dibenzyl-diaza-18-crown C-H-π interactions between the phenyl groups and ring methylene protons dominate the overall conformation of the macrocycle. [21] Cesium ions in tetrabenzo-24-crown-8 were reported to show in crystals unprecedented linear η^2-acetonitrile and even dichloromethane ligation.[22] In the solid state often disorder and occurrence of several conformations is encountered, as for example in a 1,10-diaza-18-crown-6 complex.[23]

The presence of sulfur atoms in calixarenes, involving only exchanges of methylene groups, lead to large differences in structural and chemical behavior; thus the thiacalix[4]arenes seem to prefer a 1,2-alternate conformation instead of the usual cone structure.[24] Calix[4]arenes themselves can show unusual 1,3-alternate conformations, depending on substitution.[25] On the other hand other calix[4]arenes were found to retain their cone conformation independent of substitution in solid state, in solution and on complexation with alkali metal ions.[26] Mobile calix[4]arene ligands with pendent proton-ionizable groups show different conformations as function of the metal ion; thus Li^+ salts prefer the cone conformation, while for the Na^+ and K^+ salts more than two conformations are observed, and with Cs^+ and Rb^+ ions partial cones.[27] With ketocalixarenes only the use of chiral solvents has made it possible to detect by NMR otherwise silent ring-inversion process, with unusual 1,3-alternate conformations.[28] A resorcinol based calixcrown ether crystallizes in two equally populated conformations, with unusual O-C-C-O and C-C-O-C geometry.[29] Tetraacetylcalix[4]pyrrole was shown to complex in DMSO selectively fluoride vs. acetonitrile, with dramatic effect of water present in the DMSO solution only at lower anion concentrations.[30]

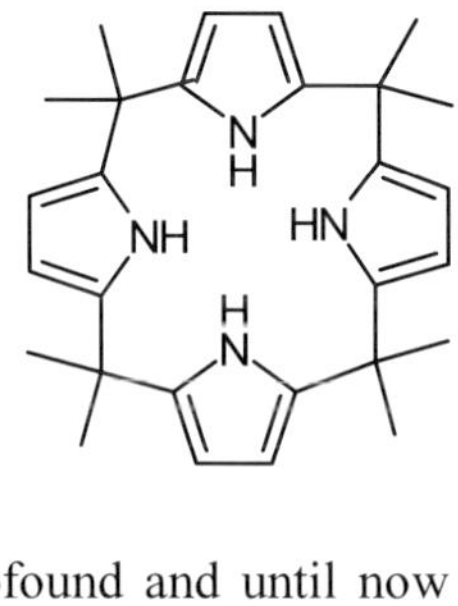

That solvents can have a profound and until now often little understood effect on selectivity and affinity of macrocylic hosts is illustrated with the complexation of halide anions with calix[4]pyrrole (Table 1).[31]

TABLE 1. Dependence of complex stabilities of a calix[4]pyrrole with halide anions on solvent; (i: results from different addition modes)

Halide	lgK in CH_2Cl_2	lgK in DMSO	lgK in MeCN	lgK in MeCN	ΔH in MeCN	ΔH in MeCN	TΔS in MeCN
F^-	4.23	3.02	5.15	6.21	34/43 i	41	8
Cl^-	2.54	3.01	4.90	4.72	44/ 41 i	45	18
Br^-	1.00	-		3.66	-	31	10
Ref.	*29a*	*29b*	*29c*	*29e*	*29c*	*29e*	*29e*

The impressive selectivity for e.g. fluoride over chloride drops from a factor of fifty in dichloromethane to zero in DMSO, and is also small in acetonitrile. Calorimetric measurements show, that the complexation is dominated by enthalpy, but influenced also by entropy disadvantage which is larger for Cl^- than for F^- (it should be noted that several published data differ significantly, also as function of addition modes). Obviously, the prediction of structures and stabilities, but in particular of selectivities in different environment is still a demanding challenge.

4. Thermodynamics / Enthalpic vs. Entropic Contributions

Thermodynamics of complex formation, discussed in detail by Schmidtchen in the present volume, hold many surprises.[32] In comparing enthalpic with entropic parameters one should not forget that ΔG and ΔS depend on the chosen units for relative concentration, which can be either mol/l, or can be given in dimensionless mole fractions; in consequence the usual partition of absolute numbers for ΔH and ΔS becomes to some degree arbitrary (see e.g. ref. 2d, p. 24, p. 106). In addition, complexation is usually characterized by sizeable changes of heat capacity, making the thermodynamic partitions temperature-dependent.

It should be remembered here that the stability increase with macrocycles in comparison to open chain analogs – the so-called macrocyclic effect – is **not** generally due to the commonly discussed entropy advantage. This is obvious from the small differences between complexes of e.g. with either triethylenetetramine or the macrocycle cyclam, which with water as solvent for Cu^{2+} are TΔΔS = -0.8 kJ/mol with a large ΔΔG = -19 kJ/mol, and even TΔΔS = -7.1 kJ/mol with ΔΔG = -16 kJ/mol for Ni^{2+}.[33] With oxygen containing ligands, on the other hand, one sees to a variable degree both enthalpic and moderate entropic advantages of the macrocyclic host.

Remarkably, enhanced binding by the chelate effect is occasionally seen not in free energies of complexation: the association constants of e.g. amino acids with sulfonatocalix[4]arenes are quite close the ones observed with phenol-4-sulfonic acid as single host unit ! Only ΔH shows an advantage of the macrocycle with ΔH = 30 compared to 1 kJ/mol, compensated by a much smaller entropy disadvantage with the smaller monomeric host, which changes from e.g. TΔS = -12 kJ/mol to favorable +12 kJ/mol.[34]

Entropy driven complexation is usually believed to be the hallmark of hydrophobic interactions, but associations in water with cyclodextrin and other strongly lipophilic macrocycles even with hydrophobic guest ligands are invariably driven by enthalpy, most often with large adverse entropy contributions. Strongly enthalpy driven complexations in water were observed with a cyclophane bearing for the sake of solubility external positive charges as host, and 1,4-disubstituted benzene derivatives or steroids as guest. With a similar but more lipophilic cyclophane calorimetric measurements were possible also in other solvents, and showed again a dominating enthalpic driving force, with small positive entropy effects only in benzene or DMF as solvent.[35] In contrast, associations involving either hydrophilic ions or strongly polar interactions, are often endothermic and dominated by favorable entropy contributions, brought about mainly by liberation of solvent molecules from the solvation shell.[2d,36] Complexes of an electroneutral cyclohexapeptide host with anions in methanol-water mixtures are enthalpically as well as entropically favored.[37]

Electroneutral combinations are often favored by enthalpic against entropic contributions, however with noticeable exceptions. Thus, crown ether 18-C-6 complexes with e.g. free amines in methanol show stability constants around lgK= 2.5 ± 0.1, quite independent of the amine structure, but very sizeable differences in ΔH, ranging from 1 or 2 kJ/mol for secondary or tertiary amines to 30 kJ/mol for primary amines; with surprisingly favorable entropic $T\Delta S$ values of up to 13 kJ/mol. Protonated amines show increased lg K values of up to 4.4, mainly due to enthalpic advantages, and not. as one might expect for the more polar combinations, due to entropic effects, although these are also sizeable.[38]

Calorimetric measurements with e.g. urea and various crown ethers, both representing electroneutral, yet polar compounds, showed even in methanol favorable $T\Delta S$ values between 12 to 13 kJ/mol, with ΔH of only 1 to 2 kJ/mol; as expected complexation with the ionic guanidinium salt shows increased values of $T\Delta S$ = 19 to 21 kJ/mol.[39] The experimentally found $T\Delta\Delta S$ differences between urea and guanidinium as guests are found to agree with entropies of fusion[40] for 1 or two methanol molecules ($T\Delta S$ = 4.1 kJ/mol each) which are liberated upon complexation. Surprisingly, association of zwitterionic amino acids and peptides with anionic sulfonatocalix[4]arenes in water [41,42] are favored by enthalpic and disfavored by entropic contributions, with negligible influence of the macrocycle ring size on the affinities.[34]

In water, related sulfonatocalixarene complexes with e.g. benzylammonium salts showed by NMR also a complexation mode without ion pairing, with the phenyl moiety inserted in the calix cone, as well as a conformation allowing a salt bridge; the observed thermodynamics were more in line with the absence of dominating ion pair interactions (Figure 5).[43]

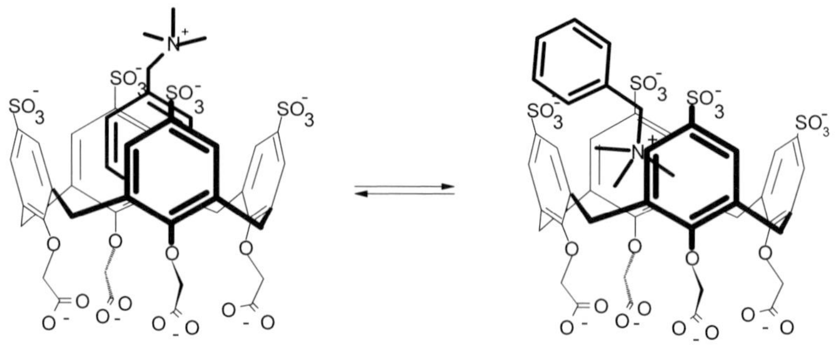

Figure 5. Two conformations in a calixarene complex (ref.42).

Metal complexes also can show contradictive behavior: with benzo-15-crown-5 the complexation of lanthanide perchlorates is entropy-driven in acetonitrile, while the complexation of lanthanide nitrates with corresponding disubstituted derivatives is primarily enthalpy-driven, with small entropic differences.[44]

Obviously, reliable predictions for thermodynamics even for long-established crown ether complexes are a far-reaching goal. Few efforts have been made until now to correlate quantitatively the often dominating complex entropies with independent parameters, such as with entropies of fusion for the used solvent molecules,[40] and known values for solvation and desolvation of the partner molecules.[45] Restriction of conformational mobility within host or guest molecules has experimentally[46] been shown to hinder complex formation less than anticipated earlier on the basis of entropic terms by loss of rotational freedom.[47] Free energies of complexation are usually better suited for stability predictions than enthalpy or entropy parameters, also as a result of the well known enthalpy-entropy compensation;[48] in particular ΔG values often show additivity of the many non-covalent binding contributions present in a macrocycle.[49]

5. *Van der Waals* or Hydrophobic Effects ?

Interactions between lipophilic partners in aqueous media are often ascribed to hydrophobic forces, although they may as well be due to *van der Waals*-type interactions. The cyclophane **CP-a** with X= $^{+}NR_2$ (R = Me or H) was found to bind saturated compounds with much smaller affinity than aromatic guest molecules; only after removing as in **CP-b** the positively charged nitrogen centers of the host the reverse is seen, as expected by the higher lipophilicity of aliphatic guest molecules (Figure 6).[50] Similarly, the cyclophane **CD-a** binds guest molecules containing positively charged nitrogen centers with remarkable affinities, with significantly increased complex stability if the host walls are made completely aromatic (**CD-b**) instead of aliphatic.[51] These observations initiated the discovery of the now well-established cation-π-interactions in many organic and biological complexes,[52] which until then were known only from e.g. mass-spectroscopic studies of K^+-benzene complexes in the gas state.[53]

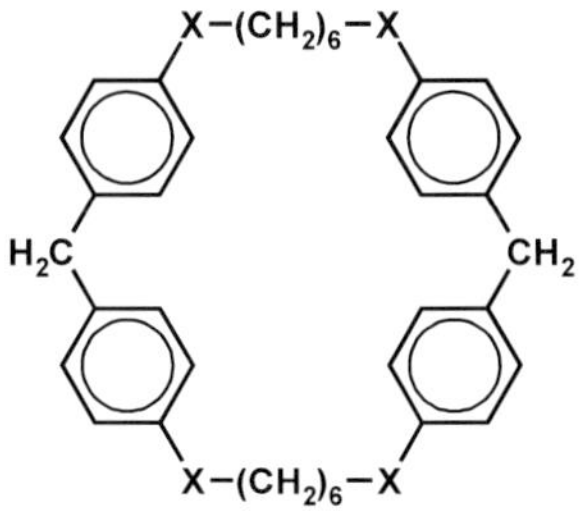

CP

CP-a: X = $^{+}NMe_2$

CP-b: X =

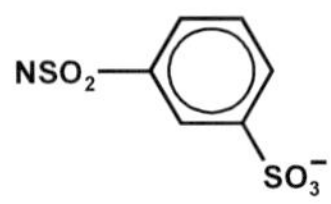

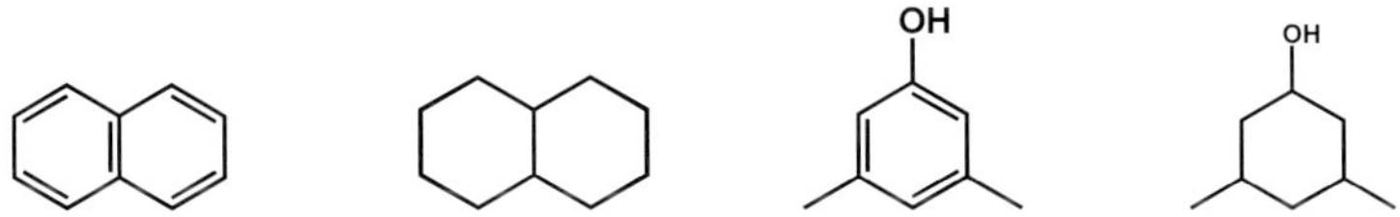

CP-a:	17	7	11	~3
CP-b:	13	-	6.5	~4

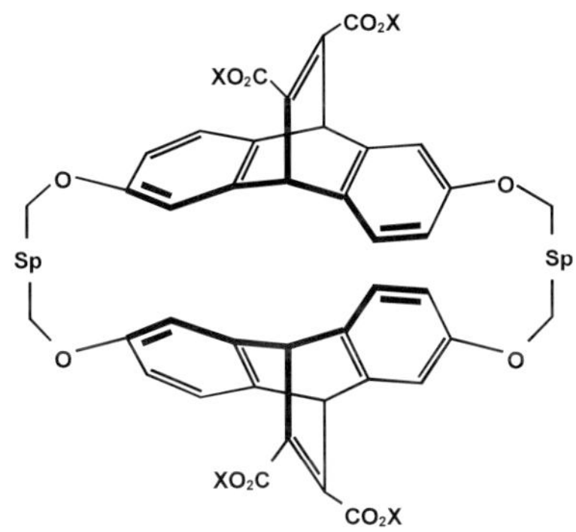

CD-a: Sp = 1,4-CH_2-Ph-CH_2, X = Cs
CD-b: Sp = 1,4-Cyclohexyl, X = Cs

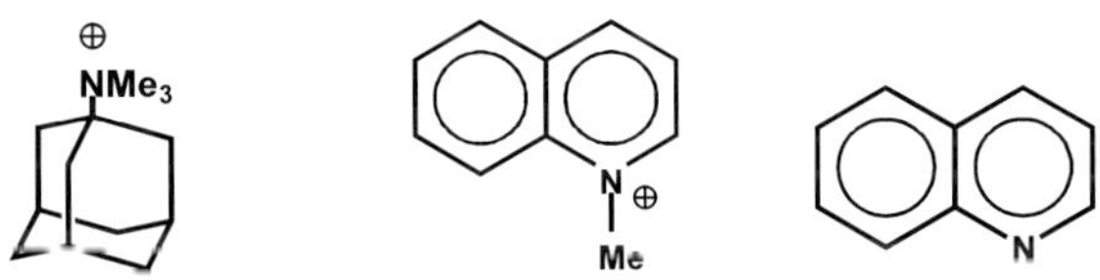

with **CD-a:** X = Cs^+	28	32	22	(in D_2O)
with **CD-b:** X = Me	9	14.5	-	(in $CDCl_3$)

Figure 6. Evidence for the cation-π effect in cyclophane complexes (all complexation free energies ΔG [kJ/mol], in water).

A dissection of hydrophobic from *van der Waals* interactions is possible by comparing the affinity of saturated with more polarizable guest molecules with suitable host compounds. Using water-soluble porphyrin host compounds it has been found, that saturated systems such as 1,4-cyclohexane dicarboxylates bind only by ion pairing to the porphyrin macrocycle, which bears positive charges at the periphery. In contrast, the almost isosteric 1,4-benzene-dicarboxylate binds by orders of magnitude better. In addition, many substituents at benzoic acid as guest lead to significant complexation stability increases, essentially as function of their polarizibility, whereas e.g. a methyl or isopropyl group has only a negligible effect (Figure 7).[54] Larger hydrophobic contributions have reported for associations between porphyrins and peptides;[55] however, they can be due to formation of 1:2 complexes and to simultaneous dispersive interactions with the backbone amide groups.

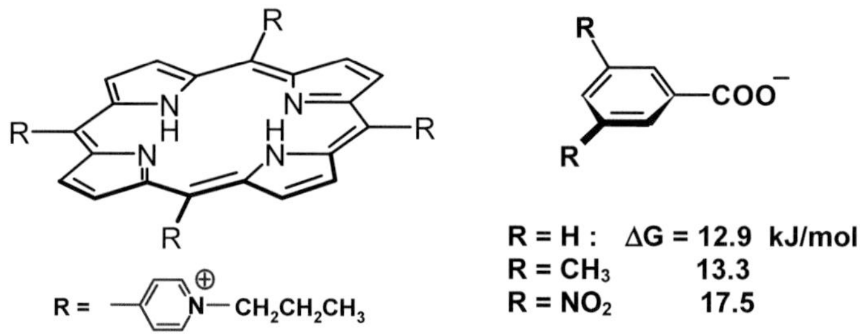

Figure 7. Complexation free energies ΔG of the porphyrin macrocycle with benzoic acid derivatives, showing the absence of hydrophobic effects with R = CH_3, and the large effect of polarizable groups such as R = NO_2.

The observed free energy changes are additive and demonstrate, that many interactions summarized as hydrophobic are in fact of dispersive nature.[56] The hydrophobic effect itself may have quite different origins. As mentioned above the classic hydrophobic effect in terms of the Frank-Evans picture of water liberation is in conflict with many calorimetric data of macrocyclic complexes such as cyclodextrins, showing the absence of entropic driving contributions.[35,36] The so-called non-classical hydrophobic effect has its origin in cohesive forces between solvent molecules, which for water are particularly large.[57, 35] Release of solvent molecules out of a macrocyclic cavity upon complexation with a guest molecule inside allows the liberated solvent molecules not only more freedom (the classical, entropic effect), but also more solvent-solvent interactions, leading to the more often observed enthalpic gain.

6. Acknowledgements

The work of my able collaborators in Saarbrücken was supported by the Deutsche Forschungsgemeinschaft, the A.v. Humboldt foundation and the Fonds der Chemischen Industrie. I appreciate valuable suggestions by Professor A. Yatsimirksi, UNAM Mexico.

7. References and Notes

[1] Popper, K. R. (**1959**, new edition: London, Routledge, 2002) *The Logic of Scientific Discovery.*

[2] a) J.-M. Lehn, *Supramolecular Chemistry. Concepts and Perspectives,* Wiley VCH Weinheim etc **1995;** b) F. Vögtle, *Supramolecular Chemistry: An Introduction,* Wiley **1993;** c) P. D. Beer, P. A. Gale, D. K. Smith, Supramolecular Chemistry (Oxford Chemistry Primers, 74) Oxford University Press; d) H.-J. Schneider and A. Yatsimirski, *Principles and Methods in Supramolecular Chemistry,* Wiley Chichester etc, **2000;** e) J. W. Steed and J. L. Atwood, *Supramolecular Chemistry,* Wiley New York etc, **2000;** f) *Comprehensive Supramolecular Chemistry*, Vol. 1-11; J. M. Lehn, J. L. Atwood, J. E. D. Davies, D. D. MacNicol, D. D. MacNicol and F. Vögtle, eds., Pergamon/Elsevier Oxford etc, **1996**.

[3] Stetter, H. and Roos, E. E. (1955): Macrocyclic ring systems. II. Bis[N,N'-Alkylenebenzidine], *Chem.Ber.* ***88***, 1390-1395.

[4] Hilgenfeld and R.; Saenger, W. (1982) Stetter's Complexes are no Intramolecular Inclusion Compounds, *Angew. Chem. Int. Ed. Engl.* ***21***, 1690-1701.

[5] Wald, P.and Schneider, H.-J.; unpublished results.

[6] Ciampolini, M., Dapporto, P., and Nardi, N. (1979): Structure and properties of some lanthanoid(III) perchlorates with the cryptand 4,7,13,16,21,24-hexaoxa-1,10-diazabicyclo[8.8.8]hexacosane, *Dalton Trans,* 974-977; Hart, F. A., Hursthouse, M. B., Abdul Malik, K. M., and Moorhouse, S. (1978): X-ray crystal structure of a cryptate complex of lanthanum nitrate, *Chem. Commun.* 549-50.

[7] Shestakova, A.K., Chertkov, V.A., Schneider, H.-J. (2000) Structures and equilibria involving the [222] cryptand and europium ions, *Tetrahedron Lett.* ***41***, 6753-6756.

[8] Szumna, A., and Jurczak, J. (2001): Unusual encapsulation of two anions in the cavity of neutral macrocyclic octalactam, *Helv. Chim. Acta* **84**, 3760-3765.

[9] See e.g. Gokel, G. W., Leevy, W. M., and Weber, M. E. (2004): Crown ethers: Sensors for ions and molecular scaffolds for materials and biological models, *Chem. Rev.* ***104***, 2723-2750.

[10] Grootenhuis, P. D. J., and Kollman, P. A. (1989): Crown ether-neutral molecule interactions studied by molecular mechanics, normal mode analysis, and free energy perturbation calculations. Near quantitative agreement between theory and experimental binding free energies, *J. Am. Chem. Soc.* ***111***, 4046-4051.

[11] Wipff, G. (1992): Molecular Modeling Studies on Molecular Recognition - Crown Ethers, Cryptands and Cryptates - from Static Models in Vacuo to Dynamic Models in Solution, *J. Coord. Chem.* ***27***, 7-37; and chapter in this volume.

[12] Schatz, J. (2004): Recent application of ab initio calculations on calixarenes and calixarene complexes. A review, *Coll. Czech. Chem. Comm.* ***69***, 1169-1194.

[13] Dang, L. X. (1995): Mechanism and Thermodynamics of Ion Selectivity in Aqueous- Solutions of 18-Crown-6 Ether - a Molecular-Dynamics Study, *J. Am. Chem. Soc.* ***117***, 6954-60.

[14] Timko, J. M.; Moore, S. S.; Walba, D. M.; Hiberty, P. C.; and Cram, D. J., (1977) Structural units that control association constants between polyethers and tert-butylammonium salts, *J. Am. Chem. Soc.*, ***99***, 4207-4219.

[15] Raevsky, O. A., Solovev, V. P., Solotnov, A. F., Schneider, H. J., and Rüdiger, V. (1996): Conformation of 18-crown-5 and its influence on complexation with alkali and ammonium cations, *J. Org. Chem.* ***61***, 8113-8116.

[16] Thus, it has been found with complexes of vancomycin with a mutated peptidoglycan that exchange of a single amide group by an ester group reduces the affinity constant by a factor of 1000; this has been ascribed to various structural changes accompanying this single point mutation; see Walsh, C. T., Fisher, S. L., Park, I. S., Prahalad, M., and Wu, Z. (1996): Bacterial resistance to vancomycin: five genes and one missing hydrogen bond tell the story, *Chem. & Biol.* ***3***, 21-8; McComas, C. C., Crowley, B. M., and Boger, D. L. (2003): Partitioning the Loss in Vancomycin Binding Affinity for D-Ala-D-Lac into Lost H-Bond and Repulsive Lone Pair Contributions, *J. Am. Chem. Soc.* ***125***, 9314-9315.

[17] Schneider, H. J., Rüdiger, V., and Raevsky, O. A. (1993): The incremental description of host-guest complexes: free energy increments derived from hydrogen bonds applied to crown ethers and cryptands, *J. Org. Chem.* ***58***, 3648-3653.

[18] Solov'ev, V. P., Strakhova, N. N., Kazachenko, V. P., Solotnov, A. F., Baulin, V. E., Raevsky, O. A., Rüdiger, V., Eblinger, F., and Schneider, H. J. (1998): Steric and stereoelectronic effects in aza crown ether complexes, *Eur. J. Org. Chem.* 1379-1389.

[19] Meadows, E. S., De Wall, S. L., Barbour, L. J., Fronczek, F. R., Kim, M. S., and Gokel, G. W. (2000): Structural and dynamic evidence for C-H···O hydrogen bonding in lariat ethers: Implications for protein structure, *J. Am. Chem. Soc.* ***122***, 3325-3335

[20] Grant, G. J., Shoup, S. S., Baucom, C. L., and Setter, W. N. (2001): Complexation and conformational analysis studies of 11-membered ring crown trithioethers, *Inorg. Chim. Acta* ***317***, 91-102.

[21] Evans, D. J., Junk, P. C., and Smith, M. K. (2002): Intramolecular C-H center dot center dot center dot π interactions influence the conformation of N,N'-dibenzyl-4,13-diaza-18-crown-6 molecules, *New J. Chem.* ***26***, 1043-1048.
[22] Bryan, J. C., Kavallieratos, K., and Sachleben, R. A. (2000): Unusual ligand coordination for cesium, *Inorg. Chem.* ***39***, 1568-1572.
[23] Chekhlov, A. N. (2000): Crystal structure of the 1,10-diaza-18-crown-6 complex with calcium thiocyanate, *Russ. J. Coord. Chem.* ***26***, 157-161.
[24] Lhotak, P. (2004): Chemistry of thiacalixarenes, *Eur. J. Org. Chem*, 1675-1692.
[25] Simaan, S., Agbaria, K., Thondorf, I., and Biali, S. E. (2003): Conformation of syn- and anti-phenylquinazoline calix[4]arene diethers, *New J. Chem.* ***27***, 236-238.
[26] Creaven, B. S., Deasy, M., Gallagher, J. F., McGinley, J., and Murray, B. A. (2001): Unusual cone conformation retention in calix[4]arenes, *Tetrahedron* ***57***, 8883-8887.
[27] Talanov, V. S., Hwang, H. S., and Bartsch, R. A. (2001): Unusual conformational control of mobile mono- and diionizable calix[4]arene ligands by alkali metal cations, *Perkin Trans. 2*, 1103-1108
[28] Seri, N., Simaan, S., Botoshansky, M., Kaftory, M., and Biali, S. E. (2003): A conformationally flexible tetrahydroxycalix[4]arene adopting the unusual 1,3-alternate conformation, *J. Org. Chem.* ***68***, 7140-7142.
[29] Buchanan, G. W., Laister, R. C., and Yap, G. P. A. (2000): Stereochemical investigations of resorcinol based crown ethers: X-ray crystal structure, molecular mechanics and NMR studies of benzo-13-crown-4 ether, *J. Mol. Struct.* ***523***, 261-268.
[30] Woods, C. J., Camiolo, S., Light, M. E., Coles, S. J., Hursthouse, M. B., King, M. A., Gale, P. A., and Essex, J. W. (2002): Fluoride-selective binding in a new deep cavity calix[4]pyrrole: Experiment and theory, *J. Am. Chem. Soc.* ***124***, 8644-8652.
[31] a) Gale, P. A., Sessler, J. L., Kral, V., and Lynch, V. (1996): Calix[4]pyrroles: Old yet new anion-binding agents, *J. Am. Chem. Soc.* 118, 5140-5141; b) Camiolo, S., and Gale, P. A. (2000): Fluoride recognition in 'super-extended cavity' calix[4]pyrroles, *Chem. Commun.* 1129-1130; c) Schmidtchen, F. P. (2002): Surprises in the energetics of host-guest anion binding to calix[4]pyrrole, *Org. Lett.* ***4***, 431-434; d) Danil de Namor, A. F. D., and Shehab, M. (2003): Selective recognition of halide anions by calix[4]pyrrole: A detailed thermodynamic study, *J .Phys. Chem. B* ***107***, 6462-6468; e) Sessler, J. L., Camiolo, S., and Gale, P. A. (2003): Pyrrolic and polypyrrolic anion binding agents, *Coord. Chem. Rev.* ***240***, 17-55.
[32] Danil de Namor, A. F. D., Cleverley, R. M., and Zapata-Ormachea, M. L. (1998): Thermodynamics of calixarene chemistry, *Chem. Rev.* ***98***, 2495-2525.
[33] Compiled data see ref . 2d, p. 15.
[34] Buschmann, H. J., Mutihac, L., and Schollmeyer, E. (2003): Complexation of some amino acids and peptides by p- sulfonatocalix[4] arene and hexasodium p-sulfonatocalix[6] arene in aqueous solution, *J. Incl. Phenom. Macrocycl. Chem.* ***46***, 133-137; see also Douteau-Guevel, N., Perret, F., Coleman, A. W., Morel, J. P., and Morel-Desrosiers, N. (2002): Binding of dipeptides and tripeptides containing lysine or arginine by p-sulfonatocalixarenes in water: NMR and microcalorimetric studies, *Perkin Trans. 2* 524-532.
[35] Meyer, E. A., Castellano, R. K., and Diederich, F. (2003): Interactions with aromatic rings in chemical and biological recognition, *Angew. Chem. Int. Ed. Engl.* ***42***, 1210-1250.
[36] Schneider, H.-J. (2004), (a) "Hydrophobic Effects"; (b) "Van der Waals Effects" in: *Encyclopedia of Supramolecular Chemistry*, Atwood, J. L.; Steed, J. W., eds; Marcel Dekker, Inc., New York; (c) Southall, N. T., Dill, K. A., and Haymet, A. D. J. (2002): A view of the hydrophobic effect, *J. Phys. Chem. B* ***106***, 521-533, and references cited therein.
[37] Kubik, S., Kirchner, R., Nolting, D., and Seidel, J. (2002): A molecular oyster: A neutral anion receptor containing two cyclopeptide subunits with a remarkable sulfate affinity in aqueous solution, *J. Am. Chem. Soc.* ***124***, 12752-12760.
[38] Buschmann, H. J., Mutihac, L., and Jansen, K. (2001): Complexation of some amine compounds by macrocyclic receptors, *J. Incl. Phenom. Macrocycl. Chem.* ***39***, 1-11.
[39] Buschmann, H. J., Dong, H., Mutihac, L., and Schollmeyer, E. (1999): Complex formation between crown ethers and urea and some guanidinium derivatives in methanol, *J. Thermal Anal. Calor.* ***57***, 487-491.
[40] Hefter, G., Marcus, Y., and Waghorne, W. E. (2002): Enthalpies and entropies of transfer of electrolytes and ions from water to mixed aqueous organic solvents, *Chem. Rev.* ***102***, 2773-2835; and references cited therein.
[41] Arena, G., Contino, A., Gulino, F. G., Magri, A., Sansone, F., Sciotto, D., and Ungaro, R. (1999): Complexation of native L-α-aminoacids by water soluble calix[4]arenes, *Tetrahedron Lett.* ***40***, 1597-1600.
[42] Kalchenko, O. I., Perret, F., Morel-Desrosiers, N., and Coleman, A. W. (2001): A comparative study of the determination of the stability constants of inclusion complexes of p-sulfonatocalix [4]arene with amino acids by RP-HPLC and H-1 NMR, *Perkin Trans. 2* 258-263.
[43] Arena, G., Casnati, A., Contino, A., Lombardo, G. G., Sciotto, D., and Ungaro, R. (1999): Water-soluble calixarene hosts that specifically recognize the trimethylammonium group or the benzene ring of aromatic

ammonium cations: A combined H-1 NMR, calorimetric, and molecular mechanics investigation, *Chem. Eur. J.* ***5***, 738-744.

[44] Liu, Y., Zhang, H. Y., Bai, X. P., Wada, T., and Inoue, Y. (2000): Molecular design of crown ethers. 21. Synthesis of novel double-armed benzo-15-crown-5 lariats and their complexation thermodynamics with light lanthanoid nitrates in acetonitrile, *J. Org. Chem.* ***65***, 7105-7109; see also Danil de Namor, A. F. D., Chahine, S., Jafou, O., and Baron, K. (2003): Solution thermodynamics of lanthanide-cryptand 222 complexation processes, *J. Coord. Chem* ***56***, 1245-1255; Israeli, Y., Bonal, C., Detellier, C., Morel, J. P., and Morel-Desrosiers, N. (2002): Complexation of the La(III) cation by p-sulfonatocalix[4]arene - A La-139 NMR study, *Canad. J. Chem.* ***80***, 163-168.

[45] Buschmann, H. J., and Schollmeyer, E. (2000): The interpretation of thermodynamic data for the complex formation of cations with crown ethers and cryptands. Part I: The reaction entropy, *J. Incl. Phenom. Macrocycl. Chem.* ***38***, 85-97.

[46] Eblinger, F.; Schneider, H.-J. (1988), Stabilities of Hydrogen-Bonded Supramolecular Complexes with Variable Numbers of Single Bonds. Attempts to Quantify a Dogma in Host-Guest Chemistry, *Angew. Chem. Int. Ed. Engl.* ***37***, 826-829; Hossain, M. A. S., and Schneider, H. J. (1999): Flexibility, association constants, and salt effects in organic ion pairs: How single bonds affect molecular recognition, *Chem. Eur. J.* ***5***, 1284-1290.

[47] Jencks, W. P. (1981): On the attribution and additivity of binding energies, *Proc. Nat. Acad. Sci. USA* ***78***, 4046-50.; Searle, M. S., and Williams, D. H. (1992): The cost of conformational order: entropy changes in molecular associations, *J. Am. Chem. Soc.* ***114***, 10690-7, and references cited therein.

[48] see e.g. a) Liu, L., and Guo, Q. X. (2001): Isokinetic relationship, isoequilibrium relationship, and enthalpy-entropy compensation, *Chem. Rev.* ***101***, 673-695; b) Liu, Y., Han, B. H., and Chen, Y. T. (2000): The complexation thermodynamics of light lanthanides by crown ethers, *Coord. Chem. Rev* ***200***, 53-73; Rekharsky, M. V., and Inoue, Y. (1998): Complexation thermodynamics of cyclodextrins, *Chem. Rev.* ***98***, 1875-1917; Terekhova, I. V., Lapshev, P. V., and Kulikov, O. V. (2003): Thermodynamics of α-cyclodextrin complexation with bases of nucleic acids and their derivatives, *Russian J. Coord. Chem.* ***29***, 73-75.

[49] Schneider, H. J. (1994): Linear Free-Energy Relationships and Pairwise Interactions in Supramolecular Chemistry, *Chem. Soc. Rev.* ***23***, 227-234.

[50] Schneider, H.-J.; Blatter, T.(1988) Modification of Hydrophobic and Polar Interaction by Charged Groups in Synthetic Host-Guest Compounds, *Angew. Chem., Int. Ed. Engl.* ***27***, 1163-4; Schneider, H.-J.; Blatter, T.; Simova, S.; Theis, I. (1989): Large Binding Constant Differences between Aromatic and Aliphatic Substrates in Positively Charged Cavities Indicative of Higher Order Electrical Effects, *Chem.Commun.* 580-1.

[51] Shepodd, T. J.; Petti, M. A.; Dougherty, D. A. (1988) Molecular recognition in aqueous media: donor-acceptor and ion-dipole interactions produce tight binding for highly soluble guests, *J. Am. Chem. Soc.* ***110***, 1983-1985.

[52] Ma, J. C., and Dougherty, D. A. (1997): The cation-π interaction, *Chem. Rev.* ***97***, 1303-1324.

[53] Sunner, J., Nishizawa, K., and Kebarle, P. (1981): Ion-solvent molecule interactions in the gas phase. The potassium ion and benzene, *J. Phys. Chem.* ***85***, 1814-20.

[54] Schneider, H. J., Tianjun, L., Sirish, M., and Malinovski, V. (2002): Dispersive interactions in supramolecular porphyrin complexes, *Tetrahedron* ***58***, 779-786.

[55] Huffman, D. L., and Suslick, K. S. (2000): Hydrophobic Interactions in Metalloporphyrin-Peptide Complexes, *Inorg. Chem.* ***39***, 5418-5419.

[56] Liu, T. J., and Schneider, H. J. (2002). Additivity and quantification of dispersive interactions-from cyclopropyl to nitro groups: Measurements on porphyrin derivatives, *Angew. Chem. Int. Ed. Engl.* ***41***, 1368-1370.

[57] Jencks, W.P. (**1989**) *Catalysis in Chemistry and Enzymology*, McGraw Hill, New York.

19.
CALORIMETRY: AN INDISPENSIBLE TOOL IN THE DESIGN OF MOLECULAR HOSTS

FRANZ P. SCHMIDTCHEN
Technical University of Munich, Department of Chemistry, 85747 Garching, Germany

1. Introduction

The design of molecular host compounds is at the heart of supramolecular chemistry. In essence we strive to construct and assemble covalently bound sets of atoms that are capable of performing certain functions better than any other ensemble under given conditions. In order to approach this goal we look out for conceptual ideas that in any generality will not reflect the entirety of the true or putative facets of the real system but will rather be a xylographic projection of the bare necessities supposed to govern the desired properties. We seek a model, a faithful yet simple representation of the original that undoubtedly must contain a bias, but by virtue of the reduction in complexity enables the physical realization of the underlying idea. Even though the model might describe a systems behaviour in one particular situation it is a consequence of its fictitious nature that it may totally fail in another. Learning about the limitations of a model is as important for success in molecular design as is the parent concept, because it helps to avoid futile efforts and false reasoning.

2. On the Role of Calorimetry

A powerful tool in this respect is calorimetry since the measurement of heat effects addresses the most fundamental result of an interaction between chemical species. Whenever there is an interaction between non-ideal particles there is an energetic exchange which is characteristic in sign and magnitude of the microscopic event within the system and is communicated to the outside (ΔH). In all real cases relevant in macroscopic chemistry the determination of the enthalpy of a supramolecular process does not allow direct back tracing to identify and quantify the molecular origin, because the observable is the composite result of myriads of basic events encompassing all components of the system and not just the species of interest to the investigator. On the basis of averaging over truly huge numbers of individual species and events the measurable outcome is highly reproducible and thus constitutes a reliable platform to build any real application upon.

Karsten Gloe (ed.), Macrocyclic Chemistry: Current Trends and Future Perspectives, 291–302.
© 2005 *Springer. Printed in the Netherlands.*

For weak interactions as they occur in host-guest-binding it is important to realize a conceptual difference between the experimental value that is based on the weighted average ensemble of species and the merging of this ensemble into one more or less representative structure that is then commonly used in the delineation of interaction types between the binding partners.

This virtual shrinking in applying a model condenses and removes many of the quite subtle features of the binding partners that are certainly relevant in their mutual interaction and are present in somewhat disguised form in the experimental result but cannot be accounted for and in principle cannot even be recognized from model reasoning unless dedicated measures are taken. Thus, detecting and respecting the limits of applicability of a model is of paramount necessity.

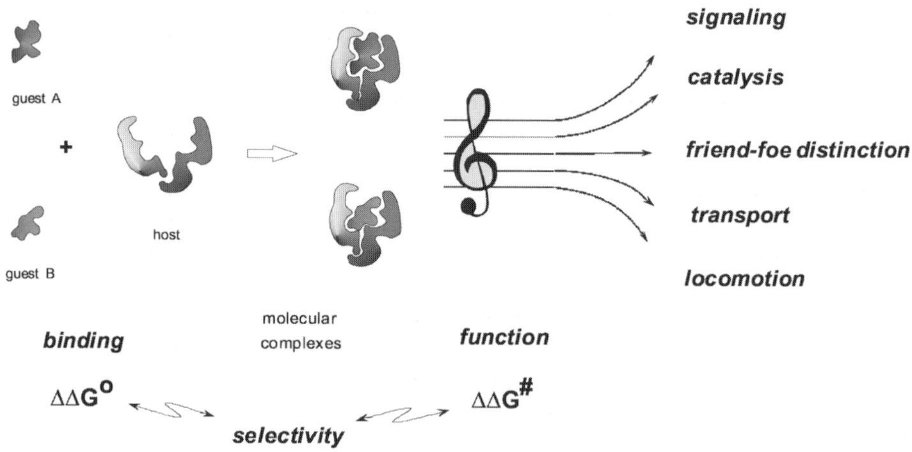

Figure 1. The selectivity of molecular recognition addresses two steps: The equilibrium binding between the host and the competing guest species A and B and their subsequent conversion. Either step may be dominant in selectivity generation.

In spite of the lack of immediate expressiveness in terms of the molecular events calorimetry is an invaluable aid in the elucidation of molecular recognition as it poses blunt restrictions on any explanatory attempt. However, in order to tap its usefulness one prerequisite must be met that is not self-evident in molecular recognition processes at large and is even constitutively absent in rapidly changing systems as they occur in the biological world: Reversibility! [1] To the extent that molecular recognition emerges from different rates on competing pathways the measurement of heat effects in the ground state may be of quite limited value. If instead discriminatory action (selectivity) preponderantly arises from equilibrium processes prior to the product generating step (this can be a new chemical species as resulting from catalysis, but also a physical signal, a flux across a barrier or a vectorial translation on a surface, see Figure 1) the influential factors of differential binding are open to energetic inspection. In many cases equilibrium conditions can be readily established and checked and even form the physical basis for selectivity as e.g. in two-phase extractions. In others the kinetic process is very fast and sensible to the molecular environment to yield an instantaneous response on the equilibrium status as in analyte sensing. These applications are so prominent that molecular recognition is frequently looked upon as a time and concentration independent status which can be expressed as a ratio of formation constants. This is clearly a misconception as pointed out above, because it deliberately suppresses the kinetic aspect,[2] which certainly is the dominant factor in a broad variety of applications from catalysis to membrane transport or to the friend-foe distinction in the immune response where some binding event beyond doubt is mandatory, yet, the decisive action is the removal of the antigen from the body which is subject to a series of kinetically controlled steps.

Taking the ratio of binding constants as a measure of molecular recognition is only justified to ease the comparison of different host designs since it refers to the rather odd situation when two guest species compete for the same site on the host in the *absence* of the respective competitor. The relation builds on the assumption that no complexes of higher stoichiometries than in a 1:1 host-guest-relationship are relevant in association. Furthermore, also no cooperative effects (be them positive or negative) triggered by the mere presence of the competitor should occur which is the more unlikely the more subtle and flexible (adaptable to different molecular situations) the host is. Negligence in the proper appreciation of the various caveats may lead to erroneous conclusions as was demonstrated recently in a couple of concrete cases. [3,4] In total, the comparison of binding constants yields a rather skew and distorted image of the genuine molecular recognition processes, yet the simplicity of the procedure and the availability of the binding constants renders this way the preferred mode of assessing host design.

Calorimetry can readily contribute to the production of the required data on this stage since both methodological distinct versions, differential scanning calorimetry (DSC) and isothermal titration calorimetry (ITC) provide equilibrium constants of the supramolecular events from the direct fit of the prima facie observables to a binding scheme. [5-8] There is benefit and burden combined owing to the fundamental nature of these methods that rest on heat effects. On one hand they reflect in the most encompassing way the molecular processes running all simultaneously in the system and thus they mirror the true energetics without retreat to any structural peculiarity (e. g. a chromophore or a special NMR signal) and they do so in particularly sensitive fashion (association constants in the $10^2 - 10^9$ M^{-1} range can be directly determined and stronger or weaker complexes are accessible by indirect variants) [9,10]. On the back side calorimetry also senses all non-specific interactions which makes the measurement very sensible to impurities and to the exact adjustment of the composition of third-party-ingredients (buffer, salts, solvent, counterions etc.) Naturally these influences gain weight the less specific and strong the host-guest-complexes under study are, since simple dilution, the most effective way to elude interfering complications, would also render the desired interaction unobservable.

In addition to binding constants that emerge from a fitting process itself requiring the input of a binding model that even in the case of a good fit might not resolve the ambiguities in the underlying molecular scenario [11] calorimetry yields enthalpies or heat capacities that are not readily available by any other means. Owing to the direct method of determination calorimetric enthalpies do not require model assumptions or excessive extrapolations that put the so called van't Hoff enthalpies obtained from the temperature dependence of the equilibrium constants into question [12]. Connecting enthalpies ΔH with Gibbs enthalpies ΔG via the Gibbs-Helmholtz equation ($\Delta G = \Delta H - T\Delta S$) yields the change in the system's entropy ΔS at the temperature of the measurement, a piece of information that is dramatically underappreciated in the design of artificial host compounds. As molecular chemistry traditionally was engaged in the construction and rearrangement of covalent molecular frameworks addressing enthalpy-dominated processes [13] it was safe to neglect entropic factors. This is improper in supramolecular chemistry, because here enthalpic and entropic influences frequently balance each other (enthalpy-entropy compensation [14]) undermining all rationalizations on the exclusive basis of attractive and repulsive binding interactions.

Though entropies in principle can be determined from the temperature dependence of the heat capacity starting from zero Kelvin, in all practical instances it is obtained by calculation from ΔG and ΔH. Since both state functions are available from the same calorimetric measurement e.g. by ITC, this provides also the immediate access to the relevant entropy change possessing no equally precise alternative way of determination [12].

The following examples were selected from our own work in order to illustrate the power, however, also the shortcomings of titration calorimetry in the investigation of artificial host-guest-binding. The case studies are directed to the question of solvent participation in hydrophobic complexation by α-cyclodextrin and the use of solvent isotope effects to find out about supramolecular enantiodifferentiation. The prominent guanidinium-oxoanion interaction on the other hand is taken to explore the correctness of the purported singularity of the binding mode in this widely-spread supramolecular motif.

3. Estimating Solvent Contributions to Host-guest Binding

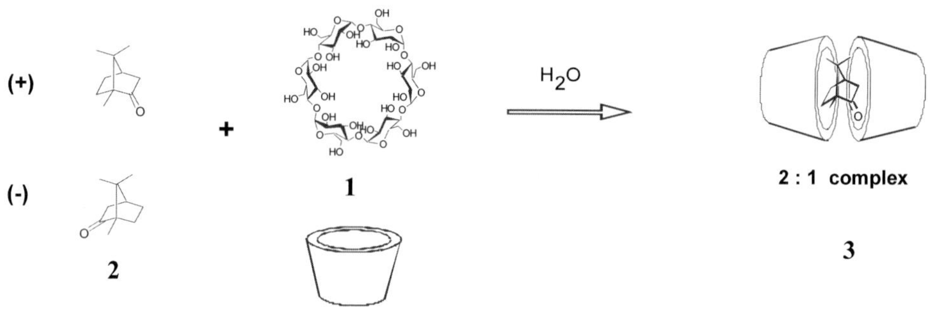

α-cyclodextrin **1** (α-CD) forms complexes with both enantiomers of camphor **(+)2, (-)2** in water. Chromatographic and NMR-analysis [16,17] gave evidence of 2:1 stoichiometric complexes which result in a highly synergistic process, the second step yielding a 10000-fold higher affinity constant than the initial 1:1 binding. Based on NMR-data [17] and the comparison to the complexation of other guest molecules by α-CD the camphor guest is occluded in a cavity prepared by the toroidal host compounds. This mechanism requires that most if not all water molecules solvating the host and guest contact surface must be stripped off on guest encapsulation. Since the host-guest interface area, however, not the rim region of the host molecules mediating the host-host contacts is almost entirely hydrophobic, it seems safe to apply a peculiarity of hydrophobic hydration: The free enthalpy of hydration is unaffected on switching the solvent from light to heavy water (a solvent isotope effect, SIE). This experimental observation arises from an almost exact compensation of the change in enthalpy by a counteracting change in entropy (solvent isotope effect: $\Delta G^0_{H_2O} - \Delta G^0_{D_2O} = \Delta\Delta G^0_{SIE} \sim 0$; $\Delta\Delta H^0_{SIE} \sim T\Delta\Delta S^0_{SIE}$).

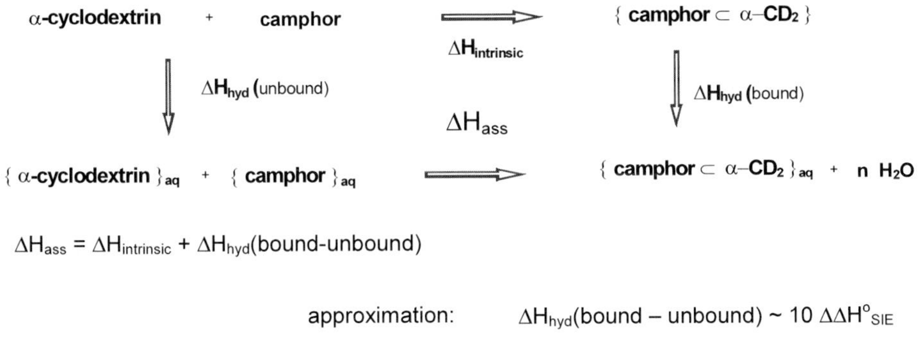

Figure 2. Thermodynamic cycle to relate the observable ΔH_{ass} to the change in solvation ΔH_{hyd}(bound-unbound) and the direct mutual interaction $\Delta H_{intrinsic}$.

It is well known from hydration studies in the gas phase that the hydrogen bond in D_2O is about 10% stronger than in H_2O resulting in a distinct differential heat effect when a hydrophobic surface is desolvated from light versus heavy water.[18] In as far as the total interaction enthalpy between water molecules emerge solely from hydrogen bonding (which is a good approximation) the measurement of the solvent isotope effect allows the estimation of the enthalpy contribution of solvent reorganisation to the total observable enthalpy change ΔH_{ass}.[19] The thermodynamic cycle (Figure 2) illustrates the relations which shows the measurable quantity ΔH_{ass} as a merger of the intrinsic interactions of host and guest, $\Delta H_{intrinsic}$, that remains unaffected by a solvent change and the difference in the hydration enthalpies of the bound and unbound partners ($\Delta H_{(bound\text{-}unbound)}$). It is this contribution that is estimated as ten times the enthalpic solvent isotope effect. In the calorimetric measurement the system is well behaved (cf. Figure 3) and furnishes the following energetic parameters for the complexation of (-)-camphor at 298 K : $\Delta G^0 = -31.4$ kJ mol^{-1}; $\Delta H^0 = -68.4$ kJ mol^{-1}; $T\Delta S^0 = -37.0$ kJ mol^{-1}.

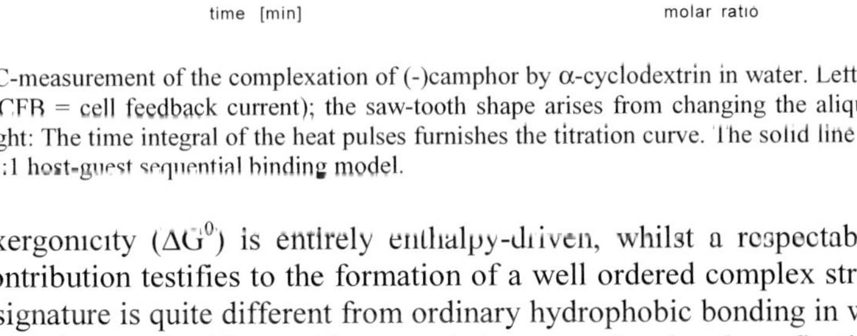

Figure 3. ITC-measurement of the complexation of (-)camphor by α-cyclodextrin in water. Left: Primary heat pulse trace (CFB = cell feedback current); the saw-tooth shape arises from changing the aliquots of titrand solution.; Right: The time integral of the heat pulses furnishes the titration curve. The solid line represents the best fit to a 2:1 host-guest sequential binding model.

The exergonicity (ΔG^0) is entirely enthalpy-driven, whilst a respectable negative entropy contribution testifies to the formation of a well ordered complex structure. The energetic signature is quite different from ordinary hydrophobic bonding in water which is commonly characterized by positive association entropies thereby reflecting the poor structural definition of the associate.

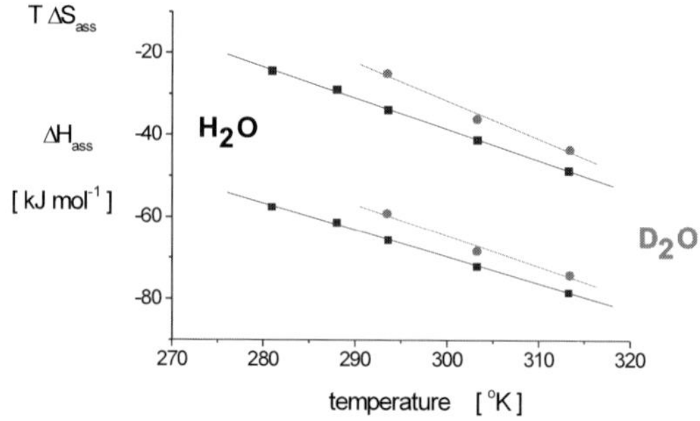

Figure 4. Energetics of (-)camphor binding to α-cyclodextrin in light (black lines) versus heavy water.

Repetition of the measurement this time in heavy water yielded the expected distinct difference with the corroboration of the initial premise that the enthalpic and entropic differences between the two solvents almost cancel (Figure 4). The difference in enthalpy, the solvent isotope effect at 298 K, was determined to be -5.2 kJ mol^{-1}. Correspondingly, the total change in the enthalpy of hydration is estimated as 10 × -5.2 = -52 kJ mol^{-1} leaving (-68-(-52)) = -16 kJ mol^{-1} for the intrinsic enthalpy of the mutual interaction of host and guest. In view of many other biological host guest systems which show similar or even more massive solvent contributions the result appears as a rule rather than an exception. Only 25% of the observable association enthalpy arises from the direct interaction and as such can operate in the enantiodifferentiation of the guest. A 75% share of solvent contribution if taken as the basis for generalization would clearly compromise any attempt in host design that exclusively focuses on a two-partner two-state binding scheme (e.g. a lock-and-key picture).

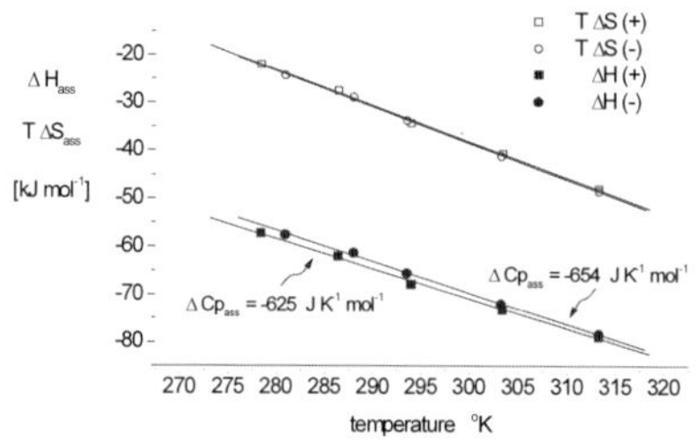

Chiral discrimination of camphor at 298 K

H_2O	(+) / (-) Camphor	D_2O
1,94	$K_{ass}(+) / K_{ass}(-)$	2,22
1,35	$\Delta\Delta G^o$ kJ mol^{-1}	2.02
1,35	$\Delta\Delta H^o$ kJ mol^{-1}	4,17
~0	$T\Delta\Delta S^o$ kJ mol^{-1}	-2,14
-29	$\Delta\Delta Cp^o$ J mol^{-1} K^{-1}	+206

Figure 5. Enantiodifferentiation of camphor by host-guest complexation with α-cyclodextrin. Left: The temperature dependence in light water; Right: comparison of the differences in the enantiodifferentiation in light versus heavy water.

Despite the small fraction of association enthalpy attributable to the intrinsic binding it suffices to allow equilibrium discrimination between the camphor enantiomers to an extent that outmatches most other enantiodifferentiations with cyclodextrins studied to date.[20] Binding of the (+)-enantiomer is better by a factor of ~ 2 than for the (-)-antipode (Figure 5) and this enantioselectivity increases with decreasing temperatures. In light water the discrimination is exclusively enthalpy based (Figure 5, right) whilst in D_2O the enantioselectivity emerges from the partial compensation of a greater difference in enthalpy that is levelled by an unfavourable entropy component. The puzzling difference in the energetic pattern is even amplified in considering the variation of the heat capacity. This parameter characterizes the change in the thermally excitable degrees of freedom. Comparing the difference in heat capacity of the diastereomeric complexes $\{(+)\mathbf{2} \subset \alpha\text{ - CD}\}$ / $\{(-)\mathbf{2} \subset \alpha\text{ - CD}\}$ (Figure 5, right) in light versus heavy water it is clear from the switch in sign and the grossly different magnitude that in the light of the small and rigid guest structures this result cannot be accounted for by a change in the host-guest binding modes alone. We are guided by the energetic signature of this interaction to assume the participation of "structural" water molecules. Such solvent molecules have been detected in biological and artificial host systems [21] and must be considered as stoichiometric ligands fortifying the structural layout. Though calorimetry due to the global response character is certainly not a first choice tool in answering structural problems, it may in some instances reveal subtleties that are quite hard to probe by any other means.

4. Evaluation of Binding Mode Diversity

Many host-guest interactions especially in polar solvents are characterized by positive association entropies. This is surprising in the light of the formal attachment of two species to make one since enthalpy is an extensive state function and thus depends on the sheer number of particles belonging to the system. The puzzling increase in disorder on binding is commonly explained [22, 23] by the release of bound solvent molecules from the interacting interfaces to the bulk. Of course, this explanation is valid only if the solvation shell is more structured than the surrounding bulk solvent. Since polar solvents, in particular the hydrogen bonding variety, possess a distinct near-neighbour structure that even prevents the free translation of particles making diffusion a very slow process one can expect only very tight binding or totally non-binding surfaces to provide enough urge to order the directly attached solvent molecules. Thus, only ionic or hydrophobic moieties of host or guest qualify as potential candidates to show this effect especially in well structured solvents like water or acetonitrile. A prominent example of biological and artificial host-guest binding is the guanidinium-oxoanion motif (Figure 6) that might serve as a touchstone to probe this hypothesis. Nature's recurrent use of the guanidinium group to bind anions in water in spite of its well proven extreme hydrophilicity [24] may signal a special virtue of this functional group that is not apparent at first sight [25]. Numerous X-ray crystal structures suggest a greatly predominant binding mode of an anionic guest in which the oxoanion is held and positioned in the way depicted in Figure 6 by joint action of the ionic charge and two parallel isodirectional hydrogen bonds suggesting an optimal complementary fit. This view was challenged by neutron diffraction studies [26] that did not detect a strongly held hydration shell in aqueous solution around the guanidinium cation. If there is only weak enthalpic hydration, the proven hydrophilicity of the guanidinium cation must arise from a strongly positive entropy of hydration meaning that guanidinium cation is a structure breaker. This conclusion is in line with guanidiniums' role as a protein denaturing agent, however, it raises a question mark for the supposition that

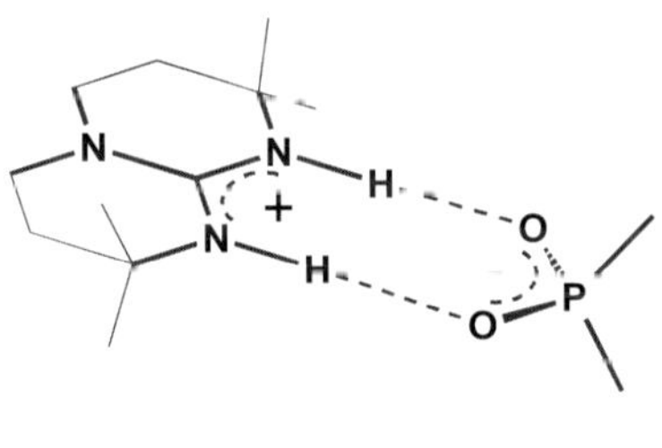

Figure 6: A sketch of the guanidinium oxoanion binding motif as it occurs in many biological and abiotic receptors.

this cation forms a uniquely dedicated host-guest structure with oxoanions. Rather one should expect the presence of an ensemble of different relative configurations of host and guest in solution that undergo very rapid interconversions as sketched in Figure 7.

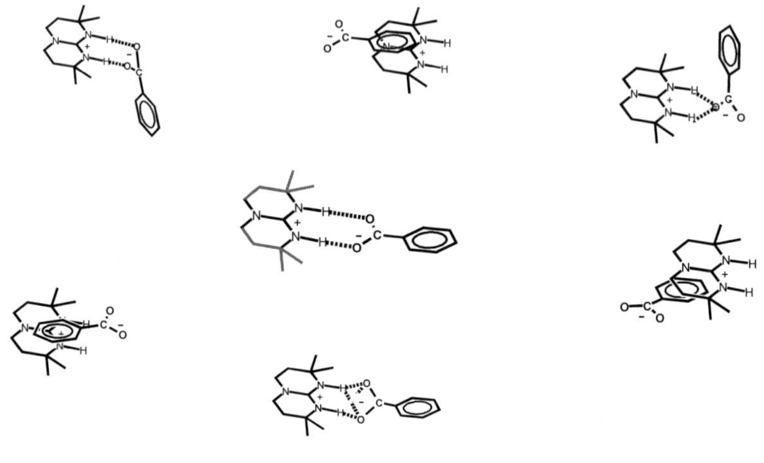

Figure 7. A sketch of the proposed solution ensemble of bicyclic guanidinium-carboxylate associated species. The configurations can also be considered as snapshots taken from an individual host-guest pair in certain time intervals, because all configurations should be rapidly interconverting.

Such a theatre would be completely compatible with the results of host-guest complexation with respect to stoichiometry and the effects observed with structural probes e.g. by NMR, yet it offers an additional rationalization for the positive entropies observed. If not one host-guest structure is formed, but a huge family of interconverting configurations that are all populated at ambient temperature because of the tiny energetic differences resulting from the levelling effect of the participating solvent molecules an increase in entropy versus the situation of the separated binding partners seems well conceivable. For many if not most applications using the guanidinium-oxoanion motif the difference between a snug-fit or fuzzy-family structural representation would go unnoticed and would be essentially irrelevant. However, there are fields of supramolecular applications where the structural integrity of the complexes is vital for success. Prominent examples encompass self-assembly and catalysis both of which cannot tolerate ambiguity of the topological and orientational layout in the host-guest complex. Thus, we are in need of some experimental confirmation on the "fuzziness" of supramolecular binding in solution which in general cannot be addressed by NMR due to the averaging of fluctuating configurations. X-ray crystallographic analyses in turn are also silent in this respect, because of their origin from a *kinetic* selection process (crystallization) and their confinement to a low entropy environment (the lattice). Calorimetry can help here if the multitude of reasons for an observable enthalpic effect can be limited. A promising way to do so is trend analysis where common patterns of response are correlated with subtle structural modifications in an ensemble of closely related compounds. One example documenting the success of this strategy refers to an attempt in designing improved affinity and directionality into the guanidinium-oxoanion interaction taking the bicyclic variant depicted in Figure 6 as the parent structure. Reasoning that supplementary hydrogen bond donor groups would help in the attraction of the anionic guest we envisaged the implementation of four *sec*-carboxamido functions, since this donor group is of proven virtue in anion binding. [27]

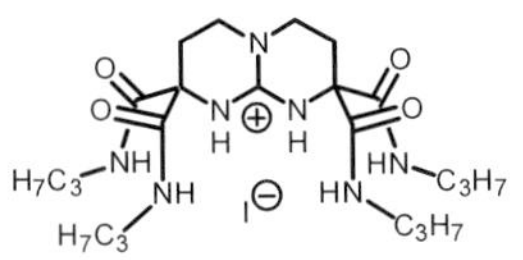

4

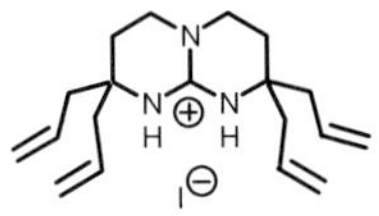

5

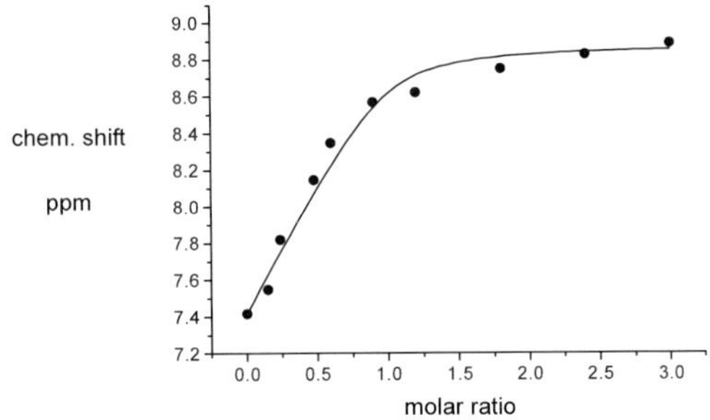

Figure 8. NMR titration of guanidinium host **1** (amide NH signal) with dihydrogenphosphate in acetonitrile at 1.0 mM, ambient temperature. The solid line represents the best fit to a 1:1 binding model.

First choice as a respective derivative was the tetracarboxamide **4**, because the introduction of the four carboxamido functions next to the ionic guanidinium moiety hampers their adequate solvation and should surface as an enhanced binding affinity due to reduced desolvation costs. Moreover, the stickiness towards the guest is spatially condensed creating a binding module that can readily be integrated into polymodular host compounds. The optimal convergence of all hydrogen bond donor groups necessitates the quite unfavourable superposition of the dipolar vectors resulting in additive, however, somewhat attenuated attractive contributions, that may even be compromised by some back bonding within the malonamide moieties. Nevertheless, stronger enthalpic interaction with oxoanionic guests are expected which must show in an augmented exothermicity relative to a tetraallyl-substituted analogue **5** that is incapable of additional hydrogen bonding. When subjected to dihydrogenphosphate complexation using NMR monitoring, host **4** displayed saturation behavior indicative of complex formation, yet the data evaluation disclosed significant deviations from the anticipated 1:1 stoichiometric model (Figure 8) suggesting the participation of higher complexes.

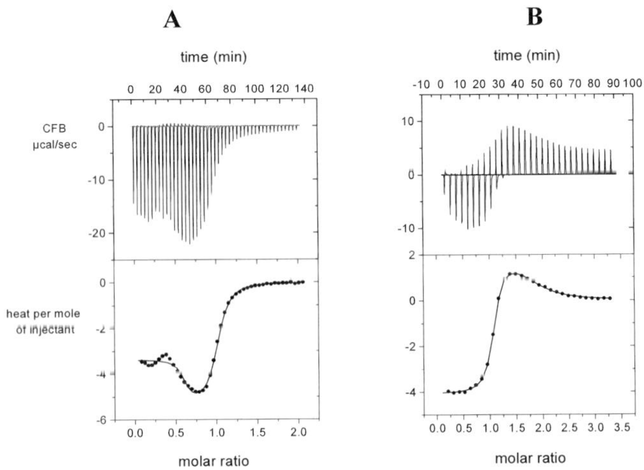

Figure 9. ITC-traces of the titration of **4** into dihydrogenphosphate (1.53 mM) in acetonitrile at 293 K (panel **A**) or adding $H_2PO_4^-$ into host **4** solution at 0.69 mM (panel **B**). The lines represent the best fit to a two-independent-site-model.

Whilst the data analysis did not allow a more concrete quantification at this stage, ITC measurements proved to be more telling. The addition of host **4** into the solution of $H_2PO_4^-$ in acetonitrile produced a heat response distinguishing various successive phases (Figure 9, panel A). Obviously, several processes happened depending on the actual stoichiometric ratio of the host and guest partners. The initial stages featuring a high excess of phosphate anions over the guanidinium host (host : guest ratio < 0.3) undoubtedly mirror higher order complexation but eluded our quantitative analysis. The stoichiometric regime with n > 0.5 was cleanly described by a binding model in which a low affinity and endothermic 1:2 host : guest binding step is taken over by an exothermic 1 : 1 complexation as soon as the host-guest ratio allows the latter complex to be formed in substantial amounts. The validity of this sequence of events was probed by inverting the titration sequence (panel B). Under these conditions the exothermic formation of the 1:1 complex precedes the endothermic 1:2 process after which the heat evolution ceases because higher complexes are not formed due to their lower affinity and the moderate excess of guest applied.

The thermodynamic state functions obtained by non-linear regression in both experiments are identical within the error limits and are considered reliable contrary to the parameters of higher complexation that suffer from dramatically increased errors and their cross correlation.

On comparing the binding energetics of the carboxamido substituted guanidinium host **4** to the tetraallyl substituted congener **5** that is incapable of supplementary hydrogen bond donation we notice a quite unexpected, yet seemingly constitutive feature of polytopic guest binding. In the entire series of oxoanions probed (Figure 10) the 1:1 complex formation constants of host **4** versus host **5** are enhanced, putatively corroborating the naïve anticipation that the sheer number of attractive interactions leads to increased binding strength and thereby determines the affinity. The inspection of the calorimetric data discloses this as an error. Throughout the series the exothermicity of binding representing the enthalpic outcome of the host-guest attractive interactions compromised by the cost of desolvation at the interacting sites is diminished in the multitopic host **4** with respect to the alkyl substituted analogue **5**. The enhanced affinity observed is exclusively due to an overwhelming increase in the entropic component of association, $T\,\Delta S$. A straightforward explanation to account for this surprising result can name the release of the supposedly well structured solvent shell solvating the polar amido functions to the bulk solvent. We note, however, that commonly solvent release correlates with the interface area of host and guest [23] which is very similar within the ensemble of oxoanions tested. The significant differences found are thus believed to originate from variations in the number and stiffness of mutual binding modes encompassing the partners. The importance of this entropic contribution has been analyzed recently.[28]

5. Conclusions

Affinity and selectivity in directional supramolecular systems are composite responses of frequently counteracting influences. The analysis of structure in many instances may then be a dull tool in their study because of the averaging between rapidly interconverting species. Here, an example from abiotic host-guest complexation demonstrates the utility of an energetic analysis how to avoid the pitfall of false reasoning that may follow from all too simple complementarity concepts.

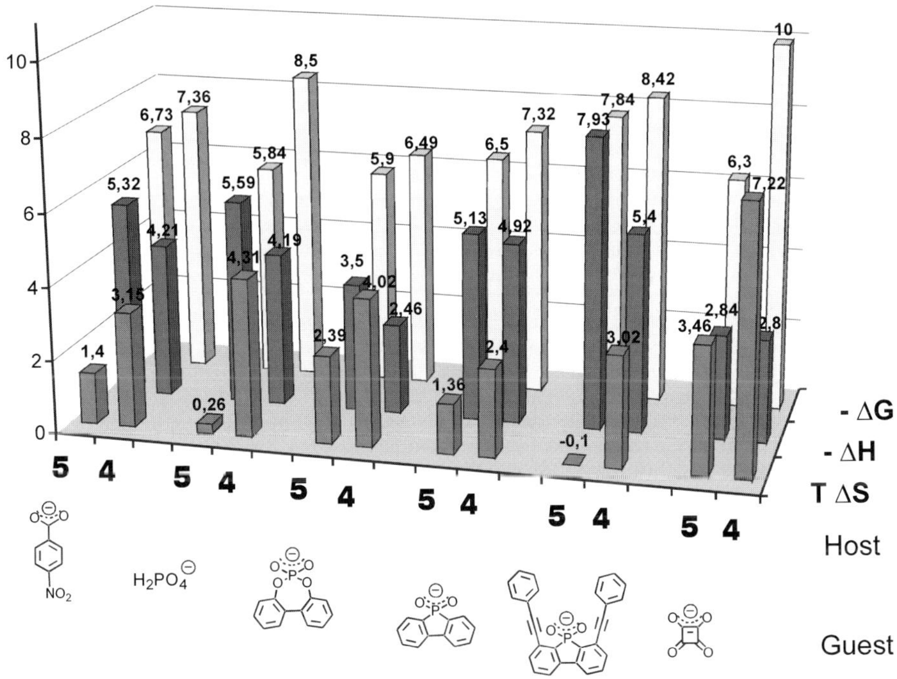

Figure 10. Comparison of the binding energetics of 1:1 stoichiometric complexes of guest anions to guanidinium hosts **4** and **5** in acetonitrile.

With the advent of modern instrumentation that furnish all pertinent thermodynamic data from a single experiment within a few hours calorimetry can help to tackle problems in host-guest binding that are inaccessible (principally or on a quantitative basis) to any other method. The power of enthalpic measurements, however, is unleashed only if put in a well-defined structural context. As a corollary, the benefit of sound structure-energy correlation on this basis calls for a more intense synthetic input.

6. Acknowledgements

This work was supported by Deutsche Forschungsgemeinschaft and Hans-Fischer-Gesellschaft, München.

7. References

[1] Schmidtchen, F.P. (2004) Thermodynamic and Kinetic Selectivity, *Encyclopedia of Supramolecular Chemistry*, (J.Atwood ed.), Marcel Dekker, New York, **DOI:** 10.1081/E-ESMC-120012656;

[2] Van Regenmortel,M.H.V. [1999) Molecular recognition in the post-reductionist era, *J.Mol.Recognit. 12*, 1-2;

[3] Huang,F., Jones,J.W., Slebodnik,C., Gibson, H.W. (2003) Ion pairing in fast-exchange host-guest systems: Concentration dependence of apparent association constants for complexes of neutral hosts and divalent guest salts with monovalent counterions, J.Am.Chem.Soc. 125, 14458-14464;

[4] Jones, J.W., Huang, F., Bryant, W.S. Gibson, H.W. (2004) A cautionary note regarding the investigation of supramolecular complexes involving secondary ammonium salts in acetone, Tetrahedron Lett. 45, 5961-5963;

[5] Stödeman, M., Wadsö, I., (1995) Scope of microcalorimetzry in the area of macrocyclic chemistry, Pure Appl. Chem 67, 1059-1068;
[6] Jelesarov, I., Bosshard, H.R. (1999) Isothermal titration calorimetry and differential scanning calorimetry as complementary tools to investigate the energetics of biomolecular recognition, *J.Mol.Recognit. 12*, 3-18;
[7] Holdgate, G.A. (2001) Making cool drugs hot: The use of isothermal titration calorimetry as a tool to study binding energetics, *BioTechniques 31*, 164-184;
[8] Ladbury, J.E. (2001) Isothermal titration calorimetry: Application to structure-based drug design, *Thermochim.Acta 380,* 209-215;
[9] Sigurskjold, B.W. (2000) Exact analysis of competition ligand binding by displacement isothermal titration calorimetry, *Anal.Biochem. 277*, 260-266;
[10] Turnbull, W.B., Daranas, A.H. (2003) On the value of C: Can low affinity systems be studied by isothermal titration calorimetry? *J.Am.Chem.Soc. 125*, 14859-14866;
[11] Schellman, J.A. (2002) Fifty years of solvent denaturation, *Biophys.Chem. 96*, 91-101;
[12] Cornish-Bowden, A. (2002) Statistical analysis of enzyme kinetic data, Oxford University Press, Oxford, p.249-268;
[13] Kolb, H.C., Finn, M.G., Sharpless, K.B. (2001) Click-chemistry: Diverse chemical function from a few good reactions, *Angew.Chem.Int.Ed.Engl. 40*, 2004-2021;
[14] Dunitz, J.D. (1995) win some,lose some: Enthalpy-entropy compensation in weak intermolecular interactions, Chem.&Biol. 2, 709-712;
[15] Schmidtchen, F.P. (2002) The anatomy of the energetics of molecular recognition by calorimetry: Chiral discrimination of camphor by α-cyclodextrin, *Chem.Eur.J. 8*, 3522-3529;
[16] Bielejewska, A., Duszcyk, K., Sybilska,D. (2001) Influence of organic solvent on the behavior of camphor and α-pinene enantiomers in reversed-phase liquid chromatography systems with α-cyclodextrin as chiral additive, *J.Chromatog. A 931*, 81-93;
[17] Dodziuk, H., Ejchart, A., Lukin, O., Vysotsky, M.O. (1999) ^{1}H and ^{13}C NMR and molecular dynamics study of chiral recognition of camphor enantiomers by α-cyclodextrin, *J.Org.Chem. 64*, 1503-1507;
[18] Chervenak, M.C., Toone, E.J. (1994) A direct measure of the contribution of solvent reorganisation to the enthalpy of binding, *J.Am.Chem.Soc. 116*, 10533-10539;
[19] Oas, T.G. Toone, E.J. (1997) Thermodynamic solvent isotope effects and molecular hydrophobicity, *Adv.Biophys.Chem. 6*, 1-52;
[20] Rekharsky, M.V., Inoue,Y. (1998) Complexation thermodynamics of cyclodextrins, *Chem.Rev. 98*, 1875-1917;
[21] Pometun, M.S., Gundusharma,U.M., Richardson, J.F., Wittebort, R.J. (2002) Solid state NMR and calorimetry of structural waters in helical peptides, *J.Am.Chem.Soc. 124*, 2345-2351;
[22] Rekharsky, M.V., Inoue, Y., Tobey,S., Metzger, A., Anslyn, E. (2002) Ion-pairing molecular recognition in water: Aggregation at low concentrations that is entropy-driven, *J.Am.Chem.Soc. 124,* 14959-14967;
[23] Deanda, F., Smith, K.M., Liu, J., Pearlman, R.S. (2004) GSSI, a general model for solute-solvent interactions. Description of the model, *Mol.Pharmaceut. 1*, 23-39;
[24] Radzika, A., Wolfenden, R. (1988) Comparing the polarities of the amino acids:Side-chain distribution coefficients between the vapor phase, cyclohexane, 1-octanol and neutral aqueous solution, *Biochemistry 27*, 1664-1670;
[25] Hauser, S.L., Johanson, E.W., Green, H.P., Smith, P.J. (2000) Aryl phosphate complexation by cationic cyclodextrins: An enthalpic advantage of guanidinium over ammonium and unusual enthalpy-entropy compensation, *Org.Lett. 2*, 3575-3578;
[26] Mason, P.E., Neilson,G.W., Dempsey, C.E., Barnes, A.C., Cruikshank, J.M. (2003) The hydration structure of guanidinium and thiocyanate ions: Implications for protein stability in aqueous solution, *Proc.Natl.Acad.Sci. 100*, 4557-4561;
[27] Bondy, C.R., Loeb, S.J. (2003) Amide based receptors for anions, *Coord.Chem.Rev. 240*, 77-99;
[28] a) Haj-Zaroubi, M., Schmidtchen, F.P. submitted, b) Jadhav,V., Schmidtchen, F.P. submitted;

20. MODELING OF MACROCYCLIC LIGAND COMPLEXES

PETER COMBA AND BODO MARTIN
Universität Heidelberg, Anorganisch-Chemisches Institut, INF 270, 69120 Heidelberg, Germany

1. Introduction

The history of the International Symposia of Macrocyclic Chemistry and this series of articles on current trends and possible future developments in the area of macrocycles indicate that applications of macrocycles, which range from thorough investigations of fundamental principles (photophysics, electrochemistry, spectroscopy, mechanistic problems, molecular recognition and chemical bonding) to applications in analytical chemistry, separation science, catalysis, biomimetics and medicine, usually have host-guest interactions, often metal complexation, as a focal point. The salient feature of macrocyclic hosts is a degree of rigidity of the ligand and the concomitant enforcement of a particular shape to the host-guest complex. In all areas mentioned the (ligand-enforced) geometry of the complex and the fit (or misfit) between various guests (ligands) and a specific host (metal ion) are of particular importance. A good host/guest fit leads to the selective stabilization of specific complexes, i.e., to metal-ion-selective complexation. A misfit between guest and host may lead to the selective destabilization of a specific assembly (energization) and therefore to a selectively enhanced reactivity. These two principles are well known from natural systems (bioavailability of metal ions, detoxification and stabilization of reactive intermediates on one hand and energization (entatic states) on the other),[1-9] and have also been discussed extensively, some times controversially, in the area of macrocyclic ligand metal complexes.[9-19]

With the importance of the shape of the complex and the fit of the guest into the host cavity, the conformational flexibility of the host (ligand), i.e., the number of isomers, their relative energies and the energy barriers between conformations, the elasticity of the ligand, i.e., the shape persistence of a particular ligand conformation and the elasticity of the structure of the complex (variation of the complex geometry with rigid ligand structures, i.e., the translation of the metal ion within a rigid ligand cavity) are decisive for successful thermodymanic and kinetic selectivity.[19] Ligand preorganization and host-guest (metal-ligand) complementarity (geometry, electronics, solvation) are the two important and well-defined principles which help to design, understand and tune selectivity.[14,19-25]

Karsten Gloe (ed.), Macrocyclic Chemistry: Current Trends and Future Perspectives, 303–325.
© 2005 *Springer. Printed in the Netherlands.*

Conformational freedom (flexibility) of a ligand and ligand preorganization with respect to a specific coordination geometry (this depends on the metal ion - on the guest molecule in a more general sense); complementarity between a ligand and a specific metal ion (elasticity in terms of geometry, the type of donors with respect to the bonding); dynamics and secondary interactions, solvation; ligand- and metal-ion-enforced distortions; the stabilization of specific oxidation and spin-states: all these are properties, where modeling is hoped to be able to give qualitative, semi-quantitative or, in favorable cases, quantitative answers.

2. Methods

The entire range of computational methods has been used to model structures, stabilities and properties of transition metal complexes (specifically those of macrocyclic ligands), including Hartree-Fock and post-Hartree-Fock methods, approximate density functional theory (DFT), semi-empirical molecular orbital approaches, QM-MM hybrid methods, empirical force-field calculations (MM) and, for specific problems, methods involving neural networks and genetic algorithms, ligand field theory (LFT, specifically the angular overlap model, AOM), molecular dynamics (MD), quantitative structure-activity or structure-property correlations (QSAR, QSPR), and data mining (statistical analysis of experimental structural data). These are also the methods found in the discussion of published data in Section 3. It is beyond the scope of this review to go into details for any of these methods or to compare them comprehensively. A number of authoritative textbooks do this in great detail,[26-28] and monographs on macrocyclic ligand chemistry often have short sections on molecular modeling.[29-31] For the most frequently used approaches, i.e., DFT[18,32] and MM[18,33] there are monographs and many reviews (see below for some of these, specifically in Section 3).

The two main methods currently used in computational and combined computational/experimental studies in the general area of transition metal coordination compounds, and specifically also with macrocyclic ligands, are DFT and MM. While DFT yields structural data, energies and molecular vibrations, as well as electronic information (the ground state wave function, spin density, charge distribution etc.), the latter is missing in MM.

Despite the fact that the awareness of experimentalists in computational chemistry has increased to a high level, that the infrastructure and experience in many research labs are present to perform detailed computational studies, that efficient and user-friendly software has become available and CPU time has become relatively cheap, for a number of reasons MM is still extensively used, specifically in the area of macrocyclic ligand coordination chemistry. (i) There are reliable and extensively validated methods and force fields available for the accurate computation of structures and conformational equilibria of transition metal complexes; (ii) full conformational analyses of even relatively small molecules require the refinement of hundreds and thousands of structures, and this often is not possible with quantum mechanical methods (similar arguments emerge for the inclusion of solvation and crystal lattices); (iii) typical questions in the area of macrocyclic ligand chemistry (see Section 1) relate to structural distortions and concomitant strain energies, where reliable results from MM can be expected; reactivities, transition state structures and general information on bonding, where MM does not necessarily yield useful information, are often not of central importance in the area of macrocyclic ligand coordination chemistry.

MM modeling of transition metal compounds has been reviewed extensively.[18,33-37] One reason for the excellent performance of MM in terms of structural modeling (and to a lesser but still astonishingly large extent for the energetics) is that there are highly parameterized and thoroughly validated force fields which are based on large data sets. Also, the assertion that MM fully neglects electrons and specific effects of bonding is not entirely correct. Rather, the type of a bond (e.g., single vs. double) is accounted for by the corresponding potential energy function and the force field parameters, both of which are selected by the atom types.[35] An elegant extension of these principles is the VALBOND force field which is based on valence bond electronic theory and computes the hybridization, followed by the selection of the pertinent force field from an atom-based parameter set.[38-42] While this approach might be difficult to apply to transition metal coordination compounds (and it never has been developed in this area) the DOMMINO force field includes a ligand field term in addition to the usual set of potential energy functions which describe bond-stretching, angle-bending, non-bonded and for other interactions.[43-46] Advantages of this latter approach are that specific electronic effects such as Jahn-Teller distortions in copper(II) complexes or spin transitions as, e.g., in nickel(II), are automatically included with a single parameter set and without additional functions or specific and not always unambiguous assignments of variable atom types.[47]

The critical and in many ways most important part of MM is the parameterization of the force field, and this is true in general for the methods discussed above, even for generic force fields (e.g. the UFF),[48,49] for adaptions of general force fields such as MM2, MM3[50] and AMBER,[51,52] and for specific force fields for transition metal compounds such as MOMEC.[53,54] Various methods to develop parameter sets have been reviewed,[55-57] and these also include automatic parameter fitting,[55,58-62] neural networks[63,64] and genetic algorithms.[63,65-68] Limits and possible pitfalls of MM, which relate to the approximations and over-simplifications of the empirical force field approach, have been discussed in detail.[18,35,69]

3. Applications

3.1 SCOPE

Small cyclic ligands such as tacn (1,4,7-triazacyclononane),[70] clathrochelates (cage ligands) such as the sarcophagines (sar=3,6,10,13,16,19-hexaazabicyclo eicosane),[31,71-73] calixarenes[30] and other "container molecules",[74] and rigid open-chained polydentate ligands such as the bispidine-type ligands (2,4-, 3,7-, 2,3,4-, 2,4,7-, 2,3,4,7-substituted 3,7-diazabicyclo nonane)[75-78] are not typical macrocycles but have rigid cavities and will also be discussed, where appropriate, in the following Sections. It is not our aim to give a comprehensive overview in terms of all systems and modeling studies which have been published in the area of macrocycles in recent years. However, the selection of published work discussed here represents examples of all areas, where the various computational methods have helped to understand macrocyclic ligand coordination chemistry.

3.2 SHAPE, DYNAMICS AND CONFORMATIONAL SPACE

Structure optimization still is a major area of molecular modeling, specifically, where no crystals for X-ray structural studies are available, where the structure in solution is of particular interest or where geometries of short-lived intermediates and exited or transition state structural data is sought. A recent highlight of an MM structure is that of the dicopper(I) complex of the macrocyclic Schiff-base ligand **L**^{1} shown in Figure 1. This complex is extremely air sensitive and, therefore, no solid state structural data was available; rapid oxygenation, followed by hydroxylation of one of the benzene rings indicates tyrosinase-like reactivity.[79] Indeed, the MM-optimized structure[80] had a Cu···Cu distance similar to that of the enzyme.[81,82] Of particular interest here is the fact that the computed structure[80] was confirmed by X-ray structural analysis six years after publication.[83,84] There are minor structural differences between the computed and experimental structures, which are believed to partially be due to the fact that the computed structure has two relatively bulky NH_3 donors as co-ligands, while in the two experimental structures there are two sterically less demanding $NCCH_3$ donors. More important is that the conformations of all four five-membered chelate rings have been predicted correctly. This helps to give credit to other published structures, where still no single crystal structural analyses are available,[85-90] and which all have been solved with the same philosophy, i.e., structure optimization coupled with the computation of a molecular property, based on the computed structure (for more examples, see Section 3.4).

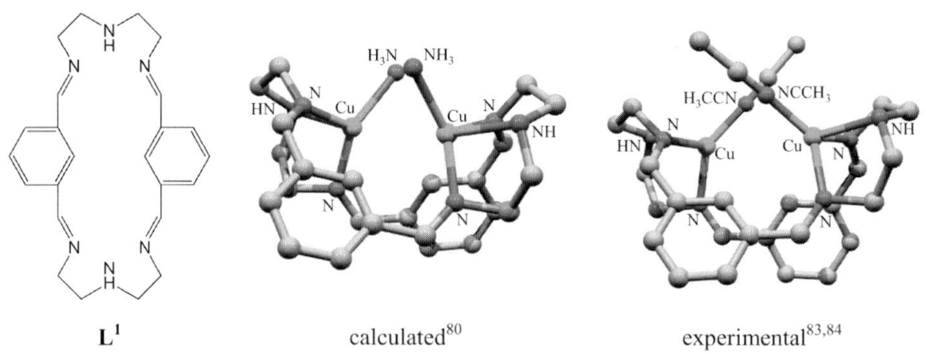

Figure 1. MM-optimized[80] and experimentally observed (X-ray)[83,84] structures of $[Cu_2L^1(Y)_2]^{2+}$; Y=NH_3 (computed) or Y=$NCCH_3$ (experimental).

For a thorough structural analysis an extensive search of the conformational space is required,[18] and this may be done with deterministic or stochastic search protocols or with MD.[91-94] It is interesting to note here that conformational analysis was the pioneering application of MM in coordination chemistry.[95,96] An accurate prediction of isomer distributions requires a force field which is well tuned with respect to the energetics, and this might be different from a structural force field.[18] Also, for the accurate prediction of isomer ratios, entropic terms and secondary interactions, such as solvation, ion-pairing, hydrogen bonding and intermolecular van der Waals interactions (see Section 3.3), need to be considered.[18] The classic example in this area is that of the three isomers of $[Co(dien)_2]^{3+}$ (dien = 3-aza-pentane-1,2-diamine).[18,97-101] Many similar studies have been performed with hexaaminecobalt(III) complexes, and the corresponding force field is accepted to be very well tuned and very reliable.[18] Isomer ratios have also been used for structure determination (comparison of computed and observed isomer distributions (NMR, HPLC, etc.) to confirm the computed structures, see also Section 3.4),[102-104] and this has been extended to other transition metal ions and donor sets, and used to evaluate materials for racemate separation,[105] metal-ion-selective extractions[106] and to determine structures in solution.[107]

For the conformational analysis of macrocyclic ligand transition metal complexes a range of methods are used. The relative energies of the various configurations and conformations are computed by MM,[108-112] semi-empirical MO theory[113] or DFT.[114] The conformational space is searched by stochastic (Monte Carlo) methods or MD[94] or on the basis of structures available from the CSD[115] which then are analyzed by MM.[116-122] MM has also been used to analyze pathways of metal ion incorporation to macrocyclic ligands.[123] MM in combination with structural studies and solution NMR spectroscopy has been used to analyze the dynamics of figure-of-eight shaped, folded, large macrocyclic ligand dicopper(I) complexes.[124-126] For the Schiff-base ligands with two N_2S_2 coordination sites and para-substituted xylylene-spacer groups there are two low energy conformations, which are interconverted via an energy barrier of approx. 50 kJ/mol (Figure 2). For the corresponding ligands with meta-substituted xylylene spacer groups there is a considerably higher energy barrier between two low energy structures.[125,126]

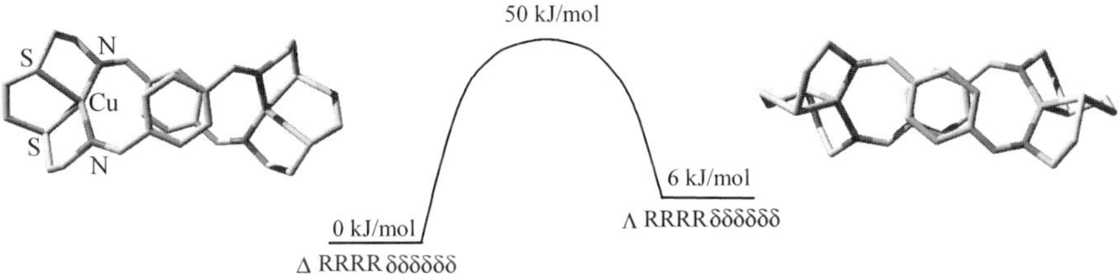

Figure 2. Experimental (Δ isomer) and computed (Λ isomer) low energy structures of the dicopper(I) complex of the macrocyclic Schiff-base ligand with para-xylylene spacer groups, and computed energy barrier (confirmed by NMR spectroscopy).[125,126]

Due to the biochemical importance of porphyrine systems their metal complexes have been modeled extensively. Metalloporphyrines are assumed to be planar but there is a relatively high degree of flexibility, and six low energy distortion modes have been described (Figure 3).[127] MM with a variety of force fields has been used to compute the structural and dynamic properties,[65,66,128-131] and these studies also include information on the orientation of the axial donors.[132,133] Reaction dynamics related to the interaction of substrates with the metal site have also been modeled,[134-136] and MM has been used to correlate structural and spectroscopic properties (primarily Raman frequencies).[137,138]

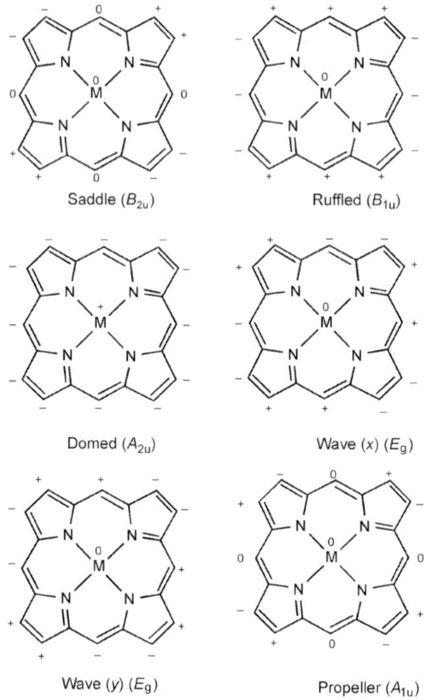

Figure 3. The common distortions of a porphyrin, written in terms of the irreducible representations of the D_{4h} point group.[66,127][*Phys. Chem. Chem. Phys.* 2002, *4*, 5878] - Reproduced with permission of the PCCP Owner Societies.

Of particular interest in terms of the shape of the coordination sphere are complexes with bispidine-type ligands (Figure 4). These ligands have a very rigid adamantane-derived structure but the corresponding coordination geometries are highly elastic (see also Section 3.5).[75-77] In the particular case of copper(II) complexes of tetra- and pentadentate bispidine-type ligands various minima on the warped rim of the Mexican hat potential energy surface could be trapped and characterized both structurally and spectroscopically, and the combination of experimental data with MM, AOM and DFT helped to understand the factors which are responsible for the stabilization/destabilization of the various structural forms.[76-78,139,140]

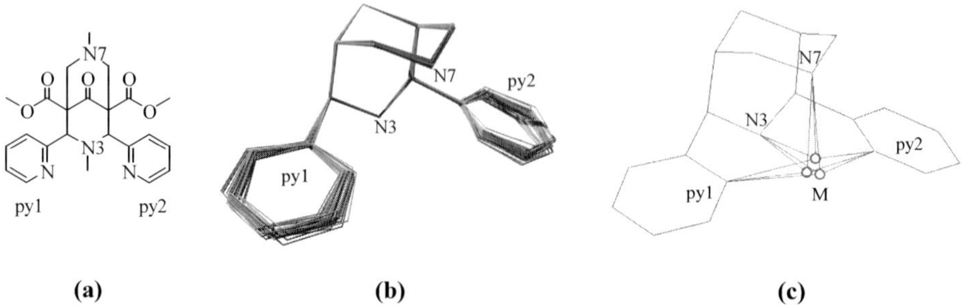

Figure 4. (b) Overlay of the ligand backbones of 40 metal complexes of the tetradentate ligand shown in (a); (c) averaged ligand structure with metal positions of three discrete clusters (Mn^{II}, Cu^{I}, Cu^{II}).[19,75]

3.3 WEAK INTERACTIONS

Weak interactions are of importance in the general areas of supramolecular chemistry,[23] self-assembly[24], molecular devices and machines,[141,142] in host-guest chemistry in general (see Sections 1 and 3.5), in stereoselective synthesis and racemate separation, and in areas which are based on solvation effects, selective interactions between liquid and solid phases and crystal lattice effects (solvent extraction, chromatography, crystal engineering and polymorph prediction). Computer modeling in these areas relies on the prediction of small energy differences (typically less than 5 kJ/mol), and these usually are the result of a large number of interactions. In addition, due to the large size of the systems to be analyzed, the search for the global and close lying local energy minima of the system requires computationally efficient methods. Therefore, predominantly force-field-based methods are used in this area. The thorough development of reliable model potentials used for the simulation of weak intermolecular forces has also been based on ab-initio calculations of small diatomic molecules,[143] and perturbation theory was used for the computation of the weak interaction energy of van der Waals complexes.[143]

There is a conceptional difference between computational force fields used for the optimization of isolated molecules (note that, generally, these are parameterized to reproduce experimental structural data and therefore do not yield "gas phase structures"[18,144]), and force fields tuned to the modeling of condensed phases, because in those attractive long range non-bonded interaction contribute to the total energy.[18,144] This has been tested in comparative studies on copper(II)-amino acid complexes.[145,146] The prediction of polymorphs and the solution of structures by powder diffraction[147] involves the generation and optimization of possible structures, the computation of lattice energies, the Rietveld refinement and, if possible, comparison with experiment.[148]

There are similar problems and methodologies in the modeling of solvent-solute interactions. In the area of solvent interactions (see also Sections 3.4 and 4) the combination of QM- or MM-based methods with MD has been used successfully in the computational modeling and rather accurate predictions of the ligand-assisted distribution of metal ions between aqueous and organic liquid phases.[149,150]

There are various methods for the prediction of stereoselectivities, both in racemate separation (thermodynamics, usually computed by force field methods) and enantioselective catalysis (reactivity, usually computed by quantum mechanics)[18]. Promising recent developments, primarily based on force field and statistical methods, but also involving QM modeling, are based on stereocartography, the computation of the chirality content and the evaluation of chirophores.[151-154]

3.4 MOLECULAR PROPERTIES

Molecular properties depend on molecular structures and, with appropriate methods and software, it is possible to compute these on the basis of structural information, established by experiment or computation (see Sections 2 and 3.2).[18,155,156] This allows for the design of new materials with given properties and often is used for structure determination (see Section 3.2 for more examples in this area). Complex stabilities will be discussed separately and in detail in Section 3.5; the application of molecular modeling in the area of isomer distributions (conformational equilibria), in combination with experimental data (usually spectroscopy and chromatography), is discussed in Section 3.2.

Clearly, the best known, most available and commonly used method in structural chemistry, which relies heavily on computation but incorrectly is generally not associated with molecular modeling, is structure determination by single crystal (or powder) X-ray diffraction (recent developments in crystal structure refinement also include QM-based methods[157,158]). The best known structural method which uses a combination of spectroscopy and computational structure optimization is the structure determination by NMR spectroscopy, which primarily is based on nuclear Overhauser effects (NOEs),[159] and which is best known in the field of protein structure determination,[160-163] where also an increasing number of metalloproteins are structurally characterized.[164,165] Structures of other biomolecules[160,166] and small molecular weight coordination compounds[167-171] have also been reported. The computation of NMR chemical shifts (and EPR spin Hamiltonian parameters) relies on accurate charge and spin distributions, and these usually are computed quantum-mechanically[172]. There are also more efficient yet accurate empirical[173] and semi-empirical methods[174] for the calculation of charge distributions and NMR chemical shifts[175,176] but, so far, these are only parameterized for a very limited range of elements in the periodic table.

The computation of molecular vibrations is possible with all methods for structure refinement which compute the Hessian matrix (for MM this is the case for optimizers based on second derivatives such as the Newton-Raphson method[18]). The computed frequencies may then be used for comparison with experimental data[90]. Recent developments in this area are novel QM-based approaches for the efficient computation of specific vibrational frequencies in large molecules.[177]

Reactivities and mechanisms, including product selectivities and catalytic cycles, are quite generally computed with quantum mechanical approaches, in the field of transition metal complexes often with DFT-based methods. There are few examples where MM has been used for the computation of bond-breaking / bond-making processes[42,46] and transition states.[178,179] In the field of macrocyclic ligand coordination chemistry the largest contribution of computational work on reactivities and mechanisms has appeared in the area of porphyrine systems, and there are numerous studies on heme-iron oxidation catalysis, primarily based on DFT.[180-182] There are also various studies based on semi-empirical MO theory and using QM-MM either with semi-empirical[183] or DFT-based modules used for the quantum-mechanical region.[184]

With respect to the various spectroscopic methods not discussed so far there are four methodologically different concepts of molecular modeling: (i) quantum-mechanical methods which result structural and electronic information; (ii) MM-based structural modeling, followed by a single-point MO-based computation of the electronic properties; (iii) spectra simulation with given electronic parameters in cases, where structural parameters (e.g., distances such as used in protein modeling, see above) are involved in the simulation; (iv) structure property correlations. The latter approach will be discussed briefly at the end of this Section (redox potentials) and in Section 4.

(i) Among the QM-based methods time-dependent DFT (TD-DFT) is gaining in popularity although, in terms of the theoretical basis and obvious problems in terms of accuracy, TD-DFT clearly is not a routine method for the computation of excitation energies of transition metal complexes.[185] TD-DFT has been used for the computation of electronic absorption and CD spectra.[186,187] A very promising method for the computation of spectroscopic properties (Mössbauer, electronic and EPR spectra), based on multireference ab-initio methods, has recently been developed.[188-191] Another interesting new approach for the computation of ligand-field-spectroscopic parameters of transition metal complexes is a recently developed DFT-based ligand field method (LF-DFT).[192]

(ii) Ligand field transitions and EPR spectra of transition metal complexes can be computed with the Angular Overlap Model, AOM,[193,194] based on a set of electronic parameters and the coordinates of the chromophore. MM-AOM provides the accurate computation of electronic dd-transitions and EPR g-values (based on first order ligand field or on MO theory the hyperfine constants A are also available).[195-197] MM-AOM has been used to compute electronic spectra of chromium(III), cobalt(III) and nickel(II) hexamines,[195] EPR g-values of low spin iron(III) complexes[196] and ligand field and EPR spectra of copper(II) tetraamines.[197] Three highlights of structure determinations by solution (and single crystal) spectroscopy, in combination with MM-AOM calculations, are shown in Figure 5. (a) The MM-AOM optimized structure of the low spin iron(III) complex of a bis-pendent-amine-tetraazamacrocycle ($[Fe(\mathbf{L}^2)]^{3+}$) is shown in Figure 5(a).[196] Three independent molecular cations of two X-ray crystal structural analyses have been known[198,199] but, due to structural disorder, the conformations (δ or λ) of the two in-plane five-membered chelate rings (three possible diastereomeric forms) were not known from crystallography, but could be analyzed unambiguously by the combination of experimental spectroscopy and MM-AOM modeling.[196] (b) Solution spectroscopy and electrochemical experiments, together with MM-AOM and in combination with MM-Redox (see below), were used to determine the structure of the pair of stable conformations of a hexamine cobalt(III) cage complex $[Co(L^3)]^{3+}$.[85,200] The two isomers have similar stability, the structure of the cobalt(III) complex of the ob_3-isomer and that of the cobalt(II) complex of the lel_3-isomer are known from crystallography and, most importantly, the spectroscopic and redox parameters of the two isomers are accurately reproduced (notable are the low energy dd transition ($^1A_1 \rightarrow {}^1T_1$) of the elusive, blue isomer which is at an unprecedentedly low energy (600 nm, 16700 cm^{-1}), and which has a very high reduction potential (+0.84 V vs. SHE).[85] (c) The cyclam-derived bis-bispidine-caped macrocycle $\mathbf{L}^4$, shown in Figure 5 (c), has the strongest ligand field observed with copper(II) tetraamines[88] (see also Section 3.5). The $[Cu(\mathbf{L}^4)]^{2+}$ and $[Cu(\mathbf{L}^4)(OH_2)_2]^{2+}$ complexes are, so far, only available in very low yield; the structure was solved by MM-AOM.[88]

(iii) The known technology to simulate EPR spectra of dipole-dipole-coupled dicopper(II) systems, based on electronic parameters, which may be available from mononuclear reference complexes or MM-AOM modeling, and a set of geometric parameters (distance and orientation of the two g-tensors of the copper(II) sites[201-203]) has led to the development of EPR-constrained molecular mechanical structure optimization (MM-EPR).[80,86,89,204-207] MM-EPR has been used to solve structures of bis-macrocyclic ligand dicopper(II) complexes and of dicopper(II) complexes of cyclic peptides. In the most recent example[207] MM EPR allowed to show that the bismacrocyclic ligand dicopper(II) complex shown in Figure 6 has a solution structure (d) which is different from that in the solid (a) and that two additional stable configurations (no experimental structures available) can also be isolated (b, c). There probably is no other method than MM-EPR, which could have solved this structural problem unambiguously and also very quickly.

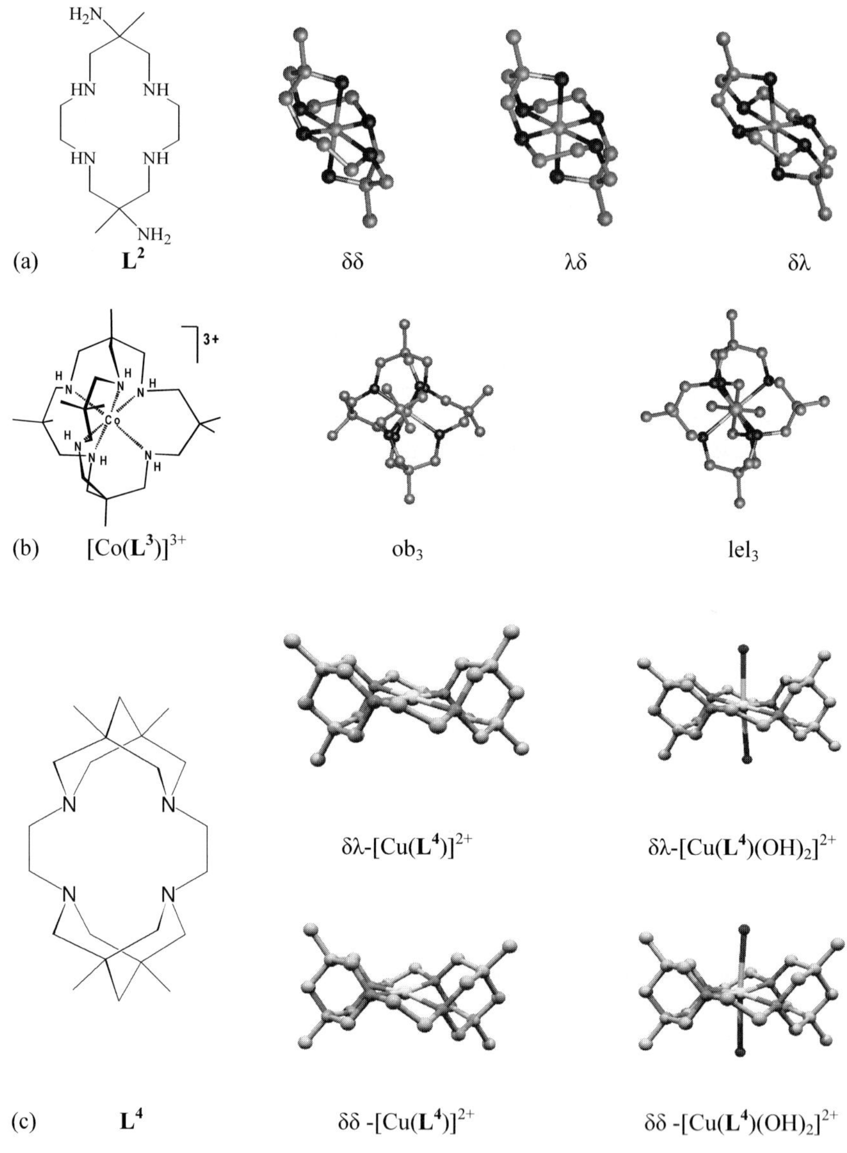

Figure 5. MM-AOM-based structures of (a) $[Fe(\mathbf{L^2})]^{3+}$ (the δλ isomer is the most stable form,[196] (b) $[Co(\mathbf{L^3})]^{3+}$ (ob_3 and lel_3 have similar stabilities[85]), (c) $[Cu(\mathbf{L^4})]^{2+}$ (yellow) and $[Cu(\mathbf{L^4})(OH_2)_2]^{2+}$ (purple).[88]

(iv) There is a geometric contribution to the stabilization of a specific oxidation state of a transition metal complex, and in the context of macrocyclic ligands (with a given donor set) this is the primary feature (geometric complementarity, see Section 1 and Section 3.5). This involves the ligand-enforced orientation of the donor groups around the metal ion (e.g., square planar vs square pyramidal vs tetrahedral), the orientation of the ligand lone pairs with respect to those of the metal ion, and the metal-donor distances. MM-based strain energy differences between the oxidized and reduced forms of complexes have been correlated with experimentally determined redox potentials, and the resulting linear fit can be used to predict redox potentials of new compounds or for structure determination (MM-Redox).[18,85,208,209] A good visualization of the correlation of the geometry with the redox potential (and with spectroscopic parameters) is given in Figure 7, where the ring sizes of tetraazamacrocyclic ligands are plotted against structural (tetrahedral twist), electrochemical and various spectroscopic parameters.[210] One of the most convincing examples of the MM-Redox approach is the pair of isomeric hexamine cobalt(III) cage complexes discussed above (see Figure 5(b)), where the conformational change leads to a difference in redox potential of over 800 mV,[85] accurately predicted by MM.

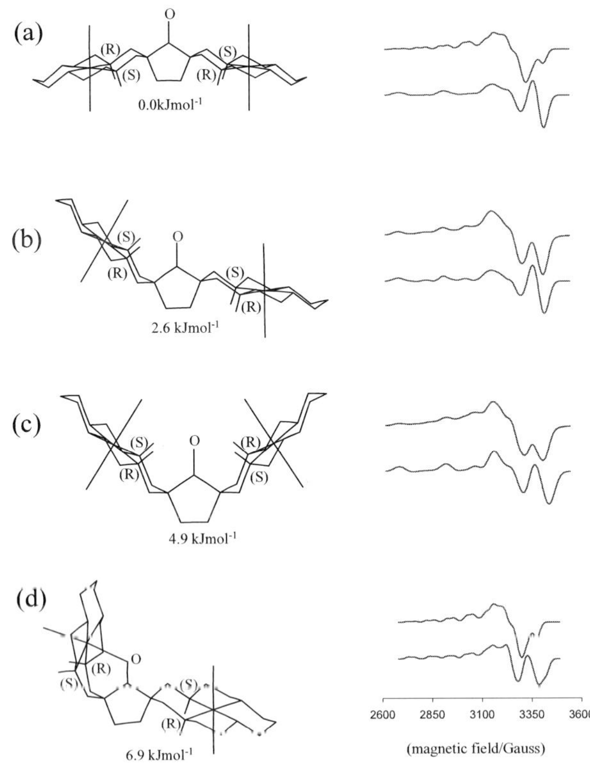

Figure 6. (a) – (c) are the computed structures and relative strain energies of the three possible (R,R,S,S)$_2$ configurations of $[Cu_2(\mathbf{L}^4)(OH_2)_4]^{4+}$, (most stable conformers; chair/chair, λ/δ), (d) is the observed chair/boat conformer of $[Cu_2(\mathbf{L}^4)(OH_2)_3]^{4+}$; corresponding observed (top) and simulated (bottom) EPR spectra; g- and A-parameters ($g_\perp$, $g_\parallel$, $A_\perp$, $A_\parallel$ in $10^{-4}cm^{-1}$): (a,b) 2.045, 2.19, 23, 200; (c) 2.045, 2.185, 23, 205; (d) 2.045, 2.185, 23, 205; 2.045, 2.13, 19, 180.

Clearly, other factors, such as solvation and electronic complemetarity also contribute to the redox potential and, therefore, MM-Redox is strictly restricted to the comparison of systems which involve identical redox couples (e.g., cobalt(III/II)), donor sets (e.g., hexamines) and solvents (e.g., water).[208] The well-known effect of solvation (solvent-dependent redox potentials) has been studied with complexes of N-alkylated cyclam derivatives[211-213] and confirmed by DFT calculations.[214] The electronics of the donor groups have been included in ligand additivity models.[215-218] However, these largely neglect steric influences. A multi-parameter fit, including geometric as well as electronic descriptors, has also been proposed[219] but this also has severe limitations.

3.5 FIT AND MISFIT

Molecular recognition in general and specifically metal-ion-selective complexation rely on the selective and strong binding of a host by a guest molecule, of a metal ion by a (macrocyclic) ligand (see also Section 3.2). Based on the fundamental lock-and-key concept[220] the complementarity of shape, size and functionality, more specifically, shape, size and electronic compatibility[221,222] between the host and the guest is generally accepted to be a major factor in molecular recognition. With ligands which exist in various conformations, high stability is observed when there is no necessity for conformational rearrangement, i.e., when the ligand is preorganized.[20-22] It follows (see Figure 8) that ligand rigidity is of importance for metal ion recognition: (i) a lack of rigidity in the sense of conformational flexibility may lead to various shapes of the host molecule with different size, i.e., to various hosts with different selectivity; (ii) a lack of rigidity in the sense of a high elasticity of a single host isomer (steepness of the individual potential energy curves in Figure 8 (b)) also reduces the selectivity.[19] These two ligand-based properties have also been described as multiple juxtapositional fixedness.[11,223] Not generally appreciated but also of importance for metal-ion-selective and strong coordination are poor solvation of the ligand[21] and a rigid coordination sphere (which does not necessarily result from a rigid ligand, see Section 3.2 and below).[19] Fit and misfit between a host and a guest are also of importance in other areas of molecular recognition, including those with supramolecular and biomolecular systems (receptor modeling, substrate binding and activation, enzyme inhibition).[9,166,224-226]

The application of molecular modeling techniques to the prediction and interpretation of metal-ion-selective complexation by macrocyclic and other ligands has been used, described and reviewed abundantly.[10,12-19,227-230] A thorough theoretical study of the energetics of complex formation needs to include the solvation (desolvation) of the metal ion, ligand and complex; a full conformational analysis of the ligand (preorganization) and complex (coordination number and geometry); and the geometric and electronic complementarity of the ligand / metal ion couple. The problem to solve is of similar complexity to that of the computation of redox potentials (see Section 3.4) and, in fact, the two problems are related,[231] and the electrochemical determination of stability constants is a common procedure. Quite generally, modeling studies concentrate on influences derived from preorganization and geometric complementarity.[15,232] In many cases these are not even well separated,[17-19] and the determination of the size, shape and elasticity of the ligand cavity and its fit to the guest (metal ion) is often the major or only problem which is addressed. Studies which include the electronics of metal-ligand binding, solvation and dynamics are available but not abundantly.[233-235]

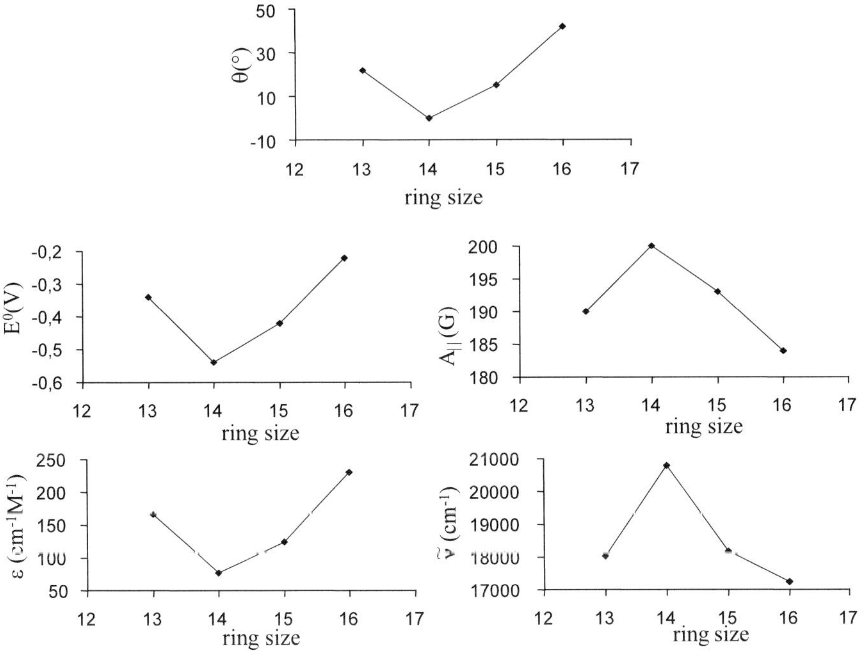

Figure 7. Correlation of the CuN_4 plane twist angle θ, the reduction potential E°, the EPR hyperfine parameter $A_{||}$, the d-d transition $\tilde{\nu}$ and its extinction coefficient ε with the macrocycle ring size.[210]

A number of approaches for the computation of cavity shapes, sizes and elasticities have been reported and discussed in detail and controversially, and the methods, applications and interpretations have been reviewed extensively.[9,10,16-19,228-230,236-238] These also include models which appreciate chelate ring sizes[12,17,239,240] and the directionality of the metal d-orbitals and the ligand lone pairs (misdirected valences).[14,19,197,241]

The ligand field splitting has been used to probe the complementarity in macrocyclic ligand complexes[10,12,18] (note that ligand field strengths have also been correlated with redox potentials,[18,242-244] and these are related to complex stabilities, see above). An interesting study in this context is that involving the set of copper(II) complexes with the three tetraazamacrocyclic ligands in Figure 9 (**L**^{5}, cyclam has secondary amines, the two reinforced ligands **L**^{6} and **L**^{7} have tertiary amines with similar nucleophilicity). Interestingly, the ligand field induced by **L**^{6} is smaller than that by **L**^{7} (540 vs. 500 nm), the unresolved dd transition of $[Cu(L^7)(OH_2)_2]^{2+}$ is at 430 nm (that of $[Cu(L^5)]^{2+}$ is at an unprecedented 390 nm), see Table 1;[19,245-247] the stability constants follow qualitatively a similar trend.[19] The conclusion, based on the computed structures (MM, full conformational analysis) and electronic spectra (AOM, see MM-AOM in Section 3.4), is that the lack of complementarity with **L**^{6} is due to misdirected valences, and this also emerges from Figure 8.[19] The bispidine-based macrocyclic ligand **L**^{7} is not only very rigid but also highly complementary with respect to copper(II).

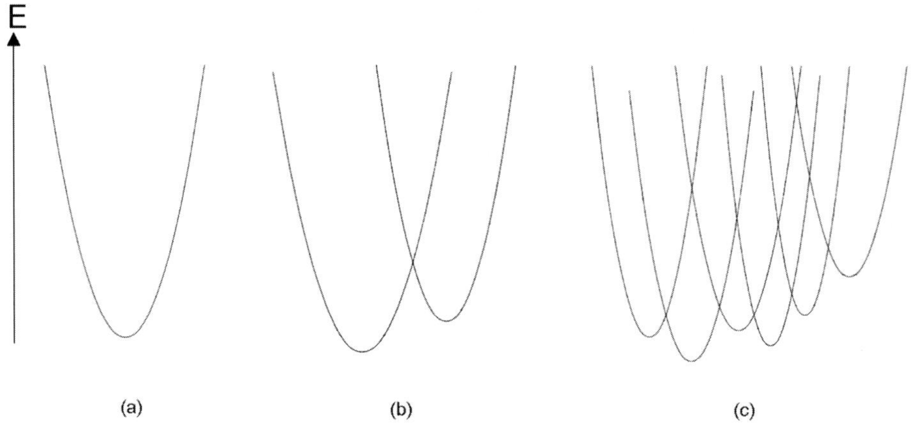

Figure 8. (a) Rigidity and selectivity (b) elasticity (different line shapes) (c) conformational flexibility.

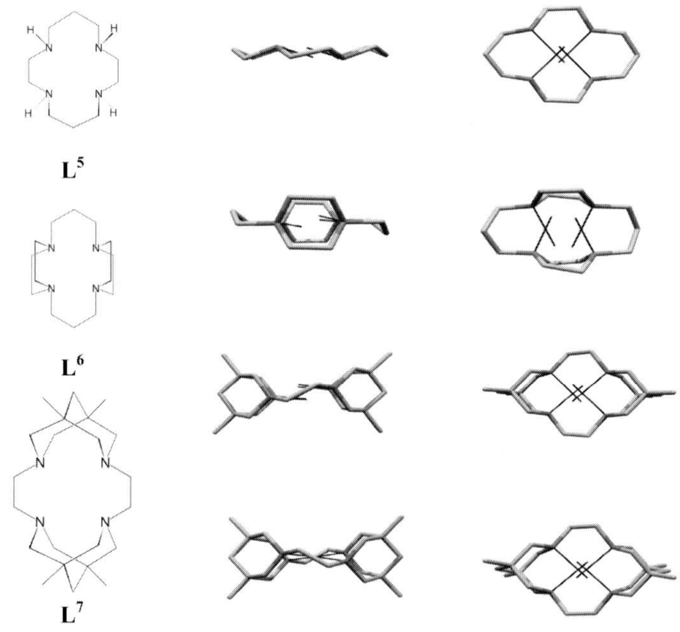

Figure 9. Visualization of the lone pair directionality (complementarity) of three tetraazamacrocyclic ligands **L**[5], **L**[6] and **L**[7].[19]

TABLE 1. Experimentally determined and computed (MM-AOM) electronic transitions in cm^{-1} (nm) of $[Cu(\mathbf{L^5})(OH_2)_2]^{2+}$, $[Cu(\mathbf{L^6})(OH_2)_n]^{2+}$ and $[Cu(\mathbf{L^7})(OH_2)_n]^{2+}$ (n=0,1,2).[88]

Compound		$E_1(xz)$	$E_2(yz)$	$E_3(xy)$	$E_4(z^2)$
$[Cu(\mathbf{L^5})(OH_2)_2]^{2+}$ (n=1,2)	exp.[248]	19900 (500)			
$[Cu(\mathbf{L^5})(OH_2)_2]^{2+}$	calc.	19200	20020	20400	13600
$[Cu(\mathbf{L^6})(OH_2)_n]^{2+}$ (n=1,2)	exp.[246]	18590 (540)			
$[Cu(\mathbf{L^6})(OH_2)]^{2+}$		18654	20330	17506	16590
$[Cu(\mathbf{L^6})(OH_2)_2]^{2+}$		21070	21600	20395	17225
$[Cu(\mathbf{L^7})(OH_2)^n]^{2+}$ in H_2O (n=1,2)	exp.	23250 (430)			~19200 (520)
λλ-$[Cu(\mathbf{L^7})(OH_2)_2]^{2+}$	calc.	23000	23380	22120	19530
λδ-$[Cu(\mathbf{L^7})(OH_2)_2]^{2+}$	calc.	23560	23640	22650	19640
$[Cu(\mathbf{L^7})]^{2+}$ in $MeNO_2$	exp.	25640 (390)			
λλ-$[Cu(\mathbf{L^7})]^{2+}$	calc.	24660	25040	23400	23580
λδ-$[Cu(\mathbf{L^7})]^{2+}$	calc.	24660	25220	23770	23770

A cavity size curve of the related hexadentate ligand is shown in Figure 9.[78] The hexadentate bispidine-based ligand $\mathbf{L^8}$ is, with respect to the bispidine (3,7-diazabicyclo[3.3.1]nonane) core, very rigid and highly preorganized; the pyridine donors in 2,3,4,7-positions are flexible (low energy C-C rotational barriers). Qualitatively similar cavity size curves have also been observed for 2,4-disubstituted tetradentate and 2,3,4- or 2,3,7-trisubstituted pentadentate ligands. A general observation, based on experimental structural data[19,75,76,78] is that, with a large variation of the metal ion size (0.1 – 0.2 Å per M-N bond), the ligand structure is basically constant, with the exception of low energy rotations of the pyridyl groups. However, the metal center has little restriction to move within and out of the ligand cavity.

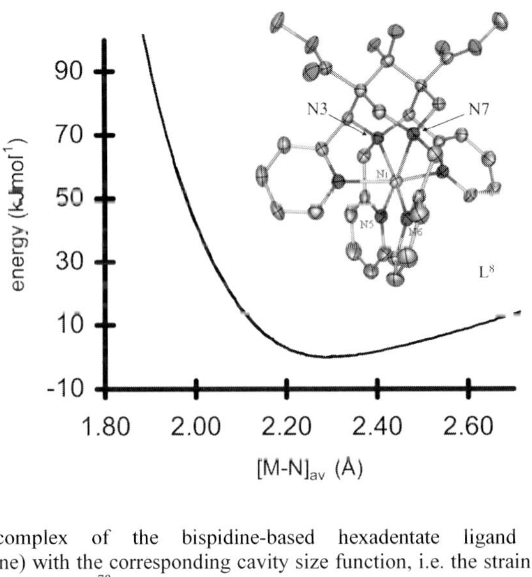

Figure 10. Ni(II) complex of the bispidine-based hexadentate ligand $\mathbf{L^8}$ (bispidine= 3,7-diazabicyclo[3.3.1]nonane) with the corresponding cavity size function, i.e. the strain energy as a function of the averaged metal donor distances.[78]

This is nicely supported by the cavity size curve (Figure 10): the high degree of elasticity of the coordination geometry is a result of the very flat tail towards longer metal donor distances. For metal ions with larger average M-N distances than approx. 2.1 Å the hexadentate ligand $\mathbf{L^8}$ is virtually unselective (note that this analysis does not include electronic effects). Complexes of small metal ions are destabilized. This is due to subtle metal-ion-enforced ligand distortions as shown in Table 2. This analysis, based on the cavity size curves and a detailed MM analysis is supported by spectroscopy, electrochemistry and potentiometrically determined stability constants (the latter are also shown in Table 2).[78]

TABLE 2. Selected structural data of $[Co(\mathbf{L^8})]^{2+}$, $[Ni(\mathbf{L^8})]^{2+}$, $[Cu(\mathbf{L^8})]^{2+}$, $[Zn(\mathbf{L^8})]^{2+}$ (distances in Å, angles in degrees).

	$[Co(\mathbf{L^8})]^{2+}$	$[Ni(\mathbf{L^8})]^{2+}$	$[Cu(\mathbf{L^8})]^{2+}$	$[Zn(\mathbf{L^8})]^{2+}$
distances Å				
$\sum_{i=1}^{6}(M-N_i)$	(12.83)	12.64 (12.62)	13.04(13.09)	13.04 (13.11)
N3•••N7	(3.06)	2.88 (3.02)	2.84(2.91)	2.90 (3.11)
strain energies kJ/mol				
E_{strain}	6.0	11.0	45.0	0.0
stability constants log K				
$M^n + L \leftrightarrows [ML]^n$	15.0	14.1	19.8	21.5

The interesting observation, unexpected on the basis of limited published data of a tetradentate derivative of $\mathbf{L^8}$ [249,250] but in full agreement with the MM analysis is that the stabilities do not follow the usual Irving-Williams behavior ($Co^{2+} < Ni^{2+} < Cu^{2+} > Zn^{2+}$)[251,252] but that $\mathbf{L^4}$ is zinc(II)-selective. This was interpreted as a combination of the steric effects discussed above, electronic effects (distorted coordination geometries, misdirected valences), which are unimportant for the d^{10} ion zinc(II), in contrast to copper(II), especially nickel(II) and smaller metal ions such as cobalt(III).[78]

4. Conclusions and Outlook

The fundamental basis, hard- and software and the education in theoretical and computational chemistry have developed to an extent where experimentalists in molecular chemistry generally have the basic knowledge, experience and tools to compute structures and properties of compounds of their interest. The increasing awareness of both the power of simple, empirical and approximate approaches, and their limits and pitfalls leads to the hope that "the danger that the cowboys of the keyboard could turn a serious pursuit into disrepute and a circus of pretty pictures" [253] is decreasing. The collection of examples in the previous Sections indicates that structures, conformational equilibria and dynamics, as well as many properties of isolated molecules, solids and systems in solutions may be thoroughly and reasonably accurately computed. The design of novel compounds with given properties is a possible aim but in reality, for a number of reasons (not least also based on difficulties with the experimental part of the studies[19]) not usually fulfilled. More often, molecular modeling is used for the interpretation of structures and properties (stabilities, reactivities and electronics) of novel compounds, and in this way is of much help to design (by further computation or pure intuition) ligands (macrocycles) and complexes with improved performance.

Specifically for experimentalists, simple methods are required, which allow to compute real-life molecules in their preferred conformation and in the medium studied in the laboratory. Within limits (discussed in Section 2 and extensively in the literature cited there) empirical force field and ligand field calculations are such approaches, specifically for macrocyclic ligand complexes. However, it is not only the enormous difference in CPU and real time and effort which have to be invested, which make empirical models appealing: there are examples where interpretations based on a combination of empirical models are at least as insightful as heavy computations on "a higher level of theory";[77,78,196,207] however, there also are many examples where simplistic models alone do not yield the required answers.[245] While one of the major advantages of empirical modeling is that full conformational analyses and dynamics of relatively large systems with solvents or crystal lattices included are accessible, the truth is that these are often neglected. While, in some examples, there might be good reasons to do this, it is clear that this will have to (and probably will) change in the future.

Methods which are attracting increasing attention are data mining which often (but not necessarily always) are based on searches in structural data bases (e.g. the CSD[115]) and concomitant statistical analyses[122,254] and structure correlations in general.[255,256] So far, less widely used, specifically in the area of macrocyclic ligand and general coordination chemistry are quantitative structure-activity and structure-property relationships (QSAR, QSPR) and related methods[150,208,257-264]. There are interesting developments in the area of QSAR, which include, apart from structural information, electronics in the description of the molecules (e.g., the electron-topological (ET) method).[265] Specifically, for transition metal compounds this seems to be a promising approach.

A large variety of methods and the corresponding computer programs for modeling of macrocyclic ligand complexes are available – the choice is dictated by the problem to be solved and, quite generally, this is not the confirmation of a known structure. However, whether modeling is used for designing novel macrocycles with interesting and useful properties or for the interpretation of challenging new experimental results, the basis or aim always is the comparison with an experimental structure or data on other experimentally determined molecular properties and, in the area of macrocycles, usually the most critical and important part is preparative chemistry. A good reason why, in most of the chapters in this book experiments and, specifically, syntheses are a central feature.

5. Acknowledgements

Our own contributions are generously supported financially, primarily by the German Science Foundation (DFG), the Volkswagenstiftung, the Alexander von Humboldt Stiftung and the University of Heidelberg. This and the creative, highly professional and hard work by coworkers, whose names appear in the references, are gratefully acknowledged.

6. References

(1) Williams, R. J. P.; Frausto da Silva, J. J. R. *The natural selection of the chemical elements*; Clarendon Press: Oxford New York, 1996.
(2) Frausto da Silva, J. J. R.; Williams, R. J. P. *The biological chemistry of the elements*; Clarendon Press: Oxford New York, 1991.
(3) Williams, R. J. P. *Bringing Chemistry to Life*; Oxford University Press: Oxford, 2000.
(4) Vallee, B. L.; Williams, R. J. P. *Biochemistry* 1968, *59*, 498.
(5) Williams, R. J. P. *Eur. J. Biochem.* 1995, *234*, 363.
(6) Williams, R. J. P. *Inorg. Chim. Acta* 1971, *5*, 137.
(7) Gray, H. B.; Malmström, B. G. *Biochemistry* 1989, *28*, 7499.
(8) Malmström, B. G. *Eur. J. Biochem.* 1994, *223*, 711.
(9) Comba, P. *Coord. Chem. Rev.* 2000, *200*, 217.
(10) Busch, D. H. *Acc. Chem. Res.* 1978, *11*, 392.
(11) Busch, D. H. *Chem. Rev.* 1993, *93*, 847.
(12) Hancock, R. D. *Progr. Inorg. Chem.* 1989, *37*, 187.
(13) Hancock, R. D. In *Crown Compounds: Toward Future Applications*; Cooper, S. R., Ed.; VCH: Weinheim, New York, 1992, p. 167.
(14) Hay, B. P.; Hancock, R. D. *Coord. Chem. Rev.* 2001, *212*, 61.
(15) Hay, B. P.; Zhang, D.; Rustad, J. R. *Inorg. Chem.* 1996, *35*, 2650.
(16) Comba, P. *Coord. Chem. Rev.* 1993, *123*, 1.
(17) Comba, P. *Coord. Chem. Rev.* 1999, *185*, 81.
(18) Comba, P.; Hambley, T. W. *Molecular Modeling of Inorganic Compounds. 2nd Edition, with a Tutorial, based on MOMEClite*; 2nd ed.; Wiley-VCH: Weinheim, 2001.
(19) Comba, P.; Schiek, W. *Coord. Chem. Rev.* 2003, *238*, 21.
(20) Cram, D. J.; Lein, G. M.; Kaneda, T.; Helgeson, R. C.; Knobler, C. B.; Maverick, E.; Trueblood, K. N. *J. Am. Chem. Soc.* 1981, *103*, 6228.
(21) Artz, S. P.; Cram, D. J. *J. Am. Chem. Soc.* 1984, *106*, 2160.
(22) Cram, D. J. *Science* 1988, *240*, 76.
(23) Lehn, J.-M. *Supramolecular Chemistry: Concepts and Perspectives*; VCH: New York Weinheim Basel, 1995.
(24) Lindoy, L. F.; Atkinson, I. M. *Self-Assembly in Supramolecular Systems in*; RSC: Cambridge, 2000.
(25) Cram, D. J.; deGrandpre, M. P.; Knobler, C. B.; Trueblood, K. N. *J. Am. Chem. Soc.* 1984, *106*, 3286.
(26) Leach, A. R. *Molecular Modelling*; Longman: Edinburgh, 1996.
(27) Jensen, F. *Introduction to Computational Chemistry*; Wiley & Sons Ltd.: Chichester, 1999.
(28) Cramer, C. J. *Essentials of Computational Chemistry*; Wiley-VCH: New York, Weinheim, 2002.
(29) Lindoy, L. F. *The chemistry of macrocyclic ligand complexes*; Cambridge University Press: Cambridge New York, 1989.
(30) Gutsche, C. D. *Calixarenes Revisited*; The Royal Society of Chemistry: Hertfortshire, UK, 1998.
(31) Voloshin, Y. Z.; Kostromina, N. A.; Krämer, R. *Clathrochelates: Synthesis, Structure and Properties*; Elsevier: Amsterdam, 2002.
(32) Koch, W.; Holthausen, M. C. *A Chemist´s Guide to Density Funcional Theory*; Wiley-VCH: Weinheim New York, 2000.
(33) Rappe, A. K.; Casewit, C. J. *Molecular Mechanics across Chemistry*; University Science Books: Sausalito, 1997.
(34) Landis, C. R.; Root, D. M.; Cleveland, T. In *Reviews in Computational Chemistry*; Lipkowitz, K. B., Boyd, D. B., Eds.; VCH Verlag: New York, Weinheim, 1995; Vol. 6, p. 73.
(35) Boeyens, J. C. A.; Comba, P. *Coord. Chem. Rev.* 2001, *212*, 3.
(36) Hay, B. P.; Clement, O. In *The Encyclopedia of Computational Chemistry*; Schleyer, E., Ed.; Wiley, 1998.
(37) Deeth, R. *Comprehensive Coordination Chemistry, Elsevier* 2003, Vol. 2.
(38) Root, D. M.; Landis, C. R.; Cleveland, T. *J. Am. Chem. Soc.* 1993, *115*, 4201.
(39) Cleveland, T.; Landis, C. R. *J. Am. Chem. Soc.* 1996, *118*, 6020.
(40) Landis, C. R.; Cleveland, T.; Firman, T. K. *J. Am. Chem. Soc.* 1998, *120*, 2641.
(41) Firman, T. K.; Landis, C. R. *J. Am. Chem. Soc.* 2001, *123*, 11728.
(42) Landis, C. R.; Firman, T. K.; Cleveland, T.; Root, D. M. In *Molecular Modeling and Dynamics of Bioinorganic Systems*; Banci, L., Comba, P., Eds.; Kluwer: Dordrecht, 1997, p. 49.
(43) Burton, V. J.; Deeth, R. J.; Kemp, C. M.; Gilbert, P. J. *J. Am. Chem. Soc.* 1995, *117*, 8407.
(44) Burton, V. J.; Deeth, R. J. *J. Chem. Soc. Chem. Commun.* 1995, 573.
(45) Deeth, R. J.; Paget, V. J. *J. Chem. Soc., Dalton Trans.* 1997, 537.
(46) Deeth, R. J.; Munslow, I. J.; Paget, V. J. In *in: Molecular Modeling and Dynamics of Bioinorganic Systems*; Banci, L., Comba, P., Eds.; Kluwer: Dordrecht, 1997, p. 77.

(47) Comba, P.; Zimmer, M. *Inorg. Chem.* 1994, *33*, 5368.
(48) Casewit, C. J.; Colwell, K. S.; Rappe, A. K. *J. Am. Chem. Soc.* 1992, *114*, 10035.
(49) Rappé, A. K.; Colwell, K. S.; Casewit, C. J. *Inorg. Chem.* 1993, *32*, 3438.
(50) Allinger, N. L.; Yuh, Y. H.; Lii, J.-H. *J. Am. Chem. Soc.* 1989, *111*, 8551.
(51) Weiner, S. J.; Kollmann, P. A.; Case, D. A.; Singh, U. C.; Ghio, C.; Alagona, G.; Profeta Jr., S.; Weiner, P. *J. Am. Chem. Soc.* 1984, *106*, 765.
(52) Cornell, W. D.; Cieplak, P.; Bayly, C. I.; Gould, I. R.; Merz, K. M.; Ferguson, D. M.; Spellmeyer, D. C.; Fox, T.; Caldwell, J. W.; Kollmann, P. A. *J. Am. Chem. Soc.* 1995, *117*, 5179.
(53) Bernhardt, P. V.; Comba, P. *Inorg. Chem.* 1992, *31*, 2638.
(54) Bol, J. E.; Buning, C.; Comba, P.; Reedijk, J.; Ströhle, M. *J. Comput. Chem.* 1998, *19*, 512.
(55) Norrby, P.-O.; Brandt, P. *Coord. Chem. Rev.* 2001, *212*, 79.
(56) Norrby, P.-O. In *in "Computational Organometallic Chemistry"*; Cundari, E. T. R., Ed.; Dekker: New York, Basel, 2001, p. 7.
(57) Comba, P.; Remenyi, R. *Coord. Chem. Rev.* 2003, *238*, 9.
(58) Dillen, J. L. M. *J. Comput. Chem.* 1992, *13*, 257.
(59) Lopis, A. S.; Glasser, L.; Marsicano, F. *QCPE Bulletin* 1997, *17*, 16.
(60) Norrby, P. O.; Liliefors, T. *J. Comput. Chem.* 1998, *19*, 1146.
(61) Lifson, S.; Warshel, A. *J. Chem. Phys.* 1968, *49*, 5116.
(62) Comba, P.; Daubinet, A.; Stephan, H. *in preparation.*
(63) Hunger, J.; Beyreuther, S.; Huttner, G.; Allinger, K.; Radelow, U.; Zsolnai, L. *Eur. J. Inorg. Chem.* 1998, 693.
(64) Strassner, T.; Busold, M. *J. Mol. Mod.* 2001, *7*, 374.
(65) Marques, H. M.; Cukrowski, I. *Phys. Chem. Chem. Phys.* 2003, 5499.
(66) Marques, H. M.; Cukrowski, I. *Phys. Chem. Chem. Phys.* 2002, *4*, 5878.
(67) Hunger, J.; Huttner, G. *J. Comput. Chem.* 1999, *20*, 455.
(68) Cundari, T. R.; Fu, W. *Inorg. Chim. Acta* 2000, *300*, 113.
(69) Lipkowitz, K. B. *J. Chem. Educ.* 1995, *72*, 1070.
(70) Stockheim, C.; Hoster, L.; Weyhermueller, T.; Wieghardt, K.; Nuber, B. *J. Chem. Soc., Dalton Trans.* 1996, 4409.
(71) Sargeson, A. M. *Coord. Chem. Rev.* 1996, *151*, 89.
(72) Sargeson, A. M. *Pure & Appl. Chem.* 1984, *56*, 1603.
(73) Sargeson, A. M. *Pure & Appl. Chem.* 1986, *58*, 1511.
(74) Cram, D. J.; Cram, J. M. *Container Molecules and their Guests*; The Royal Society of Chemistry: Cambridge, 1997.
(75) Comba, P.; Kerscher, M.; Merz, M.; Müller, V.; Pritzkow, H.; Remenyi, R.; Schiek, W.; Xiong, Y. *Chem. Eur. J.* 2002, *8*, 5750.
(76) Comba, P.; Kerscher, M. *Cryst. Eng.*, 2004, *6*, 197.
(77) Comba, P.; Martin, B.; Prikhod'ko, A.; Pritzkow, H.; Rohwer, H. *Comptes Rendus Chimie, accepted.*
(78) Bleiholder, C.; Börzel, H.; Comba, P.; Heydt, A.; Kerscher, M.; Kuwata, S.; Laurenczy, G.; Lawrance, G. A.; Lienke, A.; Martin, B.; Merz, M.; Nuber, B.; Pritzkow, H. *Inorg. Chem., accepted.*
(79) Menif, R.; Martell, A. E. *Inorg. Chem.* 1993.
(80) Comba, P.; Hambley, T. W.; Hilfenhaus, P.; Richens, D. T. *J. Chem. Soc., Dalton Trans.* 1996, 533.
(81) Karlin, K. D.; Tyeklar, Z. *Bioinorganic Chemistry of Copper*; Chapman & Hall: New York, 1993.
(82) Magnus, K. A.; Ton-That, H.; Carpenter, J. E. *Chem. Rev.* 1994, *94*, 727.
(83) Ma, H.; Allmendinger, M.; Thewalt, U.; Lentz, A.; Klinga, M.; Rieger, B. *Eur. J. Inorg. Chem.* 2002, 2857.
(84) Schindler *Inorg. Chem.* 2003, *42*, 1430.
(85) Comba, P.; Sickmüller, A. F. *Angew. Chem.* 1997, *109*, 2089.
(86) Comba, P.; Cusack, R.; Fairlie, D. P.; Gahan, L. R.; Hanson, G. R.; Kazmaier, U.; Ramlow, A. *Inorg. Chem.* 1998, *37*, 6721.
(87) Börzel, H.; Comba, P.; Katsichtis, C.; Kiefer, W.; Lienke, A.; Nagel, V.; Pritzkow, H. *Chem. Eur. J.* 1999, *5*, 1716.
(88) Comba, P.; Pritzkow, H.; Schiek, W. *Angew. Chem.* 2001, *113*, 2556.
(89) Bernhardt, P. V.; Comba, P.; Fairlie, D. P.; Gahan, L. A.; Hanson, G. R.; Lötzbeyer, L. *Chem. Eur. J.* 2002, *8*, 1527.
(90) Bukowski, M.; Comba, P.; Limberg, C.; Merz, M.; Que Jr., L.; Wistuba, T. *Angew. Chem.* 2004, *116*, 1303.
(91) Saunders, M.; Houk, K. N.; Wu, Y.-D.; Still, W. C.; Lipton, M.; Chang, G.; Guida, W. C. *J. Am. Chem. Soc.* 1990, *112*, 1419.
(92) Cundari, T. R.; Fu, W. *J. Mol. Struct. (Theochem)* 1998, *425*, 51.
(93) Csiki, C.; Zimmer, M. *J. Biomol. Struct. Dyn.* 1999, *17*, 121.
(94) Bartol, J.; Comba, P.; Melter, M.; Zimmer, M. *J. Comput. Chem.* 1999, *20*, 1549.

(95) Mathieu, J.-P. *Ann. de Phys.* 1944, *19*, 335.
(96) Corey, E. J.; Bailar Jr., J. C. *J. Am. Chem. Soc.* 1959, *81*, 2620.
(97) Keene, F. R.; Searle, G. H. *Inorg. Chem.* 1974, *13*, 2173.
(98) Bond, A. M.; Keene, F. R.; Rumble, N. W.; Searle, G. H.; Snow, M. R. *Inorg, Chem.* 1978, *17*, 2847.
(99) Dwyer, M.; Searle, G. H. *J. Chem. Soc., Chem. Commun.* 1972, 726.
(100) Yoshikawa, Y. *Bull. Chem. Soc. Jpn.* 1976, *49*, 159.
(101) Bond, A. M.; Hambley, T. W.; Snow, M. R. *Inorg. Chem.* 1985, *24*, 1920.
(102) Comba, P.; Hambley, T. W.; Zipper, L. *Helv. Chim. Acta* 1988, *71*, 1875.
(103) Comba, P.; Maeder, M.; Zipper, L. *Helv. Chim. Acta* 1989, *72*, 1029.
(104) Comba, P.; Hörmann, A.; Martin, L. L.; Zipper, L. *Helv. Chim. Acta* 1990, *73*, 874.
(105) Bernhardt, P. V.; Comba, P.; Gyr, T.; Várnagy, K. *Inorg. Chem.* 1992, *31*, 1220.
(106) Comba, P.; Goll, W.; Nuber, B.; Várnágy, K. *Eur. J. Inorg. Chem.* 1998, 2041.
(107) Comba, P.; Jakob, H.; Nuber, B.; Keppler, B. K. *Inorg. Chem.* 1994, *33*, 3396.
(108) Thöm, V. J.; Fox, C. C.; Boeyens, J. C. A.; Hancock, R. D. *J. Am. Chem. Soc.* 1984, *106*, 5947.
(109) Connelly, P. J.; Billo, E. J. *Inorg. Chem.* 1987, *26*, 3224.
(110) Adam, K. R.; Atkinson, I. M.; Antolovich, M.; Brigden, L. G.; Lindoy, L. F. *J. Mol. Struct.* 1994, *323*, 223.
(111) Hambley, T. W. *J. Chem. Soc., Dalton Trans.* 1986, 565.
(112) Comba, P.; Luther, S. M.; Maas, O.; Pritzkow, H.; Vielfort, A. *Inorg. Chem.* 2001, *40*, 2335.
(113) Adam, K. R.; Atkinson, I. M.; Lindoy, L. F. *J. Mol. Struct. (THEOCHEM)* 1996, *384*, 183.
(114) Adam, K. R.; Atkinson, I. M.; Lindoy, L. F. *Inorg. Chem.* 1997, *36*, 480.
(115) Allen, F. H. *Acta Cryst.* 2002, *B58*, 380.
(116) Raithby, P. R.; Shields, G. P.; Allen, F. H. *Acta Cryst.* 1997, *1353*, 241.
(117) Raithby, P. R.; Shields, G. P.; Allen, F. H. *Acta Cryst.* 1997, *1353*, 478.
(118) Allen, F. H.; Raithby, P. R.; Shields, G. P. *Struct. Chem.* 1997, *8*, 385.
(119) Cooper, C.; Zimmer, M. *Struct. Chem.* 1999, *10*, 17.
(120) Bakaj, M.; Zimmer, M. *J. Mol. Struct. (THEOCHEM)* 1999, *508*, 59.
(121) Donnelly, M. A.; Zimmer, M. *Inorg. Chem.* 1999, *38*, 1650.
(122) Zimmer, M. *Coord. Chem. Rev.* 2001, *212*, 133.
(123) Csiki, C.; Norenberg, K. M.; Shoemaker, C. M.; Zimmer, M. In *Molecular Modeling and Dynamics of Bioinorganic Systems*; Banci, L. C., P., Ed.; Kluwer Academic Publishers: Dordrecht, Boston, London, 1997, p. 105.
(124) Comba, P.; Fath, A.; Hambley, T. W.; Kühner, A.; Richens, D. T.; Vielfort, A. *Inorg. Chem.* 1998, *37*, 4389.
(125) Comba, P.; Kühner, A. *Eur. J. Inorg. Chem.* 1999, 509.
(126) Comba, P.; Kühner, A.; Peters, A. *J. Chem. Soc., Dalton Trans.* 1999, 509.
(127) Jentzen, W.; Simpson, M. C.; Hobbs, J. D.; Song, X.; Ema, T.; Nelson, N. Y.; Medforth, C. J.; Smith, K. M.; Veyrath, M.; Mazzanti, M.; Ramasseul, R.; Marchon, J.-C.; Takeuchi, T.; Goddard, W. A.; Shelnutt, J. A. *J. Am. Chem. Soc.* 1995, *117*, 11085.
(128) Munro, O. Q.; Bradley, J. C.; Hancock, R. D.; Marques, H. M.; Marsicano, F. W.; Wade, P. W. *J. Am. Chem. Soc.* 1992, *114*, 7218.
(129) Munro, O. Q.; Marques, H. M.; Debrunner, P. G.; Mohanrao, K.; Scheidt, W. R. *J. Am. Chem. Soc.* 1995, *117*, 935.
(130) Zimmer, M.; Crabtree, R. H. *J. Am. Chem. Soc.* 1990, *112*, 1062.
(131) Sparks, L. D.; Medforth, C. J.; Park, M.-S.; Chamberlain, J. R.; Ondrias, M. R.; Senge, M. O.; Smith, K. M.; Shelmutt, J. A. *J. Am. Chem. Soc.* 1993, *115*, 581.
(132) Sato, M.; Walker, F. A.; Raitsimring, A. M.; Walters, W. P.; Dolata, D. P.; Debrunner, P. G.; Scheidt, W. R. *J. Am. Chem. Soc.* 1994, *116*, 7760.
(133) Abraham, R. J.; Margden, I. *Tetrahedron* 1992, *48*, 7489.
(134) Kottalam, J.; Case, D. A. *J. Am. Chem. Soc.* 1988, *110*, 7690.
(135) Kollman, P. A.; Grootenhuis, P. D. J.; Lopez, M. A. *Pure Appl. Chem.* 1989, *61*, 593.
(136) Hancock, R. D.; Weaving, J. S.; Marques, H. M. *J. Chem. Soc.,Chem. Commun.* 1989, 1176.
(137) Shelnutt, J. A.; Medforth, C. J.; Berber, M. D.; Barkigia, K. M.; Smith, K. M. *J. Am. Chem. Soc.* 1991, *113*, 4077.
(138) Medforth, C. J.; Senge, M. O.; Smith, K. M.; Sparks, L. D.; Shelmutt, J. A. *J. Am. Chem. Soc.* 1992, *114*, 9859.
(139) Comba, P.; Hauser, A.; Kerscher, M.; Pritzkow, H. *Angew. Chem.* 2003, *115*, 4675.
(140) Comba, P.; Laorden, C. L. d.; Pritzkow, H. *in preparation.*
(141) Balzani, V.; Venturi, M.; Credi, A. *Molecular Devices and Machines*; Wiley-VCH: Weinheim, 2003.
(142) Schliwa, M. *Molecular Motors*; Wiley-VCH: Weinheim, 2003.
(143) Price, S. L. In *Implications of Molecular and Materials Structure for New Technologies*; Howard, J. A. K., Allen, F. Eds., Ed.; Kluwer Academic Publishers, Dorderecht: Boston, London, 1999.

(144) Comba, P. In *Intermolecular Interactions*; Gans, W., Boeyens, J. C. A., Eds.; Plenum Press: New York, 1998.
(145) Sabolovic, J.; Rasmussen, K. *Inorg. Chem.* 1995, *34*, 1221.
(146) Sabolovic, J.; Liedl, K. R. *Inorg. Chem.* 1999, *38*, 2764.
(147) Gavezzotti, A. *Theoretical aspects and computer modeling of the molecular solid state*; John Wiley & Sons: Chichester New York, 1997.
(148) Leusen, F. J. J.; Wilke, S.; Verwer, P.; Engel, G. E. In *Implications of Molecular and Materials Structure for New Technologies*; Comba, P., Ed.; Kluwer Academic Publishers, Dorderecht: Boston, London, 1999.
(149) Boehme, C.; Wipff, G. *Chem. Eur. J.* 2001, *7*, 1398.
(150) Yoshizuka, K. *Anal. Sciences* 2004, *20*, 761.
(151) Lipkowitz, K. B.; Gao, D.; Schefzick, S.; Katzenelson, O. *J. Am. Chem. Soc.* 1999, *121*, 5559.
(152) Gao, D.; Schefzick, S.; Lipkowitz, K. B. *J. Am. Chem. Soc.* 1999, *121*, 9481.
(153) Lipkowitz, K. B. *Acc. Chem. Res.* 2000, *33*, 555.
(154) Lipkowitz, K. B.; D'Hue, C. A.; Sakamoto, T.; Stack, J. N. *J. Am. Chem. Soc.* 2002, *124*, 14255.
(155) Comba, P. *Comm. Inorg. Chem.* 1994, *16*, 133.
(156) Comba, P. In *Implications of Molecular and Materials Structure for New Technologies*; Howard, J. K. A., Allen, F. H., Shields, G. P., Eds.; Kluwer: Dordrecht, 1999.
(157) Ryde, U.; Nilsson, K. *J. Am. Chem. Soc.* 2003, *125*, 14232.
(158) Ryde, U.; Nilsson, K. *J. Mol. Struct. (THEOCHEM)* 2003, *632*, 259.
(159) Neuhaus, D.; Williamson, M. P. *The nuclear overhauser effect in structural and conformational analysis*; VCH: Weinheim New York, 1989.
(160) Wüthrich, K. *NMR of proteins and nucleic acids*; John Wiley & Sons: New York Chichester, 1986.
(161) Wüthrich, K. *NMR in structural biology*, 1995.
(162) Wüthrich, K. *Acc. Chem. Res.* 1989, *22*, 36.
(163) Wüthrich, K. *Science* 1984, *243*, 45.
(164) Bertini, I.; Luchinat, C.; Parigi, G. *Solution NMR of Paramagnetic Molecules*; Elsevier: Amsterdam, 2001.
(165) Bertini, I.; Rosato, A. In *Molecular Modeling and Dynamics of Bioinorganic Systems*; Banci, L., Comba, P., Eds.; Kluwer: Dordrecht, 1997.
(166) Hambley, T. W.; Jones, A. R. *Coord. Chem. Rev.* 2001, *212*, 35.
(167) Albinati, A.; Lianza, F.; Berger, H.; Pregosin, P. S.; Rüegger, H.; Kunz, R. W. *Inorg. Chem.* 1993, *32*, 478.
(168) Ruegger, H.; Kunz, R. W.; Amman, C. J.; Pregosin, P. S. *Magnetic Resonance in Chemistry* 1991, *29*, 197.
(169) Bunel, S.; Ibarra, C.; Moraga, E.; Blasko, A.; Bunton, C. A. *Carbohydr. Res.* 1993, *244*, 1.
(170) Won, H.; Olson, K. D.; Hare, D. R.; Wolfe, R. S.; Kratky, C.; Summers, M. F. *J. Am. Chem. Soc.* 1992, *114*, 6880.
(171) Beyreuther, S.; Hunger, J.; Cunskis, S.; Diercks, T.; Frick, A.; Planker, E.; Huttner, G. *Eur. J. Inorg. Chem.* 1998, 1641.
(172) Kaupp, M.; Bühl, M.; Malkin, V. G. *Calculation of NMR and EPR Parameters*; Wiley-VCH: Weinheim, 2004.
(173) Rappé, A. K.; Goddard III, W. A. *J. Phys. Chem.* 1991, *95*, 3358.
(174) Sternberg, U., Koch, F.-T.; Möllhoff, M. *J. Comput. Chem.* 1994, *15*, 524.
(175) Sternberg, U.; Witter, R. *J. Comp. Chem.* 2002, *23*, 298.
(176) Sternberg, U.; Pries, W. *THEOCHEM* 2001, *544*, 181
(177) Reiher, M.; Neugebauer, J. *Phys. Chem. Chem. Phys.* 2004, *6*, 4621.
(178) Eksterowicz, J. E.; Houk, K. N. *Chem. Rev.* 1993, *93*, 2439.
(179) Norrby, P.-O.; Rasmussen, T., Haller, J., Strassner, T.; Houk, K. N. *J. Am. Chem. Soc.* 1999, *121*, 10186.
(180) Meunier, B., de Visser, S. P., Shaik, S. *Chem. Rev.* 2004, *104*, 3947.
(181) Shaik, S.; van Eldik, R.; Clark, T. *J. Biol. Inog. Chem.* 2004, *9*, 661.
(182) Selcuki, R.; van Eldik, R. *Inorg. Chem.* 2004, *43*, 2828.
(183) Beierlein, F.; Lanig, H.; Schürer, G.; Horn, A. H. C.; Clark, T. *Mol. Phys.* 2003, *101*, 2469.
(184) Ryde, U. *Curr. Opin. Chem. Biol.* 2003, *1*, 136.
(185) Ziegler, T. *J. Chem. Soc., Dalton Trans.* 2002, 642.
(186) Autschbach, J.; Jorge, F. E.; Zieger, T. *Inorg. Chem.* 2003, *42*, 2867.
(187) Autschbach, J.; Jorge, F. E.; Zieger, T. *Inorg. Chem.* 2003, *43*, 8902.
(188) Neese, F. *J. Chem. Phys.* 2003, *119*, 9428.
(189) Neese, F. *Inorg. Chim. Acta* 2002, *337C*, 181.
(190) Neese, F. *Curr. Opin. Chem. Biol.* 2003, *7*, 125.

(191) Fujita, K.; Schenker, R.; Gu, W.; Brunold, T. C.; Cramer, S. P.; Riordan, C. G. *Inorg. Chem.* 2004, *43*, 3329.
(192) Atanasov, M. A.; Daul, C. A.; Ranzy, C. *Chem. Phys. Lett.* 2003, *367*, 737.
(193) Larsen, E.; La Mar, G. N. *J. Chem. Educ.* 1974, *51*, 633.
(194) Schäffer, C. E. *Pure Appl. Chem.* 1970, 361.
(195) Bernhardt, P. V.; Comba, P. *Inorg. Chem.* 1993, *32*, 2798.
(196) Comba, P. *Inorg. Chem.* 1994, *33*, 4577.
(197) Comba, P.; Hambley, T. W.; Hitchman, M. A.; Stratemeier, H. *Inorg. Chem.* 1995, *34*, 3903.
(198) Bernhardt, P. V.; Comba, P.; Hambley, T. W.; Lawrance, G. A. *Inorg. Chem.* 1991, *30*, 942.
(199) Stratemeier, H.; Hitchman, M. A.; Comba, P.; Bernhardt, P. V.; Riley, M. J. *Inorg.Chem.* 1991, *30*, 4088.
(200) Geue, R. J.; Hanna, J.; Höhn, A.; Qin, C. J.; Ralph, S. F.; Sargeson, A. M.; Willis, A. C. In *Electron Transfer Reactions*; Issied, S. S., Ed.; American Chemical Society: New York, 1997.
(201) Smith, T. D.; Pilbrow, J. R. *Coord. Chem. Rev.* 1974, *13*, 173.
(202) Wang, D.; Hanson, G. R. *J. Magn. Reson. A* 1995, *117*, 1.
(203) Wang, D.; Hanson, G. R. *Appl. Magn. Reson.* 1996, *11*, 401.
(204) Bernhardt, P. V.; Comba, P.; Hambley, T. W.; Massoud, S. S.; Stebler, S. *Inorg. Chem.* 1992, *31*, 2644.
(205) Comba, P.; Hilfenhaus, P. *J. Chem. Soc., Dalton Trans.* 1995, 3269.
(206) Comba, P.; Gavrish, S. P.; Hay, R. W.; Hilfenhaus, P.; Lampeka, Y. D.; Lightfoot, P.; Peters, A. *Inorg. Chem.* 1999, *38*, 1416.
(207) Comba, P.; Kerscher, M.; Lampeka, Y. D.; Lötzbeyer, L.; Prikhod'ko, A. I.; Pritzkow, H.; Tsymbal, L. *Inorg. Chem.* 2003, *42*, 3387.
(208) Comba, P.; Sickmüller, A. F. *Inorg. Chem.* 1997, *36*, 4500.
(209) Comba, P.; Jakob, H. *Helv. Chim. Acta* 1997, *80*, 1983.
(210) Comba, P. *Coord. Chem. Rev.* 1999, *182*, 343.
(211) Golub, G.; Cohen, H.; Meyerstein, D. *J. Chem. Soc., Chem. Commun.* 1992, 397.
(212) Golub, G.; Cohen, H.; Paoletti, P.; Bencini, A.; Messori, L.; Bertini, I.; Meyerstein, D. *J. Am. Chem. Soc.* 1995, *117*, 8353.
(213) Bernhardt, P. V. *J. Am. Chem. Soc.* 1997, *119*, 771.
(214) Clark, T.; Hennemann, M.; van Eldik, R.; Meyerstein, D. *Inorg. Chem.* 2002, *41*, 2927.
(215) Picket, C. J.; Pletcher, D. *J. Organomet. Chem.* 1975, *102*, 327.
(216) Bursten, B. E.; Green, M. R. *Inorg. Chem.* 1988, *35*, 393.
(217) Lever, A. B. P. *Inorg. Chem.* 1990, *29*, 1271.
(218) Lever, A. B. P. *Inorg. Chem.* 1991, *30*, 1980.
(219) Sima, J.; Labud, J. *Comments Inorg. Chem.* 1988, *27*, 2496.
(220) Fischer, E. *Chem. Ber.* 1984.
(221) Rebek, J. J. *Prog. Mol. Recog.* 1988, 222.
(222) Lehn, J.-M. *J. Inclusion Phenom.* 1988, *6*, 353.
(223) Busch, D. H.; Farmery, K.; Goedken, V.; Katrich, V.; Melnyk, A. C.; Sperati, C. R.; Tokel, N. *Adv. Chem. Ser.* 1971, *100*, 44.
(224) Vedani, A.; Dobler, M.; Dollinger, H.; Hasselbach, K.-M.; Birke, F.; Lill, M. A. *J. Med. Chem.*
(225) Zimmer, M. *Chem. Rev.* 1995, *8*, 2629.
(226) Doucet, J.-P.; Weber, J. *Computer-Aided Molecular Design*; Academic Press: London, 1996.
(227) Drummond, L. A.; Hendrick, K.; Kanagasundaram, M. J. L.; Lindoy, L. F.; MacPartlin, M.; Tasker, P. A. *Inorg. Chem.* 1982, *21*, 3923.
(228) Henrick, K.; Tasker, P. A.; Lindoy, L. F. *Prog. Inorg. Chem.* 1985, *33*, 1.
(229) Lindoy, L. F. *Progr. Macrocycl. Chem. (Izatt, Ed.)* 1986, *33*, 53.
(230) Martin, L. Y.; DeHayes, L. J.; Zompa, L. J.; Busch, D. H. *J. Am. Chem. Soc.* 1974, 4046.
(231) Buckingham, D. A.; Sargeson, A. M. In *Chelating Agents and Metal Chelats*; Dwyer, F. P., Mellor, D. P., Eds.; Academic Press: New York, 1964, p. 237.
(232) Comba, P.; Fath, A.; Kühner, A.; Nuber, B. *J. Chem. Soc., Dalton Trans.* 1997, 1889.
(233) Troxler, L.; A. Dedien; Hutschka, F.; Wipff, G. *J. Mol. Struct. (THEOCHEM)* 1998, *431*, 151.
(234) Boehme, C.; Wipff, G. *Inorg. Chem.* 1999, *38*, 5734.
(235) Boehme, C.; Wipff, G. *Chem. Eur. J.* 2001, 1398.
(236) Drew, M. G. B.; Rice, D. A.; bin Silong, S.; Yates, P. C., *Dalton Trans.* 1986, 1081.
(237) Hancock, R. D. *J. Chem. Soc., Dalton Trans.* 1986, 2505.
(238) Comba, P.; Okon, N.; Remenyi, R. *J. Comput. Chem.* 1999, *20*, 781.
(239) Hancock, R. D. *Acc. Chem. Res.* 1990, *23*, 253.
(240) Hay, B. P.; Rustad, J. R. *Supramolecular Chemistry* 1996, *6*, 383.
(241) Comba, P.; Hambley, T. W.; Ströhle, M. *Helv. Chim. Acta.* 1995, *78*, 2042.
(242) Curtis, N. J.; Lawrance, G. A.; Sargeson, A. M. *Aust. J. Chem.* 1983, *36*, 1327.
(243) Ventur, D.; Wieghardt, K.; Nuber, B.; Weiss, J. *Z. anorg. allg. Chem.* 1987, *551*, 33.

(244) Hambley, T. W. *Inorg. Chem.* 1988, *27*, 2496.
(245) Comba, P.; Lienke, A. *Inorg. Chem.* 2001, *40*, 5206.
(246) Chapman, J.; Ferguson, G.; Gallagher, J. F.; Jennings, M. C.; Parker, D., *J. Chem. Soc, Dalton Trans.* 1992, 345.
(247) Fabbrizzi, L.; Micheloni, M.; Paoletti, P. *J. Chem. Soc., Dalton Trans.* 1979, 1581.
(248) Fabbrizzi, L.; Micheloni, M.; Paoletti, P. *J. Chem. Soc , Dalton Trans.* 1979, 389.
(249) Hosken, G. D.; Hancock, R. D. *J. Chem. Soc., Chem. Commun.* 1994, 1363.
(250) Hosken, G. D.; Allan, C. C.; Boeyens, J. C. A.; Hancock, R. D. *J. Chem. Soc., Dalton Trans.* 1995, 3705.
(251) Irving, H.; Williams, R. J. P. *J. Chem. Soc.* 1953, 3192.
(252) Irving, H.; Williams, R. J. P. *Nature* 1948, *162*, 746.
(253) Boeyens, J. C. A. In *Fundamental priciples of molecular modeling*; Gans, W., Amann, A., Boeyens, J. C. A., Eds.; Plenum Press: New York, 1996, p. 1.
(254) Harris, S. E.; Pascual, I.; Orpen, A. G. *J. Chem. Soc., Dalton Trans.* 2001, 2996.
(255) Bürgi, H.-B.; Dunitz, J. D. *Structure Correlation, Vol.1*; VCH: Weinheim New York, 1994.
(256) Bürgi, H.-B.; Dunitz, J. D. *Structure Correlation, Vol. 2*; VCH: Weinheim New York, 1994.
(257) Hansch, C.; Fujita, T. *J. Am. Chem. Soc.* 1964, *86*, 1616.
(258) Oprea, T. I.; Waller, C. L. In *Reviews in Computational Chemistry*; Lipkowitz, K. B., Boyd, D. B., Eds.; VCH Verlag: New York Weinheim, 1997; Vol. 11, p. 127.
(259) Vedani, A.; Dobler, M. *J. Med. Chem.* 2002, *45*, 2139.
(260) Hancock, R. D.; McDougall, G. J.; Marsicano, F. *Inorg. Chem.* 1979, *18*, 2847.
(261) Hay, B. P.; Rustad, J. R.; Hostetler, C. J. *J. Am. Chem. Soc.* 1993, *115*, 11158.
(262) Comba, P.; Gloe, K.; Inoue, T.; Stephan, H.; Yoshizuka, K. *Inorg. Chem.* 1998, *37*, 3310.
(263) Selzer, P.; Gasteiger, J.; Thomas, H.; Salzer, R. *Chem. Eur. J.* 2000, *6*, 920.
(264) Hay, B. P.; Dixon, D. A.; Vargas, R.; Garza, J.; Raymond, K. N. *Inorg. Chem.* 2001, *40*, 3922.
(265) Bersuker, I. B.; Dimoglo, A. S. In *Reviews in Computational Chemistry*; Lipkowitz, K. B., Boyd, D. B., Eds.; Wiley-VCH: Weinheim New York, 1991; Vol. 2, p. 423.

21.
SIMULATIONS OF THE DYNAMICS OF 18-CROWN-6 AND ITS COMPLEXES: FROM THE GAS PHASE TO AQUEOUS INTERFACES WITH SC-CO_2 AND A ROOM-TEMPERATURE IONIC LIQUID

A. CHAUMONT, R. SCHURHAMMER,
P. VAYSSIÈRE AND G. WIPFF
Laboratoire MSM, Institut de Chimie,
4, rue B. Pascal, 67 000 Strasbourg, France

1. Introduction

There have always been tight connections between supramolecular and theoretical chemistry. In the early seventies, quantum chemists introduced the "supermolecule" approach to analyze the strength and nature of the non -covalent interactions which assemble simple systems like the water dimer or water-ion adducts in the gas phase, and presumably larger host-guest complexes in solution.[1] Among the many macro(poly)cyclic hosts developed experimentally,[2-4] 18-crown-6 ("18C6") turned out to be the most studied by theoreticians. Indeed, in spite of its simplicity and relative small size, 18C6 is a highly versatile molecule which acts as a selective host for small ions and molecules and displays fundamental features of molecular recognition: preorganization and macrocyclic effect, flexibility, induced fit upon ligand binding and solvent dependent recognition properties.[5-7] In the early eighties, it was still too computer time demanding to study real host-guest complexes by quantum mechanical (QM) *ab initio* methods, and most of the studies employed empirical force field approaches, in which the system is represented by soft charged spheres (atoms) connected by harmonic springs (bonds). The first simulations were of molecular mechanics type, i.e. based on energy minimized "static" models in the gas phase.[8] Later on, insights into the motions of 18C6 and its cation complexes were obtained from harmonic dynamics (normal modes of vibration), or classical molecular dynamics (MD) simulations, also employing a force field representation of the potential energy.[9] The simulated time scales evolved with the computer resources, from about 0.1 ns for the first simulations *in vacuo*, to about 10 ns today for complexes in a solvent environment represented explicitly at the molecular level. The possibility to calculate thermodynamic properties via statistical perturbation theory coupled with MD simulations was a major breakthrough in the field of molecular recognition for host-guest complexation in solution[10, 11] and has been developed for 18C6 complexes.[12-14]

Karsten Gloe (ed.), Macrocyclic Chemistry: Current Trends and Future Perspectives, 327–348.
© 2005 *Springer. Printed in the Netherlands.*

Meanwhile, cation - crown complexes were fully optimized by *ab initio* QM techniques in the gas phase, allowing to investigate the effect of electronic reorganization (mainly charge transfer + polarization effects) on the recognition properties.[15, 16] QM/MM hybrid methods which combine a QM representation of the ion coordination sphere with a MM representation of the more remote species, have been tested on 18C6/ cation complexes.[17, 18] Typical applications of "computational approaches in supramolecular chemistry" up to 1994 can be found in ref. [19] There are also reviews on more specialized fields like calixarenes.[20, 21]

In this paper, we present three recent applications of computer modeling, again focusing on 18C6. The first one is mainly methodological and concerns the application of Car-Parrinello molecular dynamics (CP-MD) on "supermolecules" involving 18C6, i.e. the 18C6-H_2O adduct. It will be shown that a dynamic picture of the system solves the apparent conflicting results on hydrogen binding patterns obtained by IR spectroscopy, and solid-state or classical MD or Monte Carlo results. The two other applications, based on classical MD simulations, address the question of solvation and liquid-liquid extraction of the Sr^{2+} cation by 18C6 to "green solvents", namely supercritical-CO_2 (SC-CO_2) and a room temperature ionic liquid, for which the "dry" *versus* "humid" states will be compared, focusing on what happens at the aqueous interface.

2. Methods

2.1 CLASSICAL MOLECULAR DYNAMICS

An introduction to the modeling methods can be found in refs. [22, 23]. The classical MD simulations reported here were performed with the modified AMBER software,[24] in which the potential energy consists of harmonic deformations of bond and angles, dihedral energies, plus non-bonded interactions represented by a sum of pair wise additive coulombic and van der Waals contributions:

$$U = \sum_{bonds} K_r (r - r_{eq})^2 + \sum_{angles} K_\theta (\theta - \theta_{eq})^2 + \sum_{dihe} \sum_{n} V_n (1 + \cos n\phi)$$
$$+ \sum_{i<j} \frac{q_i q_j}{R_{ij}} - 2\varepsilon_{ij} \left(\frac{R_{ij}^*}{R_{ij}} \right)^6 + \varepsilon_{ij} \left(\frac{R_{ij}^*}{R_{ij}} \right)^{12}$$

One important feature is the representation of ion interactions by a non-covalent model, which allows for changes in its coordination sphere during the dynamics, while MM3-derived covalent models cannot.[25] The parameters for 18C6 and its hydrates are from ref. [26] and we used ESP derived atomic charges of Kollman *et al.* [27] on the crown (q_O = -0.404, q_C = 0.244 and q_H = -0.021 e). The H_2O and CO_2 molecules were represented with the TIP3P [28] and Murthy *et al.* [29] three point models, respectively. The Sr^{2+} ion parameters were fitted from its free energy of hydration.[30] For the study of the water / SC-CO_2 interfacial system, more details are given in ref. [31]. The corresponding non-bonded interactions were calculated with a 15 Å cutoff and a reaction field correction for long range electrostatics, while for the ionic liquid simulations, the electrostatics was calculated with a 12 Å cutoff plus Ewald correction. After energy minimization and stepwise equilibration of each system, MD was run in the (N,V,T) ensemble, typically for 1 - 10 ns. The methodology and parameters used for the MD simulations in the

[BMI][PF$_6$] ionic liquid are described in refs. [32-34] The solvent box contains about 300 BMI$^+$ PF$_6^-$ ions for the dry liquid, plus 300 H_2O molecules for the humid liquid. Several distinguishing features of these ionic liquid compared to water systems should be noted. (i) As concerns the force field, there is presently limited experience and validations, compared to the many studies on water, particularly as far as the consistency between the solvent and solute parameters, and the possible importance of many-body, charge transfer and polarization effects are concerned. (ii) The ionic nature of the liquids requires an adequate treatment of long range electrostatics (with e.g. Ewald summation methods), which may be more critical than with liquids based on neutral components. (iii) The relaxation times of the solvent and solutes are several orders of magnitude slower in the ionic liquid than in water, thus rising the important issue of adequate sampling of the different configurations of the system.

2.2 CP-MD CALCULATIONS

While classical MD methods assume a fixed electrostatic representation (atomic charges) of the system, QM methods focus on the electron distribution within the Born-Oppenheimer approximation, i.e. with a given configuration of the nuclei. During the last few years, it has become feasible to perform *ab initio* molecular dynamics, in which the forces between nuclei are determined "on the fly" from QM calculations based on the density-functional theory (Car-Parrinello molecular dynamics "CP-MD"). CP-MD simulations have been successfully employed in the study of many ions in solution,[35-37] and gas phase water clusters. In contrast to classical MD, they allow for bond-making or bond-breaking processes during the dynamics.[38-43] The CP-MD results presented here were obtained using the density-functional based Car-Parrinello scheme [44] as implemented in the CPMD program.[45] The BLYP functional combination was employed, together with norm-conserving pseudopotentials generated according to the Martins and Troullier procedure [46] and transformed into the Kleinman-Bylander form (BLYP MT).[47] Periodic boundary conditions were imposed using cubic supercells with box length of 15 Å. Kohn-Sham orbitals were expanded in plane waves up to a kinetic energy cutoff of 80 Ry. The CP-MD simulations were performed starting from the lowest-energy structure, using a fictitious electronic mass of 600 a.u. and a time step of 0.121 fs. The simulations were performed in the N,V,E ensemble over 2 - 10 ps at ca. 300(±30) K. More details are given in refs. [48, 49]

3. Results

3.1 18-CROWN-6 AND ITS HYDRATE: THE MERRY-GO-ROUND DYNAMICS OF H_2O OVER THE D_{3D} CROWN

Figure 1. Hydration of the D_{3d} crown (X-ray structure of the 18C6.4H_2O, from refs. 56,57). Orthogonal views. Only 2 H_2O molecules are shown on the left side for clarity.

The 18C6 crown is very flexible, as shown by the diversity of structures found in the solid state,[50] and by sampling simulations in the gas phase [51] or in solution.[52, 53] Two forms, respectively of C_i and D_{3d} symmetry, are however particularly important: the C_i form adopted by the uncomplexed crown in the solid state has no cavity, while D_{3d} displays a "circular cavity" in which K^+ fits nicely, leading to the K^+ recognition by the crown in solution. In 1984, Monte Carlo calculations predicted that dissolution of the crystal (C_i form) in water would lead to conformational change to D_{3d}, due to specific hydration patterns.[54] The predicted D_{3d} hydration scheme involved centro-symmetrically related 2+2 H_2O molecules (Figure 1). There are two H_2O molecules at each face of the crown, one in $O_1 \cdots O_7$ bridging position, and the second also bridging over the O_4 oxygen of the crown and the $O(H_2O)$ oxygen. These predictions were confirmed by X-ray crystallography on crystals of 18C6 containing H_2O molecules,[55-57] by Raman or IR spectroscopy of 18C6 hydrates [58-61] and by more refined MD simulations of 18C6 in water.[62] The bridging hydration of the crown seems however to be inconsistent with the IR spectroscopy results in humid CCl_4 [63] or in humid SC-CO_2 solutions.[64] According to these IR studies, H_2O mainly binds monodentate to the crown ("linear" HO-H...O_{18C6} hydrogen bonds, see Figure 2) while a smaller amount, whose proportion increases at lower temperatures, binds bidentate.[63] Which form of 18C6 is present in solution cannot be inferred from these IR studies.

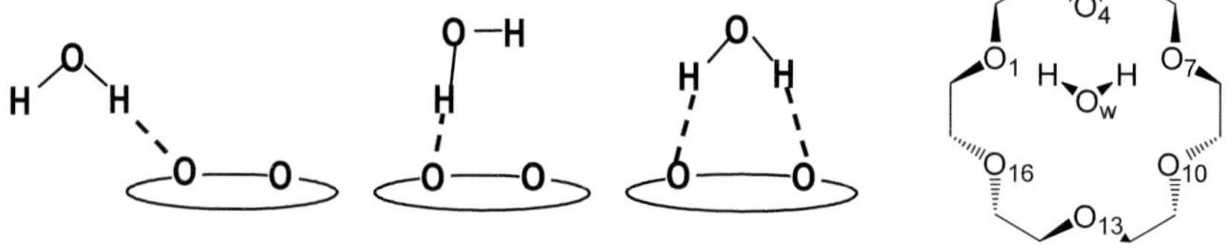

Figure 2. Schematic representation of single-*out*, single-*in* (monodentate) and bridging (bidentate) coordination of H_2O to 18C6, and of the D_{3d} bridging hydrate.

Recent QM and CP-MD investigations on 18C6 and its hydrates solve this apparent contradiction between IR observations and the large occurrence of bridging hydration of 18C6.[49] First, high level QM DFT calculations at the BLYP/6-311G++(3df,3pd) // BLYP/6-31G* level reveal that the C_i and D_{3d} forms are quasi iso-energetic in the gas phase ($\Delta E = 0.1$ kcal/mol) as found with AMBER calculations ($\Delta E = 0.3$ kcal/mol), but contrasts with previously reported MP2 calculations which used nonoptimized structures ($\Delta E = 17$ kcal/mol in favor of C_i).[65]

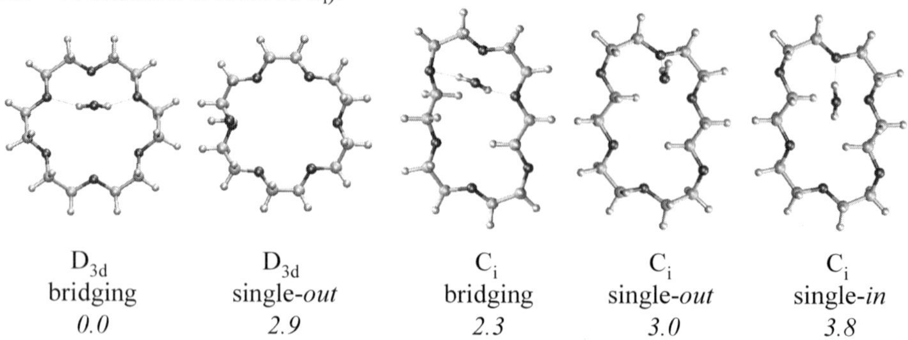

Figure 3. Optimized structures of single-*out*, single-*in* (monodentate) and bridging (bidentate) hydrates of 18C6, and relative energies (kcal/mol).

Several forms of the 18C6-H_2O adduct were then modeled, based on the C_i *versus* D_{3d} crown, where H_2O is either bridging or makes a "single" hydrogen bond with the crown, turned "inside" or "outside" the crown (Figure 3). They were found to follow the order of relative energies: C_i-*single-in* > C_i-*single-out* > D_{3d}-*single-out* > C_i-*bridging* > D_{3d}-*bridging* (see Figure 3), thus confirming the highest stability of the D_{3d}-bridging hydration and that, for a given conformer of the crown, bridging is more stable than "single" hydrogen bonding. The calculated binding energy of bridging water (6.1 kcal/mol for the D_{3d} crown, and 3.8 kcal/mol for the C_i form) also favors the D_{3d} crown. As mentioned above, this D_{3d}-bridging "static" structure of the hydrate is not consistent, however, with IR data. We notice that the two hydrogen bond distances formed by this H_2O molecule are relatively long (2.12 and 2.16 Å, respectively) and not equal, which suggests that the D_{3d} crown is somewhat too big to accommodate the H_2O guest, and this will have consequences for the dynamics of the 18C6-H_2O adducts.

The CP-MD trajectories of the 18C6-H_2O "supermolecule" were then calculated for the C_i and D_{3d} crowns starting with one bridging water molecule. The latter dissociated from the C_i crown in less than 1 ps, indicating that this complex is too weakly bound to compete with the kinetic energy at 300 K. This contrasts with the hydrate of the D_{3d} crown, which remained bound when the dynamics was pushed up to 10 ps, indicating that it is kinetically stable.

The D_{3d} crown -H_2O adduct reveals very interesting dynamics features, in which the motions of the water molecule and the crown are coupled. Typical snapshots are shown in Figure 4 and the time evolution of O¨O distances is shown in Figure 5. First, it should be noted that the crown is never of exact D_{3d} symmetry but deforms in such a way that opposite O_i¨O_{i+7} distances fluctuate (between about 4.6 and 5.6 Å), which prevents optimal bridging interactions with the H_2O molecule. As a result, the latter is most often "monodentate", with one proton anchored to one O_i oxygen (i = 1, 7, 13) on top of the crown, while its other proton is loosely bound to the O_{i+7} or the O_{i-7} oxygens, at too large distances (> 2.5 Å), however, to form real hydrogen bonds. From time to time, this free H_{H2O} proton "jumps" towards and anchors to one of the crown oxygens, while the other H_{H2O} proton becomes "free". As a result, the H_2O molecule undergoes a merry-go-round rotation motion over the crown during the dynamics. Due to its shape, the D_{3d} crown "catalyzes" this process during which quadrupole (18C6) / dipole (H_2O) interactions are retained, while the C_i form cannot, thus also explaining why the bridging hydrate dissociates from the C_i crown. As a result of the dynamics, the lifetime of the "bridging" hydration is much shorter than the "linear" hydration. Among the 10 000 saved structures, only 15 % simultaneously display two HO-H¨O "hydrogen bonds" shorter than 2.5 Å. Molecules are always in motion and, given the threefold symmetry of the crown, one can make an analogy with "waltzing": the H_2O molecule is "dancing" over the crown, sometimes hesitating between on foot or the other, between one direction or the other, and has therefore statistically one foot (hydrogen) "on the ground", and the other one lifted. The slower the music (reduced temperature), the slower is the dance, and the larger is the probability of seeing the "two feet on the ground" (bridging hydration). This is what is seen by IR spectroscopy in weakly humid apolar media like CCl_4 or CO_2.

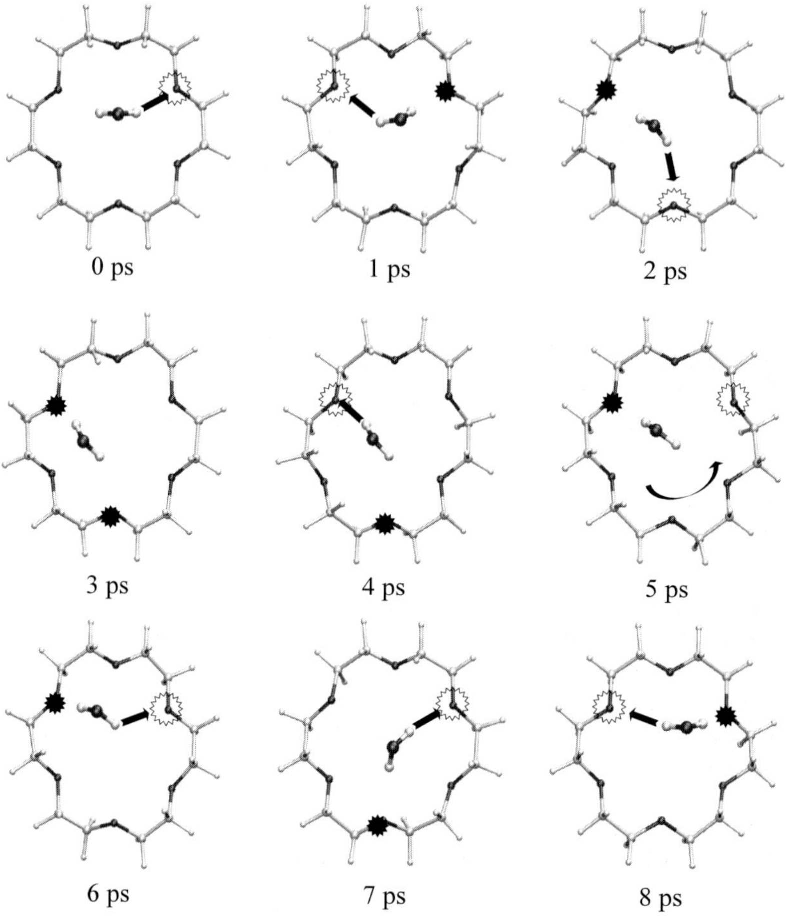

Figure 4. CP-MD dynamics of the 18C6 D_{3d} monohydrate. Selected structures with schematic representation of the H_2O anchoring and motions

It should be noted that in aqueous solution, as well as in solid state structures of hydrates, the D_{3d} crown is generally tetra-hydrated, and that the bridging H_2O molecule is prevented from "waltzing" by another, cooperatively bound, H_2O molecule (Figure 2). This CP-MD study thus demonstrates the importance of dynamics in host - guest complexes, and solves the apparent contradiction between different binding modes "observed" at different time scales in somewhat different environments. The recent CP-MD study of the $H_3O^+ \subset 18C6$ "supermolecule" provides another example of how the host's topology "catalyzes" the motions of the guest.[48] As pointed out previously,[9] dynamics coupling of the host and guest motions may have deep consequences in the field of molecular recognition in supramolecular chemistry, as well as in biology.

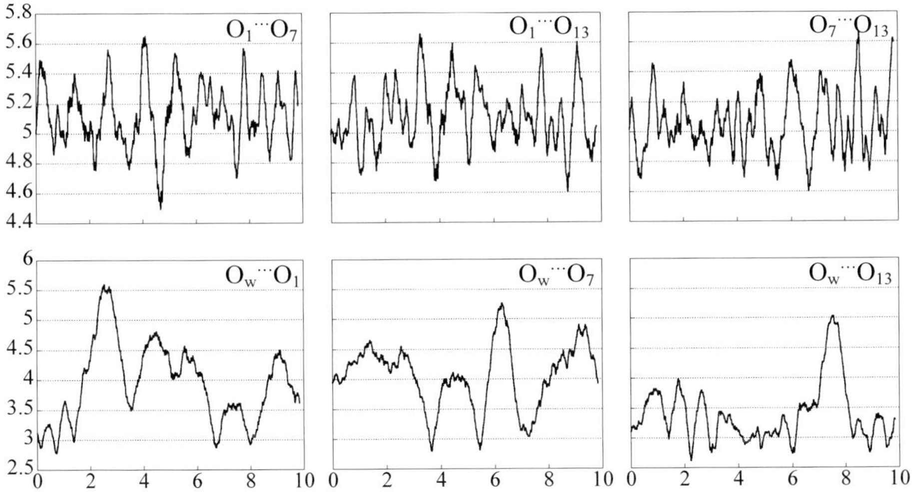

Figure 5. CP-MD dynamics of the 18C6 D_{3d} monohydrate. $O_{18C6}\cdots O_{18C6}$ and $O_{18C6}\cdots O_{H2O}$ distances (Å) as a function of time (ps). See Figure 2 for atom labels.

On the methodological side, it is gratifying to note, once these computer time consuming CP-MD calculations have been performed, that rapid classical dynamics AMBER calculations on the 18C6-H_2O hydrate essentially yield similar qualitative conclusions: instability of the C_i hydrate, and "waltzing dynamics" of H_2O over the D_{3d} crown, which also validates early and subsequent force field studies,[8] also pointing out the limitations of static views, even obtained by sophisticated quantum calculations.

3.2 THE STRONTIUM EXTRACTION BY 18C6 TO SC-CO_2: IMPORTANCE OF INTERFACIAL PHENOMENA AND COUNTERIONS

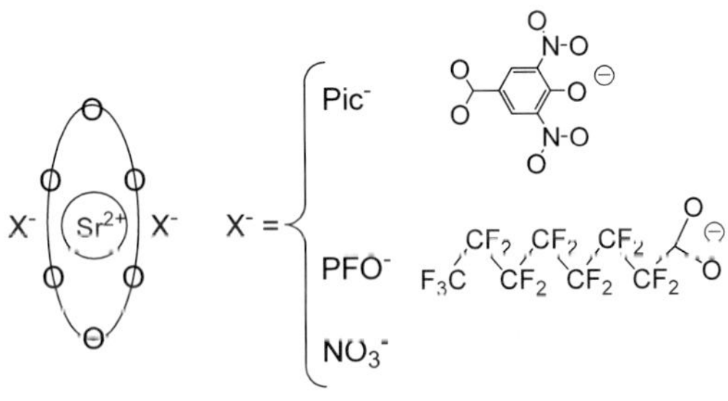

Figure 6. The simulated Sr⊂18C6$(X)_2$ complexes (X^- = Pic^-, PFO^-, NO_3^-).

There are strong analogies between metal ion extraction by complexant molecules to classical organic liquids (e.g. haloalkanes, aromatics) and to supercritical fluids like SC-CO_2, thus leading to promising developments in the context of "green chemistry".[66-69] In this section, we focus on the ion extraction by crown ethers, which is generally conducted with lipophilic counterions (e.g. phenolate derivatives, fatty acids) in classical conditions, and with more CO_2-philic fluorinated anions to SC-CO_2.[70, 71] More specifically, we wanted to understand why the perfluorooctanoate anion (noted PFO^-) promotes the Sr^{2+} extraction to SC-CO_2,[70, 72] while the picrate anion (noted Pic^-), widely used in classical extraction does not. We thus simulated a concentrated solution of Sr⊂18C6$(X)_2$ complexes, with X^- = Pic^- *versus* PFO^- as counterions (see Figure 6) at the water / SC-CO_2 interface. The interface is presumably the place where hydrophobic extractants and anions (initially in the "receiving phase") and metallic ions (initially in water) meet, recognize each other and form a complex hydrophobic enough to be extracted. Computer simulations contributed to our understanding of "what happens at the interface",[73] and highlighted the importance of interfacial phenomena in ion extraction. The first simulations at classical interfaces showed that extractant molecules like calixarenes, crown ethers, cryptands phosphoryl-containing ligands are highly surface active, be they uncomplexed or complexed,[74-78] as do most of hydrophobic ions, even lacking an amphiphilic topology.[79] Later on, MD simulations pointed to the analogies between aqueous interfaces with classical and SC-CO_2 phases.[80, 81] A recently reported paper allows us to understand why PFO^- anions are more effective than Pic^- anions in promoting the extraction of Sr⊂18C6^{2+} to SC-CO_2.[26]

Figure 7 shows the interfacial CO_2 / water landscape of concentrated solutions of strontium complexes at the end of the simulation, and reveals the importance of interfacial phenomena for both Sr⊂18C6$(PFO)_2$ and Sr⊂18C6$(Pic)_2$ complexes. None of them sits in water, as anticipated for these complexes which instead concentrate at the interface. This can be rationalized from the fact that, at the interface, they still enjoy significant attractions with the aqueous phase (-117 and -202 kcal/mol, respectively on the average per complex, according to an energy component analysis) without having to pay for a high "cavitation energy" in water. There are however remarkable counterion effects on the interfacial distributions. With Pic^- as counterions, all complexes are trapped at the interface. An important driving force is the tendency of Pic^- anions to aggregate via Pic^- /Pic^- hydrophobic π–stacking interactions (Figure 7). According to MD studies,[82] these anions self-assemble in aqueous environments and at aqueous interfaces [79] and these interactions also contribute to the supramolecular 2D-arrangements formed by the Sr⊂18C6$(Pic)_2$ complexes at the interface. In the context of liquid-liquid extraction, this is deleterious, however, as the resulting two dimensional film forms a kind of "interfacial crust" which prevents the interface crossing by the complexes and the ligands.

The case of Sr⊂18C6$(PFO)_2$ complexes differs, as the PFO^- counterions are amphiphilic in nature and their carboxylate head coordinates to strontium, while their perfluoroalkyl chain cannot and do not attract each other, thus preventing self-aggregation of the complexes. The affinity of these fluorinated chains for CO_2 contributes to solubilize the strontium complexes. As a result, the Sr⊂18C6$(PFO)_2$ complexes sitting at the interface are "diluted", while the others migrated from the interface to SC-CO_2: 40 % are found at more than 8 Å from the interface, and can thus be considered as "extracted" at 305 K.

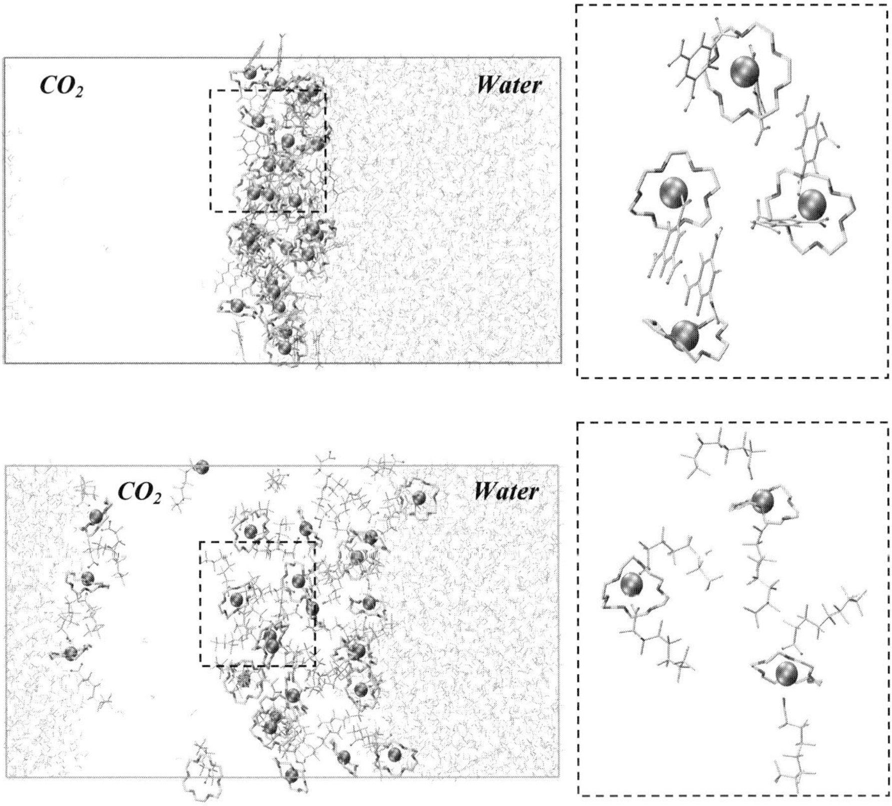

Figure 7. Distribution of 25 Sr⊂18C6(Pic)$_2$ complexes *(top)* and 25 Sr⊂18C6(PFO)$_2$ *(bottom)* complexes at the CO_2 / water interface. *Left:* snapshots after 1.2 ns of dynamics at 305 K, where CO_2 is not shown for clarity. *Right:* zoom on typical complexes. A color version is given in the Appendix.

The uncomplexed forms of the SrX_2 salts display marked analogies with the strontium complexes of 18C6: two-dimensional self-assembling of Sr(Pic)$_2$ species at the interface, and diluted monolayer of PFO^- anions, neutralized by the Sr^{2+} counterions.[31] In real extraction systems, the PFO^- anions stem from the PFOH acid, or from salts (e.g. quaternary ammonium$^+$ PFO^-). Like 18C6 itself, these species are surface active and compete with the strontium complexes to saturate the interface, leading to an equilibrium between adsorbed and extracted species, thus pointing to the importance of interfacial phenomena in ion extraction and, in the context of supramolecular chemistry, to the importance of self-assembling processes. Beyond simple molecules or "supermolecules" of host-guest type, it is thus important to study concentration and collective effects at interfaces. Depending on the concentration of ions, extractants, "synergy agents" (e.g. counterions, co-extractants, solvent modifiers) and on the water - "oil" ratio, the "interface" may evolve from a well-defined "flat" abrupt zone of 15- 20 Å width, to a mixed "third phase" between aqueous and CO_2 phases, to micelles or microemulsions, and important microscopic insights into these systems can be gained by simulation studies.[83, 84]

3.3 THE STRONTIUM EXTRACTION BY CROWN ETHERS TO A ROOM - TEMPERATURE IONIC LIQUID. SOLVATION IN DRY VERSUS "HUMID" FORMS OF THE LIQUID, AND PARTITIONING AT THE AQUEOUS INTERFACE

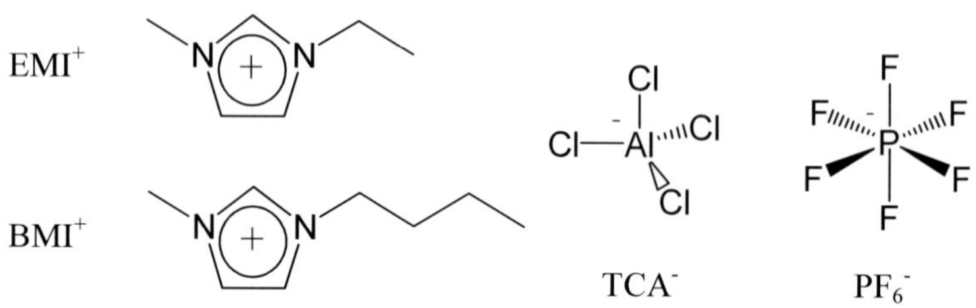

Figure 8. Ionic components of RTILs.

When mixed with water, room temperature ionic liquids "RTILs" based on an organic cation (e. g. alkyl-substituted phosphonium$^+$, -pyridinium$^+$, -imidazolium$^+$) and hydrophobic anions (e.g. PF_6^-, $[CF_3SO_2]_2N^-$) form distinct phases and can be used for liquid-liquid ion extraction purposes. Recent applications involve the extraction of actinide and lanthanide cations by phosphoryl-containing ligands, and of alkali and alkaline-earth cations by crown ethers or calixarenes.[85-89] These liquids, although inherently ionic in nature, display interesting analogies with classical organic solvents used for liquid-liquid extraction purposes. Due to their low volatility, non-flammability and presumed chemical stability (see however [90]) they can be regarded as "green solvents".[91-94] Their properties can be tailored experimentally via an appropriate design of the cation and anion components or substituents, but their microscopic properties are yet poorly known. This led us to simulate by classical molecular dynamics RTILs like [EMI][TCA] that is miscible with water, and [BMI][PF_6] that is not at the macroscopic level (see Figure 8 for definitions). We investigated the solvation of uncomplexed lanthanide (La^{3+}, Eu^{3+}, Yb^{3+}) and uranyl UO_2^{2+} cations [32, 95] in these "neat" liquids, also looking at the effect of Lewis basicity (added Cl^- anions) and solvent humidity [96] on the nature and solvation of the complexes.[33, 34]

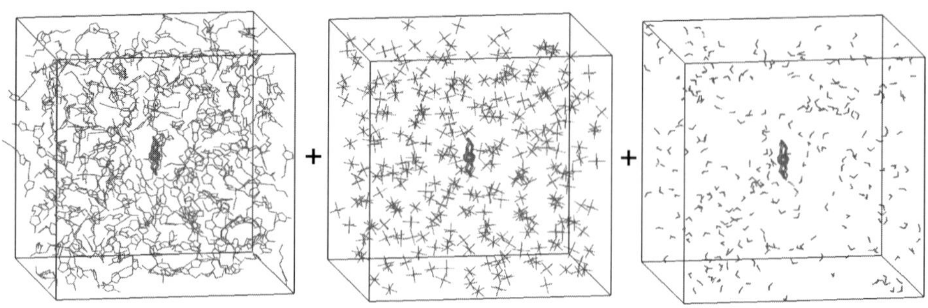

Figure 9. The Sr⊂18C6(NO_3)$_2$ complex in humid [BMI][PF_6][H_2O] solution, showing the BMI^+, PF_6^- and H_2O components side by side instead of superposed for clarity.

In this section, we focus on the Sr^{2+} extraction by crown ethers to [BMI][PF_6], as this system has been recently investigated by experiment, revealing that extraction to RTILs is several orders of magnitude more efficient than to classical organic solvents.[97-100] We thus investigate the solvation of 18C6 free, comparing its C_i and D_{3d} forms, as well as two forms of its strontium complex, namely $Sr\subset18C6^{2+}$ without counterions and $Sr\subset18C6(NO_3)_2$ with neutralizing bidentate nitrate counterions. Indeed, according to EXAFS studies, the strontium coordination to the nitrates is retained upon dissolution of the $Sr\subset18C6(NO_3)_2$ complex in the RTIL, but is likely to be lost when the complex is extracted from water, presumably because the cation coordinates to water molecules.[101, 102] We also compare two forms of the RTIL: the dry $[BMI][PF_6]$ form, and the "humid" $[BMI][PF_6][H_2O]$ form which contains one H_2O molecule per BMI^+ PF_6^- ion pair, and is thus somewhat sursaturated in water.[103] Figure 9 represents the simulated humid solvent box with one 18C6 strontium complex.

3.3.1 Solvation of the 18C6 free Crown. The C_i to D_{3d} Conversion

The simulations in the dry and in the humid RTIL started with 12 crowns (C_i form) "diluted" in the solvent box. The crowns rapidly underwent a conformational change to D_{3d}, showing that this structure is more stable in the two forms of the RTIL. Typical snapshots at the end of the dynamics (1.5 ns) and radial distribution functions (see Figure 10) reveal the importance of solvent cations.

In the dry RTIL, about one BMI^+ cation is, on the average, "complexed" at each face of 18C6, pointing its butyl or its methyl chain towards the crown's cavity. This is consistent with the well-known affinity of 18C6 for positively charged guests. The C_i form of 18C6 has no cavity and cannot therefore enjoy such stabilizing interactions, as confirmed by an energy component analysis on the C_i (constrained geometry) and D_{3d} monomers: both display attractive interaction energies with the dry RTIL, but these are $\approx$20 kcal/mol stronger for the D_{3d} than for the C_i form, thus explaining the solvation-induced conformational change of the crown.

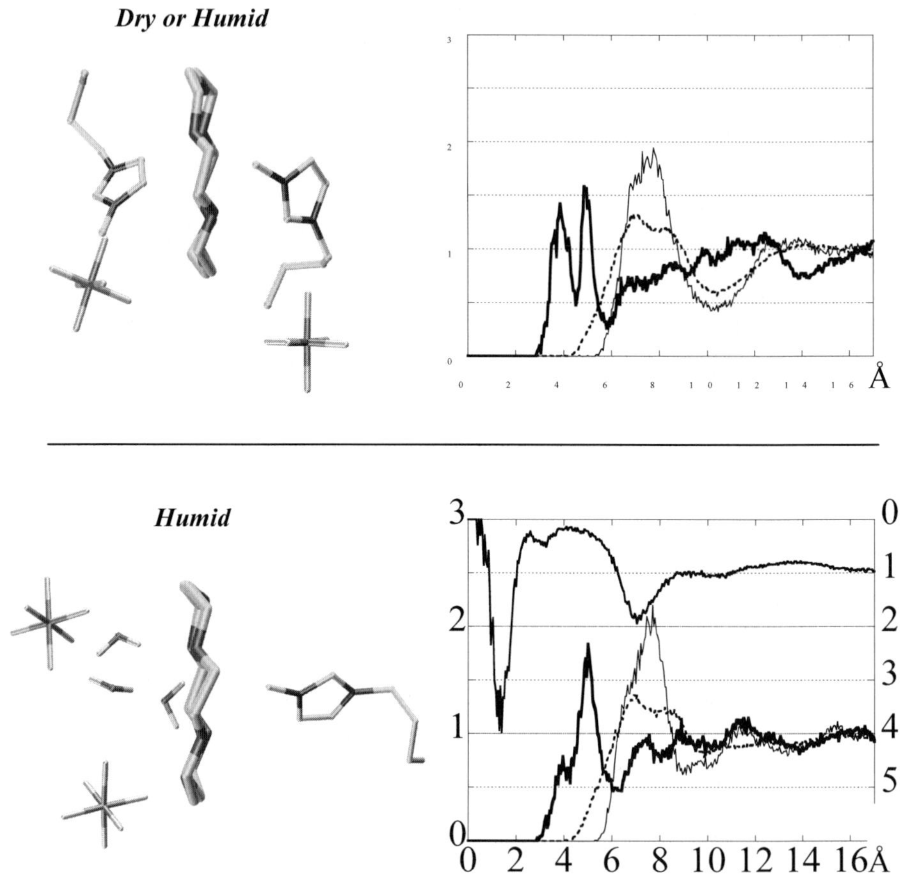

Figure 10. Solvation of twelve 18C6 crowns "diluted" in dry *versus* humid [BMI][PF_6] solutions. Typical snapshots of the "first shell" solvent molecules and radial distribution functions "RDFs" around the center of the crown: 18C6···BMI (*bold*), 18C6···F (*dotted*), 18C6···P (*plain*) and Sr···OH_2 (inversed ordinate).

In the humid RTIL, somewhat different solvation patterns are observed. Ten of the twelve crowns are solvated as in the dry liquid, with about one BMI^+ cation at each face, while the two others are solvated by 1 BMI^+ + 1 H_2O, or by 1+1 H_2O molecules (Figure 10). Interestingly, the "first shell" H_2O molecules sit in bridging position as described in section 1, and are further stabilized by second shell interactions with other H_2O, or BMI^+ species. In the two forms of the liquid, there is no solvent anion at less than 7 Å from the center of the crown, thus confirming the importance of the solvent cations and, to a lesser extent, of water. Comparison of the two forms of the RTIL shows that one 18C6 interacts somewhat more with the humid than with the dry liquid (-64 and -58 kcal/mol, respectively).

3.3.2 Solvation of two Forms of the Strontium Complex: $Sr\subset 18C6^{2+}$ (no Counterion) and $Sr\subset 18C6(NO_3)_2$ with bidentate Counterions

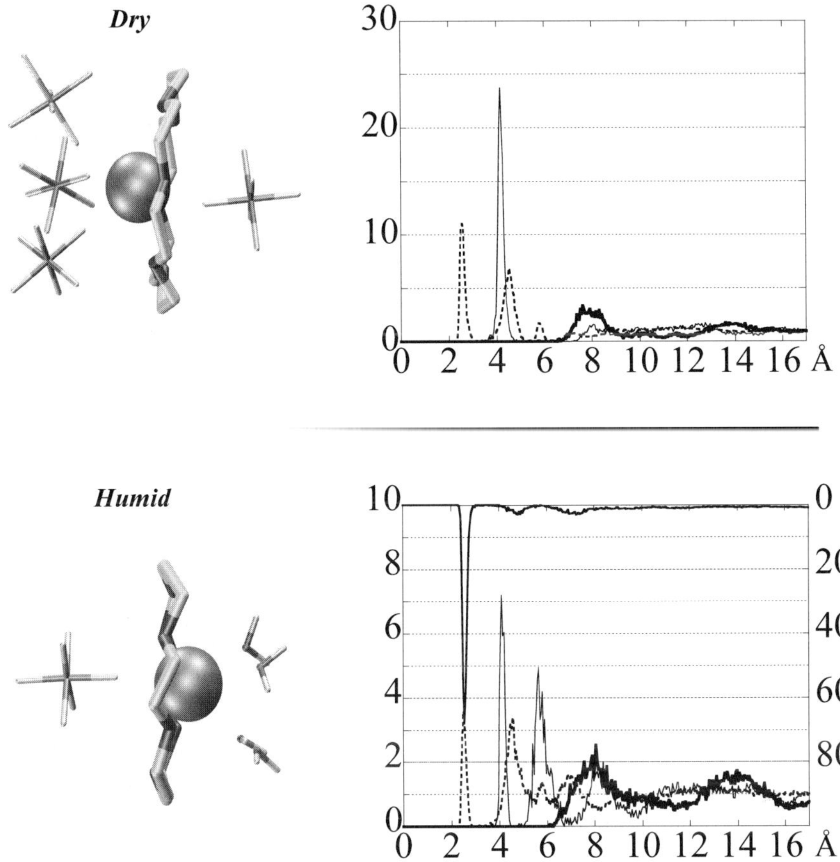

Figure 11. Solvation of the $Sr\subset 18C6^{2+}$ complex (no counterions) in dry *versus* "humid" [BMI][PF_6] solution. Left: typical snapshots at the end of the dynamics. Right: RDFs Sr···BMI (bold), Sr···P (plain) Sr···F (dotted) and Sr···OH_2 (inversed ordinate).

Two forms of the strontium complex, one without counterions and one with bidentate nitrate counterions, were simulated for 2 ns in the dry versus humid liquid. For the $Sr\subset 18C6^{2+}$ charged complex, two models were compared, i.e. one with remote (> 15 Å) Cl^- counterions, and one "without counterions", thus removing two BMI^+ solvent ions to maintain the system neutral. The two neutralization procedures led to similar solvation patterns, revealing the importance of solvent humidity. In the dry RTIL, the complexed Sr^{2+} cation is further coordinated to 1+2, or 1+3 PF_6^- anions, i.e. a monodentate one at one face and two or three bidentate ones at the other face of the crown (Figure 11). A modelbuilt $Sr\subset 18C6(PF_6)_4^{2-}$ complex with 2+2 symmetrically coordinated PF_6^- anions also spontaneously evolved to an 1+2 coordination during the dynamics, which hints for an equilibrium between 1+2 and 1+3 solvation patterns. These anions are surrounded by BMI^+ cations, at more than 7 Å from Sr^{2+}.

In the humid RTIL, the strontium solvation is completely different and dominated by water, as seen from two independent simulations. First, in Sr⊂18C6^{2+} (with remote Cl^- anions) Sr^{2+} captured 3 H_2O + 1 PF_6^- molecules in its inner sphere (Figure 11). These water ligands are further hydrogen bonded to second shell H_2O molecules or to BMI^+ cations. Another simulation started with the 2+2 coordinated Sr⊂18C6$(PF_6)_4^{2-}$ complex as above, but the four PF_6- ligands dissociated during the dynamics and were replaced by 1+4 H_2O molecules. The substitution of negatively charged PF_6^- ligands by neutral H_2O ones upon moistening of the RTIL is, according to an energy component analysis, beneficial: the Sr⊂18C6^{2+} interacts better with the humid than with the dry RTIL (-390 kcal/mol and -330±10 kcal/mol, respectively).

The Sr⊂18C6$(NO_3)_2$ complex with bidentate counterions

When the two NO_3^- anions lock Sr^{2+} at the center of the crown, the cation is shielded from the solvent. In the dry liquid, the nitrates remained mostly bidentate during the dynamics, but adopted a less symmetrical arrangement, in order to further coordinate one PF_6^- monodentate anion on the average, and the complex was surrounded by BMI^+ cations at about 7 Å from strontium (Figure 12). In the humid liquid, the two nitrates became monodentate. When they were constrained being bidentate, Sr^{2+} coordinated one H_2O molecule, on the average (Figure 12) leading to a total coordination number of 11 oxygens (6 from the crown, 4 from the nitrates, 1 from H_2O), i.e. one more than suggested by EXAFS studies on the extracted complex.[102] One also finds H_2O molecules at 4 - 7 Å, which solvate the nitrate anions, and display therefore repulsive interactions with Sr^{2+}. These likely contribute to make the complex somewhat "hydrophobic", a prerequisite in the context of liquid-liquid extraction. In terms of energetics, the Sr⊂18C6$(NO_3)_2$ complex with bidentate counterions is better solvated in the humid than in the dry RTIL: its interaction energies with the solvent are -133 and -107±15 kcal/mol, respectively.

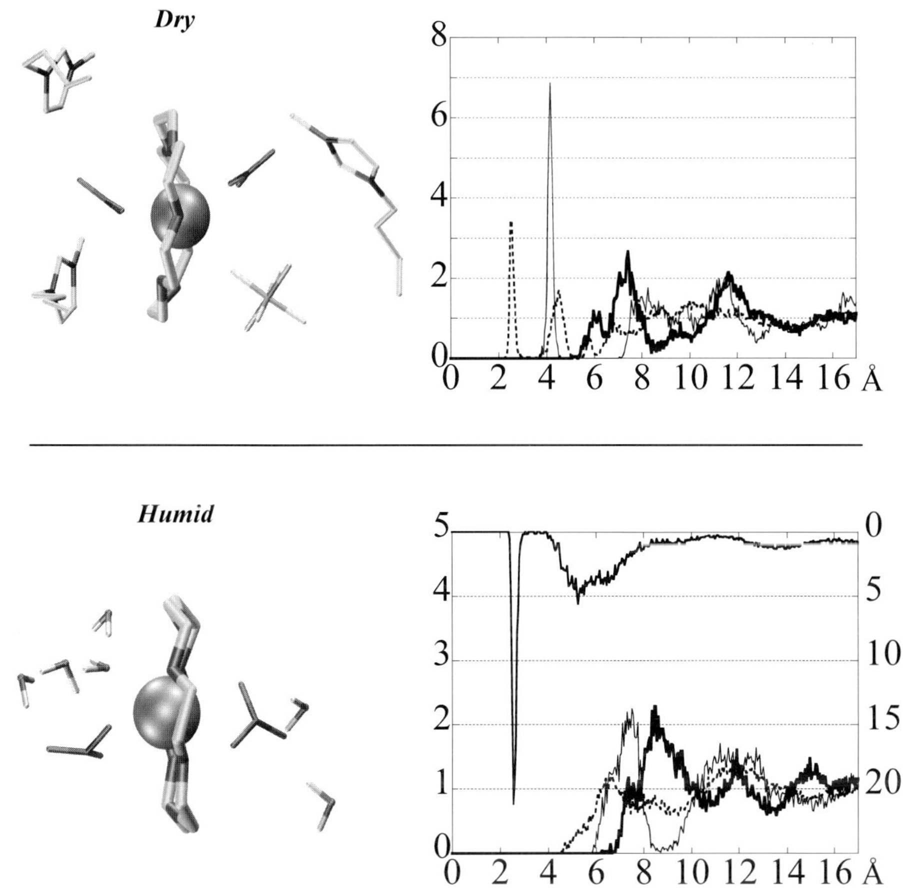

Figure 12. Solvation of the $Sr\subset 18C6(NO_3)_2$ complex with coordinated nitrates in dry *versus* "humid" $[BMI][PF_6]$ solution. Left: typical snapshots at the end of the dynamics. Right: RDFs Sr···BMI (bold), Sr···P (plain) Sr···F (dotted) and $Sr\cdots OH_2$ (inversed ordinate).

3.3.3 The 18C6 Molecules and their Strontium Complexes at the Ionic Liquid / Water Interface

We initiated MD simulations on the interfacial behavior of the 18C6 molecules, in order to gain insights into the nature of the interface and on the distribution of the crowns and of their complexes, and to compare with the results obtained at the "classical" aqueous interface with chloroform.[104] The interface was first modelbuilt, as in the case of classical interfaces, starting with adjacent "cubic" boxes of the two liquids, and the solutes were immersed near the interface. After equilibration and 4 ns of dynamics at 300 K, important solvent mixing was observed (Figure 13 *top*), as found in a previous work,[96] suggesting that the BMI^+ and PF_6^- ions represented with +1 and -1 charges are somewhat too hydrophilic. We thus decided to scale down the BMI^+ and PF_6^- charges by a factor 0.9, which also somewhat mimics the anion to cation charge transfer. For the neat interface after 9 ns of dynamics, this was found to lead to molar fractions of about 0.26 for water in the RTIL, and of 0.0023 for the ionic liquid in water, in good agreement with experimental solubilities.[105] We note that the resulting interface (see

Figure 13) is broader, with more inter-solvent mixing than at aqueous interface with, e.g. chloroform [106] or dichloroethane,[107] and this have deep consequences as far as the mechanism of M^{n+} cation extraction is concerned. In classical extraction, the solvent molecules are neutral and the neutrality of the receiving organic phase is generally achieved by co-extracting anions from the aqueous phase, or by exchanging *n* protons of acidic extractants with M^{n+}. Extraction to RTILs may proceed via another mechanism, i.e. by exchanging *n* solvent (e.g. BMI^+) cations of the RTIL with M^{n+}, and the simulated interfacial landscape observed with $[BMI][PF_6]$ is clearly consistent with such an ion exchange mechanism. When the solvent cations become more hydrophobic (e.g. by using longer alkyl imidazolium substituents), the interface becomes narrower with less solvent mixing, and extraction rather proceeds via an anion exchange mechanism.
The uncomplexed 18C6 was simulated for 4 ns, starting with a 3x4 grid of C_i crowns initially equally shared between the aqueous and the RTIL sides of the interface. Figure 13 shows the resulting distribution after 4 ns of dynamics at 300 K. Interestingly, most of the crowns sit now on the RTIL-side of the interface. In particular, those which were on the water-side diffused to the RTIL. All crowns rearranged from C_i to a D_{3d}-type of conformation, as seen in the pure or humid RTIL. Most of them are finally solvated by BMI^+ cations, which point their N-alkyl chains to the center of the crown, as in the humid RTIL. The better solvation of 18C6 by the RTIL cations, compared to water, presumably explains why 18C6 moves to the RTIL side of the interface. With classical organic solvents, the crowns cannot enjoy such stabilizing interactions and sit therefore more on the aqueous side of interface.[104]

The strontium complexes were simulated for 4 ns as a mixture of two $Sr\subset 18C6^{2+}$ (no inner sphere counterion) plus two $Sr\subset 18C6(NO_3)_2$ complexes, initially right at the interface. During the dynamics, the complexes oscillated near the interface and, after 4 ns (see Figure 14) the two neutral $Sr\subset 18C6(NO_3)_2$ ones sit right at the border between the two liquid phases, while the charged ones $Sr\subset 18C6^{2+}$ sit somewhat deeper in water. There is thus no extraction to the RTIL phase under these conditions. Not surprisingly, $Sr\subset 18C6^{2+}$ is too hydrophilic and too much hydrated to be extracted.

For $Sr\subset 18C6(NO_3)_2$, interesting aggregation between the two complexes is observed at the interface, due to bridging nitrate counterions. Such a static view may be, however, somewhat misleading as, during the last 2 ns, the two types of complexes exchange their position. For instance, at 2 ns (Figure 14), the $Sr\subset 18C6(NO_3)_2$ complexes sit more deeply in water, while the $Sr\subset 18C6^{2+}$ complexes are closer to the interface.

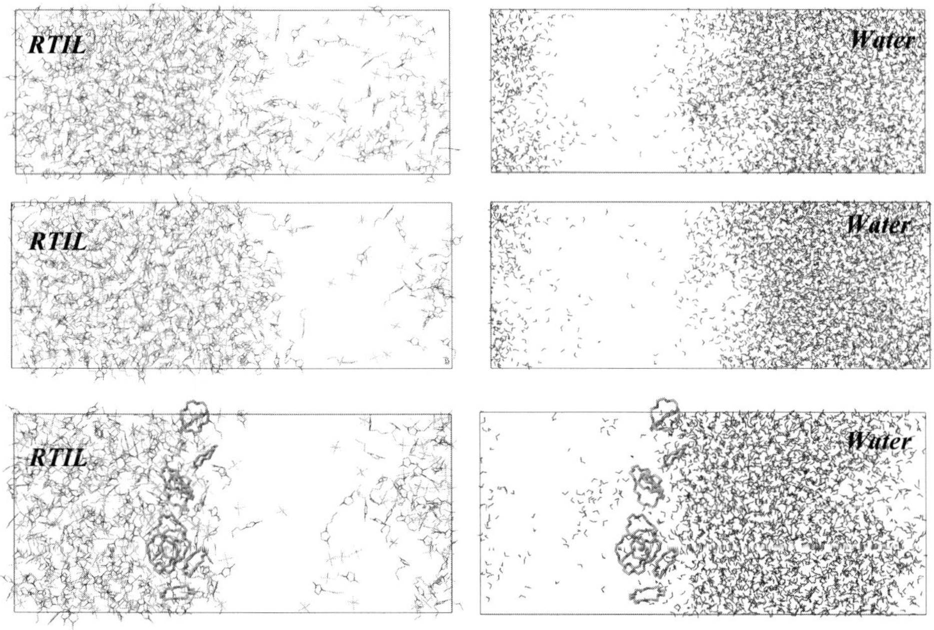

Figure 13. The [BMI][PF_6] / water neat interface *(top - charges +1/-1 after 4ns and middle - charges +0.9/-0.9 after 9ns)* and with twelve 18C6 crowns *(bottom)* after 4 ns of dynamics. Solvents are shown side by side, instead of superposed, for clarity. A color version is given in the Appendix.

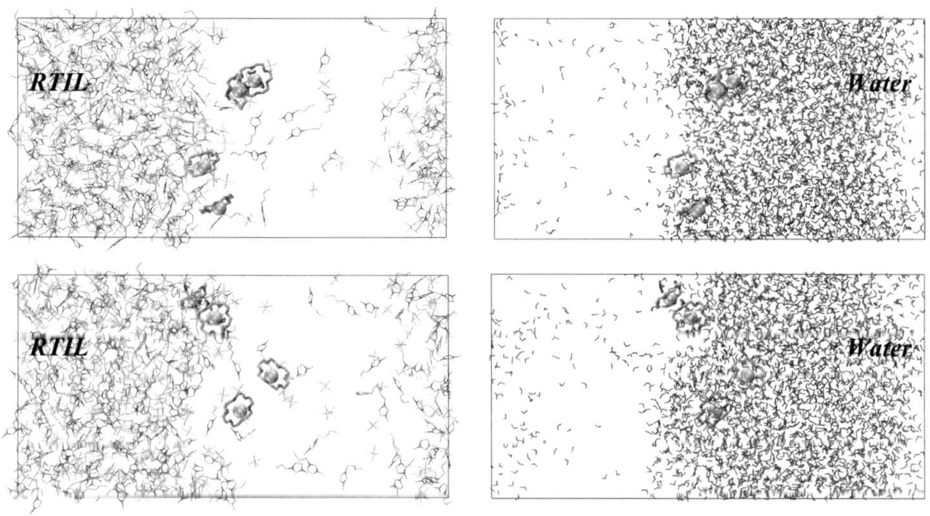

Figure 14. The [BMI][PF_6] / water interface with two $Sr\subset 18C6^{2+}$ plus two $Sr\subset 18C6(NO_3)_2$ complexes after 2 ns *(top)* and 4 ns *(bottom)* of dynamics. Solvents are shown side by side, instead of superposed, for clarity. A color version is given in the Appendix.

The interface itself is dynamics, non-planar and ill-defined, due to important local solvent mixing, and the position of the complexes with respect to the interface fluctuates much more than at classical aqueous interfaces. There is thus likely an equilibrium involving two types of positions for the strontium complex at the interface, which is consistent with cation exchange, as well as anion co-extraction mechanisms.

4. Conclusions

We have presented computer modelling studies on 18C6, focusing on the importance of dynamics at different times scales system of different sizes. The CP-MD studies in the gas phase account for the vibrations of the systems without a priori hypothesis on their harmonic character, and thus account for the coupling between atomic motions (sometimes involving bond making or bond breaking processes) and electron reorganization as a function of time. They are presently limited to small systems and short time scale (a few picoseconds), but will no doubt find important applications in supramolecular chemistry, for instance in the field of metal complexes in which the bond is hardly modelled by classical force field methods. Quantum mechanical calculations can also be "routinely" performed on real complexes involving, e.g. trivalent lanthanide or actinide ions and calixarenes, thus giving important insights into their coordination patterns in the gas phase. Ion extraction and transport has been one of the cornerstones of supramolecular chemistry, and the applications presented here illustrate the power of MD simulations to investigate complex systems in heterogeneous environments represented explicitly at the atomic level. A major issue in modelling is the sampling of the different states of the systems, and this can be presently achieved by simulations of several nanoseconds for systems of about 20 000 atoms in a day on PC-clusters. The processes of phase separation and organization to form, e.g. interfaces, micelles, microemulsions can thus be simulated, involving new media like supercritical fluids or ionic liquids. Further challenges will be to move from the micro- to the mesoscopic scale and to longer time scales, and to investigate the energetics of recognition or assembling between nano-objects. This is important for the basic, as well as technological and industrial viewpoints.

5. Acknowledgments

The authors are grateful to IDRIS, CINES, Université Louis Pasteur, and PARIS for computer resources and to Etienne Engler for assistance. PV thanks the French ministry of research for a thesis grant.

6. References

(1) Pullman, A. (1976), Progrès récents dans l'étude de la solvatation et de la fixation de cations des groupes I et II par l'approche supramoléculaire *ab initio*, *Bull. Soc. Chim. Belg.* **85**, 963-968.

(2) Lehn, J. M. (1978), Cryptates: Inclusion Complexes of Macropolycyclic Receptor Molecules, *Pure Appl. Chem.* **50**, 871-892.

(3) Lehn, J. M. (1988), Supramolecular Chemistry-Scope and Perspectives. Molecules, Supramolecules, and Molecular Devices (Nobel Lecture), *Angew. Chem. Int. Ed.* **27**, 89-112.

(4) Cram, D. J. (1988), The Design of Molecular Hosts, Guests, and Their Complexes, *Science* **240**, 760-767.

(5) Pedersen, C. J. (1967), Cyclic Polyethers and Their Complexes with Salt Metals, *J. Am. Chem. Soc.* **89**, 7017.

(6) De Jong, F., Reinhoudt, D. N. (1981) *Stability and Reactivity of Crown Ether Complexes*, Academic Press: New York.

(7) Moyer, B. A. (1996) Basic Principles of Extraction and Liquid-Liquid Systems Employing Crown Ethers and Related Metal-Ion Receptors, in Atwood, J. L., Davies, J. E. D., McNicol, D. D., Vögtle, F., Lehn, J.-M. (eds.), *Molecular Recognition: Receptors for Cationic Guests*, Pergamon, New York, pp 325-365 and references cited therein.

(8) Wipff, G., Weiner, P., Kollman, P. A. (1982), A Molecular Mechanics Study of 18-Crown-6 and Its Alkali Complexes: An Analysis of Structural Flexibility, Ligand Specificity, and the Macrocyclic Effect, *J. Am. Chem. Soc.* **104**, 3249-3258 and refs cited therein.
(9) Wipff, G., Wurtz, J. M. (1988) Dynamics Views of Macrocyclic Receptors: Molecular Dynamics Simulations and Normal Modes Analysis, in Pullman, R. (eds.), *Transport through Membranes, Carriers, Channels and Pumps*, Reidel, Dorbrecht, pp 1-26.
(10) Lybrand, T. P., McCammon, J. A., Wipff, G. (1986), Theoretical Calculation of Relative Binding Affinity in Host-Guest Systems, *Proc. Natl. Acad. Sci. USA* **83**, 833-835.
(11) Jorgensen, W. L. (1989), Free Energy Calculations: A Breakthrough for Modeling Organic Chemistry in Solution, *Acc. Chem. Res.* **22**, 184-189.
(12) Grootenhuis, P. D. J., Kollman, P. A. (1989), Molecular Mechanics and Dynamics Studies of Crown Ether-Cation Interactions: Free Energy Calculations on the Cation Selectivity of Dibenzo-18-Crown-6 and Dibenzo-30-Crown-10, *J. Am. Chem. Soc.* **111**, 2152-2158.
(13) van Eerden, J. V., Harkema, S., Feil, D. (1988), Molecular Dynamics of 18-Crown-6 Complexes with Alkali-Metal Cations: Calculation of the Relative Free Energies of Complexation, *J. Phys. Chem.* **92**, 5076-5079.
(14) Dang, L. X., Kollman, P. A. (1990), Free Energy of Association of the 18-Crown-6:K^+ Complex in Water: A Molecular Dynamics Simulation, *J. Am. Chem. Soc.* **112**, 5716-5720.
(15) Glendening, E. D., Feller, D., Thompson, M. A. (1994), An *ab initio* investigation of the structure and alkali metal cation selectivity of 18-crown-6, *J. Am. Chem. Soc.* **116**, 10657-10669.
(16) Feller, D. (1997), *Ab Initio* Study of M^+: 18-Crown-6 Microsolvation, *J. Phys. Chem. A* **101**, 2723-2731.
(17) Thompson, M. A., Glendening, E. D., Feller, D. (1994), The Nature of K^+ / Crown Ether Interactions: a Hybrid QM-MM Study, *J. Am. Chem. Soc.* **116**, 10465-10476.
(18) Thompson, M. A. (1995), Hybrid QM/MM Molecular Force Field Development for Large Flexible Molecules: a MD study of 18-crown-6, *J. Phys. Chem.* **99**, 4794-4804.
(19) Wipff, G. (1994) *Computational Approaches in Supramolecular Chemistry*, Kluwer: Dordrecht.
(20) Wipff, G. (2001) Molecular dynamics simulations of cation complexation and extraction, in Vicens, J., Asfari, Z., Harrowfield, J., Böhmer, V. (eds.), *Calixarenes-2001*, Kluwer Acad. Publishers, Dordrecht, pp 312-333 and references cited therein.
(21) van Veggel, N. R. (1999) Chap. 2: Molecular Modeling of Calixarenes and their Host-Guest Complexes, in Mandolini, L., Ungaro, R. (eds.), *Calixarenes in Action*, Imperial College Press, Singapore, pp 11-36.
(22) Allen, M. P., Tildesley, D. J. (1987) *Computer Simulation of Liquids,* in van Gunsteren, W. F. , Weiner, P. K. (eds.), Clarendon press: Oxford.
(23) Leach, A. R. (2001) *Molecular Modelling. Pinciples and applications*, Longman: Southampton.
(24) Case, D. A., Pearlman, D. A., Caldwell, J. C., Cheatham III, T. E., Ross, W. S., Simmerling, C. L., Darden, T. A., Merz, K. M., Stanton, R. V., Cheng, A. L., Vincent, J. J., Crowley, M., Ferguson, D. M., Radmer, R. J., Seibel, G. L., Singh, U. C., Weiner, P. K., Kollman, P. A. (1997), AMBER5, *University of California, San Francisco.*
(25) Hay, B. P., Rustad, J. R., Hostetler, C. J. (1993), Quantitative Structure-Stability Relationship for K^+ Ion Complexation by Crown Ethers. A MM and *ab initio* Study, *J. Am. Chem. Soc.* **115**, 11158-11164.
(26) Vayssière, P., Wipff, G. (2003), Ethers, Crown ethers and 18-crown-6 - K^+ Complexes at a Water / SC-CO_2 Interface: a Molecular Dynamics Study, *Phys. Chem. Chem. Phys.* **5**, 127-135.
(27) Singh, U. C., Kollman, P. A. (1984), An Approach to Computing Electrostatic Charges for Molecules, *J. Comput. Chem.* **5**, 129-145.
(28) Jorgensen, W. L., Chandrasekhar, J., Madura, J. D., Impey, R. W., Klein, M. L. (1983), Comparison of Simple Potential Functions for Simulating Liquid Water, *J. Chem. Phys.* **79**, 926-936.
(29) Murthy, C. S., K. Singer, McDonald, I. R. (1981), Interaction Site Models for Carbon Dioxyde, *Mol. Phys.* **44**, 135-143.
(30) Åqvist, J. (1990), Ion-Water Interaction Potentials Derived from Free Energy Perturbation Simulations, *J. Phys. Chem.* **94**, 8021-8024.
(31) Vayssière, P., Wipff, G. (2003), Importance of Counterions in Alkali and Alkaline Earth Cation Extraction by 18-Crown-6: Molecular Dynamics Studies at the Water / SC-CO_2 Interface, *Phys. Chem. Chem. Phys.* **5**, 2842-2850.
(32) Chaumont, A., Engler, E., Wipff, G. (2003), Uranyl and Strontium Nitrate Solvation in Room-Temperature Ionic Liquids. A Molecular Dynamics Investigation, *Inorg. Chem.* **42**, 5348-5356.
(33) Chaumont, A., Wipff, G. (2004), M^{3+} Lanthanide Chloride Complexes in "Neutral" Room-Temperature Ionic Liquids. A Theoretical Study, *J. Phys. Chem A* **108**, 3311-3319.
(34) Chaumont, A., Wipff, G. (2004), Solvation of Uranyl(II), Europium(III) and Europium(II) Cations in "Basic" Room-Temperature Ionic Liquids. A Theoretical Study, *Chem. Eur. J.* **10**, 3919-3930.

(35) Saiz, L., Klein, M. L. (2002), Computer Simulation Studies of Model Biological Membranes, *Acc. Chem. Res.* **35**, 482-489.
(36) Jungwirth, P., Tobias, D. J. (2002), Ions at the Air / Water Interface, *J. Phys. Chem. B* **106**, 6361-6373.
(37) Martinez, J. M., Pappalardo, R. R., Marcos, E. S. (1999), First principles Ion-Water interaction potentials for highly charged monoatomic ions. Computer simulations of Al^{3+}, Mg^{2+}, and Be^{2+} in water, *J. Am. Chem. Soc.* **121**, 3175-3184.
(38) Zheng, G., Irle, S., Elstner, M., Morokuma, K. (2004), Quantum Chemical Molecular Dynamics Model Study of Fullerene Formation from Open-Ended Carbon Nanotubes, *J. Phys. Chem. A* **108**, 3182-3194.
(39) Raugei, S., Klein, M. L. (2002), Hydrocarbon Reactivity in the Superacid SbF_5/HF: an *ab initio* Molecular Dynamics Study, *J. Phys. Chem. B* **106**, 11596-11605.
(40) Chen, B., Ivanov, I., Park, J. M., Parrinello, M., Klein, M. L. (2002), Solvation Structure and Mobility Mechansims of OH^{-}: A Car-Parrinello Molecular Dynamics Investigation of Alkaline Solutions, *J. Phys. chem. B* **106**, 12006-12016.
(41) Marx, D., Tuckerman, M. E., Hutter, J., Parrinello, M. (1999), The nature of the Hydrated excess proton in water, *Nature* **397**, 601-604 and references cited therein.
(42) Boese, A. D., Chandra, A., Martin, J. M. L., Marx, D. (2003), From *ab initio* quantum chemistry to molecular dynamics: The delicate case of hydrogen bonding in ammonia, *J. Chem. Phys.* **119**, 5965-5980.
(43) Pagliai, M., Raugei, S., Cardini, G., Schettino, V. (2003), Car-Parrinello Molecular Dynamics on the SN_2 reaction Cl^- + CH_3Br in water, *J. Mol. Struct. (THEOCHEM)* **630**, 141-149.
(44) Car, R., Parrinello, M. (1985), Unified Approach for Molecular Dynamics and Density-Functional Theory, *Phys. Rev. Lett.* **55**, 2471-2474.
(45) Hutter, J., Alavi, A., Deutsch, T., Bernasconi, M., Goedecker, S., Marx, D., Tuckerman, M., Parrinello, M. (1995-1999), CPMD Version 3.3a, *Max-Planck-Institut für Festkörperforschung and IBM Research Laboratory*.
(46) Troullier, N., Martins, J. L. (1991), *Phys. Rev. B* **43**, 1993-2006.
(47) Kleinman, L., Bylander, D. M. (1982), *Phys. Rev. Lett.* **48**, 1425-1428.
(48) Bühl, M., Wipff, G. (2002), Hydronium Ion Complex of 18-Crown-6: Where Are the Protons? A Density Functional Study of Static and Dynamic Properties, *J. Am. Chem. Soc.* **124**, 4473-4480.
(49) Schurhammer, R., Vayssière, P., Wipff, G. (2003), 18-Crown-6 and its Hydrates: Bridging, but Versatile Hydrogen Bond. A Theoretical Study of Static and Dynamic Properties, *J. Phys. Chem. A* **107**, 11128-11138 and references cites therein.
(50) Dobler, M. (1981) *Ionophores and their Structures,* Wiley Interscience: New York.
(51) Sun, Y., Kollman, P. A. (1992), Conformational Sampling and Ensemble Generation by MD Simulations: 18-crown-6 as a Test Case, *J. Comput. Chem.* **12**, 33-40.
(52) Straatsma, T. P., McCammon, J. A. (1989), Treatment of Rotational Isomers in Free Energy Calculations. II. Molecular Dynamics Simulation of 18-Crown-6 in Aqueous Solution as an Example of Systems with Large Numbers of Rotational Isomeric States, *J. Chem. Phys.* **91**, 3631-3637.
(53) Troxler, L., Wipff, G. (1994), Conformation and Dynamics of 18-crown-6, Cryptand 222, and their Cation Complexes studied by MD Simulations, *J. Am. Chem. Soc.* **116**, 1468-1480.
(54) Ranghino, G., Romano, S., Lehn, J. M., Wipff, G. (1985), Monte-Carlo Study of the Conformation-Dependent Hydration of the 18-Crown-6 Macrocycle, *J. Am. Chem. Soc.* **107**, 7873-7877.
(55) Nordlander, E. H., Burns, J. H. (1986), Structure of a Hydrogen-bonded Lattice Coumpound between Phosphoric Acid and 18-Crown-6: $H_3PO_4.0.5(C_{12}H_{24}O_3).3H_2O$, *Inorg. Chim. Acta* **115**, 31-36.
(56) Albert, A., Mootz, D. (1997), Formation and Crystal Structures of the Hydrates of 18-Crown-6, *Z. Naturforsch.* **52b**, 615-619.
(57) Mootz, D., Albert, A., Schaefgen, S., Stäben, D. (1994), 18-crown-6 and water: crystal structure of a binary hydrate, *J. Am. Chem. Soc.* **116**, 12045-12046.
(58) Matsuura, H., Fukuhara, K., Ikeda, K., Tachikake, M. (1989), Crystalline Complex of 18-Crown-6 with Water and a Phase Diagram of the Binary System as Studied by Raman Spectroscopy, *Chem. Commun.* 1814-1816.
(59) Fukuhara, K., Tachikake, M. (1995), Raman Spectroscopic Study of the Hydrates of 18-Crown-6, *J. Phys. Chem.* **99**, 8617-8623.
(60) Patil, K. J., Pawar, R. B. (1999), Near IR spectral Studies for investigating the Hydration of 18-crown-6 in Aqueous Solutions, *J. Phys. Chem. B* **103**, 2256-2261.
(61) Patil, K. J., Pawar, R. B. (2003), Near IR spectral investigations of hydration of 18-crown-6 in HDO-D_2O solutions at different temperatures, *Spectrochim. Acta* **59**, 1289-1297.
(62) Wipff, G., Troxler, L. (1994) MD Simulations on Synthetic Ionophores and on their Cation Complexes: Comparison of Aqueous / non-Aqueous Solvents, in Wipff, G. (eds.), *Computational Approaches in Supramolecular Chemistry*, Kluwer, Dordrecht, pp 319-348.

(63) Bryan, S. A., Willis, R. R., Moyer, B. A. (1990), Hydration of 18-Crown-6 in Carbon Tetrachloride. Infrared Spectral Evidence for an Equilibrium between Monodentate and Bidentate Forms of Bound Water in the 1:1 Crown-Water Adduct, *J. Phys. Chem.* **94**, 5230-5233.

(64) Rustenholtz, A., Fulton, J. L., Wai, C. M. (2003), An FT-IR Study of Crown Ether-Water Complexation in Supercritical CO_2, *J. Phys. Chem. A* **107**, 11239-11244.

(65) von Szentpaly, L., Shamovsky, I. L. (1994), Consistent point charges at the 18-crown-6 atoms from correlated *ab initio* calculations, *J. Mol. Struct. (THEOCHEM)* **305**, 249-260.

(66) Lin, Y., Smart, N. G., Wai, C. M. (1995), Supercritical Fluid Extraction of Uranium and Thorium from Nitric Acid Solutions with Organophosphorus Reagents, *Environ. Sci. Technol.* **29**, 2706-2708.

(67) Phelps, C. L., Smart, N. G., Wai, C. M. (1997), Past, Present and Possible Future of Supercritical Fluid Extraction Technology, *J. Chem. Ed.* **73**, 1163-1168.

(68) Lin, Y., Smart, N. G. (2003) Supercritical Fluid Extraction of Actinides and Heavy Metals for Environmental Cleanup: A Process Development Perspective, in Gopalan, A.S., Wai, C.M., Jacobs, H.K. (eds.), *Supercritical Carbon Dioxide*, Washington, pp 23-35.

(69) Erkey, C. (2000), Supercritical Carbon Dioxide Extraction of Metals from Aqueous Solutions: a Review, *J. Supercrit. Fl.* **17**, 259-287.

(70) Mochizuki, S., Wada, N., Smith, R. L., Inomata, H. (1999), Perfluorocarboxylic acid Counter ion Enhanced Extraction of Aqueous Alkali Metal Ions with Supercritical Carbon Dioxide, *Analyst* **124**, 1507-1511.

(71) Mochizuki, S., Wada, N., Smith, R. L., Inomata, H. (1999), Supercritical Fluid Extraction of Alkali Metal Ions using Crown Ethers with Perfluorocarboxylic Acids from Aqueous Solution, *Anal. Commun.* **36**, 51.

(72) Wai, C. M., Kulyako, Y., Yak, H.-K., Chen, X., Lee, S.-J. (1999), Selective Extraction of Strontium with Supercritical fluid Carbon Dioxide, *Chem. Commun.* 2533-2534.

(73) Watarai, H. (1993), What's Happening at the Liquid-Liquid Interface in Solvent Extraction Chemistry?, *Trends in Analytical Chemistry* **12**, 313-318.

(74) Wipff, G., Lauterbach, M. (1995), Complexation of Alkali Cations by Calix[4]crown Ionophores: Conformation and Solvent Dependent Na^+ / Cs^+ Binding Selectivity: A MD FEP study, *Supramol. Chem.* **6**, 187-207.

(75) Berny, F., Muzet, N., Schurhammer, R., Troxler, L., Wipff, G. (1998) MD Simulations on Ions and Ionophores at a Liquid-Liquid Interface: from Adsorption to Recognition, in Tsoucaris, G. (eds.), *Current Challenges in Supramolecular Assemblies, NATO ARW Athens*, Kluwer Acad. Publishers, Dordrecht, pp 221-248.

(76) Baaden, M., Berny, F., Muzet, N., Troxler, L., Wipff, G. (2000) Interfacial Features of Assisted Liquid-Liquid Extraction of Uranyl and Cesium Salts: a Molecular Dynamics Investigation, in Lumetta, G., Rogers, R., Gopalan, A. (eds.), *Calixarenes for Separation. ACS Symposium Series 757*, ACS, Washington DC, pp 71-85.

(77) Baaden, M., Berny, F., Muzet, N., Schurhammer, R., Troxler, L., Wipff, G. (2000) Separation of Cs^+ and UO_2^{2+} cations by liquid-liquid extraction: computer simulations of the demixing of water / oil binary solutions, in Davies, C. (eds.), *Euradwaste 1999: Radioactive Waste management Strategies and Issues*, European Commission, Brussels, pp 519-522.

(78) Berny, F., Muzet, N., Troxler, L., Wipff, G. (1999) Simulations of liquid-liquid interfaces: A key border in Supramolecular Chemistry, in Ungaro, R., Dalcanale, E. (eds.), *Supramolecular Science: where it is and where it is going*, Kluwer Acad. Publishers, Dordrecht, pp 95-125 and references cited therein.

(79) Berny, F., Schurhammer, R., Wipff, G. (2000), Distribution of Hydrophilic, Amphiphilic and Hydrophobic Ions at a Liquid-Liquid Interface: a Molecular Dynamics Investigation, *Inorg. Chim. Acta* **300-302**, 384-394.

(80) Schurhammer, R., Berny, F., Wipff, G. (2001), Importance of Interfacial Phenomena in Assisted Ion Extraction by Supercritical CO_2: a Molecular Dynamics Investigation, *Phys. Chem. Chem. Phys.* **3**, 647-656.

(81) Baaden, M., Schurhammer, R., Wipff, G. (2002), Molecular Dynamics Study of the Uranyl Extraction by Tri-n-butylphosphate (TBP): Demixing of Water/"Oil"/TBP Solutions with a Comparison of Supercritical CO_2 and Chloroform, *J. Phys. Chem. B* **106**, 434-441.

(82) Troxler, L., Harrowfield, J. M., Wipff, G. (1998), Do Picrate Anions "Attract Each Other" in Solution ? Molecular Dynamics Simulations in Water and in Acetonitrile Solutions, *J. Phys. Chem. A* **102**, 6821-6830.

(83) Coupez, B., Wipff, G. (2004), The Synergistic Effect of Cobalt-dicarbollide Anions on the Extraction of M^{3+} Lanthanide Cations by Calix[4]arene-CMPO ligands: a Molecular Dynamics Study at the Water-"oil" Interface, *C. R. Chimie Acad. Sc. Paris*, in press.

(84) Coupez, B., Boehme, C., Wipff, G. (2003), Importance of Interfacial Phenomena and Synergistic Effects in Cation Extraction by Dithiophosphinic Ligands: a Molecular Dynamics Study, *J. Phys. Chem. B* **107**, 9484-9490.

(85) Visser, A. E., Rogers, R. D. (2003), Room-Temperature ionic Liquids: New Solvents for *f*-element Separations and Associated Solution Chemistry, *J. Solid State Chem.* **171**, 106-113.
(86) Dai, S., Shin, Y. S., Toth, L. M., Barnes, C. E. (1997), Comparative UV-vis Studies of Uranyl Chloride Complex in two Basic Ambient Temperature Melt Systems: the Observation of Spectral and Thermodynamic Variations induced via Hydrogen Bonding, *Inorg. Chem.* **36**, 4900-4902.
(87) Visser, A. E., Swatloski, R. P., Griffin, S. T., Hartman, D. H., Rogers, R. D. (2001), Liquid/Liquid Extraction of Metal ions in Room Temperature Ionic Liquids, *Separ. Sc. Technol.* **36**, 785-804.
(88) Visser, A. E., Swatlosli, R. P., Reichert, W. M., Mayton, R., Sheff, S., Wierzbicki, A., Davies, J., Rogers, R. D. (2002), Task-Specific Ionic Liquids Incoporating Novel Cations for the Extraction of Hg^{2+} and Cd^{2+}: Synthesis, Characterization and Extraction Studies, *Environ. Sci. Technol.* **36**, 2523-2529.
(89) Luo, H., Dai, S., Bonnesen, P. V., III, A. C. B., Holbrey, J. D., Bridges, N. J., Rogers, R. D. (2004), Extraction of Cesium Ions from Aqueous Solutions Using Calix[4]arene-bis(*tert*-octylbenzo-crown-6) in Ionic Liquids, *Anal. Chem.* **76**, 3078-3083.
(90) Swatloski, R. P., Holbrey, J. D., Rogers, R. D. (2003), Ionic Liquids are not always Green: Hydrolysis of 1-butyl-3-Methylimidazolium Hexafluorophosphate, *Green Chem.* **5**, 361-363.
(91) Welton, T. (1999), Room-Temperature Ionic Liquids. Solvents for Synthesis and Catalysis, *Chem. Rev.* **99**, 2071-2083.
(92) Wasserscheid, P., Welton, T. (2002) *Ionic Liquids in Synthesis*, Wiley-VCH (eds.), Weinheim.
(93) Huddleston, J. G., Willauer, H. D., Swatloski, R. P., Visser, A. E., Rogers, R. D. (1998), Room Temperature Ionic Liquids as novel media for "clean" Liquid-Liquid Extraction, *Chem. Commun.* 1765-1766.
(94) Rogers, R. D., Seddon, K. R. (2003), Ionics Liquids Solvents of the Future ?, *Science* **302**, 792.
(95) Chaumont, A., Wipff, G. (2003), Solvation of M^{3+} lanthanide cations in Room-Temperature Ionic Liquids. A Molecular Dynamics Investigation, *Phys. Chem. Chem. Phys.* **5**, 3481-3488.
(96) Chaumont, A., Wipff, G. (2004), Solvation of Uranyle$^{(II)}$, Europium$^{(III)}$ Cations and their Chloro complexes in a Room-Temperature Ionic Liquid. A Theoretical Study of the effect of Solvent Humidity, *Inorg. Chem.* **43**, in press.
(97) Dai, S., Ju, Y. H., Barnes, C. E. (1999), Solvent Extraction of Strontium Nitrate by a Crown Ether using Room-temperature Ionic Liquids, *Dalton Trans.* 1201-1202.
(98) Visser, A. E., Swatloski, R. P., Reichert, W. M., Griffin, S. T., Rogers, R. D. (2000), Traditional Extractants in nontraditional Solvents: Groups 1 and 2 Extraction by Crown Ethers in Room Temperature Ionic Liquids, *Ind. Eng. Chem. Res.* **39**, 3596-3604.
(99) Dietz, M. L., Dzielawa, J. A. (2001), Ion-Exchange as a Mode of Cation Transfer into Room-temperature Ionic Liquids containing Crown ethers: Implications for the Greeness" of Ionic liquids as Diluents in Liquid-Liquid Extraction, *Chem. Commun.* 2124-2125.
(100) Chun, S., Dzyuba, S. V., Bartsch, R. A. (2001), Influence of Structural Variation in Room-Temperature Ionic Liquids on the Selectivity and Efficiency of competitive Alkali Metal Salt Extraction by a Crown Ether, *Anal. Chem.* **73**, 3737-3741.
(101) Dietz, M. L., Dzielawa, J. A., Laszak, I., Young, B. A., Jensen, M. P. (2003), Influence of Solvent Structural Variations on the Mechanism of facilitated Ion Transfer into Room-Temperature Ionic Liquids, *Green Chem.* **5**, 682-685.
(102) Jensen, M. P., Dzielawa, J. A., Rickert, P., Dietz, M. L. (2002), EXAFS Investigations of the Mechanism of Facilitated Ion Transfer into a Room-Temperature Ionic Liquid, *J. Am. Chem. Soc.* **124**, 10664-10665.
(103) Anthony, J. L., Maginn, E. J., Brennecke, J. F. (2002), Solubilities and Thermodynamic Properties of Gases in the Ionic Liquid 1-*n*-Butyl-3-methylimidazolium Hexafluorophosphate, *J. Phys. Chem. B* **106**, 7315-7320.
(104) Troxler, L., Wipff, G. (1998), Interfacial Behaviour of Ionophoric Systems: Molecular Dynamics Studies on 18-crown-6 and its Complexes at the Water-Chloroform Interface, *Anal. Sci.* **14**, 43-56.
(105) Anthony, J. L., Maginn, E. J., Brennecke, J. F. (2001), Solution Thermodynamics of Imidazolium-Based Ionic Liquids and Water, *J. Phys. Chem. B* **105**, 10942-10949.
(106) Wipff, G., Engler, E., Guilbaud, P., Lauterbach, M., Troxler, L., Varnek, A. (1996), Free Calixarene and Cryptand Ionophores at the Water / Chloroform interface, *New J. Chem.* **20**, 403-417.
(107) Schweighofer, K. J., Benjamin, I. (1995), Transfer of Small Ions Across the Water/1,2-Dichloroethane Interface, *J. Phys. Chem.* **99**, 9974-9985.

22. APPLICATION OF MACROCYCLIC LIGANDS TO ANALYTICAL CHROMATOGRAPHY

JOHN D. LAMB AND JOSEPH S. GARDNER
Department of Chemistry and Biochemistry, Brigham Young University, Provo, UT 84602, U.S.A.

1. Introduction

As recognized from the time of their discovery, one of the most intriguing features of macrocyclic hosts is the selectivity with which they bind specific guest species. Cation, anion, and neutral guests have been selectively sequestered by macrocycles of widely varying design and size. This selectivity has prompted the incorporation of macrocyclic hosts into a variety of practical separations systems, including but not limited to, solvent extraction, liquid membranes, capillary electrophoresis, and chromatography. Such separations systems typically fall into three categories: batch processes designed to recover and concentrate desired species, e.g., the recovery of precious metals; purification processes designed to remove undesirable species, such as the purification of water; and analytical methods designed to analyze species of interest, such as the determination of metal ions in water. Related to this last category is the incorporation of macrocycles into other types of analytical techniques which do not rely on separations, such as luminescent sensors and ion selective electrodes. Several reviews detail the broad range of applications of macrocyclic ligands to analytical applications in general, many of which are referenced below.

This review chapter focuses on research in one important separations-based analytical technique, chromatography. Specifically, we summarize recent advances in the application of macrocyclic ligands to analytical chromatography for the determination of ions and neutral species. We have organized the review by macrocycle type, beginning with synthetic macrocycles, and followed by a series of naturally-occurring macrocycles and their analogs, including porphyrins, cyclodextrins, and macrocyclic antibiotics. Most of the review describes the direct incorporation of macrocycles into the stationary or mobile phases of high performance chromatographic systems. However, brief mention is also made of analyte concentration methods which exploit macrocyclic ligand selectivity and which are based on chromatographic principles.

Karsten Gloe (ed.), Macrocyclic Chemistry: Current Trends and Future Perspectives, 349–363.
© 2005 *Springer. Printed in the Netherlands.*

2. Synthetic Macrocyclic Ligands

The era of synthetic macrocycles effectively got its start with the work of Pedersen, who reported the synthesis and cation-binding characteristics of the crown ethers in 1967. [1] Not long thereafter, the selectivity of macrocycles was first applied to a chromatographic separation when Cram's group [2-4] reported using optically pure (*RR*)-tetranaphthyl-22-crown-6 in the chloroform mobile phase for the separation of cationic amino acid isomers. During subsequent years, a host of investigators has made steady progress in advancing the use of macrocyclic ligands in chromatographic systems, including the separation and analysis of cations and anions, as well as of organic species. This progress is detailed up to the early 1990's in the chapter by Lamb and Smith in *Comprehensive Supramolecular Chemistry* [5], some highlights of which are summarized below. The major emphasis of this chapter is on work performed since that review.

2.1 ADSORBED TO THE STATIONARY PHASE

A successful approach in applying macrocycles to chromatography has involved the adsorption of hydrophobic macrocycles to reversed phased chromatographic packings. This approach avoids the complications inherent in attempting to covalently attach the ligand to the packing material. Kimura and coworkers [6,7] pioneered the adsorption of hydrophobic crown ethers and cryptands onto silica-based reversed-phase HPLC packings for the effective separation of alkali metal cations. Around the same time, Shinbo et al. [8] achieved highly efficient separations of amino acid enantiomers using a chiral crown ether adsorbed onto a reversed phase packing. Daicel Chemical Industries markets an HPLC column [9] which contains a chiral crown adsorbed to silica gel for the rapid and efficient analysis of chiral amines, amino alcohols and up to 25 amino acids.

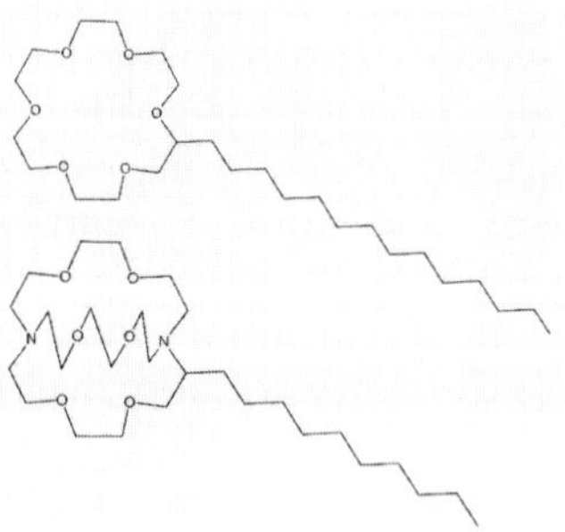

Figure 1. Structures of hydrophobically substituted 18-crown-6 (above) and 2.2.2 (below) adsorbed onto ion chromatography polymer resins for ion separations.

Lamb's group has adsorbed the hydrophobic crown ethers and cryptands shown in Fig 1 onto reversed phase chromatography packings for application to ion chromatography in the analysis of cations and anions. A brief introduction to ion chromatography (IC) is in order to lay the foundation for a description of this work.

The term "ion chromatography" has taken on a specific meaning in analytical chemistry: it refers to high performance liquid chromatographic analysis of ions, as first described by Small et al. in 1975. [10] The key to their innovation was the development not only of high efficiency column packings for cation and anion separations, but also of a universal form of detection (conductimetry) which was made practical by an innovative "suppressor" technology. The suppressor currently employs an ion-permeable membrane to remove the background conductivity of the ionic eluent. In the case of anion separations, which employ a basic eluent, after separation has been achieved on the column, protons pass through the membrane into the eluent stream to neutralize the eluent. In the case of cation separations, the opposite acid-base reaction is used. Such IC systems were the first to achieve true HPLC analysis of ions and have become a standard tool now found in almost all analytical labs.

Lamb and Drake [11] introduced the use of macrocycles in IC when they described using column-adsorbed macrocycle-cation complexes for the IC analysis of anions in aqueous solution in 1989. [12] Since the macrocycle-cation complexes are labile in water solution, the anion capacity of the resulting columns was variable, according to the type and concentration of metal cation in the eluent. This variable capacity of the column opened the door to a new type of gradient separation, which they called a "capacity gradient". In typical chromatography, when a gradient is carried out to more rapidly elute late-eluting peaks, it is the eluent strength which is modified over time. But in capacity gradient IC, it is the capacity of the column which varies over the course of the chromatogram, achieved by switching from one metal cation to another with a different affinity for the macrocycle. In Fig 2, a capacity gradient is used to achieve good separation among a series of anions of widely varying hydrophilicity. Without a gradient, the late peaks in these separations would emerge (if at all) at extremely long times in very broad, unusable peaks.

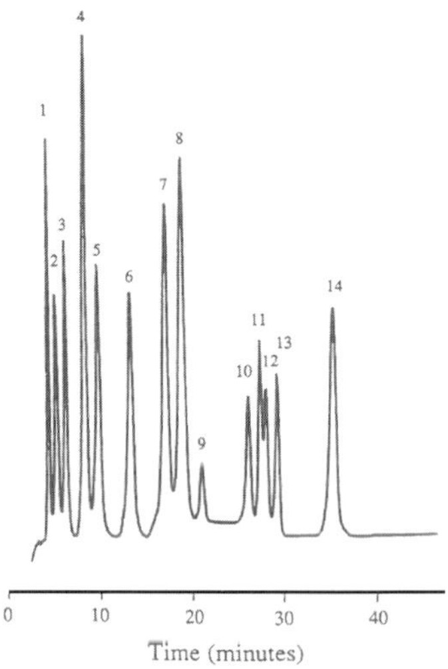

Figure 2. Ion chromatographic separation of a series of anions on polymer-based column with adsorbed decyl-2.2.2 using gradient capacity from sodium hydroxide to lithium hydroxide aqueous eluent. Anions: 1) fluoride 2) acetate 3) chloride 4) nitrite 5) bromide 6) nitrate 7) sulfate 8) oxalate 9) chromate 10) iodide 11) phosphate 12) phthalate 13) citrate 14) thiocyanate (from refs. 13,14)

The gradient capacity in each case is achieved by changing from an eluent cation of high affinity for the cryptand adsorbed to the polystyrene column to one of low affinity. This gradient can be achieved either by the gradual substitution of one ion for another, or by a step substitution. [15] Furthermore, organic modifiers may be added to the eluent to adjust elution times and peak resolution. [16] High resolution separations of carbohydrate oligomers of increasing chain length was achieved using gradients of this type. [17] This approach is especially effective for the determination of fluoride ion in environmental samples. [18]

Capacity gradients can be achieved in another way in IC: by changing the temperature of the macrocycle-based column. Since the reaction of cryptands with metal cations is typically exothermic, raising the temperature reduces the degree of complexation. Based on this concept, gradient separations of anions can be achieved using a 2.2.2 column when the temperature is raised to 80 °C during the separation. [19]

In addition to their application to liquid chromatography, macrocycles have been applied to gas chromatographic separations as well. Kartsova et al. [20] performed a systematic study of a series of crown ethers and cryptands adsorbed onto GC stationary phases. The influence on the polarity and selectivity of the stationary phases of the type and number of heteroatoms, conformational lability of the cavity, the presence of substituents, and the concentration of macrocycle was studied with respect to the separations of various classes of organic compounds. Phases containing mixtures of two macrocycles were found to be most promising.

In addition to crown ethers and cryptands, macrocycles of more complex structure have been adsorbed to stationary phase materials to effect separations. For example, the lipophilic calix[4]resorcinarene shown in Figure 3 was adsorbed to ODS silica by Petraszkiewicz [21] for HPLC separations of pyrimidine bases and substituted phenols. Furthermore, Sokoliess et al. [22] used six calix[n]arene phases (n=4, 6, 8) and a calix[4]resorcinarene columns to separate a number of organic species and their cis-trans isomers using organic modifiers in the eluent. Selectivities were influenced by macrocycle ring size as well as the degree of substitution on the upper rim with p-tert-butyl groups. The same group also reported the use of bonded columns. [23] Calix[4]arene derivatives were used by Yu et al. [24] to separate aromatic isomers. A recent review summarizes the use of calixarenes as stationary phases in gas, liquid, and capillary chromatography techniques. [25] They note that one of the attractive features of these macrocycles for chromatographic applications is their ability not only to sequester substrates in the native cavity, but also the ease of their derivatization. Specifically, additional substituents and functional groups are easily appended at the macrocycle rims to act as independent binding sites or to adjust cavity interactions.

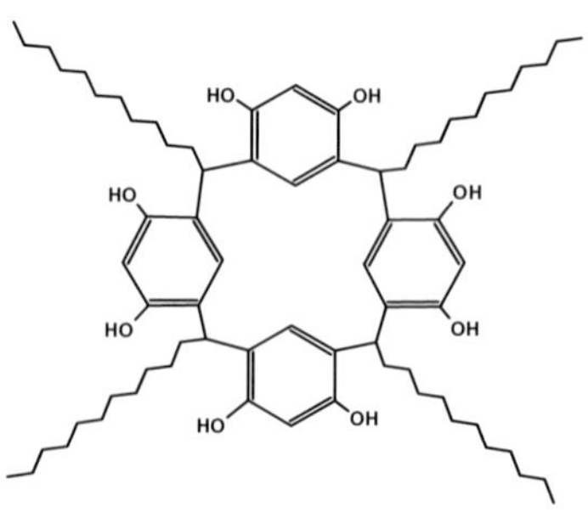

Figure 3. Structure of hydrophobically substituted calix[4]resorcinarene adsorbed onto silica-based resin for HPLC separations (see ref. 21).

2.2 BONDED TO THE STATIONARY PHASE

In recent years, the chemistry described above using adsorbed crown ethers and cryptands in IC has been incorporated into commercially available polymer-based columns for anion analysis by Dionex Corporation. Specifically, 18-crown-6 was bonded to the polymeric substrate of the traditional cation separator column which originally contained only phosphonate and carboxylate cation exchange sites to produce a new column, the CS15, which facilitated the analysis of ammonium ion in the presence of high concentrations of sodium ion. Addition of 18-crown-6 as a third cation exchange moiety improved resolution between some cation peaks. [26-29] In another column, cryptand 2.2.2 was covalently bonded to polystyrene/divinylbenzene packing materials. The column performed both for cation and anion separations in fashion very similar to comparable columns in which the macrocycle was adsorbed to the substrate, as described in the section above. The extra stability gained by bonding the macrocycle to the substrate did not detract from the separations, nor from the ability to carry out capacity gradient separations. The separation of a series of anions by this column using a capacity gradient is shown in Figure 4. [30,31].

The 2.2.2 column is especially advantageous in the determination of fluoride ion, which in traditional IC often elutes so early as to be masked by the injection peak. [32] In similar fashion, Tsai and Shih [33] derivatized polystyrene/divinylbenzene resin with cryptand 2.2.2 for the ion chromatographic separation of cations or anions.

In the early 1980's, the groups of Blasius and of K. Kimura pioneered the grafting of macrocycles to silica particles for use in chromatographic separations of cations. [34-37] Lauth and Gramain [38,39] were subsequently able to effect good anion separations in water samples using benzo-18-crown-6 derivatized silica stationary phases and a common cation. Unlike polymer-based column particles, these stationary phase materials have not found practical application in ion chromatography because of their strong sensitivity to acid/base degradation.

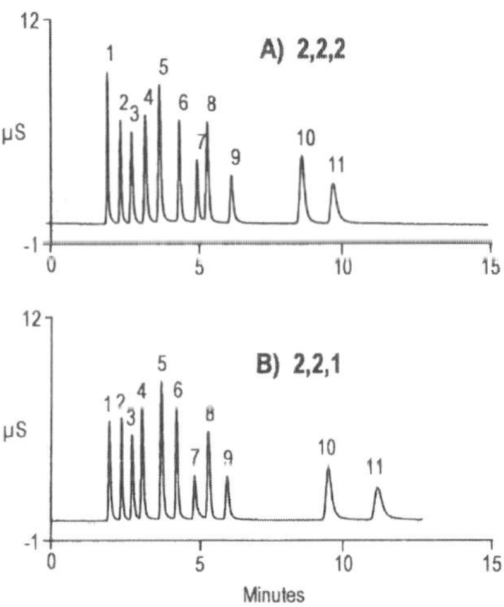

Figure 4. Ion chromatographic separations of a series of anions on polymer-based column with covalently bonded (A) 2.2.2 and (B) 2.2.1. using gradient capacity step changes from one alkali metal cation to another. Anions: 1) fluoride 2) chloride 3) nitrite 4) bromide 5) nitrate 6) sulfate 7) thiosulfate 8) phosphate 9) iodide 10) thiocyanate 11) perchlorate (from ref. 31)

As an example of the latter application, the group of Izatt and of IBC Advanced Technologies has applied chromatographic materials based on silica-bound macrocycles to the separation and concentration of chemical species, especially metal cations, for subsequent analysis by various methods. [40-42] Heavy metal ions such as Pb^{2+} and precious metals can be concentrated from the ppm to the ppt range with high selectivity using these materials. In a recent example, Talanova et al. [43] report the preparation of dibenzo-18-crown-6 covalently grafted onto polystyrene. The structure is shown in Figure 5.

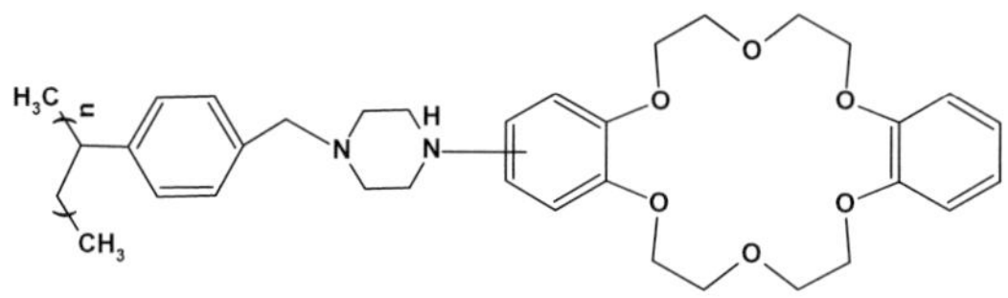

Figure 5. Recent example of crown ether bonded to polymer substrate (see ref. 43)

The group of Alexandratos has reported attaching a variety of macrocyclic ligands, such as crown ethers and calixarenes, to polymer supports for chromatographic separations. The calixarenes were substituted with phosphorus-containing ligands for selective cation coordination. [44,45] Although these materials were not specifically prepared for high performance chromatographic separations, they have the potential for application in sample pretreatment and concentration. And the same principles could be used to synthesize high performance chromatographic packings. A recent review summarizes the strategies available to the synthetic chemist for covalently bonding macrocycles to organic polymers. [46]

Among more complex macrocycles, Li et al. [47-52] reported the preparation and characterization of stationary phases incorporating calixarenes or calix-crowns bonded to silica. With individual columns, high selectivity was observed in the separation of alkylated aromatics, aromatic carboxylic acids, sulfonamides, nucleosides, and water-soluble vitamins. In other work, Sokoliess et al. [53] have characterized calixarene- and resorcinarene-bonded stationary phases similar to those described in the previous section of this chapter. And Huai et al. [54] used an end-capped p-tert-butyl-calix[4]arene-bonded silica phase for HPLC separation of a number of organic compounds. Resorcinarenes have also found application in GC. [55-57] Recently, exotic macrocycles have been used in capillary electrochromatography, as reported by Gong et al. [58]

2.3 POLYMERIZED AS STATIONARY PHASE MATERIALS

In addition to post-functionalizing polymers by bonding the macrocycle to the pre-formed polymer backbone, macrocycles can be incorporated into polymer matrices by direct polymerization of the macrocycle, either by a step-growth mechanism or a chain-growth mechanism. [46] Polymeric crown ether stationary phases were pioneered by Blasius et al. [34, 59-62] These resins were used to separate both cations (including protonated amines) with a common anion, and anions with a common cation in high performance mode. More recent examples of incorporating macrocycles into polymer chains (shown in Figure 6) display metal cation selectivities which typically echo the selectivity of the native macrocycle. The hydrophilicity of the resin is a strong determinant of loading capacities and overall ionic affinities. These latter materials have largely been applied to preparative chromatography.

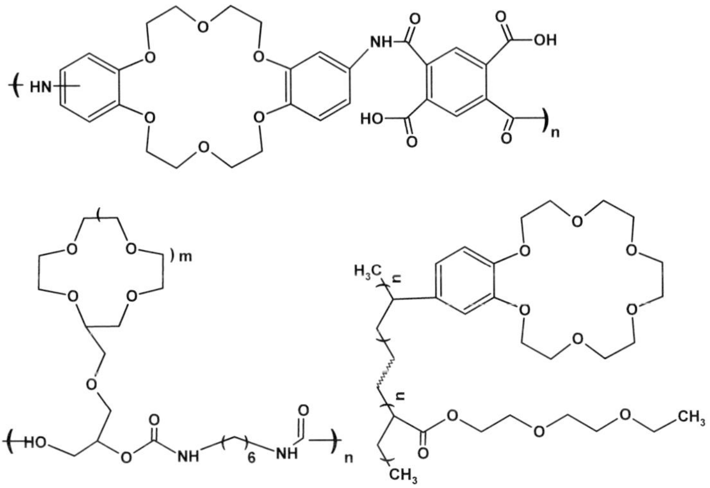

Figure 6. Recent examples of polymerized crown ethers (see ref. 46)

For our purposes here, the problem with resins formed by co-polymerization of the macrocycle is that they are often soft and unable to endure the pressures needed for HPLC applications. Furthermore, adsorption kinetics are slow and band broadening more significant when exchange sites are distributed throughout the polymer, rather than on the surface. However, these resins present good candidates for sample preparation applications where separation efficiency is not critical.

2.4 DISSOLVED IN THE MOBILE PHASE

Although the majority of reports of macrocycles in analytical chromatography have involved ligand association with the stationary phase, their use as mobile phase constituents has also been investigated. Lamb and Drake [11] showed that addition of water-soluble crown ethers to the mobile phase altered the retention of alkali metal cations on an underivatized reversed phase column. Nakagawa et al. [63-66] also used crown ether-containing mobile phases in the separation of protonated amines, amino acids and peptides, and β-lactam antibiotics.

Interest in this mode of introducing macrocycles into the chromatographic separation has increased in recent years. For example, Ohta and Tanaka [67] used mobile phase 18-crown-6 on a silica get column for the determination of common mono- and divalent cations. Läubli and Kampus [68] demonstrated how mobile phase 18-crown-6 improves the separation of NH_4^+ and Na^+ on a column containing carboxylate groups. Dumont and Fritz [69] explained how the addition of 18-crown-6 improved peak shape of some cations. Okada [70] showed how mobile phase 18-crown-6 on an aminopropylated silica gel ($Si—NH_3^+NO_3^-$) could be used to alter the retention times of several anions. Finally, Jensen and Joppert [71] described how the use of mobile phase 18-crown-6 with a low capacity column based on carboxylate and phosphonate functional groups could cause K^+ to elute after the divalent cations.

Using a macrocycle as a mobile phase additive is only practical for simple, inexpensive, relatively low-toxicity macrocycles such as 18-crown-6. A recent example of the benefits to be derived from addition of 18-crown-6 to an ion chromatographic eluent is found in work published by Lamb's group. [72] Specifically, in separating alkali, alkaline earth and amine cations by IC, it was found that addition of the macrocycle to the mobile phase could be used to adjust the retention time of ammonium cation so that ammonium ion could be determined under concentration ratios of 60,000:1 Na^+ to NH_4^+ (see Figure 7). This method has specific application in the determination of ammonia produced by nitrogenase—analysis time and sample size are considerably reduced over traditional wet chemical methods.

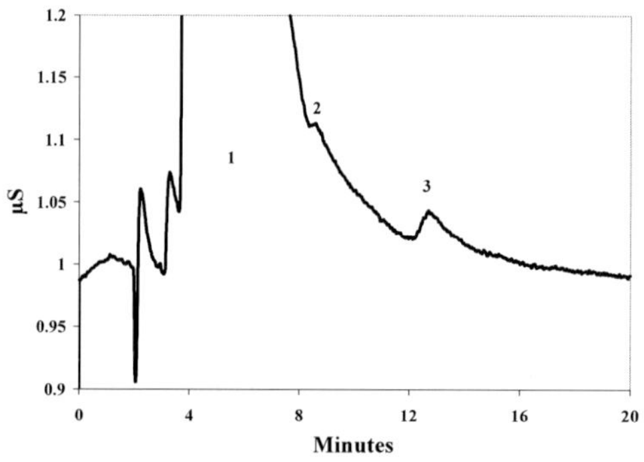

Figure 7. Ion chromatographic separation of ammonium ion (3) from sodium (1) and calcium (2) ions with a concentration ratio of sodium to ammonium of 60,000:1. Column: Dionex CS15; eluent: sulfuric acid and 18-crown-6. (from ref. 32)

Another interesting application of mobile phase 18-crown-6 in IC involves the analysis of perchlorate in water samples. [73] The U.S. Environmental Protection Agency has focused recently on perchlorate as a toxin of interest, proposing to reduce the acceptable level in drinking water from the current 4 ppb to 1 ppb. Analytical methods which can reliably determine perchlorate at this low level are few. The only currently accepted method is an IC method using a separator column based on standard quaternary amine exchange sites. [74] Lamb's group has recently found that an 18-crown-6 based mobile phase coupled with a polystyrene reversed phase column can be used to determine perchlorate at this level, as in the chromatogram in Figure 8.

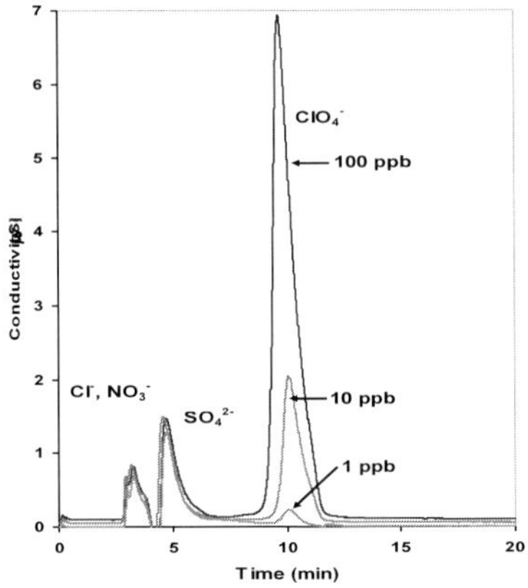

Figure 8. Analysis of perchlorate ion at 100, 10 and 1 ppb concentrations in water using 18-crown-6 in an aqueous mobile phase and a polystyrene reversed phase column (from ref. 75)

Several papers have reported use of a weak acid/crown ether eluent in ion-exclusion/cation exchange chromatography with conductimetric detection on a weakly acidic cation-exchange column to effect the simultaneous determination of both cations and anions. Resolution was significantly improved when 18-crown-6 was present in the eluent. Detector response was positive for anions and negative for cations. [76-80] A sample chromatogram is shown in Figure 9.

Cyclam has also been used in the mobile phase for HPLC analysis. For example, Colgan et al. [81] added the macrocycle to a reversed phase analysis of the drug tenidap to reduce peak tailing and improve reproducibility.

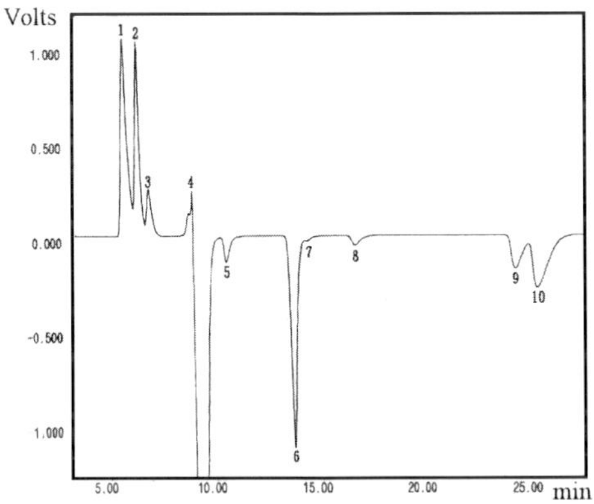

Figure 9. Analysis of anions and cations in river water using tartaric acid/18-crown-6/methanol-water eluent with a carboxylated polyacylate stationary phase in the protonated form. Ions: 1) sulfate 2) chloride 3) nitrate 4) eluent dip 5) unknown 6) sodium 7) ammonium 8) potassium 9) magnesium 10) calcium (from ref. 80)

3. Naturally Occurring Macrocycles

3.1 PORPHYRINS

Porphyrins and their analogs have been investigated for inclusion as mobile phase modifiers and in novel stationary phases which offer unusual resolution in HPLC, as summarized in a number of reviews. [82-86] For example, porphyrins have been used as complexing agents in HPLC for transition metal determination. An obstacle to metal ion determination is the slow kinetics of complexation, which leads to significant band broadening. Porphyrin complexes serve as exchange sites in immobilized metal ion affinity chromatography (IMAC). In a very recent example of IMAC work, Mifune et al. [87] report the preparation of HPLC silica stationary phases modified with metal-porphines and metal-phthalocyanines for the separation of πp-electron-rich polyaromatic hydrocarbons (PAH's). Similarly, Kibbey et al. [88] reported a tetraphenylporphyrin-based stationary phase for HPLC separation of PAH's; and when the ligands were metalated with tin(IV) or indium(III), the stationary phase was useful in the separation of aromatic sulfonates and aromatic carboxylates. Stationary phases containing metal-complexed porphyrin analogs have been used in the HPLC separation of anions, including common anions, aromatic carboxylates, and sulfonates. [88-89] Sessler's group used a silica gel-bound sapphyrin as a stationary phase for the HPLC separation of anions, as well as adenosine, AMP, ADP, and ATP. [90] As is true for several other types of macrocycles, recent reports describe the inclusion of porphyrin analogs and their metal complexes in electrochromatographic systems. [91]

3.2 CYCLODEXTRINS

Biologically derived cyclodextrins and their synthetically derivatized analogs form inclusion complexes with a host of guest molecules, often with unusually high enantioselectivity. For this reason, they make good candidates for use in chromatographic separations and analysis of many species, including positional isomers, analogs containing different functional groups, homologs and enantiomers. The latter are of particular interest to those working with pharmaceuticals. Applications of α-, β-, and γ-cyclodextrins to thin layer chromatography, gas chromatography, HPLC, and supercritical fluid chromatography have been extensively reviewed in a number of recent, highly informative sources, including a 2003 Royal Society monograph and a 2004 book on chiral separations. [92-99] The vast majority of anantiomeric separations have been performed using β-cyclodextrin and its many derivatives.

In gas chromatography, the addition of bulky substituents at the primary C6-hydroxy groups improves solubility in the polysiloxane coating and affects access of substrates to the ligand cavity. GC columns coated with modified cyclodextrins have been used for the enantiomeric analysis of essential oils, flavors and fragrances, ingredients in alcoholic beverages, clinically-important species, terpenoids, pheromones, enzymatic reaction ingredients, pesticides, atmospheric ingredients, and pharmaceuticals. [98] Sometimes these methods include multidimensional chromatography performed in two or more stages, with an initial cut to separate components of interest (by GC or LC), followed by the separation of enantiomers using the cyclodextrin column. A mixed chiral stationary phase for GC has been described containing modified resorcinarene and β-cyclodextrin bonded to polysiloxane for the separation of enantiomers of apolar hydrocarbons and polar amino acids. [100]

In LC, cyclodextrins have been used in both the mobile and stationary phases, although the latter application predominates using bonded phases. Both normal and reversed phase separations are possible. With polar mobile phases, the cyclodextrin cavity hosts and retains hydrophobic analytes; but when large concentrations of organic solvent are used in the mobile phase, solvent molecules occupy the cavity—inclusion of analytes is suppressed and hydrogen bonding and dipole-dipole interactions with analytes predominate. Applications to electrochromatography have also been described. [96]

3.3 MACROCYCLIC ANTIBIOTICS

Macrocyclic antibiotics such as teicoplanin and vancomycin have been used in chiral stationary phases separations of amino acids, drugs, and other species using HPLC and other separations methods. These applications have been reviewed in a number of recent sources, including a 2004 monograph on chiral separations. [97, 101-107]

4. References

1. Pedersen, C. J. (1967) Cyclic Polyethers and their Complexes with Metal Salts, *J. Am. Chem. Soc.* **89**, 2495-2496.
2. Kaplan, L., Sousa, L. R., Hoffman, D. H., and Cram, D. J. (1974) Total optical resolution of amino esters by designed host-guest relations in molecular complexation, *J. Am. Chem. Soc.* **96**, 7100-7101.
3. Dotsevi, G., Sogah, Y., and Cram, D. J. (1975) Chromatographic Optical Resolution Through Chiral Complexation of Amino Ester Salts by a Host Covalently Bound to Silica Gel, *J. Am. Chem. Soc.* **97**, 1259-1261.
4. Dotsevi, G., Sogah, Y., Cram, D. J. (1976) Total Chromatographic Optical Resolutions of Amino Acid and Ester Salts Through Chiral Recognition by a Host Covalently Bound to Polystyrene Resin, *J. Am. Chem. Soc.* **98**, 3038-3041.
5. Lamb, J. D. and Smith, R. G. (1996) Application of Macrocyclic Ligands to High-Performance Ion Analysis, in D. Reinhoudt (ed.), *Comprehensive Supramolecular Chemistry* **10**, Pergamon, Oxford, pp. 79-112.
6. Kimura, K., Harino, H., Hayata, E., and Shono, T. (1986) Liquid chromatography of alkali and alkaline-earth metal ions using octadecylsilanized silica columns modified in situ with lipophilic crown ethers, *Anal. Chem.* **58**, 2233-2237.
7. Kimura, K., Hayata, E., and Shono, T. (1984) Convenient, efficient crown ether-containing stationary phases for chromatographic separation of alkali metal ions: dynamic coating of highly lipophilic crown ethers on octadecylsilanized silica, *Chem. Commun.*, 271-272.
8. Shinbo, T., Yamaguchi, T., Nishimura, K., and Sugiura, M. (1987) Chromatographic Separation of Racemic Amino Acids by Use of Chiral Crown Ether-coated Reversed-phase Packings, *J. Chromatogr.* **405**, 145-53
9. see the website http://www.chiraltech.com/New/Cat04.pdf
10. Small, H., Stevens, T. S., and Bauman, W. C. (1975) Novel Ion Exchange Chromatographic Method Using Conductimetric Detection, *Anal. Chem.* **47**, 1801-1809.
11. Lamb, J. D. and Drake, P. A. (1989) Chemically-Suppressed Anion Chromatography Based On Macrocycle Cation Complexation, *J. Chromatogr.* **482**, 367-380 .
12. Weiss, J. and Jensen, D. (2003) Modern Stationary Phases for Ion Chromatography, *Analyt. Biooanalyt. Chem.* **375**, 81-98.
13. Phillip A. Drake (1990), Master's Thesis: "Anion Chromatography with Variable Capacity Separators Based on Macrocyclic Complexes," Brigham Young University.
14. Robert G. Smith (1993), PhD Dissertation: "Anion Separations Using Macrocycle-Based Ion Exchange Columns," Brigham Young University.

15. Lamb, J. D. and Smith, R. G. (1994) Use of Step Gradients on Different Polymetric Substrates in the Separation of Anions by Macrocycle-based Ion Chromatography, *J. Chromatogr. A* **671**, 89-94.
16. Lamb, J. D., Smith, R. G., and Jagodzinski, J. (1993) Anion Chromatography With a Crown Ether-based Stationary Phase and an Organic Modifier in the Eluent, *J. Chromatogr.* **640**, 33-40, .
17. Niederhauser, T. L., Halling, J., Polson, N. A., and Lamb, J. D. (1998) High-performance Anion-exchange Chromatographic Separations of Carbohydrates on a Macrocycle-based Stationary Phase with Eluents of Relatively Low pH and Concentration, *J.Chromatogr. A* **804**, 69-77.
18. Lamb, J. D., Eatough, D. J., and Smith, R. G. (1993) Determination of Fluoride in Atmospheric Samples Using Macrocycle-based Ion Chromatography, *Int. J. Envir. Analyt. Chem.* **56**, 317-325.
19. Lamb, J. D. and Smith, R. G. (1992) A Comparison of Gradient Capacity Anion Chromatography Using Macrocycles D-2.2.2 and D-2.2.1 in Constant or Variable Temperature Mode, *Talanta* **39**, 923-930.
20. Kartsova, L. A., Markova, O. V., Amel'chenko, A. I., Ostryanina, N. D. (2000) Macrocycles as Components of Gas-Chromatographic Phases, *J. Analyt. Chem.* **55**, 270-279.
21. Pietraszkiewicz, O. and Pietraszkiewicz, M. (1999) Separation of Pyrimidine Bases on a HPLC Stationary RP-18 Phase Coated with Calix[4]resorcinarene, *J. Incl. Phen.* **35**, 261-270.
22. Sokoliess, T., Menyes, U., Roth, U., and Jira, T. (2002) Separation of Cis- and Trans-isomers of Thioxanthene and Dibenz[b,e]oxepin Derivatives on Calixarene- and Resorcinarene-bonded High-performance Liquid Chromatography Stationary Phases, *J. Chromatogr. A* **948**, 309-319.
23. Sokoliess, T., Opolka, A., Menyes, U., Roth, U., and Jira, T. (2002) Separation of Racemic Drugs on Chiral Resorcinarene-bonded HPLC-columns, *Pharmazie* **57**, 589-590.
24. Yu, X., Fang, H., Lin, L., Han, H., and Wu, C. (2001) Study of the Gas Chromatographic Properties of Calix[4]arene Derivatives Containing Different Substituents on the Upper Rim, *Chromatographia* **53**, 519-524.
25. Milbradt, R. and Böhmer, V. (2001) Calixarenes as stationary phases, in Asfari, Z. (ed.), *Calixarenes,* Kluwer Academic Publishers, Dordrecht, pp. 663-676.
26. Rey, M. A., Pohl, C. A., Jagodzinski, J. J., Kaiser, E. Q., and Riviello, J. M. (1998) A New Approach to Dealing with High-to-low Concentration Ratios of Sodium and Ammonium Ions in Ion Chromatography, *J. Chromatogr. A* **804**, 201-209.
27. Pohl, C. A., Rey, M. A., Jensen, D., and Kerth, J. (1999) Determination of Sodium and Ammonium Ions in Disproportionate Concentration Ratios by Ion Chromatography, *J. Chromatogr. A* **850**, 239-245.
28. Rey, M. A., and Pohl, C. A. (1996) Novel Cation-exchange Stationary Phase for the Separation of Amines and of Six Common Inorganic Cations, *J. Chromatogr. A* **739**, 87-97.
29. Jensen, D. and Joppert, G. (1998) Advances in Analytical Cation Chromatography, *LaborPraxis* **22**, 60-67.
30. Woodruff, A., Pohl, C. A., Bordunov, A., and Avdalovic, N. (2002) Adjustable-capacity Anion-exchange Separator, *J. Chromatogr. A* **956**, 35-41.
31. Woodruff A., Pohl, C. A., Bordunov, A., and Avdalovic, N. (2003) Environmental Applications of a Cryptand Adjustable-capacity Anion-exchange Separator, *J. Chromatogr. A* **997**, 33-39.
32. Richens, D. A. (2001), Master's Thesis: "Use of Macrocyclic Ligands in Mobile and Stationary Phases for Ion Separation by Ion Chromatography," Brigham Young University.
33. Tsai, S. S. and Shih, J. S. (1998) Application of the bifunctional switch-type cryptand ion chromatographic stationary phase for ion separation, *Sep. Sci. Tech.* **33**, 1407-1421.
34. Blasius, E., Janzen, K. P., Klein, W., Klotz, H., Nguyen, V. B., Nguyen-Tien, T., Pfeiffer, R., Scholten, G., and Simon, H. (1980) Preparation, Characterization and Application of Ion Exchangers with Cyclic Polyether Anchor Groups, *J. Chromatogr.* **201**, 147-166.
35. Kimura, K., Nakajima, M., and Shono, T. (1980) Poly(crown ether)-Modified Silica for Stationary Phase of Liquid Chromatography, *Analyt. Lett.* **13**, 741-750.
36. Nakajima, M., Kimura, K., and Shono, T. (1983) Liquid Chromatography of Alkali and Alkaline Earth Metal Salts on Poly(benzo-15-crown-5)- and Bis(benzo-15-crown-5)-modified Silicas, *Anal. Chem.* **55**, 463-467.
37. Kimura, K., Nakajima, M., and Shono, T. (1985) Characterization of Poly(crown ether)-Modified Silicas, *J. Polymer Sci., Polymer Chemistry Edition* **23**, 2327-2331.
38. Lauth, M. and Gramain, P. (1987) Chromatographic Separations of Alkali and Alkaline Earth Metal Cations and Some Anions on Benzo-18-crown-6-modified Silicas, *J. Chromatogr.* **395**, 153-158.
39. Lauth, M. and Gramain, P. (1985) Ion Chromatographic Separation on Silica Grafted with Benzo-18-C-6 Crown Ether, *J. Liq. Chromatogr.* **8**, 2403-2415.
40. Izatt, R. M., Bradshaw, J. S., Bruening, R. L., and Bruening, M. L. (1994) Solid Phase Extraction of Ions of Analytical Interest using Molecular Recognition Technology, *Amer. Lab* **26**, 28C-28M.

41. Izatt, N., Bruening, R., Krakowiak, K., and Izatt, S. (2000) Contributions of Professor Reed M. Izatt to Molecular Recognition Technology: From Laboratory to Commercial Application, *Ind. Eng. Chem. Res.* **39,** 3405-3411.
42. Goken, G. L., Bruening, R. L., Krakowiak, K. E., and Izatt, R. M. (1999) Metal-ion Separations Using SuperLig or AnaLig Materials Encased in Empore Cartridges and Disks, *ACS Symposium Series* **716**, 251-259.
43. Talanova, G. G., Yatsimirskii, K. B., and Kravchenko, O. V. (2000) Peculiarities of Dipotassium Palladium Tetrachloride (K_2PdCl_4) and Dipotassium Platinum Tetrachloride (K_2PtCl_4) Complexation with Polymer-supported Dibenzo-18-crown-6, *Ind. Eng. Chem. Res.* **39**, 3611-3615.
44. Alexandratos, S. D. and Natesan, S. (2000) Coordination Chemistry of Phosphorylated Calixarenes and their Application to Separations Science, *Ind. Eng. Chem. Res.* **39**, 3998-4010.
45. Alexandratos, S. and Natesan, S. (2001) Synthesis and Ion-Binding Affinities of Calix[4]arenes Immobilized on Cross-Linked Polystyrene, *Macromolecules* **34**, 206-210.
46. Alexandratos, S. and Stine, C. (2004) Synthesis of Ion-selective Polymer-supported Crown Ethers: A Review, *React. Funct. Polymers* **60**, 3-16.
47. Li, L., Da, S., Feng, Y., and Liu, M. (2004) Preparation and Characterization of a P-tert-butyl-calix[6]-1,4-benzocrown-4-bonded Stationary Phase for Liquid Chromatography, *J. Chromatogr. A* **1040**, 53-61.
48. Li, L., Liu, M., Da, S., and Feng, Y. (2004) High Performance Liquid Chromatography of Aromatic Carbonxylic Acids on p-tert-butyl-calix[8]arene-bonded Silica Gel Stationary Phase, *Talanta* **62**, 643-648.
49. Li, L., Da, S., Feng, Y., and Liu, M. (2004) Study on the Chromatographic Behavior of Water-soluble Vitamins on P-tert-butyl-calix[8]arene-bonded Silica Gel Stationary Phase by HPLC, *Talanta* **64**, 373-379.
50. Li, L., Liu, M., Da, S., and Feng, Y. (2004) Studies on the Chromatographic Behavior of Nucleosides and Bases on P-tert-butyl-calix[8]arene-bonded Silica Gel Stationary Phase by HPLC, *Talanta* **63**, 433-441.
51. Li, L., Da, S., Feng, Y., and Liu, M. (2004) Preparation and Evaluation of a New Calix[4]arene-Bonded Stationary Phase for HPLC, *J.Liq. Chromatogr. Rel. Techn.* **27**, 2167-2188.
52. Li, L., Da, S., Feng, Y., and Liu, M. (2004) Preparation and Characterization of a New P-tert-butyl-calix[8]arene-bonded Stationary Phase for High-performance Liquid Chromatography, *Analyt. Sci.* **20**, 561-564.
53. Sokoliess, T., Schonherr, J., Menyes, U., Roth, U., Jira, T. (2003) Characterization of calixarene- and resorcinarene-bonded stationary phases I. Hydrophobic interactions. *J. Chromatogr. A* **1021**, 71-82.
54. Huai, Q. Y., Zhao, B., and Zuo, Y. M. (2004) Preparation and Evaluation of an End-capped P-tert-butyl-calix[4]arene-bonded-silica Stationary Phase for Reversed-Phase High-performance Liquid Chromatography, *Chromatographia* **59**, 637-645.
55. Pfeiffer, J. and Schurig, V. (1999) Enantiomer Separation of Amino Acid Derivatives on a New Polymeric Chiral Resorc[4]arene Stationary Phase by Capillary Gas Chromatography, *J. Chromatogr. A* **840**, 145-150.
56. Zhang, H., Ling, Y., Dai, R., Wen, Y., Fu, R., and Gu, J. (1997) Two Alkyl Resorcarenes as New Stationary Phases for Isomer Separation by GC, *Chem. Lett.* **3**, 225-226.
57. Zhang, H., Ling, Y., Dai, R., Wen, Y., Fu, R., Gu, J., and Zhang, S. (1997) Resorcarene Derivative Used As a New Stationary Phase for Capillary Gas Chromatography, *J.Chromatogr. A* **787**, 161-169.
58. Gong, Y., Lee, H. (**2003**) Application of Cyclam-Capped ß-Cyclodextrin-Bonded Silica Particles as a Chiral Stationary Phase in Capillary Electrochromatography for Enantiomeric Separations, *Anal. Chem.* **75**, 1348-1354.
59. Blasius, E. and Janzen, K. P. (1985) Ion Chromatography and Catalysis of Organic Reactions Using Polymers with Cyclic Polyethers as Anchor Groups, *Israel J.Chem.* **26**, 25-34.
60. Blasius, E., Janzen, K. P., Luxenburger, H., Nguyen, V. B., Klotz, H., and Stockemer, J. (1978) Application of Exchangers with Crown Ethers as Anchor Groups in Analytical and Preparative Chemistry, *J.Chromatogr.* **167**, 307 320.
61. Blasius, E., Janzen, K. P., and Zender, J. (1985) Silica gels coated or substituted with cyclic polyethers for ion chromatography, *Fresenius' Z. Analyt. Chem.* **320**, 435-438.
62. Blasius, E., Janzen, K. P., and Zender, J. (1986) The Mechanism of Anion Separation with Silica Gel Coated with Crosslinked poly(dibenzo-18-crown-6), *Fresenius' Z. Analyt. Chem.* **325**, 126-128.
63. Iwagami, S., and Nakagawa, **T.** (1986) Liquid Chromatography with Crown Ether c-containing Mobile Phases VIII. Retention behavior of Amino compounds in cation-exchange chromatography. *J. Chromatogr.* **369**, 49-58.

64. Nakagawa, T., Shibukawa, A., Kaihara, A., Itamochi,T., Tananka, H. (1986) Liquid Chromatography with Crown Ether-containing Mobile Phases VI. Molecular Recognition of Amino Acids and Peptides. *J. Chromatogr.* **353**, 399-408.
65 Nakagawa, T., Shibukawa, A., and Uno, T. (1982) Liquid Chromatography with Crown Ether-containing Mobile Phases II. Retention behavior of β-Lactam Antibiotics in Reversed-phase High-performance Liquid Chromatography, *J. Chromatogr.* **239**, 695-706.
66. Nakagawa, T., Murata, H., Shibukawa, A., Murakami, K., and Tanaka, H. (1985) Liquid Chromatography with Crown Ether-containing Mobile Phases V. Effect of Hydrophobicity and Cavity Size of the Crown Ether on Retention of Amino Compounds in Reversed-phase High-performance Liquid Chromatography, *J. Chromatogr.* **330**, 43-53.
67. Ohta, K. and Tanaka, K. (1999) Simultaneous Determination of Common Mono- and Divalent Cations in Natural Water Samples by Conductimetric Detention Ion Chromatography with an Unmodified Silica Gel Column and Oxalic Acid/18-crown-6 as Eluent, *Anal. Chim. Acta.* **381**, 265-273.
68. Laeubli, M. and Kampus, B. (1995) Selectivity Enhancement on a Poly (butadiene-maleic acid) Coated Cation Phase Induced by Ethylene Oxide-based Complexing Agents, *J.Chromatogr. A* **706**, 103-107.
69. Dumont, P. J. and Fritz, J. S. (1995) Ion chromatographic separation of alkali metals in organic solvents. *J.Chromatogr. A* ***706***, 149-158.
70. Okada, T. (1997) Nonaqueous Anion–exchange Chromatography II. Changing Anion-exchange Selectivity by Resin Surface Complex Formation of Crown Ethers, *J. Chromatogr. A* **758**, 29-35.
71. Jensen, D. and Joppert, G. (1998) Advances in Analytical Cation Chromatography, *LaborPraxis* **22**, 60-67.
72 Richens, D. A., Peterson, S., Simpson, D., McGinn, A., and Lamb, J. D. (2003) Use of Mobile Phase 18-Crown-6 To Improve Peak Resolution Between Mono- and Divalent Metal and Amine Cations in Ion Chromatography, *J. Chromatogr. A* **1016**, 155-164.
73. Urbansky, E. T. (ed.) (2000) *Perchlorate in the Environment*, Environmental Science Research **57**, Kluwer Academic Publishers, New York, NY.
74. Hautman, D. P., Munch, D. J., Eaton, A. D., and Haghani, A. W. (1999) *EPA Method 314.0, Determination of Perchlorate in Drinking Water using Ion chromatography, Revision 1.0*, US Environmental Protection Agency, Cincinnati, Ohio.
75. Lamb, J. D., Simpson, D., Jensen, B., and Peterson, Q. (2004), 18-crown-6 as an Eluent Additive for the Ion Chromatographic Determination of Perchlorate in Drinking Water, in preparation.
76. Ikedo, M., Mori, M., Xu, Q., Taoda, H., Sato, S., Hu, W., and Tanaka, K. (2003) Simultaneous Determination of Anions and Cations in Raw Material of Ceramic Glaze by Ion-exclusion/cation-exchange Chromatography with a Weakly Acidic Cation-exchange Resin, *Bunseki Kagaku* **52**, 1173-1180.
77. Ding, M., Tanaka, K., Hu, W., Hasebe, K., and Haddad, P. R. (2001) Simultaneous Ion-exclusion Chromatography and Cation-exchange Chromatography of Anions and Cations in Environmental Water Samples on a Weakly Acidic Cation-exchange Resin by Elution with Pyridine-2,6-dicarboxylic Acid, *Analyst* **126**, 567-570.
78. Ding, M., Tanaka, K., Hu, W., Haddad, P. R., and Miyanaga, A. (2001) Simultaneous Determination of Anions and Cations in Environmental Samples by Using a Weakly Acidic Cation-exchange Column with 2, 6-pyridinedicarboxylic Acid, Sulfosalicylic acid, and 18-crown-6 as Eluent, *J. Liq. Chromatogr. Rel. Techn.* **24**, 3105-3117.
79. Tanaka, K., Ohta, K., Haddad, P. R., Fritz, J. S., Miyanaga, A., Hu, W., Hasebe, K., Lee, K. P., and Sarzanini, C. (2001) High-performance Ion-exclusion/cation-exchange Chromatography of Anions and Cations in Acid Rain Waters on a Weakly Acidic Cation-exchange Resin, *J. Chromatogr. A* **920**, 239-245.
80. Kwon, S., Lee, K., Tanaka, K., and Ohta, K. (1999) Simultaneous Determination of Anions and Cations by Ion-exclusion Chromatography-Cation-Exchange Chromatography with Tartaric Acid/18-Crown-6 as Eluent, *J. Chromatogr. A* **850**, 79-84.
81. Colgan, S., Hammen, P., Knutson, K., and Bordner, J. (1996) Investigation of Cyclam-Containing Mobile Phases for the Liquid Chromatographic Analysis of Tenidap, *J. Chromatogr. Sci.* **34**, 111-114.
82. Biesaga, M., Pyrzynska, K., and Trojanowicz, M. (2000) Porphyrins in Analytical Chemistry. A Review, *Talanta* **51**, 209-224.
83. Zolotov, Yu. A. (ed.) (1997) Macrocyclic Compounds in Analytical Chemistry, *Chemical Analysis* **43,** Winefordner, J. D. (series ed.), John Wiley & Sons, New York.
84. Shi, Z. and Fu, C. (1997) Porphyrins as Ligands for Trace Metal Analysis by High-performance Liquid Chromatography, *Talanta* **44**, 593-604.

85. Uvarova, M. I., Brykina, G. D., and Shpigun, O. A. (2000) Porphyrins and Phthalocyanines in High-performance Liquid Chromatography, *J. Analyt. Chem.* **55**, 910-925.
86. Menon, S. K., Jogani, S. K., and Agrawal, Y. K. (2000) Macrocyclic Schiff Bases and their Analytical Applications, *Rev. Analyt. Chem.* **19**, 361-412.
87. Mifune, M., Minato, K., Kitamura, Y., Okazaki, K., Iwado, A., Akizawa, H., Haginaka, J., Motohashi, N., and Saito, Y. (2004) Electron Interaction of PAHs with Anion-exchange Silica Gels Modified with Anionic Metal-porphine and -Phthalocyanine Derivatives as HPLC Stationary Phase for Preparative Column in Organic Solvents, *Talanta* **63**, 1035-1038.
88. Kibbey, C. E. and Meyerhoff, M. E. (1993) Preparation and Characterization of Covalently Bound Tetraphenylporphyrin-silica Gel Stationary Phases for Reversed-phase and Anion-exchange Chromatography, *Anal. Chem.* **65**, 2189-2196.
89. Biesaga, M., Stolarczyk, E., Pyrzynska, K., and Trojanowicz, M. (2002) Retention of Anions on Silica-based Metalloporphyrin Stationary Phases, *Analyt. Sci.* **18**, 151-154.
90. Iverson, B., Thomas, R., Kral, V., and Sessler, J. (1994) Molecular Recognition of Anionic Species by Silica Gel Bound Sapphyrin, *J. Am. Chem. Soc.* **116**, 2663-2664.
91. Charvatova, J., Kasicka, V., Barth, T., Deyl, Z., Mikski, I., and Kral V. (2003) Separation of Structurally Related Peptides by Open-tubular Capillary Electrochromatography Using (Metallo) Porphyrins as the Adsorbed Stationary Phase, *J. Chromatogr. A* **1009**, 73-80.
92. Cserhati, T. and Forgacs, E. (2003) *Cyclodextrins in Chromatography*, RSC Chromatography Monographs, Loughborough, UK.
93. Mitchell, C. and Armstrong, D. (2004) Cyclodextrin-based Chiral Stationary Phases for Liquid Chromatography: A Twenty-year Review, *Meth. Molecular Bio.* **243**, 61-112.
94. Mosinger, J., Tomankova, V., Nemcova, I., and Zyka, J. (2001) Cyclodextrins in Analytical Chemistry, *Analyt. Lett.* **34**, 1979-2004.
95. Schneiderman, E. and Stalcup, A. (2000) Cyclodextrins: A Versatile Tool in Separation Science, *J. Chromatogr. B* **745**, 83-102.
96. Shpigun, A., Ananieva, I., Budanova, N., and Shapovalova, E. (2003) Use of Cyclodextrins for Separation of Enantiomers, *Russ. Chem. Rev.* **72**, 1035-1054.
97. Ward, T. (2000) Chiral Separations, *Anal. Chem.* **72,** 4521-4528.
98. Schurig, V. (2002) Chiral Separations Using Gas Chromatography, *Trends Anal. Chem.* **21,** 647-661.
99. Atwood, J. L., Davies, J. E. D., MacNicol, D. D., and Vogtle, F. (Eds) (1996) Comprehensive Supramolecular Chemistry, Vol. 3 "Cyclodextrins", Pergamon, Oxford.
100. Ruderisch, A., Pfeiffer, J., and Schurig, V. (2003) Mixed Chiral Stationary Phase Containing Modified Resorcinarene and Beta-Cyclodextrin Selectors Bonded to a Polysiloxane for Enantioselective Gas Chromatography, *J. Chromatogr.* **994**, 127-135.
101. Gubitz, G. and Schmid, M. G. (eds.) (2004) Chiral Separations: Methods and Protocols, *Methods in Molecular Biology* **243,** Humana Press, Totowa, New Jersey.
102. Dungelova, J., Lehotay, J., and Rojkovicova, T. (2004) HPLC Chiral Separations Utilizing Macrocyclic Antibiotics–A Review, *Chemia Analityczna* **49**, 1-17.
103. Dungelova, J., Lehotay, J., Rojkovicova, T., and Cizmarik, J. (2003) Chiral Separations of Some Drugs on the Basis of Macrocyclic Antibiotics in HPLC, SFC and CEC, *Ceska a Slovenska Farmacie* **52**, 119-125.
104. Ward, T. J. and Farris, A. B. (2001) Chiral separations using the macrocyclic antibiotics: a review. *J. Chromatogr. A* **906**, 73-89.
105. Hui, F. (2004) High Performance Liquid Chromatography and Capillary Electrophoresis Chiral Recognition Mechanisms Using Glycopeptide Macrocyclic Antibiotics as Selectors, *Fenxi Huaxue* **32**, 964-968.
106. Aboul-Enein, H. and Ali, I. (2002) Optimization Strategies for HPLC Enantioseparation of Racemic Drugs Using Polysaccharides and Macrocyclic Glycopeptide Antibiotic Chiral Stationary Phases, *Farmaco* **57**, 513-529.
107. Rojkovicova, T., Lehotay, J., and Cizmarik, J. (2001) Enantiomeric Separations of Drugs Using Macrocyclic Antibiotics, *Ceska a Slovenska Farmacie* **50**, 166-172.

23.
LIGAND DESIGN FOR BASE METAL RECOVERY

P. A. TASKER AND V. GASPEROV
School of Chemistry, The University of Edinburgh, Edinburgh, EH9 3JJ, U.K.

1. Introduction

Historically, base metals have been recovered from primary sources mainly by *pyrometallurgy* which "produces metals or intermediate products directly from the ore by use of high-temperature oxidative or reductive processes".[1] The sequence of unit operations: *concentration, separation, reduction* and *refining* [2-4] usually only makes use of organic ligands in flotation processes in which the finely ground mineral particles are separated from silicaceous materials on the basis of their different surface activities.[5] Separation by froth flotation is frequently employed to concentrate sulfide ores using *collectors* with soft, typically sulfur, donors which preferentially bind to the ore particles and increase their hydrophobicity, transferring them to an oil-based froth. Approximately two billion tonnes of material per year is currently treated by froth flotation.[6] The nature of the chemi- and physi-sorption shown by these collector and related activators, depressants and dispersants falls outside the scope of this review and has been very effectively discussed by others.[5, 7, 8] The nature of the downstream processing of the ore concentrates by roasting and smelting result in destruction of the organic ligands. In contrast, ligands used in extractive *hydrometallurgy* are largely recovered and recycled.[9]

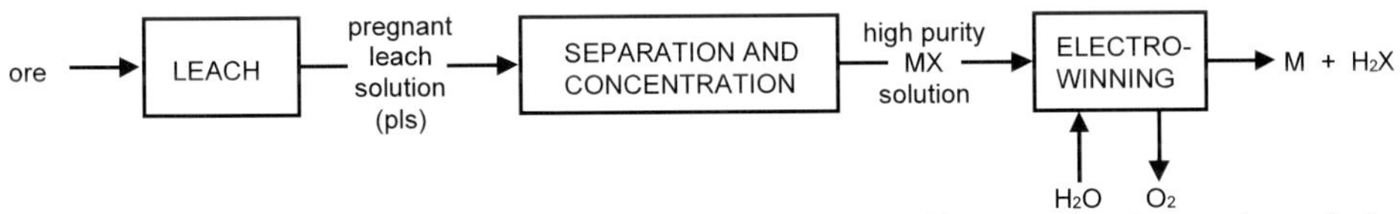

Figure 1. A simplified flowsheet for the hydrometallurgical recovery of base metals using metal complexing agents.

Karsten Gloe (ed.), Macrocyclic Chemistry: Current Trends and Future Perspectives, 365–382.
© 2005 *Springer. Printed in the Netherlands.*

There has been a significant growth since the mid-1980s in the use of hydrometallurgy to recover base metals. This has come about as a result of the proven robustness of technology based on solvent extraction in the nuclear industries, [10-12] and in the very successful heap/leaching/solvent extraction/electrowinning recovery of copper [13-16] which has stimulated the development of new leaching and separation processes for a wide range of base metals.[9] Once contents of the ore have been transferred to an aqueous medium it is possible to use organic reagents to effect the separation and concentration of the target metal (see the flowsheet in Figure 1). The design of ligands with the appropriate "strength" and selectivity to interface with the front-end (leaching) and the back-end (reduction to generate pure metal) of such a flowsheet presents challenging targets for the coordination chemist.

For base-metal production the separation technology most commonly employed is solvent extraction because this can be operated in continuous rather than batch processing on a very large scale. Solvent extraction using conventional mixer-settlers is inefficient when used to treat very dilute aqueous feeds because an impracticably large volume of the aqueous phase must be contacted with an organic extractant to achieve concentration along the flowsheet. For dilute feeds solid state supported reagents (ion exchange resins) are preferred although these generally have low metal capacities and thus very large quantities of resin are required for bulk recovery of base-metals. This review will focus on the use of complexing agents in solvent extraction for the primary recovery of base metals. In such processes, even though the reagents are usually recycled with high efficiency, their cost is a major factor in determining commercial viability. As a consequence, no macrocyclic reagents are currently used in commercial operations. However, the principles of ligand design which have been established from studies of macrocyclic chemistry [17-19] underpin the development of extractants which have the appropriate strength and selectivity to achieve "separation" and "concentration" in Figure 1.

2. The Design of Solvent Extractants for Base Metal Recovery

As organic solvent extractants have been applied for many years [20, 21] in qualitative and quantitative analysis of metals it is logical to use similar classes of molecules as a starting point for the development of industrial reagents, particularly those which are required to show high selectivity of phase transfer. However, other requirements of extractive hydrometallurgy differ considerably; the very high distribution coefficients needed for quantitative analysis are often inappropriate for processes in which the metal has to be stripped efficiently and the reagent recycled. In the laboratory, the chemistry of the feed solution will normally be under control of the analyst. The pH can be adjusted using buffers, masking agents can be added and generally reagent costs and toxicities are not major concerns. In contrast, in hydrometallurgy the extractant must meet the requirements of the front-end (leaching) and the back-end (electro-winning) of a flowsheet (Figure 1), and with minimal adjustment of the composition of aqueous process streams must give high mass transport efficiency and show high stability to chemical degradation by oxidation or hydrolysis under what are frequently fairly aggressive conditions.[9]

High boiling inert diluents such as kerosenes are preferred for large scale industrial operations. Consequently, the extractants are usually functionalized with large, often branched, alkyl groups to ensure high solubilities in such media and negligible solubility of the ligand or metal complexes in the aqueous phase. High solubility in such low polarity diluents is only possible if metal-containing complexes are charge-neutral. The four methods for achieving this can be classified [1, 9] as follows: Extraction by *cation exchange* involves the formation of an electrically neutral complex in which the desired metal cation displaces another cation (most commonly a proton) from a complexing agent.

$$M^{n+} + nLH_{(org)} \rightleftharpoons [ML_n]_{(org)} + nH^+$$

When the extraction involves the release of a proton, from an organic acid, LH, the equilibrium is dependent on the pH of the aqueous phase and the pH associated with 50% loading of the reagent (the $pH_{½}$) defines the "strength" of a reagent at a stated concentration for a defined composition of an aqueous feed.

Extraction by *anion exchange* is characterised by the transfer of an anionic metal complex from the aqueous phase, displacing an anion from the organic medium. Such reactions are commonly used for transport of chlorometallate complexes, e.g.

$$FeCl_4^- + [R_4N]Cl_{(org)} \rightleftharpoons [R_4N][FeCl_4]_{(org)} + Cl^-$$

Often the cationic component of the ion pair is generated by the protonation of an organic base, e.g.

$$FeCl_4^- + HCl + R_3N_{(org)} \rightleftharpoons [R_3NH][FeCl_4]_{(org)} + Cl^-,$$

and consequently extraction and stripping are dependent on both the pH and the chloride concentration of the aqueous phase.

Extraction by *solvation* involves the displacement of some or all of the water molecules in the coordination sphere of a metal complex by neutral organic donors, conferring high solubility in water-immiscible solvents. *Solvating* extractants are typically ethers, ketones, or neutral phosphorus(V) molecules containing P=O units, e.g. the extraction of uranium(VI) from nitrate solutions by tri-*n*-butyl phosphate (TBP).

$$UO_2(NO_3)_2 + 2TBP_{(org)} \rightleftharpoons [UO_2(NO_3)_2(TBP)_2]_{(org)}$$

Extraction by *physical distribution* involves the transfer of a discrete molecular entity from the aqueous phase to an inert solvent and has little relevance to extractive metallurgy.

Whilst such a classification [1, 9] is useful because it indicates the different types of chemical changes which can occur at the metal centres during phase transfer, it oversimplifies the situation in many cases and fails to indicate the importance of the outer sphere coordination chemistry and of solvation effects in general on the free energies of extraction. Alternative classifications of extraction processes which take better account of these have been presented recently by the Moyer Group.[22] The importance of such supramolecular effects in the design of reagents will be stressed in the examples below.

The *anion exchange* and *solvation* mechanisms make it possible to transport metal salts across flowsheets. Stripping a loaded chlorometallate solution of an anion exchanger back into an aqueous phase which contains a low chloride concentration liberates a metal chloride, e.g.

$$[R_4N][FeCl_4]_{(org)} \rightleftharpoons [R_4N]Cl_{(org)} + FeCl_3$$

and effectively transports $FeCl_3$.

Table 1. Examples of commercial solvent extractants for metals.

Reaction Classification	Type of Reagent	Examples
cation exchange	carboxylic acids	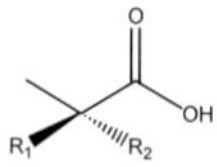 versatic acid R_1 = n-hexyl R_2 = n-octyl
	aryl sulfonic	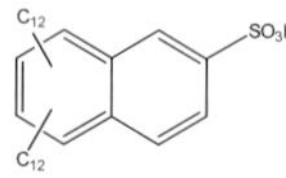 HDDNS didodecylnapthalenesulfonic acid
	alkyl phosphoric	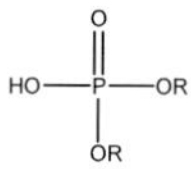 D2EHPA R = $CH_2CH(C_2H_5)C_4H_9$ 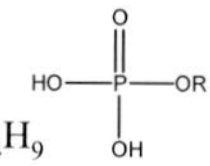 MEHPA R = $CH_2CH(C_2H_5)C_4H_9$
	phosphinic	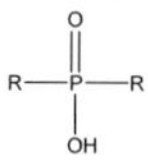 Cyanex 272 R = $CH_2CH(CH_3)CH_2C_4H_9$
	phosphonic	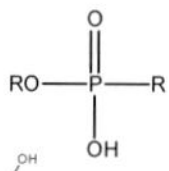 Ionquest 801 R = $CH_2CH(C_2H_5)C_4H_9$
	hydroxyoximes	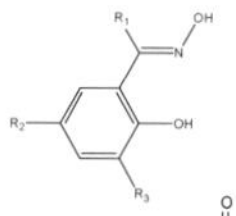 R_1 = H, C_6H_5 R_2 = C_9H_{19}, $C_{12}H_{25}$ R_3 = H, Cl
	β-diketones	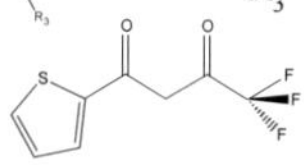 HTTA thenoyltrifluoroacetone
anion exchange	primary amines	Primene 81R, *t*-C_{12}-C_{14}
	secondary amines	Adogen 283, Adogen 283D, di-*i*-octyl
	tertiary amines	Alamine 336, Adogen 364, tri-C_8-C_{10}
	quarternary	Adogen 364, Aliquat 336, methyl, tri-C_8-C_{10}
	guanidines	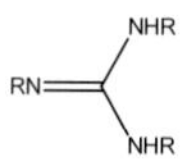 LIX79
solvating extractants	neutral phosphorus(V) molecules	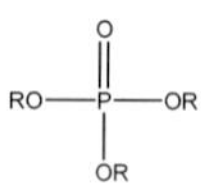 TBP R = C_4H_9, TOPO C_8H_{17}
	esters	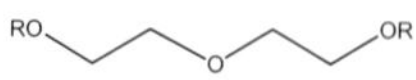 Butex R = C_4H_9
	ketones	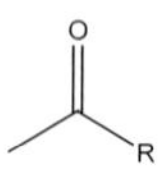 MIBK R = C_4H_9

Similarly uranyl nitrate is moved across a circuit by stripping a solvating reagent,

$$[UO_2(NO_3)_2(TBP)_2]_{(org)} \rightleftharpoons UO_2(NO_3)_2 + 2TBP_{(org)}$$

into an aqueous solution containing nitrate at relatively low concentrations.

The advent of anion recognition has opened up the possibility of developing ditopic ligands for extraction of *metal salts,*

$$M^{n+} + nX^- + L_{(org)} \rightleftharpoons [MLX_n]_{(org)}$$

which have separated binding sites for a metal cation and its attendant anion(s).[23] In principle the cation and anion can then be separately stripped for downstream processing (section 5).

3. Macrocycles in Extractive Metallurgical Research and Development

The high costs associated with multistep syntheses of macrocyclic extractants have limited their application in commercial recovery of base metals from primary sources. They have been more seriously considered for the recovery and recycling of precious metals (see *Section 3.2*) and for use as "polishing agents" to remove harmful impurities present in process streams. Solid-state supported crown ethers and related ligands are offered commercially for a wide range of such applications. The ligand design and synthesis for such systems which show very high selectivities has been underpinned by fundamental work on the thermodynamics and kinetics of metal complex formation by groups at the Universities of Utah and Sydney.[24, 25] The commercial sensitivity associated with many of the metal-recovery systems offered by companies means that it is often not possible to identify the nature of the ligating groups present.

Macrocyclic reagents have proved particularly useful in understanding the ligand design requirements for efficient solvent extraction processes. The increased stability of macrocyclic complexes over their open chain counterparts and the ability to tune their selectivity (by controlling the type, number and disposition of the donor atoms, using variations of ring size, the flexibility/rigidity of the backbone and the substituents on the inner great ring) has underpinned many investigations of metal solvent extraction. Polyethers have been most extensively studied, [21, 26-30] focusing particularly on the extraction of alkali and alkaline earth metals, [21] and lanthanides and actinides.[31-37] The following chapter in this volume by B. A. Moyer et al. deals specifically with the recovery of these metals. Aza macrocycles have been used most in transport of a wide variety of transition and post-transition metals, [21, 26, 38-45] whilst thioethers have attracted interest for the recovery of precious metals or other "soft" metal ions such as Cu^+, and Hg^{2+}.[46, 47] The ability to append a variety of functional groups to calixarenes has facilitated the development of extractants showing selectivity for particular alkali and alkaline earth as well as transition and heavy metal ions.[40, 48]

Whilst it is unlikely that most of the macrocyclic extractants studied will be used in bulk recovery processes on cost grounds, the principles of supramolecular and coordination chemistry that have emerged are of great consequence in the design of "strong" and selective new commercial reagents.

3.1 SYNERGISM AND SELF-ASSEMBLY

In the context of solvent extraction, synergism is the enhancement that is observed when two reagents, $\mathbf{L}^1$ and $\mathbf{L}^2$, in combination extract more than they would separately under otherwise similar conditions.[20]

$$D_{\mathrm{M}}(\mathbf{L}^1 + \mathbf{L}^2) > D_{\mathrm{M}}(\mathbf{L}^1) + D_{\mathrm{M}}(\mathbf{L}^2)$$

The increase in extraction efficiency, evaluated by the distribution coefficients D_{M}, is often considered to be the result of an increase in the overall lipophilicity of a system. Synergism has been studied systematically using four types of extractant combinations: 1) chelating extractant/neutral extractant; 2) acidic extractant/neutral extractant; 3) two neutral extractants; 4) two chelating extractants, with the observed synergistic effect generally decreasing from class 1 to 4.[49]

The majority of the reported synergistic combinations involve crown ethers.[49] Synergism has been observed mainly in the extraction of alkali and alkaline earth metals where crown ethers have been used in combination with hydrophobic organic acids such as alkyl phosphoric, alkyl sulfonic or alkyl carboxylic acid extractants.[30] It has also been reported for combinations of crown ethers and chelating agents such as thenoyltrifluoroacetone which have been used successfully for the extraction of alkali and alkaline metal ions as well as Mn^{2+}, Co^{2+}, Zn^{2+} and Cd^{2+}.[49] This combination has also been used in the extraction of lanthanides.[49] Neutral ligands such as tri-*n*-octylphosphine oxide and *p*-*tert*-butylcalix[4]arene have also been used with crown ethers although the observed synergism is slight.[49]

Few reports give an explanation for the origins of the synergism at a molecular level although the thermodynamic basis of the beneficial effects of admixtures has been discussed.[20, 49] It has been suggested that a combination of extractants ensures the saturation of the coordination sphere, thus increasing the lipophilicity of the complex.[20, 49] Such an explanation accounts for the efficient synergistic extraction by either O- or S-donor macrocycles in the presence of carboxylic, phosphoric or sulfonic acids and is supported by X-ray structure determinations of some mixed-ligand complexes. In the Mn^{2+} complex of di-*tert*-butylnapthelenesulfonate and cyclohexano-15-crown-5, $[Mn(C_{18}H_{23}O_3S)(C_{14}H_{26}O_5)]\ C_{18}H_{23}O_3S$, the cation is bound to the five oxygen atoms from the crown ether, one from a sulfonate anion and one from water molecule which form a pentagonal bipyramidal arrangement.[50, 51] A second sulfonate ion is hydrogen bonded to the water molecule.

In some systems it has been possible in synergistic mixtures of extractants to identify assembled ligand-packages in the absence of metal cations which contain complete donor sets needed for metal complexation.[52, 53] NMR titrations provide a useful method of establishing the stoichiometry of metal-free ligand assemblies.[52] In some cases crystal structure determinations of adducts of carboxylic acids and azamacrocycles show these to be salts in which the transfer of a proton to an amino group results in close association of the carboxylate with the macrocycle donor set, based on a combination of electrostatic and hydrogen bonding.

The assembly of 4-*tert*-butylbenzoic acid and cyclam (**A**, Figure 2) [54] has a planar arrangement of the macrocycle's N_4 donor set with a *trans* disposition of the carboxylates very similar to that observed in the structure of the nickel complex, [Ni(cyclam)(benzoate)$_2$], (**B**, Figure 2).[55] Formation of the nickel complex from the assembly and extraction of the nickel ion can then be represented by the "pH-swing" equilibrium

$$Ni^{2+} + [(LH_2)(RCO_2)_2]_{(org)} \rightleftharpoons [NiL(RCO_2)_2]_{(org)} + 2H^+$$

in which two protons are displaced from the centre of the assembly and the hydrophobic ligand packages $[(LH_2)(RCO_2)_2]_{(org)}$ effectively function as cation exchange reagents (section 2).

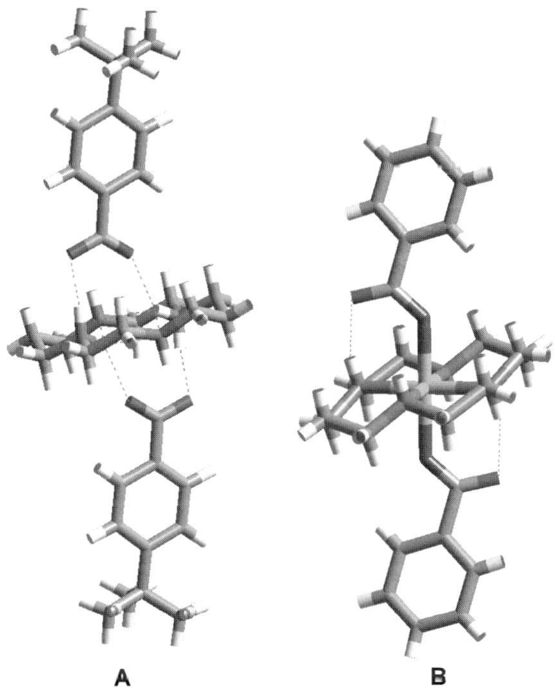

Figure 2. Structures of [(cyclamH$_2$)(4-*t*-butylbenzoate)$_2$], (**A**) and [Ni(cyclam)(benzoate)$_2$], (**B**).[54, 55] A coloured version is given in the Appendix.

Very similar assemblies are proposed to explain the synergism of extraction of Cu^{2+}, Co^{2+}, Ni^{2+} and Zn^{2+} by the macrocycles shown in Figure 3 in combination with either carboxylic or phosphinic acids.[53, 56]

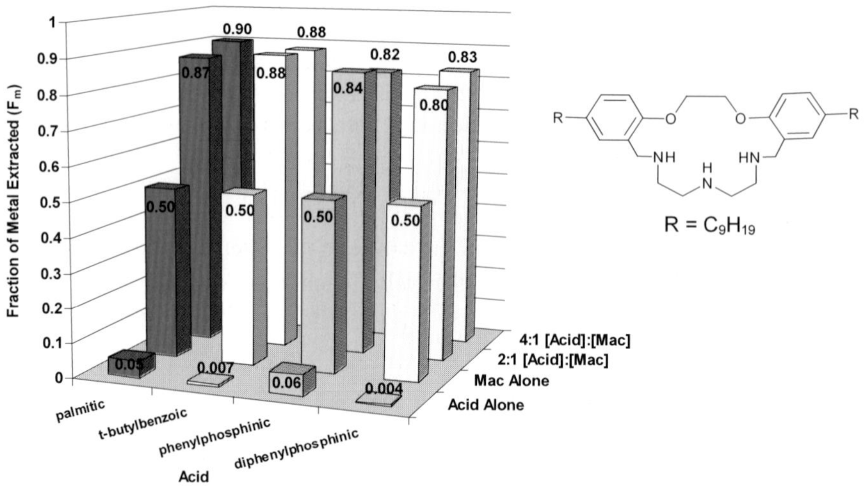

Figure 3. Synergistic extraction of Cu^{2+} from a 10^{-3} M aqueous nitrate solution, pH 5.0, by mixtures of a lipophilic N_3O_2-macrocycle (10^{-3} M) and a carboxylic or phosphinic acid in chloroform.[53]

The high development costs associated with the design, synthesis, manufacture, toxicity-testing and registration of new organic compounds provide a strong driving force for the development of synergistic combinations of existing reagents rather than investment in new types of complexing agents.[9, 57] The recent expansion of supramolecular chemistry and the development of techniques for determining the stability and structures of organic assemblies in solution should underpin a more rational approach to the development of synergistic mixtures of extractants based on molecular design.

3.2 PRECIOUS METAL RECOVERY

α-Hydroxy-ketoximes, phenolic-oximes, dialkylsulfides, esters of pyridine mono and di-carboxylic acids, alkyl derivatives of 8-hydroxyquinoline, trialkylamines, alkyl derivatives of aniline, aliphatic ethers and ketones have all been evaluated as solvent extractants for precious metals.[9] The high metal values make the application of macrocyclic ligands more potentially viable but there is no evidence that they are used yet on any scale in PMG recovery.

Calixarenes with pendant sulfur donors show a high affinity for Au^{3+}.[58, 59] Sulfur-containing macrocycles, including calixarenes, with pendant sulfur and nitrogen donors also show very high selectivity toward Ag^{+} over other cations.[9] An investigation of mono- and tri-nucleating mixed donor (nitrogen and sulfur) macrocycles has revealed their ability to transport Ag^{+} exclusively from solutions containing Co^{2+}, Ni^{2+}, Cu^{2+}, Zn^{2+}, Pb^{2+} and Cd^{2+}. Macrocyclic tetrathioethers have also been reported to be selective for Pt^{2+}. While extraction of $Pt^{2+}$is reportedly very slow, enhanced extraction is observed with the addition of thiourea.[60]

4. "*Pseudo*-macrocyclic" Systems in Extractive Metallurgy

A feature of the use of very low polarity diluents such as the kerosenes used in solvent extraction of base metals is that it favours the formations of secondary bonds based on hydrogen-bonding components or strong dipoles present in ligands or in the outer coordination spheres of metal ions. Such secondary bonding is probably responsible for many of the "modifier" and "synergism" phenomena observed by extractive metallurgists (see *section 3.1*).[61] In this section, we deal with intra-complex hydrogen-bonding which leads to the formation of *pseudo*–cyclic ligand assemblies which have a significant effect on the strength and selectivity of extraction of metal cations. One of the best examples of this is the 14-membered *pseudo*-macrocycle formed by the phenolic oxime copper extractants.

Both the aldoximes, shown in Figure 4, and related ketoximes form *pseudo*-macrocyclic complexes.[62] Those with divalent cations, VO^{2+}, $CoNO^{2+}$, Ni^{2+}, Cu^{2+}, Zn^{2+} and Pd^{2+}, have been shown to have approximately planar $N_2O_2^{2-}$ donor sets in solid state structures. The metal cation lies at the centre of these donor sets in the 4-coordinate Ni^{2+}, Cu^{2+} and Pd^{2+} complexes but is displaced to one side in square-pyramidal $VO^{2+}$and $CoNO^{2+}$ complexes.[62]

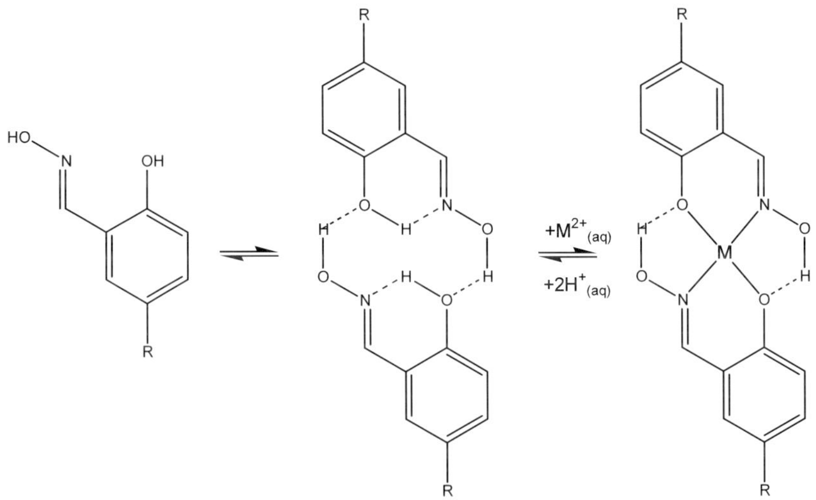

Figure 4. The pre-assembly of bis-salicylaldoxime 14-membered *pseudo*-macrocyclic extractant in non-polar solvents.

The *pseudo* macrocyclic structure is also present in the solid state forms of some free ligands.[62] Overall, these are less planar than their 4-coordinate metal complexes, having a step conformation to reduce the repulsion between the central phenolic protons. Nevertheless, they are apparently well organized for complex formation and donor cavity sizes [63] give a good fit for divalent copper.[62]

The cyclic structures are retained in solution [16, 61, 64] and such dimers must be considered in modelling the extraction equilibria for the recovery of copper using the commercial extractants shown in Table 1.[61, 65]

$$(LH)_{2(org)} + M^{2+} \rightleftharpoons ML_{2(org)} + 2H^{+}$$

In commercial operations which now account for nearly one third of the world's copper extraction,[15] reagents can be selected or combined to meet the "strength" requirements of particular circuits.[16] Ketoximes are slightly weaker extractants than aldoximes and are often used when the pregnant leach solution is warm or has a pH of 1.8 or above.[15] The aldoximes are stronger and are preferred when processing feeds with higher Cu content or lower temperature and pH. It is also common practice to add "modifiers" to the aldoxime extractants to generate weaker formulations which match the requirements of a particular circuit. Most of such modifiers, e.g. 4-alkylphenols, "fatty" alcohols and esters (Y below), have the potential to form hydrogen-bonds to monomeric, dimeric or other forms of the extractant in kerosene solutions, e.g.

$$(LH)_{2(org)} + 2Y_{(org)} \rightleftharpoons 2LHY_{(org)} \quad \text{or}$$

$$(LH)_{(org)} + Y_{(org)} \rightleftharpoons LHY_{(org)} \quad \text{etc,}$$

and the modifier then competes with metal cations for the donor functionalities, depressing metal-loading. The reagent suppliers' ability to formulate reagents in this way is crucial to commercial success and represents a good example of the practical application of supramolecular chemistry.

The favourable intra-molecular hydrogen-bonding between the oxime proton and the phenol oxygen atom will reduce or overcome ligand:ligand repulsion enthalpy terms normally involved in the formation of bis-chelate complexes. The integrity of this head-to-tail macrocyclic assembly is preserved in *cis*-octahedral complexes which are formed in the presence of α,ω-diamines.[66, 67]

$$2LH_{(org)} + diamine_{(org)} + Ni^{2+} \rightleftharpoons [NiL_2(diamine)]_{(org)} + 2H^{+}$$

The addition of lipophilic 1,2-diamines to the extractant formulation enhances nickel extraction.[66, 67] The solid state structures of a series of diamine adducts [NiL_2(diamine)] all contain the *trans*-oxime isomer in Figure 5 in which the *pseudo*-macrocycle is folded about the N_{imino}-N_{imino} diagonal. This allows both of the oxime OH to phenolate hydrogen-bonds to remain intact, and permits the N and O donors to form longer bonds to the larger high-spin Ni^{2+} ion than would be possible with the flat $N_2O_2^{2-}$ donor set.

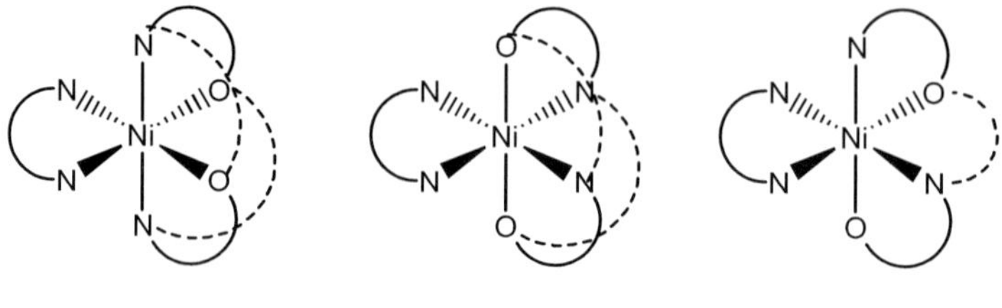

trans N oxime *trans* O phenol All *cis*

Figure 5. Isomers of [NiL_2diamine] complexes labelled to show differences in the orientation of the donors in the two salicylaldimato chelates (L^-).

Inter-ligand hydrogen bonding is also very significant in the mode of action of most of the commercial phosphorus(V) acid extractants containing at least one P-OH or P-SH group (examples are listed in Table 1). Such ligands are strongly associated in the non-polar solvents, the dimeric form (**A**) predominating. The 8-membered rings in these dimers are preserved when complexes (**B**) are formed in the presence of excess extractant.[1, 9] The large bite angle defined by these eight-membered chelate rings favours tetrahedral coordination geometry.[68]

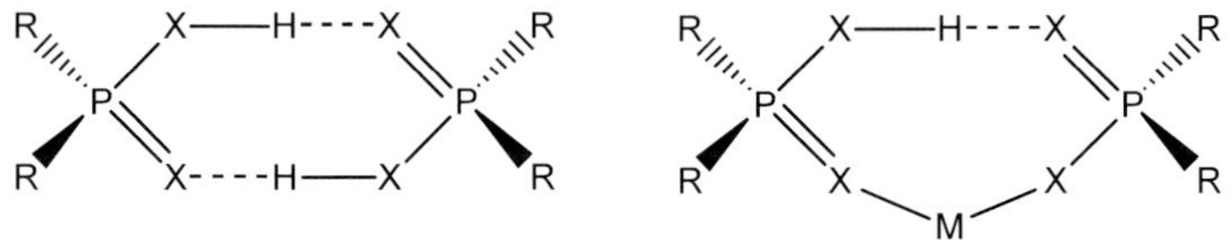

A **B**

Figure 6. The dimers formed by phosphorus(V) acids in non-polar solvents, (**A**) and the structurally related *pseudo*-chelates (**B**) formed on extraction of metal cations.

As a consequence, the selectivity of extraction of first transition series dications does not follow the Irving Williams order when these reagents are used in base metal recovery. The bis(2-ethylhexyl)ester of phosphoric acid (D2EHPA) shows [1] a preference: $Zn^{2+} > Cr^{2+} > Mn^{2+} > Fe^{2+} > Co^{2+} > Ni^{2+} \sim V^{2+}$ which is exploited in the recovery of zinc from primary sources.[69] M^{2+} ions which form tetrahedral complexes and M^{3+} ions which show a preference for octahedral donor sets give neutral complexes with 4 : 1 and 6 : 1 D2EHPA : metal stoichiometries respectively,

$$4LH_{(org)} + M^{2+} \rightleftharpoons [ML_2(LH)_2]_{(org)} + 2H^+$$

$$6LH_{(org)} + M^{3+} \rightleftharpoons [ML_3(LH)_3]_{(org)} + 3H^+$$

These stoichiometries allow the 8-membered *pseudo*-chelate rings (**B** in Figure 6) to be preserved and provide the complex unit with a very hydrophobic exterior.

D2EHPA has been used to effect the separation of Co^{2+} from Ni^{2+}, reflecting the greater ease by which the former adopts tetrahedral geometry.[1] D2EHPA is a poor extractant for Ni^{2+} as this shows a preference for a *pseudo*-octahedral structure in which two axial sites are occupied either by fully protonated LH molecules, or by fully protonated $(LH)_2$ dimers or by water molecules, depending on extractant concentration and the equatorial plane contains two 8-membered *pseudo*-chelates (**B** in Figure 6). Selectivity of extraction of Co^{2+} from Ni^{2+}-containing feeds can be enhanced by using phosphonic acid and phosphinic acid reagents such as Ionquest 801 and Cyanex 272 (Table 1).[70-72] This has been ascribed to the destabilisaton of the octahedral Ni^{2+} complexes which results from the greater steric bulk of the ligands with alkyl groups attached directly to the phosphorus atom rather than through an oxygen atom in alkoxyl unit.[68]

Ligand assembly in the metal coordination sphere is also implied by the unusual stoichiometries of extraction shown when hydrophobic carboxylic acids are used in hydrocarbon diluents.[1, 9]

Distribution data for the extraction of Ni^{2+} by Versatic acid (Table 1) are consistent with a generalized reaction, [73, 74]

$$Ni^{2+} + pH_2O + (1+m/2)[HA]_{2(org)} \rightleftharpoons 1/x[NiA_2.mHA.pH_2O]_{x(org)} + 2H^+$$

At low Ni-loading a monomeric species $[NiA_2.4HA]$ predominates. Dinuclear complexes and hydrated species are found at higher Ni-loadings.[73, 74] It is possible that 8-membered or other *pseudo*-chelate rings could be formed in the hydrocarbon phase in a similar manner to that show in Figure 6, enhancing the lipophilicity of the complexed package. Alternatively, or additionally, the hydrogen bonding arrangements between undeprotonated carboxylic acids and carboxylate ligands could lead to more effective solvation of the complex and enhanced stability in a hydrocarbon phase.

The secondary bonding between ligands and the resulting assembly processes which characterize the modes of action of most of the commercial extractants for base metals appear to be essential to their efficacy in commercial operations. They lead to increased "strength" and selectivity of metal-extraction, essentially by enhancements of thermodynamic stability analogous to those involved in the chelate and macrocyclic effects. Consequently the development of new reagents, or formulations of reagents, to achieve *concentration* and *separation* in flowsheets such as Figure 1 will be substantially improved if the principles of supramolecular chemistry are employed. These principles apply particularly to the design of systems to transport both a metal cation and its attendant anion(s), see below.

5. Reagents to Transport Metal Salts

The systems described above all result in the transport of metal cations across a metal-recovery circuit. In many cases this leads to very good materials balances in metal-recovery, especially when the circuit uses acid-leaching of the ore followed by solvent extraction using an organic acid (LH). The extraction then releases protons back into the aqueous phase, regenerating the acid needed for leaching. This underpins the very successful copper recovery operations outlined in Figure 7 in which copper oxide in the crude ore is essentially split into its component elements with the consumption of only electrical power.

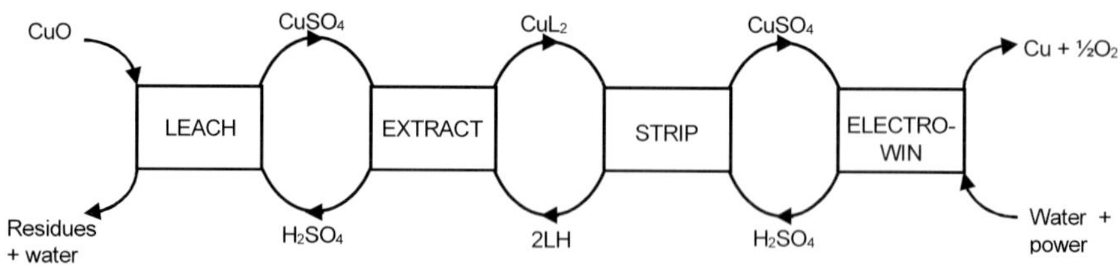

$CuO + H_2SO_4$	$\rightleftharpoons$	$CuSO_4 + H_2O$	Leach
$CuSO_4 + 2LH_{(org)}$	$\rightleftharpoons$	$CuL_{2\,(org)} + H_2SO_4$	Extraction
$CuL_{2\,(org)} + H_2SO_4$	$\rightleftharpoons$	$CuSO_4 + 2LH_{(org)}$	Strip
$CuSO_4 + H_2O$	$\rightleftharpoons$	$Cu + ½ O_2 + H_2SO_4$	Electrowin
CuO	$\rightleftharpoons$	*Cu + ½ O_2*	*Overall*

Figure 7. A simplified flowsheet and materials balance for the recovery of copper from oxidic and transition ores by heap leaching, solvent extraction and electrowinning.[9]

When the leaching step does not consume acid it is not possible to obtain such excellent materials balances using the commercial cation exchange reagents LH. This is the case when processing sulfidic ores, the most common source of base metals, and until recently these have largely been recovered by *pyro*metallurgy. There has been significant investment in the development of technology to leach sulfidic ores.[9] In the CUPREX process outlined in Figure 8, oxidative leaching with ferric chloride generates elemental sulfur, avoiding the liberation of SO_2, and generates a pregnant leach solution containing high concentrations of metal chlorides and chlorometallate anions.[75] *Solvating extractants* such as esters of pyridine carboxylic acids can then be used to transport $CuCl_2$ across the circuit.[76-79]

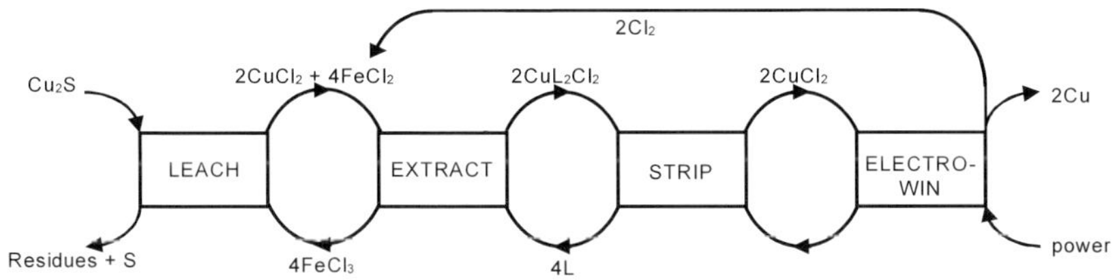

$4FeCl_3 + Cu_2S$	$\rightleftharpoons$	$4FeCl_2 + 2CuCl_2 + S\downarrow$	Leach
$2CuCl_2 + 4L$	$\rightleftharpoons$	$2CuL_2Cl_{2\,(org)}$	Extraction at high Cl^- activity
$2CuL_2Cl_{2\,(org)}$	$\rightleftharpoons$	$2CuCl_2 + 4L_{(org)}$	Stripping at low Cl^- activity
$2CuCl_2$	$\rightleftharpoons$	$2Cu + 2Cl_2$	Electrowinning
$4FeCl_2 + 2Cl_2$	$\rightleftharpoons$	$4FeCl_3$	Leach regeneration
Cu_2S	$\rightleftharpoons$	*$2Cu + S$*	*Overall*

Figure 8. A simplified flowsheet and materials balance for the recovery of Cu from sulfidic ores by oxidative chloride leaching, using "solvating" extractants involving formation of dichloroCu(II) complexes.

In these processes chlorometal complexes are formed in the organic phase and loading and stripping is largely controlled by variation of the chloride concentration in the aqueous phase. Processing sulfate-based feed solutions obtained by microbial or pressure leaching with oxygen by a similar mechanism is problematic because sulfate does not coordinate strongly to base metal cations. The hydrophilicity of sulfate also contributes to its poor extractability into organic media.[80] To overcome these problems ditopic ligands which contain a 'tailor-made' binding site for sulfate and a separate site for the metal cation (Figure 9) have been developed.[81]

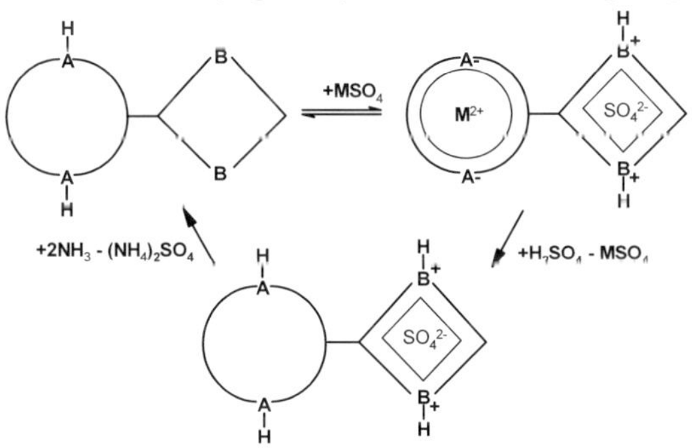

Figure 9. A schematic representation of the loading of a metal(II) sulfate into the zwitterionic form of a ditopic reagent and sequential stripping of the cation and sulfate by pH-adjustment.[81]

Relatively simple reagents containing a "salen"-like metal binding unit with protonatable pendant tertiary amine groups to accommodate the sulfate dianion have been developed to establish proof-of-concept,[82] and the incorporation of the metal cation in the salen cavity has been shown to template the sulfate-binding site and enhance extraction (Figure 10).[83]

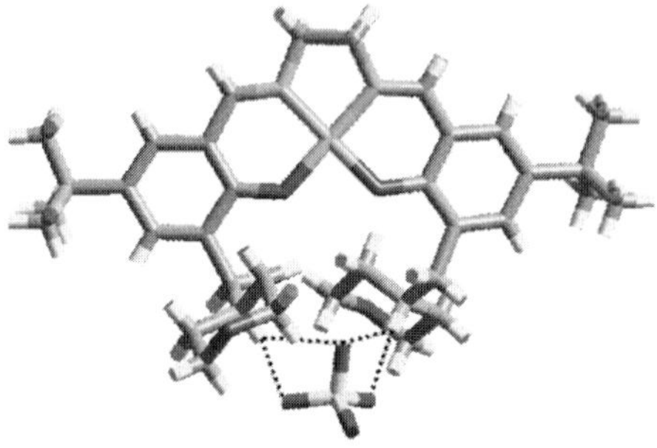

Figure 10. A $NiSO_4$ complex of a bis-morpholinomethyl substituted "salen" ligand showing the binding of sulfate by a combination of electrostatic forces and hydrogen bonds to the protonated morpholine groups. A coloured version is given in the Appendix.

6. Conclusions

Hydrometallurgical methods for the recovery of base metals are becoming increasingly preferred over the more traditional pyrometallurgical processes and the mining industry has invested considerably in the development of new leaching technology to transfer metal values from mineral deposits to aqueous feed streams.[9] The downstream *separation* and *concentration* of metal values require new technologies and present challenging targets for coordination chemists and those involved in ligand design. One of the major challenges is to identify relatively inexpensive reagents which are highly selective in transporting metal ions or metal salts, but are very robust and can be efficiently stripped and recycled. Whilst macrocyclic ligands are not yet used in any large scale operations, the development of macrocyclic chemistry over the last 30 years and, more recently, of supramolecular chemistry provides important guidelines on how to achieve high selectivity through organization of donor sets and recognition of anions. This review, while not comprehensive, has attempted to demonstrate that a feature of extraction into the hydrocarbon solvents used in extractive metallurgy is that secondary bonding between ligands, especially hydrogen bonding, can have a very significant effect on assembly of ligating packages both before and after binding to the metal ion/salt. As a consequence it should now be possible to approach the design of "strong" and selective reagents using the principles of supramolecular chemistry.

7. Acknowledgements

We would like to thank the EPSRC UK for funding for VG under the "Greener Technologies" initiative and Mr James Davidson for assistance in the preparation of the manuscript.

8. References

1. Nicol, M. J., Fleming, C. A., and Preston, J. S. (1987) Application to Extractive Metallurgy., in *Comprehensive Coordination Chemistry*, Wilkinson, G., Gillard, R. D., and McCleverty, J. A., Editors. Pergamon Oxford. p. 779-842.
2. Habashi, F. (1997) *Handbook of Extractive Metallurgy*, ed. Habashi, F. Vol. 3. Wiley-VCH Verlag GmbH, Weinheim
3. Burkin, A. R. (2001) *Chemical Hydrometallurgy - Theory and Principles,* 1st ed. Imperial College, London
4. Swaddle, T. W. (1996) *Inorganic Chemistry: An Industrial and Environmental Perspective*. Academic Press, San Diego
5. Klimpel, R. R. (1999) A review of sulfide mineral collector practice, *Advances in Flotation Technology, [Proceedings of the Symposium \"Advances in Flotation Technology\" held at the SME Annual Meeting], Denver, Mar. 1-3, 1999*, 115-127.
6. Fuerstenau, D. W. (1995) Where we are in flotation chemistry after 70 years of research, *Proceedings of the International Mineral Processing Congress, 19th, San Francisco, 1995*, **3**, 3-17.
7. Nagaraj, D. R. (1997) Development of new flotation chemicals, *Transactions of the Indian Institute of Metals*, **50**(5), 355-363.
8. Grano, S. R., Prestidge, C. A., and Ralston, J. (1997) Solution interaction of ethyl xanthate and sulfite and its effect on galena flotation and xanthate adsorption, *Int. J. Miner. Process.*, **52**(2-3), 161-186.
9. Tasker, P. A., Plieger, P. G., and West, L. C. (2004) Metal complexes for hydrometallurgy and extraction, *Comprehensive Coordination Chemistry II*, **9**, 759-808.
10. Singh, H. and Gupta, C. K. (2000) Solvent extraction in production and processing of uranium and thorium, *Miner. Process. Extr. Metall. Rev.*, **21**(1-5), 307-349.
11. Burns, C. J., Neu, M. P., Boukhalfa, H., Gutowski, K. E., Bridges, N. J., and Rogers, R. D. (2004) The actinides, *Comprehensive Coordination Chemistry II*, **3**, 189-345.
12. Cotton, S. (2004) Scandium, yttrium, and the lanthanides, *Comprehensive Coordination Chemistry II*, **3**, 93-188.
13. Davenport, W. G. (1999) Copper extraction from the 60's into the 21st century. in *Copper 99-Cobre 99 International conference*. Phoenix.
14. Avecia (2000-2001) Continued evolution of copper SX reagents. Avecia, Blackley Manchester. p. 2-5.
15. Kordosky, G. A. (2002) Copper recovery using leach/solvent extraction/electrowinning technology: forty years of innovation, 2.2 million tonnes of copper annually. in *International Solvent Extraction Conference, Cape Town, South Africa, Mar. 17-21, 2002.*
16. Szymanowski, J. (1993) *Hydroxyoximes and Copper Hydrometallurgy*. CRC Press, Boca Raton
17. Lindoy, L. F. (1992) *The Chemistry of Macrocyclic Ligand Complexes*. Cambridge University Press, Cambridge, UK
18. Gokel, G. W. (1991) *Crown Ethers and Cryptands*. The Royal Society of Chemistry, Cambridge, UK
19. Melson, G. A. (1979) *Coordination Chemistry of Macrocyclic Compounds*. Plenum Press, New York, U.S.A
20. Rydberg, J., Musikas, C., Choppin, G. R., (eds.) (1992) *Principles and Practices of Solvent Extraction,* Marcel Dekker, New York.
21. Beklemishev, M. K., Dmitrienko, S. G., and Isakova, N. V. (1997) Solvent extraction of metals with macrocyclic reagents and its analytical applications, *Chemical Analysis (New York)*, **143** (Macrocyclic Compounds in Analytical Chemistry), 63-208.
22. Moyer, B. A., Bonnesen, P. V., Chambliss, C. K., Haverlock, T. J., Marchand, A. P., Chong, H.-S., McKim, A. S., Krishnudu, K., Ravikumar, K. S., Kumar, V. S., and Takhi, M. (2001) Use of cage-functionalized macrocycles and fluorinated alcohols in the liquid-liquid extraction of NaOH and other sodium salts, *ACS Symposium Series*, **778** (Nuclear Site Remediation), 114-132.
23. Coxall, R. A., Lindoy, L. F., Miller, H. A., Parkin, A., Parsons, S., Tasker, P. A., and White, D. J. (2003) Solvent extraction of metal sulfates by zwitterionic forms of ditopic ligands, *Dalton Trans.* (1), 55-64.
24. Bradshaw, J. S., Izatt, R. M., Savage, P. B., Bruening, R. L., and Krakowiak, K. E. (2000) The design of ion selective macrocycles and the solid-phase extraction of ions using molecular recognition technology. A synopsis, *Supramol. Chem.* **12**(1), 23-26.
25. Dudler, V., Lindoy, L. F., Sallin, D., and Schlaepfer, C. W. (1987) An oxygen-nitrogen donor macrocycle immobilized on silica gel. A reagent showing high selectivity for copper(II) in the presence of cobalt(II), nickel(II), zinc(II) or cadmium(II), *Aust. J. Chem.* **40**(9), 1557-1563.
26. Bradshaw, J. S., Izatt, R. M., Bordunov, A. V., Zhu, C. Y., and Hathaway, J. K. (1996) Crown ethers, *Comprehensive Supramolecular Chemistry*, **1**, 35-95.
27. Yakshin, V. V. (2002) Structural and Chemical Features of Extraction with Crown Ethers, *Russ. J. Coord. Chem.*, **28**(10), 697-705.

28. Moyer, B. A., Bonnesen, P. V., Delmau, L. H., Haverlock, T. J., Kavallieratos, K., and Levitskaia, T. G. (2002) Toward recognition of the anion in the extraction of alkali metal salts by crown ethers and calixarenes, *International Solvent Extraction Conference, Cape Town, South Africa, Mar. 17-21, 2002*, 299-306.
29. Bartsch, R. A., Ivy, S. N., Lu, J., Huber, V. J., Talanov, V. S., Walkowiak, W., Park, C., and Amiri-Eliasi, B. (1998) Metal ion extraction by lariat ethers with "tunable" proton-ionizable groups, *Pure Appl. Chem.*, **70**(12), 2393-2400.
30. McDowell, W. J. (1988) Crown ethers as solvent extraction reagents: where do we stand?, *Sep. Sci. Technol.*, **23**(12-13), 1251-1268.
31. Wai, C. M. (1997) Separation of lanthanides and actinides with ionizable crown ethers, *Recent Progress in Actinides Separation Chemistry, Proceedings of the Workshop on Actinides Solution Chemistry, WASC '94, Tokai, Japan, Sept. 1-2, 1994*, 81-94.
32. Meguro, Y., Kitatsuji, Y., and Yoshida, Z. (1997) Synergistic extraction of actinides(III) and lanthanides(III) with thenoyltrifluoroacetone and crown ethers, *Recent Progress in Actinides Separation Chemistry, Proceedings of the Workshop on Actinides Solution Chemistry, WASC '94, Tokai, Japan, Sept. 1-2, 1994*, 53-71.
33. Meguro, Y., Kitatsuji, Y., Kimura, T., and Yoshida, Z. (1995) Separation factors in synergistic ion-pair extraction of lanthanide(III) with crown ether and b-diketone, *Kidorui*, **26**, 380-381.
34. Mathur, J. N. and Khopker, P. K. (1988) Use of crown ethers as synergists in the solvent extraction of trivalent actinides and lanthanides by 1-phenyl-3-methyl-4-trifluoroacetyl-5-pyrazolone, *Solvent Extr. Ion Exch.*, **6**(1), 111-124.
35. King, R. B. and Heckley, P. R. (1974) Lanthanide nitrate complexes of some macrocyclic polyethers, *J. Am. Chem. Soc.*, **96**(10), 3118-3123.
36. Buenzli, J. C. G. and Wessner, D. (1984) Rare earth complexes with neutral macrocyclic ligands, *Coord. Chem. Rev.*, **60**, 191-253.
37. Vanura, P. (1990) Extraction of trivalent lanthanides and actinides in the presence of crowns, polyethylene glycols and other polyoxonic compounds, *Chem. Listy*, **84**(2), 141-163.
38. Dong, Y., Lawrance, G. A., Lindoy, L. F., and Turner, P. (2003) Macrocyclic ligand design. Interaction of a series of successively N-benzylated derivatives of 1,4,8,11-tetraazacyclotetradecane (cyclam) with copper(II) and nickel(II), *Dalton Trans.* (8), 1567-1576.
39. Kasahara, K. and Endo, M. (2001) Solvent extraction of metal ions with acyclic and macrocyclic Schiff bases, *Anal. Sci*, **17**(Suppl.), a339-a342.
40. Yordanov, A. T. and Roundhill, D. M. (1998) Solution extraction of transition and post-transition heavy and precious metals by chelate and macrocyclic ligands, *Coord. Chem. Rev.*, **170**, 93-124.
41. Zolotov, Y. A., Morosanova, E. I., Dmitrienko, S. G., Formanovskii, A. A., and Ivanov, G. V. (1984) Solvent extraction of metals with macrocyclic compounds. II. The use of 19-membered compounds with donor systems 2N,2O and 2N,3O, *Mikrochim. Acta*, **3**(5-6), 399-408.
42. Handel, H., Muller, F. R., and Guglielmetti, R. (1983) Lipophilic tetraazamacrocycles: application to liquid-liquid extraction of metal ions, *Helv. Chim. Acta*, **66**(2), 514-521.
43. Fujiwara, M., Nakajima, Y., Matsushita, T., and Shono, T. (1985) Preparation and characterization of copper(II) and nickel(II) complexes with novel tetraaza-macrocyclic Schiff bases, *Polyhedron*, **4**(9), 1589-1594.
44. Zolotov, Y. A., Ionov, V. P., Niz'eva, N. V., and Formanovskii, A. A. (1984) Use of macrocyclic Schiff bases for the selective extraction and spectrophotometric determination of copper, *Doklady Akademii Nauk SSSR*, **277**(5), 1145-1148 [Chem.].
45. Bell, T. W. and Sahni, S. K. (1991) Bridged pyridine hosts, *Inclusion Compd.*, **4**, 325-390.
46. Saito, K., Masuda, Y., and Sekido, E. (1983) Liquid-liquid extraction of metal ions by the thiacrown compound 1,4,8,11-tetrathiacyclotetradecane, *Anal. Chim. Acta*, **151**(2), 447-455.
47. Sekido, E., Saito, K., Naganuma, Y., and Kumazaki, H. (1985) Liquid-liquid extraction of some class B metal ions with thiacrown ether 1,4,8,11-tetrathiacyclotetradecane, *Anal. Sci*, **1**(4), 363-368.
48. Reddy, M. L. P. and Francis, T. (2001) Recent advances in the solvent extraction of mercury(II) with calixarenes and crown ethers, *Solvent Extraction and Ion Exchange*, **19**(5), 839-863.
49. Bond, A. H., Dietz, M. L., and Chiarizia, R. (2000) Incorporating Size Selectivity into Synergistic Solvent Extraction: A Review of Crown Ether-Containing Systems, *Ind. Eng. Chem. Res.*, **39**(10), 3442-3464.
50. Burns, J. H. and Lumetta, G. J. (1991) Complexes of manganese and zinc di-tert-butylnaphthalenesulfonate with cyclohexano-15-crown-5 ether toluene solvate, *Acta Crystallogr., Sect. C: Cryst. Struct. Commun.*, **C47**(10), 2069-2073.
51. Burns, J. H. and Kessler, R. M. (1987) Structural and molecular mechanics studies of bis(dibutylphosphato)aquastrontium-18crown-6 and analogous alkaline-earth-metal complexes, *Inorg. Chem.*, **26**(9), 1370-1375.

52. Adam, K. R., Atkinson, I. M., Farquhar, S., Leong, A. J., Lindoy, L. F., Mahinay, M. S., Tasker, P. A., and Thorp, D. (1998) Tailoring molecular assemblies for metal ion binding, in *Pure Appl. Chem.* p. 2345-2350.
53. Byriel, K. A., Gasperov, V., Gloe, K., Kennard, C. H. L., Leong, A. J., Lindoy, L. F., Mahinay, M. S., Pham, H. T., Tasker, P. A., Thorp, D., and Turner, P. (2003) Host-guest assembly of ligand systems for metal ion complexation; synergistic solvent extraction of copper(II) ions by N3O2-donor macrocycles and carboxylic or phosphinic acids, *Dalton Trans.* (15), 3034-3040.
54. Adam, K. R., Antolovich, M., Atkinson, I. M., Leong, A. J., Lindoy, L. F., McCool, B. J., Davis, R. L., Kennard, C. H. L., and Tasker, P. A. (1994) On the nature of the host-guest interaction between cyclam and 4-tert-butylbenzoic acid - a system preassembled for metal complex formation, *Chem. Commun.* (13), 1539-1540.
55. Lindoy, L. F., Mahinay, M. S., Skelton, B. W., and White, A. H. (2003) Ligand assembly and metal ion complexation: Syntheses and X-ray structures of Ni(II) and Cu(II) benzoate and 4-tert-butylbenzoate complexes of cyclam, *Journal of Coordination Chemistry*, **56**(14), 1203-1213.
56. Gasperov, V., Gloe, K., Leong, A. J., Lindoy, L. F., Mahinay, M. S., Stephan, H., Tasker, P. A., and Wichmann, K. (2002) Synergistic solvent extraction of Co(II), Ni(II), Cu(II) and Zn(II) based on supramolecular assemblies, *International Solvent Extraction Conference, Cape Town, South Africa, Mar. 17-21, 2002*, 353-359.
57. Flett, D. S. (1999) New reagents or new ways with old reagents, *J. Chem. Technol. Biotechnol.*, **74**(2), 99-105.
58. Yordanov, A. T. and Roundhill, D. M. (1998) Electronic absorption spectroscopy for the investigation of the solution binding of gold, palladium, mercury and silver salts to the lower rim substituted calix[4]arenes 25,26,27,28-(2-N,N-dimethyl- dithiocarbamoylethoxy)calix[4]arene and 25,26,27,28- (2-methylthioethoxy)calix[4]arene, *Inorg. Chim. Acta*, **270**(1,2), 216-220.
59. Belhamel, K., Dzung, N. T. K., Benamor, M., and Ludwig, R. (2002) Extraction of gold from hydrochloric acid solutions by calixarenes. in *International Solvent Extraction Conference, Cape Town, South Africa, Mar. 17-21, 2002.*
60. Saito, K., Taninaka, I., Yamamoto, Y., Murakami, S., and Muromatsu, A. (2000) Liquid-liquid extraction of platinum(II) with cyclic tetrathioethers, *Talanta*, **51**(5), 913-919.
61. Szymanowski, J. (1995) Physicochemical modification of extractants, *Crit. Rev. Anal. Chem.*, **25**(3), 143-194.
62. Smith, A. G., Tasker, P. A., and White, D. J. (2003) The structures of phenolic oximes and their complexes, *Coord. Chem. Rev.*, **241**(1-2), 61-85.
63. Henrick, K., Tasker, P. A., and Lindoy, L. F. (1985) The specification of bonding cavities in macrocyclic ligands, *Prog. Inorg. Chem.*, **33**, 1-58.
64. Szymanowski, J. (1989) Surfactants in Solution, in *Surfactants in Solution*, Mittal, B. K. L. and Lindman, B., Editors. Plenum Press New York.
65. Russell, J. H. and Rickel, R. L. (1990) Modeling copper extraction from acidic solutions using P5100, PT5050, and LIX 84, *Solvent Extr. Ion Exch.*, **8**(6), 855-873.
66. Hultgren, V. M., Beddoes, R. L., Collison, D., Helliwell, M., Atkinson, I. M., Garner, C. D., Lindoy, L. F., and Tasker, P. A. (2001) Formation of folded complexes retaining intramolecular H-bonding in the extraction of nickel(II) by phenolic oxime and aliphatic diamine ligands, *Chem. Commun.* (6), 573-574.
67. Jones, V. M. (1997) PhD Thesis. University of Manchester.
68. Zhu, T. (2002) A structure parameter characterizing the steric effect of organophosphorous acid extractants, *International Solvent Extraction Conference, Cape Town, South Africa, Mar. 17-21, 2002*, 203-207.
69. Cole, P. M. and Sole, K. C. (2002) Solvent extraction in the primary and secondary processing of zinc, *International Solvent Extraction Conference, Cape Town, South Africa, Mar. 17-21, 2002*, 863-870.
70. Cole, P. M. (2002) The introduction of solvent-extraction steps during upgrading of a cobalt refinery, *Hydrometallurgy*, **64**(1), 69-77.
71. Thakur, N. V. and Mishra, S. L. (1998) Separation of Co, Ni and Cu by solvent extraction using di-(2-ethylhexyl) phosphonic acid, PC 88A, *Hydrometallurgy*, **48**(3), 277-289.
72. Maljkovic, D. and Lenhard, Z. (2002) The effect of organophosphoric extractant concentration and initial phase volume ratio on cobalt(II) and nickel(II) extraction, *International Solvent Extraction Conference, Cape Town, South Africa, Mar. 17-21, 2002*, 982-987.
73. Preston, J. S. (1985) Solvent-Extraction of Metals by Carboxylic-Acids, *Hydrometallurgy*, **14**(2), 171-188.
74. Jaaskelainen, E. and Paatero, E. (2000) Characterisation of organic phase species in the extraction of nickel by pre-neutralized Versatic 10, *Hydrometallurgy*, **55**(2), 181-200.
75. Dalton, R. F., Diaz, G., Price, R., and Zunkel, A. D. (1991) The Cuprex Metal Extraction Process - Recovering Copper from Sulfide Ores, *JOM-J. Miner. Met. Mater. Soc.*, **43**(8), 51-56.

76. Dalton, R. F., Price, R., Hermana, E., and Hoffmann, B. (1987) Separation Process in Hydrometallurgy, in *Separation Process in Hydrometallurgy*, Davies, G. A., Editor. Ellis Horwood Ltd. Chichester. p. 466.
77. Cote, G., Jakubiak, A., Bauer, D., Szymanowski, J., Mokili, B., and Poitrenaud, C. (1994) Modeling of extraction equilibrium for copper(II) extraction by pyridinecarboxylic acid esters from concentrated chloride solutions at constant water activity and constant total concentration of ionic or molecular species dissolved in the aqueous solution, *Solvent Extr. Ion Exch.* , **12**(1), 99-120.
78. Szymanowski, J. (1996) Copper hydrometallurgy and extraction from chloride media, *J. Radioanal. Nucl. Chem.-Artic.*, **208**(1), 183-194.
79. Soldenhoff, K. H. (1987) Solvent-Extraction of Copper(Ii) from Chloride Solutions by Some Pyridine Carboxylate Esters, *Solvent Extr. Ion Exch.* , **5**(5), 833-851.
80. Hofmeister, F. (1888) *Schmeidebergs Arch. Exp. Pathol. Pharmakol.*, **23**, 247.
81. White, D. J., Laing, N., Miller, H., Parsons, S., Tasker, P. A., and Coles, S. (1999) Ditopic ligands for the simultaneous solvent extraction of cations and anions, *Chemical Communications (Cambridge)* (20), 2077-2078.
82. Galbraith, S. G., Henderson, D. K., Miller, H. A., Plieger, P. G., Tasker, P. A., Smith, K. J., and West, L. C. (2003) Development of extractants to transport metal salts in base metal recovery processes, *Hydrometallurgy 2003, Proceedings of the International Symposium honoring Professor Ian M. Ritchie, 5th, Vancouver, BC, Canada, Aug. 24-27, 2003*, **1**, 941-954.
83. Galbraith, S. G., Plieger, P. G., and Tasker, P. A. (2002) Cooperative sulfate binding by metal salt extractants containing 3-dialkylaminomethylsalicylaldimine units, *Chem. Commun.* (22), 2662-2663.

24. USE OF MACROCYCLES IN NUCLEAR-WASTE CLEANUP: A REAL-WORLD APPLICATION OF A CALIXCROWN IN CESIUM SEPARATION TECHNOLOGY

BRUCE A. MOYER, JOSEPH F. BIRDWELL, JR.,
PETER V. BONNESEN AND LAETITIA H. DELMAU
Oak Ridge National Laboratory
P. O. Box 2008, Oak Ridge, TN 37831-6119, U.S.A.

1. Introduction

Crown ethers and more recently calixarenes have long been thought to have potential application in the separation of the fission products ^{137}Cs and ^{90}Sr from nuclear waste [1–8]. Through the 1980s, however, practical extraction systems proved elusive. Although much research toward this end had demonstrated a wealth of creative chemistry, fundamental problems related to insufficient extractive strength and selectivity and lack of an efficient means to reverse the extraction stood as barriers to progress. In this review, a solution to the problem of extracting cesium from alkaline nuclear waste will be presented based on calixcrown chemistry originally reported by European investigators starting in 1994 [9–16]. This chemistry was extended and developed toward the specific technology now called the Caustic-Side Solvent Extraction (CSSX) process [17,18] for the separation of cesium from the legacy high-level wastes stored in underground tanks at U.S. Department of Energy (USDOE) sites [19–21]. The technology has been designed and successfully demonstrated for near-future application at the Savannah River Site (SRS) [22–24], but because of the similarity in waste compositions, it is expected that the CSSX process stands as a well-suited candidate for application to Hanford wastes [25–27] as well. The science background and technology development pertaining to the CSSX process have been described previously in various publications [28–48] and open-literature reports (not all of which are cited here) [48–61]. Thus, in this contribution, we will describe the overall technology itself, both in terms of the underlying chemistry and the engineering that makes the use of CSSX practical. As such, this article is intended to serve as an example of how macrocyclic chemistry can be adapted for an industrial use to meet the demanding requirements of real-world systems. Below, the CSSX process is introduced via a brief historical summary of its development, followed by a description of solvent extraction as applied to nuclear problems and a description of the chemistry and engineering of CSSX.

Karsten Gloe (ed.), Macrocyclic Chemistry: Current Trends and Future Perspectives, 383–405.
© 2005 *Springer. Printed in the Netherlands.*

2. Historical Background

In the 1990s, needs emerged at multiple sites in the USDOE complex for separations technologies applicable to the problem of fission-product removal from high-level waste (HLW) [19–27]. The motivation for concentrating high-activity fission-products, especially ^{137}Cs, is based on cost and handling issues aimed at (a) minimizing the amount of transuranic waste volume due to limited disposal space and high disposal costs, and (b) minimizing the volume of waste that requires vitrification due to the relatively high costs associated with production of this waste form.

The most urgent needs were identified at the Hanford Site and SRS, where respectively 55 and 37 million gallons of alkaline HLW had accumulated from defense-related reprocessing activities. Other than the roughly 10–15% of the waste mass that consists of mostly metal hydroxide sludge, the bulk of the waste, often referred to as salt waste, can be described as a mixture of sodium hydroxide, sodium nitrate, and other soluble salts whose radionuclide content is dominated by ^{137}Cs.

In accord with the general need to divert the ^{137}Cs to the HLW stream destined for vitrification and geologic disposal [20], several technologies were developed and demonstrated for removal and concentration of the ^{137}Cs from the alkaline waste. Among these, the most promising were initially either based on ion exchange with crystalline silicotitanate (CST) [62] or cation-exchange resins [63–66] or based on precipitation with tetraphenylborate [23,24]. As the CSSX process did not appear until the late 1990s, ion exchange with SuperLig [66] and precipitation with tetraphenylborate, by what is referred to as the In-Tank-Precipitation (ITP) process [23,24], became the baseline technologies for ^{137}Cs removal at respectively Hanford and the SRS. CSSX was fundamentally not possible until late 1994, when European researchers reported the first extractants possessing sufficiently high selectivity to remove cesium from a 10^4–10^6 higher background concentration of sodium [9–16]. Thereupon, a viable solvent-extraction alternative became possible, with its attendant advantages in throughput, liquid handling, and low radionuclide holdup. Given the high selectivity of the new extractants, called calixarene-crown ethers or simply calixcrowns, and the ability to strip them with essentially water, it was realized that a process could likely be developed that would require little adjustment to the waste feed, produce little secondary waste, and deliver a concentrated and nearly pure cesium stream ideal for vitrification in borosilicate glass. The intervening period through 2002 entailed finding and demonstrating an optimal solvent system [37,43,49] and flowsheet [32,35,41] that would possess these advantageous qualities. The extractant that was adopted, called BOBCalixC6, is shown in Figure 1; it will be described in greater detail below. Batch tests on waste simulants and real Hanford wastes demonstrated that rudimentary extraction and stripping was possible [42], spurring further solvent and flowsheet development. In 1998, the first flowsheet was developed and demonstrated on simulated SRS HLW [51,67], and an improved solvent [43,50] and flowsheet [41] were demonstrated on both simulated [41] and real [32] SRS HLW in 2000 and 2001.

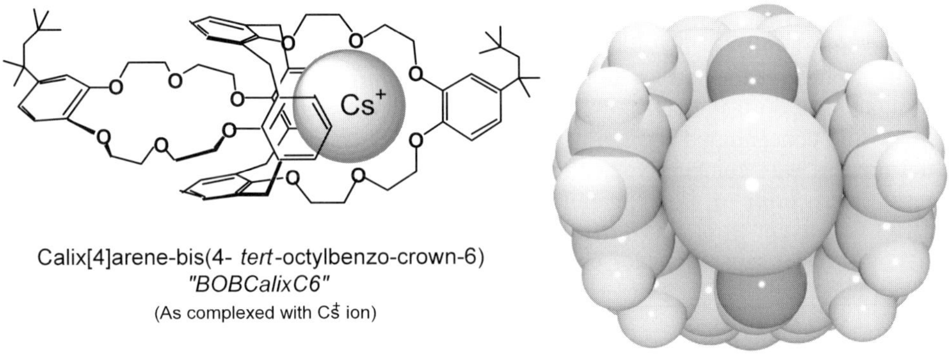

Figure 1. Calixcrown extractant adopted for CSSX, as complexed with Cs^+ ion. Left: chemical drawing of the complex. Right: a space-filling view of part of a crystal structure of the model complex Cs_2Calix[4]arene-bis(benzo-crown-6)$(NO_3)_2$•$3CHCl_3$ [69] showing the good fit of the Cs^+ ion inside the calixarene cavity; the crown ether atoms have been removed for clarity.

In 2001, the SRS announced its choice of CSSX as the baseline cesium-removal technology over small tank precipitation (a small-scale version of the ITP process) and ion exchange with CST for its Salt Waste Processing Facility (SWPF) to go into operation in 2010 [22]. An optimized solvent system, model, and flowsheet were developed and demonstrated in 2001 and 2002 [37,49], and a modular concept was developed by ORNL in 2003 [68]. Thus, the past decade has seen the emergence and maturation of a powerful new technology based on a macrocyclic cation receptor designed to function in solvent extraction to meet the critical need of the USDOE for a means of cleanly separating Cs from alkaline tank waste.

3. Generic Advantages of Solvent Extraction

Key Advantages of SX
High throughput
High selectivity
All liquid handling
Continuous processing
Stage-wise design flexibility

Although solvent extraction in acid-side processing has long been established as a standard technology in the nuclear industry [70–76], precedent for caustic-side solvent extraction is a recent development. On the acid side, introduction of solvent extraction with REDOX and then PUREX processes five decades ago greatly reduced waste generation in reprocessing compared with previous precipitation technology [26]. Other recognized advantages of solvent extraction included improved selectivity, ability to handle only liquid streams, continuous processing, stage-wise operation, and high throughput. These same advantages were later exploited in processes developed to treat HLW, as in TRUEX, SREX, and DIAMEX, to name a few [70]. SREX is notable in that it demonstrated the technical and economic feasibility of employing a relatively expensive extractant, the crown ether di(*tert*-butylcyclohexano)-18-crown-6 (DtBuCH18C6), for removal of the fission product ^{90}Sr from HLW [77]. Russian researchers demonstrated processes for removing ^{90}Sr from HLW using dicyclohexano-18-crown-6 [5] and ^{137}Cs from HLW using the crown ether, dibenzo-21-crown-7 [4,5]. On the caustic side, a close analog of SREX called SRTALK was developed for removal of technetium as pertechnetate from a Hanford simulant [78–80].

SRTALK was the first example of a solvent extraction process applied to caustic HLW waste. It employed the same crown ether as SREX, DtBuCH18C6, and was demonstrated using a 12-stage cascade of centrifugal contactors to achieve a decontamination factor (DF) of 10.7, meeting the process goal, and to produce a stream of 10-fold concentrated sodium pertechnetate [78]. Economic calculations showed that the process was competitive with ion exchange using a commercial anion exchange resin. Thus, prior to the appearance of CSSX, there was ample precedent for use of effective solvent-extraction process technology for nuclear separations in general, for fission-product separation, for caustic-side feeds, and for exploiting the unprecedented performance of designer extractants like crown ethers in an economical manner.

One key property of all of the processes named above is the fact that extraction is driven by the mass-action effect of high nitrate (CSSX, SREX, TRUEX, etc.) or sodium (SRTALK) concentration in the aqueous feed. That means that stripping can be effected in principle with water, which greatly reduces waste production and chemical consumption. This concept is illustrated in the chemical cycle as depicted for CSSX in Figure 2. In actual practice, a simple solvent-extraction flowsheet employs a water-immiscible solvent that flows counter-current to the aqueous phases that enter and exit the cascade at defined stages [81]. In the extraction section, the aqueous phase is the feed to be treated, whereupon the solvent becomes loaded with the desired metal and also lesser amounts of other extractable species. Often these other extractable species are scrubbed out of the loaded solvent by a scrub solution that leaves the bulk of the target metal in the solvent. The solvent leaving the scrub section is then stripped with an aqueous solution that removes the target metal in a concentrated stream. The solvent may then be contacted with a wash solution that removes traces of solvent degradation products or other troublesome species, whereupon the solvent can be reused, ideally thousands of times. The flowsheet actually used in CSSX will be described in greater detail below.

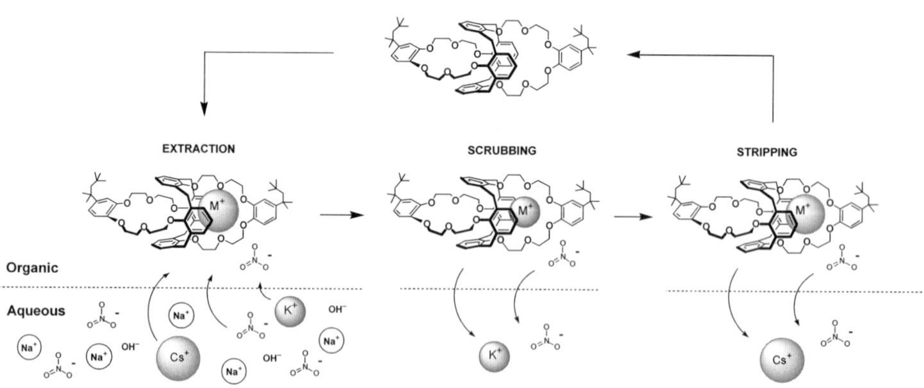

Figure 2. Chemistry occurring in three sections of a simple flowsheet in solvent extraction using CSSX as an example. In the extraction step, essentially all of the Cs^+ and a minor fraction of the K^+ ions in the waste are extracted (i.e., $M^+ = Cs^+$ or K^+). The K^+ ions are removed in scrubbing as shown, while the calixcrown-Cs^+ complex effectively remains in the solvent so that a pure cesium nitrate product is obtained on stripping.

4. CSSX Solvent Composition

Key Features of the CSSX Solvent
High selectivity, good D_{Cs} values
Reversible: can be cycled many times
Commercial availability of components
Robust to degradation
Degradation products wash out
Low solubility loss to the aqueous phase
Resists effects of impurities

The CSSX process utilizes a novel solvent made up of four components: calix[4]arene-bis-(4-*tert*-octylbenzo-crown-6) known as BOBCalixC6 as extractant; a lipophilic fluorinated alcohol, 1-(2,2,3,3-tetrafluoropropoxy)-3-(4-*sec*-butylphenoxy)-2-propanol known as Cs-7SB, as diluent modifier; tri-*n*-octylamine as a suppressor of impurity effects; and the isoparaffinic diluent Isopar L, a mixture of branched hydrocarbons with an average chain-length of 12 carbons. Figure 3 shows the composition of the solvent as currently optimized for the SWPF application at the SRS [37,49]. The chemistry of the solvent is well understood, with regards to both its fundamental properties and its performance under process conditions. All of the components are commercially available, and efficient synthetic and purification procedures have been worked out [17,18,37]. Thus, these key components may be obtained from multiple chemical suppliers capable of specialty synthesis.

The chemistry of calixcrown extractants has advanced to maturity in the past decade, lending a great deal of confidence in CSSX from this perspective. Many structural variations of calixcrowns have been reported in the literature, and extensive investigations of the applicability of such extractants to separation of cesium and other metals from acidic HLW have taken place in Europe [6]. As shown in Figure 1, calix[4]arene-crown-6 compounds like BOBCalixC6 possess the appropriate crown bridging length (i.e., 6 oxygen atoms in the bridge) and calixarene ring size (i.e., 4 phenol units) to create an excellent cavity for Cs^+ ion, giving rise to high selectivity. It may be noted that BOBCalixC6 possesses two cavities, only one of which is expected to be occupied by a Cs^+ cation at any time under normal CSSX operating conditions [45]. Although the high selectivity of the calixcrown extractant is essential for removing the trace concentration of Cs^+ ions (ca. 3×10^{-4} M) from a sea of Na^+ ions (5–7 M), the strength of the extraction is on the order of 10- to 100-fold higher than is attainable with conventional crown ethers [45]. We have attributed this strength to a combination of preorganization of the calixarene cavity and weak or absent solvation of the cavity [39]. The strong extraction strength translates to minimal concentration of BOBCalixC6 (7 mM) needed to achieve a useful cesium distribution ratio (see below), which in turn minimizes the cost of the solvent.

A crystal structure of BOBCalixC6 [82] was made possible by the fact that the compound can be crystallized from certain solvents (Figure 4). As so often is the case for crown ethers, the flexible polyether chain twists in upon itself, and the cavity does not exist until a guest species such as Cs^+ ion is present.

BOBCalixC6 was chosen as the preferred extractant for the CSSX solvent on the basis of synthetic accessibility, solubility characteristics, and selectivity. Calix[4]arene-crown-6 compounds may be synthesized with one or two crown loops, and both mono- and bis-crown calixes have good Cs^+ extraction strength [9–16].

Cesium Extractant *@ 0.007 M*	***BOBCalixC6*** **Calix[4]arene-bis(4-*tert*-octylbenzo-crown-6)**
• Economical synthesis • Selective - rejects sodium • Strong - effective at low concentration • Good radiation and chemical stability	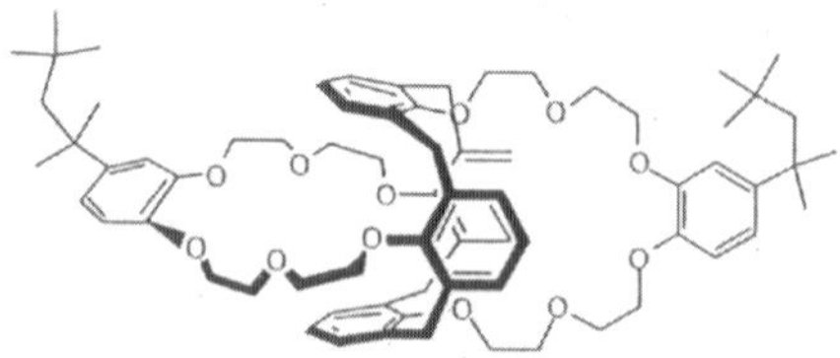
Diluent Modifier *@ 0.75 M*	***Cs-7SB*** **1-(2,2,3,3-Tetrafluoropropoxy)-3-(4-*sec*-butylphenoxy)-2-propanol**
• Increases Cs extraction • Increases extractant solubility • Excellent stability characteristics • Benign degradation products wash out	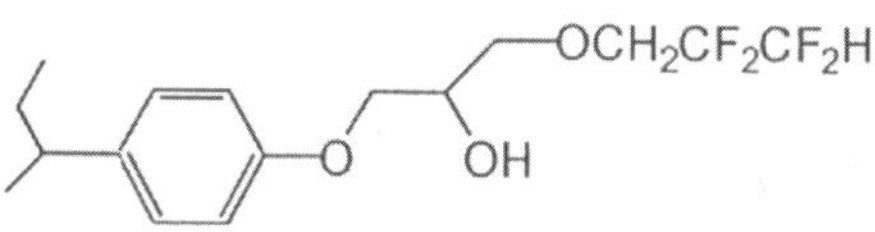
Suppressor *@ 0.003 M*	***TOA*** **Tri-*n*-octylamine**
• Suppresses impurity effects • Suppresses ion-pair dissociation • Improves and stabilizes stripping	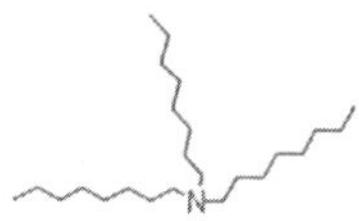
Diluent	***Isopar L***
• Promotes good hydraulics	Branched avg. 12-carbon aliphatic

Figure 3. CSSX baseline solvent for the Salt Waste Processing Facility (SWPF) application at the Savannah River Site (SRS) [49].

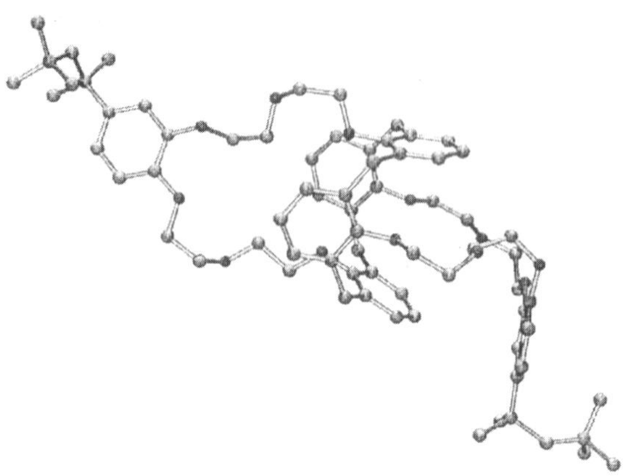

Figure 4. Crystal structure of BOBCalixC6 [82].

Because the bis-crown calixes require fewer synthetic steps, they offer an economic advantage, other factors being equal. Selectivity considerations led us to prefer a benzo ring substituent, which confers a lower affinity for Na^+ ions without sacrifice of Cs^+ affinity [46]. The branched alkyl substituent on each benzo group was originally considered to be capable of both increasing the solubility of the calixcrown in alkane diluents and boosting lipophilicity. BOBCalixC6 indeed has a demonstrated lipophilicity great enough to render its loss to the process aqueous phases negligible over the course of annual use, but accurate measurements showed that its actual solubility is only slightly more than 7 mM in the CSSX solvent shown in Figure 3 [49,50]. This solubility limit is not considered an issue for process purposes, as a good cesium extraction strength is still obtained. Nucleation and crystallization of BOBCalixC6 from slightly supersaturated solvent was noted to be exceedingly slow [50], and thus, even excursions of BOBCalixC6 into the supersaturated state do not represent a risk. Nevertheless, if desired this limitation can be removed by employing the alternative calixcrown calix[4]arene-*bis*[4-(2-ethylhexyl)benzo-crown-6], or BEHBCalixC6, which possesses 2-ethylhexyl groups in place of the *tert*-octyl groups on BOBCalixC6 [30]. The 2-ethylhexyl groups confer solubility greater than 50 mM and otherwise have no apparent effect on the extraction behavior.

Extensive tests of thermal and radiation stability of the CSSX solvent under expected operating conditions consistently showed negligible degradation of BOBCalixC6 [33,49–51,60,84]. The only apparent degradation detected was slow nitration of the benzo group at elevated nitric acid concentrations and temperature [50,84], conditions not encountered in the CSSX flowsheet. However, BOBCalixC6 would therefore be inappropriate for cesium removal from nitric acid solutions. In that case, other calixcrowns not possessing an alkylbenzo group would be preferred [44].

A product of years of basic and applied research at ORNL, the solvent modifer Cs-7SB plays a critical role in boosting the extraction power of BOBCalixC6 and promoting adequate solubility of all solvent species [18,37]. Without the modifier, the calixcrown is both poorly soluble in the preferred aliphatic diluent and incapable of extracting cesium. Careful analysis of solvation phenomena in ion-pair extraction by crown ethers led to the conclusion that the hydrogen-bond donor (HBD) ability of the solvent medium strongly influences the overall extraction power of the crown ether by specific solvation of the co-extracted anion [83]. Accordingly, we found alcohols to be effective modifiers and were able to model the enhancement in cesium extraction strength in terms of the formation of organic-phase species of the form $CsCalix(NO_3)_2{\bullet}(ROH)_n$ [28]. The extraction strength of BOBCalixC6 correlated with the HBD ability of the modified solvent as measured by the solvatochromic E_T parameter [40]. Therefore, it was advantageous to boost the solvating power of the modifier by employing electron-withdrawing fluorinated substituents in proximity to the –OH group. A forerunner of Cs-7SB possessing a 1,1,2,2-tetrafluoroethoxy group proved unstable under simulated process conditions [18,37,51]. Replacing this group with a 2,2,3,3-tetrafluoropropoxy group conferred the needed stability.

Tests consistently showed the alkylphenoxy platform to be effective and versatile for modifier design, and the starting materials are economical. The *sec*-butyl group in particular was chosen to balance the needs for resistance to third-phase formation, lowest loss to the aqueous phase, and greatest ease of washing out degradation products. The ability to wash out degradation products was ultimately a decisive design factor. Although tests showed that the alkylphenoxy modifiers possess very good stability with application of heat or radiation within the range of expected operating conditions, slight degradation to give the respective alkylphenol was detected [49–51]. Although our original choice was to employ a *tert*-octyl group, which conferred very good solubility behavior, low loss to the aqueous phase, and good economics, the 4-*tert*-octylphenol breakdown product distributes weakly to alkaline aqueous solutions and therefore cannot be efficiently washed out. To prevent buildup of alkylphenol in the solvent, an alkyl substituent with fewer carbon atoms was needed, but this compromised the solubilizing power of the modifier, leading to greater tendency to form a third phase. Among the choices, Cs-7SB was found to be the best compromise. Stability tests [33,49–51,60,84] showed Cs-7SB to have adequate radiation and thermochemical stability overall. Negligible absolute degradation of the Cs-7SB inventory may be expected in annual use, and the traces of degradation products, such as *sec*-butylphenol, are readily washed out of the solvent. The lipophilicity of Cs-7SB is good, and only minor losses to the aqueous phase are expected in annual use [49]. Though originally available only from ORNL, Cs-7SB has been recently commercialized.

The tri-*n*-octylamine (TOA) is present in the solvent to improve stripping performance by suppressing the deleterious effect of lipophilic anion impurities [31,49–51,85]. Since stripping is carried out under acidic conditions, the amine in the ammonium form provides an available concentration of cations in the solvent to act as counterions for traces of surfactant anions, dibutylphosphate, or other lipophilic anions that might be present. The effect of anions is therefore suppressed, allowing cesium nitrate to be stripped. During extraction of cesium, the alkaline conditions will render the amine neutral, in which form its effect on extraction equilibria will be negligible. Figure 5 illustrates the function of the amine in this regard. The well-known commercial tertiary amine extractant TOA works well and was adopted at 3 mM concentration [49].

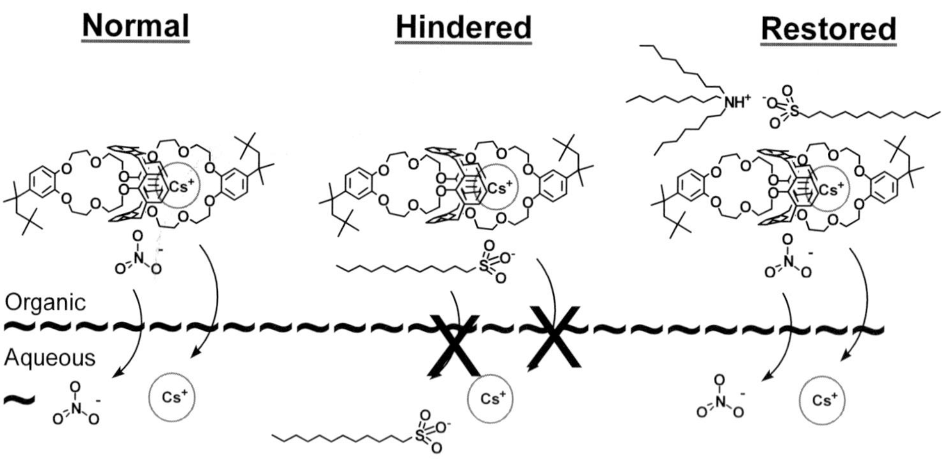

Figure 5. Stripping under three conditions: normal, hindered by the presence of a lipophilic anion, and restored by addition of tri-*n*-octylammonium cation. The tri-*n*-octylamine in the solvent exists in its protonated form under stripping conditions, where it provides charge balance for traces of lipophilic anions such as surfactants or dibutylphosphate, allowing cesium nitrate to be stripped. Under extraction conditions, the amine is in the neutral form and has negligible effect.

An isoparaffinic diluent, Isopar L, available from ExxonMobil Chemical Company, was chosen for the CSSX solvent. Isopar L is a blend of C_{10} to C_{12} branched alkanes with a distillation range of 185–211 °C, a viscosity of 1.6 centipoise at 25 °C, a specific gravity of 0.765–0.772 g/mL at 60 °F (15.6 °C), and a thermal closed-cup flash point of at least 60 °C (data courtesy of ExxonMobil Chemical Company). Aliphatic diluents are desirable because they improve hydraulics (low density and viscosity), have excellent stability, and dissolve negligibly in water. Straight-chain hydrocarbons would be preferred for maximum radiation stability, but use of the equivalent straight-chain diluent Norpar 12, for example, entails a slightly greater tendency to third-phase formation. Radiation degradation effects were found to be minimal in any case for SRS waste, as the radiation dose expected for the solvent is modest (ca. 100 krad/yr [33]). Hanford wastes tend to have lower levels of cesium on average, and accordingly, radiation effects are not considered a problem either for potential Hanford applications.

5. Flowsheet Design

Given the high performance of the CSSX solvent, engineering flowsheets can be designed to meet varied needs, including exceedingly demanding ones. The flowsheet designed to treat SRS feed and to deliver a stream of nearly pure cesium nitrate with a decontamination factor (DF) of 40,000 (99.9975% removal) and a concentration factor (CF) of 15 [35] is shown in Figure 6. Demonstration-phase tests on simulated waste feed and on actual waste (both supernatant liquid and dissolved salt cake) at the SRS showed the CSSX process to be capable of meeting these stringent goals [32,86,87].

Designed to operate by a "nitrate-swing" principle, the chemistry occurring in the flowsheet is simplistically depicted above in Figure 2. Batch-contacting results in Table 1 show a relatively high cesium distribution ratio (D_{Cs}) in extraction, moderate in scrubbing, and low in stripping. Generally, simulant testing was shown to be a valid predictor of performance by direct comparison of D_{Cs} values obtained using simulants with those obtained with actual waste [34].

TABLE 1. Performance of optimized CSSX solvent in extract-scrub-strip (ESS) batch tests[a]

D_{Cs}						
Extract	Scrub #1	Scrub #2	Strip #1	Strip #2	Strip #3	Strip #4
14.16	1.14	1.35	0.115	0.079	0.091	0.053

[a]Average SRS simulant was employed for extraction (O:A = 1:3), 50 mM HNO_3 for scrubbing (O:A = 5:1), and 1 mM HNO_3 for stripping (O:A = 5:1). All contacts were made at 25 °C. Fresh aqueous phase was used for each contact (cross-current contacting). Data from reference [49].

In the flowsheet (Figure 6), the solvent is contacted with the alkaline waste feed stream in a series of centrifugal contactors configured for countercurrent flow, resulting in the transfer of the cesium to the solvent phase. The high aqueous nitrate concentration drives the extraction, forming organic-phase complexes consisting of a Cs^+ ion bound inside the calixcrown and an associated nitrate anion. The high alkalinity also drives the extraction by what is thought to be deprotonation of the modifier (i.e., cation exchange), as supported by equilibrium modeling [36,52–54]. The existence of dual extraction mechanisms contributes to the ability of the solvent to accommodate varied feeds. Following the extraction step, the solvent is scrubbed with 0.05 M nitric acid to remove other soluble salts, mainly sodium and potassium, and to acid-condition the solvent. These salts would otherwise be stripped in the first strip stage, thereby raising the aqueous nitrate concentration, inhibiting cesium stripping, and adding to the mass of material to be vitrified.

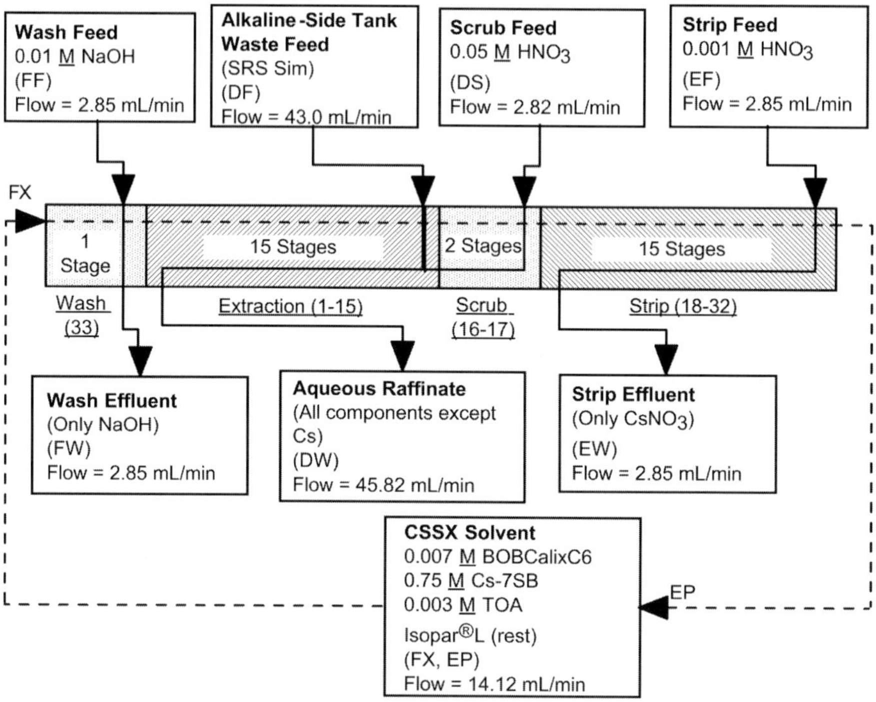

Figure 6. CSSX demonstration flowsheet for SRS feed giving DF ≥ 40,000 and CF = 15 [35].

The scrub solution joins the aqueous waste feed stream to constitute the aqueous phase in the extraction section of the contactor cascade; its low flow rate and low electrolyte concentration have little effect on the extraction section. The scrubbed solvent passes to the strip section of the cascade, where it is contacted with 1 mM nitric acid to transfer the cesium to the aqueous phase. This concentration of nitric acid was chosen to minimize the D_{Cs} for best stripping efficiency, while maintaining sufficient acidity to keep the amine protonated and sufficient ionic strength for adequate coalescence.

The stripped solvent is then recycled to the extraction section after first passing through a solvent-wash stage, shown in Figure 6 as stage #1. The wash stage employs 0.01 M NaOH to remove any rogue anions such as surfactants, *sec*-butylphenolate (degradation product of Cs-7SB), and possibly other anionic components of the feed (e.g., dibutylphosphate) that persist in the solvent through scrubbing and stripping. The solvent is designed to be recycled thousands of times and retain its integrity for at least one year of use, which is necessary for both environmental and economic reasons. In the first simulant test in centrifugal contactors in 2001 [88], a test of 71 h duration, the solvent was recycled 42 times. This and subsequent tests on real waste have repeatedly shown the CSSX process to be robust and to meet SRS processing goals [32,86,87].

The flowsheet has three aqueous effluents. The first is the decontaminated aqueous raffinate stream. For economic reasons, entrained solvent in the raffinate must be recovered; test results show that coalescers are practical and effective [89]. In the baseline SWPF, the raffinate stream will be transferred to the Saltstone Facility, where it will be disposed of in a cementitious low-activity waste form called saltstone. The second effluent is the mildly acidic aqueous strip solution containing the cesium concentrated by a factor of 15. Consisting of a few millimolar $CsNO_3$ in 0.001 M HNO_3, the strip effluent is an ideal feed for vitrification, one of the key advantages of the CSSX process. The third effluent is the NaOH effluent from the solvent-wash stage.

A modular CSSX flowsheet concept has been designed to increase the flexibility of deployment [68]. The ORNL concept was developed in specific response to a need to expedite deployment of cesium treatment technologies at the SRS. The primary design criteria for the SRS system were attainment of a cesium decontamination factor of 40 with a cesium concentration factor of 15. This was achieved using a flowsheet similar to that of Figure 6, but with stages and flow-rates configured to meet the less stringent goals. The design throughput was 3 million gallons per year, based on 90% system availability. The process was configured to be housed in three, standard-size shipping containers that would be delivered to the deployment site for integration with existing facilities, with the treatment equipment already installed inside the modules.

6. Secondary Waste Generation

CSSX chemistry features minimal consumption of chemicals and generation of secondary waste. The fact that stripping may in principle be effected by contacting the loaded solvent with water provides an ideal means of closing the solvent cycle, from the perspective of both secondary-waste generation and vitrification of the cesium product stream. Some compromises to the ideal situation were required, however, to effect scrubbing, ensure good phase coalescence in stripping, and minimize buildup of impurities in the solvent over repeated cycles. As discussed above, scrubbing with dilute nitric acid represents a minor chemical addition to the bulk waste stream. To effect good phase coalescence, some aqueous electrolyte is generally required in practical solvent-extraction systems, and thus, a very dilute nitric acid stripping solution is employed, virtually of no consequence in terms of a burden to vitrification. Likewise, washing the solvent with a slow stream of 10 mM NaOH is quite acceptable. Thus, the CSSX process will effectively generate two secondary waste streams: (a) a dilute caustic effluent stream produced by washing the solvent as it is recycled through the solvent extraction cascade and (b) the spent solvent. The caustic wash waste is generated continuously during cascade operation. Since this solution contacts the solvent only after stripping, it will contain extremely low levels of cesium, and can be combined with the treated waste feed (i.e., the raffinate from extraction) for final disposal. Despite continuous washing during use, the solvent will periodically require replacement on a batch basis. The total volume of solvent in a full-scale system capable of processing waste at an annual rate of a few million gallons is estimated to be on the order of a few hundred liters. While the spent solvent may contain minor concentrations of radionuclides, it will not contain significant levels of transuranics. Due to the infrequent and batchwise nature of solvent disposal, installation of a dedicated solvent destruction process is not practical. It is expected that spent solvent can be immobilized for disposal using any of a number of additives that have been tested and found acceptable for disposal of organic wastes at the SRS [90].

7. Equipment Design

Centrifugal Contactor
High-throughput Small footprint Minimizes solvent degradation Minimizes solvent inventory Proven in nuclear industry

The CSSX process is best performed using centrifugal solvent-extraction contactors. One imperative when employing an expensive extractant is that the extractant must be "turned over" very rapidly. This requires both rapid chemical kinetics and fast phase contacting. It was found by simple batch contacting experiments by hand vortexing that extraction takes place on the order of seconds [51]. Thus, for fast solvent cycling a contacting device that can generate a dispersion with a huge interfacial area (20 μm droplets) and coalesce the dispersion in a matter of seconds is needed. Such a device is the centrifugal contactor [91,92], which was initially developed for use in actinide recovery (PUREX processing) as an evolution of mixer-settlers. Centrifugal contactors utilize force generated by centrifugal acceleration to obtain highly efficient separation of dispersed, immiscible liquid phases. Feed solutions to the contactor are introduced into a narrow vertical annulus between the stationary contactor housing and a moving hollow rotor, where shear forces exerted on the solution promote rapid formation of a dispersion (Figure 7). It is within this region of the contactor that mass transfer occurs. As the dispersion flows downward inside the housing, it enters the rotor through an opening located in the bottom end of the rotor at the centerline. The configuration of the rotor is such that the liquid is supported along the rotor wall as the dispersion separates into light and heavy phases under the influence of centrifugal force. Channels inside the upper end of the rotor direct the separated phases from the rotor into collection troughs that are machined into the stationary housing. Withdrawal nozzles are connected to these collection troughs, and are configured to facilitate connection of multiple contactors in a cascade arrangement with countercurrent flow.

Because of its ease of operation, rapid attainment of steady state, high mass-transfer rate, rapid coalescence, and compact size, the centrifugal contactor is the preferred device for solvent-extraction processing. A key advantage of centrifugal contactors for radioactive processing is the short contact time between organic solvent and aqueous phases, typically less than 10 seconds, thereby minimizing solvent degradation due to radiolysis and extending solvent life. A second key advantage of centrifugal contactors is the low solvent holdup and rapid turnover, which translates into a low solvent inventory and acceptable capital cost. Finally, centrifugal contactors require a small horizontal and vertical footprint, minimizing facility capital costs and allowing modular design. In nuclear plants, footprint is the most important capital-cost determinant.

Contactor-based solvent extraction systems have been operated successfully at several USDOE sites in various applications involving radioactive materials. Designs incorporating features for remote maintenance and operation are easily implemented. While initial multi-stage testing of CSSX was carried out using 2-cm minicontactors [32,35,41,67,78,86–88], tests of 5.5-cm contactors [56,58,68] have validated the expectation that scale-up may be accomplished in a straightforward manner.

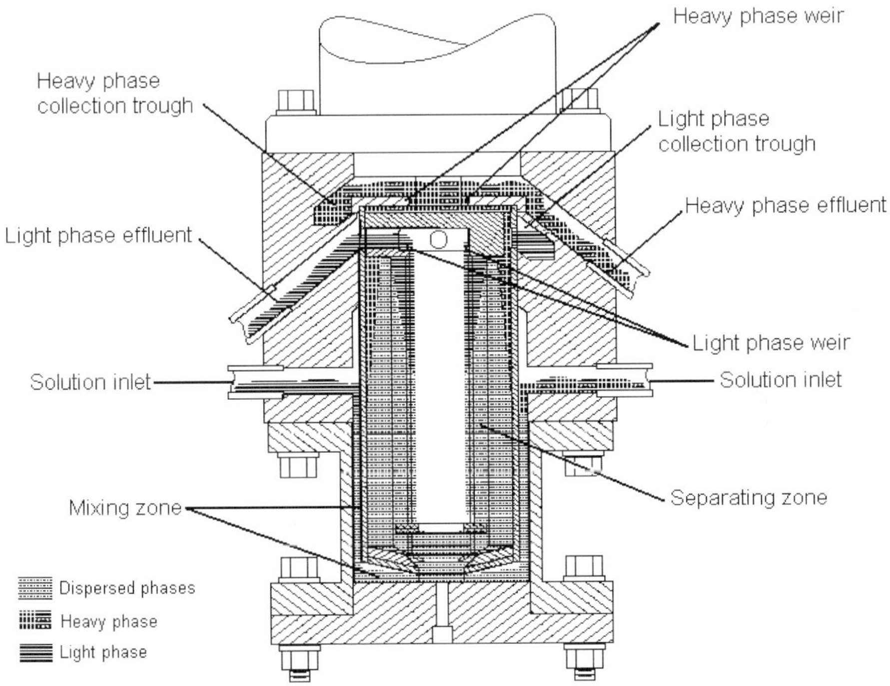

Figure 7. Cross-section of a centrifugal contactor.

8. Issues Related to Further Development and New Applications of CSSX

8.1 GENERIC CRITERIA

As new types of applications emerge for macrocyclic extractants such as found in CSSX, future development and R&D opportunities may be expected. Accordingly, a set of generic criteria have been identified pertaining to the key properties of a solvent-extraction system, specifically focusing on CSSX, that must be assured for proper functioning of the process over long-term operation, including the ability to demonstrate the flowsheet on both simulants and real feed streams. Chemical development issues of CSSX in particular have been discussed in an available report [49].

Four generic criteria must be addressed for development of any solvent extraction process in the nuclear or hydrometallurgical industries. Namely, the solvent must maintain its integrity over time and use. The process must accept wide feed variation, and the solvent components cannot be lost at an appreciable rate. The solvent must possess a relatively high affinity and selectivity for the component that is to be recovered from the feed stream. Also, the opposing phases in the process must be immiscible or very nearly so in order to effect efficient phase separation. Finally, the prototype process must undergo successful demonstration on the feed types expected and under the conditions expected, meeting the specified processing requirements.

8.2 SOLVENT INTEGRITY

Elements of solvent integrity include resistance to a) third-phase formation, b) formation of emulsions and cruds, c) radiation and chemical degradation, and d) buildup of impurities that do not readily wash out. Although emulsions have been observed in contactor tests, it has been concluded that this phenomenon is a matter of proper contactor design and operation parameters [57]. Cruds have not been observed in CSSX tests, though this phenomenon tends to be ubiquitous in long-term operation in any solvent-extraction process. Radiation degradation and thermal/chemical degradation have been studied in extensive tests under the conditions of SRS "average" salt waste [33,37,49-51,59,60,84]. All results have indicated that the solvent will withstand stresses at the extreme of temperature (35 °C constant with spikes to 60 °C), all expected aqueous components (including numerous potential trace catalysts), and radiation (15-y dose) without consequence to performance for at least one year, even without solvent washing [49,50]. It was shown that the traces of degradation products that could be identified (fluoride ion and *sec*-butylphenol for the modifier and dioctylamine for TOA) would be washed out by the process streams and wash stage. No evidence was found that minor aqueous components present in the waste build up in the solvent. Traces of anionic species that might be extracted are both neutralized by the TOA suppressor and washed out in the wash stage.

Resistance to third-phase formation is generally regarded as a development issue in solvent extraction, because of the preferred use of aliphatic diluents, which are naturally poor solvents for the polar extractants and complexes that must be held in solution. For CSSX in particular, we are demanding that the solvent system accept ion pairs consisting of the Cs^+-calixcrown complex and associated counter-anion. Tests show that this is generally not a problem, apparently because of the nature of this complex and because of the relatively low concentration of complex formed, cesium being at trace levels in the waste. However, the concentrations of K^+ ion in the waste can be significant, especially at Hanford, and the resulting high loading of potassium has been linked to third-phase formation in previous tests [50]. Third-phase formation is defined as the splitting of the solvent into two phases. It is basically a solubility phenomenon wherein certain extraction complexes exceed their solubility limit, falling out of solution in the form of a heavy, viscous liquid phase that collects usually at the interface. The heavy solvent phase is rich in some extracted component (e.g., potassium and cesium in this case) together with extractant, modifier, diluent, and water molecules. The presence of a third phase is unacceptable, as it would seriously impair contactor performance and flowsheet chemistry.

The CSSX solvent exhibits good resistance to third-phase formation. Tests have indicated that the temperature at which the third phase appears decreases rapidly as BOBCalixC6 concentration decreases and as Cs-7SB concentration increases [49]. The current CSSX solvent was formulated in part to increase resistance to third-phase formation and accordingly has a lower BOBCalixC6 concentration and a higher Cs-7SB concentration than previous formulations [37,50,51]. From the reported results, it may be expected that a third phase would not form above 10 °C when the current CSSX solvent is contacted with the SRS average waste composition at its maximum potassium concentration (50 mM). Since the solvent is highly loaded with potassium under these conditions, approaching its loading plateau, further increase in aqueous potassium concentration is not expected to entail a significant risk.

8.3 EXTRACTION ROBUSTNESS TO FEED VARIATION

As discussed above, aqueous feed variation presents a generic issue for flowsheet design in order to handle the worst-case aqueous feed. Fortunately, the performance of the strip section is independent of the waste feed to the extraction section, because trace aqueous species other than cesium nitrate are scrubbed out of the solvent prior to stripping. Batch tests at the SRS [34,93] demonstrated good performance for seven types of feeds, including supernatant liquid and dissolved salt cake, and D_{Cs} values were closely predicted by the computer model [52–54]. Hanford wastes present even greater variability, including waste types high in competing K^+ ion concentration or organic complexants, for example. Cesium was extracted from two real Hanford supernatant tank wastes and stripped with 10 mM nitric acid [42]. The DSSF waste represents a severe case, since extraction of K^+ ion (at ca. 1 M in the waste) competes significantly with Cs^+ ion extraction. Values of D_{Cs} around 2 were obtained, considered low but usable for flowsheet design. By comparison, excellent extraction (D_{Cs} = 20) was obtained from the CC waste, which contains low levels of K^+ ion but a high concentration of organic complexants. Excellent results were also obtained in extraction of cesium from Hanford B-110 sludge wash solution [37].

Extraction D_{Cs} values have been shown to be affected most strongly by the potassium concentration in the feed, temperature variation, nitrate concentration, and hydroxide concentration [41,49–51,53,54]. Flowsheet design parameters include O/A ratios, number of extraction stages, and possibly temperature control, and the computer model developed for CSSX extraction behavior [53,54] may be employed to estimate the values of D_{Cs} for various waste compositions at 25 °C. The model remains to be expanded to cover expected changes in some compositional variables (especially high K^+ concentrations), temperature, and concentrations of solvent components.

Selectivity is not considered an issue regardless of the expected feed composition. A survey of the elements present in the average SRS simulant, including Al, Cr, Cu, Fe, Mo, Hg, Ag, Pb, Pd, Rh, Ru, Zn, Sn, Cl, and F, could not detect any buildup in the previous CSSX solvent on limited cycles, nor did these elements appear in the scrub or strip stages in significant concentrations [43,50]. Five metals not included in the full simulant recipe, namely U, Np, Pu, Sr, and Tc, were not significantly extracted by the earlier CSSX solvent nor did they affect extraction or stripping of cesium. In tests of the optimized CSSX solvent [49], certain organic anions like dibutylphosphate, present in the waste in trace amounts, were found to partition weakly into the solvent on extraction, remaining in the solvent through stripping, ultimately being washed out efficiently by the NaOH wash stage. Trace Ag, Cr, Cu, Hg, Mn, Mo, Pb, Pd, Rh, Ru, Sn, and Zn were not extracted; traces of Al, Ca, Sr, and Fe were weakly detected in the optimized solvent after extraction but were found to be scrubbed out. The bulk metals Na and K were also extracted as expected, and these also were scrubbed out. Technetium as pertechnetate anion was extracted very weakly from the simulant (D_{Tc} = 0.038 or 1% extracted) and, like dibutylphosphate, remained in the solvent in the acidic scrub and strip steps, subsequently to be washed out in the NaOH wash step.

8.4 SOLVENT LOSS RATE

All solvent components have very high partition ratios, as demonstrated on average SRS waste simulant [49–51]. The annual processing losses estimated from partition data under the SRS conditions were <4% of BOBCalixC6, <27% Cs-7SB, and <9% of TOA [49]. These values are upper limits, because the partition ratios were higher than could be accurately measured. Thus, there is no basis to believe that reagent loss by partitioning to the aqueous phase poses an issue in normal applications (e.g., SRS and Hanford).

In the absence of a solvent-recovery method, entrainment is expected to be the major solvent-loss factor in all solvent-extraction applications [94], including CSSX [89], potentially amounting to several hundred ppm of the aqueous effluent. Solvent loss is known to be the economic determinant in most commercial solvent-extraction systems [94], and its mitigation is essential for expensive extractants such as BOBCalixC6. Indeed, CSSX would be economically prohibitive if the solvent entrained in the aqueous raffinate destined for bulk vitrification is not recovered for reuse. Test data show that 90% of the entrainment loss of CSSX solvent may be recovered by fibrous coalescers [89]. Because economics might be further improved by further reducing entrainment losses, it may be beneficial to examine other possible techniques, such as centrifugation.

Diluent evaporation and accompanying increase in solvent density, viscosity, and component concentrations has been documented [55]. Evidence of diluent loss during the CSSX demonstration and downselect indicates that such losses will not be large, but diluent makeup will be needed. Solvent density measurement is a convenient indicator, as it is a sensitive and easily determined solvent property.

Stability tests indicated that the CSSX solvent performance would meet SRS processing requirements for more than a year under maximum bounding conditions of radiation, chemical, and thermal stress [33,37,49–51,59,60,84]. Losses of BOBCalixC6, Cs-7SB, and TOA were significant only when bounding conditions were greatly exceeded (e.g., for a 160-year radiation dose or contacting for many weeks at 61 °C).

8.5 VERIFICATION OF PROCESS CONCEPT

In addition to issues related to the fundamental chemistry of the process, there are two general, engineering-based aspects of system performance that must be verified to ensure successful process operation in any new application of solvent extraction: phase separation performance and efficient transfer of solute(s) between phases.

Efficient phase separation is critical, since cross-phase contamination has an inherently adverse effect on mass-transfer efficiency. In addition, carryover of solvent in aqueous effluent streams results in loss of solvent from the process, impacting process economics. Phase separation is affected by several physicochemical factors, including the viscosities and densities of the opposing bulk phases and the interfacial tension of the two-phase system. All of these properties contribute to the dimensionless dispersion number, which describes the tendency of two dispersed phases to separate [95]. The separation residence time determines the relationship between separator size and separator throughput. Extensive property determinations were made during prior development and testing of candidate CSSX solvent formulations [49].

Phase separation in all solvent extraction systems, including CSSX, is greatly enhanced by the application of relatively high centrifugal forces that are generated in the separating zone of centrifugal solvent extraction contactors. Typical acceleration fields applied in engineering- and production-scale contactors exceed 300 times the acceleration due to gravity. Consequently, dispersions that do not separate readily by gravity settling, (i.e., solution pairs having high viscosities or having nearly equal densities) can be separated at efficiencies exceeding 99%.

As part of prior CSSX process demonstration testing, the ability of engineering-scale centrifugal contactors to efficiently separate the phases present in the extraction, scrubbing, and stripping sections of the baseline CSSX flowsheet was evaluated over ranges of phase ratios [58]. In all cases, carryover of opposing phases in contactor effluents was so low as to be undetectable. However, it is generally recommended for new feed types that phase-separation evaluations be included in development work.

In addition to phase-separation performance, mass-transfer efficiency in solvent-extraction applications is controlled by the extent to which interfacial area between phases is maximized (thereby enhancing solute transfer) and by the ability to maintain the dispersed condition long enough to accommodate the kinetics of the solvation and/or coordination processes. During demonstration of CSSX technology for application to SRS waste, mass-transfer performance in single, engineering-scale centrifugal contactor stages was demonstrated under extraction and stripping conditions [58] using simulated waste solution. In addition, multi-stage mass-transfer performance was confirmed using a cascade of laboratory-scale contactors to process actual SRS waste solution [96]. However, because the flow ratios in any new CSSX flowsheet may differ from those in the SRS CSSX process, and because these differences affect interfacial area and residence time, demonstration and quantification of mass transfer performance in lab- and engineering-scale contactors are desirable prior to final design of a process system for deployment. Experience with contactors of various sizes indicates that mass-transfer efficiencies obtained in contactors having rotor diameters of 4-cm and greater are scaleable to larger units. Efficiencies obtained in smaller contactors tend to be lower than those in larger devices; consequently, flowsheets designed based on lab-scale performance are overly conservative.

9. Conclusions

From the perspective of the macrocyclic chemistry community, the development of the CSSX process represents validation of the value of basic research in recognition and inclusion phenomena. The treatment of nuclear waste is one of the greatest technical challenges of our times. It involves tremendous complexity, extreme requirements, harsh conditions, high cost, and low tolerance for failure. It puts to the test the pervasive assumption in our community that macrocyclic chemistry offers the potential to meet demanding situations. The long path through fundamental research, technology development, demonstration, plant design, and final implementation for any application subjects the elegant macrocycle to conditions and expectations that could not have been conceived at the time of the initial synthesis and characterization. Having ushered one system through this maze, we may observe that the fundamental understanding of the macrocycle is but one part of the total understanding needed to develop a fully functional system for real-world needs. Its response to every chemical variable and condition comes into play and must be understood well enough that the behavior of the system is predictable within a defined envelope of operating parameters. For the specific application involving cesium removal from alkaline nuclear waste, macrocyclic chemistry has offered a number of possible approaches. The CSSX process, with its advantages summarized below, stands as one such example that has now been taken through a rigorous demonstration phase.

The CSSX process features an ability to handle many types of reprocessing wastes and even has an acid-side variant that may be applicable to INEEL waste. Laboratory batch tests have shown applicability of CSSX for the range of expected SRS wastes, and contactor tests have been successfully conducted on both SRS supernatant and SRS dissolved salt cake. Batch tests on Hanford wastes, both real and simulated, lend encouragement for a direct application at Hanford. Because CSSX uses a nitrate swing principle, a nearly pure, concentrated cesium nitrate product is produced upon stripping with 0.001 M HNO_3. This product stream is ideal for direct vitrification or other end processing. Very little secondary waste is generated, as dilute scrubbing and washing solutions are used at low flow rates. The solvent is reusable for at least a year. The flexibility of solvent-extraction flowsheet design allows the CSSX flowsheet to be easily configured for different feed types and processing requirements. Centrifugal contactors feature small footprints, allowing modular designs as well as fixed-plant applications. Centrifugal contactors are easily maintained, and recovery from process interruptions is rapid. These contactors also minimize solvent inventory, which makes it possible to employ expensive extractants. Unlike ion exchange, there is no significant inventory of radionuclides. Extracted ^{137}Cs is immediately stripped, minimizing radiation damage to the solvent, lowering shielding requirements, and enhancing radiation safety. CSSX chemistry is well understood. An equilibrium model has already been implemented and used in conjunction with flowsheet-design codes and performance prediction. The solvent has been commercialized.

10. Acknowledgments

We thank our many colleagues for their varied contributions to the CSSX chemistry and flowsheet development and demonstration. We thank contributors at Oak Ridge National Laboratory, Kim Anderson, Charles F. Baes, Jr., Jeffrey C. Bryan, Nancy L. Engle, Tamara J. Haverlock, Leon N. Klatt, Douglas D. Lee, Tatiana G. Levitskaia, Mike Maskarinec, Richard A. Sachleben, Frederick V. Sloop, Jr., Roger D. Spence, and Bruce A. Tomkins; Argonne National Laboratory, Scott B. Aase, Hassan A. Arafat, Cliff Conner, J. R. Falkenberg, Ralph A. Leonard, Candido Pereira, Monica C. Regalbuto, and George F. Vandegrift; Savannah River Technology Center, Seth G. Campbell, Stephen L. Crump, Michael A. Norato, Reid A. Peterson, Robert A. Pierce, Kenneth J. Rueter, Paul L. Rutland, Major C. Thompson, D. Douglas Walker, Thomas L. White, and William R. Wilmarth; Pacific Northwest National Laboratory, Harry Harmon and Gregg J. Lumetta; and Idaho National Engineering and Environmental Laboratory, Jack D. Law and Terry A. Todd. We are indebted to Radu Custelcean for internal review of the manuscript and Sandra L. Solands for preparation of the manuscript. This research and development, up to and including the CSSX demonstration, was sponsored by the following offices, all within the U.S. Department of Energy: 1) Division of Chemical Sciences, Geosciences, and Biosciences, Office of Basic Energy Sciences, Office of Science; 2) the Efficient Separations and Processing Crosscutting Program, Office of Science and Technology, Office of Environmental Management; 3) Environmental Management Science Program, Office of Science; and 4) the Office of Project Completion, Office of Environmental Management. Oak Ridge National Laboratory is managed and operated for the U. S. Department of Energy by UT-Battelle, LLC. under contract number DE-AC05-00OR22725.

11. References

1. Schulz, W. W. and Bray, L. A.: Solvent Extraction Recovery of Byproduct Cs-137 and Sr-90 from HNO_3 Solutions—a Technology Review and Assessment, *Sep. Sci. Technol.* **22** (2&3) (1987), 191–214.
2. Horwitz, E. P. and Schulz, W. W.: Solvent Extraction in the Treatment of Acidic High-Level Liquid Waste, Where Do We Stand? In *Metal-Ion Separation and Preconcentration*, Bond, A. H., Dietz, M. L., and Rogers, R. D. (Eds.), ACS Symp. Ser., Vol. 716, American Chemical Society, Washington, DC, 1999, pp. 20–50.
3. Dozol, J.-F., Casas i Garcia, J., and Sastre, A. M.: Applications of Crown-Ethers to Cesium and Strontium Removal from Marcoule Reprocessing Concentrate, in *New Separation Chemistry Techniques for Radioactive Waste and Other Specific Applications*, Cecille, L., Casarci, M., and Pietrelli, L. (Eds.), Elsevier Applied Science, Amsterdam, 1991, pp. 173–185.
4. Egorov, N. N., Zakharov, M. A., Lazarev, L. N., Lyubtsev, R. I., Nikiforov, A. S., Strakhov, M. V., and Filippov, E. A.: New Solutions to the Problem of Handling Long Lived Radionuclides. *Soviet Atomic Energy* **72**(2) (1992), 148–150.
5. Filippov, E. A., Dzekun, E. G., Nardova, A. K., Mamakin, I. V., Gelis, V. M., and Milyutin, V. V.: Application of Crown Ethers and Ferrocyanide-Based Inorganic Material for Cesium and Strontium Recovery from High-Level Radioactive Wastes, Proc. Waste Management '92, Tucson, Arizona, U.S.A., Mar. 1–5, 1992, pp. 1021–1025.
6. Dozol, J.-F., Bohmer, V., McKervey, A., Lopez Calahorra, F., Reinhoudt, D., Schwing, M.-J., Ungaro, R., and Wipff, G.: New Macrocyclic Extractants for Radioactive Waste Treatment, Ionizable Crown Ethers and Functionalized Calixarenes, Report EUR-17615, European Communities, Luxembourg, 1997.
7. Blasius, E. and Nilles, K.-H.: The Removal of Cesium from Medium-Active Waste Solutions. I. Evaluation of Crown Ethers and Special Crown-Ether Adducts in the Solvent Extraction of Cesium, *Radiochim. Acta* **35** (1984), 173–182.
8. Gerow, I. H. and Davis, M. V.: The Use of 24-Crown-8s in the Solvent Extraction of $CsNO_3$ and $Sr(NO_3)_2$, *Sep. Sci. Technol.* **14**(5) (1979), 395–414.
9. Ungaro, R., Casnati, A., Ugozzoli, F., Pochini, A., Dozol, J.-F., and Hill, C., Rouquette, H.: 1,3-Dialkoxycalix[4]arenecrowns-6 in 1,3-Alternate Conformation, Cesium-Selective Ligands that Exploit Cation-Arene Interactions, *Angew. Chem., Int. Ed Engl.* **33**(14) (1994), 1506–1509.
10. Asfari, Z., Wenger, S., and Vicens, J.: Calixcrowns and Related Molecules. *J. Inclusion Phenom. Mol. Recognit. Chem.* **19** (1994), 137–148.
11. Hill, C., Dozol, J. F., Lamare, V., Rouquette, H., Eymard, S., Tournois, B., Vicens, J., Asfari, Z., Bressot, C., Ungaro, R., Casnati, A.: Nuclear Waste Treatment by Means of Supported Liquid Membranes Containing Calix-Crown Compounds, *J. Incl. Phenom. Mol. Recogn. Chem.* **19** (1994), 399–408.
12. Asfari, Z., Bressot, C., Vicens, J., Hill, C., Dozol, J.-F., Rouquette, H., Eymard, S., Lamare, V., Tournois, B.: Doubly Crowned Calix[4]arenes in the 1,3-Alternate Conformation as Cesium-Selective Carriers in Supported Liquid Membranes. *Anal. Chem.* **67** (1995), 3133–3139.
13. Asfari, Z., Bressot, C., Vicens, J., Hill, C., Dozol, J. F., Rouquette, H., Eymard, S., Lamare, V., Tournois, B.: Cesium Removal From Nuclear Waste Water by Supported Liquid Membranes Containing Calix-bis-crown Compounds, in *Chemical Separations with Liquid Membranes*, ACS Symp. Ser., Vol. 642, American Chemical Society, Washington, 1996, pp. 376–390.
14. Asfari, Z., Wenger, S., Vicens, J.: Calixcrowns and Related Molecules, *Pure Appl. Chem.* **67**(7) (1995), 1037–1043.
15. Lamare, V., Bressot, C., Dozol, J. F., Vicens, J., Asfari, Z., Ungaro, R., and Casnati, A.: Selective Extraction of Cesium at Tracer Level Concentration from a Sodium Nitrate Solution With Calix-crowns. Molecular Modeling Study of the Cs^+/Na^+ Selectivity, *Sep. Sci. Technol.* **32** (1997), 175–191.
16. Dozol, J. F., Simon, N., Lamare, V., Rouquette, H., Eymard, S., Tournois, B., De Marc, D., Macias, R. M.: A Solution for Cesium Removal from High-Salinity Acidic or Alkaline Liquid Waste, The Crown Calix[4]arenes, *Sep. Sci. Technol.* **34** (1999), 877–909.
17. Moyer, B. A., Bonnesen, P. V., Sachleben, R. A., Presley, D. J.: "Solvent and Process for Extracting Cesium from Alkaline Waste Solutions," U. S. Patent 6,174,503, Jan. 16, 2001.
18. Bonnesen, P. V., Moyer, B. A., Sachleben, R. A.: "Fluoro-alcohol Phase Modifiers and Process for Cesium Solvent Extraction." U. S. Patent 6,566,561, May 20, 2003.
19. Research Needs for High-Level Waste Stored in Tanks and Bins at U.S. Department of Energy Sites, National Research Council, National Academy Press, 2001.
20. *Nuclear Wastes, Technologies for Separations and Transmutation*, National Research Council, National Academy Press, Washington, D.C., 1996, pp. 192, 193.
21. Linking Legacies, USDOE Report DOE/EM-0319, 1997.

22. U.S. Department of Energy Record of Decision, Savannah River Site Salt Processing Alternatives, Federal Register, Vol. 66, No. 201, pp. 52752–52756, October 17, 2001.
23. Mow Lee, L. and Kilpatrik, L. L.: A Precipitation Process for Supernatant Decontamination, Report DP-1636, E. I. DuPont de Nemours & Co., Savannah River Laboratory, Aiken, SC, 1982.
24. Walker, D. D., Barnes, M. J., Crawford, C. L., Peterson, R. A., Swingle, R. F., and Fink, S. D.: In-Tank Precipitation with Tetraphenylborate, Recent Progress and Research Results, in *Science and Technology for Disposal of Radioactive Tank Wastes*, Schultz, W. W., and Lombardo, N. J. (Eds.), Plenum Press, New York, 1998, pp. 219–230.
25. Bunker, B., Virden, J., Kuhn, B., and Quinn, R.: Nuclear Materials, Radioactive Tank Wastes, in *Encyclopedia of Energy Technology and the Environment*, John Wiley & Sons, Inc., New York, 1995, pp. 2023–2032.
26. Hunter, J. R.: in Proceedings of the First Hanford Workshop on Separation Science and Technology, Richland, WA, July 23–25, 1991, Report PNL-SA-21775, Pacific Northwest National Laboratory, Richland, WA, U.S.A., May, 1993, pp. I.5–I.10.
27. Gephart, R. E. and Lundgren, R. E.: *Hanford Tank Clean up: A Guide to Understanding the Technical Issues*, Report PNL-10773, Pacific Northwest National Laboratory, Richland, WA, 1995.
28. Delmau, L. H., Lefranc, T. J., Bonnesen, P. V., Bryan, J. C., Presley, D. J., and Moyer, B. A.: Fundamental Studies Regarding Synergism Between Calix[4]arene-bis(*tert*-octylbenzo-crown-6) and Alcohol Modifiers in the Solvent Extraction of Cesium Nitrate, *Solvent Extr. Ion Exch.*, in press.
29. Delmau, L. H., Bonnesen, P. V., Herlinger, A. W., and Chiarizia, R.: Aggregation Behavior of Solvent Modifiers for the Extraction of Cesium from Caustic Media, *Solvent Extr. Ion Exch.*, in press.
30. Engle, N. L., Bonnesen, P. V., Tomkins, B. A., Haverlock, T. J., and Moyer, B. A.: Synthesis and Properties of Calix[4]arene-bis[4-(2-ethylhexyl)benzo-crown-6], a Cesium Extractant with Improved Solubility, *Solvent Extr. Ion Exch.* **22**(4) (2004), 611–636.
31. Delmau, L. H., Bonnesen, P. V., and Moyer, B. A.: A Solution to Stripping Problems Caused by Organophilic Anion Impurities in Crown-Ether Based Solvent Extraction Systems, A Case Study of Cesium Removal from Radioactive Wastes, *Hydrometallurgy* **72**(1,2) (2004), 9–19.
32. Norato, M. A., Beasley, M. H., Campbell, S. G., Coleman, A. D., Geeting, M. W., Guthrie, J. W., Kennell, C. W., Pierce, R. A., Ryberg, R. C., Walker, D. D., Law, J. D., and Todd, T. A.: Demonstration of the Caustic-side Solvent Extraction Process for the Removal of Cs-137 from Savannah River Site High Level Waste, *Sep. Sci. Technol.* **38**(12,13) (2003), 2647–2666.
33. White, T. L., Peterson, R. A., Wilmarth, W. R., Norato, M. A., Crump, S. L., and Delmau, L. H.: Stability Study of Cs Extraction Solvent, *Sep. Sci. Technol.* **38**(12,13) (2003), 2667–2683.
34. Wilmarth, W. R., Mills, J. T., Dukes, V. H., Beasley, M. C., Coleman, A. D., Diprete, C. C., and Diprete, D. P.: Caustic-side Solvent Extraction Batch Distribution Coefficient Measurements for Savannah River Site High-level Wastes, *Sep. Sci. Technol.* **38**(12,13) (2003), 2637–2645.
35. Leonard, R. A., Aase, S. B., Arafat, H. A., Conner, C., Chamberlain, D. B., Falkenberg, J. R., Regalbuto, M. C., and Vandegrift, G. F.: Experimental Verification of Caustic-side Solvent Extraction Removal of Cesium from Tank Waste, *Solvent Extr. Ion Exch.* **21**(4) (2003), 505–526.
36. Delmau, L. H., Baes, C. F., Jr., Bostick, D. A., Haverlock, T. J., and Moyer, B. A.: Solvent Extraction System Modeling Using the Program SXFIT, in *Leaching and Solution Purification,* Hydrometallurgy 2003—Fifth International Conference, Vol. 1, Proceedings of the 2003 International Symposium on Hydrometallurgy, Vancouver, Canada, August 24–27, 2003, Young, C. A., Alfantazi, A. M., Anderson, C. G., Dreisinger, D. B., Harris, B., and James, A. (Eds.), TMS (The Minerals, Metals, and Materials Society), Warrendale, PA, 2003, pp. 969–981.
37. Bonnesen, P. V., Delmau, L. H., Moyer, B. A., and Lumetta, G. J.: Development of Effective Solvent Modifiers for the Solvent Extraction of Cesium from Alkaline High-Level Tank Waste, *Solvent Extr. Ion Exch.* **21**(2) (2003), 141–170.
38. Bonnesen, P. V., Delmau, L. H., Haverlock, T. J., Levitskaia, T. G., Sachleben, R. A., Sloop, F. V., Jr., and Moyer, B. A.: Science to Applications, Development of the Caustic-Side Solvent Extraction Process for Cesium Removal from Savannah River Waste, Proc. 9th Biennial Internat. Conf. Nuclear and Hazardous Waste Management (Spectrum 2002), Aug. 4–8, 2002, Reno, Nevada.
39. Bryan, J. C., Chen, T., Levitskaia, T. G., Haverlock, T. J., Barnes, C. E., and Moyer, B. A.: Solvation of Calix[4]arene-bis-crown-6 Molecules, *J. Inclusion Phenom.* **42**(3–4) (2002), 241–245).
40. Duchemin, C. R., Engle, N. L., Bonnesen, P. V., Haverlock, T. J., Delmau, L. H., and Moyer, B. A.: Solvatochromic Solvent Polarity Measurements of Alcohol Solvent Modifiers and Correlation with Cesium Extraction Strength, *Solvent Extr. Ion Exch.* **19**(2001), 1037–1058.
41. Leonard, R. A., Conner, C., Liberatore, M. W., Sedlet, J., Aase, S. B., Vandegrift, G. F., Delmau, L. H., Bonnesen, P. V., and Moyer, B. A.: Development of a Solvent Extraction Process for Cesium Removal from SRS Tank Waste, *Sep. Sci. Technol*. **36**(5&6) (2001), 743–766.
42. Moyer, B. A., Bonnesen, P. V., Delmau, L. H., Haverlock, T. J., Sachleben, R. A., Leonard, R. A., Conner, C., and Lumetta, G. J.: Solvent Extraction of Tc and Cs from Alkaline Nitrate Wastes, in

Solvent Extraction for the 21st Century, M. Cox, M. Hidalgo, and M. Valiente (Eds.), Society of Chemical Industry, London, 2001, Proc. Internat. Solvent Extraction Conference (ISEC '99), Barcelona, Spain, July 11–16, 1999, Vol. 2, pp. 1365–1369.

43. Bonnesen, P. V., Delmau, L. H., Moyer, B. A., and Leonard, R. A.: A Robust Alkaline-side CSEX Solvent Suitable for Removing Cesium from Savannah River High Level Waste, *Solvent Extr. Ion Exch.* **18**(6) (2000), 1079–1107.

44. Bonnesen, P. V., Haverlock, T. J., Engle, N. L., Sachleben, R. A., and Moyer, B. A.: Development of Process Chemistry for the Removal of Cesium from Acidic Nuclear Waste by Calix[4]arene-Crown-6 Ethers, in *Calixarene Molecules for Separations*, G. J. Lumetta, R. D. Rogers, and A. S. Gopalan (Eds.), ACS Symposium Series, Vol. 757, American Chemical Society, Washington, D. C., 2000, Chap. 3, pp. 26–44.

45. Haverlock, T. J., Bonnesen, P. V., Sachleben, R. A., and Moyer, B. A.: Analysis of Equilibria in the Extraction of Cesium Nitrate by Calix[4]arene-bis(t-octylbenzo-crown-6) in 1,2-Dichloroethane, *J. Incl. Phenom. Mol. Recognit. Chem.* **36** (2000), 21–37.

46. Sachleben, R. A., Bonnesen, P. V., Descazeaud, T., Haverlock, T. J., Urvoas, A., and Moyer, B. A.: Surveying the Extraction of Cesium Nitrate by 1,3-Alternate Calix[4]arene Crown-6 Ethers in 1,2-Dichloroethane, *Solvent Extr. Ion Exch.* **17** (1999), 1445–1459.

47. Thuéry, P., Nierlich, M., Bryan, J. C., Lamare, V., Dozol, J.-F., Asfari, Z., and Vicens, J.:. Crown Ether Conformations in 1,3-Calix[4]arene Bis-crown Ethers, Crystal Structures of a Caesium Complex and Solvent Adducts and Molecular Dynamic Simulations, *J. Chem. Soc., Dalton Trans.* (1997) 4191–4202.

48. Haverlock, T. J., Bonnesen, P. V., Sachleben, R. A., and Moyer, B. A.: Applicability of a Calixarene-Crown Compound for the Removal of Cesium from Alkaline Tank Waste, *Radiochim. Acta.* **76** (1997), 103–108.

49. Delmau, L. H., Birdwell, J. F., Jr., Bonnesen, P. V., Foote, L. J., Haverlock, T. J., Klatt, L. N., Lee, D. D., Leonard, R. A., Levitskaia, T. G., Maskarinec, M. P., Moyer, B. A., Sloop, F. V. Sloop, Jr., and Tomkins, B. A.: Caustic-Side Solvent Extraction, Chemical and Physical Properties of the Optimized Solvent, Report ORNL/TM-2002/190, Oak Ridge National Laboratory, Oak Ridge, TN, October, 2002.

50. Moyer, B. A., Alexandratos, S. D., Bonnesen, P. V., Brown, G. M., Caton, J. E., Jr., Delmau, L. H., Duchemin, C. R., Haverlock, T. J., Levitskaia, T. G., Maskarinec, M. P., Sloop, F. V., Jr., and Stine, C. L.: Caustic-Side Solvent Extraction Chemical and Physical Properties, Progress in FY 2000 and FY 2001, Report ORNL/TM-2001/285, Oak Ridge National Laboratory, Oak Ridge, TN, February 2002.

51. Bonnesen, P. V., Delmau, L. H., Haverlock, T. J., and Moyer, B. A.: Alkaline-Side Extraction of Cesium from Savannah River Tank Waste Using a Calixarene-Crown Ether Extractant, Report ORNL/TM-13704, Oak Ridge National Laboratory, Oak Ridge, TN, December 1998.

52. Delmau, L. H., Haverlock, T. J., Sloop, F. V., Jr., and Moyer, B. A.: Caustic-Side Solvent Extraction, Prediction of Cesium Extraction from Actual Wastes and Actual Waste Simulants, Report ORNL/TM-2003/011, Oak Ridge National Laboratory, Oak Ridge, TN, February 2003.

53. Delmau, L. H., Bostick, D. A., Haverlock, T. J., and Moyer, B. A.: Caustic-Side Solvent Extraction, Extended Equilibrium Modeling of Cesium and Potassium Distribution Behavior, Report ORNL/TM-2002/116, Oak Ridge National Laboratory, Oak Ridge, TN, May 2002.

54. Delmau, L. H., Haverlock, T. J., Levitskaia, T. G., Sloop, F. V., Jr., and Moyer, B. A.: Caustic-Side Solvent Extraction Chemical and Physical Properties, Equilibrium Modeling of Distribution Behavior, Report ORNL/TM-2001/267, Oak Ridge National Laboratory, Oak Ridge, TN, December 2001.

55. Lee, D. D., Birdwell, J. F., Jr., Bonnesen, P. V., and Tomkins, B. A.: Density Changes in the Optimized CSSX Solvent System, Report ORNL/TM 2002/204, Oak Ridge National Laboratory, Oak Ridge, TN, October 2002.

56. Birdwell, J. F., Jr., and Anderson, K. K.: Evaluation of Mass Transfer Performance for Caustic-Side Solvent Extraction of Cesium in a Conventional 5-cm Centrifugal Contactor, Report ORNL/TM-2001/278, Oak Ridge National Laboratory, Oak Ridge, TN, January 2002.

57. Birdwell, J. F., Jr., Anderson, K. K., and Hobson, D. E.: Investigation of Emulsion Formation in Solvent Washing in the Caustic-Side Solvent Extraction (CSSX) Process, Report ORNL/TM-2002/126, Oak Ridge National Laboratory, Oak Ridge, TN, June 2002.

58. Birdwell, J. F., Jr., and Anderson, K. K.: Evaluation of 5-cm Centrifugal Contactor Hydraulic and Mass Transfer Performance for Caustic-Side Solvent Extraction of Cesium, Report ORNL/TM-2001/137, Oak Ridge National Laboratory, Oak Ridge, TN, August 2001.

59. Birdwell, J. F., Jr., and Cummings, R. L.: Irradiation Effects on Phase-Separation Performance Using a Centrifugal Contactor in a Caustic-Side Solvent Extraction Process, Report ORNL/TM-2001/91, Oak Ridge National Laboratory, Oak Ridge, TN, August 2001.

60. Spence, R. D., Delmau, L. H., Klatt, L. N., Sloop, F. V., Jr., Bonnesen, P. V., and Moyer, B. A.: Batch-Equilibrium Hot-Cell Tests of Caustic-Side Solvent Extraction (CSSX) with SRS Simulant Waste and

Internal ^{137}Cs Irradiation, Report ORNL/TM-2001/49, Oak Ridge National Laboratory, Oak Ridge, TN, September, 2001.

61. Harmon, H., Schlahta, S, Wester, D., Walker, J., Fink, S., and Thompson, M.: Tanks Focus Area Savannah River Site Salt Processing Project Research and Development Summary Report, Report TFA-0105, Pacific Northwest National Laboratory, Richland, WA, May 2001.
62. Miller, J. E., Brown, N. E., Krumhansl, J. L., Trudell, D. E., Anthony, R. G., and Philip, C. V.: Development and Properties of Cesium Selective Crystalline Silicotitanate (CST) Ion Exchangers for Radioactive Waste Applications, in *Science and Technology for Disposal of Radioactive Tank Wastes*, Schultz, W. W. and Lombardo, N. J. (Eds.), Plenum Press, New York, 1998, pp. 269–286.
63. Samanta, S. K. and Misra, B. M.: Ion Exchange Selectivity of a Resorcinol-Formaldehyde Polycondensate Resin for Cesium in Relation to Other Alkali Metal Ions, *Solvent Extr. Ion Exch,* **13**(3) (1995), 575–589.
64. Bibler, J. P., Wallace, R. M., and Bray, L. A.: Testing a New Cesium-Specific Ion Exchange Resin for Decontamination of Alkaline High-Activity Waste, Report WSRC-RP-89, Westinghouse Savannah River Co., Aiken, SC, 1989.
65. Samanta, S. K., Ramaswamy, M., and Misra, B. M.: Studies on Cesium Uptake by Phenolic Resins, *Sep. Sci. Technol.* **27**(2) (1992), 255–267.
66. Hassan, N. M., McCabe, D. J., King, W. D., Hamm, L. L., and Johnson, M. E.: Ion exchange removal of cesium from Hanford tank waste supernates with SuperLig (R) 644 resin, *J. Radioanal. Nucl. Chem.* **254**(1) (2002), 33–40.
67. Leonard, R. A., Conner, C., Liberatore, M. W., Sedlet, J., Aase, S. B., and Vandegrift, G. F.: Evaluation of an Alkaline-Side Solvent Extraction Process for Cesium Removal from SRS Tank Waste Using Laboratory-Scale Centrifugal Contactors, Report ANL-99/14, Argonne National Laboratory, Argonne, IL, August, 1999.
68. Birdwell, J. F., Jr., Counce, R. M., and Slater, C. O.: Conceptual Design of a Simplified Skid-Mounted Caustic-Side Solvent Extraction Process for Removal of Cesium from Savannah River Site High-Level Waste, Report ORNL/TM-2004/59, Oak Ridge National Laboratory, Oak Ridge, TN, May 2004.
69. Thuery, P., Nierlich, M., Bressot, C., Lamare, V., Dozol, J. F., Asfari, Z., and Vicens, J.: Crystal and molecular structures of binuclear caesium complexes with 1,3-calix[4]-bis-crown-6 and 1,3-calix[4]-bis-benzo-crown- 6, *J. Inclusion Phenom. Mol. Recognit. Chem.* **23**(4) (1995), 305–312.
70. Horwitz, E. P.: Solvent Extraction in the Treatment of Acidic High-Level Liquid Waste, Where Do We Stand? In *Metal Ion Separation and Preconcentration. Progress and Opportunities,* ACS Symposium Series, Vol. 716, Bond, A. H., Dietz, M. L., and Rogers, R. D. (Eds.), American Chemical Society, Washington, D.C., 1999, pp. 20–50.
71. *Chemical Pretreatment of Nuclear Waste for Disposal*, Schulz, W. W. and Horwitz, E. P. (Eds.), Plenum, New York, 1994.
72. *Light Water Reactor Nuclear Fuel Cycle*, Wymer, R. G. and Vondra, B. L. (Eds.), CRC, Boca Raton, 1981.
73. Musikas, C. and Schulz, W. W.: Solvent Extraction in Nuclear Science and Technology, in *Principles and Practices of Solvent Extraction*, Rydberg, J., Musikas, C., and Choppin, G. R. (Eds.), Marcel Dekker, New York, 1992, pp. 413–448.
74. Danesi, P. R.: Solvent Extraction in the Nuclear Industry, in *Developments in Solvent Extraction*, Alegret, S. (Ed.), Ellis Horwood, Chichester, 1988, pp. 188–214.
75. Nash, K. L., Barrans, R. E., Chiarizia, R., Dietz, M. L., Jensen, M. P., Rickert, P. G., Moyer, B. A., Bonnesen, P. V., Bryan, J. C., and Sachleben, R. A.: Fundamental Investigations of Separations Science for Radioactive Materials, *Solvent Extr. Ion Exch.* **18**(4) (2000), 605–631.
76. Todd, T. A., Law, J. D., Herbst, R. S., Lumetta, G. J., and Moyer, B. A.: Treatment of Radioactive Wastes Using Liquid-Liquid Extraction Technologies—Fears, Facts, and Issues, *Proc. Waste Management 00,* Tucson, AZ, March 2000.
77. Dietz, M. L., Horwitz. E. P., and Rogers, R. D.: (1995) Extraction of Strontium from Acidic Nitrate Media Using a Modified Purex Solvent, *Solvent Extr. Ion Exch.* 13, 1–17.
78. Leonard, R. A., Conner, C., Liberatore, M. W., Bonnesen, P. V., Presley, D. J., Moyer, B. A., and Lumetta, G. J.: Developing and Testing an Alkaline-Side Solvent Extraction Process for Technetium Separation from Tank Waste, *Sep. Sci. Technol.* **34**(6&7) (1999), 1043–1068.
79. Bonnesen, P. V., Moyer, B. A., Haverlock, T. J., Armstrong, V. S., and Sachleben, R. A.: Removal of Technetium from Alkaline Waste Media by a New Solvent Extraction Process, in *Emerging Technologies in Hazardous Waste Management VI*, D. W. Tedder and F. G. Pohland (Eds.), American Academy of Environmental Engineers, Annapolis, MD, 1996, pp. 245–262.
80. Moyer, B. A., Sachleben, R. A., and Bonnesen, P. V.: "Process for Extracting Technetium from Alkaline Solutions," U. S. Patent 5,443,731, August 22, 1995.
81. Leonard, R. A.: Design Rules for Solvent Extraction, *Solvent Extr. Ion Exch.* **17**(3) (1999), 597–612.
82. Bryan, J. C. and Bonnesen, P. V.: unpublished results.

83. Marcus, Y., Pross, E., and Hormadaly, J.: Anion Solvation Properties of Protic Solvents. 2. Salt Distribution Study, *J. Phys. Chem.* **84** (1980), 2708–2715.
84. Bonnesen, P. V., Sloop, F. V., Jr., and Engle, N. L.: Stability of the Caustic-Side Solvent Extraction (CSSX) Process Solvent, Effect of High Nitrite on Solvent Nitration, ORNL Report ORNL/TM-2002/115, Oak Ridge National Laboratory, Oak Ridge, TN, July 2002.
85. Delmau, L. H., Van Berkel, G. J., Bonnesen, P. V., and Moyer, B. A.: Improved Performance of the Alkaline-Side CSEX Process for Cesium Extraction from Alkaline High-Level Waste Obtained by Characterization of the Effect of Surfactant Impurities, Report ORNL/TM-1999/209, Oak Ridge National Laboratory, Oak Ridge, TN, 1999.
86. Norato, M. A., Cambell, S. G., Crowder, M. L., Geeting, M. W., Kessinger, G. F., Pierce, R. A., and Walker, D. D.: High-Level Waste Demonstration of the Caustic-Side Solvent Extraction Process with Optimized Solvent in the 2-cm Centrifugal Contactor Apparatus Using Tank 37H/44F Supernate, Report WSRC-TR-2002-00243, Rev. 0, Westinghouse Savannah River Company, Aiken, SC, November 1, 2002.
87. Norato, M. A., Fink, S. D., Fondeur, F. F., Kessinger, G. F., Pierce, R. A., and Walker, D. D.: Demonstration of Caustic-Side Solvent Extraction with Optimized Solvent in the 2-cm Centrifugal Contactor Apparatus Using Dissolved Salt Cake from Tank 37H, Report WSRC-TR-2002-00307, Westinghouse Savannah River Company, Aiken, SC, September 3, 2002.
88. Leonard, R. A., Aase, S. B., Arafat, H. A., Chamberlain, D. B., Conner, C., Regalbuto, M. C., and Vandegrift, G. F.: Multi-day Test of the Caustic-Side Solvent Extraction Flowsheet for Cesium Removal from a Simulated SRS Tank Waste, Report ANL-02/11, Argonne National Laboratory, Argonne, IL, January 15, 2002.
89. Pereira, C., Arafat, H. A., Falkenberg, J. R., Regalbuto, M. C., and Vandegrift, G. F.: Recovery of Entrained CSSX Solvent from Caustic Aqueous Raffinate Using Coalescers, Report ANL-02/34, Argonne National Laboratory, Argonne, IL, November 2002.
90. C. A. Langton: *PUREX Organic Solidification*, Report WSRC-TR-2002-00570, Rev. 0, Savannah River Technology Center, Aiken, SC, December, 2002.
91. Leonard, R. A., Chamberlain, D. B., and Conner, C.: Centrifugal Contactors for Laboratory-Scale Solvent Extraction Tests, *Sep. Sci. Technol.* **32**(1–4) (1997), 193–210.
92. Leonard, R. A., Bernstein, G. J., Ziegler, A. A., and Pelto, R. H.: Annular Centrifugal Contactors for Solvent Extraction, *Sep. Sci. Technol.* **15**(4) (1980), 925–943.
93. Wilmarth, W. R., Healy, D. P., Wheeler, D. J., Mills, J. T., Dukes, V. H., and D. P. DiPrete: *Caustic Side Solvent Extraction Batch Distribution Measurements for SRS High Level Waste Samples and Dissolved Saltcake,* Report WSRC-TR-2002-00336, Rev. 1, Westinghouse Savannah River Company, Aiken, SC, December 5, 2002.
94. Ritcey, G. M. and Ashbrook, A. W.: *Solvent Extraction, Principles and Applications to Process Metallurgy*, Elsevier, New York, 1984, Part I.
95. Leonard, R. A.: Solvent Characterization Using the Dispersion Number, *Sep. Sci. Technol.* **30**(7–9) (1995), 1103–1122.
96. Campbell, S. G., Geeting, M. W., Kennell, C. W., Law, J. D., Leonard, R. A., Norato, M. A., Pierce, R. A., Todd, T. A., Walker, D. D., and Wilmarth, W. R.: Demonstration of Caustic-Side Solvent Extraction with Savannah River Site-High-Level Waste, Report WSRC-TR-2001-00223, Rev. 0, Westinghouse Savannah River Company, Aiken, SC, April 19, 2001.

25.

TEXAPHYRIN CONJUGATES. PROGRESS TOWARDS SECOND GENERATION DIAGNOSTIC AND THERAPEUTIC AGENTS

WEN-HAO WEI[a], MARK E. FOUNTAIN[a], JONATHAN L. SESSLER[a], DARREN J. MAGDA[b], ZHONG WANG[b] AND RICHARD A. MILLER[b]

[a] *Department of Chemistry & Biochemistry, The University of Texas at Austin, 1 University Station, A5300 Austin, TX 78712-0165 U.S.A.*

[b] *Pharmacyclics Incorporated, 995 East Arques Avenue, Sunnyvale, CA 94085-4521 U.S.A.*

1. Introduction

Cancer, despite significant research efforts, remains a major health issue and the second leading cause of death, exceeded only by heart disease [1]. In both, cancer and heart disease, the ability to image diseased tissues, the use of localizing therapeutic agents, and the elaboration of therapeutic mechanisms has resulted in useful treatment options. The localization and fluorescence of porphyrins in tumors was seen as early as 1924 [2]. From this early foundation, a mixture of hematoporphyrin oligomers, (Photofrin®), was developed as a photodynamic therapy agent (PDT) for clinical treatment of tumors in the 1960's [3]. The medical usefulness of Photofrin® is unfortunately hampered by the high light absorption of biological tissues at the excitation wavelength, and a nearly unacceptable level of skin phototoxicity. As a consequence, a number of so-called second generation photosensitizers have received attention in recent years [4,5].

Texaphyrin (Fig. 1) was first reported in the late 1980's [6]. The texaphyrins are pentaaza Schiff-base expanded porphyrins that are known to form stable 1:1 complexes with many metal cations, including those of the trivalent lanthanide series. Peripheral ring substitution with alcohol and PEG-like functionalities in **1** and **2** endows the compounds with sufficient water solubility for clinical use [7,8].

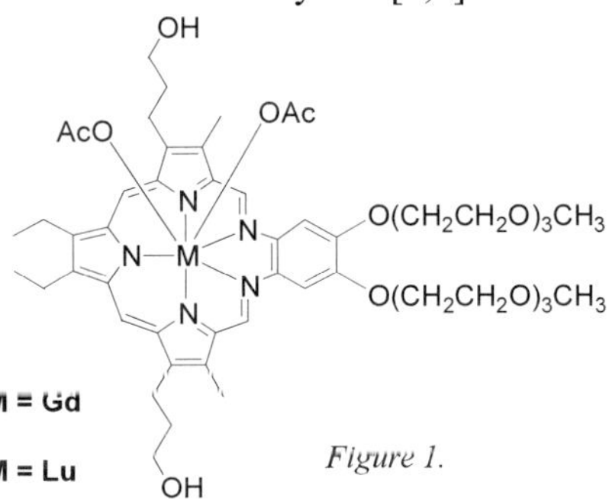

1 M = Gd

2 M = Lu

Figure 1.

Karsten Gloe (ed.), Macrocyclic Chemistry: Current Trends and Future Perspectives, 407–425.
© 2005 *Springer. Printed in the Netherlands.*

The gadolinium(III) complex, **1** (motexafin gadolinium; MGd; Xcytrin®), has been extensively studied as a potential MRI detectable anticancer agent [9-17]. Both clinical and preclinical studies have confirmed that this agent is taken up into, and retained in, tumors with high specificity [12-15]. Likewise, the lutetium(III) complex (motexafin lutetium, MLu; Antrin®; **2**) is currently being tested clinically as a photosensitizer for the PDT treatment of age-related macular degeneration, atherosclerotic plaque indications, peripheral vascular, coronary artery disease, and cancer [11,18-21].

Although structurally and spectrally similar, the differing redox and photophysical properties of the various lanthanide(III) texaphyrin complexes allow their use in such disparate areas as PDT, direct cancer treatment, and both X-ray and chemotherapy enhancement protocols. Both compounds **1** and **2** generate reactive oxygen species (ROS), albeit via mechanistically distinct pathways (vide infra). The ROS are thought to be responsible, at least in part, for the observed biological activity of MGd and MLu [22].

Promising as these agents are, efforts are underway to develop systems with improved therapeutic efficacy. Towards this end several conjugates with known chemotherapeutic agents are being prepared. To date, conjugates containing L-buthionine-S,R-sulfoximine (glutathione synthesis inhibitor), nitroimidazole (hypoxic radiation sensitizer), doxorubicin (DNA synthesis inhibitor), cisplatin (DNA modifier), and methotrexate (folate inhibitor), have been prepared [23]. Related systems, containing covalently linked fluorophores have also been synthesized. This chapter provides an update concerning these ongoing synthetic efforts and places them in the context of other texaphyrin-related studies.

2. Chemistry

2.1 INITIAL SYNTHETIC EFFORTS

The preparation of the basic texaphyrin ring starts from the diformyl tripyrrane fragment **5** (Scheme 1), an intermediate that in turn is prepared in three steps from the simple monomeric pyrroles **3** and **4**.

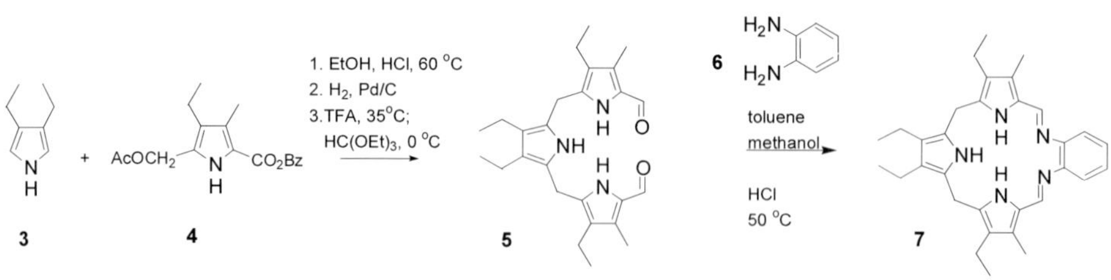

Scheme 1. Synthesis of sp^3 texaphyrin.

An acid catalyzed double Schiff-base condensation of **5** with *o*-phenylenediamine, **6**, then gives the cyclic non-aromatic sp^3 texaphyrin ring **7**, a macrocyclic product that can be viewed as being an "expanded porphyrinogen" [24]. The sp^3 form can be oxidized with ferrocenium ion to give the corresponding aromatic derivative, namely the metal-free sp^2 form **8** (Scheme 2) [25].

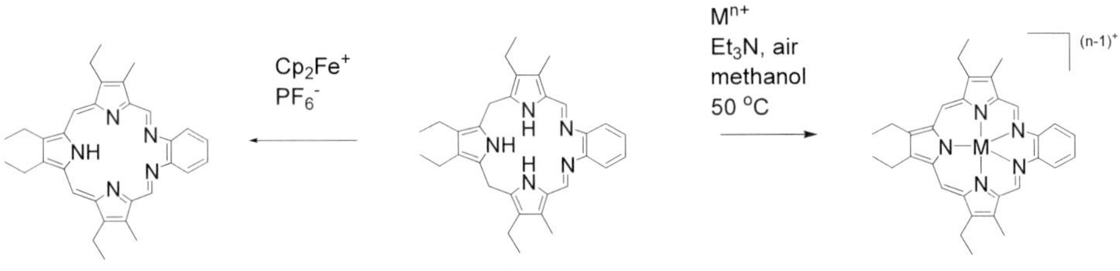

8 sp^2-texaphyrin **7** sp^3-texaphyrin **9** sp^2-metallotexaphyrin

Scheme 2. Synthesis of sp^2-texaphyrin and sp^2-metallotexaphyrin.

Alternatively, reaction of **7** with the appropriate metal salt in the presence of base and air gives the stable aromatic (sp^2) texaphyrin metal complex **9** in a one-step oxidation and metalation procedure [6,7,20,26-28].

The cadmium(II) complex corresponding to **9** (M = Cd; n = 2) was the first texaphyrin made [6]. This aromatic expanded porphyrin was found to differ substantially from various porphyrin complexes and it was noted that its spectral and photophysical properties were such that it might prove useful as a PDT agent. However, it was also appreciated that the poor aqueous solubility and inherent toxicity of this particular metal complex would likely preclude its use in vivo [29-31]. Nonetheless, the coordination chemistry of texaphyrins such as **9** was soon generalized to allow for the coordination of late first row transition metal (Mn(II), Co(II), Ni(II), Zn (II), Fe(III)) and trivalent lanthanide cations [26]. This, in turn, opened up several possibilities for rational drug development. For instance, the Mn(II) texaphyrin complex was found to act as a peroxynitrite decomposition catalyst [32] and is being studied currently for possible use in treating amyotrophic lateral sclerosis. This work, which is outside the scope of this review, has recently been summarized by Crow [33].

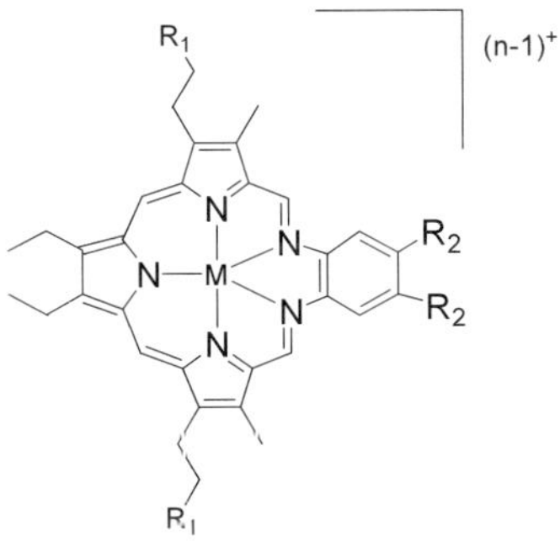

9 $R_1 = R_2 = H$
10 $R_1 = CH_2OH$,
$R_2 = (OCH_2CH_2)_3OCH_3$

Figure 2. Generalized metallo-texaphyrin structure.

Of the lanthanide complexes, the gadolinium(III) and lutetium(III) complexes (abbreviated in a general sense as Gd-Tex and Lu-Tex, respectively) were immediately recognized as being the most interesting. The interest in the Gd(III) species reflects the fact that gadolinium(III) is the single most paramagnetic monoatomic cation known. The realization that this cation forms a stable texaphyrin complex thus led to the consideration that it might display MRI signal enhancing properties [26,28,34]. By contrast, Lu(III) is diamagnetic and this fact, coupled with the ability of texaphyrins to absorb strongly in the >700 nm spectral region, led to the proposal that Lu(III)-containing texaphyrin complexes might be useful as PDT photosensitizers [30,35].

2.2 SYNTHESIS OF CURRENT CLINICAL CANDIDATES

Although the spectral and redox properties of this initial set of lanthanide(III) texaphyrin complexes appeared suitable for various medical uses, the alkyl substituents present in the original series (cf. generalized structure **9**, Fig. 2) made them largely insoluble in water. The alkyl substitution on the diformyl-tripyrrane was thus modified to give the bis-hydroxypropyl derivative. The phenyl ring was also modified to allow for PEG-like substitution. This afforded a new series of texaphyrins with generalized structure **10**, as shown in Figure 2. Fortunately, within this series the Gd(III) and Lu(III) complexes, structures **1** and **2** shown in Figure 1, proved sufficiently water soluble for use in vivo. It has thus been these species that have been carried forward for preclinical and clinical studies [26,35-37].

3. Summary of Current Clinical Studies

Motexafin gadolinium, MGd, utilized both as a single agent and in combination therapy approaches is either currently in or has been involved in clinical trials for several indications. For the lead indication, involving the use of MGd as an adjuvant in the radiation therapeutic treatment of metastatic brain cancers, a first Phase III trial has been completed and a definitive, Phase III international trial (termed “SMART”) is ongoing in patients with brain metastases from non-small cell lung cancer [15,38]. Additionally, several Phase I and II trials, involving the study of MGd as a single agent in the treatment of low-grade lymphoma, multiple myeloma [39], chronic lymphocytic leukemia and renal cell cancer are ongoing. Data derived from these trials provide support for the notion that MGd is well tolerated when administered at the established clinical dose levels (e.g., 10 days at 5 mg/kg per day in the case of the Phase III trial) and serve to highlight the fact that good uptake and retention in tumors is achieved as judged by MRI scans (vide infra) [13,40].

Motexafin lutetium is also in clinical trials, having completed several Phase I and II studies. The completed trials are for the photodynamic treatment of recurrent breast cancer [41,42], light-based treatment of choroidal neovascularization, [43] and the photoangioplastic reduction of atherosclerotic plaque in peripheral [19,44] and coronary arterial disease [20,45]. On the basis of these studies, MLu (Antrin® Phototherapy) is currently being developed for the treatment of atherosclerotic plaque. Additionally, the National Cancer Institute is testing MLu for the PDT based treatment of prostate [46,47] and cervical cancers.

4. Biolocalization Studies

One of the salient features of MGd is that it contains a highly paramagnetic Gd(III) center. This has provided the added benefit in clinical work of allowing neoplastic lesions to be imaged by MRI [9,28,48]. These studies have also served to establish inter alia that MGd localizes well to tumors. In the case of MLu, which displays stronger fluorescence than MGd, localization in tumors and abnormal cells has repeatedly been demonstrated in experiments. The mechanisms by which this texaphyrin complex in particular and porphyrins in general, are selectively concentrated in tumors and abnormal tissues is still not completely understood. Nonetheless, it is appreciated that macroscopic tumor targeting as well as sub-cellular localization plays a significant role in the activity of (at least) PDT agents [49-53]. Moreover, the specifics of porphyrin localization at the sub-cellular (organelle) level have been shown to be correlated with the charge, polarity, and lipophilicity of the compound [54,55]. The main targets for many porphyrin type agents are the mitochondria and lysosomes, organelles where the production of singlet oxygen and other ROS leads to oxidative stress, and, generally, apoptosis mediated cell death. UV-Vis, Raman, and fluorescence spectroscopy have allowed in situ detection of texaphyrins in both cancer cells and various sub-cellular organelles. Intact cell studies, using interferometric Fourier fluorescence microscopy and high sensitivity charge coupled device (CCD) cameras, have demonstrated a gradual uptake of MGd, MLu, and other texaphyrins. Localization is primarily in the lysosomes and endoplasmic reticulum, with lesser distribution to other organelles such as the mitochondria [56-60]. On a more macroscopic level, the uptake of MLu in atherosclerotic plaque has been demonstrated in vivo [21,61-64].

5. Redox Chemistry and Therapeutic Mechanism

The structurally similar, but magnetically distinct, lanthanide(III) texaphyrin complexes, MLu and MGd both generate reactive oxygen species (ROS), albeit via different mechanisms. Photoirradiation of MLu causes excitation from the singlet ground state to the triplet state (Fig. 3).

Photodynamic Therapy

$$\text{Sens. }(S_o) \xrightarrow[\text{2. ISC}]{\text{1. }h\nu} \text{Sens. }(T_1)$$

$$\text{Sens. }(T_1) + {}^3O_2 \longrightarrow \text{Sens. }(S_o) + {}^1O_2$$

$$\text{Biomolecule} + {}^1O_2 \longrightarrow \text{Products}$$

Figure 3. The PDT singlet oxygen toxicity mechanism.

In aerobic environments, relaxation of this latter excited state can be coupled to up-conversion of normal triplet oxygen to singlet oxygen. The high reactivity of singlet oxygen with biomolecules is thought to account for much of the effect, as well as toxicity, displayed by this and other photosensitizers under conditions of photoexcitation. Thus, while the mode of action of MLu is not new, it does have several potential advantages when considered as a photosensitizer for treating vascular disease. These include, 1) its long wavelength absorption (which reduces interference from endogenous chromophores such as heme) and 2) its ability to localize well in plaque (a key prerequisite to selectivity). The high quantum yield for singlet oxygen production is also considered a positive attribute [65]. Mechanistically, X-ray therapy differs from PDT in that the irradiation of water generates hydroxyl radical among other products. Under normal conditions of clinical application, it is this species (OH·) that is considered the active cytotoxin. However, radiation also leads to oxidative stress, making irradiated cells more susceptible to agents that can produce other reactive oxygen species (ROS) or block the synthesis of so-called reducing metabolites like ascorbate, glutathione, or NADP(H). Such considerations are important in the case of texaphyrins. Indeed, MGd is thought to mediate its effect, at least in part, as the result of its ability to produce ROS, while concurrently depleting electron rich species, such as ascorbate or glutathione that can protect against the cytotoxic effects of ionizing radiation. The actual mechanism of MGd radiation sensitization is likely to be quite complex, and may well involve more than one apoptosis triggering pathway. Nonetheless, as implied above and detailed in Figure 4, one postulated mechanism for MGd sensitization is that electron transfer from a reducing metabolite (e.g., ascorbate) serves to produce a one-electron reduced MGd radical cation ($MGd^{+\cdot}$) which then reacts with molecular oxygen, to give superoxide as the initial product. Rapid disproportionation in conjunction with chemically or biologically enhanced conversion then gives hydrogen peroxide, a known apoptosis triggering agent [66]. The net result is additional cell damage beyond that produced by the hydroxyl radical arising from ionizing radiation [10,67,68]. Because the aromatic MGd complex (a dication) is regenerated in the course of the electron transfer from $MGd^{+\cdot}$ to O_2, this reaction sequence can play a catalytic role in the overall process of electron transfer from a biological reductant (e.g., ascorbate) to oxygen. Since the net result is the depletion of species, such as ascorbate and glutathione, that can protect against the effects of ionizing radiation (as well as the production of ROS) this process is potentially quite potent as a drug development paradigm. It has thus come to be termed "futile redox cycling" (Fig. 4), [66,69].

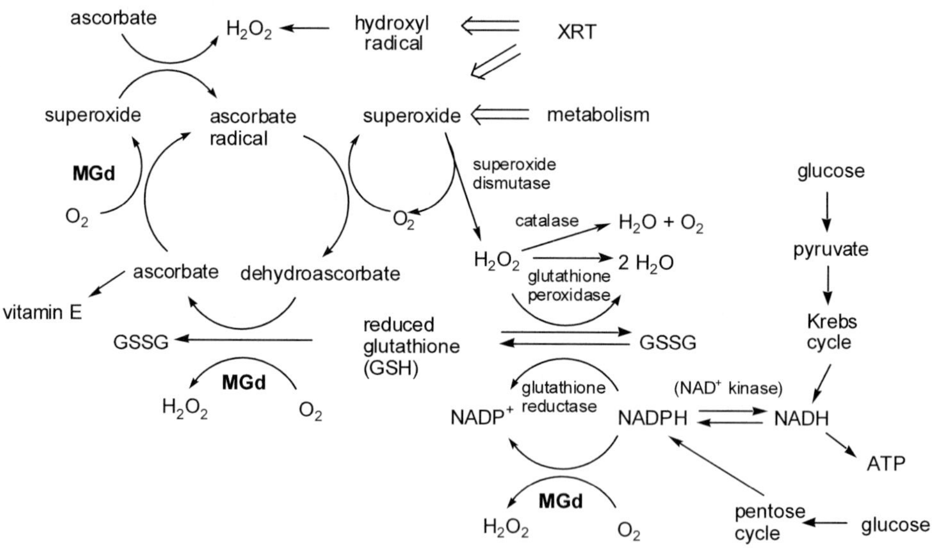

Figure 4. Postulated schematic showing the mechanism for the metabolic effects of MGd under X-ray sensitization conditions.

6. Synthesis of Conjugates

To improve the therapeutic efficacy of MGd, several conjugates are being developed. These are elaborated texaphyrin derivatives wherein the texaphyrin core is linked to various "active species," such as electron deficient species, hypoxic radiosensitizers, fluorophores, or therapeutic agents with alternate modes of action. The result of this conjugation process could be MGd analogues that either enhance the futile redox coupling process or which serve to deliver known cytotoxins to cancerous sites more effectively. In the case of the fluorophore linked systems, the result could be agents with improved diagnostic potential.

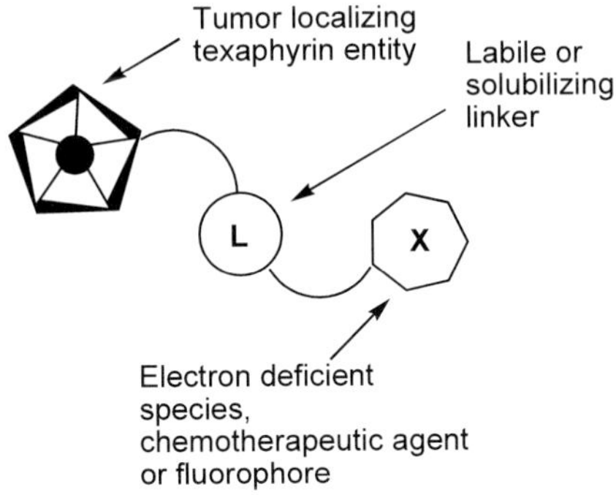

Figure 5. Generalized approach to conjugate construction.

As illustrated schematically in Figure 5, conjugation could be effected either through direct covalent attachment, or through solubilizing linkers. In the case of an appended cytotoxic agent it is possible that the conjugation process could end up interfering with activity (by, e.g., targeting an active species to a different sub-cellular site than is required for action). Because of this, efforts are being made to develop conjugates that incorporate so-called fusible linkers, i.e., tethers that will undergo cleavage under biological conditions [23,70].

6.1 CONJUGATES BASED ON REDOX ACTIVE GROUPS

As noted above, the X-ray sensitization of cells using MGd has been investigated extensively and appears to operate at least in part in an oxygen dependent manner, with cytotoxic activity that is substantially enhanced upon the addition of ascorbate to the media [69]. The cytotoxic activity of MGd can also be enhanced with L-buthionine-S,R-sulfoximine (BSO), an irreversible inhibitor of an enzyme involved in glutathione biosynthesis (γ-glutamylcysteine synthetase), that serves to suppress the synthesis of ROS-protective glutathione. Presumably, this activity is responsible for the in vitro radiation enhancing effect of BSO [66,71].

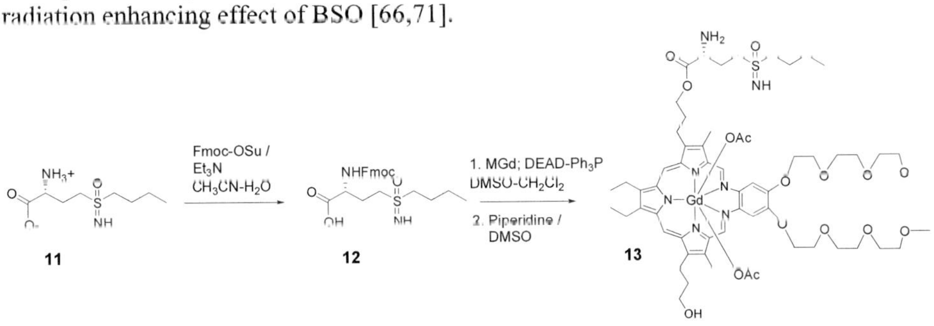

Scheme 3. Synthesis of MGd-BSO conjugate. DEAD = diethyl azodicarboxylate; Fmoc = 9-fluorenylmethoxycarbonyl.

This potential mechanistic synergy made BSO an early candidate for conjugation to MGd. Thus, FMOC protection of the amine group of BSO **11** was carried out to give the corresponding carboxylic acid **12**. Subsequent Mitsunobu coupling to MGd and deprotection afforded the desired conjugate **13** in 35% yield (Scheme 3).

Nitroimidazoles have long been recognized for their potential as radiation sensitizers [72]. These species are easy to reduce and have been used off-label as adjuvants for radiation therapy; however, their lack of tumor localizing specificity and their high toxicity has tended to limit their widespread clinical use [72-76]. Given the enhancement a combined texaphyrin nitroimidazole system might be expected to display, complex **14,** a texaphyrin nitroimidazole conjugate, was targeted for synthesis. It was prepared in good yield by conjugating 2-nitroimidazole **15** to MGd **1** under Mitsunobu conditions (Scheme 4).

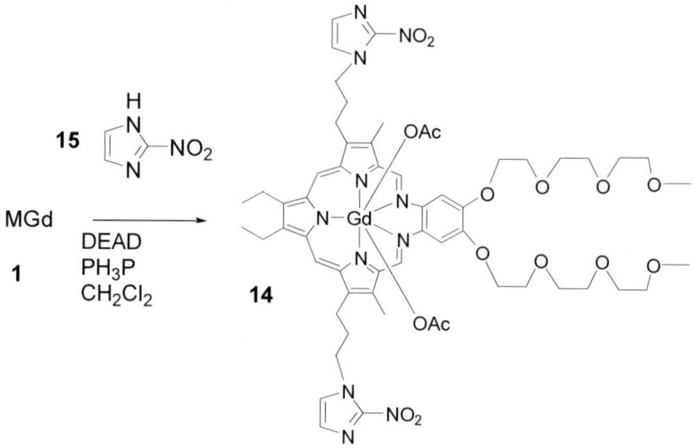

Scheme 4. Synthesis of bis-nitroimidazole-MGd conjugate 14.

6.2 CONJUGATES BASED ON KNOWN CHEMOTHERAPEUTIC AGENTS

Doxorubicin (Adriamycin®) is an important chemotherapeutic agent. However, it suffers from dose-limiting cardiomyopathy. It thus represents an attractive building block for the preparation of an MGd conjugate. In particular, the easy-to-reduce quinone functionality could provide an additional site for electron capture (from, e.g., NAD(P)H), which in turn could aid in the production of ROS [77-87].

Consistent with such a proposal is the finding that administering doxorubicin in conjunction with MGd, **1**, leads to an enhanced effect in vivo. Thus, an effort was made to attach this "active agent" to a Gd(III) texaphyrin [88]. The resulting synthesis is shown in Scheme 5. It involves protection of one alcohol of MGd as the acetate ester. The other free hydroxyl is then converted to the activated succimidyl carbonate (giving intermediate **16**), and then coupled to doxorubicin **17** to give the conjugate **18.**

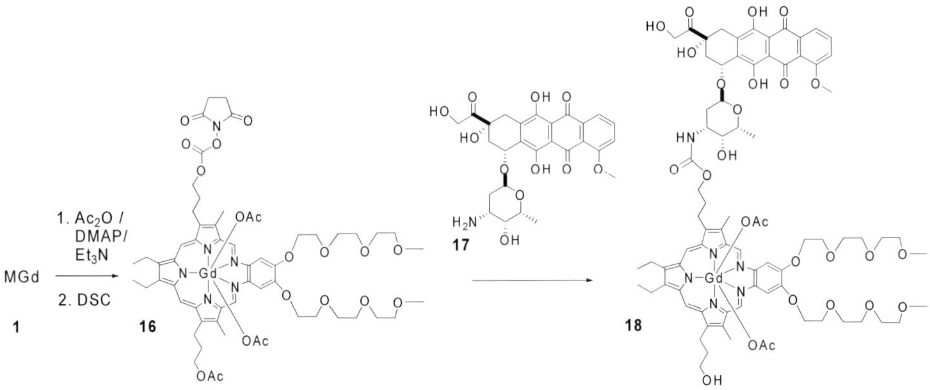

Scheme 5. Synthesis of MGd-doxirubicin conjugate **18**. Here, DMAP = dimethylaminopyridine; DSC = N,N'-disuccinimidyl carbonate.

Cisplatin, is one of the most widely used of all cancer chemotherapeutics. It displays high activity against a variety of tumors. However, it is also plagued by toxicity, and displays little if any in the way of inherent tumor localization [89,90]. Because of these limitations, cisplatin represents an attractive target for conjugation. A significant advantage of cisplatin-MGd conjugates would be a possible reduction in the kidney toxicity normally associated with cisplatin [91,92]. This reduction in toxicity, it is proposed, could arise from an ability to lower the administered dose levels by using a cancer targeting conjugate, or from possible changes in clearance pathway that arise as a result of being tethered to the texaphyrin core. There is another potential benefit associated with a putative texaphyrin cisplatin conjugate. This benefit has its origins in the fact that cisplatin, after cellular uptake, undergoes hydrolysis of the reactive chlorides, to give the therapeutically active "diaquo" species. These latter species react with DNA, forming G-G crosslinks. In a conjugate where cisplatin is bound to a MGd core, this targeting could allow for increased levels of texaphyrin derived ROS within or near the DNA backbone, thus potentially enhancing DNA damage and apoptosis triggering [91,93-95].

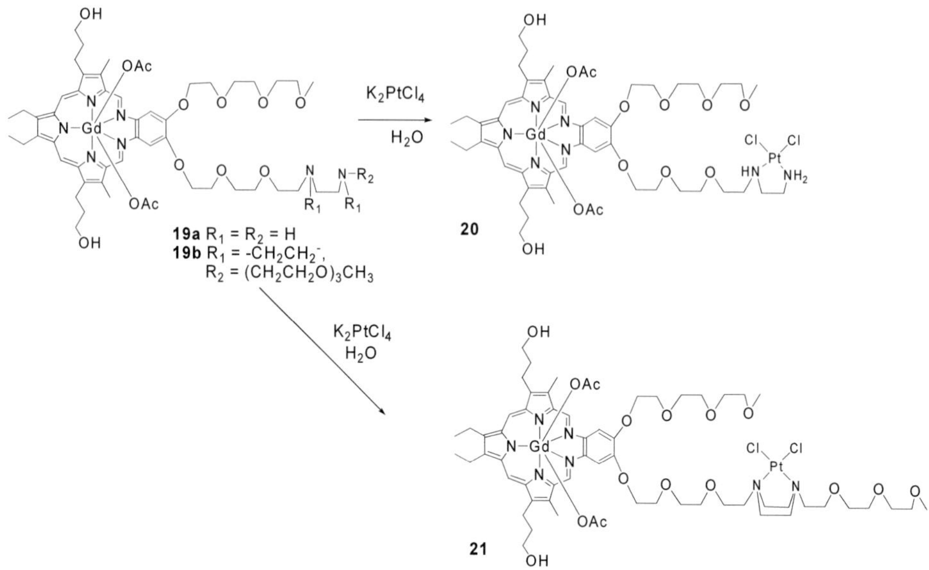

Scheme 6. Synthesis of MGd-cisplatin type conjugate.

To date, two cisplatin-type texaphyrin conjugates have been prepared (Scheme 6). The first of these, **20**, was synthesized by treating a Gd(III) texaphyrin diamine conjugate **19a** with K_2PtCl_4 under aqueous conditions. Unfortunately, the resulting product, **20**, proved too insoluble in both aqueous and non-aqueous media to allow for characterization by methods other than UV-Vis spectroscopy and mass spectrometry. The second conjugate, **19b**, incorporates increased PEG substitution which was included in the design in an effort to improve the solubility characteristics. Sadly, like its congener **20**, complex **21** proved too insoluble to permit analysis by any means other than UV-Vis spectroscopy and mass spectrometry. Current efforts are therefore focused on producing platinum-containing texaphyrin conjugates with improved aqueous solubility.

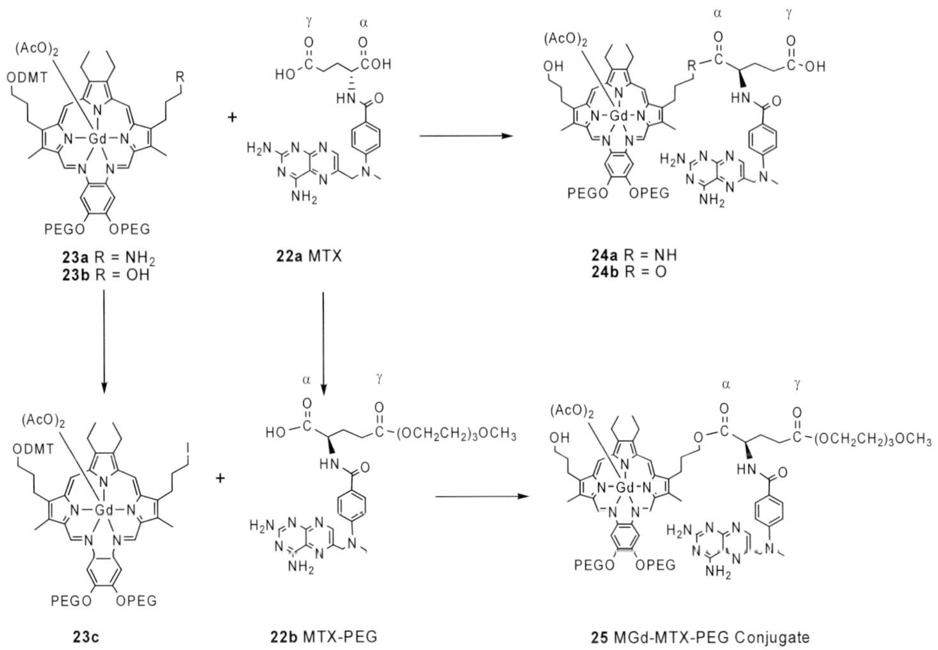

Scheme 7. Synthesis of MGd-MTX conjugates.

Recently, a set of water soluble texaphyrin conjugates that rely on the use of methotrexate (MTX, **22**), rather than cisplatin, as the active chemotherapeutic group has been prepared (Scheme 7) [23]. The MTX conjugates may be more suitable for use in vitro and in vivo. This is because methotrexate bears diacid functionality and is thus inherently water soluble; however, therapeutically the α-carboxyl group must be free in order to retain activity. Mechanistically, MTX is an anti-folate type cell cycle inhibitor, poorly taken up by cells through the folate transport system [96,97]. Tumors are known to rapidly develop resistance to MTX. A significant resistance mechanism in multi-drug resistant tumor cells is lowered MTX uptake. Conjugation to MGd may provide a means of overcoming this type of resistance. To date, three MTX-MGd conjugates, **24a**, **24b**, and **25** have been prepared. In the case of the amide **24a**, the first key steps involved protection of one of the hydroxyls present in MGd as the dimethoxytrityl (DMT) ether followed by conversion of the other hydroxyl to the corresponding phthalimide by a Mitsonobu reaction. Methylamine deprotection then gave the amine **23a**. Finally, direct DCC mediated coupling of this latter intermediate to MTX, followed by deprotection of the DMT ether, gave the amide-linked product **24a**. In the case of the ester-linked conjugate **24b**, protection of one hydroxyl group, also as the dimethoxytrityl (DMT) ether, gave the alcohol **23b**. Direct DCC mediated coupling to MTX gave the product **24b** after DMT deprotection. The PEG solubilized conjugate **25** was also synthesized from the PEG mono-ester of MTX **22b**. In this case, the DMT protected MGd alcohol **23b** was converted to the corresponding iodide **23c** through a Mitsonobu reaction. Direct coupling to the MTX-PEG monoester and deprotection then gave the solubilized conjugate **25**. In all cases, a mixture of the α and γ isomers was obtained in the coupling steps and in the MTX esterification. This mixture was found to favor the γ reaction products slightly.

In preliminary studies, A549 lung cancer cells were treated with the ester conjugate **24b** or the amide conjugate **24a** for 8 or 24 h, and analyzed for cellular proliferation after 3 days. Non-conjugated methotrexate, MGd, and the combination of these were also tested for comparison. Only MTX-Gd-Tx ester **24b** led to significant inhibition of cellular proliferation after 8 h treatment. After 24 h treatment, free methotrexate was more active than the ester conjugate at the lower concentrations tested. Amide conjugate **24a** displayed modest activity after 24 h treatment. We interpret these findings as being consistent with an altered mode of cellular uptake of methotrexate by conjugation to the texaphyrin moiety, and with the greater stability of the amide relative to the ester linkage in the biological milieu.

These results are also consistent with the finding that the ester product **24b** was seen to hydrolyze slowly in aqueous solution, whereas the amide conjugate **24a** was stable. However these results could also reflect an alternate uptake mechanism. Support for this latter explanation comes from the finding that alkylation of the γ site of MTX is thought to lead to enhanced uptake [98]. Previously, tumor localizing monoclonal antibody conjugates containing methotrexate were prepared by Umemoto, containing both fusible and non-fusible linkers. In mouse mammary tumor cell studies, the fusible conjugate had an IC_{50} of 0.043 μM, compared to 0.021 μM for MTX. Compared to the fusible conjugate, the non-cleavable conjugate showed similar uptake patterns, but a 100-fold reduction in the potency of the drug over the fusible conjugate. This result from the literature coincides with those inferred from our own studies of **23b** and **24b** and thus supports the conclusion that the choice of linker is an important one [99,100].

6.3 CONJUGATES BASED ON FLUOROPHORES

The use of fluorescence imaging techniques to show the location of texaphyrins in abnormal cells (tumors, lesions, plaques) and in animal models is a potentially attractive means of obtaining detailed localization information. Metallotexaphyrins, although displaying a low fluorescence quantum yield, and although subject to photobleaching under strong illumination, may nonetheless be visualized in the 750 nm spectral region using CCD cameras. Use of this technique allows differentiation of the tumor from surrounding tissues, as has been demonstrated in mice using both MGd and MLu. MLu was also used to effect selective visualization of early stage lesions in a hamster cheek pouch model. In these studies the fluorescence centered at 758 nm was monitored [58-60].

The availability of additional fluorescent texaphyrin marker systems would allow these kinds of studies to be extended and expanded. This would be particularly useful in the case of the gadolinium(III) derivatives, since these paramagnetic species are inherently less fluorescent than MLu and related diamagnetic texaphyrin complexes. The selection of fluorophores with higher quantum efficiency and lower photobleaching (than MGd or MLu) would allow localization of the conjugate using techniques such as fluorescence and confocal laser microscopy (CFM). It is possible that the development of these "dual fluorophore" conjugates would permit the study of photoelectronic/redox phenomena using techniques such as fluorescence resonance electron transfer (FRET), sub-cellular movement using fluorescence recovery after photobleaching (FRAP), and allow dual wavelength imaging to be used to define localization. Confirmatory studies using MRI would also be possible in the case of the Gd(III) systems.

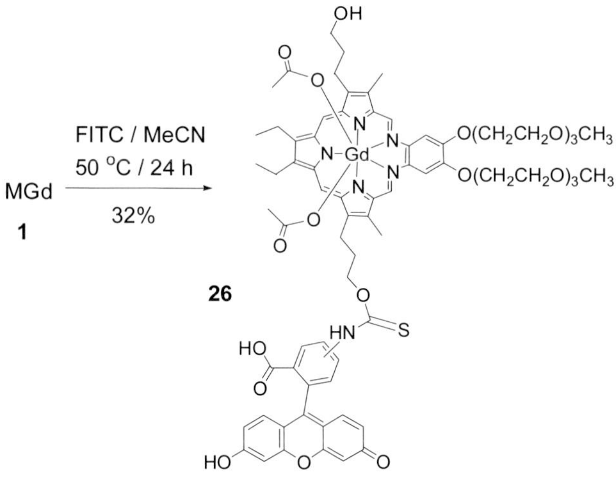

Scheme 8. Synthesis of MGd-FITC conjugate.

Additionally, fluorophores are available that are cationic and lipophilic, raising the possibility of sub-cellular targeting of the conjugates and/or detailed studies of structure-function relationships.

With the above considerations in mind, several MGd fluorophores were prepared. The fluorophore FITC could be coupled to MGd directly by heating in MeCN. This provided conjugate **26** (Scheme 8) in 32% yield as a mixture of the 5(6) isomers. The resulting conjugate displays low solubility in aqueous media, which may limit its usefulness. Accordingly, the fluorescein and rhodamine 6G conjugates **27a** and **27b**, respectively, were targeted for synthesis (Scheme 9). They were prepared from the solubilized-dinitrobenzene fluorophores **28a** and **28b**, respectively. After reduction, cyclization with dialdehyde **29** to the sp^3 texaphyrin and oxidative metal insertion, the respective fluorophore conjugates were obtained.

All three MGd-fluorophores could be observed in imaging studies with Jurkat-T lymphoma cells as a result of the fluorescence produced by the marker fluorophores (Fig. 6). The cells displayed punctate localization predominantly, as would be expected for retention that is primarily in cellular vesicles or lysosomes. Importantly, fluorescence from the MGd core could not be seen under standard CFM conditions. This means that the paramagnetic Gd(III) center does not quench the fluorescence of the appended fluorophore. It also highlights the fact that these new conjugates are potentially far more effective as detectable targeting agents then MGd alone.

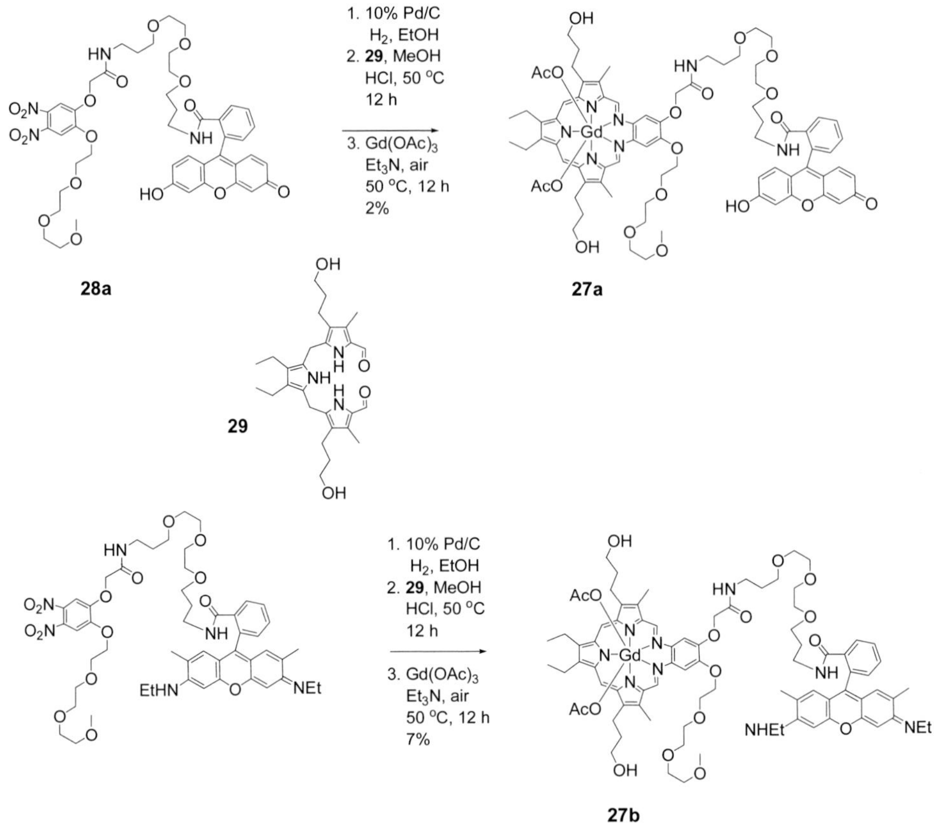

28b

Scheme 9. Synthesis of MGd-fluorophore conjugates.

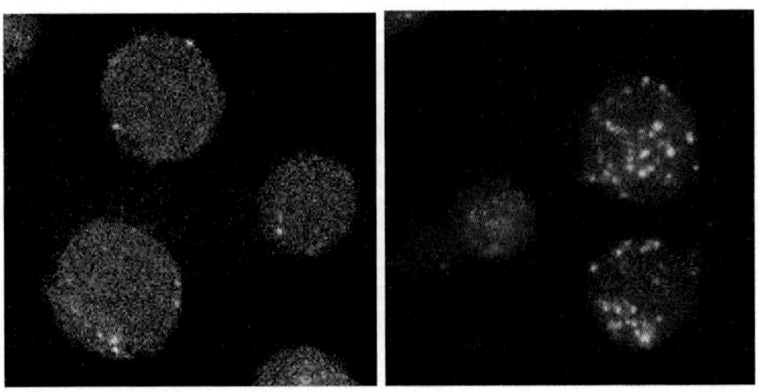

a. **b.**

Figure 6. Control (a.) and Fluorescein-MGd conjugate (b.) by CFM.

7. Future Prospects and Directions

An increasing number of preclinical and clinical studies are attesting to the unique biolocalization, safety, potential therapeutic utility, and imaging ability of MGd and MLu. The ability to produce ROS upon photoirradiation (MLu) or in the presence of reducing metabolites via "futile redox cycling" (MGd), coupled with their documented biolocalization ability, endows these texaphyrin complexes with several perceived advantages relative to more classic and generally less specific cytotoxic chemotherapeutic agents. The texaphyrins, being small molecules in the biological sense, have a further advantage in that they may be "custom built" for a needed application. One way of doing this involves the construction of so-called conjugates wherein a texaphyrin core is linked to either some other active species or to a fluorescent marker subunit. As discussed in this chapter, the linkage used to effect the tethering may be labile or nonlabile such that bond scission either occurs or does not occur in various biological milieus. In both cases, the well-recognized ability of texaphyrins to localize to, e.g., rapidly proliferating tissues is considered advantageous since it may allow for the site-selective delivery of cytotoxic agents while mitigating their potentially adverse effects, either by lowering the required dosage levels or by altering the pathways for elimination. In a similar vein, the attachment of a fluorophore is considered potentially useful since it may permit a normally poorly fluorescent species, such as MGd, to become the foundation for the construction of new tissue selective, fluorescence-based diagnostic agents.

The range of texaphyrin conjugates prepared to date serves to illustrate some of the structural diversity that is possible within the context of this general paradigm. It also serves to highlight some of the synthetic approaches that are making the construction of such systems feasible. While the chemistry of texaphyrin conjugates is still in its early days, it is gratifying to note that the MGd-MTX ester conjugate **24b**, containing a fusible linkage, shows an observable improvement in efficacy over MTX in cell studies at early time points. Such improvements are not seen in the case of the corresponding amide-linked conjugate **24a**. Several of the MGd-fluorophore conjugates, notably **26**, **27a**, and **27b**, were found to be readily visible as the result of their built-in fluorescent tags. Further studies with such systems should allow, at least in principle, the importance of charge, polarity, and lipophilicity on sub-cellular localization to be probed in a detailed manner. These same kinds of studies are also expected to provide important insights into which kinds of linkers are most appropriate for any given drug delivery task or biological targeting application. As such, it is hoped that the conjugates prepared to date and discussed in this chapter will help provide a firm foundation for the rational design of yet-improved texaphyrin-type therapeutic and diagnostic agents.

8. Acknowledgment

This work was supported in part by a grant from the National Institutes of Health (CA 68682 to J.L.S.).

9. References

[1] American Cancer Society (2004) Cancer Facts and Figures-2004, American Cancer Society, Atlanta.

[2] Bonnett, R. (2000) Chemical aspects of photodynamic therapy, Gordon and Breach Science Publishers, Amsterdam.

[3] Sternberg, E.D. and Dolphin, D. (1998) Porphyrin-based Photosensitizers for Use in Photodynamic Therapy, Tetrahedron 54, 4151-4202.

[4] Leunig, M. et al. (1993) Tumour localisation kinetics of photofrin and three synthetic porphyrinoids in an amelanotic melanoma of the hamster, Br. J. Cancer 68, 225-234.

[5] Paquette, B. and Van Lier, J.E. (1992) Phthalocyanines and related compounds: Structure-activity relationships, Photodyn. Ther. 145-156.

[6] Sessler, J.L. et al. (1988) An "expanded porphyrin": the synthesis and structure of a new aromatic pentadentate ligand, J. Am. Chem. Soc. 110, 5586-5588.

[7] Sessler, J.L. et al. (1993) Synthesis and structural characterization of lanthanide(III) texaphyrins, Inorg. Chem. 32, 3175-3187.

[8] Sessler, J.L. et al. (1996) New texaphyrin-type expanded porphyrins, Pure Appl. Chem. 68, 1291-1295.

[9] Young, S.W. et al. (1996) Gadolinium(III) texaphyrin: a tumor selective radiation sensitizer that is detectable by MRI, Proc. Nat. Acad. Sci. U S A 93, 6610-6615.

[10] Miller, R.A. et al. (1999) In vivo animal studies with gadolinium(III) texaphyrin as a radiation enhancer, Int. J. Radiat. Oncol., Biol., Phys. 45, 981-989.

[11] Sessler, J.L. and Miller, R.A. (2000) Texaphyrins: new drugs with diverse clinical applications in radiation and photodynamic therapy, Biochem. Pharmacol. 59, 733-739.

[12] Rosenthal, D.I. et al. (1999) A phase I single-dose trial of gadolinium texaphyrin (Gd-Tex), a tumor selective radiation sensitizer detectable by magnetic resonance imaging, Clin. Cancer Res. 5, 739-745.

[13] Carde, P. et al. (2001) Multicenter phase Ib/II trial of the radiation enhancer motexafin gadolinium in patients with brain metastases, J. Clin. Oncol. 19, 2074-2083.

[14] Mehta, M.P. et al. (2002) Lead-in phase to randomized trial of motexafin gadolinium and whole-brain radiation for patients with brain metastases: centralized assessment of magnetic resonance imaging, neurocognitive, and neurologic end points, J. Clin. Oncol. 20, 3445-3453.

[15] Mehta, M.P. et al. (2003) Survival and neurologic outcomes in a randomized trial of motexafin gadolinium and whole-brain radiation therapy in brain metastases, J. Clin. Oncol. 21, 2529-2536.

[16] Sessler, J.L. et al. (1993) Gadolinium(III) texaphyrin: a novel MRI contrast agent, J. Am. Chem. Soc. 115, 10368-10369.

[17] Young, S.W. et al. (1994) Preclinical evaluation of gadolinium(III) texaphyrin complex. A new paramagnetic contrast agent for magnetic resonance imaging, Invest Radiol. 29, 330-338.

[18] Mody, T.D. and Sessler, J.L. (2001) Texaphyrins: a new approach to drug development, Journal of Porphyrins and Phthalocyanines 5, 134-142.

[19] Chou, T.M. et al. (2002) Photodynamic therapy: applications in atherosclerotic vascular disease with motexafin lutetium, Catheterization and cardiovascular interventions : official journal of the Society for Cardiac Angiography & Interventions 57, 387-394.

[20] Kereiakes, D.J. et al. (2003) Phase I drug and light dose-escalation trial of motexafin lutetium and far red light activation (phototherapy) in subjects with coronary artery disease undergoing percutaneous coronary intervention and stent deployment: procedural and long-term results, Circulation 108, 1310-1315.

[21] Yeung, P.K.F. (2001) Motexafin lutetium (Pharmacyclics), IDrugs 4, 351-359.

[22] Sessler, J.L. et al. (2001) Probing the reactivity of the radiation sensitizer motexafin gadolinium (Xcytrin) and a series of lanthanide(III) analogues in the presence of both hydroxyl radicals and aqueous electrons, J. Porphyrins Phthalocyanines 5, 593-599.

[23] Magda, D.J. et al. (2004) Synthesis of texaphyrin conjugates, Pure Appl. Chem. 76, 365-374.

[24] Sessler, J.L. et al. (1987) Synthesis and crystal structure of a novel tripyrrane-containing porphyrinogen-like macrocycle, J. Org. Chem. 52, 4394-4397.

[25] Hannah, S. et al. (2001) Synthesis of a Metal-Free Texaphyrin, Org. Lett. 3, 3911-3914.

[26] Sessler, J.L. et al. (1994) Texaphyrins: Synthesis and Applications, Acc. Chem. Res. 27, 43-50.

[27] Sessler, J.L. et al. (1988) The coordination chemistry of planar pentadentate "porphyrin-like" ligands, Comments Inorg. Chem. 7, 333-350.

[28] Mody, T.D. et al. (2001) Texaphyrins: synthesis and development of a novel class of therapeutic agents, Prog. Inorg. Chem. 49, 551-598.

[29] Harriman, A. et al. (1989) Metallotexaphyrins: a new family of photosensitisers for efficient generation of singlet oxygen, Chem. Commun. 314-316.

[30] Sessler, J.L. et al. (1991) Tripyrroledimethine-derived ("texaphyrin"-type) macrocycles: potential photosensitizers which absorb in the far-red spectral region, Proc. SPIE-The International Society for Optical Engineering 1426, 318-329.

[31] Ehrenberg, B. et al. (1992) Spectroscopy, photokinetics and cellular effect of far-red and near infrared absorbing photosensitizers, Proc. SPIE-The International Society for Optical Engineering 1645, 259-263.

[32] Shimanovich, R. et al. (2001) Mn(II)-Texaphyrin as a Catalyst for the Decomposition of Peroxynitrite, J. Amer. Chem. Soc. 123, 3613-3614.

[33] Crow, J. (in press) in: Medicinal Inorganic Chemistry, ACS Symposium Series, (Sessler, J.L. et al., Eds.).

[34] Hannah, S. et al. (2002) Late First-Row Transition-Metal Complexes of Texaphyrin, J. Am. Chem. Soc. 124, 8416-8427.

[35] Young, S.W. et al. (1996) Lutetium texaphyrin (PCI-0123): a near-infrared, water-soluble photosensitizer, Photochem. Photobiol. 63, 892-897.

[36] Geraldes, C.F. et al. (1995) Nuclear magnetic relaxation dispersion studies of water-soluble gadolinium(III)-texaphyrin complexes, J. Magn. Reson. Imag. 5, 725-729.

[37] Sessler, J.L. et al. (1996) Water soluble texaphyrin metal complex preparation, US Pat. 5,569,759.

[38] Meyers, C.A. et al. (2004) Neurocognitive function and progression in patients with brain metastases treated with whole-brain radiation and motexafin gadolinium: results of a randomized phase III trial, J. Clin. Oncol. 22, 157-165.

[39] Anonymous (2004) Motexafin gadolinium: gadolinium (III) texaphyrin, gadolinium texaphyrin, Gd-Tex, GdT2B2, PCI 0120, Drugs in R&D 5, 52-57.

[40] Rosenthal, D.I. et al. (2000) Reversible renal toxicity resulting from high single doses of the new radiosensitizer gadolinium texaphyrin, Department of Radiation Oncology, University of Pennsylvania Medical Center, Philadelphia 19104-4283, USA, 593-598.

[41] Dimofte, A. et al. (2002) In vivo light dosimetry for motexafin lutetium-mediated PDT of recurrent breast cancer, Lasers Surg. Med. 31, 305-312.

[42] Renschler, M.F. et al. (1998) Photodynamic therapy trials with lutetium texaphyrin (Lu-Tex) in patients with locally recurrent breast cancer, Proceedings of SPIE-The International Society for Optical Engineering 3247, 35-39.

[43] Woodburn, K.W. et al. (2002) Photodynamic therapy for choroidal neovascularization: a review, Retina 22, 391-405.

[44] Rockson, S.G. et al. (2000) Photoangioplasty for human peripheral atherosclerosis: Results of a phase I trial of photodynamic therapy with motexafin lutetium (antrin), Circulation 102, 2322-2324.

[45] Yeung, A. (2004) Vulnerable Plaque: Pathophysiology, Detection and Therapeutic Intervention, Cardiovascular Research Foundation 16th Annual Scientific Meeting of Transcatheter Cardiovascular Therapeutics (TCT), Washington, DC.

[46] Griffin, G.M. et al. (2001) Preclinical evaluation of motexafin lutetium-mediated intraperitoneal photodynamic therapy in a canine model, Clin. Cancer Res. 7, 374-81.

[47] Hsi, R.A. et al. (2001) Photodynamic therapy in the canine prostate using motexafin lutetium, Clin. Cancer Res. 7, 651-660.

[48] Viala, J. et al. (1999) Phases IB and II multidose trial of gadolinium texaphyrin, a radiation sensitizer detectable at MR imaging: preliminary results in brain metastases, Radiology 212, 755 759.

[49] Kessel, D. et al. (2003) Localization and photodynamic efficacy of two cationic porphyrins varying in charge distribution, Photochem. Photobiol. 78, 431-435.

[50] Kessel, D. (2002) Relocalization of cationic porphyrins during photodynamic therapy, Photochem. Photobiol. Sci. 1, 837-840.

[51] Jori, G. et al. (1984) In Vitro and In Vivo Studies on the Interaction of Hematoporphyrin and its Dimethyl Ester with Normal and Malignant Cells in: Porphyrin Localization and Treatment of Tumors, Progress in Clinical and Biological Research, Vol. 170, pp. 471-482 (Doiron, D.R. and Gomer, C.J., Eds.) Liss, NY.

[52] El-Far, M. and Pimstone, N. (1984) Porphyrin: Protoporphyrin: Plasma Protein Interaction - The Metabolic Basis for Tumor Localization of Hematoporphyrin Derivative - A Preliminary Report

in: Porphyrin Localization and Treatment of Tumors, Progress in Clinical and Biological Research, Vol. 170, pp. 657-660 (Doiron, D.R. and Gomer, C.J., Eds.) Liss, NY.

[53] Ruck, A. and Diddens, H. (1996) Upake and Subcellular Distribution of Photosensitizing Drugs in Malignant Cells in: The Fundamental Basis of Phototherapy (Honigsman, H.J., G.; Young, A., Ed.) OEMF, Milano.

[54] Vorndran, C. et al. (1995) New fluorescent calcium indicators designed for cytosolic retention or measuring calcium near membranes, Biophys. J. 69, 2112-2124.

[55] Peng, Q. et al. (1996) Correlation of subcellular and intratumoral photosensitizer localization with ultrastructural features after photodynamic therapy, Ultrastruct. Path. 20, 109-129.

[56] Synytsya, A. et al. (2003) In vitro interaction of macrocyclic photosensitizers with intact mitochondria: a spectroscopic study, Biochim. Biophys. Acta 1620, 85-96.

[57] Synytsya, A. et al. (2004) Biodistribution assessment of a lutetium(III) texaphyrin analogue in tumor-bearing mice using NIR Fourier-transform Raman spectroscopy, Photochem. Photobiol. 79, 453-460.

[58] Woodburn, K.W. et al. (1997) Localization and efficacy analysis of the phototherapeutic lutetium texaphyrin (PCI-0123) in the murine EMT6 sarcoma model, Photochem. Photobiol. 65, 410-415.

[59] Woodburn, K.W. (2001) Intracellular localization of the radiation enhancer motexafin gadolinium using interferometric Fourier fluorescence microscopy, J. Pharm. Exper. Ther. 297, 888-894.

[60] Zellweger, M. et al. (2000) Fluorescence pharmacokinetics of Lutetium Texaphyrin (PCI-0123, Lu-Tex) in the skin and in healthy and tumoral hamster cheek-pouch mucosa, J. Photochem. Photobiol., B 55, 56-62.

[61] Hayase, M. et al. (2001) Photoangioplasty with local motexafin lutetium delivery reduces macrophages in a rabbit post-balloon injury model, Cardiovascular Research 49, 449-455.

[62] Woodburn, K.W. et al. (1996) Phototherapy of cancer and atheromatous plaque with texaphyrins, Journal of clinical laser medicine & surgery 14, 343-8.

[63] Woodburn, K.W. et al. (1996) Selective uptake of texaphyrins by atheromatous plaque, Proceedings of SPIE-The International Society for Optical Engineering 2671, 62-71.

[64] Woodburn, K.W. et al. (1997) Photoeradication and imaging of atheromatous plaque with texaphyrins, Proc. SPIE-The International Society for Optical Engineering 2970, 44-50.

[65] Sessler, J.L. et al. (1997) Biomedical applications of lanthanide(III) texaphyrins. Lutetium(III) texaphyrins as potential photodynamic therapy photosensitizers, J. Alloys Compd. 249, 146-152.

[66] Magda, D. et al. (2001) Redox cycling by motexafin gadolinium enhances cellular response to ionizing radiation by forming reactive oxygen species, Int. J. Radiat. Oncol. Biol. 51, 1025-1036.

[67] Sessler, J.L. et al. (1999) One-Electron Reduction and Oxidation Studies of the Radiation Sensitizer Gadolinium(III) Texaphyrin (PCI-0120) and Other Water Soluble Metallotexaphyrins, J. Phys. Chem. A 103, 787-794.

[68] Sessler, J.L. et al. (2001) Pulse Radiolytic Studies of Metallotexaphyrins in the Presence of Oxygen: Relevance of the Equilibrium with Superoxide Anion to the Mechanism of Action of the Radiation Sensitizer Motexafin Gadolinium (Gd-Tex2+, Xcytrin), J. Phys. Chem. B 105, 1452-1457.

[69] Magda, D. et al. (2002) Motexafin gadolinium reacts with ascorbate to produce reactive oxygen species, Chem. Commun. 2730-2731.

[70] Chari, R.V.J. (1998) Targeted delivery in chemotherapeutics: tumor activated prodrug therapy, Adv. Drug Delivery Rev. 31, 89-104.

[71] Meister, A. (1995) Mitochondrial changes associated with glutathione deficiency, Biochim. Biophys. Acta 1271, 35-42.

[72] Hall, E.J. (1994) Radiobiology for the Radiobiologist, J. B. Lippincott Co., Philadelphia.

[73] Brown, J.M. et al. (1981) SR-2508: a 2-nitroimidazole amide which should be superior to misonidazole as a radiosensitizer for clinical use, Int. J. Radiat. Oncol. Biol. 7, 695-703.

[74] Adams, G.E. et al. (1980) Toxicity of nitro compounds toward hypoxic mammalian cells in vitro: dependence on reduction potential, J. Nat. Cancer Inst. 64, 555-560.

[75] Stratford, I.J. et al. (1981) Cytotoxic Properties of 4-Nitroimidazole (NSC 38087): A Radiosensitizer of Hypoxic Cells In Vitro, Br. J. Cancer 44, 109-116.

[76] Simpson, J.R. et al. (1982) Large Fraction Radiotherapy Plus Misonidazole for Treatments of Advanced Lung Cancer: Report of a Phase I/II Trial, Int. J. Radiation Oncology Biol. Phys. 8, 303-308.

[77] Berlin, V. and Haseltine, W.A. (1981) Reduction of adriamycin to a semiquinone-free radical by NADPH cytochrome P-450 reductase produces DNA cleavage in a reaction mediated by molecular oxygen, J. Biol. Chem. 256, 4747-4756.

[78] Doroshow, J.H. (1983) Effect of anthracycline antibiotics on oxygen radical formation in rat heart, Cancer Res. 43, 460-472.

[79] Yeh, G.C. et al. (1987) Adriamycin resistance in human tumor cells associated with marked alterations in the regulation of the hexose monophosphate shunt and its response to oxidant stress, Cancer Res. 47, 5994-5999.

[80] Lau, D.H. et al. (1989) Association of DNA cross-linking with potentiation of the morpholino derivative of doxorubicin by human liver microsomes, J. Nat. Cancer Inst. 81, 1034-1038.

[81] Rumyantseva, G.V. et al. (1989) Hydroxyl radical generation and DNA strand scission mediated by natural anticancer and synthetic quinones, FEBS letters 242, 397-400.

[82] Sinha, B.K. et al. (1989) Adriamycin activation and oxygen free radical formation in human breast tumor cells: protective role of glutathione peroxidase in adriamycin resistance, Cancer Res. 49, 3844-3848.

[83] Lewis, A.D. et al. (1992) Role of cytochrome P-450 from the human CYP3A gene family in the potentiation of morpholino doxorubicin by human liver microsomes, Cancer Res. 52, 4379-4384.

[84] Feinstein, E. et al. (1993) Dependence of nucleic acid degradation on in situ free-radical production by Adriamycin, Biochemistry 32, 13156-61.

[85] Gao, J.-P. et al. (1993) The role of reduced nicotinamide adenine dinucleotide phosphate in glucose- and temperature-dependent doxorubicin cytotoxicity, Cancer Chemother. Pharmacol. 33, 191-196.

[86] Minotti, G. et al. (1995) Secondary alcohol metabolites mediate iron delocalization in cytosolic fractions of myocardial biopsies exposed to anticancer anthracyclines: novel linkage between anthracycline metabolism and iron-induced cardiotoxicity, J. Clin. Invest. 95, 1595 605.

[87] Lau, D.H. et al. (1994) Metabolic conversion of methoxymorpholinyl doxorubicin: from a DNA strand breaker to a DNA cross-linker, Br. J. Cancer 70, 79-84.

[88] Miller, R.A. et al. (2001) Motexafin gadolinium: A redox active drug that enhances the efficacy of bleomycin and doxorubicin, Clin. Cancer Res. 7, 3215-3221.

[89] Lottner, C., Bart, K.-C., Bernhart, G., Brunner, H. (2002) Hematoporphyrin-Derived Soluble Porphyrin-Platinum Conjugates with Combined Cytotoxic and Phototoxic Antitumor Activity, J. Med. Chem. 45, 2064-2078.

[90] Zhang, J.-G., Lindup, E. (1994) Cisplatin Nephrotoxicity: Decreases in Mitochondrial Protein Sulphydryl Concentration and Calcium Uptake by Mitochondria from Rat Renal Cells, Biochem. Pharmacol. 47, 1127-1135.

[91] Pasini, A. and Zunino, F. (1987) New Cisplatin Analogues - On the Way to Better Antitumor Agents, Angew. Chem. Int. Ed. Engl. 26, 615-625.

[92] Di Francesco, A.M. et al. (2002) Cellular and molecular aspects of drugs of the future: Oxaliplatin, Cell. Molec. Life Sciences 59, 1914-1927.

[93] Ohndorf, U.-M. et al. (1999) Basis for recognition of cisplatin-modified DNA by high-mobility-group proteins, Nature 399, 708-712.

[94] Dolling, J.A. et al. (1999) Cisplatin-modification of DNA repair and ionizing radiation lethality in yeast, Saccharomyces cerevisiae, Mutation Res. 433, 127-136.

[95] Legendre, F. and Chottard, J.-C. (1999) Kinetics and Selectivity of DNA-Platination in: Cisplatin: Chemistry and Biochemistry of a Leading Anticancer Drug, pp. 223-245 (Lippert, B., Ed.) Verlag Helvitica Chimica Acta, Zurich.

[96] Messman, R.A. and Allegra, C.J. (2001) Antifolates in Cancer Chemotherapy and Biotherapy in: Cancer chemotherapy and biotherapy : principles and practice, pp. 139 (Chabner, B.A. and Longo, D.L., Eds.) Lippincott Williams & Wilkins, Philadelphia.

[97] Piper, J.R. et al. (1982) Syntheses of alpha- and gamma-substituted amides, peptides, and esters of methotrexate and their evaluation as inhibitors of folate metabolism, J. Med. Chem. 25, 182-187.

[98] Wright, J.E. et al. (1993) Methotrexate and gamma-tert-butyl methotrexate transport in CEM and CEM/MTX human leukemic lymphoblasts, Biochem. Pharmacol. 46, 871-6.

[99] Endo, N. et al. (1987) Target-selective cytotoxicity of methotrexate conjugated with monoclonal anti-MM46 antibody, Cancer Immunol. Immunother. 25, 1-6.

[100] Umemoto, N. et al. (1989) Preparation and in vitro cytotoxicity of a methotrexate-anti-MM46 monoclonal antibody conjugate via an oligopeptide spacer, Int. J. Cancer 43, 677-654.

Appendix: Coloured Figures of Chapters 3, 7, 8, 11, 12, 17, 21, 23, 24

Karsten Gloe (ed.), Macrocyclic Chemistry: Current Trends and Future Perspectives, 427–440.
© 2005 *Springer. Printed in the Netherlands.*

3. TOWARDS FUNCTIONAL MACROCYCLES: SELF-ASSEMBLY AND TEMPLATE STRATEGIES (pp. 37-52)

C. A. Schalley, H. T. Baytekin and B. Baytekin

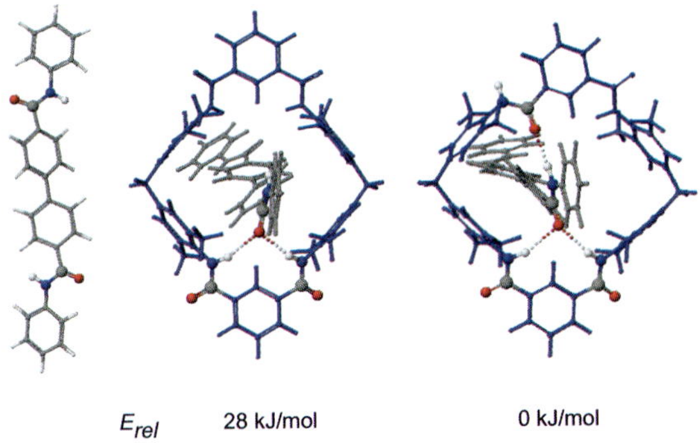

Figure 4. Structures of an axle (left), its complex with "oxygens out" conformer of the tetralactam ring (middle) and the complex with the most stable conformer.

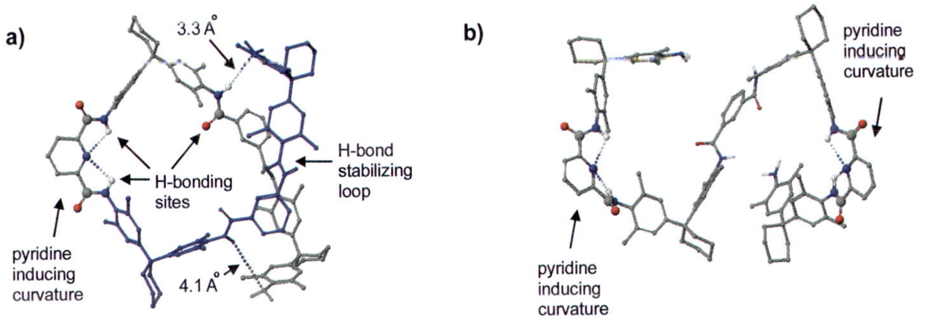

Figure 7. a) The lowest energy conformation of the intermediate **14** which promotes knot formation. b) The reversed sequence of adding the diacyl chlorides **2** and **6** into the reaction does not produce the helical intermediate; knot formation does not occur.

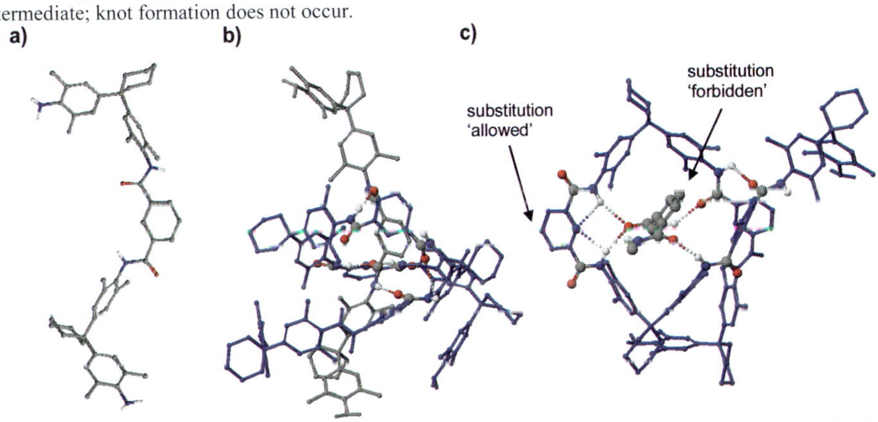

Figure 8. a) Lowest energy conformation of the extended diamine axle **13** that threads into the helix shaped intermediate **14**. b) Side view of the intermediate complex: the ends that will react are in close vicinity of each other. c) The "inner" part of the intermediate can not stand any bulkiness that is created by using bulkier isophthaloyl chlorides.

7. SENSING, TEMPLATION AND SELF-ASSEMBLY BY MACROCYCLIC LIGAND SYSTEMS (pp. 105-119)

P. D. Beer and W. W. H. Wong

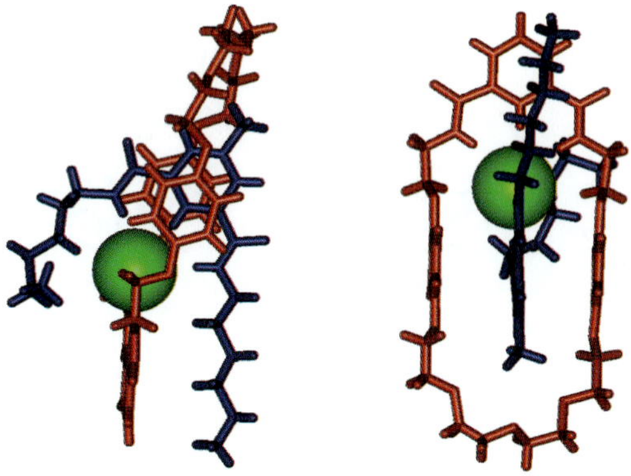

Figure 8. Two views of the structure of the pyridinium-chloride pseudorotaxane.

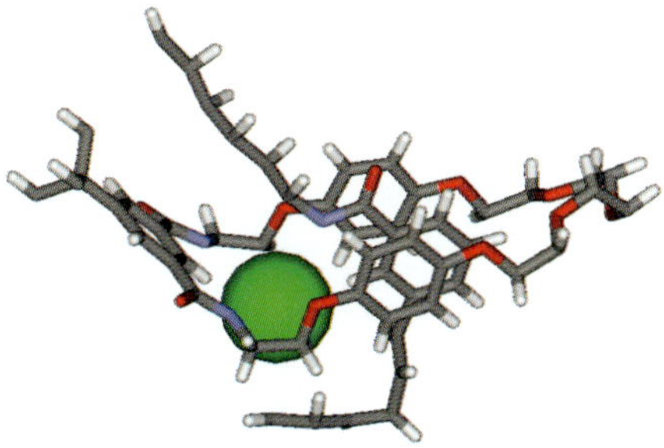

Figure 9. Structure of the nicotinamide-chloride pseudorotaxane.

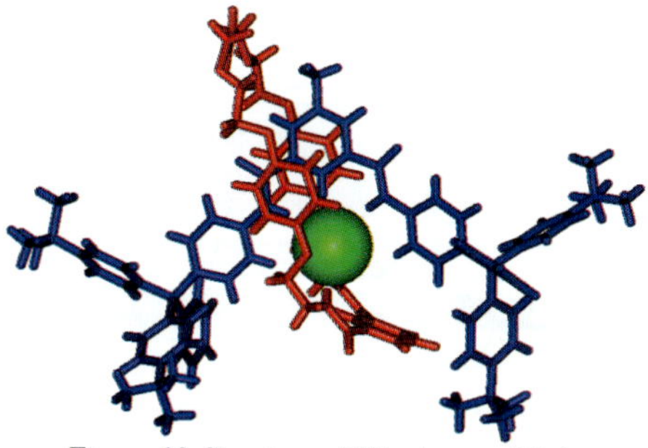

Figure 10. Structure of [2]rotaxane **(12a)**.

8. SIGNALLING REVERSIBLE ANION BINDING IN AQUEOUS MEDIA (pp. 121-136)

R. Dickens and D. Parker

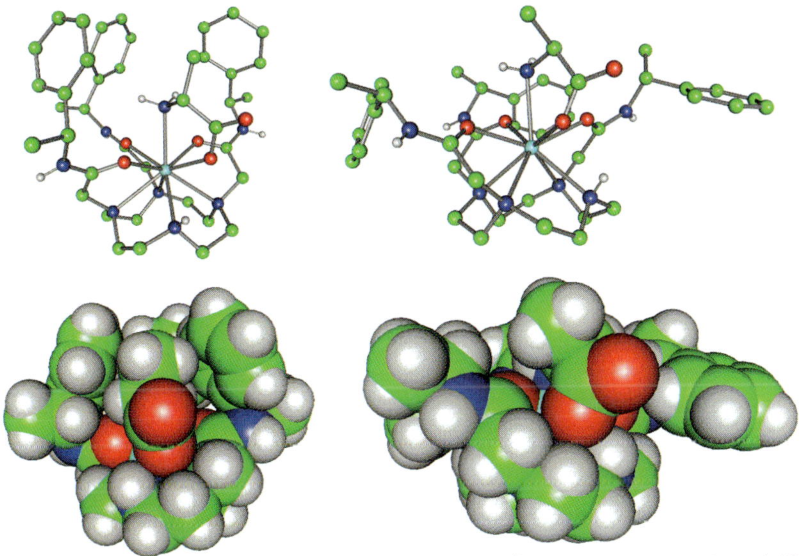

Figure 6. Crystal structures of the ternary complexes of [Yb.**1a**]$^{3+}$ with (*R*)-alanine (left) and (*S*)-alanine (right). Note the axial and equatorial conformations of the pendent side chains, respectively, and the opposite chirality of the macrocyclic ring and pendent arm configuration.

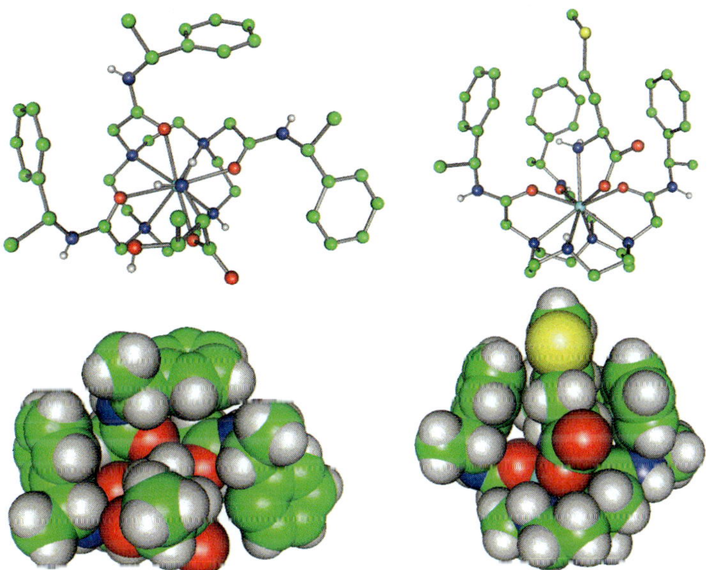

Figure 7. Crystal structures of the ternary complexes of [Yb.**1a**]$^{3+}$ with (*S*)-threonine (left) and (*S*)-methionine (right). The methinione side chain is enclosed in a tight hydrophobic pocket formed by the axial arrangement of the pendent arms.

11. STRUCTURAL ASPECTS OF HALIDES WITH CRYPTANDS (pp. 173-188)

MD. A. Hossain, S. O. Kang and K. Bowman-James

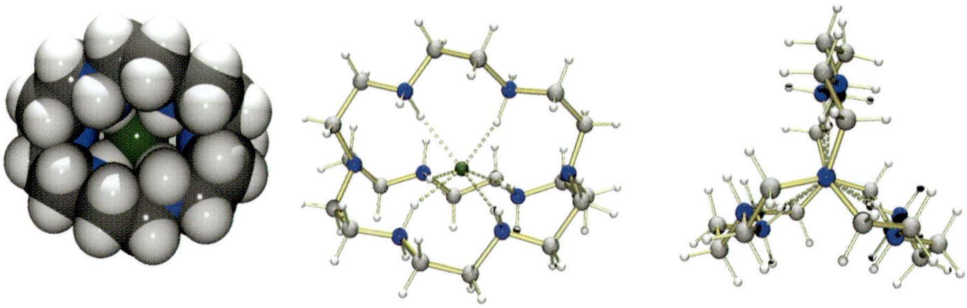

A **B** **C**

Figure 3. Crystal structure of the fluoride complex of **2** showing the encapsulated fluoride in the space filling model (**A**), in the perspective side view (**B**), and in the view down the pseudo-threefold axis (**C**).

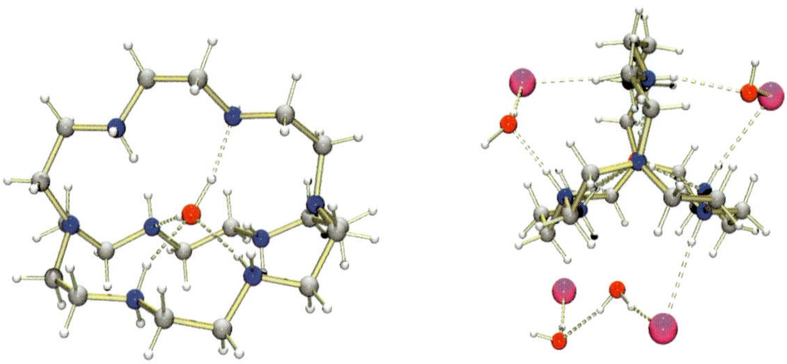

A **B**

Figure 6. Crystal structure of the iodide complex of **2** showing encapsulated water as viewed from the side (**A**), and additionally the external iodides and waters as viewed down the pseudo-threefold axis (**B**).

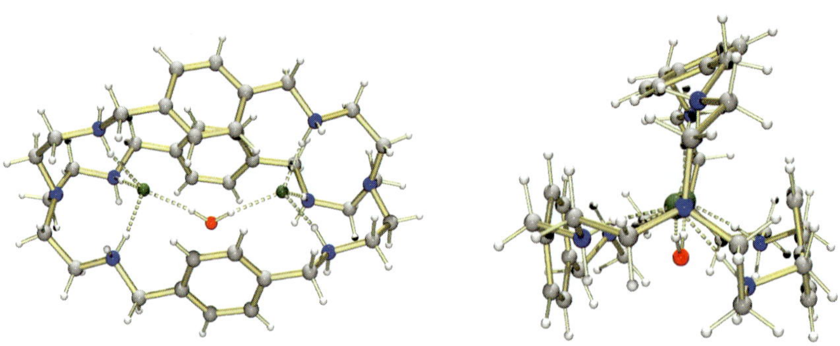

A **B**

Figure 10. Crystal structure of the fluoride cascade complex of **6** showing the side view (**A**) and view down the pseudo-threefold axis (**B**).

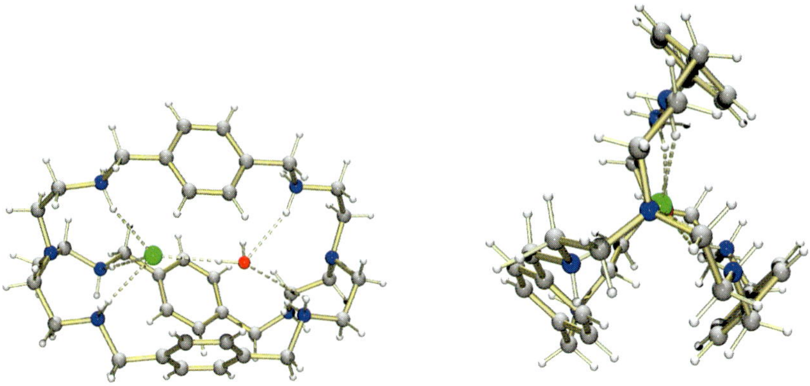

A **B**

Figure 11. Crystal structures of the complex of **6** with chloride showing the side view (**A**) and view down the pseudo-threefold axis (**B**).

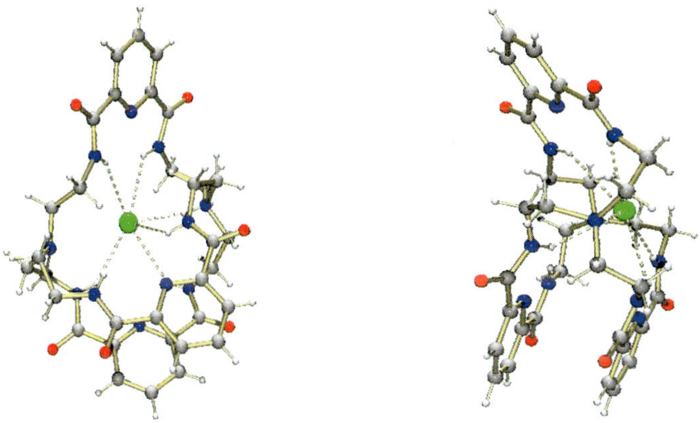

A **B**

Figure 18. The crystal structures of [H**7**(Cl)] showing the side (**A**) and end-on (**B**).

12. OXOANION SELECTIVITY WITH PROTONATED AZACRYPTATE HOSTS: THE INFLUENCE OF HYDRATION ON STRUCTURE AND STABILITY (pp. 189-201)

J. Nelson, V. McKee and R. M. Town

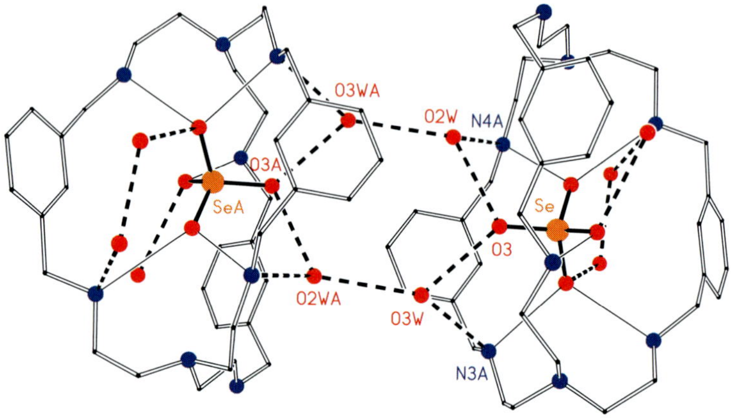

Figure 2. Selenate cryptate of H_6R3Bm^{6+}.

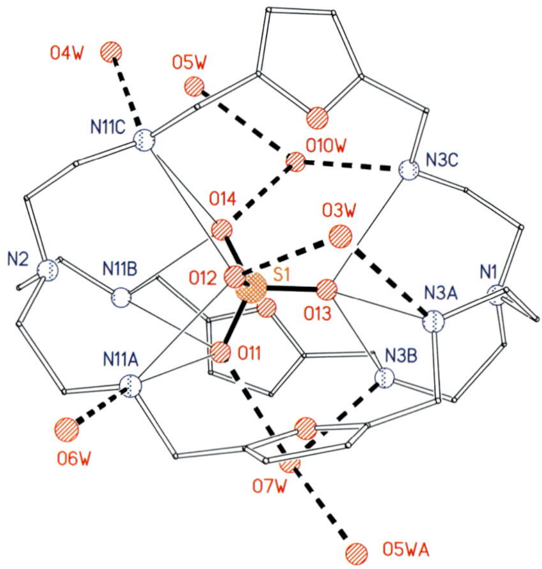

Figure 3. Chromate cryptate of H_6R3Bm^{6+}.

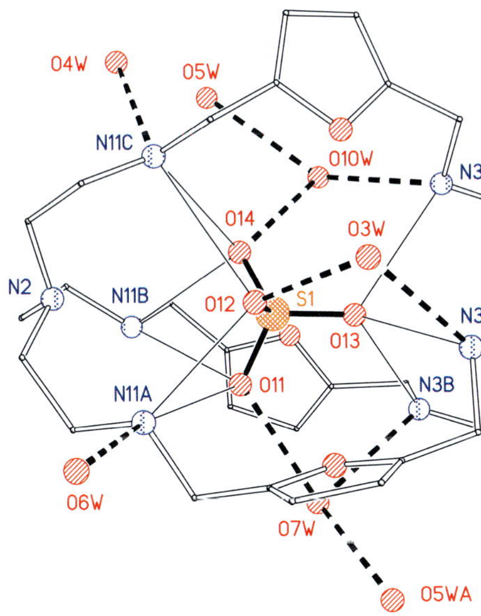

Figure 4. Sulfate cryptate of H_6R3F^{6+}.

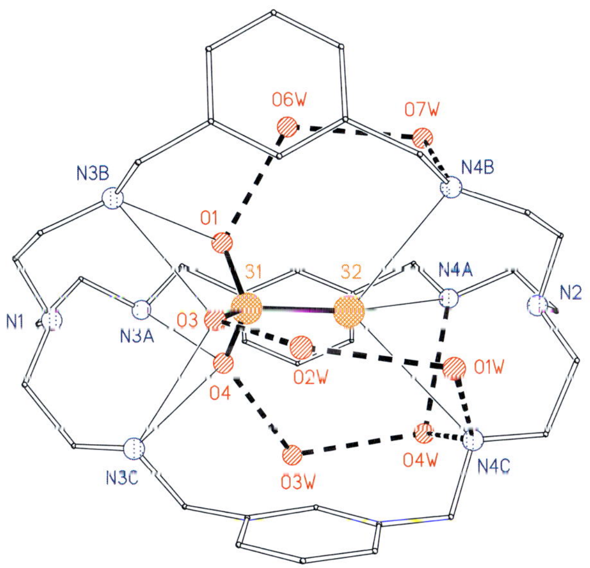

Figure 5. Thiosulfate cryptate of H_6R3Bm^{6+}.

17. RECOGNITION OF CYTOCHROME *c* BY TETRAPHENYLPORPHYRIN-BASED PROTEIN SURFACE RECEPTORS (pp. 267-275)

R. K. Jain, L. K. Tsou and A. D. Hamilton

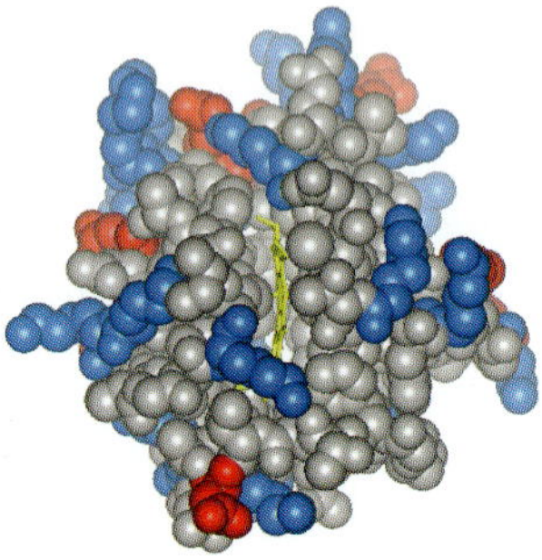

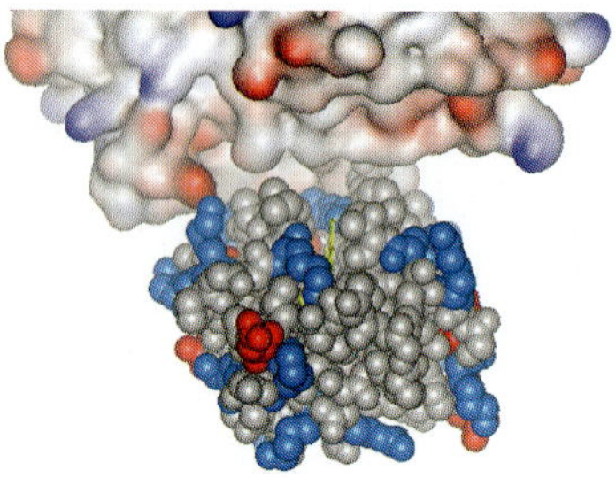

Figure 1. (a) X-ray crystal structure of horse-heart ferricytochrome *c*.[8] All protein atoms are shown in the C.-P.-K. form, while the heme group is shown in the stick form. All Arg and Lys residues are colored blue, while Glu and Asp are colored in red, to contrast the destribution of the most ionizable side chains. (b) The X-ray crystal structure of horse heart ferricytochrome *c* in complex with horse cytochrome *c* peroxidase (ccp).[9] The peroxidase is shown as a molecular surface model, with blue regions depicting positive and red representing negative electrostatic potential. Note the cluster of negative potential on ccp that surrounds the contact interface.

21. SIMULATIONS OF THE DYNAMICS 18-CROWN-6 AND ITS COMPLEXES: FROM THE GAS PHASE TO AQUEOUS INTERFACES WITH SC-CO2 AND A ROOM-TEMPERATURE IONIC LIQUID (pp. 327-348)

A. Chaumont, R. Schurhammer, P. Vayssière and G. Wipff

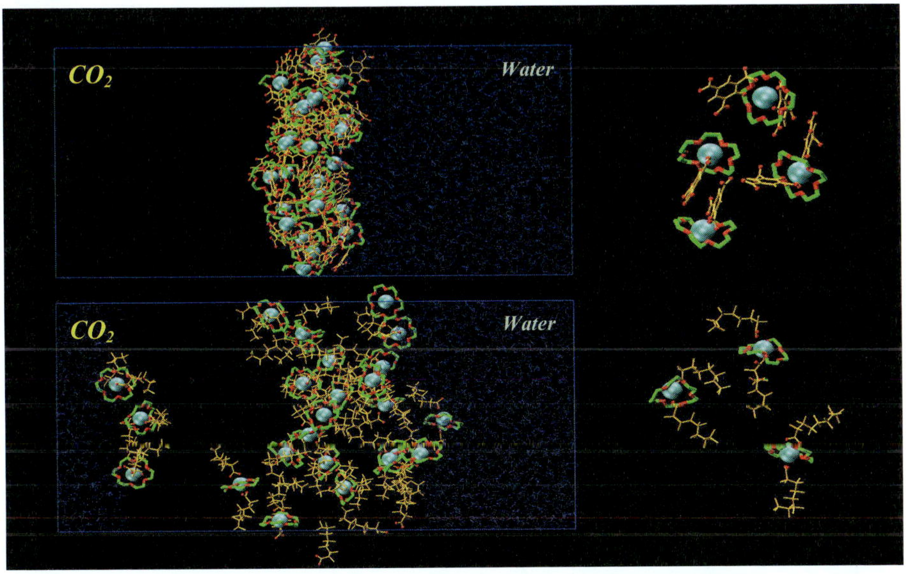

Figure 7. Distribution of 25 Sr⊂18C6(Pic)$_2$ complexes (top) and 25 Sr⊂18C6(PFO)$_2$ (bottom) complexes at the CO_2 / water interface. Left: snapshots after 1.2 ns of dynamics at 305 K, where CO_2 is not shown for clarity. Right: zoom on typical complexes.

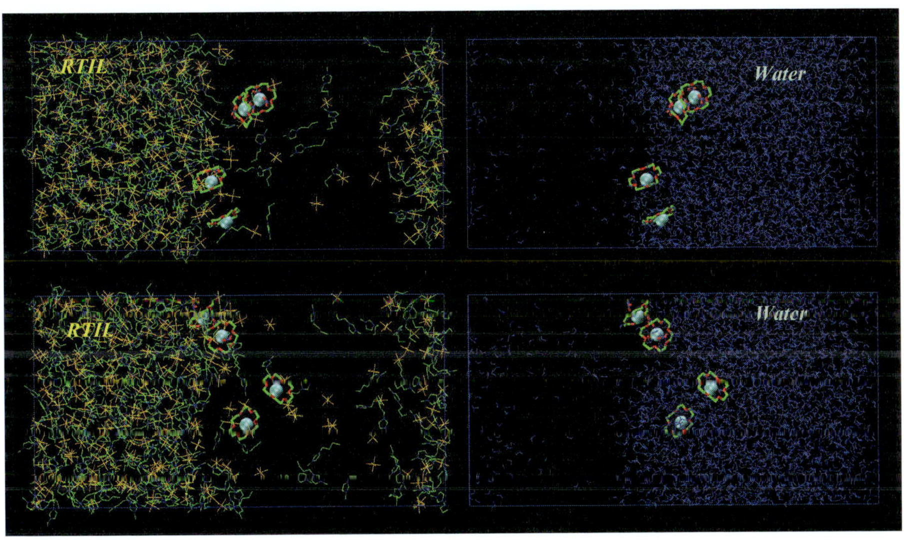

Figure 14. The [BMI][PF$_6$] / water interface with two Sr⊂18C6^{2+} plus two Sr⊂18C6(NO$_3$)$_2$ complexes after 2 ns (top) and 4 ns (bottom) of dynamics. Solvents are shown side by side, instead of superposed, for clarity.

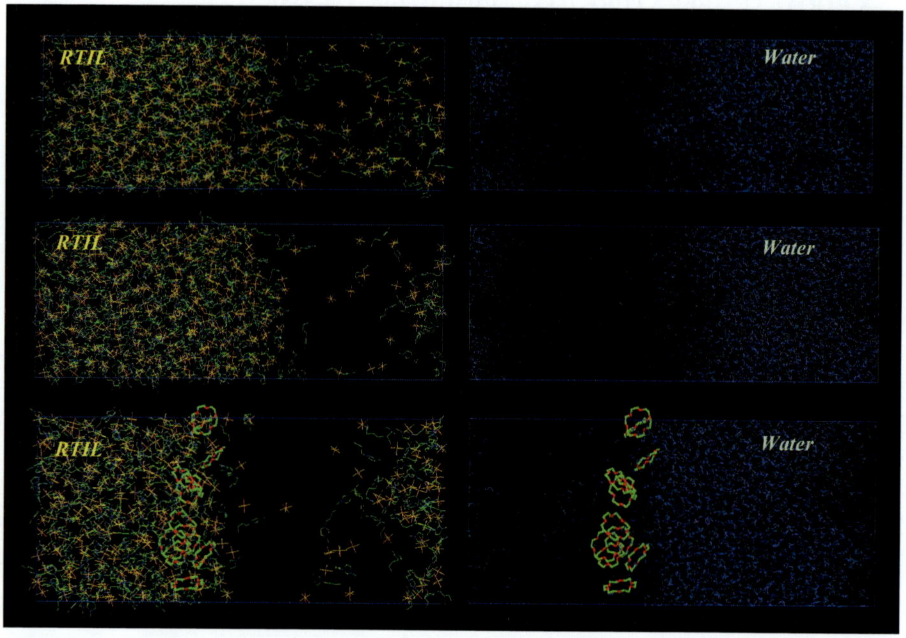

Figure 13. The $[BMI][PF_6]$ / water neat interface (top - charges +1/-1 after 4ns and middle - charges +0.9/-0.9 after 9ns) and with twelve 18C6 crowns (bottom) after 4 ns of dynamics. Solvents are shown side by side, instead of superposed, for clarity.

23. LIGAND DESIGN FOR BASE METAL RECOVERY (pp. 365-382)

P. A. Tasker and V. Gasperov

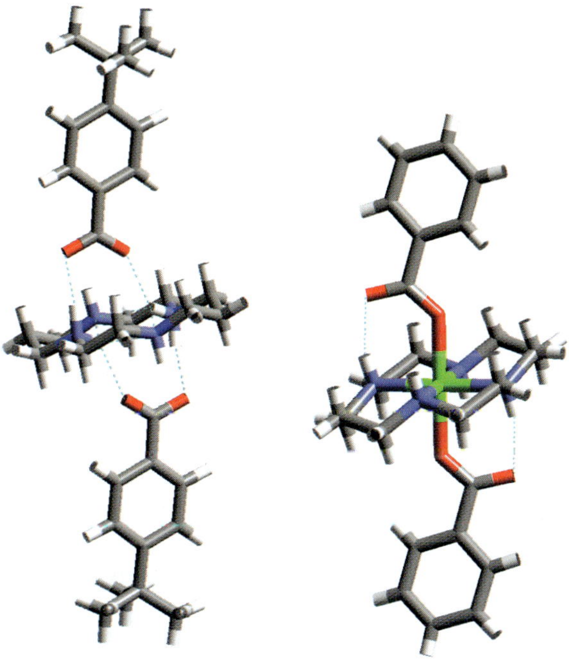

A B

Figure 2. Structures of [(cyclamH_2)(4-*t*-butylbenzoate$)_2$], (**A**) and [Ni(cyclam)(benzoate$)_2$], (**B**).[54, 55]

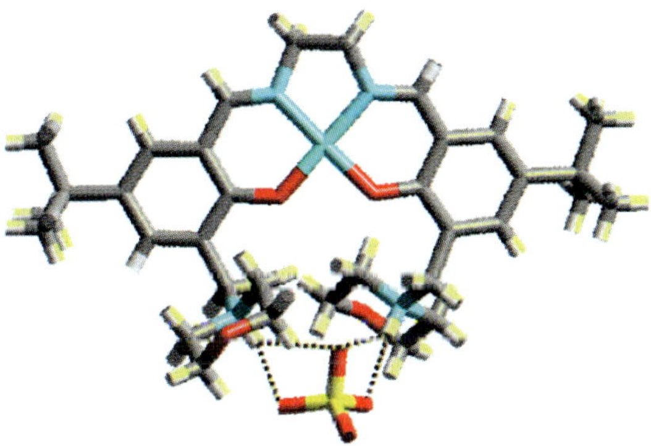

Figure 10. A $NiSO_4$ complex of a bis-morpholinomethyl substituted "salen" ligand showing the binding of sulfate by a combination of electrostatic forces and hydrogen bonds to the protonated morpoline groups.

24. USE OF MACROCYCLES IN NUCLEAR-WASTE CLEANUP: A REAL-WORLD APPLICATION OF A CALIXCROWN IN CESIUM SEPARATION TECHNOLOGY (pp. 383-405)

B. A. Moyer, J. F. Birdwell, JR., P. V. Bonnesen and L. H. Delmau

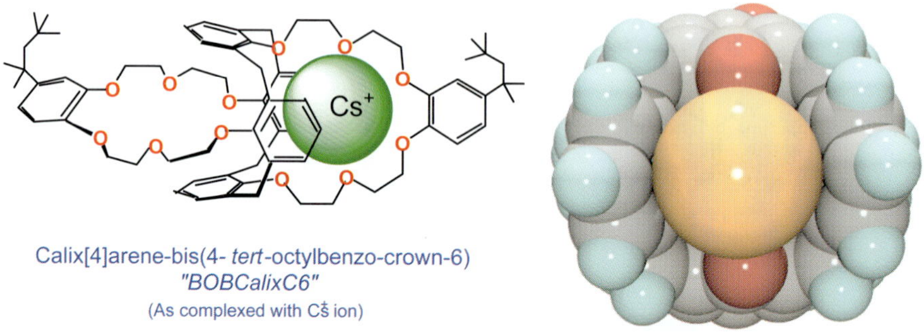

Figure 1. Calixcrown extractant adopted for CSSX, as complexed with Cs^+ ion. Left: chemical drawing of the complex. Right: a space-filling view of part of a crystal structure of the model complex Cs_2Calix[4]arene-bis(benzo-crown-6)$(NO_3)_2$•$3CHCl_3$ [69] showing the good fit of the Cs^+ ion inside the calixarene cavity; the crown ether atoms have been removed for clarity.

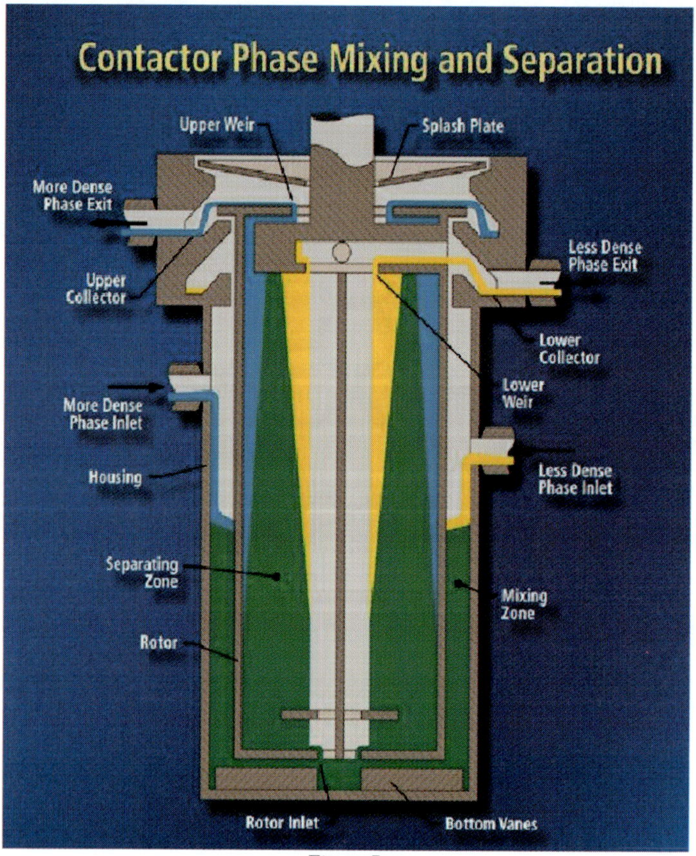

Figure 7.

INDEX

Karsten Gloe (ed.), Macrocyclic Chemistry: Current Trends and Future Perspectives, 441–450.

© 2005 *Springer. Printed in the Netherlands.*